Birkhäuser

Autonomic Systems

The AUTONOMIC SYSTEMS book series provides a platform of communication between academia and industry by publishing research monographs, outstanding PhD theses, and peer-reviewed compiled contributions on the latest developments in the field of autonomic systems.

It covers a broad range of topics from the theory of autonomic systems that are researched by academia and industry. Hence, cutting-edge research, prototypical case studies, as well as industrial applications are in the focus of this book series. Fast reviewing provides a most convenient way to publish latest results in this rapid moving research area.

The topics covered by the series include (among others):

- self-* properties in autonomic systems (e.g. self-management, self-healing)
- architectures, models, and languages for building autonomic systems
- trust, negotiation, and risk management in autonomic systems
- theoretical foundations of autonomic systems
- applications and novel computing paradigms of autonomic systems

For further volumes:
www.springer.com/series/8123

Organic Computing — A Paradigm Shift for Complex Systems

Christian Müller-Schloer
Hartmut Schmeck
Theo Ungerer
Editors

Birkhäuser

Editors
Christian Müller-Schloer
Institute for Systems Engineering
Leibniz Universität Hannover
Appelstr. 4
30167 Hannover
Germany
cms@sra.uni-hannover.de

Theo Ungerer
Department of Computer Science
University of Augsburg
Universitätsstr. 6a
86159 Augsburg
Germany
theo.ungerer@informatik.uni-augsburg.de

Hartmut Schmeck
Institute AIFB
Karlsruhe Institute of Technology
Kaiserstr. 89
76133 Karlsruhe
Germany
hartmut.schmeck@kit.edu

1998 ACM Computing Classification: C [Computer Systems Organization]; C.0 [General]: System architectures; C.2.4 [Distributed Systems]; C.3 [Special-purpose and Application-based systems]: Real-time and embedded systems; C.4 [Performance of Systems]; D.2 [Software Engineering]; L.2.9 [Robotics]; J.2 [Physical Sciences and Engineering]

ISBN 978-3-0348-0129-4 e-ISBN 978-3-0348-0130-0
DOI 10.1007/978-3-0348-0130-0

Cover design: deblik, Berlin

Printed on acid-free paper

Springer Basel AG is part of Springer Science+Business Media

www.birkhauser-science.com

Preface

It is not the question whether self-organising systems will arise, but how they will be designed and controlled.

One of the most fascinating endeavours in research consists of starting a new area of research, based on a visionary concept posing a range of challenges which can only be met by a collaborative effort. The priority research programme ("Schwerpunktprogramm") SPP 1183 "Organic Computing" of the German Research Foundation (DFG) has been such an inspiring opportunity. This compendium provides an overview of the various results which have been achieved over a period of six years.

About eight years ago a presentation of a colleague on future trends for research in Artificial Intelligence at a "Perspectives Seminar" at Dagstuhl inspired us to initiate a series of workshops on challenges for future research in Computer Engineering, looking at a time frame of ten to fifteen years. The combination of a large collection of ideas led to a joint position paper of the Gesellschaft für Informatik (GI) and the Informationstechnische Gesellschaft (ITG) on a new research area which we called *Organic Computing*. Motivated by the increasing number of intelligent devices in our vicinity being aware of their environment, capable to interact, and relying on cooperation to satisfy their functional objectives we identified the urgent necessity of a new paradigm for system design. Such a paradigm should enable future systems to carry out certain tasks on their own, they should have properties like self-organisation, self-healing, self-configuration, self-protection, self-optimisation, and they should be able to adapt reasonably to changing requirements of their operating environment. In particular, they should adjust to human needs in a trustworthy way and should allow for explicit interference in case of undesired effects of self-organisation, leading to the—somewhat contradictory—requirement of *controlled self-organisation*. Considering the numerous examples of self-organisation in nature, it was obvious to look at the mechanisms behind the various ways of generating complex but robust behaviour from simple interactions in order to exploit bio-inspired concepts for the design of a new generation of technical application systems. In this way the notion of Organic Computing is associated with two different meanings: the need for intelligent systems showing "organic" behaviour and the potential use of bio-inspired concepts in their construction.

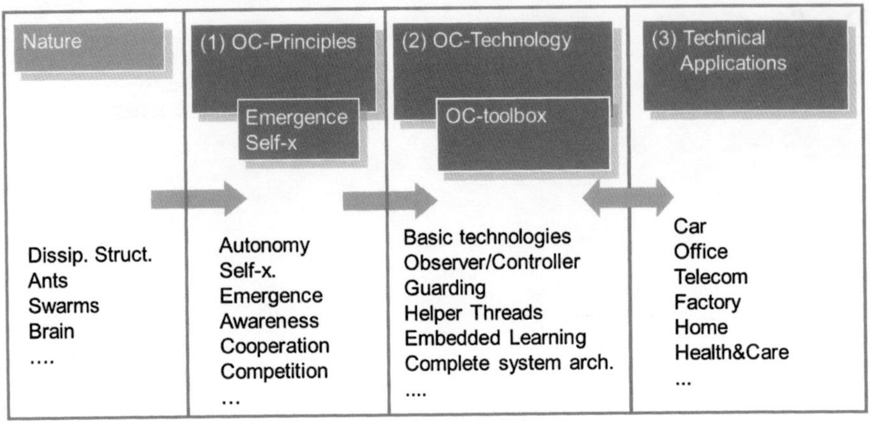

Fig. 1 Topical structure of the priority programme SPP 1183 "Organic Computing"

The concepts of Organic Computing are closely related to other research initiatives like ubiquitous and pervasive computing addressing the abundance of intelligent devices and the penetration of almost every aspect of our lives with information technology. The necessity of providing complex information processing and data management systems with self-* properties has also been the driving force of the Autonomic Computing Initiative by IBM using the biological principle of the autonomic (nervous) system as their paradigm for system design. Furthermore, independent from our initiative, Christoph von der Malsburg had used the notion of Organic Computing for his visionary concept of designing information processing systems inspired by principles of molecular biology and neuro-physiology.

Based on the ideas of the position paper our proposal to start a priority programme on Organic Computing was approved by the German Research Foundation in early 2004. The call for proposals for the first of the three two-year phases addressed the need for fundamental research on the effects of emergence due to self-organisation and on the design of system architectures and tools supporting the concepts of Organic Computing while keeping in mind the specific requirements of technical applications (see Fig. 1). In July 2005, the SPP 1183 "Organic Computing" started with 18 projects which had been selected by an international evaluation committee from 59 high quality submissions.

Semi-annual colloquia and various mini workshops supported a regular exchange of information between the projects and served to initiate and support cooperative work on various topics of joint interest like *self-organisation and emergence*, *architectures*, *applications*, *design methods*, and *bio-inspired methods*. In this way, a common understanding of essential concepts for the design of Organic Computing systems developed, which was deepened during the successive phases 2 and 3 of this programme. The programme's third and final phase emphasised the need for a clear demonstration of the achievements of this fascinating research motivated by a visionary concept for future system architectures.

This compendium now combines presentations of major results from all of the projects that are or had been part of the priority programme. The contributions start with an overview of our work on *Theoretical Foundations* characterising essential properties of Organic Computing systems, looking in-depth at different aspects of emergence as a result of self-organisation, studying formal approaches to specify correct behaviour of adaptive systems, and highlighting ways of transferring behavioural patterns of biological systems (in particular of societies of ants) into Organic Computing systems. This is complemented with an assessment of the Organic Computing research methodology from a philosophical point of view.

One of the central objectives of the priority programme was to establish a toolbox of concepts and methods which are essential for the design and operation of Organic Computing systems. The chapters on *Methods and Tools*, *Learning*, and *Architectures* present an overview of the broad range of fundamental concepts that evolved within the research programme. Learning, in particular, is an indispensible part of Organic Computing, addressing the challenge to develop adequate behaviour even in unanticipated situations. As stated in the call for proposals, the potential and relevance of the new techniques and system architectures of Organic Computing should be demonstrated with respect to demanding application scenarios. Therefore, the chapter on *Applications* reports on the work on Organic Computing in areas like traffic, robotics, sensor networks, image processing, and parallel computing.

The compendium is concluded with an *Outlook* on research that originated from or was inspired by this priority programme. Finally, we address a number of insights and lessons learnt combined with potential future research questions—following the more general observation that "good research answers some questions while—and this is more important—posing new ones."

Acknowledgement

The initiators and participants of the SPP 1183 "Organic Computing" would like to express their deepest gratitude to the German Research Foundation and to all those who contributed to the formation of this priority programme and who were responsible for providing us with the necessary financial support. In particular, we would like to thank the members of the international evaluation committee for their careful assessment and selection of the proposals. We are grateful to the DFG staff (in particular Dr. Gerit Sonntag and Dr. Valentina Damerow) for their valuable advice and continuing support. The funding instrument of priority programmes has proven to be highly effective in establishing and supporting new research communities.

This Compendium would not have been possible without the help of many more supporters. They have contributed to the organisational side of the preparation process, and they have done so in addition to the project work necessary for their research. This translates into many extra hours! Especially we would like to thank the Editorial Office with Sven Tomforde, Holger Prothmann and Lei Liu for their great work. The chapter editors Bernhard Sick, Hans-Ulrich Heiß and Wolfgang Reif have supported us in soliciting, inspecting and summarising single chapters of this book.

For quality control each article was reviewed by two members of the Review Team. We thank all of them for spending the necessary hours during the Christmas break 2010! We thank Barbara Hellriegel from Birkhäuser Verlag for her support and the smooth interface to the production. And last but not least we want to thank the authors for delivering high quality research and professional articles—and for keeping the tight deadlines.

Hannover, Germany
Karlsruhe, Germany
Augsburg, Germany

Christian Müller-Schloer
Hartmut Schmeck
Theo Ungerer

Contents

Review Team

Name	Surname	Affiliation
Gerrit	Anders	Universität Augsburg, Institut für Software und Systems Engineering
Jürgen	Becker	Karlsruhe Institute of Technology (KIT), Institute for Information Processing Technologies
Michael	Beigl	Karlsruhe Institute of Technology (KIT), Pervasive Computing Systems
Andreas	Bernauer	Eberhard Karls Universität Tübingen, Wilhelm-Schickard Institut für Informatik, Lehrstuhl Technische Informatik
Abdelmajid	Bouajila	Technische Universität München, Lehrstuhl für Integrierte Systeme
Jürgen	Branke	University of Warwick, Warwick Business School
Uwe	Brinkschulte	J.-W. Goethe Universität Frankfurt, Institut für Eingebettete Systeme
Werner	Brockmann	University of Osnabrück, Institute of Computer Science
Bastian	Degener	Universität Paderborn, Heinz Nixdorf Institute
Peter	Dittrich	Friedrich-Schiller-University Jena, Institute of Computer Science, Bio Systems Analysis Group
Rolf	Ernst	Technische Universität Braunschweig, Institut für Datentechnik und Kommunikationsnetze
Sandor	Fekete	Technische Universität Braunschweig, Institute of Operating Systems and Computer Networks

Dietmar	Fey	University of Erlangen-Nürnberg, Department of Computer Science 3
Dominik	Fisch	University of Applied Sciences Deggendorf, Computationally Intelligent Systems Lab
Stefan	Fischer	Universität zu Lübeck, Institute of Telematics
Jörg	Hähner	Leibniz Universität Hannover, Institut für Systems Engineering, System und Rechner Architektur
Hans-Ulrich	Heiß	Technische Universität Berlin, Fakultät für Elektrotechnik und Informatik, Institut für Telekommunikationssysteme
Jörg	Henkel	Karlsruhe Institute of Technology (KIT), Institute of Computer Science and Engineering
Andreas	Herkersdorf	Technische Universität München, Lehrstuhl für Integrierte Systeme
Wolfgang	Karl	Karlsruhe Institute of Technology (KIT), Institute of Computer Science and Engineering
Paul	Kaufmann	University of Paderborn, Faculty for Electrical Engineering, Computer Science and Mathematics, Department of Computer Science
Bernd	Kleinjohann	University of Paderborn, C-LAB
Erik	Maehle	University of Lübeck, Institute of Computer Engineering
Friedhelm	Meyer auf der Heide	Universität Paderborn, Heinz Nixdorf Institute
Martin	Middendorf	Universität Leipzig, Institut für Informatik
Sanaz	Mostaghim	Karlsruhe Institute of Technology (KIT), Institute AIFB
Gero	Mühl	Universität Rostock, Fakultät für Informatik und Elektrotechnik, Institut für Informatik
Christian	Müller-Schloer	Leibniz Universität Hannover, Institut für Systems Engineering, System und Rechner Architektur
Florian	Nafz	Universität Augsburg, Institut für Software und Systems Engineering
Marco	Platzner	University of Paderborn, Faculty for Electrical Engineering, Computer Science and Mathematics, Department of Computer Science

Franz	Rammig	Universität Paderborn, Heinz Nixdorf Institut
Wolfgang	Reif	Universität Augsburg, Institut für Software und Systems Engineering
Wolfgang	Rosenstiel	Eberhard Karls Universität Tübingen, Wilhelm-Schickard Institut für Informatik, Lehrstuhl Technische Informatik
Christian	Schindelhauer	Albert-Ludwigs-Universität Freiburg, Fakultät für Angewandte Wissenschaften, Institut für Informatik, Rechnernetze und Telematik
Hartmut	Schmeck	Karlsruhe Institute of Technology (KIT), Institute AIFB
Hella	Seebach	Universität Augsburg, Institut für Software und Systems Engineering
Bernhard	Sick	University of Applied Sciences Deggendorf, Computationally Intelligent Systems Lab
Florian	Siefert	Universität Augsburg, Institut für Software und Systems Engineering
Walter	Stechele	Technische Universität München, Lehrstuhl für Integrierte Systeme
Jan-Philipp	Steghöfer	Universität Augsburg, Institut für Software und Systems Engineering
Jürgen	Teich	University of Erlangen-Nuremberg, Department of Computer Science 12 (Hardware-Software-Co-Design)
Dirk	Timmermann	Universität Rostock, Institut für Angewandte Mikroelektronik und Datentechnik
Sven	Tomforde	Leibniz Universität Hannover, Institut für Systems Engineering, System und Rechner Architektur
Theo	Ungerer	Universität Augsburg, Institut für Informatik, Lehrstuhl für Systemnahe Informatik und Kommunikationssysteme
Arno	Wacker	Universität Duisburg-Essen, Fakultät für Ingenieurwissenschaften, Fachgebiet Verteilte Systeme
Torben	Weis	Universität Duisburg-Essen, Fakultät für Ingenieurwissenschaften, Fachgebiet Verteilte Systeme
Stefan	Wildermann	University of Erlangen-Nuremberg, Department of Computer Science 12 (Hardware-Software-Co-Design)

Rolf	Würtz	Ruhr-Universität Bochum, Institut für Neuroinformatik
Johannes	Zeppenfeld	Technische Universität München, Lehrstuhl für Integrierte Systeme
Tobias	Ziermann	University of Erlangen-Nuremberg, Department of Computer Science 12 (Hardware-Software-Co-Design)

Projects

Acronym	*Project name* **project leader(s), affiliation(s)**	Chapter reference
ASoC	*Architecture and Design Methodology for Autonomic Systems on Chip* Wolfgang Rosenstiel, Universität Tübingen Andreas Herkersdorf, Technische Universität München	3.2, 4.7, 5.3
AutoNomos	*A Distributed and Self-Regulating Approach for Organizing a Large System of Mobile Objects* Sandor Fekete, Technische Universität Braunschweig Stefan Fischer, Universität zu Lübeck	2.7, 5.2
CHEMORG	*The Bio-Chemical Information Processing Metaphor as a Programming Paradigm for Organic Computing* Peter Dittrich, Friedrich-Schiller-Universität Jena	1.4, 2.6
DodOrg	*Digital On-Demand Computing Organism for Real-Time Systems* Jürgen Becker, Karlsruher Institut für Technologie Uwe Brinkschulte, Universität Frankfurt am Main Jörg Henkel, Karlsruher Institut für Technologie Wolfgang Karl, Karlsruher Institut für Technologie	2.3, 4.3, 4.4
EmergentRadio	*Emergent strategies to optimise collaborative transmission schemes* Michael Beigl, Karlsruher Institut für Technologie	5.4
	Energy Aware Self Organized Communication in Complex Networks Dirk Timmermann, Universität Rostock	5.6

EPOC	*Embedded Performance Analysis for Organic Computing* Rolf Ernst, Technische Universität Braunschweig	4.6
ESLAS	*A Modular Approach for Evolving Societies of Learning Autonomous Systems* Franz J. Rammig, Universität Paderborn Bernd Kleinjohann, Universität Paderborn	3.5, 5.9
Functional Knowledge Fusion	*On-line Fusion of Functional Knowledge within Distributed Sensor Networks* Bernhard Sick, Universität Passau	1.3, 3.3
	Learning to look at humans Rolf P. Würtz, Ruhr-Universität Bochum	3.6
Marching Pixels	*Generic Emergent Computing in Chip Architectures* Dietmar Fey, Friedrich-Schiller-Universität Jena	2.4, 5.10
MODOC	*Model-Driven Development of Self-Organizing Control Applications* Hans-Ulrich Heiß, Technische Universität Berlin Gero Mühl, Universität Rostock Jan Richling, Technische Universität Berlin Arno Wacker, Universität Duisburg-Essen Torben Weis, Universität Duisburg-Essen	2.1
MOVES	*Multi-Objective Intrinsic Evolution of Embedded Systems* Marco Platzner, Universität Paderborn	2.5
OCμ	*Organic Computing Middleware for Ubiquitous Environments* Theo Ungerer, Universität Augsburg Wolfgang Trumler, Siemens AG	4.2
OCCS	*Observation and Control of Collaborative Systems* Hartmut Schmeck, Karlsruher Institut für Technologie Christian Müller-Schloer, Leibniz Universität Hannover Jürgen Branke, University of Warwick Jörg Hähner, Leibniz Universität Hannover	1.1, 1.2, 3.1
ORCA	*Organic Fault-Tolerant Control Architecture for Robotic Applications* Erik Maehle, Universität Lübeck Werner Brockmann, Universität Osnabrück	3.4, 4.5, 5.7

OrganicBus	*Organic Self-organizing Bus-Based Communication Systems* Jürgen Teich, Universität Erlangen-Nürnberg	5.5
OTC	*Organic Traffic Control* Christian Müller-Schloer, Leibniz Universität Hannover Hartmut Schmeck, Karlsruher Institut für Technologie Jürgen Branke, University of Warwick Jörg Hähner, Leibniz Universität Hannover	4.1, 5.1
SAVE ORCA	*Formal Modeling, Safety Analysis, and Verification of Organic Computing Applications* Wolfgang Reif, Universität Augsburg	1.5, 2.2
Smart Teams	*Local Distributed Strategies for Self-Organizing Robotic Exploration Teams* Friedhelm Meyer auf der Heide, Universität Paderborn Christian Schindelhauer, Universität Freiburg	5.8
SOS-TC	*Organisation and Control of Self-Organising Systems in Technical Compounds* Martin Middendorf, Universität Leipzig	1.6

Contributors

Name	Surname	Affiliation
Ahmad	Al-Homsy	University of Lübeck, Institute of Computer Engineering
Florian	Allerding	Karlsruhe Institute of Technology (KIT), Institute AIFB
Gerrit	Anders	Universität Augsburg, Institut für Software und Systems Engineering
Behnam	Banitalebi	Karlsruhe Institute of Technology (KIT), Pervasive Computing Systems
Birger	Becker	Karlsruhe Institute of Technology (KIT), Institute AIFB
Jürgen	Becker	Karlsruhe Institute of Technology (KIT), Institute for Information Processing Technologies
Ralf	Behnke	Universität Rostock, Institut für Angewandte Mikroelektronik und Datentechnik
Michael	Beigl	Karlsruhe Institute of Technology (KIT), Pervasive Computing Systems
Andreas	Bernauer	Eberhard Karls Universität Tübingen, Wilhelm-Schickard Institut für Informatik, Lehrstuhl Technische Informatik
Abdelmajid	Bouajila	Technische Universität München, Lehrstuhl für Integrierte Systeme
Jürgen	Branke	University of Warwick, Warwick Business School
Oliver	Bringmann	Forschungszentrum Informatik Karlsruhe

Uwe	Brinkschulte	J.-W. Goethe Universität Frankfurt, Institut für Eingebettete Systeme
Werner	Brockmann	University of Osnabrück, Institute of Computer Science
Arne	Brutschy	Université Libre de Bruxelles, IRIDIA – CODE
Rainer	Buchty	Karlsruhe Institute of Technology (KIT), Institute of Computer Science and Engineering
Andreas	Buschermöhle	University of Osnabrück, Institute of Computer Science
Emre	Cakar	Leibniz Universität Hannover, Institut für Systems Engineering, System und Rechner Architektur
Bastian	Degener	Universität Paderborn, Heinz Nixdorf Institute
Peter	Dittrich	Friedrich-Schiller-University Jena, Institute of Computer Science, Bio Systems Analysis Group
Konrad	Diwold	Universität Leipzig, Institut für Informatik
Sebastian	Ebers	Universität zu Lübeck, Institute of Telematics
Thomas	Ebi	Karlsruhe Institute of Technology (KIT), Institute of Computer Science and Engineering
Adam	El Sayed Auf	University of Lübeck, Institute of Computer Engineering
Rolf	Ernst	Technische Universität Braunschweig, Institut für Datentechnik und Kommunikationsnetze
Sandor	Fekete	Technische Universität Braunschweig, Institute of Operating Systems and Computer Networks
Dietmar	Fey	University of Erlangen-Nürnberg, Department of Computer Science 3
Dominik	Fisch	University of Applied Sciences Deggendorf, Computationally Intelligent Systems Lab
Stefan	Fischer	Universität zu Lübeck, Institute of Telematics
Nugroho	Fredivianus	Karlsruhe Institute of Technology (KIT), Institute AIFB

Arne	Garbade	Universität Augsburg, Institut für Informatik, Lehrstuhl für Systemnahe Informatik und Kommunikationssysteme
Marcus	Geimer	Karlsruhe Institute of Technology (KIT), Institute AIFB
Karl-Erwin	Großpietsch	University of Lübeck, Institute of Computer Engineering
Mathias	Gutmann	Karlsruhe Institute of Technology (KIT), Institut für Philosophie
Jörg	Hähner	Leibniz Universität Hannover, Institut für Systems Engineering, System und Rechner Architektur
Hans-Ulrich	Heiß	Technische Universität Berlin, Fakultät für Elektrotechnik und Informatik, Institut für Telekommunikationssysteme
Horst	Hellbrück	University of Applied Sciences Lübeck, Electrical Engineering and Computer Science
Björn	Hendriks	Technische Universität Braunschweig, Institute of Operating Systems and Computer Networks
Jörg	Henkel	Karlsruhe Institute of Technology (KIT), Institute of Computer Science and Engineering
Andreas	Herkersdorf	Technische Universität München, Lehrstuhl für Integrierte Systeme
Sebastian	Holzapfel	Universität Duisburg-Essen, Fakultät für Ingenieurwissenschaften, Fachgebiet Verteilte Systeme
Jens	Hülsmann	University of Osnabrück, Institute of Computer Science
Björn	Hurling	Leibniz Universität Hannover, Institut für Systems Engineering, System und Rechner Architektur
Martin	Jänicke	University of Applied Sciences Deggendorf, Computationally Intelligent Systems Lab
Janmartin	Jahn	Karlsruhe Institute of Technology (KIT), Institute of Computer Science and Engineering
Bojan	Jakimovski	University of Lübeck, Institute of Computer Engineering
Alexander	Jungmann	University of Paderborn, C-LAB

Edgar	Kalkowski	University of Applied Sciences Deggendorf, Computationally Intelligent Systems Lab
Wolfgang	Karl	Karlsruhe Institute of Technology (KIT), Institute of Computer Science and Engineering
Paul	Kaufmann	University of Paderborn, Faculty for Electrical Engineering, Computer Science and Mathematics, Department of Computer Science
Timo	Kautzmann	Karlsruhe Institute of Technology (KIT), Institute AIFB
Barbara	Kempkes	Universität Paderborn, Heinz Nixdorf Institute
Rolf	Kiefhaber	Universität Augsburg, Institut für Informatik, Lehrstuhl für Systemnahe Informatik und Kommunikationssysteme
Bernd	Kleinjohann	University of Paderborn, C-LAB
Florian	Kluge	Universität Augsburg, Institut für Informatik, Lehrstuhl für Systemnahe Informatik und Kommunikationssysteme
Marcus	Komann	Friedrich-Schiller-Universität Jena, Institute of Computer Science, Chair for Computer Architecture and Communication
David	Kramer	Karlsruhe Institute of Technology (KIT), Institute of Computer Science and Engineering
Stephan	Krannich	University of Lübeck, Institute of Computer Engineering
Peter	Kreyssig	Friedrich-Schiller-University Jena, Institute of Computer Science, Bio Systems Analysis Group
Marek	Litza	University of Lübeck, Institute of Computer Engineering
Raphael	Maas	University of Lübeck, Institute of Computer Engineering
Erik	Maehle	University of Lübeck, Institute of Computer Engineering
Naoki	Matsumaru	Friedrich-Schiller-University Jena, Institute of Computer Science, Bio Systems Analysis Group
Daniel	Merkle	University of Southern Denmark, Department of Mathematics & Computer Science

Friedhelm	Meyer auf der Heide	Universität Paderborn, Heinz Nixdorf Institute
Martin	Middendorf	Universität Leipzig, Institut für Informatik
Jörg	Mische	Universität Augsburg, Institut für Informatik, Lehrstuhl für Systemnahe Informatik und Kommunikationssysteme
Moez	Mnif	Leibniz Universität Hannover, Institut für Systems Engineering, System und Rechner Architektur
Sanaz	Mostaghim	Karlsruhe Institute of Technology (KIT), Institute AIFB
Gero	Mühl	Universität Rostock, Fakultät für Informatik und Elektrotechnik, Institut für Informatik
Christian	Müller-Schloer	Leibniz Universität Hannover, Institut für Systems Engineering, System und Rechner Architektur
Florian	Nafz	Universität Augsburg, Institut für Software und Systems Engineering
Moritz	Neukirchner	Technische Universität Braunschweig, Institut für Datentechnik und Kommunikationsnetze
Mathias	Pacher	J.-W. Goethe Universität Frankfurt, Institut für Eingebettete Systeme
Helge	Parzyjegla	Universität Rostock, Fakultät für Informatik und Elektrotechnik, Institut für Informatik
Benjamin	Pfundt	University of Erlangen-Nürnberg, Department of Computer Science 3
Marco	Platzner	University of Paderborn, Faculty for Electrical Engineering, Computer Science and Mathematics, Department of Computer Science
Franz	Rammig	Universität Paderborn, Heinz Nixdorf Institut
Benjamin	Rathgeber	Karlsruhe Institute of Technology (KIT), Institut für Philosophie
Marc	Reichenbach	University of Erlangen-Nürnberg, Department of Computer Science 3
Wolfgang	Reif	Universität Augsburg, Institut für Software und Systems Engineering
Willi	Richert	University of Paderborn, C-LAB

Jan	Richling	Technische Universität Berlin, Fakultät für Elektrotechnik und Informatik, Institut für Telekommunikationssysteme
Urban	Richter	Karlsruhe Institute of Technology (KIT), Institute AIFB
Nils	Rosemann	University of Osnabrück, Institute of Computer Science
Wolfgang	Rosenstiel	Eberhard Karls Universität Tübingen, Wilhelm-Schickard Institut für Informatik, Lehrstuhl Technische Informatik
Michael	Roth	Universität Augsburg, Institut für Informatik, Lehrstuhl für Systemnahe Informatik und Kommunikationssysteme
Jakob	Salzmann	Universität Rostock, Institut für Angewandte Mikroelektronik und Datentechnik
Alexander	Scheidler	Université Libre de Bruxelles, IRIDIA – CODE
Julia	Schmitt	Universität Augsburg, Institut für Informatik, Lehrstuhl für Systemnahe Informatik und Kommunikationssysteme
Christian	Schindelhauer	Albert-Ludwigs-Universität Freiburg, Fakultät für Angewandte Wissenschaften, Institut für Informatik, Rechnernetze und Telematik
Sebastian	Schlingmann	Universität Augsburg, Institut für Informatik, Lehrstuhl für Systemnahe Informatik und Kommunikationssysteme
Hartmut	Schmeck	Karlsruhe Institute of Technology (KIT), Institute AIFB
Arnd	Schröter	Technische Universität Berlin, Fakultät für Elektrotechnik und Informatik, Institut für Telekommunikationssysteme
Christian	Schuck	Karlsruhe Institute of Technology (KIT), Institute for Information Processing Technologies
Hella	Seebach	Universität Augsburg, Institut für Software und Systems Engineering
Enrico	Seib	Universität Rostock, Fakultät für Informatik und Elektrotechnik, Institut für Informatik
Ralf	Seidler	University of Erlangen-Nürnberg, Department of Computer Science 3

Bernhard	Sick	University of Applied Sciences Deggendorf, Computationally Intelligent Systems Lab
Florian	Siefert	Universität Augsburg, Institut für Software und Systems Engineering
Stephan	Sigg	Karlsruhe Institute of Technology (KIT), Pervasive Computing Systems
Walter	Stechele	Technische Universität München, Lehrstuhl für Integrierte Systeme
Jan-Philipp	Steghöfer	Universität Augsburg, Institut für Software und Systems Engineering
Steffen	Stein	Technische Universität Braunschweig, Institut für Datentechnik und Kommunikationsnetze
Tareq	Syed	Karlsruhe Institute of Technology (KIT), Institut für Philosophie
Jürgen	Teich	University of Erlangen-Nuremberg, Department of Computer Science 12 (Hardware-Software-Co-Design)
Christopher	Tessars	Technische Universität Braunschweig, Institute of Operating Systems and Computer Networks
Dirk	Timmermann	Universität Rostock, Institut für Angewandte Mikroelektronik und Datentechnik
Sven	Tomforde	Leibniz Universität Hannover, Institut für Systems Engineering, System und Rechner Architektur
Theo	Ungerer	Universität Augsburg, Institut für Informatik, Lehrstuhl für Systemnahe Informatik und Kommunikationssysteme
Alexander	von Renteln	J.-W. Goethe Universität Frankfurt, Institut für Eingebettete Systeme
Arno	Wacker	Universität Duisburg-Essen, Fakultät für Ingenieurwissenschaften, Fachgebiet Verteilte Systeme
Thomas	Walther	Ruhr-Universität Bochum, Institut für Neuroinformatik
Matthäus	Wander	Universität Duisburg-Essen, Fakultät für Ingenieurwissenschaften, Fachgebiet Verteilte Systeme
Axel	Wegener	Universität zu Lübeck, Institute of Telematics

Sebastian	Weis	Universität Augsburg, Institut für Informatik, Lehrstuhl für Systemnahe Informatik und Kommunikationssysteme
Torben	Weis	Universität Duisburg-Essen, Fakultät für Ingenieurwissenschaften, Fachgebiet Verteilte Systeme
Mathias	Werner	Technische Universität Chemnitz, Fakultät für Informatik, Professur Betriebssysteme
Stefan	Wildermann	University of Erlangen-Nuremberg, Department of Computer Science 12 (Hardware-Software-Co-Design)
Michael	Wittke	Leibniz Universität Hannover, Institut für Systems Engineering, System und Rechner Architektur
Micaela	Wünsche	Karlsruhe Institute of Technology (KIT), Institute AIFB
Rolf	Würtz	Ruhr-Universität Bochum, Institut für Neuroinformatik
Johannes	Zeppenfeld	Technische Universität München, Lehrstuhl für Integrierte Systeme
Tobias	Ziermann	University of Erlangen-Nuremberg, Department of Computer Science 12 (Hardware-Software-Co-Design)

Chapter 1
Theoretical Foundations

Starting out into a visionary direction of research requires a solid foundation. Organic Computing claims to reduce the risks of operating multitudes of intelligent devices in environments with some intrinsic uncertainty about potential changes with respect to behavioural requirements. The goal of an engineer is the design of technical systems that will show a dependable behaviour in compliance with specific requirements. The concepts of self-organisation and evolution in response to environmental effects have to be looked at from a fundamental point of view. Therefore, fundamental research has been an essential part of this Priority Programme (SPP). In particular, the call for proposals for the first phase of the programme stated "Insights about the behaviour of natural and artificial complex systems shall be used to open up the necessary degrees of freedom for self-organised behaviour of technical systems. This requires projects on the theory of complex systems, investigating the possibilities to establish goal-oriented emergent processes and, in particular, looking at the problems of security and stability of self-organised technical systems. Methods are needed for controlling emergent behaviour, i.e. to direct a system towards desired emergent behaviour and to detect and prevent undesirable emergence."

Several projects of the SPP have addressed these fundamental aspects. Selected results of this research are presented in the contributions of this chapter. Chapter 1.1 is a reprint from a journal (ACM Transactions on Autonomous and Adaptive Systems, vol. 5, no. 3, 2010) where H. Schmeck et al. provide a characterisation of essential properties of Organic Computing systems which is a prerequisite for arguing about the achievements of Organic Computing in comparison to classical systems engineering. *Adaptivity* and *self-organisation* as well as *robustness* and *flexibility* deal with the response of a system to certain changes or disturbances that might occur within the system or in its environment. To argue about self-organisation the authors introduce a *degree of autonomy* resembling the relative amount of external control compared to internal control actions. This is used to characterise the notion of *controlled self-organisation* as the ability of a system to run safely without external control over long periods of time while allowing for external interference whenever necessary.

The decision whether to interfere with a *System under Observation and Control* depends on a continuous observation and analysis of the current system status with respect to the occurrence of emergent effects due to local interactions of the system components. In Chap. 1.2 (reprinted from the proceedings of the 2006 IEEE Mountain Workshop on Adaptive and Learning Systems), M. Mnif and C. Müller-Schloer introduce the notion of "Quantitative Emergence" as a measurement of emergent effects by looking at the entropy values of essential system parameters. Significant changes of these entropy values indicate potential changes of the degree of order in the system which are typically associated with emergent phenomena. This is extended in Chap. 1.3 by D. Fisch et al. who present "Divergence measures as a generalised approach to quantitative emergence". Here, emergence is regarded as an unexpected or unpredictable change of the distribution underlying the observed samples which is measured by assessing the divergence between the density functions underlying the different observations.

While many authors consider emergent phenomena as being mainly indicators of unexpected global effects due to local interactions, P. Dittrich and P. Kreyssig consider "Emergent Control" in Chap. 1.4 as an alternative strategy of controlling large scale systems. That means they consider the explicit design of local behaviour which should allow satisfying the global requirements as an effect of emergence, in contrast to the strategy of feedback control which is continually influencing the system based on some global view of the system behaviour. Their approach is motivated by their research on chemical programming and chemical organisation theory.

The control of Organic Computing systems is also the topic of Chap. 1.5 by F. Nafz et al. They propose a "Restore-invariant approach" for "Constraining self-organisation through corridors of correct behaviour". The invariants are specifying corridors of acceptable behaviour which are closely related to the notion of "acceptance space" introduced in Chap. 1.1. This is combined with an approach for a self-organised restoration of invariants in case of violations by forming coalitions for local reconfiguration.

Whereas all of these contributions are considering the design and operation of technical systems, A. Scheidler et al. analyse the behaviour of societies of ants as a source of inspiration for designing Organic Computing systems. In Chap. 1.6 they are presenting a self-organised task distribution system for mobile agents inspired by the spatial distribution of work in ant societies, transfer the behaviour of certain house-hunting ants into collective decision-making in Organic Computing systems and finally show how to exploit pheromones in the design of a self-organised production network that automatically sorts items with respect to their types while they are moving through the network.

The chapter is concluded with a contribution from a different discipline: In Chap. 1.7, M. Gutmann et al. assess the concepts of Organic Computing from a philosophical point of view. They discuss the phenomenological meaning of essential notions connected to Organic Computing. In particular, they see Evolutionary Robotics as closely related to Organic Computing and distinguish the evolutionary from the engineering paradigm in robotic design, they analyse the role of self-organisation and conclude that Organic Computing extends Evolutionary Robotics by connecting the evolutionary with the self-organisation paradigm, thus providing a new model-theoretic perspective and a promising framework for engineering practice.

The contributions of this chapter clearly show that the projects of the SPP on Organic Computing have succeeded in getting some fundamental insights into the effects of self-organisation as observable in nature and in technical systems. This has led to several different generic approaches to the design and operation of Organic Computing systems as addressed in more detail in the remaining chapters of this volume. Nevertheless, although M. Gutmann concludes that Organic Computing provides a new model-theoretic perspective and a promising framework for engineering practice, there is still a need for exploring the relationship between local interactions and their resulting global effects by further fundamental research.

Hartmut Schmeck
Institute AIFB, Karlsruhe Institute of Technology (KIT), Bldg. 05.20,
76128 Karlsruhe, Germany
email: hartmut.schmeck@kit.edu

Chapter 1.1
Adaptivity and Self-organisation in Organic Computing Systems

Hartmut Schmeck, Christian Müller-Schloer, Emre Çakar, Moez Mnif, and Urban Richter

Abstract Organic Computing (OC) and other research initiatives like Autonomic Computing or Proactive Computing have developed the vision of systems possessing life-like properties: they self-organise, adapt to their dynamically changing environments, and establish other so-called self-x properties, like self-healing, self-configuration, self-optimisation, etc. What we are searching for in OC are methodologies and concepts for systems that allow to cope with increasingly complex networked application systems by introduction of self-x properties and at the same time guarantee a trustworthy and adaptive response to externally provided system objectives and control actions. Therefore, in OC we talk about *controlled self-organisation.*

Although the terms *self-organisation* and *adaptivity* have been discussed for years, we miss a clear definition of self-organisation in most publications, which have a technically motivated background.

In this article, we briefly summarise the state of the art and suggest a characterisation of (controlled) self-organisation and adaptivity that is motivated by the main objectives of the OC initiative. We present a system classification of robust, adaptable, and adaptive systems and define a degree of autonomy to be able to quantify how autonomously a system is working. The degree of autonomy distinguishes and measures external control which is exerted directly by the user (*no autonomy*) from internal control of a system which might be fully controlled by an Observer/Controller architecture that is part of the system (*full autonomy*). The quantitative degree of autonomy provides the basis for characterising the notion of controlled self-organisation. Furthermore, we discuss several alternatives for the design of organic systems.

Keywords Adaptation · Adaptivity · Observer/controller architecture · Robustness · Flexibility · Self-organisation · System classification · Configuration space

---

C. Müller-Schloer et al. (eds.), *Organic Computing — A Paradigm Shift for Complex Systems*, 5–37
Autonomic Systems, DOI 10.1007/978-3-0348-0130-0_1, © Springer Basel AG 2011

1 Introduction

Today, we are struck by the trend of increasing complexity in the design, development, and maintenance of technical systems. Organic Computing (OC), like other initiatives such as IBM's Autonomic Computing [17] or Proactive Computing [40], postulates the necessity of a paradigm shift in the design of future technical applications, see e.g. [32]. As stated there:

> *"It is not the question whether self-organised and adaptive systems will arise but how they will be designed and controlled."*

Obviously, the classical top-down design process has attained its limits. At the same time a full bottom-up design seems to be not feasible. As often, the solution has to be in the middle: On the one hand we have to endow our systems with more degrees of freedom allowing them to self-organise in order to be able to *adapt* to potentially changing environmental conditions and external goals and constraints, and on the other hand we have to make sure that self-organisation does not result in unanticipated and undesired behaviour. Therefore, we are not interested in fully autonomous systems. At last instance, human beings should always have the control over the technical systems in their vicinity.

Hence, in OC we search for concepts to achieve *controlled self-organisation* as a new design paradigm. At first glance this seems to be a paradox, but it is necessary to cope in an acceptable way with the degrees of freedom required by the process of self-organisation.

Endowing technical systems with the property of self-organisation may result in a—sometimes unanticipated—emergent behaviour exhibiting a certain degree of order. These phenomena are well known from nature (e.g. a flock of birds, the foraging behaviour of ants, or a shoal of fish). The effects of emergence can be positive as well as negative. In [20, 23] emergence is defined as *self-organised order* based on the assumption that it can be observed in terms of patterns in space and/or over time. They use a quantitative measure of emergence based on a decrease of descriptional complexity (i.e. entropy) caused by a self-organising process. While this definition proves to be practically usable in technical systems in order to quantify the emergence of higher degrees of order, it does not suffice to characterise the essential attributes and mechanisms of (controlled) self-organisation and adaptivity.

Within the German Priority Programme on OC [7] a generic architectural concept for the design and analysis of OC systems has been developed, the observer/controller architecture [2]. The self-organising *system under observation and control* (SuOC), which constitutes some productive system, will be endowed with a higher level of governance consisting of an *observer* and a *controller* (cf. Fig. 1). The observer monitors the underlying system by sampling the state and the properties of the different components and reports an aggregated quantified context (i.e. a description of the currently observed situation) to the controller. The controller evaluates this context with respect to a given objective function and takes appropriate control actions whenever it is necessary to influence the underlying system to meet the system goal. This loop of observing and controlling has to guarantee that the behaviour

Fig. 1 Observer/Controller architecture

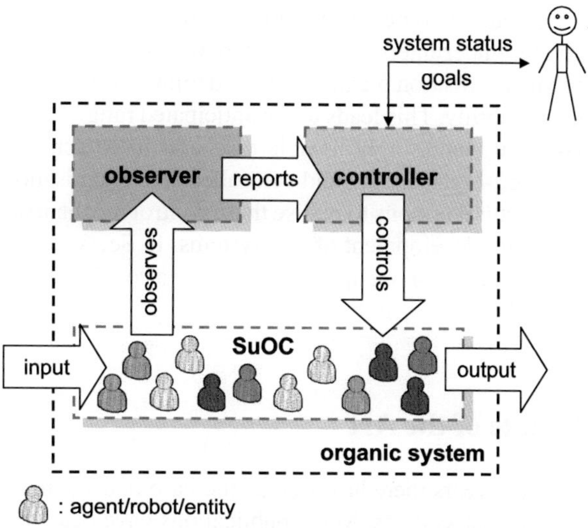

of the SuOC stays within the external constraints. In particular, explicit control actions should not be necessary unless a deviation from the desired behaviour has been detected or predicted, i.e. the SuOC would run autonomously as long as it behaves well and satisfies the requirements as specified by the developer or the user.

This generic Observer/Controller architecture bears similarities with concepts and techniques known from other scientific disciplines like control theory, mechanical engineering [13, 25], or autonomic computing [17, 38]. However, OC emphasises the crucial difference from these disciplines that the resulting architecture is not a fully autonomous element. Rather, it is driven by external goals, and it reports its system status to the user (or to some higher level object). The user may intervene explicitly, e.g. by changing some system objectives or by initiating directly some control actions. Furthermore, the underlying technical system need not have a rigid structure. Instead, it will consist of many entities interacting with each other and changing their behaviour and interconnection structure dynamically in response to influences from its environment.

An abstract view of the generic Observer/Controller loop will serve as our model for characterising the notions of adaptivity and self-organisation as a basis for defining controlled self-organisation in technical application systems. Our goal is to get significantly beyond the frequently followed attitude of using the terms self-organisation and adaptivity in an almost magical sense based on some black box model. Despite of the long-term existence of these terms, in the context of technical applications based on advanced information processing systems we still lack a precise quantitative definition as a basis for a common understanding of these crucial concepts.

In this article we will present some ideas towards this end. Section 2 summarises some previous related work concerning the technical usage and understanding of self-organisation. The essential part of a definition of self-organisation lies in the term *self*. In Sect. 3 we illustrate this using a simple example. Section 4 provides definitions of some basic terms describing a system aiming at a certain target behaviour under internal and/or external control and influenced by environmental disturbances. Section 5 extends these definitions with the terms robustness, adaptivity, and flexibility. This leads to the anticipated range of system classifications in Sect. 6 based on increasingly flexible mechanisms to achieve different degrees of autonomy, self-organisation, and controlled self-organisation. Section 7 introduces some architectural options to realise these control mechanisms and outlines a roadmap for the future development of OC systems. In Sects. 8 and 9 we discuss some remaining limitations of our approach and give a conclusion including an outlook to further research.

2 State of the Art

In recent years there has been an increase in academic research in self-organisation of technical systems. Many publications have dealt with the term self-organisation and its formalisation by using for example entropy-based approaches.

A mathematical model based on Shannon's entropy is proposed by Shalizi et al. in [34, 35]. In this context, the self-organisation process is characterised as the increase in the amount of information needed for predicting the system dynamics in the future. The method claims that the increase in statistical complexity of the system without external intervention points to the increase of self-organisation in this system.

Also, Heylighen argues that the statistical entropy can be used in determining the degree of self-organisation of a given system (cf. [14, 15]). He defines a system sitting in an attractor in state space as self-organised, since the system can no longer reach other states outside the attractor, which leads to a decrease in statistical entropy (uncertainty) and so to an increase in order (self-organisation). In addition, in [11] Gershenson and Heylighen discuss the importance of an observer in determining the degree of self-organisation using the statistical entropy.

Similarly, in [26], Parunack and Brueckner discuss an entropy model for self-organisation based on the Kugler-Turvey model, which suggests that the key to reduce disorder in a system is to couple it to another one in which the disorder increases. The proposed model is demonstrated on a system, which is composed of a macro-level coupled with the micro-level. Information entropy is used to determine the degree of self-organisation on each level.

In [42, 43] Wright proposes a method to measure self-organisation, which is based on the property that the entropy can be viewed as a function of the attractor's dimension in the state space. In this context, a self-organised system is defined to have an attractor with an intermediate dimension value.

The notion of *organisation information* is proposed in [27], which is also an information theoretical characterisation of self-organisation. In this context, a system is called self-organising, if organisation information increases during its dynamics.

Another approach to quantify self-organisation without using the notion of entropy is proposed by Lucas [19]. He introduces the notion of a self-organisation factor, which is the ratio of the non-attractor volume to the attractor volume in state space. He claims that the self-organisation factor can be used to measure the degree of self-organisation present in the system.

In [22] Mühl et al. discuss the formal classification of self-organised technical systems. They propose a *"classification of systems that models self-organising systems as a subclass of self-managing systems. The classification builds upon a definition of adaptive systems"* based on a concept of adaptivity by Zadeh [44] and defines a hierarchy of self-manageable, self-managing, and self-organising systems. Following their definition a system can be called self-organising *"if it is (i) self-managing [the system adapts to its environment without outside control], (ii) structure-adaptive [the system establishes and maintains a certain kind of structure (e.g. spatial, temporal), providing the system's primary functionality], and (iii) employs decentralised control [the system has no central point of failure]."* Following this definition strictly, OC systems could not be called self-organising, since one of the essential requirements of OC is to leave at least the opportunity for external intervention, using the generic Observer/Controller paradigm (cf. Fig. 1) which should always allow for controlled self-organisation. Nevertheless, the classification by Mühl et al. provides valuable insights. Its major shortcoming is its purely qualitative nature, which does not adequately grasp the necessity of at least sporadic external intervention.

Interesting in our context is the view of DeWolf and Holvoet who propose the possibility of self-organisation without emergence as well as vice versa: In [5] De Wolf and Holvoet discuss the relationship of emergence and self-organisation. They introduce the noun *emergent* for the result of a process—in contrast to the process itself—which leads to a certain macroscopic pattern: "A system exhibits emergence when there are coherent emergents at the macro-level that dynamically arise from the interactions between the parts at the micro-level. Such emergents are novel w.r.t. the individual parts of the system. Properties, behaviour, structure, or patterns, for instance, can be emergents." Self-organisation is defined as "... a dynamical and adaptive process where systems acquire and maintain structure themselves, without external control." Structure can be spatial, temporal, or functional. Self-organisation and emergence are seen as emphasising different characteristics of a system. Both can, according to the authors, exist in isolation or together. From an OC viewpoint, these two concepts seem to be highly related: In intelligent technical systems, emergence typically is the result of an adaptive and self-organising (in the broadest sense) process with many components. Self-organisation without emergent behaviour is possible, e.g. as the result of a predefined interaction of only a few components. In [6], De Wolf et al. describe an industrial application example of a self-organising emergent system—an automated guided vehicle warehouse transportation system—where (amongst others) an entropy measure is applied to measure the system-wide behaviour of a self-organising emergent system. In our view, self-organisation is possible without emergence if only a small number of components is involved. On the other hand, emergence as a macroscopic effect always depends on some microscopic mechanism without external control, i.e. on self-organisation.

Most of the cited literature provides only a basic understanding whether a system can be called self-organised. Also, there is a lack of architectural patterns for developing self-organised technical systems. Of course, to be able to exploit the advantages of self-organisation in technical systems, we need an appropriate formal model. But in addition to this we also need concepts and strategies to leave appropriate degrees of freedom for self-organised system behaviour and to cope with systems that can adapt their structure and behaviour in a self-organised way.

3 Is It Self-organising or Not?

For illustration purposes we use as an example a so-called *ordering game* to explain our intuition about self-organisation. The ordering game is based on a simple system, which consists of a group of sticks. In spite of their simple nature, these sticks can be viewed as agents, having some properties, e.g. length and colour (cf. Fig. 2(a)), and a local knowledge about their environment. For initialisation, the sticks and their properties are arranged randomly. The goal of this game is to have the agents rearranged with respect to some objectives resulting in different global patterns. For example, these objectives could be:

(O_1) Have the agents arranged in ascending or descending order from left to right with respect to their lengths (cf. Fig. 2(b) and 2(c)).

(O_2) Have the agents arranged into clusters with the same colour (cf. Fig. 2(d)).

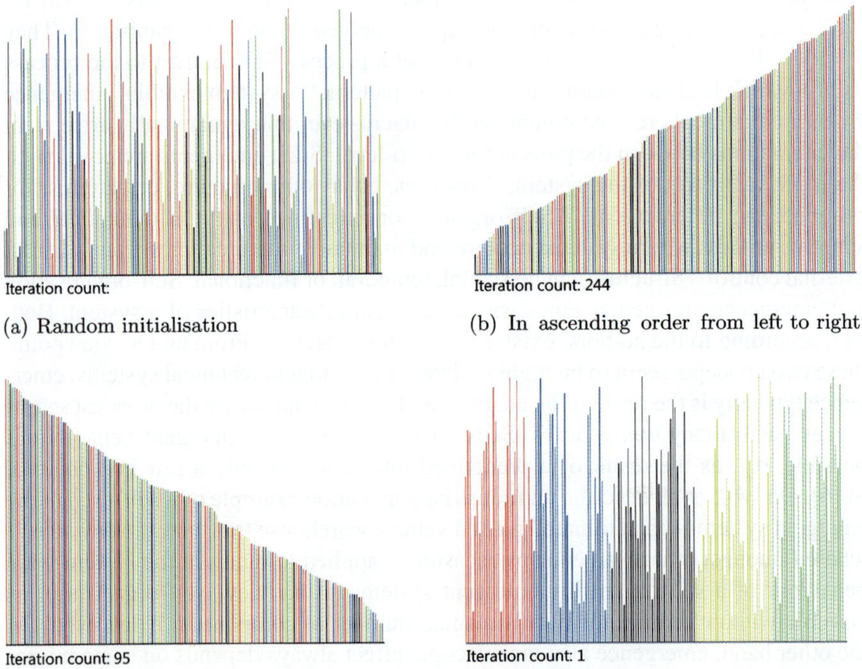

(a) Random initialisation

(b) In ascending order from left to right

(c) In descending order from left to right

(d) Arranged into coloured clusters

Fig. 2 Ordering game with different objective functions

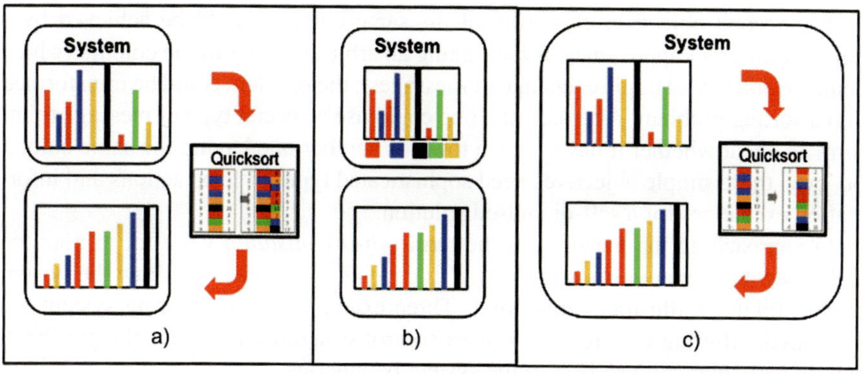

Fig. 3 Example: Is it self-organising? (**a**) No self-organisation because the reorganisation is accomplished by a sorter outside the system. (**b**) Strong self-organisation due to active agents acting autonomously and without central control. (**c**) The sorter is defined to be part of the system. There is no outside control, but the inside control is centralised. We call it (weakly) self-organised

These objectives are easily accomplished with a centralised approach to control. In this case, we could use an external global observer, which collects the relevant information about each agent (position, length, and colour) and sends it to an external global controller. The controller could use any standard algorithm to achieve the given global goal, e.g. quicksort for sorting the agents into ascending/descending order. Then it could advise the agents to move to their new positions. Hence, all the *intelligence* resides on the global level (see Fig. 3(a)). This obviously precludes a self-organisation process, because the agents do not act *autonomously*. Self-organisation or autonomic behaviour would mean that the system should manage to satisfy the objectives without or with a minimum of external intervention. This requires the definition of a local objective function on the agent level, i.e. the conversion of the global objective into a local one (or into many, possibly different, local objectives). In our example, this means to define local rules for each agent so that they can self-organise into ascending/descending order with respect to their lengths or into clusters with respect to their colours.

For the objective (O_1), it is clear how to define these rules: Each agent would determine the length of its right-hand neighbour and decide with respect to the given objective (ascending/descending), whether they have to exchange their positions. Obviously, this leads to a synchronisation issue, which must be solved locally by the agents. The resulting method would resemble an asynchronous version of the well-known *odd-even transposition sort* [18]. In Fig. 3(b) these local actions are indicated by coloured boxes below the sticks. Every agent has its local strategy/rule set and global behaviour emerges from distributed interactions. One should note that the choice of the distributed sorting algorithm is closely related to the interconnection structure (or to the communication capabilities) of the agent system: More sophisticated and much faster algorithms like odd-even merge sort or bitonic merge sort would require non-local comparison-exchange operations leading to corresponding communication requirements.

The second objective, clustering of the same colours, could be achieved by a range of different approaches. By assigning an arbitrary order to the colours (which could be done by a simple exchange of messages), the problem could be transformed into a sorting problem. Alternatively, one could use a special type of message to inform an agent whether it has to move towards another one having the same colour. So, even these simple objectives need sophisticated types of local actions and information exchanges for a self-organised solution.

In approaching the meaning of the term *self-organisation* we want to note the well-known fact that the meaning of *self* is dependent on where we draw the system border. Figure 3 illustrates this point. Three examples of an ordering system are compared, differing with respect to their control structures, the internal algorithms, and, in particular, with respect to the system boundaries.

In the first example (cf. Fig. 3(a)), the sorter is not part of the system and acts outside of the system as an external controller. Therefore, the system is certainly not self-organising. However, it could be called self-organising (or, to act autonomously), if the Observer/Controller were defined to be a part of the system (cf. Fig. 3(c)). In this case, an external observer (outside the system) would see that the system achieves a goal without any outside intervention (except for transmitting the global objective). In Sect. 6.6 this type of autonomous behaviour with central control is called *weak self-organisation*. Finally (Fig. 3(b)), the situation where the sorting or clustering processes are carried out without the assistance of a central control, truly resembles autonomous or strongly self-organised behaviour as defined in Sect. 6.6.

Hence, systems can be self-organising (i.e. no external intervention), extrinsically organised (all control is exerted from the outside), or somewhere in between (i.e. a certain amount of control is exerted from the outside, some control from the inside). In order to determine this *somewhere* the kind of self-organisation must be characterised quantitatively to enable a systematic investigation of various design patterns. Furthermore, in the following discussion of self-organisation, we separate *self* from *organisation*, since these are two different aspects. The former leads to the definition of a *degree of autonomy*, which is determined by the ratio of outside vs. inside control. The latter has to do with a characterisation of different structural properties of the system. In the following we elaborate on these ideas in more detail. This quantitative characterisation of the autonomy of a system is a major contribution of this article.

4 System Description

We start with a rather abstract description of a system that will be used for illustrating our concepts. It is guided by our generic Observer/Controller architecture that is depicted in Fig. 1. The productive system or system under observation and control (SuOC) is called here simply S. It is assumed to transform an input $\mathbf{x}$ into an output $\mathbf{y}$ (cf. Fig. 4). The domains of $\mathbf{x}$ and $\mathbf{y}$ are not specified explicitly, since their nature is not essential for the purpose of this article. $\mathbf{x}$ contains all the input parameters of

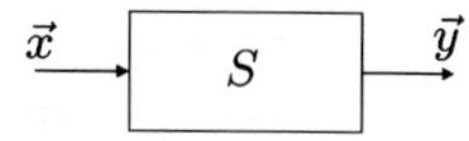

Fig. 4 System description

the intended functional behaviour of S. $\mathbf{y}$ contains all the output parameters that are controlled by S. Furthermore, we assume that $\mathbf{x}$ and $\mathbf{y}$ are accessible to an external observer (the *user* of the system). From the viewpoint of Organic Computing, S is the SuOC [30].

$\mathbf{x}$ and $\mathbf{y}$ are input and output vectors of the productive system S which has to react on a short time scale to changes of $\mathbf{x}$. A feedback-loop (see also Figs. 7 and 8) influences this behaviour by subtracting (negative feedback) the actual output value from the desired output value $\mathbf{y}_\Omega$ deriving a modified input $\mathbf{u}$ in order to minimise the difference between $\mathbf{y}$ and $\mathbf{y}_\Omega$. But, for a system S to become adaptable it has to provide an additional control input $\mathbf{c}$. This will be discussed in more detail in Sect. 6. For the time being, let us start with a system S with just an input vector $\mathbf{x}$ and an output vector $\mathbf{y}$. S is assumed to be a structured system, consisting of (many) elements and links, which interconnect these elements. Therefore, the *system structure* is viewed to be an attributed graph, the nodes of which are the elements, and its edges correspond to the links between the elements. The *attributes* are associated to the system (*global attributes*) or to the edges and elements (*local attributes*) giving information like location, performance, storage capacity, communication bandwidth, reliability etc. A *re-organisation* of S adds or removes nodes and/or edges from the graph or modifies the attributes.

In addition, an adequate system model needs some notion of an *environment*. This will subsume everything that is external to S, including input and output of S, and certain parameters (or attributes) which might have some (possibly disturbing) effect on S.

Summarising, there is a multitude of parameters and attributes which have to be included in an adequate description of a system and its behaviour. Although one could argue extensively about the distinction between a system, its environment and its inputs and outputs (see e.g. [41]), we are summarising all of them into the *state* of a system, i.e. we are using a (simplified) state space notion to argue about system behaviour.

The following aspects will be used to describe a system S with respect to its behaviour under varying environmental conditions:

1. At any given time t, the system is in a *state* $\mathbf{z}(t)$. If there are n attributes used to describe the state of S, $\mathbf{z}(t)$ is a vector in n-dimensional state space Z.
2. The performance of the system is evaluated with respect to some *evaluation criteria* (also called *objectives*) $\eta_1, \ldots, \eta_k$ mapping the system state into the set of real numbers. For simplicity, we assume that in an *ideal state* all the criteria would evaluate to zero. This set of ideal states (also called *target space*) is denoted by Z_Ω. Typical examples of such evaluation criteria could be
   (i) (for the ordering game example with increasing order): "Let r denote the number of local misorderings—i.e. the number of locations i where the length of stick no. $i+1$ is smaller than the length of stick no. i; then reduce r to zero!"

(ii) (for a scenario with distributed workers having to work on a number of tasks): "Keep the differences in the number of tasks per worker as small as possible."

(iii) (for a scenario where a vehicle has to travel from location A to location B): "Let t_{travel} be the travel time from A to B and t_{opt} the optimal realistic travel time, then minimise $(t_{travel} - t_{opt})$." An additional objective conflicting with the first one could be to minimise the fuel consumption of the vehicle. In such a multi-objective situation, it will not be possible to get optimal values for each objective but to decide for appropriate pareto-optimal parameter settings.

These examples show that the target space may be quite large (although, in some cases, it might be a unique state $\mathbf{z}_{\Omega}$). Furthermore, the evaluation might involve rather complex operations.

3. There exist internal or external *control mechanisms CM*, which allow to control the behaviour of the system by setting some attributes of the system and of its environment to specific values. Such a control mechanism will evaluate the current situation of the system and initiate adequate control actions. For example, in the ordering game, the control mechanism has to initiate a sorting process whenever necessary. Quite often, it is assumed that it is not possible to control environmental parameters. But, as e.g. in traffic control, a speed limit could be viewed to be an environmental parameter which might be modified by some external control unit. As mentioned in Sect. 1, in the generic architecture of organic systems, the control mechanism *CM* will consist of an observer and a controller having a number of standard components as described in detail in [2]. Although, for the purposes of this article, we are usually referring to the control mechanism *CM* as a central entity, in reality, it might be distributed or have a multi-level structure (as e.g. in the autonomous version of the ordering game where every stick needs a local *CM* for performing the comparison-exchange operations).
4. The system state $\mathbf{z}(t)$ is called *acceptable* if an *acceptance criterion* or *threshold* θ on $\mathbb{R}^k$ is satisfied, i.e. $\theta(\eta_1(\mathbf{z}(t)), \ldots, \eta_k(\mathbf{z}(t)))$ is true. As a special case, θ might refer to a vector of threshold values, which should not be violated (i.e. surpassed) by the evaluation criteria. The set of all acceptable states is also called the *acceptance space*. Obviously, the target space is a subset of the acceptance space. In our example scenarios, one might tolerate a certain degree of disorder before it is necessary to sort again, or (in the travel scenario) the driving performance could be acceptable as long as a certain upper limit on fuel consumption is not exceeded.
5. The system might be disturbed by some external or internal influences. Such a *disturbance* δ changes the state $\mathbf{z}(t)$ into some state $\delta(\mathbf{z}(t))$ and, consequently, the criteria values of the system change from $\boldsymbol{\eta}$ $(= (\eta_1(\mathbf{z}(t)), \ldots, \eta_k(\mathbf{z}(t))))$ to $\boldsymbol{\eta} + \mathbf{d}$ with $\eta_i(\mathbf{z}(t)) + d_i = \eta_i(\delta(\mathbf{z}(t)))$. Such a disturbance could be the sudden relocation of a stick (in the ordering game) or a failure of some component (e.g. in the travel scenario, a vehicle could get a flat tire or a road might be blocked by an accident). This notion of "disturbance" will also include changes in the evaluation criteria, like e.g. a change from an ascending to a descending ordering objective or from sorting w.r.t. length to grouping w.r.t. colour.

In the following sections we shall characterise certain crucial properties of organic (respectively self-organising and or adaptive) systems based on the notions outlined in this section.

5 Robustness and Adaptivity

An essential requirement on system performance is the capability to adapt reasonably to changes in environmental parameters or in system objectives. In particular, a common objective is the capability to maintain a required behaviour or functionality in spite of a certain range of parameter variations. The standard notion for this is *robustness*. Different from this, the requirement to modify the behaviour because of certain changes of parameter values or of objectives would correspond to the notion of *flexibility*. Obviously, both notions crucially depend on the *adaptivity* of the system.

Robustness has different meanings depending on the context. Typical definitions include *the ability of a system to maintain its functionality even in the presence of changes in their internal structure or external environment* [3], or *the degree to which a system is insensitive to effects that have not been explicitly considered in the design* [37]. In engineering, robust design generally means that the design is capable of functioning correctly, (or, at the very minimum, not failing completely) under a large range of conditions. It is also often related to manufacturing tolerances, and the corresponding literature is immense, see e.g., the works by Taguchi [39]. In scheduling, robustness of a plan generally means that it can be executed and will lead to satisfying results despite changes in the environment [33], while in computing, robustness is often associated with fault tolerance (e.g., [16]). In the context of multi-agent systems, the notion of robustness is of key importance. Therefore, it has been and still is an active area of research, as can be seen, for example, at the AAMAS series of conferences and associated workshops. An approach based on transactions is advocated by Nimis and Lockemann in [24].

Analogous to robustness, the term *flexibility* has manifold meanings in different domains and is measured according to the aspects that are specific to that application domain. Various metrics are proposed to measure the flexibility of manufacturing systems [1, 12, 36], programming paradigms, architecture styles, and design patterns in the area of software engineering [10] or different reconfigurable hardware architectures in the hardware domain [4].

In the following, we suggest a formalisation of these concepts which is based on the terms introduced in the previous section and directed towards a characterisation of Organic Computing systems.

As indicated in Fig. 5 a system S should always try to reach the target space or at least an accepted state such that no explicit external control action is necessary. If a disturbance δ changes the system state $\mathbf{z}(t)$ to a position $\delta(\mathbf{z}(t))$ outside of the acceptance space, the control mechanism has to intervene to guide S back into the acceptance space. Such a correction (e.g. via reconfiguration of S) is possible only

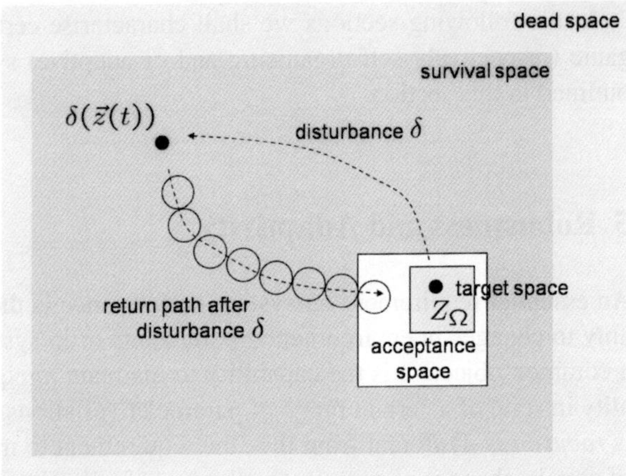

Fig. 5 Simplified representation of system state spaces

for a certain subset of states, the survival space. Any disturbance δ moving $\mathbf{z}(t)$ outside the survival space will be lethal for S. This leads to the following definition:

Definition 1 The *survival space* consists of the maximal range of states that S may (temporarily) assume such that there exists a sequence of control actions leading the system back into its acceptance space. A system leaving its survival space will be damaged permanently, it has entered the *dead space*.

As is obvious from the definition, the acceptance space is part of the survival space (since an empty sequence of control actions is sufficient). Large survival spaces require powerful and flexible control mechanisms, which in turn will benefit from a large configuration space (see below) of S.

The remaining set of states outside of the survival space is called the *dead space*, resembling the fact that the control mechanism is incapable of restoring acceptable system behaviour.

This hierarchy of state spaces is used to characterise the notion of robustness: Intuitively, we would call a system more robust if it has a large number of states that do not lead to a reduced performance or to undesired behaviour. Formally we could define:

Definition 2 Let D be a nonempty set of disturbances.

(a) A system S is called *strongly robust* with respect to D, iff all the disturbances $\delta \in D$ are mapping the target space into itself (i.e. they are mapping ideal states into ideal states).
(b) A system S is called *weakly robust* with respect to D, iff all the disturbances $\delta \in D$ are mapping the target space into the acceptance space (i.e. they are mapping ideal states into acceptable states).
(c) The (degree of) robustness of a system increases with the size of the set of disturbances.

Example An integrated circuit with automotive specification will function correctly from $-30°C$ *to* $+50°C$. *Within this temperature range, there is no control action necessary, the system remains in an ideal state, and it is strongly robust with respect to changes in temperature that do not go beyond the specified range.*

Obviously, strong robustness means that the system will continue to satisfy the evaluation criteria even under certain parameter changes, whereas weakly robust systems might show an acceptable deviation from an ideal behaviour.

While these definitions are referring to the complete sets of ideal or acceptable states, one could define analogous notions of robustness with respect to single states only. This would allow to define different degrees of robustness for individual states. In an inverse approach, one could also define the notion of critical states:

Definition 3 A system S is in a *critical state* with respect to a set of disturbances D, iff every disturbance in D will transform the current (acceptable) state of S into an inacceptable state.

The notions of robustness rely on the capability of the system S to show acceptable behaviour even under the influence of certain changes in attribute values caused by some type of disturbance. A more general requirement would address the capability to return into an acceptable state after some disturbing event. This leads to the following definitions of adaptable and adaptive systems:

Definition 4 Let D be a non empty set of disturbances.

(a) A system is called *adaptive* with respect to a set of disturbances D, iff it is capable to move into the acceptance space after any of the disturbances $\delta \in D$ without needing external control. That means, if δ transforms state $\mathbf{z}(t)$ of S into state $\mathbf{z}' = \delta(\mathbf{z}(t))$ then, after some time interval $\Delta t \geq 0$ the state $\mathbf{z}(t + \Delta t)$ of S will be acceptable.
(b) A system is called *adaptable* with respect to D, iff explicit (external) control actions are possible and necessary to achieve the movement into the acceptance space. The behaviour of an adaptable system S can be modified from the outside via *control inputs* $\mathbf{c}$ by (1) changing parameters and/or (2) changing its structure (elements and links).

This notion of adaptivity is synonymous with the notion of *self-stabilisation* introduced by E.W. Dijkstra [8]. As is obvious from our definition, every (weakly or strongly) robust system is also adaptive (since, in a robust system, the time interval Δt equals zero).

Although we did not specify how an adaptive system manages to move into an acceptable state, we assume that in order to achieve adaptable and/or adaptive behaviour, control inputs $\mathbf{c}$ are triggered (such as switching on a cooling device or lowering the clock frequency) whenever S enters a state outside of its acceptance space (but still inside its survival space, cf. Fig. 5). These control inputs (which are also called *control actions*) have to come from an external source (the *user*) in the case of an adaptable system and will be triggered by the internal control mechanism *CM* in the case of an adaptive system.

Hence, adaptive systems increase their effective robustness with the help of the internal control mechanism, which modifies the values of some parameters of the system or of the environment and thus influences the system behaviour or structure. However, this might involve a temporary deviation from the acceptance space and thus leads to a decrease in system quality.

The *quality* of an adaptive system may be measured by determining

1. the distance of S from an ideal state (which may be defined as the norm of $\eta(\mathbf{z}(t))$ in $\mathbb{R}^k$, e.g. a maximum norm), or, alternatively, the distance from an acceptable state,
2. the time it takes to move S from the present state in survival space back into the acceptance space,
3. the time it takes to move S into an ideal state, or
4. the maximum deviation of the system state from an ideal state (or the sum of all such deviations) on the path from an initial state in the survival space into the acceptance space.

In general, we would like to distinguish two possible reasons for changes in the state of S:

1. The system state $\mathbf{z}(t)$ changes due to a change of the system (e.g. broken component) or a change of the environment (disturbance δ). If the system continues to be acceptable, this corresponds to the common understanding of a *robust system*.
2. The evaluation and acceptance criteria change. This would modify the target space and the acceptance space. If a system is able to cope with such changes in its behavioural specification, it is called a *flexible system*.

Case (1) above is also shown in Fig. 5: After a state change caused by an external or internal disturbance δ the system returns to its acceptance space (weak robustness) or target space (strong robustness) guided by its Observer/Controller. Case (2) is illustrated by Fig. 6: The state vector remains unchanged but the acceptance and target space change. Formally, for the Observer/Controller mechanism there is no difference to case (1): They just have to minimise the difference between current state and target state. In reality, though, flexibility will require different system capabilities than robustness.

After having characterised these different types of adaptive behaviour we now focus on a system classification with respect to different types of control mechanisms.

6 System Classification

As outlined in Sect. 5 internal or external control actions are responsible for the capability of a system to adjust its behaviour in response to changes in system or environmental attributes such that the acceptance criterion is eventually met. In this

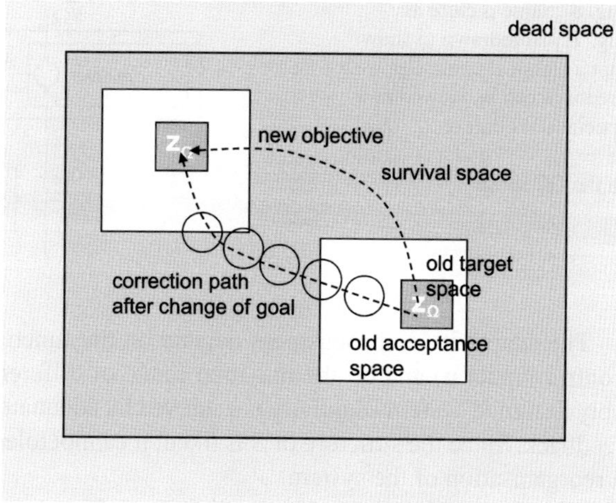

Fig. 6 Illustrating the notion of flexibility

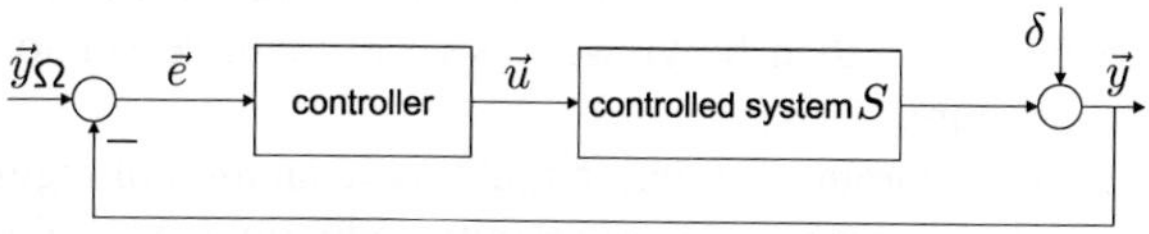

Fig. 7 Classical feedback control loop

section we are suggesting a system classification with respect to different ways of exerting control, starting with a simple, classical feedback control system and ending with a characterisation of self-organising systems.

6.1 Classical Feedback Control Loop System

In classical control theory, a control loop evaluates (*observes*) the output $\mathbf{y}$ of the system, which might be modified by a disturbance δ. It computes its deviation from an ideal output $\mathbf{y}_\Omega$, which refers to the output parameter values in an ideal state $\mathbf{z}_\Omega$. This deviation (or error) $\mathbf{e}$ is used to compute a control input $\mathbf{u}$ for the controlled system S, as shown in Fig. 7.

Please note that the system S still has only one input, i.e. the regular input (called $\mathbf{x}$ in Fig. 4). This input is modified now into the controlled input $\mathbf{u}$ (correction signal) such that S produces $\mathbf{y} = \mathbf{y}_\Omega$. Since, in this example, S is not yet adaptable, there is no other input channel available.

Hence, in this case, the evaluation of a system merely consists of a distance calculation in the state space (or, even more restricted, in the output space) with respect to a specific target state. The computation of an adequate input is based on a mathematical model of the behaviour of S (which might be some lincar model or some set of differential equations). The structure of S is assumed to be fixed, i.e. it cannot be changed by control actions. In addition, there might be an acceptance threshold value $\theta \geq 0$ specifying some tolerance with respect to deviations from the target state.

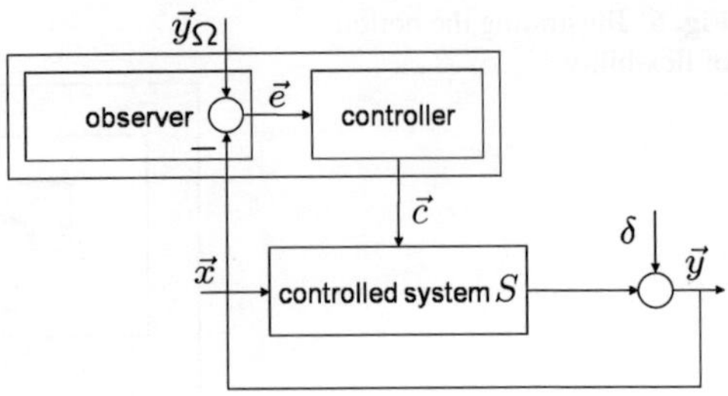

Fig. 8 Same picture as Fig. 7, just redrawn to show that a classical controller with feedback can be viewed as a specialised observer ($\mathbf{e} := \mathbf{y}_\Omega - \mathbf{y}$) and a controller in the OC sense

The adaptivity and robustness depend on the functionality of S (its response to control inputs $\mathbf{u}$) and on the maximal range of different values for $\mathbf{u}$. In a reasonably designed system adaptivity is achieved by adequately using negative or positive feedback. Since the structure of S is fixed, it cannot tolerate disturbances δ requiring a reorganisation of the system.

If we assume that S provides explicit control input separate from the regular input $\mathbf{x}$, we can redraw Fig. 7: As shown in Fig. 8, the classical feedback control loop system could then be viewed as a very simple version of an Observer/Controller architecture.

It is important to mention that this separation of regular input from control is also crucial with respect to complexity reduction: A disabled Observer/Controller will prevent the system from adaptation but not from working correctly on the productive level—as long as the disturbances remain tolerable. Arguing the other way: If we adhere to a design principle which separates regular from control input, the added complexity of observers and controllers does not increase the vulnerability of the system since a failure of observers or controllers will not prevent the system from working but just from adapting.

6.2 *Configuration Space*

In Organic Computing systems, the control mechanisms are much more general than in the classical setting described above. First of all, we have to specify the range of possibilities for influencing the SuOC. This is described by the notion of a configuration space representing the capability of S to be changed (passively!), i.e. it is determined by the parameters of the system that can be changed directly by control actions. This set of parameters may be rather small (e.g. referring to a description of the system objectives only) but could also extend to all the components of the system structure (the nodes and edges of the system graph and their attributes) and to further parameters which might be associated with the system (as for example a speed limit in traffic systems). A control input $\mathbf{c}$ selects some configuration (or, phrased differently, modifies the values of controllable attributes). In the following we assume that S is an adaptable system.

Definition 5 A *configuration* of the system S is determined by the values of a collection of system or environmental attributes which are open to be modified by

Fig. 9 Control possibilities of the adaptable system S and the adaptive system S'

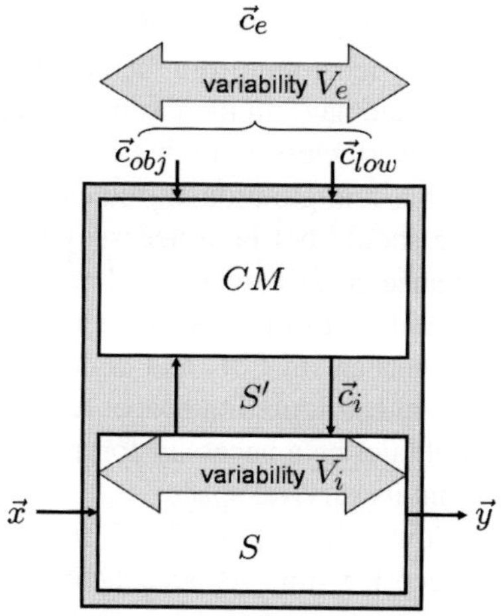

control actions. These attributes will also be called *configuration attributes*. The *configuration space* is the set of configurations that S can take. The *variability* of the configuration space is measured by $V = \log_2(\textit{number of configurations})$.

Obviously, V corresponds to the number of bits necessary to address all the different configurations. The designer of a system has to specify explicitly which of the system and environmental attributes will be configuration attributes. Typically, this will include structural attributes of S. Furthermore, any evaluation and acceptance criteria that are used inside of S will belong to this set of attributes.

The control mechanism *CM* is responsible for selecting specific configurations of S. Therefore, every control action influences the values of a subset of the configuration attributes. We denote the number of bits of a specific control action $\mathbf{c}$ by $v(\mathbf{c})$.

As mentioned already, the system S is assumed to be adaptable. Therefore, the control mechanism *CM* has the task to direct the system back into the acceptance space whenever a violation of the acceptance threshold is detected or predicted. Hence, a successful control mechanism will result in an adaptive system S' which is the combination of S and *CM* as indicated in Fig. 9. Now, with respect to S' we can distinguish two configuration spaces:

1. The *internal* configuration space is the configuration space of S, its variability is denoted by V_i.
2. The *external* configuration space is the configuration space of S', its variability is denoted by V_e.

Correspondingly, we have internal and external control actions $\mathbf{c}_i$ and $\mathbf{c}_e$, respectively. Attributes of the external configuration space will always include the evaluation and acceptance criteria. Furthermore, it might include some configuration attributes of the internal configuration space, in order to allow for direct influence of the *user* on *internal* attributes of the system S.

An important design objective would be to reduce the variability of the configuration space going from the internal to the external level, i.e. in a multi-level design we should have an increasing degree of abstraction going from lower to higher levels. Nevertheless, it might be reasonable to provide direct higher-level control access to some configuration attributes of lower levels. Furthermore, in an adaptive system there should not be a necessity to specify anything more than evaluation and acceptance criteria by external control actions $\mathbf{c}_e$. Therefore, we distinguish between external control inputs $\mathbf{c}_{obj}$, specifying system objectives as a directive for the control mechanism *CM*, and control actions $\mathbf{c}_{low}$ directly related to attributes of the internal configuration space. According to our definition of adaptive systems, there should not be a necessity for any control actions of type $\mathbf{c}_{low}$ as long as S remains within its survival space.

Now, referring to Fig. 5 and Fig. 9 again, *CM* influences (controls) S via $\mathbf{c}_i$ to keep it within its acceptance space by selecting a sequence of control actions $\mathbf{c}_1, \mathbf{c}_2, \ldots$ from its *behavioural repertoire* B. *CM* knows the evaluation criteria and the acceptance criteria as given by the higher-level controller. Ideally, it should be able to keep S within the acceptance space without additional intervention from a higher level. However, if acceptance criteria are violated for too long, the higher level controller might intervene by sending further external control actions of type $\mathbf{c}_{low}$ to influence directly some of the configuration attributes of the SuOC. Such an intervention is undesirable since it decreases the degree of autonomy of S' (see below).

The control mechanisms *CM* are not restricted to the type of a central observer/controller architecture, even if Fig. 9 may suggest this impression. This will be addressed in Sect. 7.3.

In general, we would say that a system S is operating *autonomously*, if it does not need any external control (except for the specification of system objectives). Hence, an adaptive system could operate autonomously as long as it does not leave its survival space (or, more precisely, as long as there are no other disturbances than specified).

Furthermore, an adaptive system keeps working even with its control mechanism (temporarily) not working, but possibly with a reduced degree of robustness.

Finally, there is no(!) reasonably designed technical system without evaluation and acceptance criteria (which, in particular, define Z_Ω and the acceptance space)! Sometimes these criteria might be hidden or built-in at design time, sometimes they can be changed at runtime. On the other hand control actions of type $\mathbf{c}_{low}$ might be non-existent (as in the case of a fully autonomous system).

6.3 Limitations of Adaptivity

A crucial objective for the design of organic systems is to get the utmost adaptivity for its operation in order to reduce the necessity for external interference as far as possible, even, if the system is operating under dynamically changing environmental conditions. But, in the anticipated highly complex application systems we have to be aware of certain limitations of the actual adaptivity of a system S' (assuming again the generic structure depicted in Fig. 9).

In particular, the adaptivity of S' may be limited by

(a) a limited configuration space of S and
(b) a limited behavioural repertoire B of CM.

Limitation (a) may be caused by an inappropriate design, i.e. the necessity for including additional attributes of the system or of the environment into the configuration space has not been seen at design time. There might also be some system specific reasons that prohibit any external modifying access to certain structural attributes of the system. In these cases, it does not seem reasonable to expect a possibility to extend the configuration space at runtime of the system.

What we need is a new design process which takes into account certain configuration options already at design time (e.g. a switch between different modes). Alternatively, we could open the whole design space for run-time modifications, e.g. in terms of a full VHDL description of a circuit which can be loaded at runtime. It must be clear, however, that any increase of the configuration space might also increase the failure probability.

Limitation (b), however, might occur quite often in systems having a very large configuration space but only a limited behavioural repertoire. In particular, it might be a complex optimisation problem to find the most suitable configuration for satisfying the system objectives, as e.g. in traffic light control, where you have to determine the most appropriate values for a number of control parameters of an adaptive controller system (cf. [31]). In this example, the quality of a particular parameter setting will be determined either off-line by a simulation or on-line by observing and analysing the system behaviour. Whenever the control mechanism *CM* or the user detect that the currently available range of control actions is not sufficient to get acceptable behaviour, it should be possible to extend the behavioural repertoire with new, more appropriate control actions. Definitely, it would be desirable to have a system that is capable of generating the necessary modifications of its behavioural repertoire in an autonomous way. This will be addressed in the following subsection.

6.4 Learning

In this subsection we briefly address the aspect of learning in adaptive and self-organising systems.

In general, a system has the capability to learn, if it can improve autonomously its response to input values from some set X. That means, there are time values t_1 and $t_2 > t_1$ such that for any $t > 0$ the response to an input from the set X at time $t_2 + t$ has a higher quality than the response to the same input at time t_1.

This learning capability requires some *learning mechanism LM* which may modify the behaviour of the system S' by

(a) changing the values of some attributes of the system S or of its environment, or
(b) changing the behavioural repertoire B of the control mechanism CM.

An interesting mechanism for learning by modifying parameters of the environment is the stigmergic use of pheromones by ant colonies. This aggregation of individual experiences combined with some degree of evaporation leads to the amazing capability of constructing shortest path ant roads even in a dynamically changing topography. This has inspired a whole range of new design patterns for optimisation algorithms and, with particular relevance for Organic Computing systems, for the self-organised establishment of efficient communication patterns in dynamically changing networks like sensor networks or ad-hoc networks (cf. [9]).

For case (b), speaking in terms of Fig. 9, this means: The learning mechanism modifies CM, e.g. by increasing the fitness of some rules in CM, which have been successful in the past, or by inserting new rules to cope more adequately with the observed system situations. This would lead eventually to a selection of the most adequate rules and the usage of their parameters for the modification of S via the control actions $\mathbf{c}_i$. Such a LM can be implemented e.g. with a combination of a classifier system for online learning and an evolutionary algorithm running offline in a simulated environment for generating new rules.

The design of an organic traffic control system may serve as an example for the second type of learning method (cf. [31]): A classifier system for selecting parameter settings for a traffic light controller (in the real traffic system S) is using on-line learning by associating a fitness value to classifier rules based on the performance of their parameter settings in real traffic situations, combined with off-line learning which produces new classifier rules for inadequately handled traffic situations by using a genetic algorithm which generates new rules of some minimum quality level by evaluating their performance in a traffic simulator. In this way, the learning mechanism manages to improve the system performance on known traffic situations and it is also capable of generating adequate responses (i.e. control actions) to previously unknown traffic situations.

6.5 Degree of Autonomy

One of the major objectives of this article is to develop a quantitative measure of the degree of autonomy of technical application systems. Whereas in autonomic computing the ultimate goal is to have a system that can operate reliably without any external intervention (i.e. fully autonomous), we might concede a limited degree of autonomy for an Organic Computing system, since it should always allow for (externally) controlled self-organisation.

In this subsection we shall use the relation between external versus internal control actions (as defined in Sect. 6.2) to characterise the degree of autonomy.

Again, we assume a system S' having the structure depicted in Fig. 9, i.e. consisting of a SuOC S and a control mechanism CM. S' has a certain degree of autonomy, i.e. self-control, exerted by some CM. The origin of the control actions $\mathbf{c}_i$ determines how autonomous a system is:

1. External origin of $\mathbf{c}_i$, i.e. it equals some $\mathbf{c}_{low}$: direct control by an external (e.g. human) operator, no autonomy.
2. Internal origin of $\mathbf{c}_i$: S is fully controlled by CM which is part of the system S'. Full autonomy.
3. Intermediate: control actions of type $\mathbf{c}_{low}$ and $\mathbf{c}_i$ are both used to control S.

It is a goal of autonomous systems to hide the internal control $\mathbf{c}_i$ from the external controller or user of S' at least partially. As defined above, the external configuration space of S' is determined by the collection of configuration attributes and its corresponding variability V_e. Furthermore, the control actions $\mathbf{c}_e$ have a maximum number of bits $\max v(\mathbf{c}_e)$. Obviously, we may assume that $V_e \geq \max v(\mathbf{c}_e)$. Analogously, the configuration space of S is determined by the internal configuration attributes and their corresponding variability V_i.

The autonomy of the system S' may now be characterised by the *complexity reduction*

$$R = V_i - V_e \tag{1}$$

If the system has been designed with respect to the objective of having increasingly higher levels of abstraction, R will be a positive value. A negative value would indicate that the control mechanism leads to an additional complexity of the system which would be an undesirable situation which, nevertheless, may be unavoidable sometimes. In particular, in order to provide a large range of potential system adaptations at runtime, the external configuration space might be very large.

Definition 6 The (*static*) *degree of autonomy* of system S' is defined by

$$\alpha = \frac{R}{V_i} = \frac{V_i - V_e}{V_i} \tag{2}$$

The value of α will be at most one (if V_e equals zero). In this case, there is no external variability, i.e. there is no possibility to modify any attributes of S' by external control actions. Therefore, such a system would be called *fully autonomous*. The only way to interact with such a system would be to influence the input values $\mathbf{x}$ of S.

If the value of α is zero, the internal and the external variabilities are the same, which indicates that there is no reduction in complexity. If α drops below zero, we have a situation where the external configuration space contains more controllable attributes than the internal configuration space. This might happen, if all the lower level attributes are available externally, and there are additional configuration attributes referring to the control mechanism CM. In this case, the system S could be controlled completely by external control actions, which would be no autonomy at all. But, as is obvious from the definition, a low value of α is merely indicating a potential loss of autonomy, since it does not take into account, whether during any time interval any external control actions have been used actually.

Therefore, in order to characterise the actual degree of autonomy (or, the actual degree of external control) we should consider the number of bits that have been actually used in any control actions over some time period $[t_1, t_2]$.

Definition 7 Let $v_e(t)$ and $v_i(t)$ be the number of bits of the external and internal control actions at time t.

(a) The *dynamic complexity reduction* r in S' over some time interval $[t_1, t_2]$ is defined to be

$$r = \int_{t_1}^{t_2} (v_i(t) - v_e(t))dt \tag{3}$$

(b) The *dynamic degree of autonomy* β in S' over some time interval $[t_1, t_2]$ is defined to be

$$\beta = \frac{\int_{t_1}^{t_2} (v_i(t) - v_e(t))dt}{\int_{t_1}^{t_2} v_i(t)dt} \tag{4}$$

This dynamic degree of autonomy quantifies the relative amount of external control that has been used during a specific time interval. Hence, a system could exhibit a fully autonomous behaviour over some time period while being completely controlled by external control actions at a different time. This definitely allows for a much more accurate characterisation of the autonomy of a system, based on the actual control flow. In particular, in the case of organic systems, we might have a fully autonomous behaviour over some time period, but might use adequate control actions whenever the self-organised behaviour of the system does not satisfy the external constraints.

6.6 Self-organising Systems

After having characterised the (dynamic) degree of autonomy of a system, we may start to look at the so-called *self-*-properties*, as the degree of autonomy would correspond to the *degree of self*. In particular, in this article we want to characterise self-organisation in technical application systems.

Intuition suggests that a self-organising system is a multi-element system (consisting of m elements, $m > 1$), which needs no external control to operate and to restructure itself (i.e. it has a high degree of autonomy). As mentioned in Sect. 2, most known characterisations of self-organisation agree on this. Furthermore, a common assumption is that the control mechanism *CM* is distributed over the elements. In natural systems, such a distributed control mechanism might be difficult to localise which, sometimes, leads to almost magical connotations of self-organisation.

Obviously, in a controllable technical system there must always be a control mechanism. It can be centralised (one *CM*), distributed over the m elements (m *CMs*) or distributed over a hierarchy of *CMs*. In such a hierarchical structure, there actually could be more than m control mechanisms, distributed over the different levels of the hierarchy. Figure 10 shows three examples of Observer/Controller architectures, from fully centralised to totally decentralised. The most interesting architecture seems to be the multi-level hierarchical scheme with regional Observer/Controllers (Fig. 10(c)).

We could define a *degree of self-organisation* by counting the number of *CMs* ($= k$) in relation to the number of elements m of the system $(k : m)$. An adaptive system with one centralised *CM* could then be denoted as $(1 : m)$, an adaptive system with full distribution of *CMs* as $(m : m)$. Definitely, an $(m : m)$ system may be called self-organised. It is a matter of taste whether we want to call a system with lower degrees of self-organisation still self-organised. Hence, in the following definition we are not referring to any degree of emerging order in the behaviour of the system S (as done e.g. by Heylighen, Parunak, and Wright (cf. Sect. 2)) but on the degree of distribution of its embedded control mechanism which guarantees the adaptivity of the system.

Definition 8 Let S be an adaptive system consisting of m elements $(m > 1)$ with large degrees of autonomy (α and β) and fully or partially distributed k control mechanisms *CM* $(k \geq 1)$ leading to a degree of self-organisation of $(k : m)$.

(a) S is called *strongly self-organised*, if $k \geq m$, i.e. the degree of self-organisation is at least $(m : m)$.
(b) S is called *self-organised*, if $k > 1$, i.e. it has a medium degree of self-organisation $(k : m)$.
(c) S is called *weakly self-organised*, if $k = 1$, i.e. there is a central control mechanism and the degree of self-organisation is $(1 : m)$.

The weak definition would mean that any adaptive system is also called self-organised. The authors of this article tend to use the medium definition. Speaking about adaptive and self-organising systems in an exact way requires to indicate (1) the degree of autonomy and (2) the degree of self-organisation $(k : m)$.

Above we have introduced a static and a dynamic degree of autonomy (α, β). How are these degrees defined in case of a (partially) distributed observer/controller architecture according to Fig. 10? Firstly, they are well defined for any subsystem with just one observer/controller and one SuOC. Aggregation of several O/C-controlled subsystems with just one higher-level O/C (Fig. 10(c)) is handled by adding all the subsystem control inputs. This holds for the static as well as the dynamic case. For a multi-level O/C hierarchy, the degrees of autonomy can be determined for all subtrees.

Here, a remark seems to be necessary: In the literature about self-organising systems (see e.g. [22]) you sometimes find the additional requirement that a self-organising system should be *structure-adaptive*. This is not explicitly stated in our definition, but it may be present in an implicit form: For example, if the system objectives refer to structural configuration attributes of S then, after some disturbance or after a change in the evaluation and acceptance criteria, an adaptive system will be able to re-organise its structure such that it is acceptable again.

Finally, for Organic Computing systems we emphasise the necessity to have the potential capability to influence explicitly the system behaviour, although, as long as the system performs acceptably, it should operate as autonomously as possible and it should be (strongly) self-organising. For example, it should be possible to modify objectives and evaluation criteria using control actions $\mathbf{c}_{obj}$ and to interfere directly with system operations using control actions $\mathbf{c}_{low}$, in particular in the case of disturbances δ going beyond the anticipated set of disturbances D. Now, using the notions of static and dynamic degrees of autonomy, we suggest the following characterisation of controlled self-organisation:

Definition 9 A self-organised system S' (composed of an SuOC S and a control mechanism CM) allows for *controlled self-organisation* iff

(i) it operates at a low static degree of autonomy $\alpha < 1$ (i.e. there is at least the option for external control of the system via control actions of type $\mathbf{c}_{obj}$ and $\mathbf{c}_{low}$) and
(ii) it operates at a high dynamic degree of autonomy β as long as the system state $\mathbf{z}$ stays in the acceptance space (i.e. after an initial statement of the objectives via $\mathbf{c}_{obj}$ the system works acceptably well without external control).

This characterisation requires that the system is adaptive (which means that it is capable of returning autonomously into an acceptable state after certain disturbing events), but provides the option to the external user to redefine certain objectives (by using a control input $\mathbf{c}_{obj}$) or to interfere directly with the operation of system S by using a control action of type $\mathbf{c}_{low}$. The simplest example of the latter would be a control action to switch off the system.

7 Architectures for Controlled Self-organisation

As mentioned above, it is the main goal of OC to design and build controllable self-organising technical systems with life-like properties. To achieve these goals it is obvious that system components will need some degrees of freedom to adapt to a dynamically changing environment. As a consequence, the whole system may develop an unexpected emergent behaviour, which might be positive as well as negative. But there are additional drawbacks. In order to adapt to new situations and to learn new logic and behaviour, a technical system has to be allowed to make errors, and it may react more slowly. Also, system behaviour might become less predictable. These emergent phenomena are often characterised as "*the whole is more than the sum of its parts*", where the study of the local behaviour of the individual components reveals little about the global system-wide behaviour.

Following the definition by [23], we define emergence as self-organised order, and in order to balance creative self-organised bottom-up processes and top-down control, we have proposed a generic Observer/Controller architecture [30], which supports controlled self-organisation: It allows to minimise higher-level control in favour of self-control while keeping overall control. Except for the requirement of controlled self-organisation and the explicit distinction between on-line and off-line learning this is quite similar to Autonomic Computing's MAPE (monitor, analyse, plan, and execute) cycle [38].

In the generic Observer/Controller architecture as mentioned already in the introduction and depicted in Fig. 1, the control mechanism *CM* consists of an observer and a controller, and it has an appropriate interface for observing attributes of the system *S* (which is the SuOC) and influencing the system configuration. Furthermore, there is an explicit interface to an upper-level entity (the *user*) which provides system objectives and which may interfere with the system by external control actions based on the reported system status. This Observer/Controller architecture is supposed to be the basic architecture of organic systems. The tasks of its two major components may be summarised as follows:

- *Role of the observer:* Sensing the system state, aggregating and analysing the observed data, calculation of invariants, prediction of future behaviour, storage of relevant patterns in memory for later usage, and comparison of new patterns with stored ones. The result of this process is a collection of situation indicators that are passed on to the controller. The specific operation profile of the observer is determined by a *model of observation*, which may be modified by the controller.
- *Role of the controller:* Generate (or select) an adequate reaction to the presented situation indicators, and, if necessary to intervene, send control action $\mathbf{c}_i$ to S. Adjust the model of observation whenever necessary. Update and adapt the behavioural repertoire for internal control by an appropriate combination of on-line and off-line learning, send system status information to the higher level entity and react adequately to external control actions.

Although the observer and the controller look like centralised components, they should be viewed in a more general way. There is a range of possible realisations, where the functionality of the Observer/Controller architecture is more or less distributed over the elements of the system under observation and control. This is addressed in the next subsection.

7.1 Architectural Options

As mentioned in [2] the generic Observer/Controller architecture has to be customised to different scenarios by adapting the various components. There are at least three main architectural options (see Fig. 10):

(a) Central: One Observer/Controller for the whole system. There exists exactly one *CM*.
(b) Decentral/distributed: One Observer/Controller for each subsystem. *CM* consists of m cooperating partial *CM* functionalities distributed over the m nodes (or elements) of the system.
(c) Multi-level/regional/hierarchical: One Observer/Controller for each subsystem as well as one (or more) for the whole system. There are first-level *CMs* responsible for subsets of elements (regions) and second-level *CMs* in a second hierarchical level for groups of first-level *CMs* (and so on).

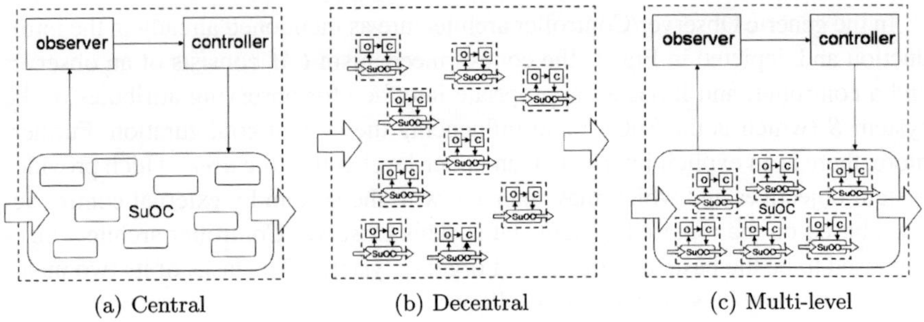

Fig. 10 Observer/Controller realisation

The central version allows for the design of an adaptive system, the behaviour of the system is controlled in a top-down way. In contrast to this, the decentral/distributed version (b) has to rely on a self-organised bottom-up establishment of the desired behaviour (i.e. it has to rely on effects of emergence), whereas the multi-level architecture (c) would provide the architectural prerequisites for controlled self-organisation which may be viewed as a flexible combination of top-down and bottom-up control.

So far, we have investigated different versions of the generic Observer/Controller architecture in elevator systems, studying the control of bunching effects [29], used it for the control of a multi-agent system [21], and utilised it as a key component in the research project Organic Traffic Control [28] where it allows to obtain an adaptive system with attractive learning capabilities for controlling individual traffic lights as well as for generating in a self-organised way progressive signal systems which are supposed to decrease traffic delays and energy consumption.

While these are just the general options, we want to emphasise that OC systems in our sense always have to be open to some external high level control, possibly by the user. We do not intend to build fully autonomous systems because we believe that we must always be able to guarantee the ultimate human control, which might be necessary to allow for spontaneous individual preferences of the user or to cope with unanticipated emerging behaviour of the system or with other kinds of unanticipated system disturbances.

7.2 Control Possibilities of OC Systems

In the previous section we quantified the degrees of autonomy and of self-organisation in adaptive technical systems and we have suggested a definition of the term *controlled self-organisation*. Now, we shall analyse and classify organic systems by the way they exhibit control of their configuration space. In the following we list different control options.

Firstly, we follow the distinction between external and internal control, or with other words: *High-level control* done by the upper level and *self-control* allocated to the internal Observer/Controller level. Assuming an SuOC consisting of a collection of cooperating agents, self-control may refer (a) to the local decision rules of the simple agents, (b) to the control of the system structure, or (c) to the control of the environment.

(a) Influencing the local decision rules of the agents modifies the local behaviour of the individuals.
(b) Influencing the system structure: We assume that the elements base their actions on local information, where *local* is defined by a neighbourhood and an interconnection network as specified in the structural attributes. Modifying this network, in particular with respect to global characteristics, will change the local behaviour of the agents and the global behaviour of the system. Also, changing the absolute number of elements influences neighbourhoods and again the behaviour of the SuOC.
(c) Influencing the environment allows indirect control of the SuOC and works only if system elements have sensors to measure and react to a modified environment, and the controller has actuators to influence the environment. (Typical examples for this would be the modification of speed limits in traffic systems or the deposit of pheromones in ant colony systems.)

Obviously, control strategy (a) has the most drastic direct influence on the behaviour of the system and requires the possibility for quite low level control actions. Strategies (b) and (c) impose increasingly less direct changes on the behaviour and attributes of the components of the SuOC and rely more on indirect effects of the control actions on the behaviour of the agents.

7.3 Roadmap to Ideal OC Systems

Sketching a roadmap for the future development of OC systems we refer to the formulas of the static and dynamic degrees of autonomy α and β and we distinguish two extreme cases.

(i) Without any complexity reduction ($R \leq 0$) or ($r \leq 0$) there is no autonomy ($\alpha \leq 0$) or ($\beta \leq 0$).
(ii) Providing no external control possibilities ($V_e = 0$) or not using external control results in full autonomy ($\alpha = 1$) or ($\beta = 1$).

Thus, α, β, and V_i span a diagram (cf. Fig. 11) illustrating the anticipated trend of OC system development: We will witness an increasing complexity of the SuOC (and a higher variability V_i), which makes it increasingly desirable to reach large complexity reductions r and hence high degrees of autonomy α or, in particular, β.

The variability of different Observer/Controller architectures will grow with the degree of distribution. This means at the same time that the task of controlling them will become increasingly difficult. The only possibility to manage this complexity will be a hierarchical approach with multiple levels of Observers/Controllers, each

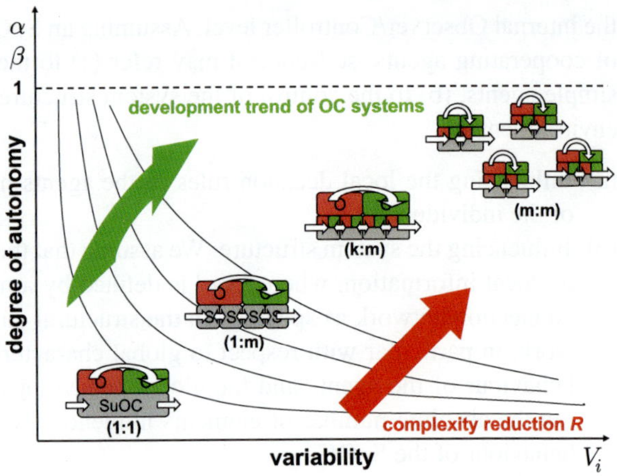

Fig. 11 Degree of autonomy (α and β) as a function of the variability V_i of the SuOC for a fixed V_e. $(1:1)$, $(1:m)$, $(k:m)$, and $(m:m)$ denote the degree of self-organisation

of them responsible for its local or regional subsystem. As shown in the diagonal in Fig. 11, as we climb up to higher Observer/Controller levels, their tasks will become less intervening and more guarding, correcting, and limiting—very much like the high-level management of a company normally will not interfere with low-level decisions, as long as they conform to the high-level directives.

8 Conclusion

In this article we focused on some of the major properties of Organic Computing systems. In technical application scenarios it is essential to build robust, flexible, or at least adaptive systems in order to reliably provide the required functionality. Similar to autonomic computing, the design of Organic Computing systems utilises autonomy or self-organisation to support adaptivity and it extends it with the notion of controlled self-organisation. The objective is to stay within or return to a target space or an acceptance space as defined by some higher-level system (the human user) with little or no external control for diverse and possibly unpredictable environments, where all kinds of disturbing events may occur. Hence, the value of a system increases if it can survive under very diverse environmental conditions while still functioning properly.

It is trivial to build an autonomous system for a low-diversity environment (i.e. only a small survival space is needed).

Example Consider a room temperature controller with $Z_\Omega = 20 \pm 2$°C *and a specified environment between* 10°C *and* 30°C. *Under standard circumstances, there is no reorganisation possible and needed to meet the system objective. However, the system may break down for environmental temperatures e.g.* >30°C *or if the air conditioner fails* (*no redundancy*).

It is relatively simple to build an adaptable system even for high-diversity environments if the task of the control mechanism *CM* is operated by an external entity (e.g. the human user).

Example Consider a car braking system on an uneven and slippery surface. The system can be modified (control parameter: pedal pressure). Therefore it is adaptable. But all the control effort comes from the outside (human user): There exists no autonomy and no (self-) adaptivity. But, if we have a modern brake-by-wire system, the human pedal pressure will be transformed into an input to an internal controller which determines the fine-grain braking actions that are adequate for the current environmental situation. Here, we would have a relatively high level of autonomy.

The challenge for system design lies in building systems,

1. which meet their target performance (stay within Z_Ω)
2. with little external control effort (high β)
3. for a large set of environmental conditions and possible disturbances (large survival space)
4. while providing the option and adequate interface for explicit external interference.

It is essential for such a system to have a built-in control mechanism (constituting the *self*). A system will become more robust if the control mechanism is distributed, i.e. its degree of self-organisation $(k : m)$ is high since this avoids a single point of failure.

Adaptive and self-organising systems will not always outperform conventional systems but they may survive under a large diversity of environmental conditions. Self-organising systems may be especially robust due to their internal distributed architecture and their structural variability.

The objective for system design should be to design controllable self-organising systems, i.e. systems, which allow for external control but have a high dynamic degree of autonomy as well. One major argument for the possibility of external control is the necessity to support the generation of trust into the dependability of a system. The topic of trustworthy systems, though, is beyond the scope of this article.

9 Outlook

In this article we have presented an approach to a quantitative definition of adaptivity and self-organisation applicable to technical systems with an explicit possibility to control their configuration space, supporting the notion of controlled self-organisation as emphasised in Organic Computing. However, several open questions remain and are subject of our future research:

- Validation: We want to check the viability of the approach by applying it to more example systems.

- Architectural classification of self-organising systems: The different options for centralisation or distribution of observers and controllers as sketched in Sect. 7 should be used to classify OC systems. We want to investigate which architectural option is preferable for which specific application scenario. This should result in design patterns determining an optimal degree of self-organisation for different system types and requirements.
- Benefit of self-organising systems: Self-organisation will result in certain desirable properties of a system (like self-x properties). On the other hand we have to spend additional effort to enable systems to self-organise and to protect us from self-organising systems running astray. It will be necessary to justify this additional cost by relating it to the achieved benefits.
- Break down global goals into local objective functions: One of the big challenges will be the *design* of controllable self-organising systems. Ultimately, it will lead to just setting the system objectives and let the system find its optimal configuration itself. This requires a systematic method to derive local rules from global goals. In the most general setting, this might even be a challenge without a satisfying answer.

References

1. Benjaafar, S., Ramakrishnan, R.: Modeling, measurement, and evaluation of sequencing flexibility in manufacturing systems. Int. J. Prod. Res. **34**, 1195–1220 (1996)
2. Branke, J., Mnif, M., Müller-Schloer, C., Prothmann, H., Richter, U., Rochner, F., Schmeck, H.: Organic Computing – Addressing complexity by controlled self-organization. In: Margaria, T., Philippou, A., Steffen, B. (eds.) Proceedings of the 2nd International Symposium on Leveraging Applications of Formal Methods, Verification and Validation (ISoLA 2006), Paphos, Cyprus, November 2006, pp. 200–206 (2006)
3. Callaway, D.S., Newman, M.E.J., Strogatz, S.H., Watts, D.J.: Network robustness and fragility: Percolation on random graphs. Phys. Rev. Lett. **85**(25), 5468–5471 (2000)
4. Compton, K.: Flexibility measurement of domain-specific reconfigurable hardware. In: Proceedings of the 2004 ACM/SIGDA 12th International Symposium on Field Programmable Gate Arrays (FPGA 2004), pp. 155–161. ACM, New York (2004)
5. De Wolf, T., Holvoet, T.: Emergence versus self-organisation: Different concepts but promising when combined. In: Brueckner, S., di Marzo Serugendo, G., Karageorgos, A., Nagpal, R. (eds.) Engineering Self-Organising Systems, Methodologies and Applications. LNCS, vol. 3464, pp. 1–15. Springer, Berlin (2005)
6. De Wolf, T., Samaey, G., Holvoet, T., Roose, D.: Decentralized autonomic computing: Analysing self-organising emergent behavior using advanced numerical methods. In: Proceedings of the Second International Conference on Autonomic Computing (ICAC 2005), pp. 52–63 (2005)
7. DFG Priority Program 1183 Organic Computing. Website. http://www.organic-computing.de/SPP (2005). Visited June 2007
8. Dijkstra, E.W.: Self-stabilizing systems in spite of distributed control. Commun. ACM **17**(11), 643–644 (1974)
9. Dorigo, M., Stützle, T.: Ant Colony Optimization. MIT Press, Cambridge (2004)
10. Eden, A.H., Mens, T.: Measuring software flexibility. IEE Proc., Softw. Eng. **153**(3), 113–125 (2006)

11. Gershenson, C., Heylighen, F.: When can we call a system self-organizing? In: Banzhaf, W., Christaller, T., Dittrich, P., Kim, J.T., Ziegler, J. (eds.) Proceedings of the 7th European Conference on Advances in Artificial Life (ECAL 2003), Dortmund, Germany. LNCS, vol. 2801, pp. 606–614. Springer, Berlin (2003)
12. Hassanzadeh, P., Maier-Speredelozzi, V.: Dynamic flexibility metrics for capability and capacity. Int. J. Flex. Manuf. Syst. **19**(3), 195–216 (2007)
13. Hestermeyer, T., Oberschelp, O., Giese, H.: Structured information processing for self-optimizing mechatronic systems. In: Araújo, H., Vieira, A., Braz, J., Encarnação, B., Carvalho, M. (eds.) Proceedings of the 1st International Conference on Informatics in Control, Automation and Robotics (ICINCO 2004), August 2004, pp. 230–237. IEEE Comput. Soc., Los Alamitos (2004)
14. Heylighen, F.: The science of self-organization and adaptivity. In: The Encyclopedia of Life Support Systems, pp. 253–280 (1999)
15. Heylighen, F., Joslyn, C.: Cybernetics and second-order cybernetics. In: Meyers, R.A. (ed.) Encyclopedia of Physical Science & Technology, 3rd edn., vol. 4, pp. 155–170. Academic Press, New York (2001)
16. Jalote, P.: Fault Tolerance in Distributed Systems. Prentice Hall, New York (1994)
17. Kephart, J.O., Chess, D.M.: The vision of autonomic computing. IEEE Comput. **1**, 41–50 (2003)
18. Knuth, D.E.: The Art of Computer Programming – Sorting and Searching, 2nd edn., vol. 3. Addison-Wesley/Longman, Amsterdam (1998)
19. Lucas, C.: Self-organizing Systems (sos) faq. http://www.calresco.org/sos/sosfaq.htm, July 2006. Frequently asked questions version 2.99, visited June 2007
20. Mnif, M., Müller-Schloer, C.: Quantitative emergence. In: Proceedings of the 2006 IEEE Mountain Workshop on Adaptive and Learning Systems (IEEE SMCals 2006), pp. 78–84 (2006)
21. Mnif, M., Richter, U., Branke, J., Schmeck, H., Müller-Schloer, C.: Measurement and control of self-organised behaviour in robot swarms. In: Lukowicz, P., Thiele, L., Tröster, G. (eds.) Proceedings of the 20th International Conference on Architecture of Computing Systems (ARCS 2007). Lecture Notes in Computer Science, vol. 4415, pp. 209–223. Springer, Berlin (2007)
22. Mühl, G., Werner, M., Jaeger, M.A., Herrmann, K., Parzyjegla, H.: On the definitions of self-managing and self-organizing systems. In: Braun, T., Carle, G., Stiller, B. (eds.) Proceedings of the KiVS 2007 Workshop: Selbstorganisierende, Adaptive, Kontextsensitive verteilte Systeme (SAKS 2007), pp. 291–301. VDE Verlag, Bern (2007)
23. Müller-Schloer, C., Sick, B.: Emergence in Organic Computing systems: Discussion of a controversial concept. In: Yang, L.T., Jin, H., Ma, J., Ungerer, T. (eds.) Proceedings of the 3rd International Conference on Autonomic and Trusted Computing (ATC 2006). LNCS, vol. 4158, pp. 1–16. Springer, Berlin (2006)
24. Nimis, J., Lockemann, P.C.: Robust multi-agent systems: The transactional conversation approach. In: Barley, M., Massacci, F., Mouratidis, H., Scerri, P. (eds.) 1st International Workshop "Safety and Security in MultiAgent Systems" (SASEMAS 2004). AAMAS, New York (2004)
25. Oberschelp, O., Hestermeyer, T., Kleinjohann, B., Kleinjohann, L.: Design of self-optimizing agent-based controllers. In: Urban, C. (ed.) Proceedings of the 3rd International Workshop on Agent Based Simulation, Passau, Germany, April 2002. SCS European Publishing House
26. Parunak, H.V.D., Brueckner, S.: Entropy and self-organization in multi-agent systems. In: Müller, J.P., Andre, E., Sen, S., Frasson, C. (eds.) Proceedings of the 5th International Conference on Autonomous Agents, Montreal, Canada, pp. 124–130. ACM, New York (2001)
27. Polani, D.: Measuring self-organization via observers. In: Banzhaf, W., Christaller, T., Dittrich, P., Kim, J.T., Ziegler, J. (eds.) Proceedings of the 7th European Conference on Advances in Artificial Life (ECAL 2003), Dortmund, Germany. LNCS, vol. 2801, pp. 667–675. Springer, Berlin (2003)

28. Prothmann, H., Rochner, F., Tomforde, S., Branke, J., Müller-Schloer, C., Schmeck, H.: Organic control of traffic lights. In: Rong, C., Jaatun, M.G., Sandnes, F.E., Yang, L.T., Ma, J. (eds.) Proceedings of the 5th International Conference on Autonomic and Trusted Computing (ATC-08). LNCS, vol. 5060, pp. 219–233. Springer, Berlin (2008)
29. Ribock, O., Richter, U., Schmeck, H.: Using organic computing to control bunching effects. In: Brinkschulte, U., Ungerer, T., Hochberger, C., Spallek, R.G. (eds.) Proceedings of the 21th International Conference on Architecture of Computing Systems (ARCS 2008). LNCS, vol. 4934, pp. 232–244. Springer, Berlin (2008)
30. Richter, U., Mnif, M., Branke, J., Müller-Schloer, C., Schmeck, H.: Towards a generic observer/controller architecture for Organic Computing. In: Hochberger, C., Liskowsky, R. (eds.) INFORMATIK 2006 – Informatik für Menschen! Lecture Notes in Informatics (LNI), vol. P-93, pp. 112–119. Köllen Verlag, Bonn (2006)
31. Rochner, F., Prothmann, H., Branke, J., Müller-Schloer, C., Schmeck, H.: An organic architecture for traffic light controllers. In: Hochberger, C., Liskowsky, R. (eds.) INFORMATIK 2006 – Informatik für Menschen! Lecture Notes in Informatics (LNI), vol. P-93, pp. 120–127. Köllen Verlag, Bonn (2006)
32. Schmeck, H.: Organic Computing – A new vision for distributed embedded systems. In: Proceedings of the 8th IEEE International Symposium on Object-Oriented Real-Time Distributed Computing (ISORC 2005), pp. 201–203. IEEE Comput. Soc., Los Alamitos (2005)
33. Scholl, A.: Robuste Planung und Optimierung – Grundlagen, Konzepte und Methoden, Experimentelle Untersuchungen. Physica-Verlag, Heidelberg (2001)
34. Shalizi, C.R., Shalizi, K.L.: Quantifying self-organization in cyclic cellular automata (2005). arXiv:nlin/0507067
35. Shalizi, C.R., Shalizi, K.L., Haslinger, R.: Quantifying self-organization with optimal predictors. Phys. Rev. Lett. **93**(11), 1–4 (2004)
36. Shuiabi, E., Thomson, V., Bhuiyan, N.: Entropy as a measure of operational flexibility. Eur. J. Oper. Res. **165**(3), 696–707 (2005)
37. Slotine, J.-J.E., Li, W.: Applied Nonlinear Control. Prentice Hall, New York (1990)
38. Sterritt, R.: Autonomic Computing. Innov. Syst. Softw. Eng. **1**(1), 79–88 (2005)
39. Taguchi, G.: Taguchi on Robust Technology Development – Bringing Quality Engineering Upstream. Am. Soc. Mechanical (1993)
40. Tennenhouse, D.: Proactive computing. Commun. ACM **43**, 43–50 (2000)
41. Weyns, D., Parunak, H.V.D., Michel, F., Holvoet, T., Ferber, J.: Environments for multiagent systems state-of-the-art and research challenges. In: Environments for Multi-Agent Systems, First International Workshop, E4MAS 2004, Revised Selected Papers. Lecture Notes in Computer Science, vol. 3374, pp. 1–47. Springer, Berlin (2005)
42. Wright, W.A., Smith, R.E., Danek, M., Greenway, P.: A measure of emergence in an adapting, multi-agent context. In: Meyer, J., Berthoz, A., Floreano, D., Roitblat, H., Wilson, S. (eds.) Proceedings of the 6th International Conference on the Simulation of Adaptive Behaviour (SAB 2000), pp. 20–27. ISAB Press (2000)
43. Wright, W.A., Smith, R.E., Danek, M., Greenway, P.: A generalisable measure of self-organisation and emergence. In: Dorffner, G., Bischof, H., Hornik, K. (eds.) Proceedings of the International Conference on Artificial Neural Networks (ICANN 2001), Vienna, Austria, August 2001. LNCS, vol. 2130, pp. 857–864. Springer, Berlin (2001)
44. Zadeh, L.A.: On the definition of adaptivity. Proc. IEEE **51**(3), 469–470 (1963)

H. Schmeck · U. Richter
Institute AIFB, Karlsruhe Institute of Technology (KIT), Bldg. 05.20, 76128 Karlsruhe, Germany

H. Schmeck
e-mail: hartmut.schmeck@kit.edu

U. Richter
e-mail: urban.richter@kit.edu

C. Müller-Schloer (✉) · E. Çakar · M. Mnif
Institute of Systems Engineering, Leibniz Universität Hannover, Appelstr. 4, 30167 Hannover, Germany
e-mail: cms@sra.uni-hannover.de

E. Çakar
e-mail: cakar@sra.uni-hannover.de

M. Mnif
e-mail: mnif@sra.uni-hannover.de

Chapter 1.2
Quantitative Emergence

Moez Mnif and Christian Müller-Schloer

Abstract Emergence can be defined as the formation of order from disorder based on self-organisation. Humans—by looking at a self-organising system—can decide intuitively whether emergence was taking place or not. To build self-organising technical systems we need to automate the recognition of emergent behaviour. In this paper we try to give a quantitative and practically usable definition of emergence. The presented theoretical approach is applied to an experimental environment, which shows emergent behaviour. An Observer/Controller architecture with emergence detectors is introduced. The proposed definition of emergence is discussed in comparison with Shannon's information theoretical approach.

Keywords Emergence · Self-organisation · Order · Entropy · Observer/controller architecture

1 Introduction

Organic Computing (OC) has become a major research activity in Germany and worldwide [6]. Its goal is the technical utilisation of emergence and self-organisation as observed in natural systems. Emergent and self-organising behaviour has been observed in nature, demonstrated in a variety of computer simulated systems in artificial life research, and it has occurred also in highly complex technical systems (like the Internet) where it has led to unexpected global functionality. Despite the importance of a rigorous description of these phenomena, the quantitative analysis of technical self-organising systems is still a rather unexplored area.

Emergence and self-organisation have been discussed by a variety of authors for many years. The most commonly accepted definition is that the whole is more than the sum of the parts. We want to avoid a new verbal definition and refer instead to a few excellent papers and books [1, 2, 9, 11].

There seem to be certain necessary ingredients for an observed phenomenon to be called "emergent": A large population of interacting elements (or agents) without central control and hence based only on local rules leads to a macroscopic behaviour which displays new properties not existent on the element-level. This macroscopic pattern is perceived as structure or order. Although the resulting order is a necessary precondition for emergence, it is not sufficient. We require that this order has been developed without external intervention—i.e. self-organised. Hence we define emergence as self-organised order.[1] An attempt to measure emergence quantitatively should therefore rely on a well-known metric for order, i.e. entropy.

Section 2 of this article proposes a method to determine the entropy of an arbitrary system based on Shannon's entropy definition. This definition relies on the selection of observable attributes, which means a subjective influence on the measured values. Section 3 introduces the notion of an observation model, which subsumes these subjective decisions. Emergence is not the same as entropy. In Sect. 4 we derive an emergence measure based on entropy and discuss its implications in comparison to Shannon's redundancy. Section 5 discusses the limitations of our approach, the relationship of redundancy and emergence, and the term "pragmatic information" as introduced by von Weizsäcker, Sect. 6 proposes an Observer/Controller architecture we are presently implementing, which includes detectors for the measurement of emergence. Section 7 presents first experimental results of emergence measurements.

2 The Measurement of Order

The meaning of order as perceived by a human[2] observer is not clear without ambiguity. A homogeneous mixture of two liquids can be regarded as "orderly" (Fig. 1, right). Applying the thermodynamic entropy, however, will result in lower entropy (i.e. higher order) for the example on the left of Fig. 1. Apparently, order depends on the selection of certain attributes by the (human) observer. If we are interested in the spatial structure we have to base our measurement on the positions of the molecules (Fig. 1, left), if we are interested in homogeneity we can use the relative distances between the molecules (Fig. 1, right). The emergence definition presented in this article is based on the statistical definition of entropy (which essentially can be explained as counting events or occurrences).

The computation of the entropy of a system S with N elements e_i is done as follows:

[1] We appreciate the discussion of the possible separation of emergence and self-organisation by T. De Wolf but claim that for practically interesting phenomena emergence always implies self-organisation.

[2] Currently the only observers who make these decisions are human designers and researchers, but eventually one could in fact imagine a system that could make these decisions based on knowledge bases and experiments with a target system (e.g. trying out a set of likely candidate attributes, etc.). The theoretical approach of this section is the basis for our observer/controller architectures discussed later in this article.

Fig. 1 Order perception: Both pictures could be perceived as high order (*left*: more structure, *right*: more homogeneity) depending on the objective of the observer

1. Select an attribute A of the system elements of S with discrete, enumerable values a_j.
2. Observe all elements e_i and assign a value a_j to each e_i. This step corresponds to a quantisation.
3. Transform into a probability distribution (by considering the relative frequency as a probability) over the attribute values a_j (i.e. a histogram) with p_j being the probability of occurrence of attribute a_j in the ensemble of elements e_i.
4. Compute the entropy according to Shannon's definition

$$H_A = -\sum_{j=0}^{N-1} p_j \cdot ld(p_j) \tag{1}$$

If the attribute values are equally distributed (all p_j equal) we will obtain the maximum entropy. Any deviation from the equal distribution will result in lower entropy values (i.e. higher order). In other words: The more structure is present (unequal distribution), the more order is measured. The unit of measurement is bit/element. So the entropy value can be interpreted as the information content necessary to describe the given system S with regard to attribute A. A highly ordered system requires a simpler description than a chaotic one.[3]

3 Observation Model

The resulting entropy value depends on two decisions of the observer: (1) Which attribute A is measured? and (2) With what resolution (or quantisation) is A measured? The quantisation determines the information content of the system description but it is not a property of the system. Neither is the selection of a certain attribute A a system property. This means that a measured entropy value is only meaningful if we know the exact observation context. This context is subsumed by the observation model.

This reflects the fact that order is not an intrinsic property of the system. Rather order depends on subjective decisions or capabilities of the observer. In living systems the sensory equipment limits the selection of observable attributes and the

[3]This reminds of the definition of Kolmogorov complexity [5].

resolution of the measurement. In addition, the brain directs attention to certain observables, which are relevant in the present situation, and masks other attributes or registers them with lower effort, i.e. lower resolution. Hence, order results from an interaction between the observer and the observed system guided by the observation model. The observation model depends on the capabilities of the sensory equipment and the utility of certain observations with regard to the purpose.

An observer might be interested in more than one attribute. In this case, we obtain a vector of entropy values $(H_A, H_B, H_C, \dots)$ with respect to attributes $A, B, C, \dots$ We could add them into a total system entropy HS. HS denotes the information content of the total system description under the given observation models. It has the drawback of hiding or averaging the single attribute entropies. Therefore we prefer the non-added entropy values.

4 Emergence

Entropy is not the same as emergence. Entropy decreases with increasing order while emergence should increase with order. As a first try we define emergence as the difference ΔH between the entropy at the beginning of some process and at the end:

$$\Delta H = H_{start} - H_{end} \tag{2}$$

In case of an increase of order this results in a positive value of ΔH. A process is called emergent if (1) $\Delta H > 0$ and (2) the process is self-organised. This definition has two problems:

1. The measurement of H depends on the observation model (or abstraction level). An observation on a higher abstraction level will lead to a lower entropy value H even when there is no change of S in terms of self-organised order.
2. Since the start condition of the system is arbitrary, ΔH represents only a relative value for the increase of order. It would be preferable to have a normalised emergence measure.

The first problem can be solved by determining the portion of ΔH, which is due to a change of abstraction level (ΔH_{view}) and subtracting it.

$$\Delta H = \Delta H_{emergence} + \Delta H_{view} \tag{3}$$

$$\Delta H_{emergence} = \Delta H - \Delta H_{view} \tag{4}$$

In other words: Equation (1) holds only if we have not changed the abstraction level or if ΔH_{view} can be determined and subtracted.

The second problem is solved by definition of an absolute reference as starting condition. Obviously this starting condition could be the state of maximum disorder with an entropy of H_{max}. H_{max} corresponds to the equal probability distribution.

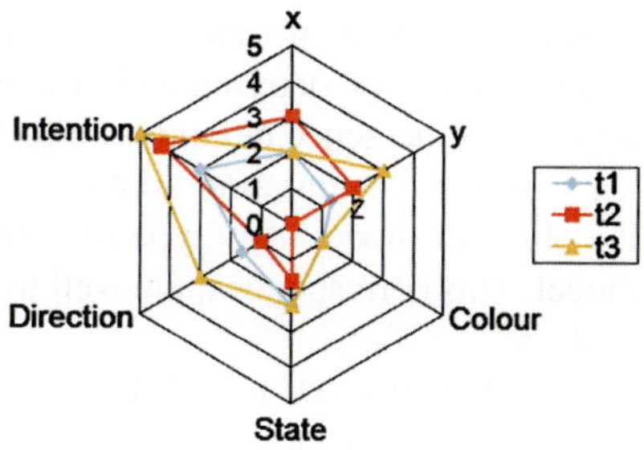

Fig. 2 Visualisation of the emergence fingerprint

This leads to the following definition:

$$\text{Emergence} \quad M = \Delta H_{emergence} = H_{max} - H - \Delta H_{view} \tag{5}$$

Absolute emergence is the increase of order due to self-organised processes between the elements of a system S in relation to a starting condition of maximal disorder.

The observation model used for both observations must be the same ($\Delta H_{view} = 0$) or ΔH_{view} must be determined and subtracted.

We can also define the relative emergence m as

$$m = \frac{H_{max} - H}{H_{max}} \quad (\text{if } \Delta H_{view} = 0) \tag{6}$$

m has a value between 0 and 1, $m = 0$ means high disorder and $m = 1$ high order.

Of course we can define—in analogy to the above discussion—attribute-specific emergence-values M_A, M_B, M_C or m_A, m_B, m_C. The vector of M_k or m_k (with k denoting the attributes of the elements of a given system) constitutes a so-called emergence fingerprint. An emergence fingerprint can be visualised e.g. as a 6-dimensional Kiviat graph (Fig. 2). It represents the "order pattern" with regard to the chosen attributes. In the example of Fig. 2, the 6 attributes x and y coordinate, status, intention (of an animat), direction of movement and colour have been chosen. The fingerprint changes over time ($t0$, $t1$, $t2$) and thus represents the development of order in the different dimensions (attributes).

5 Discussion

5.1 Limitations

The definition is not applicable if ΔH_{view} cannot be made zero or determined quantitatively. This is always the case if the macro phenomenon is totally different from the micro behaviours as seemingly in the case of the resonance frequency as an emergent property resulting from the interaction of a capacitor and an inductivity. Our quantitative definition of emergence is based on the assumption that emergent phenomena can always be observed in terms of patterns (space and/or time) consisting of large ensembles of elements. The resonance frequency of a resonant circuit does not constitute the emergent pattern but is rather a property of such a pattern. Order can also be determined in the time or frequency domain. Therefore, we can apply our emergence definition to the resonance frequency example if we observe the system behaviour after a Fourier analysis. This extends the above definition of the observer model: Any type of preprocessing can also be a part of the observer model. This corresponds quite well to the operation of the animal (and human) perception.[4]

We admit that our model does not cover the so-called 'strong emergence' definition, which demands that emergence is a phenomenon *principally* unexplainable. But this is a quite unscientific argumentation, which we reject. To the contrary, we would like to propose, that only a quantifiable phenomenon resulting in a (self-organised) increase of order, deserves to be accepted as emergence. If this definition is too restrictive, excluding some unexplainable emergent effects, we could accept that what we measure with our method is "quantitative emergence" and constitutes a certain form of emergence meaningful in technical systems. This definition leaves room for wider definitions of emergence in a more general meaning.

5.2 Redundancy and Emergence

The reader familiar with Shannon's information theory might have recognised that our definition of emergence is formally equivalent to Shannon's redundancy since redundancy

$$R = H_{max} - H \tag{7}$$

and relative redundancy

$$r = \frac{H_{max} - H}{H_{max}}. \tag{8}$$

Redundancy is a property of a message source and a code, which should be reduced as far as possible in order to utilise a given channel optimally. Emergence on the other hand is a measure of order in a system, which in many cases is highly desirable. At least, as we have seen, a high emergence value means that the description complexity of the system is low. The explanation of the apparent contradiction lies in the different points of view of a communication engineer (Shannon) and a systems engineer.

A channel is utilised optimally if it transports only information, which is new to the receiver. Shannon defines predictable information as useless or redundant. But the notion that predictable information is useless contradicts both our intuition and biological research. It is justified only in communications engineering. Shannon assumes that sender and receiver have the same semantic framework, the receiver has a-priori knowledge, he can match a received message against a known

[4]In the cochlea, the sound moves hair bundles, which respond to certain frequencies. The brain therefore reacts to preprocessed signals [10].

set of messages. A received message has a value only within this pre-defined context. The common context in a traditional technical system is given by the fact that both, sender and receiver (and the communication system in between), have been designed by the "same" engineer.

This is not true for adaptive systems (living systems or organic computing systems). Communicating adaptive systems have (1) to maintain a valid context, which might change over time, and (2) to validate new information within this context. The maintenance of context requires regular *affirmation* and a certain degree of modification. The more context the receiver has accumulated in his memory (be it genetic or acquired), the less new or current information he needs in order to recognise a certain situation and act accordingly.

This means that each animal (or animat[5]) will strive to build and maintain a dependable and stable context "database" which allows it to reduce the amount of new information to be transferred in a possibly dangerous situation and hence save valuable time and energy. This works fine as long as the environment is stable and in agreement with the memorised context database. In changing situations this agreement will be disturbed, and erroneous decisions will follow. The context information has to be updated as fast as possible.

The above discussion suggests a possible approach to separate information into two parts: *affirmation* information and *newness* information. The first part is used to affirm and/or modify the context database, the second selects between several contexts. But it is more realistic to assume that each received message is used as a whole for both purposes: It is compared to the stored contexts and selects one of them in case there is a sufficient match. A successful match results in an action (stimulus-response relationship). If there is no match or if there is a wrong match, a detrimental action will result leading to the necessity to update the context database. An update phase will require the temporary collection of more information until the animat can return to business as usual.

An animat (animal) prefers a low entropy environment (orderly and predictable, with high emergence)! On the other hand, instability (caused by random processes like mutation) is the necessary precondition for the exploration of unknown areas of the configuration space.

5.3 Pragmatic Information

The term *pragmatic information* has been introduced by Christine and Ernst von Weizsäcker, cited by Küppers in [4]. Pragmatic information means information, which has an effect on the receiver. This could be a structural change (leading to some kind of action) or the readiness of the receiver for some action ([4, p. 85]).

[5]Animats are artificial animals, e.g. robots equipped with sensors, actuators and some decision mechanism.

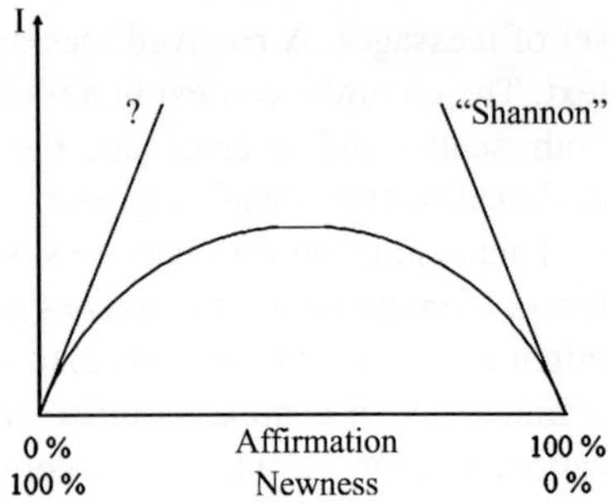

Fig. 3 Newness and affirmation [4]

Pragmatic information is determined by two aspects: Newness (German: *Erstmaligkeit*) and affirmation (German: *Bestätigung*). A qualitative curve proposed by Christine and Ernst von Weizsäcker (Fig. 3) claims that the Shannon part of the information measures newness. It is zero for a known (i.e. predicted) event and increases with increasing newness. On the other hand, for a message to be recognised within a context there must be a-priori knowledge of this context. Therefore, pragmatic information is also zero if there has been no prior communication between sender and receiver. In this case, newness = 100%. Affirmation is complementary to newness: It is zero when a message is totally new, and 100% when a message is known in advance. If there are two zero values of pragmatic information, there must be at least one maximum in between. Von Weizsäcker concludes that to achieve a maximum of pragmatic information, there must be a certain optimal combination of newness and affirmation in the communication relation between two partners.

A highly predictable system as message source (i.e. a system displaying high order = high emergence) requires a channel with low bandwidth because the transmitted information serves essentially as affirmation, which is needed less frequently. The receiver compares these messages to his context database, finds matches with a high probability and initiates corresponding actions. If affirmation is sent too frequently, it becomes useless: the channel transports redundant information (already known by the receiver). As soon as the message source changes its behaviour, the receiver needs more frequent update information in order to change his context database. The newness aspect of the messages becomes more important.

Technically speaking, there are two mechanisms in the receiver working in parallel on all received messages. On the lower "working" level, messages are compared against known probability distributions and mapped to actions. On a higher semantic level, the newness of all messages and the effectiveness of the corresponding actions are monitored. In case of inadequate results, the structure of the receiver has to be adapted by changing/extending the context database and adding new actions. This higher level is realised by (possibly multi-level) Observer/Controller architectures as shown in the next section. The lower level corresponds to the production system.

6 Observer/Controller Architecture

The objective of our work is to make emergent phenomena accessible to a practical measurement process. This is important in technical applications that have to detect emergent phenomena in order to support or to suppress them. In other projects presently run by the authors the collective behaviour of chicken in a chicken farm [3], the behaviour of cars in the environment of an intersection [7] or the synchronisation of elevators (so-called Bunching effect [3]) is of interest. We propose a generalised Observer/Controller architecture [8] (Fig. 4). The observer collects and aggregates information about the production system. The aggregated values (system indicators) are reported to the controller who takes appropriate actions to influence the production system. The observer contains several specialised detectors to calculate the system indicators from the observed raw data (Fig. 5). We are presently building emergence detectors specialised for certain attributes. The collection of attribute emergence values (the emergence fingerprint) is a part of the observation

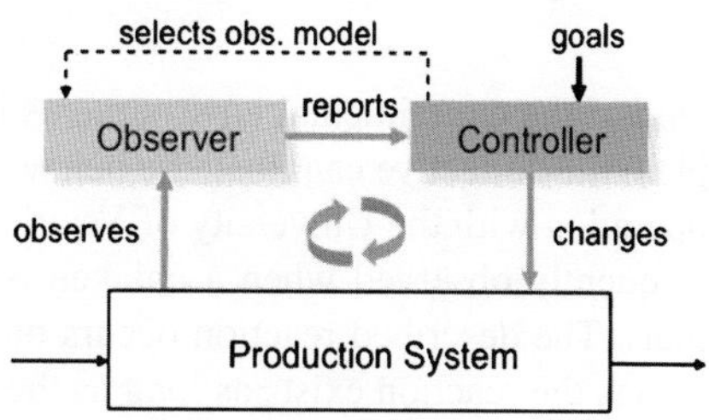

Fig. 4 Observer/controller architecture

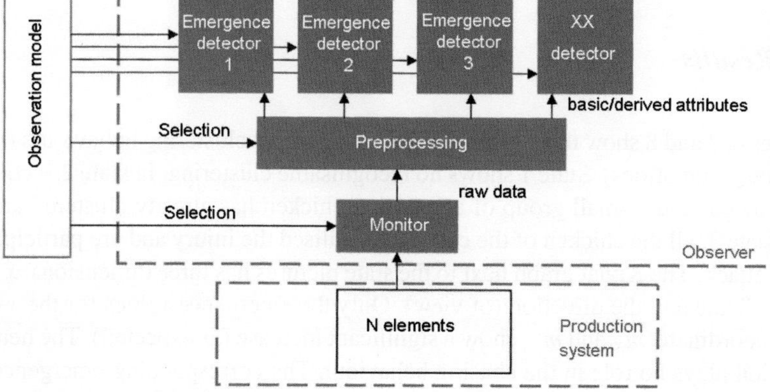

Fig. 5 Observer architecture

result as determined by the observer. The observer model influences the observation procedure, e.g. by selecting certain detectors or certain attributes of interest. The feedback from the controller to the observer directs attention to certain observables of interest in the current context.

7 Experimental Results

In this section we present first experimental results of the emergence fingerprint. The results discussed here could be part of the observer in the Observer/Controller architecture presented in Sect. 6.

7.1 Experimental Environment

One of the experimental environments is a chicken simulator, whose goal is to explain the collective cannibalistic behaviour of densely packed chicken in cages (cooperation with the University of Veterinary Medicine Hannover). This behaviour is frequently observed when a chicken is injured, and leads to a major loss of animals. The described reaction occurs only on the basis of an optical stimulus. That means the reaction exists as long as the stimulus is apparent. While simulating this behaviour, order patterns emerge in form of chicken swarms. These patterns are at present interpreted by human experts. It should be possible to classify them automatically. The emergent behaviour in this scenario is spatial, but swarms move over time. This is a case of "negative", i.e. unwanted, emergence, since the global goal is to reduce chicken death rate. The controller has to react with actions to disperse the swarms.

7.2 Results

Figures 6, 7 and 8 show three typical states of emergent clustering behaviours (taken from our simulations). State 1 shows no recognisable clustering. In state 2, a chicken is wounded, and a small group of aggressing chicken has already clustered around it. In state 3, all the chicken of the cage have realised the injury and are participating in the attack. The Kiviat graph next to the state pictures has three dimensions: x- and y-coordinate and the direction (of view). Only the emergence values for the x- and the y-coordinate, m_x and m_y, show a significant increase (as expected). The heading direction plays no role in the chasing behaviour. The corresponding emergence m_d stays very small. Figure 9 shows the overlay of the 3 states and their development over time.

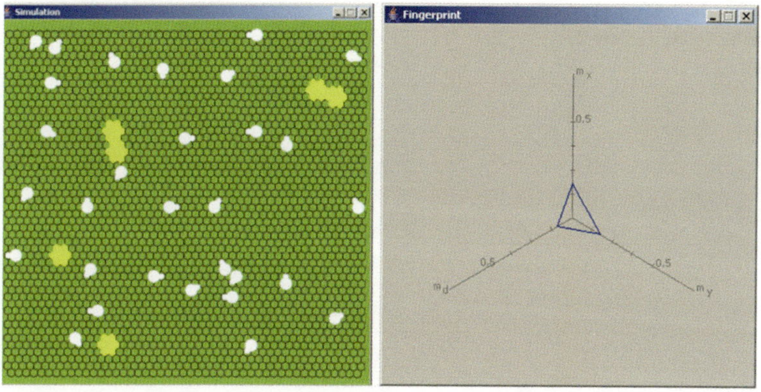

Fig. 6 Emergence fingerprint of state 1: no cluster; $m_x = 0.181$, $m_y = 0.177$, $m_d = 0.091$ (uninjured chicken: white, feeding troughs: hexagons)

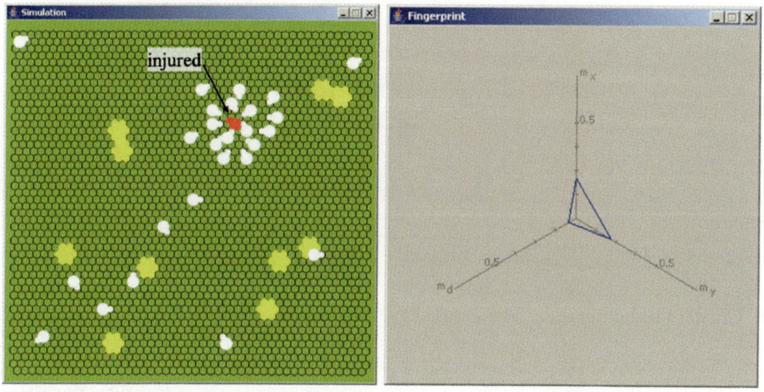

Fig. 7 Emergence fingerprint of state 2: small cluster; $m_x = 0.226$, $m_y = 0.237$, $m_d = 0.046$

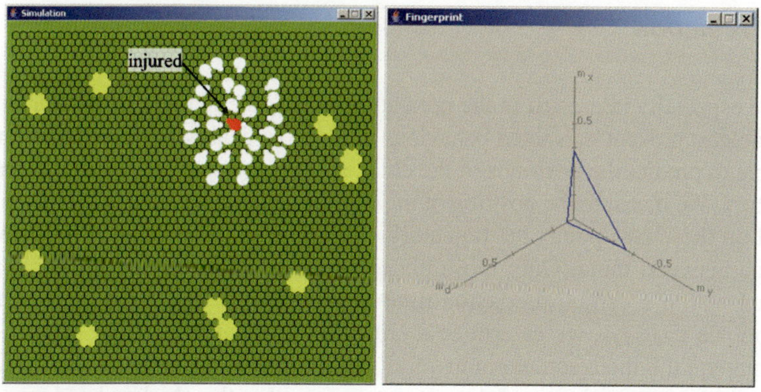

Fig. 8 Emergence fingerprint of state 3: one big cluster; $m_x = 0.359$, $m_y = 0.328$, $m_d = 0.041$

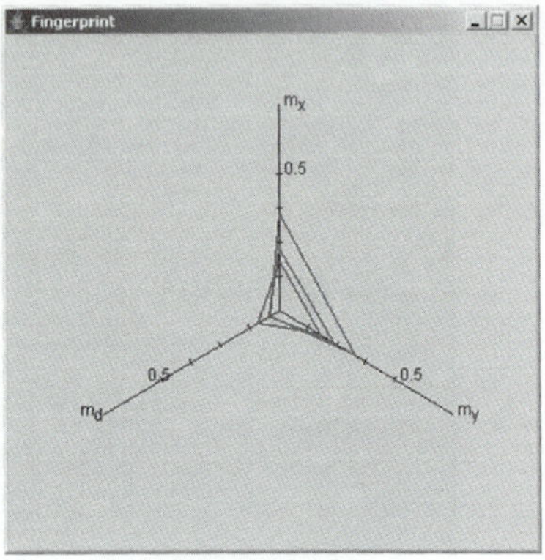

Fig. 9 Overlay of 3 fingerprints (state 1, state 2 and state 3)

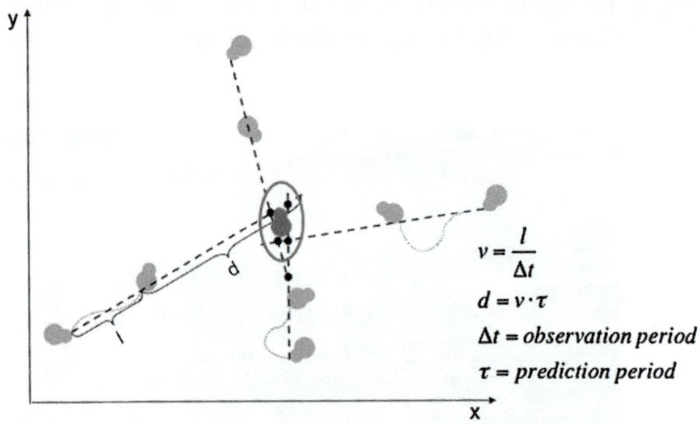

Fig. 10 Trajectory-based prediction method of chicken positions

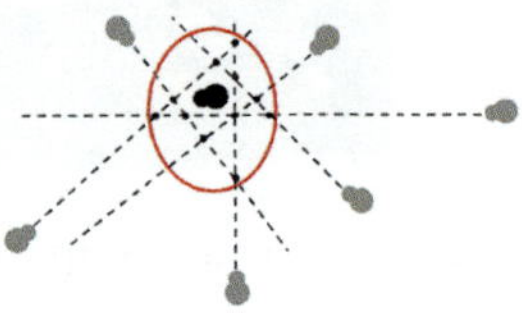

Fig. 11 Cluster prediction

7.3 Prediction

We are especially interested in the prediction of future emergent behaviour in order to be able to prevent unwanted behaviour in time. To deal with this goal, it must be possible to predict the positions of the chicken. This can be done by extrapolating a trajectory. We measure the position of the chicken at two consecutive points in time. Based on these two points the trajectory of the chicken is computed by extending the line between them. Only those positions are of practical interest that can be reached within a certain prediction time $\Delta t_{prediction}$. Using the present speed of the animals $v_{average}$, we determine a critical distance $d = v_{average} \cdot \Delta t_{prediction}$. We compute the intersection points of all trajectories within the critical distance d (Fig. 10). An accumulation of these points means that the chicken head on a point in the area, which might indicate the existence of an injured chicken (Fig. 11). This

Fig. 12 Prediction: The emergence $m_{x,prediciton}$ indicates a future cluster (m_x), which emerges about 10 time units later

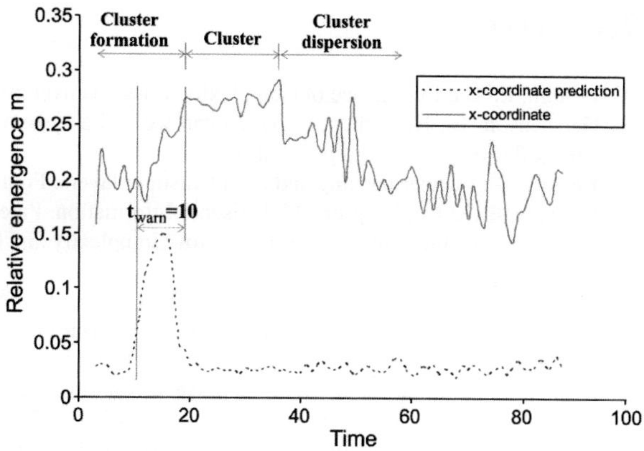

point accumulation in space can also be measured by using the emergence indicator applied to their x- and y-coordinates. Figure 12 shows the emergence values of the x-coordinate of the chicken positions and the ones of the intersection points of the trajectories. The emergence of the intersection points grows before the emergence of the actual chicken positions and can therefore be used as an early warning indicator for chicken clustering. We are currently experimenting to increase the prediction time and to reduce the effect of noise.

8 Conclusion and Outlook

We have proposed a quantitative measure of emergence based on the statistical definition of entropy and discussed it in comparison with Shannon' s information theory. While proponents of a so-called strong definition of emergence might argue that "true" emergent effects must always represent something totally new and unexpected, we claim that with our emergence definition we can measure at least some effects of the generation of order. Emergence definitions going beyond might have to live with the flaw that they principally do not lend themselves to a quantitative approach.

It is the objective of our work to make emergent effects quantitatively treatable in a technical environment. We have proposed an observer/controller architecture with special detectors for determining attribute emergence values. First experimental results obtained from a chicken simulation show the viability of the approach.

We plan to extend these detectors by preprocessing steps, using perhaps a Fourier analysis to make regular patterns like a crystal lattice treatable. The method will be applied to more technical problems like self- synchronising elevators (bunching effect) and traffic simulations.

Acknowledgements This work has been done in close co-operation with Hartmut Schmeck, Jürgen Branke, Urban Richter, Holger Prothmann (University of Karlsruhe) and Fabian Rochner (University of Hannover) within the DFG Priority Program Organic Computing. We are especially indebted to Kirstie Bellman, The Aerospace Corporation, Los Angeles, USA, for reviewing the manuscript and making valuable suggestions for improvement.

References

1. Fromm, J.: The Emergence of Complexity. Kassel University Press (2004)
2. Gershenson, C., Heylighen, F.: When can we call a system self-organizing? arXiv:nlin.AO/0303020 (2003). Informal publication
3. Keeling, L.: Feather pecking and cannibalism in layers. Poult. Int. **6** (1995)
4. Küppers, B.O.: Der Ursprung biologischer Information. Piper (1990)
5. Li, P.V.M.: An Introduction to Kolmogorov Complexity and Its Applications. Springer, Berlin (1997)
6. Müller-Schloer, C.: Organic Computing: On the feasibility of controlled emergence. In: Proceedings of the 2nd International Conference on Hardware/Software Codesign and System Synthesis, pp. 2–5 (2004)
7. Rochner, F., Müller-Schloer, C.: Adaptive decentralized and collaborative control of traffic lights. In: Dadam, P., Reichert, M. (eds.) INFORMATIK 2004 – Informatik verbindet. GI-Edition – Lecture Notes in Informatics (LNI), vol. P-51, pp. 595–599. Köllen Verlag, Bonn (2004)
8. Schöler, T., Müller-Schloer, C.: An observer/controller architecture for adaptive reconfigurable stacks. In: Beigl, M., Lukowicz, P. (eds.) ARCS. Lecture Notes in Computer Science, vol. 3432, pp. 139–153. Springer, Berlin (2005)
9. Stephan, A.: Varieties of emergence in artificial and natural systems. Z. Naturforsch. **53c**, 639–656 (1998)
10. Wikipedia: http://en.wikipedia.org/wiki/cochlea
11. Wolf, T.D., Holvoet, T.: Emergence versus self-organisation: Different concepts but promising when combined. In: Brueckner, S., Serugendo, G.D.M., Karageorgos, A., Nagpal, R. (eds.) Engineering Self-Organising Systems. Lecture Notes in Computer Science, vol. 3464, pp. 1–15. Springer, Berlin (2004).

M. Mnif · C. Müller-Schloer (✉)
Institute of Systems Engineering, Leibniz Universität Hannover, Appelstr. 4,
30167 Hannover, Germany
e-mail: cms@sra.uni-hannover.de

M. Mnif
e-mail: mnif@sra.uni-hannover.de

Chapter 1.3
Divergence Measures as a Generalised Approach to Quantitative Emergence

Dominik Fisch, Martin Jänicke, Christian Müller-Schloer, and Bernhard Sick

Abstract A technical system exhibits emergence when it has certain properties or qualities that can be termed to be irreducible in the sense that they are not traceable down to the constituent parts of the system. The article summarises three techniques for emergence detection and emergence measurement that were proposed by members of the Organic Computing community. These techniques are based on information-theoretic and probabilistic viewpoints: the discrete entropy difference discussed in detail in the previous article, the Hellinger distance which is a divergence measure for probability densities, and an iterative approach motivated by divergence measures. Advantages and drawbacks of these measures are demonstrated by means of some simulation experiments using artificial data sets. It is shown that these techniques are able to deal with different kinds of emergent phenomena such as transitions from chaos to order, concept drift, or novelty. That is, with these techniques it is possible to cover a wide range of possible applications.

Keywords Quantitative emergence · Situation awareness · Organic computing

1 Introduction

The phenomenon of *emergence* in technical systems has recently been addressed by many researchers, e.g., Abbot, Stephan, Fromm, Gabbai et al., Marzo Serugendo et al., De Wolf et al., and also the authors of this article. In [1, 2] we summarised, compared, and assessed various definitions of emergence. To give at least a quite informal definition of this term here, we can say: A technical system exhibits some kind of emergence when it has qualities or properties that are *irreducible* in the sense that they are not directly traceable down to the constituent parts of the system (for detailed definitions see [1–3]). There is an increasing demand for techniques that help to detect emergence in technical systems. Moreover, in many applications, e.g., applications of distributed, self-organising systems, it would even be advantageous to measure emergence numerically (*quantitative emergence*):

- If emergence is seen as the result of a bottom-up process in a self-organising technical system, there is a need to keep that kind of emergence under control (i.e., the behaviour of the overall system must fulfil specified requirements) and, thus, to measure emergence.

C. Müller-Schloer et al. (eds.), *Organic Computing — A Paradigm Shift for Complex Systems*, 53–66
Autonomic Systems, DOI 10.1007/978-3-0348-0130-0_3, © Springer Basel AG 2011

- If a response to detected emergence is necessary, e.g., by self-adaptation of a technical system that observes emergence in its environment, there is also a need to assess emergence numerically in order to avoid overreactions, for instance.

Various application scenarios for quantitative emergence can be found in the field of *Organic Computing*, for instance [4].

Here, we briefly summarise, compare, and assess three approaches proposed by the authors of this article in the past years: The first, proposed by Mnif and Müller-Schloer (see Chap. 1.2 of this book and [5, 6]), is based on an entropy measure. The second, suggested by Fisch, Jänicke, Sick, and Müller-Schloer in [7], is based on divergence measures. The third and most recent one, developed by Fisch, Jänicke, and Sick (see [8]), is motivated by a divergence measure but it has a lower computational complexity. While the first approach sees emergence as a change of order within a system (a kind of transition from "chaos" to "order"), the other two approaches have a more generalised view: Here, emergence is regarded as an unexpected or unpredictable change of the distribution underlying certain observations of the system (i.e., samples).

The remainder of this article begins with a discussion of some related work in Sect. 2. Section 3 briefly summarises the techniques for emergence measurement mentioned above and discusses their advantages and drawbacks. Then, some simulation experiments in Sect. 4 outline differences between these measures. Finally, Sect. 5 concludes the article by summarising the major findings.

2 State of the Art

In the literature, there are various approaches that deal with the problem of a detection or a numerical assessment of emergence. Interestingly, entropy-based measures were suggested in most cases. Divergence-based measures are quite new to this field.

In [9] and [10] (published in 1994 and 2001, respectively), emergence in time series is measured by the ability of an automaton (i.e., an ϵ-machine) to model an observed system. A process is defined to be emergent from another process if it has a greater predictive efficiency. This property is measured by relating the entropy of the automaton to the entropy of the process. An important side note can be found in an article of Gabbai et al. published in 2005 [11]: Entropy could be utilised to measure order which emerges, e.g., due to self-organisation, and, hence, to measure emergence. In the same year, De Wolf et al. proposed a discrete entropy measure to analyse self-organising emergent behaviour in decentralised systems [12]. An emergence measure based on the difference of two entropy values determined at different points in time was proposed by Mnif and Müller-Schloer in 2006 [5] for emergence measurement in self-organising systems (see also [6]). An emergence measure based on a quotient of entropy values was introduced by Holzer et al. in 2008 [13] to analyse discrete self-organising systems. Altogether, it can be stated that there was a focus on emergence measurement in discrete, self-organising systems.

Two related divergence-based emergence measures were proposed by Fisch et al. in 2010 [7] and 2011 [8], respectively. These measures evaluate probability densities that describe the state of the system, e.g., at two different points in time. An early version of the techniques set out in [8] can be found in [14].

The problem of emergence measurement is also related to problems such as novelty detection, outlier detection, or anomaly detection. References to related work can be found in [7, 8].

3 Techniques for Emergence Detection and Measurement

In this section we will briefly summarise the three techniques for emergence detection and/or emergence measurement.

3.1 Discrete Entropy Difference

To detect or to measure the emergent behaviour of a *technical system*, we observe this system (e.g., by means of sensors) and use application-specific pre-processing techniques to extract the values of *attributes* (features) from those *samples* (observations). These attribute values describe the current behaviour of the observed system. The attribute space will be modelled by a variable $\mathbf{x}$ in the following (or x in the case of a one-dimensional attribute space). Basically, $\mathbf{x}$ may consist of categorical and continuous attributes, for instance.

As outlined in the previous article of this book, Mnif and Müller-Schloer define their measure for emergence using *Shannon's entropy* measure [5]. Each attribute x of a system S is seen as a categorical random variable and $p(x)$ is the (estimated) probability of the attribute taking a certain value x. Then, the entropy of x is defined by

$$\mathrm{H}[x] = -\sum_x p(x)\,\mathrm{ld}\,p(x), \tag{1}$$

where ld is the logarithm to the base of two, as this approach is motivated by information theory. As emergence is seen as a *change of order* within S, Mnif and Müller-Schloer define emergence as the difference between an entropy at the beginning of a time interval and at the end (therefore, we call it DED: *discrete entropy difference*):

$$\mathrm{DED}[x] = -\sum_x q(x)\,\mathrm{ld}\,q(x) + \sum_x p(x)\,\mathrm{ld}\,p(x). \tag{2}$$

That is, DED compares two distributions (discrete density functions), $p(x)$ determined at an earlier point in time and $q(x)$ determined later. A process within S is called *emergent* if $\mathrm{DED}[x] > 0$ and the process is self-organised (cf. [15]). This measure may be unsatisfying for two reasons: (1) There are many attributes with *continuous values* in practical applications and (2) many applications are *multivariate*, i.e., based on several (categorical and continuous) attributes. The former problem is solved by categorisation of continuous attributes in [5]. This could be problematic as entropy measurements depend on size and position of the chosen "bins". The latter problem can be solved by averaging DED values for different attributes.

3.2 Divergence-Based Emergence Measures

A more general approach proposed by Fisch et al. in [7] does not claim that the observed process has to be self-organised. Alternatively, emergence is regarded as an *unexpected or unpredictable change of the distribution* underlying the observed samples. Then, a *divergence measure* is the obvious way to compare two density functions. Multivariate divergence measures can be defined for hybrid attribute spaces that contain continuous as well as categorical attributes. Fisch et al. suggest to use the *Hellinger distance* (Hel) [16] as an emergence measure since it obeys the triangle inequality and its values lie in the unit interval:

$$\mathrm{Hel}(p,q) = \sqrt{1 - \mathrm{BC}(p,q)}, \tag{3}$$

where BC is the *Bhattacharyya coefficient* given by

$$\mathrm{BC}(p,q) = \int \sqrt{p(\mathbf{x})\,q(\mathbf{x})}\,d\mathbf{x} \tag{4}$$

for continuous variables or by

$$\mathrm{BC}(p,q) = \sum_{x} \sqrt{p(x)\,q(x)} \tag{5}$$

for categorical ones.

There are two fundamentally different approaches to evaluate Hel, i.e., to determine values of the density functions and to compute the integral in (4) (details can be found in [7]):

- In a *non-parametric approach*, no assumptions about the functional form of the density functions are made. In the case of continuous attributes, values of the density function are determined with a *Parzen window technique*. The integral can be discretely approximated in a way that is similar to an *importance sampling* technique.
- In a *parametric approach*, assumptions are made concerning the functional form of the density functions (see [17] for details of the modelling approach). Fisch et al. use *mixture density models* with hybrid components: *Multivariate Gaussians* are taken to model the continuous variables and *multinomial distributions* are employed for the categorical ones. The categorical variables are seen as independent. The parameters of these functions are estimated with a *maximum likelihood* technique. Then, the Bhattacharyya coefficient can directly be computed from these estimates.

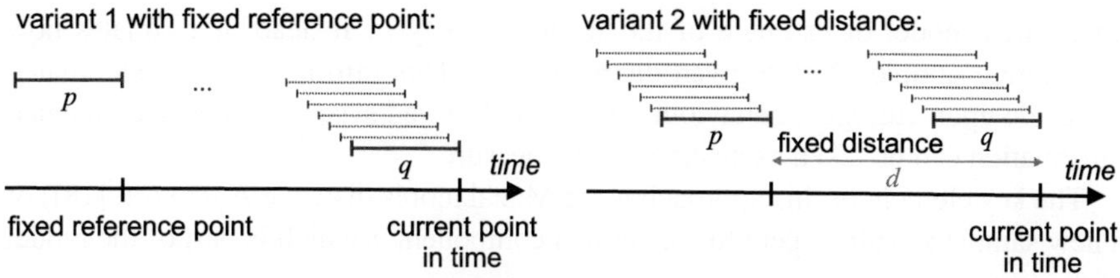

Fig. 1 Two variants of density estimation in sliding time windows

Fisch et al. suggest two ways to apply the proposed measure: It is assumed that one single observation is made at each point in time (discrete time step). Then, p and q are estimated in sliding data windows with fixed length. These windows must be long enough to estimate p and q with sufficient reliability and, at the same time, short enough to allow for the assumption that the observed processes are nearly time-invariant in these windows. There are applications where the first (earlier) time interval is fixed at a certain point in time whereas the second interval moves along the time axis with the current point in time (in an on-line application), but also applications where both windows may move along the time axis in a fixed temporal distance (see Fig. 1). Then, this distance d is an important parameter of the measurement technique. As sketched above, the densities p and q can be estimated in a non-parametric or a parametric (model-based) approach, depending on the application at hand. Hybrid approaches are viable as well.

With Hel it is not only possible to measure a "degree" of emergence if desired. With the mixture models described above we can detect processes that disappear (i.e., components that become obsolete), newly emerging processes (i.e., new components are required), or components that change their characteristics (i.e., components change their parameters such as centres or mixing coefficients). In the field of machine learning, the latter is known as *concept drift*. The problem to detect new processes is known as *novelty detection* in the literature.

Divergence-based emergence measures, however, are not free from drawbacks. First, an appropriate application-dependent window size must be defined by the user. Second, the estimation of the density in the sliding windows is (especially in the non-parametric case) computationally expensive.

3.3 Approximations of Divergence-Based Emergence Measures

The technique suggested by Fisch et al. in [8] tries to overcome the problems of the divergence-based technique described above. In [8] it is described for continuous variables only but it could basically be extended to hybrid (continuous and categorical) sets of attributes. It can be classified as a parametric approach since it assumes that the density p can be approximated with a Gaussian mixture model (GMM). The components of this GMM are multivariate Gaussians with centres $\boldsymbol{\mu}_j$ and covariance matrices $\boldsymbol{\Sigma}_j$ for a component j. However, their technique does not require an explicit model or estimate of the second density q. Instead, it estimates how well single samples "fit" a given distribution p. This information is accumulated to decide upon emergent behaviour within an observed system S. The accumulated information can be taken as an emergence measure.

The key element of this approach is the Mahalanobis distance $\Delta_j(\mathbf{x})$ (cf. [18]) of a new sample $\mathbf{x}$ with respect to a Gaussian component j which is part of the model of density p:

$$\Delta_j(\mathbf{x}) = \sqrt{(\mathbf{x} - \boldsymbol{\mu}_j)^{\mathrm{T}} \boldsymbol{\Sigma}_j^{-1} (\mathbf{x} - \boldsymbol{\mu}_j)}. \tag{6}$$

The method exploits the fact that the squared Mahalanobis distances $\Delta_j^2(\mathbf{x})$ of a Gaussian component j are χ^2-distributed. Knowing the distribution of the Mahalanobis distances, it is possible to define a hyper-ellipsoid around each centre $\boldsymbol{\mu}_j$ such that it can be expected that a certain percentage α of the samples produced by the process which is modelled by component j lies within that hyper-ellipsoid. The samples within the hyper-ellipsoids are called *inside samples* and the remaining ones are called *outside samples*. The Mahalanobis radius ρ of these hyper-ellipsoids, which depends on the chosen value of α, can be determined by means of the inverted cumulative χ^2 distribution.

Based on these hyper-ellipsoids, an emergence status s_{emg} can be defined that can be regarded as the degree of "fitness" of the trained GMM with respect to the currently observed situation. This status is updated with every new observation. It is rewarded if the observation is an inside sample and penalised otherwise. If the ratio of penalty to reward is equal to the expected ratio of inside to outside samples, s_{emg} oscillates around its initial value. In this case, the GMM fits the observed samples. If more than a percentage $1 - \alpha$ of outside samples is observed (e.g., due to a new process that is not yet covered by the GMM), s_{emg} is penalised more often and, thus, decreases.

In the case of GMM with two or more components, the situation gets more complicated as Gaussians overlap and inside samples of one component may be outside samples of another component and vice versa. Thus, it is necessary to scale the rewards and penalties in order to sustain an equilibrium. Basically, this scaling factor could be determined by approximating the volume of the overlapping regions with appropriate Monte Carlo techniques (analytic solutions are not possible in case of multivariate Gaussian components). However, it is also possible to calculate an individual scaling factor for each observation $\mathbf{x}$ as shown in [8].

With individual scaling factors, the proposed emergence detection algorithm works as set out in Algorithm 1. The computational complexity of this algorithm is $\mathcal{O}(J \cdot I^2)$ per sample, where I is the number of attributes (dimensionality of the input space) and J the overall number of components within the GMM. Note that unlike the Parzen window approach, which has a complexity of $\mathcal{O}(w \cdot I)$ per sample, this algorithm does not depend on the number of samples w in a (sliding) time window.

Details concerning the properties of this technique and its parametrisation can be found in [8].

Algorithm 1 Emergence Measure:

1. Set the percentage α of samples that are expected to be inside samples (e.g., $\alpha := 0.9$) and penalty ν_{pen} and reward ν_{rew} values with the correct ratio (e.g., $\nu_{\text{rew}} := 0.1$, $\nu_{\text{pen}} := \frac{\alpha}{1-\alpha} \cdot \nu_{\text{rew}}$).
2. For a given, new sample $\mathbf{x}_n$ at time step n, determine the set J_{in} of components for which the sample $\mathbf{x}_n$ is an inside sample and the set J_{out} of the remaining components by comparing the squared Mahalanobis distance of $\mathbf{x}_n$ to the centres to a threshold ρ:

$$J_{\text{in}} := \{j | \Delta^2_{\boldsymbol{\Sigma}_j}(\mathbf{x}_n) \leq \rho\}$$

$$J_{\text{out}} := \{j | \Delta^2_{\boldsymbol{\Sigma}_j}(\mathbf{x}_n) > \rho\}.$$

   The threshold ρ is obtained by evaluating the inverse cumulative χ^2 distribution for the user-defined value α.
3. Compute an update value for the overall emergence status of the model by summing up the scaled rewards and penalties:

$$\Delta_{\text{emg}}(\mathbf{x}_n) := \eta \cdot \left(\sum_{j \in J_{\text{in}}} \gamma_{n,j} \cdot \nu_{\text{rew}} - \sum_{j \in J_{\text{out}}} \gamma_{n,j} \cdot \nu_{\text{pen}} \right)$$

   with η being the step size controlling the reaction time (e.g., $\eta := 0.01$). The values $\gamma_{n,j}$, which are called responsibilities of a component j for a given sample $\mathbf{x}_n$, are the conditional probabilities $p(j|\mathbf{x}_n)$.
4. Compute the new emergence status with

$$s_{\text{emg}} := \min \left\{ s_{\text{init}}, s_{\text{emg}} + \Delta_{\text{emg}}(\mathbf{x}_n) \right\}.$$

   The status s_{emg}, which must be initialised appropriately (e.g., with $s_{\text{init}} := 1$), can be used to detect emergence, e.g., by checking if it sinks below a given threshold τ.

The iterative technique described in Algorithm 1 is motivated by divergence measures. At first glance, it behaves like an "inverted" divergence measure: Starting from an arbitrary point (s_{init}) this measure decreases when the corresponding Hel measure increases. The main difference, however, is that the emergence status does—unlike Hel—not converge to a (lower) bound. Instead, it decreases further and further if the system does not react on stated emergence. In this case, it is necessary to reset the emergence measure back to its initial value. Another behaviour could be realised by using the emergence status as input of a monotonically increasing, bounded function such as the hyperbolic tangent.

The key advantage of this iterative technique is the low computational complexity which allows for time-critical applications. A drawback is that it is not assumption-free concerning the distribution of samples in the input space. The approach may fail if the model assumptions are not—at least approximately—met.

4 Experimental Results

In the following, we compare the different emergence measures in three different scenarios: Chaos to order, concept drift and novelty. For *DED* and the Hellinger approaches *HelPW* (non-parametric density estimation) and *HelGMM* (parametric density estimation), we use a fixed window p and a sliding window q, each of length (size) $w = 200$ as depicted in Fig. 1 (variant 1). At each time step q is advanced by one sample, i.e., a new sample is added and the oldest sample is removed. For the χ^2-approach (parametrised with $\alpha = 0.9$, $\eta = 0.01$), only the window p is needed.

Applications of the techniques with real data sets (e.g., in the field of intrusion detection) can be found in [6–8, 17] and in Chap. 3.3 of this book.

4.1 From Chaos To Order

In this first experiment we use two artificial data sets (DS_{1a} and DS_{1b}) each consisting of 2 000 samples. The first 1 300 are generated by a uniform distribution for both data sets. The remaining 700 samples are generated for DS_{1a} by an isotropic Gaussian distribution centred at position (0, 0) with variance 0.1 and for DS_{1b} by an isotropic Gaussian distribution centred at position (1, 1) (also with variance 0.1). Thus, both data sets represent a scenario in which after 1 300 samples there is a switch from "chaos" (i.e., the uniform distribution) to "order" (i.e., the Gaussian distribution). This experimental setup is motivated by several experiments and real applications described in [5] and [6].

Snapshots of this scenario at different points in time are illustrated in Figs. 2 (DS_{1a}) and 3 (DS_{1b}). There, the horizontal and vertical lines illustrate the bins that DED uses to discretise the input space, i.e., we have 5 bins with size 2 in each dimension. Note the main difference of data sets DS_{1a} and DS_{1b}: In the first case, the Gaussian roughly "covers" a single bin in each dimension, whereas in the second case the Gaussian "covers" two bins.

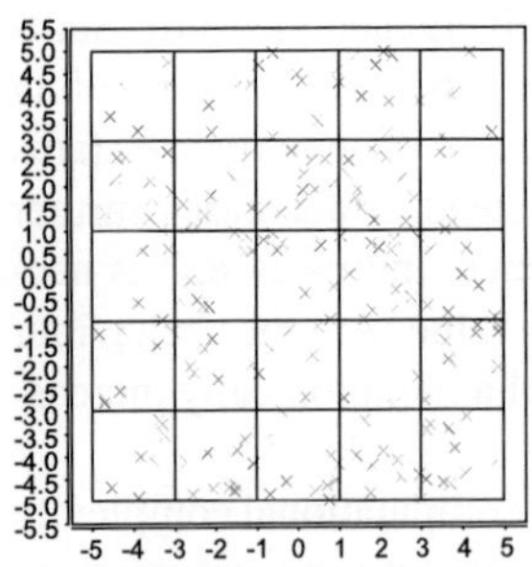

(a) Time Step 500

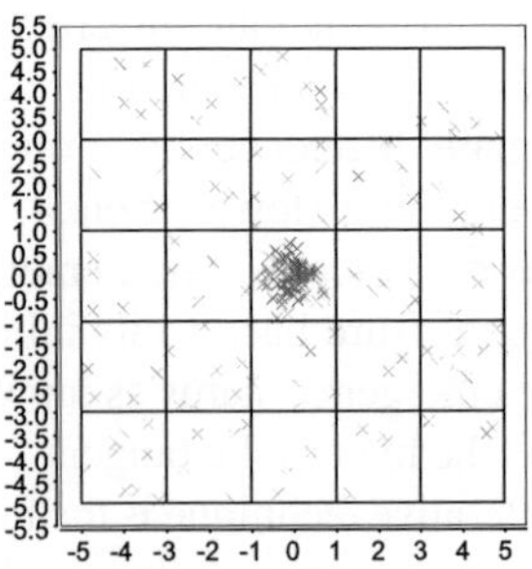

(b) Time Step 1 350

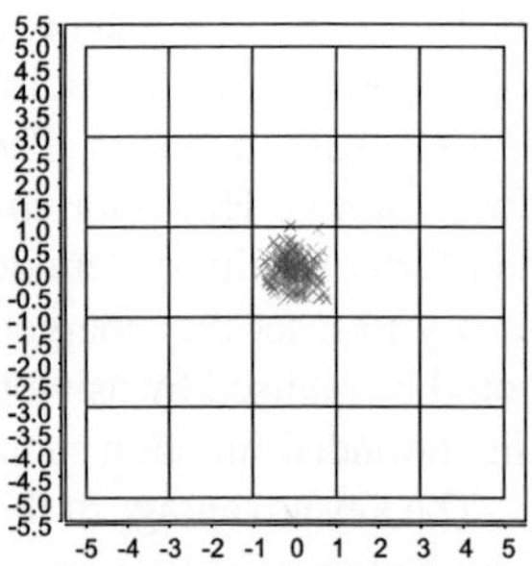

(c) Time Step 1 500

Fig. 2 Data set DS_{1a}: Distribution switch from uniform to Gaussian positioned at (0,0)

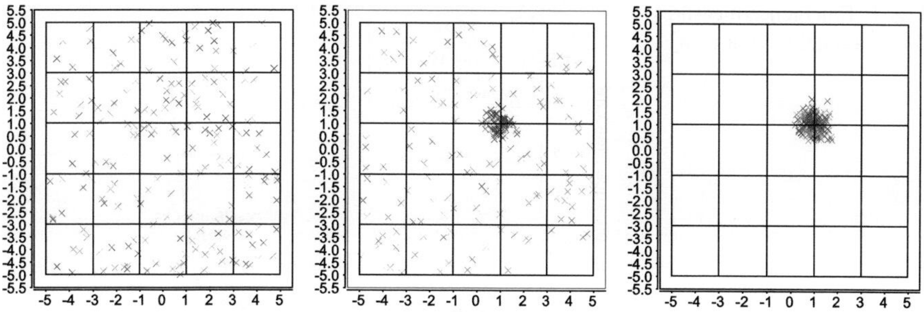

(a) Time Step 500 (b) Time Step 1 350 (c) Time Step 1 500

Fig. 3 Data set DS_{1b}: Distribution switch from uniform to Gaussian positioned at (1,1)

Trajectories of the measures are depicted in Figs. 4(a) for DS_{1a} and 4(b) for DS_{1b}. The trajectories start at time step 200 as it needs w initial samples to fill both windows such that the measures can be evaluated. First, it can be seen that DED show equal trajectories for both dimensions. This is due to the symmetry of the used data sets. Regarding the ability to quantitatively measure emergence, HelPW is superior to the remaining measures. Both data sets describe the same "type" of emergence which is accurately measured with HelPW—cf. the almost identical trajectories of HelPW in Figs. 4(a) and 4(b). DED, in contrast, highly depends on the location of the Gaussian. If it covers more than one bin as for DS_{1b}, the maximum value of DED is significantly lower than for DS_{1a}. The binning of DED, however, smooths the density estimates and, thus, has an advantage over Hel at the beginning of the scenario with the uniformly distributed samples. Here, DED measures zero emergence, whereas HelPW results in a value of about 0.2 as the window size of 200 is too small to accurately estimate the uniform distribution with a Parzen window approach. It can be stated that with both measures, a trade-off between accuracy and bin size must be effected (number of bins in the case of DED, kernel width in the case of Hel based on a Parzen window estimate). The position of the hard bins in the case of DED, however, has a major impact on the measure, whereas the position of the soft bins (Parzen windows) in the case of Hel does not have this drawback. The parametric approaches HelGMM and χ^2 clearly suffer from the heavily violated distribution assumptions. The GMM is not able to correctly model the uniform distribution and, thus, it depends on the modelling result if and to what degree emergence is detected.

4.2 Concept Drift

The scenario used in this experiment features *concept drift*, i.e., a distribution that changes gradually its characteristics. This kind of scenario can be motivated by monitoring tasks for technical systems with slowly varying (e.g., fading) parameters [19] or by click-stream analysis, for instance.

Fig. 4 Emergence measures for experiments 1a and 1b

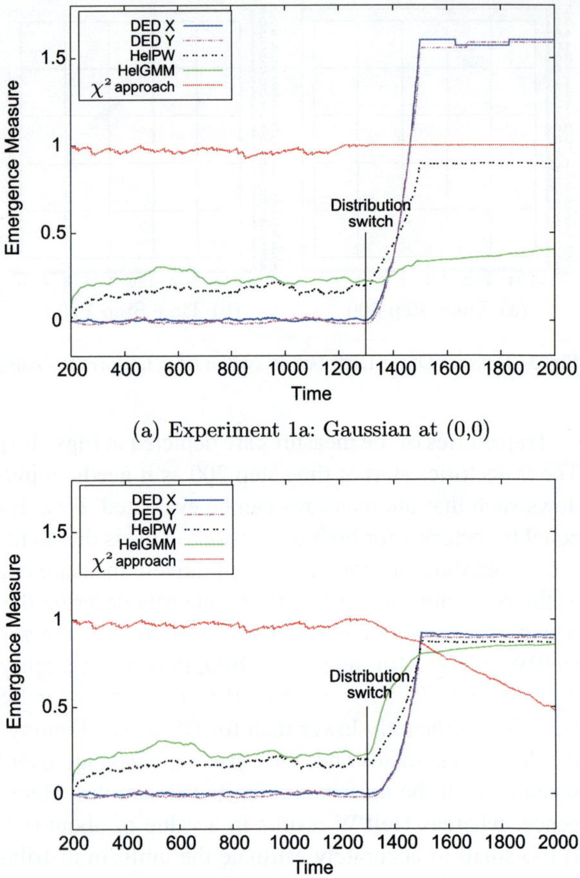

(a) Experiment 1a: Gaussian at (0,0)

(b) Experiment 1b: Gaussian at (1,1)

The scenario that will be considered here is illustrated in Fig. 5. Initially, four processes generate normally distributed samples, cf. Fig 5(a). In the course of the experiment, the process in the upper left quadrant of the input space drifts down, i.e., the centre of the corresponding component changes linearly. Snapshots of this movement at different time steps are shown in Figs. 5(b) and 5(c).

The trajectories of all measures are illustrated in Fig. 6. It can be seen that both Hellinger-based measures reflect the behaviour of the observed processes quite well, while it is much more difficult to recognise the observed concept drift from DED. There are, however, some differences between the model-based and the non-parametric estimate of Hel: The increase of HelGMM is steeper which would lead to an earlier detection of the emergent behaviour in a real application. The best result is obtained with the χ^2-approach. Nevertheless, it must be mentioned that distribution assumptions are met in this example while in real applications these assumptions hold approximately.

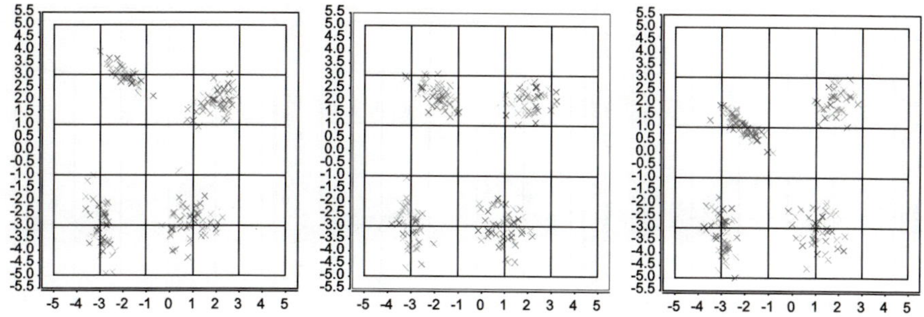

(a) Time Step 500 (b) Time Step 1 000 (c) Time Step 1 450

Fig. 5 Experiment 2: Drifting component

Fig. 6 Experiment 2: Emergence measures for the concept drift scenario

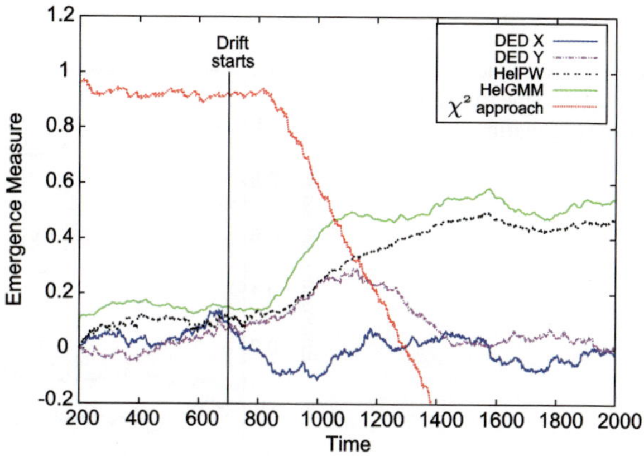

4.3 Novelty

In this final experiment we consider a scenario with a newly appearing process which is illustrated in Fig. 7. Initially, there are three processes that generate normally distributed samples, cf. Fig. 7(a). Then, at time step 1 300 a fourth process appears in the lower left quadrant of the input space as shown in Figs. 7(b) and 7(c).

Figure 8 shows the trajectories of the emergence measures. First, it can be seen that DED is unable to detect the appearance of the fourth process. Both Hellinger-based measures capture the start of the new process, however, again the model-based approach HelGMM is more rampant than HelPW. The χ^2-approach reflects the appearance of the new process most clearly.

5 Conclusion and Outlook

In this article, we summarised and compared some new techniques for a detection or a quantitative measurement of emergent phenomena in technical systems. These

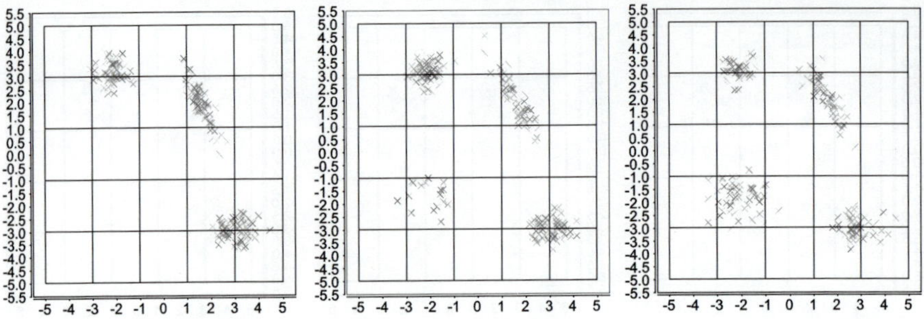

(a) Time Step 1 000 (b) Time Step 1 350 (c) Time Step 1 500

Fig. 7 Experiment 3: Novelty

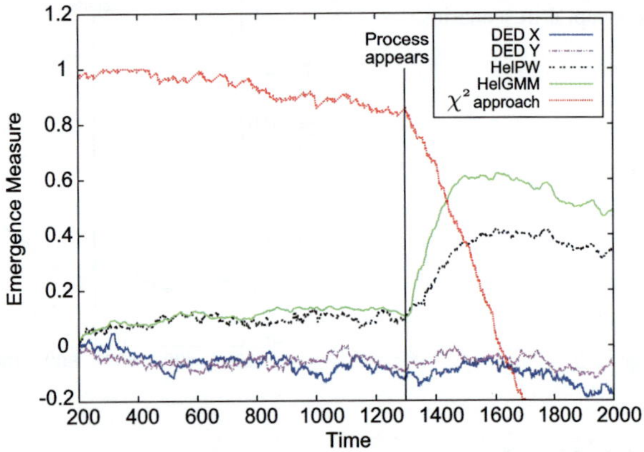

Fig. 8 Experiment 3: Emergence measures for the novelty scenario

techniques are based on either information-theoretic or probabilistic considerations. Depending on the kind of emergent phenomena—e.g., transitions from chaos to order, concept drift, novelty—and the kind of application at hand—e.g., distributed systems or self-organising systems—appropriate emergence measures must be chosen and parametrised. In our work it became obvious that a single, optimal emergence measure for all kinds of emergent phenomena does not exist. Also, it became clear that the definition of appropriate sets of attributes is the crucial issue in real applications. Pre-processing techniques for information condensing such as a data transformation based on a principal component analysis must be considered. Additionally, other types of distributions could be used.

Acknowledgement This work was supported by the German Research Foundation (DFG) under the grants SI 674/3-2, SI 674/3-3, MU 1055/9-2, and MU 1055/9-3 within the Priority Programme Organic Computing.

References

1. Müller-Schloer, C., Sick, B.: Emergence in organic computing systems: Discussion of a controversial concept. In: Yang, L.T., Jin, H., Ma, J., Ungerer, T. (eds.) Autonomic and Trusted Computing. LNCS, vol. 4158, pp. 1–16. Springer, Berlin (2006). (Proc. of the 3rd International Conference on Autonomic and Trusted Computing (ATC), Wuhan and Three Gorges, China)
2. Müller-Schloer, C., Sick, B.: Controlled emergence and self-organization. In: Würtz, R.P. (ed.) Organic Computing. Understanding Complex Systems, pp. 81–104. Springer, Berlin (2008)
3. Stephan, A.: Emergenz. Von der Unvorhersagbarkeit zur Selbstorganisation, 2nd edn. Mentis, Paderborn (2005)
4. Würtz, R.P. (ed.): Organic Computing. Understanding Complex Systems. Springer, Berlin (2008)
5. Mnif, M., Müller-Schloer, C.: Quantitative emergence. In: Proc. of the IEEE Mountain Workshop on Adaptive and Learning Systems (SMCals/06), pp. 78–84. Logan, UT (2006)
6. Mnif, M.: Quantitative Emergenz: Eine Quantifizierungsmethodik für Ordnung in selbstorganisierenden technischen Systemen. PhD thesis, Leibniz Universität Hannover, Department for Electrical Engineering and Computer Science, Hanover, Germany (2010)
7. Fisch, D., Jänicke, M., Sick, B., Müller-Schloer, C.: Quantitative emergence—a refined approach based on divergence measures. In: Proc. of the Fourth IEEE International Conference on Self-Adaptive and Self-Organizing Systems (SASO 2010), Budapest, Hungary, pp. 94–103 (2010)
8. Fisch, D., Jänicke, M., Kalkowski, E., Sick, B.: Techniques for knowledge acquisition in dynamically changing environments. ACM Transactions on Autonomous and Adaptive Systems (TAAS) (2011, accepted)
9. Crutchfield, J.: The calculi of emergence: Computation, dynamics, and induction. Physica D, 11–54 (1994)
10. Shalizi, C.: Causal architecture, complexity and self-organization in time series and cellular automata. PhD thesis, University of Wisconsin at Madison, Physics Department, Madison, WI (2001)
11. Gabbai, J.M.E., Yin, H., Wright, W.A., Allinson, N.M.: Self-organization, emergence and multi-agent systems. In: Proc. of the International Conference on Neural Networks and Brain (ICNN&B '05), vol. 3, Beijing, China, pp. 1858–1863 (2005) (Special Session 9: Self-Organization, Unsupervised Learning, Language Processing)
12. De Wolf, T., Samaey, G., Holvoet, T., Roose, D.: Decentralized autonomic computing: Analysing self-organising emergent behaviour using advanced numerical methods. In: Proc. of the Second International Conference on Autonomic Computing (ICAC 2005), Seattle, WA, pp. 52–63 (2005)
13. Holzer, R., de Meer, H., Bettstetter, C.: On autonomy and emergence in self-organizing systems. In: Hummer, K.A., Sterbenz, J.P.G. (eds.) Self-Organizing Systems. LNCS, vol. 5343, pp. 157–169. Springer, Berlin (2008). (Proc. of the Third International Workshop (IWSOS 2008), Vienna, Austria)
14. Fisch, D., Jänicke, M., Kalkowski, E., Sick, B.: Learning by teaching versus learning by doing: Knowledge exchange in organic agent systems. In: Proc. of the IEEE Symposium on Intelligent Agents (IA 2009), Nashville, TN, pp. 31–38 (2009)
15. Mühl, G., Werner, M., Jaeger, M.A., Herrmann, K., Parzyjegla, H.: On the definitions of self-managing and self-organizing systems. In: Braun, T., Carle, G., Stiller, B. (eds.) KiVS 2007 Workshop: Selbstorganisierende, Adaptive, Kontextsensitive verteilte Systeme (SAKS 2007), Bern, Switzerland, pp. 291–301. VDE Verlag (2007)
16. Le Cam, L., Lo Yang, G.: Asymptotics in Statistics, 2nd edn. Springer, Berlin (2000)
17. Fisch, D., Kastl, F., Sick, B.: Novelty-aware attack recognition—intrusion detection with organic computing techniques. In: Proc. of the 3rd IFIP Conference on Biologically-Inspired Collaborative Computing (BICC 2010) at the World Computer Congress (WCC 2010), Brisbane, Australia, pp. 242–253 (2010)

18. Bishop, C.M.: Pattern Recognition and Machine Learning. Springer, New York (2006)
19. Mauch, M.: Self reflection and adaption of intelligent systems: A combination of data stream clustering and reinforcement learning. Master's thesis, University of Passau, Department for Mathematics and Computer Science, Passau, Germany (2008)

D. Fisch · M. Jänicke · B. Sick (✉)
Computationally Intelligent Systems Lab, University of Applied Sciences Deggendorf, Daggendorf, Germany
e-mail: bernhard.sick@fh-deggendorf.de

D. Fisch
e-mail: dominik.fisch@fh-deggendorf.de

M. Jänicke
e-mail: jaenicke@fim.uni-passau.de

C. Müller-Schloer
Institute of Systems Engineering, Leibniz University Hanover, Hanover, Germany
e-mail: cms@sra.uni-hannover.de

Chapter 1.4
Emergent Control

Peter Kreyssig and Peter Dittrich

Abstract In order to control the dynamics of a system, feedback control (FC for short) is an extremely successful strategy, which is widely applied by engineers. Here we discuss a different strategy of control, called *emergent control* (EC for short), which can be found in large, distributed systems of components interacting only locally. For comparison we present a basic architecture for emergent control and two simple examples. In these examples, emergent control is achieved by a chemical computing approach. In the first example the number of objects of a particular type in a distributed system has to be kept constant. The example shows that on a macroscopic level EC and FC can display exactly the same behaviour. Hence for comparing their performance quantitatively a more refined model has to be taken into account. This model indicates a trade-off between cost and robustness. FC tends to operate at a lower cost than EC, however it also tends to instability when the system under control is large, decentralised, and/or heavily perturbed. In the second example the number of clusters in a distributed system should be controlled. The example shows how a user can "control", i.e., provide goals in EC even if the system is not tractable analytically due to highly non-linear effects.

Keywords Self-organisation · Feedback control · Distributed systems · Emergence · Seceder model · Chemical computing

1 Introduction

In order to exploit the increasing quantitative complexity of computational systems, various self-organisation principles are applied to *control* them. In general, the aim of control is to change the dynamics of a system in a desired way. This means to keep the system as close as possible to a desired attractor or, in particular, close to a stable state. The easiest way to achieve this is by adding a feedback control loop.

Here, however, we will study a strategy based on the self-organisation metaphor. *Self-organisation* refers to a phenomenon where a system becomes organised by elements of that system. That is, the system's organisation can be explained by referring to the systems components rather than to external effects. Self-organising systems are usually thermodynamically open.

C. Müller-Schloer et al. (eds.), *Organic Computing — A Paradigm Shift for Complex Systems*, 67–78
Autonomic Systems, DOI 10.1007/978-3-0348-0130-0_4, © Springer Basel AG 2011

In a technological environment a user usually wants to keep control, which results in a paradoxical situation when this demand is combined with the need for self-organisation. Here we show, how this paradox can be resolved by a strategy termed *emergent control* (cf. [4, 10, 12]). In emergent control, the controlled behaviour emerges [2] at a macro-level from many microscopic local interactions [23].

The remaining article is organised as follows. In the next section, we briefly describe the classical feedback control and sketch an architecture for emergent control. In Sect. 3 we present two examples and give an overview of techniques for achieving emergent control (Sect. 4). Finally, the difficulty of comparing EC with FC is illustrated in Sect. 5. The article concludes with a short discussion and an outlook in Sect. 6.

2 Feedback Control and Emergent Control

2.1 Feedback Control

In classical control theory a feedback controller takes measurements from the system to be controlled and uses this information to "decide" how to manipulate the system in order to achieve the desired behaviour (Fig. 1). This loop of measurement and manipulation is performed continuously. The user demands can be easily integrated by comparing them with the current system state. For example, in a *PID* (proportional integral derivative) approach the user demands, i.e., a numerical value, is simply subtracted from the measured output of the system. Then, the difference in demand and output is used to set the input of the system to be controlled. This control strategy can be found in virtually any technical system.

In passing we note that this feedback control loop is also the central explanation pattern of cybernetics [28]. Note further the relation to the *Observer/Controller (O/C) architecture*, which provides a regulatory feedback on the internal mechanism of the controlled system [20]. In contrast to a classical observer-controller approach [14] such an O/C loop can also modify how the controlled system works. Therefore, the system can still operate without the O/C on top, but looses its ability to adapt.

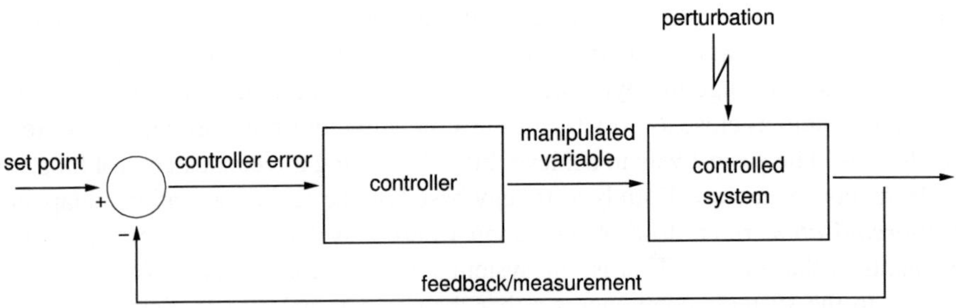

Fig. 1 Classical feedback architecture

2.2 Emergent Control

A rather different approach to the problem of controlling complex systems can be found in large, distributed systems. By that we mean systems that consist of a large amount of elementary units that only interact with a few of their fellow units. Examples for this can be found in natural (e.g., gene expression in a cell), social (e.g., the economical system) and technical (e.g., the Internet) systems as well as in many others. In these systems some focal features of feedback control like the measurement of a global state or the feedback loop do not appear. Here we can find an approach where no explicit controller in a classical sense as sketched for the feedback control above exists. A set of (usually simple) local rules is executed by the different systems components, such that the desired behaviour appears at a global scale. We can say that the global behaviour is an emergent[1] property [2, 21]. Therefore we call, as a working definition, this kind of control strategy *emergent control.*

The term emergent control appears in the literature with a similar meaning as intended here. Doursat and Ulieru [8, 27] use the term *emergent engineering* to describe a framework to realise systems that exhibit controllable self-organisation. They also mention that performance, including the self-* properties, robustness etc., cannot be measured as a numerical value yet; a difficulty we will also meet here.

From a dynamical point of view, emergent control can be seen as *programming by equilibria*, a concept suggested by Tschudin and Meyer [25]. As an example they present a self-healing communication protocol. This protocol consists of rules implying a dynamics where the desired computational behaviour is an asymptotically stable equilibrium state with a large basin of attraction. The self-healing behaviour or the generation of a computational result "emerges from the system's tendency to strive for an equilibrium" [25].

A more specialised, but also usable characterisation of emergent control systems is given in [23]: "We refer to distributed control systems based on local information as emergent control systems". Støy et al. [24] use role-based algorithms for a robot locomotion control system. The modular robots are an example for the necessity of distributed control systems. Because of this, there is a lot of work done on emergent *control of robots*. We just mention a few and point out the idea behind the control used. Tsuchiya et al. [26] define a two-tier architecture for the movement of legged robots, one for the leg motion and one for the gait pattern. Steels [22] suggests evolutionary processes for control. Digney [5] employs Q-learning for a hierarchical control structure of robots. Meeden et al. [18] train an artificial neural network with reinforcement learning to control their robot "carbot". They also experimented with the "transplantation" of the learned network from a simulator to the real robot and the other way around.

Ishiguro et al. [11] use a particular oscillator equation (inspired by slime mould behaviour) to arrange the morphology of a modular robot whose modules are con-

[1] By emergence we mean what Bedau and others call "weak emergence" [2]. In the first example in Sect. 3.1, however, we will be less strict concerning the non-linear property of the micro-macro relationship and take a linear system as a model to illustrate emergent control.

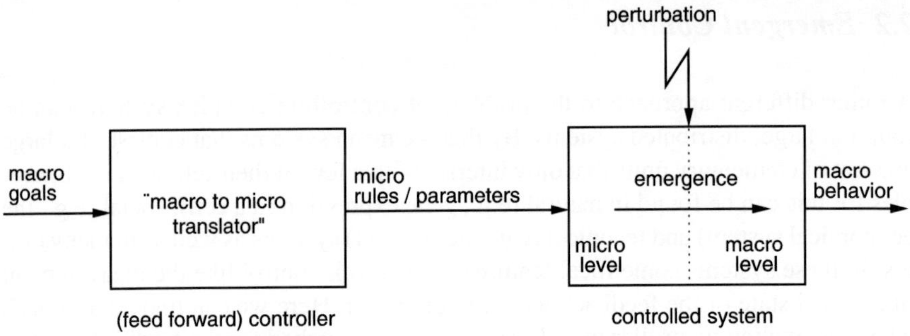

Fig. 2 Basic architecture for controlled self-organisation through emergent control. Note that there is no feedback at the macro level; except for potential user intervention as a result of the user's observation of the macro behaviour (not shown). There is especially no feedback of the macro quantity that should be controlled. There is however some kind of feedback at the micro level, which consists of many interacting elements

nected by Velcro. In a more abstract setting, emergent control can be used to coordinate the processing of distributed knowledge within a multi-agent expert system [16]. From here we envision a fascinating application area of emergent control in hybrid social systems, which consists of software agents and social actors [9, 15, 19].

A basic diagram of a possible *general architecture* for systems using emergent control is shown in Fig. 2. Most significantly there are two layers or levels to distinguish, the macro- and the micro-level. After handing the macro goals to the translator or compiler, it creates micro rules, which lead to the (emerging) macro-behaviour in the controlled productive system.

3 Examples

We present two examples. The first example is kept as simple as possible. It shows that when comparing EC with FC at macroscopic level as it is usually done in control theory by using an ODE (ordinary differential equation), their behaviour can be exactly the same. Thus refined models are required for comparison. Because the first example is linear and thus the relation between micro-level and macro-behaviour is simple, we also present a highly non-linear example. In this example, the macro-state (i.e., the number of clusters) cannot be inferred from a micro-element (i.e., a numerical value).

3.1 EC and FC Result in the Same Macro-behaviour

Let us assume that we have a system containing a number of objects of type A. And let $x(t)$ denote the number of these objects in the system at time t. The control task is to keep the number of objects at a constant value x_r provided by the user. For feedback control we assume that the system can be manipulated by adding or removing objects at a rate y. Using a linear function as a controller (i.e., a P-controller) we obtain the following ODE for the system and the controller, respectively:

$$\dot{x} = y(t) \quad \text{and} \quad y(t) = k(x_r - x(t)). \tag{1}$$

This yields the following total dynamics:

$$\dot{x} = k(x_r - x(t)) = kx_r - kx(t). \tag{2}$$

For $k > 0$ this is a linear system with one asymptotically stable fixed point at x_r, as desired. With k we can control the eigenvalues and thus the speed at which the fixed point is approached.

For emergent control there needs to be a micro-level. In our simple example, we obtain a micro-level by assuming that an object can vanish spontaneously at a certain rate k_2 (or, equivalently, with a certain probability). The micro-rules are given by the following simple chemical program:[2]

$$\xrightarrow{k_1} A \tag{3}$$

$$A \xrightarrow{k_2} . \tag{4}$$

Assuming mass-action kinetics, this chemical program can be dynamically interpreted in the following way. Within a sufficiently small time interval $d\tau$, an object of type A appears in the system with probability $k_1 \cdot d\tau$. Furthermore, each object of type A disappears with probability $k_2 \cdot d\tau$. If the number of objects is large, we can safely model the system with an ODE:

$$\dot{x} = k_1 - k_2 x(t). \tag{5}$$

Note that with $k_1 = kx_r$ and $k_2 = k$ we obtain exactly the same ODE as in our feedback control model (see Eq. (2)). Although the macroscopic ODE model is exactly the same, FC and EC differ as illustrated by Fig. 3.

In the example we see that the description of the system by ODEs only is not sufficient to describe the differences between EC and FC. Therefore we argue that a comparison is only possible with one (or several) additional refined models. Then a quantitative rather than a qualitative evaluation of properties like robustness becomes feasible. Often these refined models have to take the actual implementation into account and not only the abstracted goals. We extend this example by a toy model in Sect. 5.

---

[2]Here it is sufficient to see a chemical program just as a set of reaction rules or rewriting rules, which are equivalent to a Petri-net, see [6].

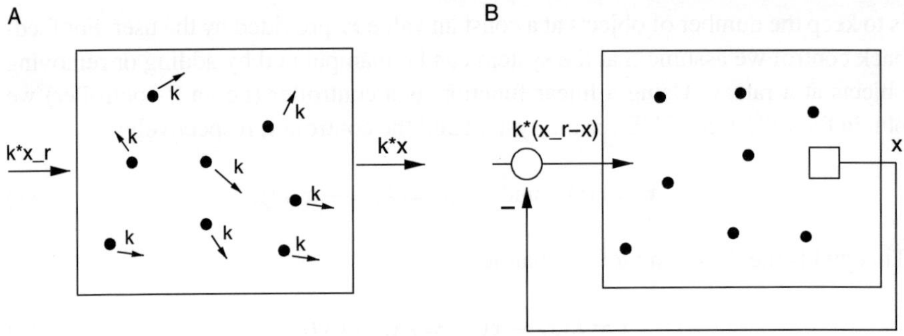

Fig. 3 Illustration of a concrete instance of emergent control (**A**) and feedback control (**B**) that have the same macro-behaviour expressed as an ordinary differential equation (cf. (5) and (2), respectively). This serves also as inspiration for the toy model shown in Sect. 5. In (**A**) every present molecule has the probability k to vanish (cumulative outflow of kx) and there is a global inflow of kx_r. In (**B**) x is measured and the inflow of molecules is regulated to $k(x_r - x)$

3.2 Emergent Control of the Number of Clusters

In the next example we assume that we have a population of objects, each characterised by a real number, which should form clusters. The aim of control in this example is to stabilise a certain amount of n clusters, while we do not care which object is part of which cluster.

To achieve this with emergent control, we assume a microscopic dynamics taken from the seceder model [7]. The seceder model is a simple individual based model that shows how a local advantage to be different gives rise to the formation of clusters. The model consists of a population of objects, which are real numbers, here. Objects reproduce and die. In a single reproduction event three objects are chosen randomly and the objects that possess the largest distance to their mean are reproduced by creating a mutated copy (offspring). Mutation is performed by adding a normally distributed random number with mean 0 and variance 1 denoted by $N(0, 1)$. The offspring replaces a randomly chosen object of the population.

Formally, the (basic) seceder model is defined as follows. The population of size M is represented by an array $P = \{P[1], \ldots, P[M]\}$ of objects $P[i] \in \mathbb{R}$. We write $P(t)$ for the population at time t, and $P(t)[i]$ for the i-th object of population $P(t)$. The population evolves over time according to the following algorithm:

```
while ¬terminate() do
    s1 := P[randomInt(1, M)]
    s2 := P[randomInt(1, M)]
    s3 := P[randomInt(1, M)]
    P[randomInt(1, M)] := f_sel(s1, s2, s3) + N(0, 1)
    t := t + 1/M
od
```

Firstly, three individuals are chosen randomly. Then a randomly chosen individual is replaced by selecting the individual with the largest distance to others and adding a random number. Finally the time counter is incremented. The

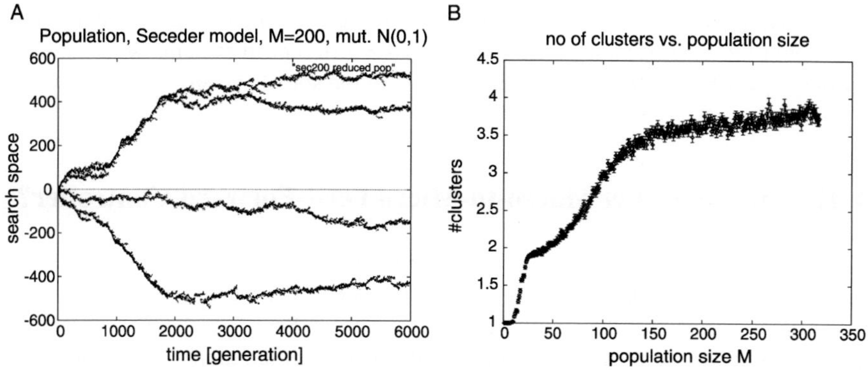

Fig. 4 Seceder model (second example). (**A**) Dynamics of the seceder model for a population size of $M = 200$. For each point in time t all individuals $P[i](t)$ are plotted on the horizontal axis. (**B**) Relation between the number of clusters and the population size M

procedure randomInt(a, b) returns a uniformly distributed random number out of $\{a, a+1, \ldots, b\}$. The selection function

$$f_{sel}(g_1, g_2, g_3) = \begin{cases} g_1 & \text{if } F_1 \geq F_2 \wedge F_1 \geq F_3, \\ g_2 & \text{if } F_2 \geq F_1 \wedge F_2 \geq F_3, \\ g_3 & \text{otherwise,} \end{cases} \quad \text{where } F_i = \left\| g_i - \tfrac{1}{3}(g_1 + g_2 + g_3) \right\|, \tag{6}$$

returns the argument that possesses the largest distance to the mean of the three arguments. Note that the seceder model can be interpreted as a chemical program with third order catalytic reaction rules and dilution flow:

$$s_1 + s_2 + s_3 \to s_1 + s_2 + s_3 + s_\mu \quad \text{with } \mu = f_{sel}(s_1, s_2, s_3), \tag{7}$$

$$s_i \to \quad \text{for all objects } i. \tag{8}$$

When running the system (e.g., initialised with, $P[i] = 0$ at $t = 0$) clusters appear spontaneously (Fig. 4). The number of clusters depends on the population size and can be also controlled by changing the tournament (e.g., using size four instead of size three).

For emergent control we need in addition to the micro-rules a *macro-micro feed-forward controller*, which maps the user demands (a number of clusters) to manipulable parameters of the micro-structure of the system. To keep things simple, we assume that this feedforward controller maps a number of groups to a number of objects. Because analytical derivation of this mapping is difficult, and in general impossible, we obtain the mapping experimentally, here. That is, we simply perform simulations for different population sizes (Fig. 4, B) and invert the experimentally determined function.

Note that there is no straightforward way how the same control objective can be achieved by classical feedback control. Measuring the number of clusters is relatively easy. But it is unclear how a feedback controller should increase and decrease the number of clusters.

Finally, it is interesting to note that in this example no microscopic entity has any clue about the macroscopic state. This could be different in systems where microscopic objects have a memory, as for example in the social system.

4 How to Construct Macro-to-Micro Feed-Forward Controller?

There are various ways to design the local or micro interactions of the system's components which achieve the emergent behaviour on the macro level.

As demonstrated for the simple chemical system in Sect. 3.1, there is sometimes the possibility to perform the deduction of local rules *manually*. The macro behaviour could then be proven mathematically. In particular, for the construction of artificial chemical systems *chemical organisation theory* can be helpful (Chap. 2.6).

Another principle is looking for and copying solutions in *nature*. A mimicking of processes in nature is also possible as for example done in swarm robotics [3], artificial hormone systems (Chap. 4.4) or ant inspired algorithms (Chap. 1.6). The Publish/Subscribe architecture (Chap. 2.1) can also be seen as imitating human behaviour.

The design of the second example in Sect. 3.2 follows a different principle. By performing experiments on a particular system (in our case the seceder model) the relationship micro to macro behaviour is established. Several experiments, i.e., executing the micro-rules to produce the associated macro behaviour, are performed. From the acquired data a general relation is deduced. This can also be described as the *inversion of an experiment* or the extrapolation from data.

Another general principle is *evolution* or more general *optimisation*. Starting from an initial population of different micro-rules and by applying mutation and selection to them, it is searched for a better macro level behaviour. An example for this approach applied to a chemical implementation of a flip-flop given by Lenser et al. [13]. Astor and Adami [1] present a method for decentralised growth of artificial neural networks using evolution and development.

The use of *scouting* or *exploration* methods is discussed by Matsumaru et al. [17]. There, systems are explored and searched for interesting behaviours. The basic idea is that an autonomous system is used, as a preliminary step, to explore the behaviour of the chemical reaction system. Then a specific aspect of the system's behaviour will be utilised for a particular computational purpose.

All the mentioned approaches are limited and, like manual programming, can only explore some regions of the space of all possible emergent control strategies.

5 Quantitative Comparison of the Performance of Emergent Control vs. Feedback Control

There is currently no satisfying theory that allows to compare the performance of EC and FC theoretically. The reason lies in the difficulty of quantifying additional features of emergent control and of self-organising systems in general, like robustness, self-optimisation, self-configuration, etc. These features are not appropriately captured by simply measuring cost and comparing the performance on this basis. We argue that the mentioned properties depend on the particular implementation. It is clear that a macro-level analysis, e.g., by ODEs like shown in the examples, is not sufficient. Therefore we suggest to use a refined model that includes a macro-level and a micro-level in order to measure performance, which includes for example robustness. Already from the toy example we presented, we can see some important differences between FC and EC.

As an example for a refined model we implemented a discrete version of the reaction vessel described in the example in Sect. 3.1. We used a 100 times 100 grid of fields, which each either carry a molecule of type A or not. We assume initially $k = 1.0$, $x_r = 0.25$ and an empty field. After 30 simulation steps we change to $x_r = 0.5$ and execute another 30 steps.

A single step in the EC mode of operation consists of the following. For each empty field a molecule of type A is produced with a probability of kx_r and each molecule of type A is erased with probability k.

The FC mode of operation is led by the idea that there is a controlled inflow at the left hand side columns of the field and a measurement at the right hand side columns of the field. Firstly, we count the molecules in a pre-defined measurement region consisting of the last 83 columns of the field. This gives an estimate for the total amount of molecules in the field. The difference between the number of molecules required (given by x_r) and the measured amount is added (or removed respectively) at the left hand side columns of the field. The molecules are subject to a diffusion process. With fixed probability each molecule is moved to a neighbouring (Moore neighbourhood) field if it is empty.

The cost can be defined to be the amount of operations on the field we have to conduct, i.e., the cost is the amount of deletion and production operations performed on the field. We remark that neither the cost for the measurement nor for the diffusion are taken into account here.

Figure 5 shows the trade-off between correctness and effort, i.e., how precise EC is in the example, but also the immense work necessary to achieve this. The EC possessed almost perfect control behaviour and response time. Whereas the FC in this case needs at least 20 steps to regulate the amount of molecules to an acceptable value, though still oscillating around the goal of 2500 molecules. After the user's interference, the setting of x_r to 0.5, the FC starts an even stronger oscillation around the goal of 5000 molecules. An interesting effect here is that the cost in the EC case depends on the amount of molecules present in the field. To summarise, FC tends to operate at a lower cost than EC, however it also tends to instability when the system under control is large and decentralised.

6 Discussion, Conclusion, and Outlook

The examples discussed here are probably not realistic. The emergent phenomenon shown in Sect. 3.1 is a simple linear accumulation of the state on the micro level,

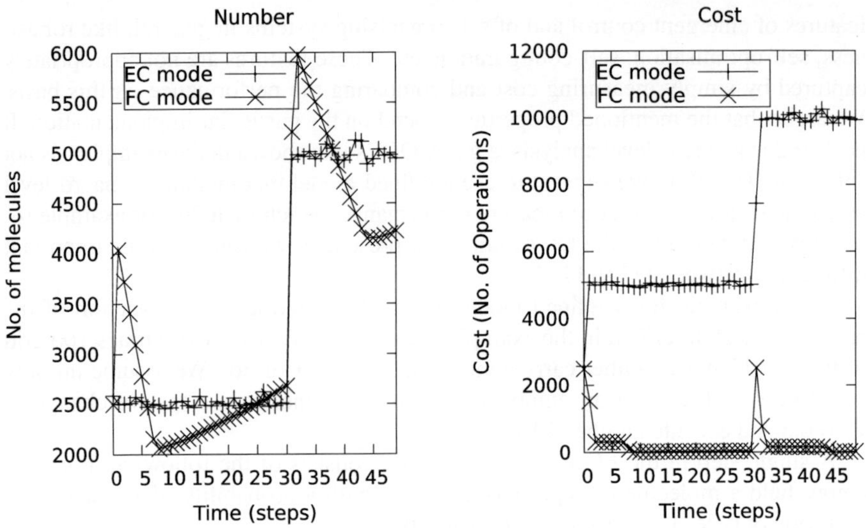

Fig. 5 Comparison EC vs FC in the toy model. The *left figure* shows the number of molecules present on the field for each step. The *right figure* shows the cost (number of modifications) needed at each step. We used the following parameter values: size of the vessel 10000, measurement on 8300 fields, $k = 1.0$, $x_r = 0.25$ (steps 0 to 30), $x_r = 0.5$ (steps 30 to 50)

i.e., it can be observed just by counting. Nevertheless Sect. 5 shows that a refined model taking into account a more realistic implementation can exhibit properties like robustness and makes it possible to quantitatively measure them. This linear relation between micro and macro level cannot be found in the seceder example (Sect. 3.2). There is not an analytical result on the relation between local rules and number of clusters known.

In our example, micro rules are changed immediately. However in an asynchronous, spatially distributed system there is almost never a direct instantaneous manipulation of the micro level. Reconfiguration can then lead to unwanted behaviour, e.g., oscillation or chaos.

Some of the methods for the construction of macro to micro translators shown in Sect. 4 are very specialised and therefore only usable in particular situations, e.g., manual design or mimicking of nature. Others, e.g., evolution or inversion, are applicable to many problems. In general there is a lack of theoretical foundation to the methods.

FC is an extremely successful strategy in a plethora of technical systems. This is also due to relatively simple architecture which has a lot of advantages. There are many cases, however, where EC seems to be preferable, because of additional qualitative properties that are missing in the FC. To make it equally successful a more abstract and reduced architecture should be beneficial. One approach is shown in Fig. 2.

EC systems are fundamentally different from FC systems. Since they cannot be described by classical feedback control loops, they require a different architectural perspective. In particular a micro-level description is necessary, since macro-level models of the dynamics are not enough for quantitative evaluation (Sect. 5). Therefore a powerful abstraction (including micro- and macro-level) of self-organisation and emergent control is needed.

One extension or addition to the presented concept is the combination of EC with FC. This can for example be simply achieved by controlling some parts of the system by FC while controlling other parts by EC. A combination with the Observer/Controller architecture (Chap. 4.1), learning algorithms, etc. is also possible.

An interesting aspect that needs to be discussed further is the inclusion of user demands, which is more difficult in EC than in FC systems. The quantitative analysis shown in Sect. 5 is not convincing yet and needs further investigation. These shortcomings relate to the observation that there is a lot more theory on emergence and self-organisation needed to understand and engineer systems and controllers using emergent phenomena resulting from the local-to-global problem.

Acknowledgement We acknowledge financial support by the German Research Foundation (DFG) under grant Di 852/4-2/3.

References

1. Astor, J.C., Adami, C.: A developmental model for the evolution of artificial neural networks. Artif. Life **6**(3), 189–218 (2000)
2. Bedau, M.A.: Downward causation and autonomy in weak emergence. Principia **6**, 5–50 (2003)
3. Bonabeau, E., Dorigo, M., Theraulaz, G.: Inspiration for optimization from social insect behaviour. Nature **406**(6791), 39–42 (2000)
4. Deneubourg, J.L., Franks, N.R.: Collective control without explicit coding: The case of communal nest excavation. J. Insect Behav. **8**(4), 417–432 (1995). doi:10.1007/BF01995316
5. Digney, B.: Learning and shaping in emergent hierarchical control systems. In: Proceedings of Space96 and Robots for Challenging Environments II (1996)
6. Dittrich, P.: Chemical computing. In: Banâtre, J.-P., Giavitto, J.-L., Fradet, P., Michel, O. (eds.) Unconventional Programming Paradigms (UPP 2004). LNCS, vol. 3566, pp. 19–32. Springer, Berlin (2005)
7. Dittrich, P., Liljeros, F., Soulier, A., Banzhaf, W.: Spontaneous group formation in the seceder model. Phys. Rev. Lett. **84**, 3205–8 (2000)
8. Doursat, R., Ulieru, M.: Emergent Engineering for the Management of Complex Situations. In: Proceedings of the 2nd International Conference on Autonomic Computing and Communication Systems (Autonomics 2008), pp. 1–10. ICST (Institute for Computer Sciences, Social-Informatics and Telecommunications Engineering) (2008)
9. Dron, J.: Social software and the emergence of control. In: Sixth International Conference on Advanced Learning Technologies, 2006, pp. 904–908. IEEE, New York (2006). ISBN 0769526322
10. Holland, O., Melhuish, C.: Stigmergy, self-organization, and sorting in collective robotics. Artif. Life **5**, 173–202 (1999)
11. Ishiguro, A., Shimizu, M., Kawakatsu, T.: Don't try to control everything!: an emergent morphology control of a modular robot. In: IEEE/RSJ International Conference on Intelligent Robots and Systems, 2004 (IROS 2004). Proceedings, vol. 1, pp. 981–985. IEEE Press, New York (2005)
12. King, M.G.: Hierarchical structure in emergent control. Sociometry **27**(1), 19–24 (1964)

13. Lenser, T., Matsumaru, N., Hinze, T., Dittrich, P.: Tracking the evolution of chemical computing networks. In: Bullock, S., Noble, J., Watson, R.A., Bedau, M.A. (eds.) Proceedings of the Eleventh International Conference on Artificial Life, pp. 343–350. MIT Press, Cambridge (2008).
14. Luenberger, D.G.: Observers for multivariable systems. IEEE Trans. Autom. Control **11**(2), 190–197 (1966). doi:10.1109/TAC.1966.1098323
15. Malsch, T.: Naming the unnamable: Socionics or the sociological turn of/to distributed artificial intelligence. Auton. Agents Multi-Agent Syst. **4**, 200–1 (2001)
16. Martínez, J.N., González Pérez, P.P.: Net of multi-agent expert systems with emergent control. Expert Syst. Appl. **14**(1–2), 109–116 (1998)
17. Matsumaru, N., Centler, F., Zauner, K.-P., Dittrich, P.: Self-adaptive scouting—autonomous experimentation for systems biology. In: Raidl, G.R., Cagnoni, S., Branke, J., Corne, D.W., Drechsler, R., Jin, Y., Johnson, C., Machado, P., Marchiori, E., Rothlauf, F., Smith, G.D., Squillero, G. (eds.) Applications of Evolutionary Computing, EvoWorkshops2004: EvoBIO, EvoCOMNET, EvoHOT, EvoIASP, EvoMUSART, EvoSTOC, Coimbra, Portugal, 5–7 Apr. 2004. LNCS, vol. 3005, pp. 52–61. Springer, Berlin (2004)
18. Meeden, L., McGraw, G., Blank, D.: Emergent control and planning in an autonomous vehicle. In: Proceedings of the Fifteenth Annual Meeting of the Cognitive Science Society, pp. 735–740. Citeseer (1993)
19. Meister, M., Schröter, K., Urbig, D., Lettkemann, E., Burkhard, H.-D., Rammert, W.: Construction and evaluation of social agents in hybrid settings: Approach and experimental results of the inka project. J. Artif. Soc. Soc. Simul. **10** (2007)
20. Müller-Schloer, C.: Organic computing—on the feasibility of controlled emergence. In: Proceedings of the 2nd IEEE/ACM/IFIP International Conference on Hardware/Software Codesign and System Synthesis, pp. 2–5. ACM, New York (2004)
21. Müller-Schloer, C., Sick, B.: Controlled emergence and self-organization. In: Würtz, R.P. (ed.) Organic Computing—Understanding Complex Systems, pp. 81–103. Springer, Berlin (2008)
22. Steels, L.: Emergent functionality in robotic agents through on-line evolution. In: Brooks, R.A., Maes, P. (eds.) Artificial Life IV: Proceedings of the Fourth International Workshop on the Synthesis and Simulation of Living Systems, pp. 8–16 (1994)
23. Støy, K.: Emergent control of self-reconfigurable robots. PhD thesis, University of Southern Denmark (2003)
24. Støy, K., Shen, W.M., Will, P.M.: Using role-based control to produce locomotion in chain-type self-reconfigurable robots. IEEE/ASME Trans. Mechatron. **7**(4), 410–417 (2002)
25. Tschudin, C., Meyer, T.: Programming by equilibria. In: Taferl, M. (ed.) 15. Kolloquium Programmiersprachen und Grundlagen der Programmierung (KPS'09), pp. 37–46 (2009)
26. Tsuchiya, K., Tsujita, K., Manabu, K., Aoi, S.: An emergent control of gait patterns of legged locomotion robots. In: Proc. of the Symposium on Intelligent Autonomous Vehicles, pp. 271–276 (2001)
27. Ulieru, M., Doursat, R.: Emergent engineering: A radical paradigm shift. In: Int. J. Auton. Adapt. Commun. Syst. (2009)
28. Wiener, N.: Cybernetics. Technol. Press, New York (1948)

P. Kreyssig · P. Dittrich (✉)
Bio Systems Analysis Group, Jena Centre for Bioinformatics and Department of Mathematics and Computer Science, Friedrich Schiller University Jena, 07743 Jena, Germany
e-mail: peter.dittrich@uni-jena.de

P. Kreyssig
e-mail: peter.kreyssig@uni-jena.de

Chapter 1.5
Constraining Self-organisation Through Corridors of Correct Behaviour: The Restore Invariant Approach

Florian Nafz, Hella Seebach, Jan-Philipp Steghöfer, Gerrit Anders, and Wolfgang Reif

Abstract Self-organisation aspects and the large number of entities in Organic Computing (OC) systems make them extremely hard to predict and analyse. However, the application of OC principles to, e.g., safety critical systems, is usually not conceivable without behavioural guarantees. In this article, a rigorous approach called the *Restore Invariant Approach* is presented, which provides a specification paradigm and a formal framework that allows to give guarantees for a system despite of self-organisation. The approach provides a method for specifying unwanted system states by constraining the system and defining a corridor of correct behaviour. Furthermore, a decentralised algorithm for monitoring and restoring the invariant based on coalition formation is presented.

Keywords Self-organisation · Formal verification · Decentralised algorithms

1 Introduction

Today, self-organisation is used to cope with the rising complexity of systems that often consist of a vast number of entities. These systems try to fulfil a global goal, often realised by a combination of local decisions and interactions with other entities. They are not influenced by an external control and they have the ability to find their system configuration on their own. This configuration is changed dynamically during runtime, depending on the current situation. These kinds of systems are desirable as self-organisation makes them highly resilient, adaptive, and robust. This makes them ideally suited for applications in which safety is a critical issue and in domains that have to deal with changing environments and unexpected disturbances. However, such domains require rigorous techniques for safety analysis or verification in order to get a system approved for deployment. Especially in automotive systems, aviation, railway, or production automation, proofs are required in order to show that the systems behave as intended.

Usually, the behaviour of a dynamic and autonomous OC-system is hard to predict and the concrete reaction to a particular situation is not necessarily deterministic. Therefore, techniques need to be developed that deal with the unforeseeable outcome of self-organisation processes. It is not feasible to specify the system by

C. Müller-Schloer et al. (eds.), *Organic Computing — A Paradigm Shift for Complex Systems*, 79–93
Autonomic Systems, DOI 10.1007/978-3-0348-0130-0_5, © Springer Basel AG 2011

explicitly listing all states that are acceptable or wanted. The *Restore Invariant Approach (RIA)*, developed by the project SAVE ORCA[1] and presented in this article, provides a different approach by specifying a corridor of correct behaviour. This is done by defining an invariant separating "good" system states from "bad" ones. The system tries to maintain this invariant as long as possible in order to stay within this corridor. The property that the system behaves correctly as long as it evolves within the corridor, is then subject to verification.

The verification techniques based on the specification of the corridor allow giving behavioural guarantees for an OC-system. However, the proof of these guarantees hinges on the correctness of the algorithms that are used for configuring the system in order to stay within the corridor. Such algorithms make use of the system's flexibility and its degrees of freedom in order to find new configurations to compensate system failures. Here, a decentralised algorithm for local reconfiguration of OC-systems based on coalition formation is presented. It uses local knowledge and does not need a central component for reconfiguration. The correctness of the reconfiguration algorithm is verified by the use of a result checker that assesses a configuration before it is enacted in the system.

This article is structured as follows: In Sect. 2 the *Restore Invariant Approach* is presented; it provides a method for the specification of OC-systems. Further, the framework for formal verification is sketched. An example scenario, which is used in the subsequent sections to illustrate the approach, is presented in Sect. 3. The definition of corridors of correct behaviour is described in Sect. 4. Then, a decentralised reconfiguration algorithm is detailed in Sect. 5. The article concludes with a summary of results and an outlook.

2 The Restore Invariant Approach

In this section the *Restore Invariant Approach (RIA)* is presented, which provides a specification technique for behavioural corridors and a core concept for treating OC-systems. The principle of the *RIA* is that the behaviour of an OC-system is specified by a predicate logic formula (called *invariant*), which defines the behavioural corridor of the system. The system's goal is to stay within this corridor as long as possible. Whenever the corridor is left, which corresponds to a violation of the invariant, the system starts a self-organisation phase and reconfigures itself in order to get back into the corridor and to make the invariant hold again. *RIA* also provides the basis for formal verification for ensuring functional correctness.

This idea of *RIA* leads to a two-layered view of an OC-system: one layer contains the functional part of the system and is often called the *system under observation and control* [6]; the other one contains the self-x intelligence and is responsible for monitoring the system and for reconfiguration in case of failures or disturbances. This is often realised with an Observer/Controller (O/C) architecture [12]. Figure 1 shows such an architecture.

[1] Formal Modelling, Safety Analysis, and Verification of Organic Computing Applications.

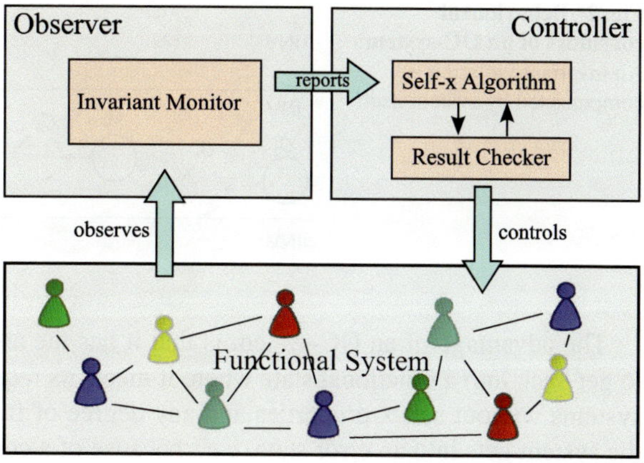

Fig. 1 Architectural view of an OC-system: An Oberserver is monitoring the functional system and reports invariant violations to the Controller. This is reconfiguring the system accordingly

The O/C layer can be either one O/C component monitoring the whole system or distributed Observer/Controllers on top of each agent, as described later in this article. The *Invariant Monitor* observes the system; whenever the specified corridor is left (i.e., the invariant is violated), it reports this and the current system state to the controller. The controller then tries to find a new configuration of the system that fulfils the invariant and leads back into the corridor. In OC-systems bio-inspired algorithms are often used. These algorithms are not necessarily sound nor complete. Therefore, a *Result Checker* component checks the output of the mechanism before forwarding it to the functional system, ensuring that only results that are consistent with the invariant are relayed.

In the following, a formal view of *RIA* is given. Further, the connection to system verification and the resulting proof obligations are presented.

2.1 A Formal View on the Restore Invariant Approach

As already mentioned above in the informal description of *RIA*, OC-systems have to deal with failures and environmental disturbances. These disturbances force the system into a state where it cannot fulfil its functionality. Therefore, two disjoint sets of states of an OC-system can be distinguished:

- a set S_{func} of *functional* states, in which the system can provide the desired functionality,
- a set S_{reconf} of *erroneous* or *reconfiguration* states in which the system cannot provide the functionality and a reconfiguration has to take place to get back into a state within S_{func}.

Formally, the execution of an OC-system is the set of all possible traces. One trace out of this set describes a sequence of states and one possible temporal evolution of the system over time.

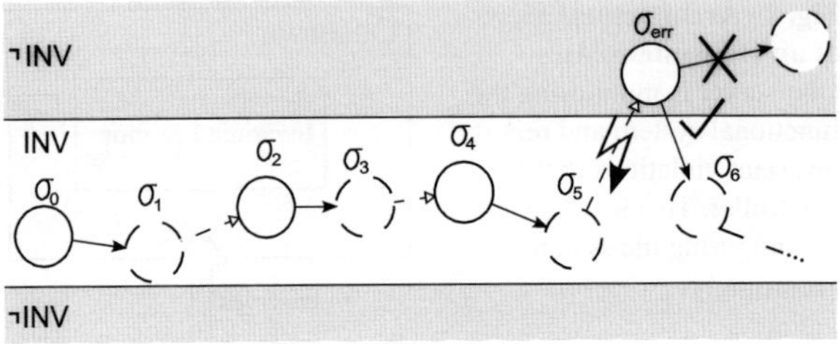

Fig. 2 Behavioural corridors of an OC-system: An invariant violation is compensated by system itself

The advantage of an OC-system is that it has the ability to compensate failures to get back into a functional state where it meets its requirements again. Traditional systems without self-x properties and any degree of freedom have traces in which the system gets into an error state, e.g., because of a component failure, and cannot recover from there.

Figure 2 shows a sample trace of a system. $\mathcal{S}_{func} = \{\sigma_0, \ldots, \sigma_6\}$ is the set of functional states that are within the behavioural corridor. All states that are not in the corridor belong to the set of reconfiguration states. Whenever a failure occurs, the system starts a self-reconfiguration and reaches a state within $\mathcal{S}_{func}$, which shows correct functional behaviour again.

This view is similar to the one proposed in Chap. 1.1. However, what is called *dead space* there is not explicitly considered. Here, states in which the system cannot successfully be reconfigured anymore are subsumed under $\mathcal{S}_{reconf}$. It is assumed that the system continuously tries to find a way to get back to the corridor, but fails to do so.

The *RIA* idea is to define the corridor of correct behaviour by defining a predicate *INV* called *invariant*, which holds in all functional states and does not hold otherwise. The predicate is called invariant as the system's goal is that this predicate holds on the entire trace. The set of states in the corridor can therefore be defined as $\mathcal{S}_{func} := \{\sigma \in \mathcal{S} | INV(\sigma)\}$. Traces consisting only of states in $\mathcal{S}_{func}$ are in some sense correct or "good" traces within the corridor and lead to expected behaviour. It is desired that an OC-system has only traces that consist of functional states or—whenever a failure occurs and it enters a state out of $\mathcal{S}_{reconf} := \mathcal{S} \setminus \mathcal{S}_{func}$—there will eventually be some state $\sigma \in \mathcal{S}_{func}$ later in the trace. This property can be formalised as a temporal logic property $\Box INV \vee \Box(\neg INV \rightarrow \Diamond INV)$. In essence, this means that the system reconfigures. The temporal operator $\Box INV$ states "*INV* holds now and will *always* hold in the future", while $\Diamond$ *INV* means "*INV* holds now or will *eventually* hold in the future". For more information about the temporal logic used here and its operators, please refer to [4].

This formula can also be used for the specification of the self-reconfiguration ($Spec_{self\text{-}x}$) mechanism of the system. The formula specifies the effect of self-organisation and its properties. Again, only properties of the result are specified, not a detailed solution. This reflects exactly the idea of the behavioural corridor, where no exact configurations are specified, just required properties.

Self-organisation after a component failure decreases redundancy, as, e.g., broken hardware components cannot be replaced or resources of an agent are not usable anymore. Nevertheless, the system's functionality can still be provided, e.g., because another component can take over. But as redundancy is not unlimited, it also means that there is some point at which restoring the invariant is not possible anymore. For a realistic (non-perfect) self-organisation, we need to modify the specification by adding a predicate Θ that states that no solution is possible anymore or some even weaker property, e.g., that no solution was found. This can be seen as the separation of *survival space* and *dead space* from Chap. 1.1. With some simplifications, the specification of the reconfiguration then looks as follows:

$$Spec_{self\text{-}x} := \Box(\neg INV \rightarrow \Diamond(INV \vee \Theta))$$

The predicate Θ can be seen as a quality predicate. Depending on how it is formulated, the used algorithms can fulfil the specification or not. For example, the weakest Θ states "Algorithm result is `found-no-solution`" whereas the strongest is $\Theta = false$ ("Reconfiguration is always possible and successful"). Usually, properties in between, like "Enough redundancy is available" or "A functional state is reachable", are used for the specification.

2.2 Behavioural Guarantees

The separation of reconfiguration and functional parts of a system and the specification of behavioural corridors by invariants which are separating 'good' from 'bad' states build the foundation for verification of functional correctness and thus for behavioural guarantees for highly dynamic OC-systems. The underlying formal foundation [9, 10] allows a separated verification approach, in which verification of the used reconfiguration mechanism can be performed independently of the verification of the system's properties by using just the specification of the self-reconfiguration instead of the actual implementation. This leads to two steps in the verification process:

- Verification of expected properties P based on the formal model of the functional part of the system SYS_{func} relative to the specification of the self-reconfiguration $Spec_{self\text{-}x}$:

$$SYS_{func} \wedge Spec_{self\text{-}x} \models P$$

- Verification of the self-reconfiguration mechanism o/c_{impl} using the invariants described in Sect. 2.1 against its specification $Spec_{self\text{-}x}$:

$$o/c_{impl} \models Spec_{self\text{-}x}$$

Verification of the Functional System

Usually, the functional system consists of several components running in parallel. A component is specified as a transition system and has trace-based semantics which is in line with the previous formal considerations. The complete functional system is the parallel composition of all components. More precisely, an interleaving semantics is used. Properties are formalised with Interval Temporal Logic (ITL), which explicitly includes the environment. This allows for a separation of system and environment. Further, an arbitrary environment can be considered without stipulating syntactic restrictions for the formulae describing the behaviour. More details on the formal semantics and the used logic can be found in [4].[2]

As OC-systems have a potentially vast number of components that are not predefined and can change during runtime, a monolithic verification of the whole system is not feasible. Therefore, a compositional method is used, in which parts of the systems are regarded separately before the analyses are combined to make statements for the complete system. This technique allows to give global guarantees by local reasoning about individual components. A compositionality theorem was proved that ensures correct lifting of the local properties to a global level and that defines the required proof obligations. The application to systems with self-x properties is presented in [10], which provides more details about compositional verification of OC-systems with *RIA*. With this approach, the verification of the functional system can be performed offline without the need of explicit consideration of the actual self-reconfiguration algorithm used.

Verification of Self-x Mechanisms by Verified Result Checking

The second step is to verify that the used self-reconfiguration mechanism is correct with respect to its specification. In OC-systems bio-inspired algorithms like genetic algorithms [7], decentralised algorithms like the one presented in Sect. 5, or learning techniques, e.g., neural networks or learning classifier systems, are often used. These algorithms do not necessarily return valid and correct results, nor are they always sound or complete. A technique to avoid direct verification of those complex algorithms and an alternative to online verification of the complete system is the use of result checking [5]. The idea is to add a short program *RC* (*Result Checker*) that checks the output of the self-reconfiguration algorithm.

Specification of **Result Checker** (*RC*)	
input	configuration (*conf*) of o/c_{impl}
output	'correct' if *Inv*(*conf*), 'incorrect', otherwise

If the calculated configuration (*conf*) of the algorithm restores the invariant, the checker just forwards the result. If *conf* violates the invariant, the result is blocked and feedback is provided to the self-reconfiguration algorithm. This reduces the

[2] Note that application of *RIA* and the general verification steps are independent of the logic used for system specification and formalisation of properties.

verification task of the O/C implementation to the formal verification of the result checker RC. This offline, a priori verification has to prove the following proof obligation:

$$RC \models Spec_{self\text{-}x}$$

If the result checker adheres to the specification of the reconfiguration process, i.e., to the properties that are required after a reconfiguration was performed, the result checker itself is formally correct and can thus detect faulty configurations.

3 Example Scenario

A running example to illustrate the application of the *RIA* is a set of self-organising resource-flow systems. Instances are, e.g., applications in production automation or logistics. The components of resource-flow systems can be described by the Organic Design Pattern (ODP), as depicted in Fig. 3. *Agents* are the main components in these systems, processing *resources* according to a given *task*. Every *agent* has several *capabilities*, which can be produce, process, or consume *capabilities*. Hence, the *task* is a sequence of *capabilities*, beginning with a *produce* capability and ending with a *consume* capability. Furthermore, each agent knows other agents it can exchange *resources* with. This is expressed in the *inputs* and *outputs* relations.

The *role* concept is introduced to define correct resource-flows through the system. Roles are assigned to agents and specify what the agent's part of the task is. A role is composed of a *precondition* telling the agent from which (*port*) it gets a resource and when the role can be applied, a sequence of capabilities (*capabilitiesToApply*) that should be applied if the role is executed, and a *postcondition* defining where the resource has to be handed over.

Agents can have multiple roles and select one when another agent wants to hand over a resource. The agent accepts resources only if the situation matches one of the role's precondition. This role is selected and the resource is processed according to

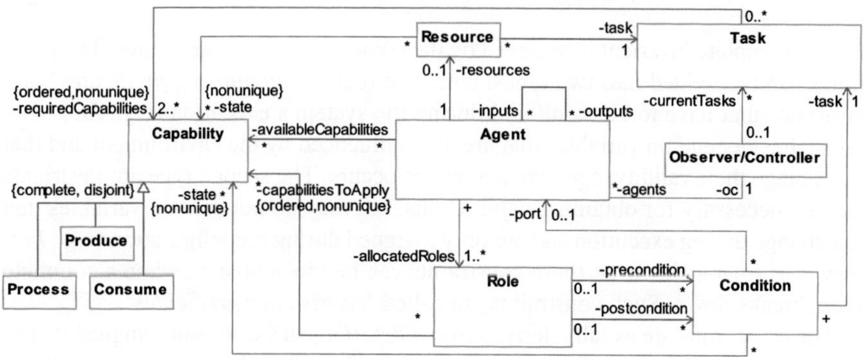

Fig. 3 Components of resource-flow systems

the capabilities defined in the role. At last, the agent tries to give the resource to the next agent, defined in the postcondition.

In this scenario, self-organisation is achieved by role allocation. In case of a failure, the system searches for a new valid role allocation. For a more elaborate model and further details on the software engineering aspects of self-organising resource-flow systems, please refer to Chap. 2.2.

4 Defining Corridors of Correct Behaviour

As described in Sect. 2, *RIA* proposes the implicit definition of correct states by specifying an invariant that defines the behavioural corridor, instead of explicitly listing all possible configurations a system may choose. The invariant can be regarded as a set of constraints on the system states and the system's configuration. In the example scenario, the invariant specifies valid role allocations with respect to the current system state. For instance, one predicate of the invariant may state that only capabilities are assigned that are available at the agent:

$$\begin{aligned} Inv_1(sys) :\Leftrightarrow\ & \forall ag \in Agents, \forall r \in ag.allocatedRoles: \\ & r.capabilitiesToApply \subseteq ag.availableCapabilities \end{aligned}$$

This predicate can be monitored locally by each agent's *Invariant Monitor*. In case an agent loses a capability, a reconfiguration is triggered. Another part is to specify the resource-flow property, which means that roles must be connected. So if one agent has a role that tells it to hand over a resource to another agent, the other agent must have a consistent role. This means that the agent is defined as one possible input.

$$\begin{aligned} Inv_2(sys) :\Leftrightarrow\ & \forall ag_{send} \in Agents, \forall r_{send} \in ag_{send}.allocatedRoles: \\ & ag_{rec} = r_{send}.postcondition.port \\ & \rightarrow \exists r_{rec} \in ag_{rec}.allocatedRoles: (r_{rec}.precondition.port = ag_{send}) \end{aligned}$$

The complete invariant is retrieved by the conjunction of all predicates. The predicates can be divided into two types. One type (called *'monitoring predicates'*) are invariants that have to be monitored during the system's execution (e.g., Inv_1), because they depend on variables that are also influenced by the environment and that can change their validity, e.g., when a failure occurs. The second type are invariants that are necessary for obtaining valid results, but they do not contain variables that can change during execution and are only assigned during reconfiguration (e.g., Inv_2 only uses roles). However, these constraints can be violated, e.g., when a complete agent breaks down. Such constraints are called *'consistency predicates'*.

The constraints are usually derived from OCL (Object Constraint Language) constraints annotated to the software engineering models. Transformation to a predicate logic formula is a straight-forward task and similar to the technique used, e.g., in [1].

The challenge in a self-reconfiguration phase is to find a valid assignment of roles with respect to the current system state. As the invariant constrains the assignments to the system variables, the reconfiguration problem can also be formalised as a constraint satisfaction problem [15]. Thus, a central reconfiguration mechanism that has global knowledge of the agents and their current state can be used to solve this problem. For an implementation, common off-the-shelf constraint solvers (e.g., Kodkod [9, 14]), genetic algorithms, or learning techniques can be used. However, a central control introduces a single point of failure as well as a bottleneck and often the agents have only local knowledge. In such situations, the problem has to be solved in a decentralised fashion. In the next section, a mechanism is presented which realises a fully decentralised Observer/Controller layer.

5 Decentralised Restoration of Invariants

The algorithm for reconfiguration presented here is characterised by three features. First, it is specialised and optimised for the class of self-organising resource-flow systems. While it is not limited to this class, it makes use of the property that roles establish a resource-flow in the system and, therefore, a topology the algorithm can exploit. Second, the algorithm is robust against failures, like partial failures of an agent (e.g., broken capabilities) or the breakdown of a complete agent. Finally and most importantly, the algorithm is decentralised and does not need global knowledge or a central control. One advantage of a local reconfiguration is that the rest of the system is not affected by a reconfiguration and therefore can continue working. Further, the algorithm is based on an asynchronous communication model.

5.1 Coalitions for Local Reconfiguration

The concept of the algorithm is to form a group of agents, called *coalitions*, that are able to restore the violated invariant locally. More detailed, the algorithm reacts to the breakdown of capabilities, the complete breakdown of an agent, and lost connections between agents. If one of those circumstances is detected, it reconfigures the system locally such that the invariant holds again for the coalition. In the following, the process of building the coalition is called *coalition formation*. The notion of coalitions and coalition formation originate in the field of multi-agent systems, where significant work on the topic was done. Particularly, this includes proposals for solving the set partitioning problem [13] or the coalition generation problem, where the challenge is to find suitable coalitions to perform a task together. In [11], a coalition is described as a goal-directed and short-lived organisation that internally coordinates its activities in order to achieve the coalition's goal. The structure of a coalition is flat, with an optional leader that represents the coalition. The coalitions presented in the following adhere to the same principles and have similar properties. The goal of a coalition is to locally restore the invariant by finding a set of agents that have every capability required for repairing the resource-flow.

5.2 Coalition Formation Strategy

As soon as an agent Ag_i recognises an invariant violation, it starts a reconfiguration by creating a new coalition C_i and becoming its leader. The objective of Ag_i is to find enough agents to be able to reconfigure the system in order to compensate the failure. In terms of the ODP, this means that processing of resources according to task τ can continue.

During the coalition formation, the algorithm exploits the existing system structure given by the current role allocation and the resource-flow defined by them, while being limited to local knowledge. Each coalition is led by one coalition member, which coordinates the coalition in order to find a set of agents that can restore the violated invariant locally. Every time a new agent is included into the coalition, the coalition's knowledge grows by the knowledge of the added agent, which consists of its allocated roles, available capabilities, and its possible inputs and outputs to other agents.

To be able to reconfigure, Ag_i must extend the coalition in order to get more degrees of freedom and increase flexibility. Therefore, the leader is asking other agents to join its coalition. If the coalition has enough members to reconfigure a continuous segment of τ, it stops acquiring further agents. This segment is called *connected task fragment(CTF)*. It is necessary to ensure that a consistent solution can be calculated and that the local solution leads to a global restoration of the invariant. Next, the agent identifies a subsequence of this segment (*task fragment to reconfigure*, *TFR*) that actually needs to be reconfigured, because not necessarily all agents that joined the coalition need new roles.

After that, leader Ag_i searches agents that are responsible for maintaining the resource-flow between the coalition and the parts of the system not involved in the reconfiguration. These agents are called *edge agents(EAg)* as they enclose the coalition with regard to the resource-flow. Edge agents are not necessarily new agents. Instead, agents that are already part of C_i can be edge agents (see Fig. 4). Agents that are not edge agents but are directly involved in the reconfiguration are called *core agents(CAg)*.

Subsequently, the resource-flow within the coalition is re-established. If this is not possible because the possible inputs or outputs do not allow a connection, new agents are recruited. These agents are called *resource-flow agents(RFAg)*. As soon as the resource-flow is established, leader Ag_i disbands the coalition. At a glance, the different agent types are defined as follows:

- *core agents(CAg)*: contains all agents of a coalition C_i that should apply at least one capability within *TFR* and all agents that were responsible for establishing the resource-flow within *TFR* before the reconfiguration started.
- *edge agents(EAg)*: consists of all agents that are responsible for a consistent interconnection to the rest of the system and whose precondition or postcondition must not change during reconfiguration.
- *resource-flow agents(RFAg)*: contains all agents that are needed to establish the resource-flow for the new role allocation and that are not already included in *CAg* or *EAg*.

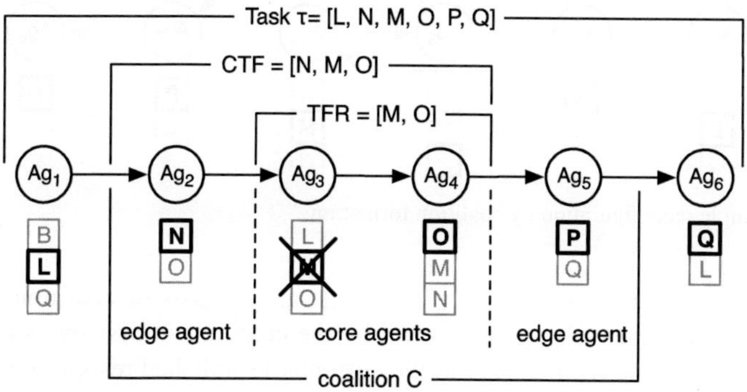

Fig. 4 Terminology and example of a coalition

The abstract example in Fig. 4 illustrates the terminology used. In the depicted situation, Ag_3 loses capability M that is needed to process resources with task τ. The resulting coalition then contains Ag_2, Ag_3, Ag_4, and Ag_5.

There are two types of reasons that trigger an agent to start a reconfiguration and form a coalition. One is an invariant violation that can be detected by one agent itself, e.g., a broken capability. The other type of invariant violation can only be detected by adjacent agents, e.g., when a complete agent breaks down. For both cases, the algorithm has a different strategy for recruiting new agents to its coalition, which is explained in the following.

5.3 Strategy for Local Variable Violation

One of the most important features for making a system more robust is reconfiguration caused by the loss of a capability, which is needed to execute an assigned role. Of course, loss of a capability diminishes redundancy and can only be compensated if there is at least one redundant capability in the system. This kind of failure violates the invariant, more precisely Inv_1 (see Sect. 4), which states that all the capabilities needed for an allocated role must be available. If a violation of the invariant on a local variable is detected, the agent starts inviting predecessors and successors with respect to the resource-flow defined by its roles.

In the situation depicted in Fig. 4, agent Ag_3 loses capability M. It creates a coalition C_3 with itself as leader. The *CTF* is $[M]$. Additionally, the agent has variable Cap_{nd} containing all needed capabilities to fulfil *CTF*, which only contains M at this moment. The goal of every leader is that the coalition's available capabilities Cap_{av} are a superset of Cap_{nd}.

In order to reach this goal and to compensate the failure, Ag_3 has to enlarge the coalition to a point where it finds an agent that has capability M. As the agents only have local knowledge, they cannot just pick the agent that has the missing capability. Instead, they have to iteratively extend their coalition until the agent that can apply M is finally found. Therefore, each leader alternately sends requests to join

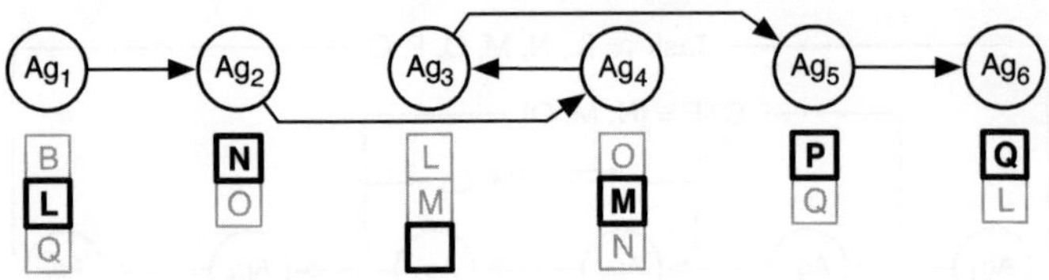

Fig. 5 Example reconfiguration by coalition formation

the coalition to agents contained in the inputs and outputs of coalition members, because these are the only agents known by the coalition. It first sends requests to its direct neighbours, with respect to the currently established resource-flow, which are in the sample scenario agents Ag_2 and Ag_4. So, first Ag_2 is asked and the coalition expands to $C_3 = \{Ag_2, Ag_3\}$. As Ag_2 has applied capability N until now, *CTF* changes to $[N, M]$. Further, Ag_2 transfers all its knowledge about inputs and outputs to Ag_3, which extends the amount of agents Ag_3 can request. Information about the available capabilities and the currently assigned roles of Ag_2 is also transferred. Because agent Ag_2 only has capabilities O and N, Ag_3 has to search further and asks Ag_4 next, which finally has the needed capability and $Cap_{nd} \subseteq Cap_{av}$ is satisfied. The situation now is: $C_3 = \{Ag_2, Ag_3, Ag_4\}$ and $CTF = [N, M, O]$.

Ag_3 now calculates a possible assignment of the capabilities and determines *TFR*, which is $[M, O]$ here. Ag_2 does not have to change its capability in order to solve the problem. The corresponding set of *core agents* is $CAg = \{Ag_3, Ag_4\}$. As the connection between reconfigured agents and non-reconfigured agents must be consistent, Ag_2 is an *edge agent* and Ag_5 is also requested to join the coalition: its role has to change because its input connection will be different afterwards. The set of *edge agents* is therefore defined as $EAg := \{Ag_2, Ag_5\}$. The coalition is now reconfigured and the new roles are allocated. Figure 5 shows the resulting capability assignment and the configured, new resource-flow. As soon as this is done, Ag_3, as the coalition's leader, disbands the coalition and processing continues.

5.4 Strategy for Complete Breakdown of an Agent

A complete breakdown of an agent violates the invariant, specifically predicate Inv_2. Because this violation cannot be detected locally by one agent, there is a slight difference in coalition formation and reconfiguration.

First of all, the breakdown must be detected, which is realised by "alive" messages exchanged by agents that are connected by the resource-flow. If, for example, agent Ag_3 in Fig. 5 completely breaks down, the agents Ag_2 and Ag_4 will detect this sooner or later as they no longer receive messages from the adjacent agent. Independently from each other, they both start a reconfiguration. As they do not know the broken agent's role and the capability it applied, they now have two objectives. Firstly, identify the capabilities that the agent has performed and, secondly, find agents that provide the missing ones. By utilising the system structure (i.e., the currently defined resource-flow) it is possible to recover the roles of the lost agent if the knowledge of both coalitions is merged. Therefore, Ag_4 needs to find the predecessor (Ag_2) and Ag_2 the successor (Ag_4) of Ag_3.

Thus, Ag_2 searches for an agent that applies a role that is applied after its own one. This can be checked because the expected incoming resource state is part of each role. In contrast to a local violation, Ag_2 does not recruit its neighbours with respect to the resource-flow. That is because one neighbour is the faulty agent Ag_3, the other one is a predecessor (Ag_1), and further search in this direction will (most likely) not return a successor. For this reason, Ag_2 prefers agents that are not direct neighbours. As soon as Ag_2 finds a successor of Ag_3, it starts to follow the resource-flow backwards until it reaches Ag_4.

Since Ag_4 also detects the violation and starts a coalition on its own, there are two coalitions in the system. When one coalition tries to recruit an agent of the other one, the coalitions clash. If this happens, the coalitions are merged and one of the former leaders becomes the new leader of the united coalition. The selection is achieved by a standard leader election algorithm.

In addition, there are other situations where coalition clashes have to be resolved. This can be the case if, for example, two different failures occur at the same time, which results in two independent coalitions. If these two coalitions meet, they are merged as in the situation above. However, in larger systems, coalitions often do not meet, for example, as failures occur on opposite ends of the system. In such a case, there are two reconfigurations in parallel without recognising each other. As they do not interfere, they fix their failures independently and, therefore, restore the invariant of the complete system.

5.5 Discussion

The algorithm can reconfigure a system after an invariant violation locally without the need of a global Observer/Controller. Further, it can deal with strongly restricted knowledge of the individual agents because it distinguishes the type of invariant violation. These properties allow the application to systems of unlimited size. Nevertheless, the size and the runtime of the coalition is closely connected to the redundancy distribution within the system. If the next available capability is far away, the resulting coalition will be larger and reconfiguration will be more complicated than when it is available in the direct neighbourhood.

Under the assumption that there are no further failures in the non-reconfigured part of the system, it can be proved that, if the invariant is restored locally within the coalition, then the invariant also holds and is restored for the complete system.

The principle of this algorithm is also applicable to other system classes. The idea is to make use of domain knowledge and the topology of the system to be able to get by with local knowledge only. In this case, the knowledge that can be derived by the currently assigned roles and the knowledge gathered by the neighbours allows to solve the reconfiguration problem locally by forming a coalition of a few agents.

The correctness of the algorithm is ensured by the verified "Result Checker" component (see Sect. 2.2) which assesses the configuration before it is enacted in the system. A more detailed view on the algorithm and discussion about some situations is presented in [2].

6 Summary and Outlook

In this article, the *Restore Invariant Approach* was presented. It provides a paradigm for the specification and analysis of OC-systems. *RIA* enables a separate treatment of self-x and functional behaviour and, therefore, allows to give behavioural guarantees despite of self-organisation. The behaviour of a system is not specified by giving explicit statements. Instead, behavioural corridors are specified to exclude unwanted behaviour. This is done by constraining the system states by so-called invariants, which the system then tries to keep satisfied.

To show the applicability of the presented theory, all results were applied to the system class of resource-flow systems and a decentralised algorithm based on local knowledge for the reconfiguration of such systems was introduced. A detailed view of the system class and a software engineering guideline based on *RIA* that allows to actually build real world systems in a repeatable and disciplined fashion is presented in Chap. 2.2.

The underlying formal framework allows to prove that a system behaves as intended and will not show undesirable behaviour without restricting its freedom to adapt to unforeseen situations. This is essential if applications in safety-critical domains should benefit from the advantages of Organic Computing techniques in the future. The framework provides methods and tools for formal specification and compositional verification of systems with self-x properties [10]. The presented technique allows the verification of systems with an unknown amount of agents and an explicit treatment of the environment. This is an important feature when considering OC-systems, which, by definition, interact with their environment. Traditional systems are developed with respect to the environment they will be deployed in, so the possible influences are known and considered during design time. OC-systems, in contrast, are able to deal with different, unforeseeable environments and therefore these considerations are shifted to run time. The complete theory is integrated into the interactive theorem prover KIV [3] and all proofs were performed with this tool.

The formal framework further allows to analyse a system with regard to the number of errors that can occur before a reconfiguration will no longer be successful. For this purpose, ADCCA [8] allows to formally analyse OC-systems for combinations of failures that can lead to a permanent violation of the invariant. Together with different metrics, the reconfiguration abilities of a system can be quantified and different systems can be compared.

References

1. Ackermann, J.: Formal description of OCL specification patterns for behavioral specification of software components. In: Workshop on Tool Support for OCL and Related Formalisms, Technical Report LGL-REPORT-2005-001, pp. 15–29, EPFL (2005)

2. Anders, G., Seebach, H., Nafz, F., Steghöfer, J.-P., Reif, W.: Decentralized reconfiguration for self-organizing resource-flow systems based on local knowledge. In: Proceedings of EASe 2011, Las Vegas, USA (2011, to appear)
3. Balser, M., Reif, W., Schellhorn, G., Stenzel, K.: KIV 3.0 for provably correct systems. In: Hutter, D., Stephan, W., Traverso, P., Ullmann, M. (eds.) Proc. Int. Wsh. Applied Formal Methods. LNCS, vol. 1641, pp. 330–337. Springer, Berlin (1999)
4. Bäumler, S., Balser, M., Nafz, F., Reif, W., Schellhorn, G.: Interactive verification of concurrent systems using symbolic execution. AI Commun. **23**(2–3), 285–307 (2010)
5. Blum, M., Kanna, S.: Designing programs that check their work. In: STOC '89: Proceedings of the Twenty-First Annual ACM Symposium on Theory of Computing, pp. 86–97. ACM, New York (1989)
6. Branke, J., Mnif, M., Müller-Schloer, C., Prothmann, H., Richter, U., Rochner, F., Schmeck, H.: Organic computing—addressing complexity by controlled self-organization. In: ISoLA, pp. 185–191 (2006)
7. Goldberg, D.E.: Genetic Algorithms in Search, Optimization, and Machine Learning, 1st edn. Addison-Wesley, Reading (1989)
8. Güdemann, M., Ortmeier, F., Reif, W.: Safety and dependability analysis of self-adaptive systems. In: Proceedings of ISoLA 2006. IEEE Comput. Soc., Los Alamitos (2006)
9. Nafz, F., Ortmeier, F., Seebach, H., Steghöfer, J.-P., Reif, W.: A universal self-organization mechanism for role-based organic computing systems. In: González Nieto, J., Reif, W., Wang, G., Indulska, J. (eds.) Autonomic and Trusted Computing. LNCS, vol. 5586, pp. 17–31. Springer, Berlin (2009)
10. Nafz, F., Seebach, H., Steghöfer, J.-P., Bäumler, S., Reif, W.: A formal framework for compositional verification of organic computing systems. In: Xie, B., Branke, J., Sadjadi, S., Zhang, D., Zhou, X. (eds.) Autonomic and Trusted Computing. LNCS, vol. 6407, pp. 17–31. Springer, Berlin (2010)
11. Rahwan, T., Ramchurn, S., Jennings, N., Giovannucci, A.: An anytime algorithm for optimal coalition structure generation. J. Artif. Intell. Res. **34**(1), 521–567 (2009)
12. Richter, U., Mnif, M., Branke, J., Müller-Schloer, C., Schmeck, H.: Towards a generic observer/controller architecture for Organic Computing. INFORMATIK 2006 – Informatik für Menschen! P-93, pp. 112–119 (2006)
13. Shehory, O., Kraus, S.: Methods for task allocation via agent coalition formation. Artif. Intell. **101**(1–2), 165–200 (1998)
14. Torlak, E., Jackson, D.: Kodkod: a relational model finder. In: Grumberg, O., Huth, M. (eds.) Tools and Algorithms for the Construction and Analysis of Systems. LNCS, vol. 4424, pp. 632–647. Springer, Berlin (2007).
15. Tsang, E.: A Glimpse of constraint satisfaction. Artif. Intell. Rev. **13**, 215–227 (1999)

F. Nafz (✉) · H. Seebach · J.-P. Steghöfer · G. Anders · W. Reif
Institute for Software & Systems Engineering, Universität Augsburg, Universitätsstr. 6a, 86159 Augsburg, Germany
e-mail: nafz@informatik.uni-augsburg.de

H. Seebach
e-mail: seebach@informatik.uni-augsburg.de

J.-P. Steghöfer
e-mail: steghoefer@informatik.uni-augsburg.de

G. Anders
e-mail: anders@informatik.uni-augsburg.de

W. Reif
e-mail: reif@informatik.uni-augsburg.de

Chapter 1.6
Ant Inspired Methods for Organic Computing

Alexander Scheidler, Arne Brutschy, Konrad Diwold, Daniel Merkle, and Martin Middendorf

Abstract In recent years social insects have been a major inspiration in the design of new computational methods. This chapter describes three examples of the application of ant-inspired methods in the domain of Organic Computing. The first example illustrates implications of theoretical findings in response-threshold models that explain division of labour in ants for Organic Computing systems. The second example outlines how principles from the house-hunting behaviour of ants can be used to organise systems that are based on reconfigurable components. The final example describes how sorting mechanisms in production networks can benefit from the indirect pheromone communication found in ants.

Keywords Ant inspired methods · Division of labour · Response-threshold models · Reconfigurable components · Sorting mechanisms in production networks

1 Introduction

Social insect colonies constitute one of the best studied examples of natural systems that possess self-x properties, e.g., self-organisation, self-healing, and self-adaptation. While single individuals are relatively simple, behaviour at the colony level is complex, highly efficient, robust, and adaptive. These properties make social insects an interesting source of inspiration for the design of new methods and systems. Typically summarised under the term Swarm Intelligence, such methods have applications in various domains such as optimisation, logistics, or robotics [1, 17].

In this article we present our work which applies self-organisation principles found in ant behaviour to Organic Computing. After a short overview we describe three examples of ant inspired Organic Computing systems in more detail.

Models of ant navigation were applied to systems consisting of a large number of relatively simple moving agents (e.g., robots). Ant movement strategies can lead to interesting emergent effects. For example, slight differences in individual behaviour can result in spatial separation at the system level [21]. However, if ant movement strategies are applied in Organic Systems the occurring emergent effects may not be desired as well. Thus, it is important to identify the origins of such negative emergence to counteract it, for example by guiding the movement of the agents through proper architectural design of the movement area [22].

C. Müller-Schloer et al. (eds.), *Organic Computing — A Paradigm Shift for Complex Systems*, 95–109
Autonomic Systems, DOI 10.1007/978-3-0348-0130-0_6, © Springer Basel AG 2011

Indeed, negative emergence is a general problem in Organic Computing systems. A common approach to overcome it is to employ a Observer/Controller architecture (see, e.g., Chap. 4.1). In this architecture observer components use sensors to detect unwanted emergent effects. If such an effect occurs, this is reported to controller components. The controller components then can send out control signals to the system to counteract the unwanted emergent effect. Swarm Controlled Emergence [15] constitutes an alternative approach for overcoming negative emergence. The main idea is to prevent negative emergence via a swarm of anti-emergence components. These components are similar to normal components, but exploit slight behavioural differences to prevent the unwanted emergent effect without directly influencing the behaviour of the normal system components.

An emergent effect on which Swarm Controlled Emergence has been tested is ant clustering [15]. In this stigmergic aggregation behaviour ants form clusters of items such as eggs or dead bodies. Individuals pick up and drop items depending on the density of items in their neighbourhood, which leads to the formation of item-clusters. Principles inspired from ant clustering have been applied, for example, to information retrieval or robotics. As clustering is a well-characterised emergent effect, we chose to test the success of our approach on preventing its occurrence, that is, we treated it in our study as a case of negative emergence. Systems with different types and fractions of anti-emergence agents were studied. The results show that Swarm Controlled Emergence can indeed be used to prevent emergent effects when a sufficient number of anti-emergent agents are present in the system.

Another example of ant behaviour that has inspired system design concerns mobile sensor networks. When watching the bustling life of an ant hill it is hard to imagine that ants take a rest. However, biologists have discovered that ants rest for quite a large fraction of their time. Surprisingly, not only single ants show alternate phases of resting and being active, but whole ant colonies exhibit synchronised activity phases. Inspired by this behaviour of ant colonies, a mechanism for self-synchronised duty-cycling in mobile sensor networks was proposed in [9]. It was shown that this approach can help to save energy and is adaptive to the amount of energy available in the single sensor nodes.

Division of labour is another behaviour of ant colonies that is particularly interesting for Organic Computing. Organic Computing systems often consist of a large number of autonomous components and thus face a problem similar to ant colonies, namely how to efficiently distribute the workload between the components of the system without using a central control. Indeed, the self-organising principles found in ants can help to organise the allocation of work in Organic Computing systems. Different behavioural models of ants have been successfully applied for the design of Organic Computing systems. For example, the observation that ants partition single tasks into subtasks that then are accomplished by different individuals has led to the design of a task partitioning method for Organic Computing systems [23]. In addition, the way in which ants distinguish nest mates from foreign ants by their odour has inspired a decentralised clustering algorithm for networks [12]. This algorithm has been employed to efficiently allocate tasks to the components of an Organic Computing system depending on the resource demands of the task [14].

A well known theoretical model that is used to explain division of labour in ants is the threshold-response model. In this model each task that has to be accomplished by the colony is associated with a stimulus value reflecting task urgency. Moreover, each agent of the colony has a personal threshold value for each task which reflects how much the agent likes to engage in this task. More exactly, the higher the stimulus value for a certain task and the lower an agent's threshold value for that task, the higher is the probability that the agent will engage in that task. We used threshold-response models, for example, to create a self-organised task allocation scheme for Organic Computing systems [16]. In [6] threshold-response models have been extended to include the spatial distribution of tasks and individuals. This is important as spatial aspects influence the division of labour in social insect colonies as well as in Organic Computing systems. In Sect. 2 of this article we describe such a spatially extended response threshold model. We are interested to know if and how the spatial distribution of tasks influences the colony's performance. Moreover, we investigate how beneficial it is for a colony to spatially rearrange its work in terms of the additional expense involved in such ongoing reorganisation. Three individual-based strategies are tested and compared regarding their adaptivity and efficiency in a dynamic task environment.

Inspired by ant house-hunting strategies, a collective decision-making process for Organic Computing systems has been proposed in [3]. The considered systems consist of two types of reconfigurable computing units: workers that execute jobs that come into the system, and scouts that are additionally responsible for the reconfiguration of all computing units in the system. In Sect. 3 of this article the proposed ant-inspired strategy is analysed experimentally and compared to a non-adaptive reference strategy. It is shown that the ant-inspired strategy leads to a collective decentralised decision process that enables the system to generate a high throughput of jobs even in dynamic environments.

A very prominent example of the successful application of self-organisation principles found in ants is Ant Colony Optimisation (ACO) [7]. ACO is a metaheuristic for solving combinatorial optimisation problems which is inspired by the foraging behaviour of ants. In ACO artificial ants use artificial pheromone to mark and reinforce paths in decision graphs that correspond to good solutions. In Sect. 4 of this article we present a network system that is also inspired by the ants' use of pheromones [20]. More precisely, variants of so-called Emergent Sorting Networks are studied. These are directed networks that consist of router agents as nodes and buffers that are located on the edges between two neighbouring agents [2]. Agents can move objects from their input buffers to their output buffers. It is assumed that these objects may differ in type and are inserted randomly into the network. With suitable rules for the agent's behaviour, the items become partially sorted in the output of the network. Pheromone-based variants of the agent behaviour are introduced and studied in different network topologies in terms of their sorting performance.

2 Spatial Organisation of Work and Response-Threshold Models

The response-threshold concept has been widely used to explain division of labour in social insects, both experimentally and analytically [4]. The threshold-reinforcement model adds learning to the threshold concept and was initially introduced to study the effect of group size on specialisation in social insect societies [10]. It was extended recently in [6] to include effects of a spatial organisation of work.

The threshold-reinforcement model consists of n individuals and m tasks. Each task T_j, $j \in \{1,\dots,m\}$ is associated with a task-specific stimulus value $S_j \geq 0$. Each individual $I_i \in \{1,\dots,n\}$ has a task-specific threshold value $\theta_{i,j}$ with $0 \leq \theta_{i,j} \leq \theta_j^{\max}$ for a maximal threshold $\theta_j^{\max} \geq 0$. An individual can either work on one of the tasks or stay idle. The probability that an idle individual I_i starts working on task T_j given m tasks is $P(X_i = j) = (1/m) \cdot (S_j{}^2/(S_j{}^2 + \theta_{ij}{}^2))$. Per timestep a working individual does $\alpha > 0$ units of work on its current task. The probability to cease working in the next timestep is $p \geq 0$. An individual that stops working is idle for at least one timestep. Hence, the maximal amount of work per task that the colony can do in one time step, on average, is $W_{\max} = n/m \cdot 1/(1+p) \cdot \alpha$ [10].

A distinctive feature of the threshold-reinforcement model is that stimulus and threshold values change dynamically. Each time step a task's T_j stimulus value changes by $S_j = S_j + \delta_j - E_j \cdot \alpha$. E_j denotes the number of individuals engaged in task T_j, and $\delta_j = D_j \cdot W_{\max}$, $0 \leq D_j \leq 1$, indicates an additive value for the stimulus in each time step. D_j constitutes the colony's demand for the task T_j per timestep. It allows to model situations with different demand conditions. For example, if $D_j = 1$, the colony must perform the amount $W_{\max}$ of work on task T_j per time step to keep the stimulus value S_j stable. The threshold values of an individual I_i also change each time step: If I_i works on task T_j then $\theta_{i,j} = \max\{\theta_{i,j} - \xi, 0\}$ (learning parameter ξ) otherwise $\theta_{i,j} = \min\{\theta_{i,j} + \phi, \theta_j{}^{\max}\}$ (forgetting parameter ϕ).

The spatially extended threshold-reinforcement model of [6] incorporates l different locations L_k, $k \in \{1,\dots,l\}$. The number of individuals at location L_k is n_k with $n = \sum_{i=1}^{l} n_i$. For simplicity, it is assumed that the number of individuals at each location is fixed. Any of the tasks can be present at each location with a certain local demand and each individual can only sense local task-stimuli. Thus, a demand D_j^k and stimulus S_j^k exists for task T_j at location L_k. Then $D_j = \sum_{k=1}^{l} D_j^k$ is the total demand for task T_j and $S_j = \sum_{k=1}^{l} S_j^k$ is the total stimulus for task T_j. Individuals only consider local stimuli when deciding which task to work on and the work of an individual contributes only to the tasks at its location. Hence, the demand and stimulus values at the different locations are independent of one another.

The stimulus of a task reflects the colony's need to work on that task. As it is correlated with the workload fulfilment of a colony it can also be seen as the accumulated amount of unprocessed work for a given task. Thus, colonies with high stimulus values have more undone work than colonies with low stimulus values. This means that stimuli can serve as a quality measure and therefore can be used to directly evaluate a system's performance. The sum of all stimulus values for all tasks at a given time step $S = \sum_{j=1}^{m} S_j$ is also called the *total stimulus*. It is assumed that the lower the total stimulus is the better is the performance of the system.

Results and Discussion

If not stated otherwise the presented results are averaged over 50 simulation runs using the following parameter values: $l = 2$, $m = 2$, $n = 100$, $p = 0.2$, $D_j = 1$, $\xi = 3.5$, $\phi = 4.0$, $\theta_{i,j}^{\max} = 20$ for $i, j \in \{1, 2\}$. Note that there is an initialisation phase (see [5] for more details) of length $t_{init} = 10000$ steps. The maximum age of an individual is $a_{\max} = 1000$. Such a system with two locations and two tasks could for example model an ant nest with two brood chambers where it is necessary to feed the larvae and to take care of the eggs.

Effect of Demand Distribution

Firstly, the effect of various demand distributions on the two locations is studied. Figure 1(a) depicts the average stimulus sum on the two locations for this case. The task distribution at both locations can be seen as a representation of how spatially separated the tasks are. In a setup with strong spatial separation each task is only present at one location. At the other extreme, when the task demands are not spatially separated the tasks are equally present at each location. As can be seen in Fig. 1(a) the system is able to keep the total stimulus at the lowest level in a scenario where both tasks are spatially separated. In this case the individuals at each location can focus on a single task only, since demand and therefore also the stimulus value for the other task is zero. Spatially separated tasks thus seem to increase a colony's performance. This leads to a second question, namely whether a colony can afford to maintain a spatial separation when there is a tendency in the system for tasks to spatially overlap, as is the case in many real systems. Figure 1(b) depicts the average total stimulus for different numbers of active individuals under different demand distributions D_j^k. It can be seen that in an environment where tasks are unequally distributed over locations (i.e., $D_1^1 = 1$) the colony can keep the total stimulus value at a lower level even when some of the individuals remain idle, compared to a situation where all individuals are active but the tasks are equally distributed over the

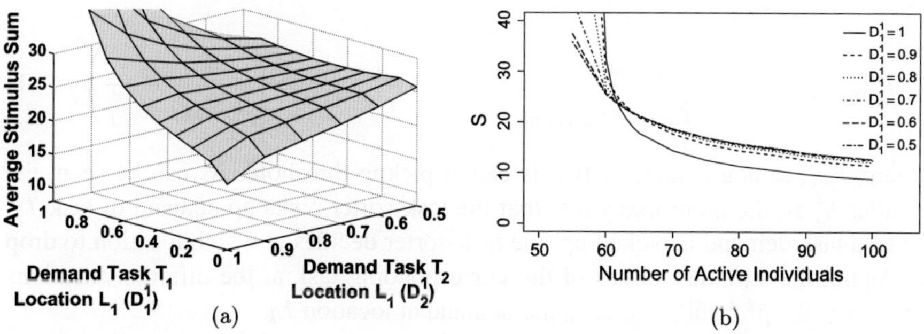

Fig. 1 (a) Average total stimulus under different demand distributions D_j^k for both tasks on both locations. (b) Average total stimulus for different number of active individuals under different demand distributions D_j^k

locations. Such idle individuals might be used in a colony to keep the tasks spatially separated (e.g., to move larvae that are in the wrong chamber in their brood chamber).

Demand Redistribution as a Third Task

To investigate a system's performance in dynamic environments it is assumed here that at each time step a certain fraction f_{cease} of each task demand D_j^k ceases at each location. The ceased demand is fed back into the system (to keep D_j for each task j). Let $f_{re\text{-}enter}^{k,j}$ denote the fraction of the vanished demand $f_{cease} * D_j$ placed on location k. At each location the task demands will slowly converge to the distribution given by $f_{re\text{-}enter}^{k,j}$ with the convergence speed depending on f_{cease}.

An individual-based task sorting mechanism is implemented in the system as a third task. As the amount of work needed for a certain task at a location is given by its demand, individuals performing task-sorting (called task-sorters) will pick up the fraction of work they could do on a given task and redistribute it to the other location. Given the sorting task T_{j+1}, it is suggested that the stimulus for this task on location L_k is given by $S_{j+1}^k = S_{j+1,\max} \times (\min_{i=1}^{j} S_i^k / \max_{i=1}^{j} S_i^k)$ where $S_{j+1,\max}$ is a parameter that denotes the maximal sorting stimulus. Two different ways of task redistribution are tested here:

i) *Deterministic redistribution.* Here a task-sorter knows which task belongs to which location and tries to shift misplaced demands accordingly. More formally: each task T_j is assigned to a location L_k. When executing the sorting task each task-sorter at a location L_k tries to pick up a demand for a task that is not assigned to this location and places it on the designated location.

ii) *Stigmergic sorting.* In nature, this sorting strategy is based on the density of brood items in an ant's vicinity, e.g., eggs or larvae. In our system, the probability for an unladen task-sorter to pick a task is higher the lower the task's stimulus value. The probability that a task-sorter at location L_k will pick up demand from task T_j is given by

$$P_{j,pickup\_from}^k = \frac{P_{j,pickup}^k}{\sum_{i=1}^m P_{i,pickup}^k}, \quad \text{with} \quad P_{j,pickup}^k = \left(\frac{T_{pickup}}{T_{pickup} + S_j^k}\right)^2 \tag{1}$$

where T_{pickup} is a parameter that is called pickup threshold. As can be seen, the smaller S_j^k is, the more likely it is that the task-sorter picks up demand of task T_j. Once a task demand is picked up, the task-sorter decides at which location to drop it. Again, the stimulus values of the corresponding task at the different locations determine the probability to drop the demand at location L_k

$$P_{j,drop\_at}^k = \frac{P_{j,drop}^k}{\sum_{h=1}^l P_{j,drop}^h}, \quad \text{with} \quad P_{j,drop}^k = \left(\frac{S_j^k}{S_j^k + T_{drop}}\right)^2 \tag{2}$$

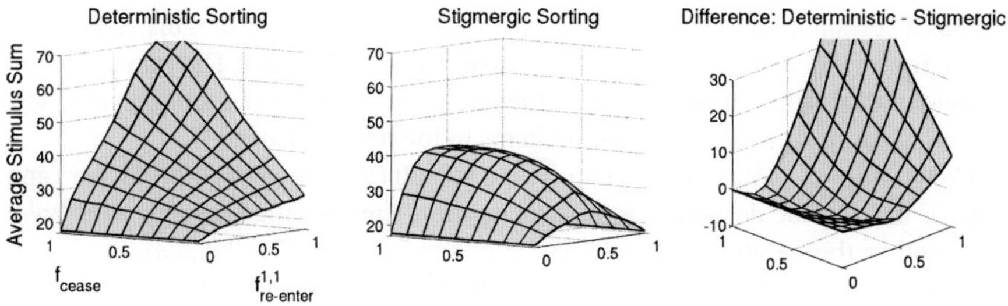

Fig. 2 Total average stimulus values and total average stimulus differences for different sorting methods under different cease f_{cease} and rearrangement fractions $f^{k,j}_{re\text{-}enter}$

where T_{drop} is a parameter that is called the drop threshold.

Figure 2 depicts the behaviour of the system with a sorting task under different cease fractions f_{cease} and rearrangement fractions $f^{k,j}_{re\text{-}enter}$. It can be seen that such a system is able to cope with strong changes in external task demand i.e. $f_{cease}, f^{k,j}_{re\text{-}enter}$. It can dynamically adjust its workforce to different situations. Its sorting abilities are limited by the fact that each task is encountered with the same likelihood (thus the maximum probability to start sorting is 0.3). Hence, the system runs into trouble when the amount of demand that needs to be rearranged per time step is larger then the maximal amount of demand that can be rearranged on average per time step given by $n/m \cdot 1/(1+p) \cdot W_{pickup}$.

3 Learning from House-Hunting Ants: Collective Decision-Making in Organic Computing Systems

Temnothorax albipennis is an ant species that has small workers and lives in small colonies. As it tends to build its nests in structurally unstable places such as in crevices and under rocks, the ant has to emigrate frequently [11]. Due to its peculiar properties, the emigration behaviour of *T. albipennis* has been studied thoroughly [11, 19]. The details of the emigration process are described in the following and thereafter we discuss how it can be applied to Organic Computing (see also [3]).

When emigration becomes necessary, scout ants start to search the surrounding environment for a new nest site. Upon finding a candidate site, a scout starts to assess the site according to several criteria (e.g., size and darkness). If the scout considers the site to be superior to the current nest, it tries to get a "second opinion" by guiding other scouts to the candidate nest site. The guidance is accomplished by *tandem-running*, which means that one ant teaches another ant the route by leading it while keeping close physical contact to allow bidirectional feedback. As soon as the other scout reaches the candidate nest site, it assesses it and, if it considers it to be good, starts recruiting other scouts as well. Scouts delay recruitment to a candidate nest site by a time that is inversely proportional to the perceived quality of the site. This behaviour ensures that better nests will attract scouts faster. As soon as a certain number of ants prefer a certain candidate nest site, the scouts will switch from tandem-running to a faster transportation behaviour. The transportation behaviour allows the scouts to migrate the rest of the colony in a more efficient way by simply carrying workers and brood items to the new nest site.

The number of scouts that are required to switch to the migration behaviour (and thus for making the decision for the specific nest site) is called *quorum threshold*. By adapting the quorum threshold to the colony's needs, *T. albipennis* is able to make a compromise between the accuracy and speed of its decisions. A low quorum threshold leads to fast but error-prone decisions.

3.1 Model of the Organic Computing System

The model for the Organic Computing system used in our study consists of a set of computing units. These units are connected by a simple interconnection network that only allows one-to-one communication; broadcast operations are not possible. Units can communicate with any other single unit at any time. As no complicated communication is employed, no cost is associated with communication. It is assumed that there are m different job types in the system. Additionally, it is assumed that the system is always saturated with jobs. All units work in parallel and each unit can only work on a single job at any time step. The purpose of the system is to maximise the total throughput. Throughput is measured as the total number of jobs processed by the system per time step.

Each computing unit consists of s slices (see also [13]). Every slice can be independently configured for exactly one type of jobs. Hence, the set of slices of a unit is partitioned with respect to the different types of jobs. This partition is called the configuration of the unit. Each unit has to be able to serve every type of job, i.e., for every type of job at least one slice has to be configured. As the purpose of the system is to maximise the throughput of jobs, configurations which result in higher throughput are considered to be superior to others.

Each job has a certain run time that depends on the number of slices a unit has configured for it. Processing speed does not necessarily increase linearly with processing resources (e.g., because of caches or verification mechanisms). The function which defines the run time for each job type depending on the number of configured slices is referred to as *run time function*.

The set of all possible configurations is called configuration space (see Fig. 3 for two example configuration spaces). The size of the configuration space is $\binom{s-1}{m-1}$ and depends on the number of slices and number of job types. In the worst case, the size of the configuration space increases factorially with linearly increasing parameters. The configuration space is strongly characterised by the run times of the underlying job types; for example job types which have a linear run time function result in spaces where configurations are ordered linearly by their quality. In order to study the impact of different underlying run time functions on the system, configuration spaces were classified by the behaviour of the function used to create the distribution

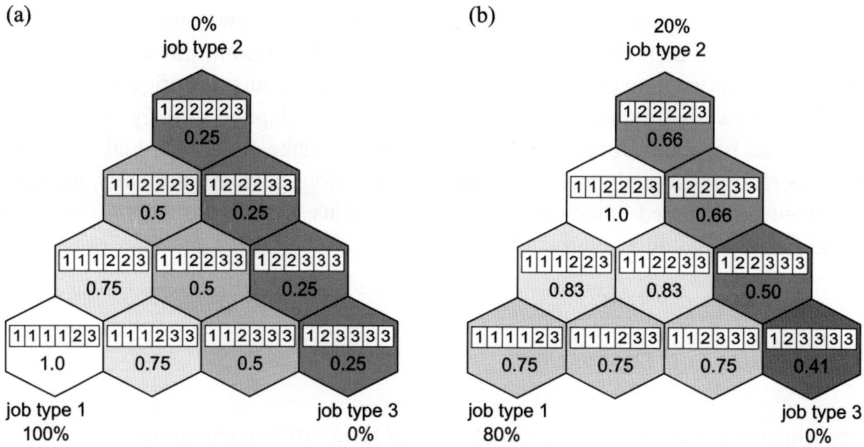

Fig. 3 Example configuration spaces with $s = 6$ and $m = 3$. Each cell denotes a configuration with configured slices per job type and its relative quality (number and cell colour). The percentages shown next to each job type are the distributions of the job types in the system. (**a**) The underlying run time function is linear, thus the configurations are ordered by their quality. The configuration in the *lower left* corner is the optimum. (**b**) Non-linear (exponential) run time function. The optimal configuration gives more slices to job type 2 than the job distribution would indicate

Table 1 Classification of underlying run time functions

Name	Behaviour with increasing number of slices	Example
constant	constant	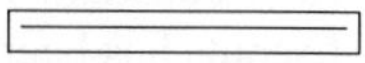
linear	linearly decreasing	
monotone	monotonically decreasing	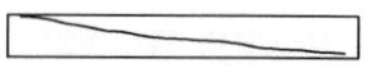
polygonal	decreasing with local maxima	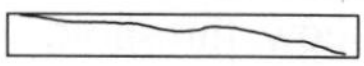
exponential	exponentially decreasing	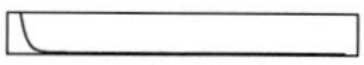
increasing	increasing	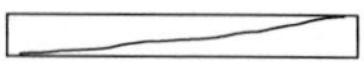
random	run times are chosen randomly	

of run times in the system. All classes of the run time functions that have been studied in this article are given in Table 1.

It is assumed that two types of computing units exist in the system: units that are only able to execute jobs, and units that can execute jobs and are additionally able to reconfigure themselves or other units. The former are referred to as *worker units*, whereas the latter are called *scout units*. The ratio between scouts and workers is assumed to be fixed. A worker unit can only change configuration with the help of a scout unit. It is the task of the scout units to find and evaluate new configurations.

We study two types of reconfiguration strategies. The *adaptive* strategy is inspired by the house-hunting behaviour of the ant *T. albipennis* described above. In the proposed system, scouts explore the configuration space, trying to agree on a new configuration using the ant's method of decision making. The required quorum threshold is adapted depending on the quality of the new configurations. Two limits have been defined, l_u and l_l, between which the quorum threshold is scaled linearly. Upon deciding on a new configuration, all the scouts reconfigure the workers to this new configuration. In order to evaluate the adaptive strategy, a *reference strategy* has been introduced. In this strategy, scouts make only individual decisions while performing a random search in the configuration space. Additionally, workers are evenly distributed among the scouts. The scouts reconfigure theirs associated workers upon finding a better configuration.

3.2 Results and Discussion

Experiments are conducted in a simulated and time-discrete environment. Each parameter is evaluated using 20 experimental runs. The metric we use in the experiments is a *fitness* metric f which is defined as the ratio of the average system throughput to optimal system throughput. An optimal system exhibits a fitness of $f = 1$.

One of the most characteristic features of the model presented is the distinction between scout and worker units. Therefore, first the impact of the scout-worker ratio on the system's fitness is studied. The optimal partition of the units found experimentally is when approx. 20% of the units act as scouts. Below this value, the fitness drops rapidly. On the other hand, above this value the fitness remains nearly constant. This behaviour is analogous to the real ants, where the percentage of scouts has been found to be 20–30% [18].

In order to study the impact of the quorum threshold scaling on the adaptive strategy, we vary the scaling limits l_l and l_u as shown in Fig. 4(a). As the results show, threshold scaling strongly affects the fitness of the system. An upper limit of

(a)

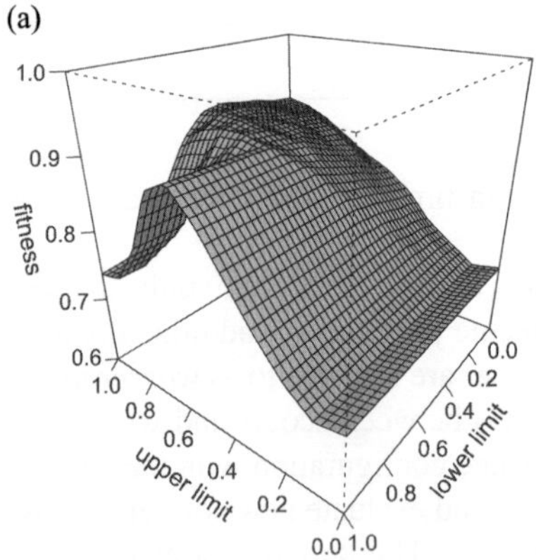

(b)

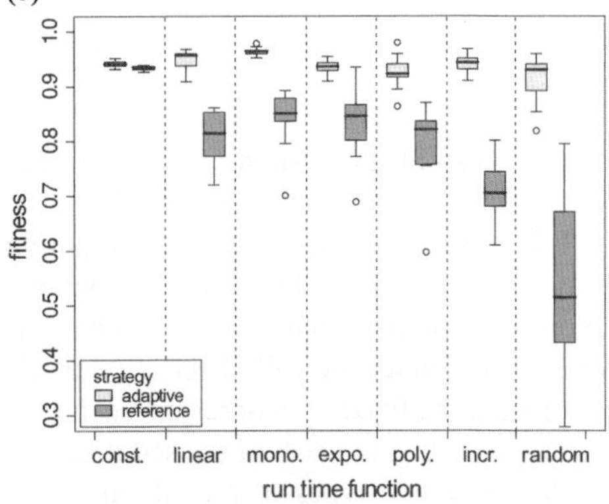

Fig. 4 (**a**) Fitness of the adaptive strategy for different quorum threshold scaling limits l_l and l_u. Limits from 0.0 to 1.0 increased by 0.1. (**b**) Fitness of the two strategies compared on different classes of configuration spaces (see Table 1)

less than 0.6 yields too many bad decisions, thus reducing the fitness substantially. The system does not react as sensitively to changes of the lower limit, although the results show that a lower limit over 0.5 results in a quorum threshold that requires too many units to decide unanimously. Having a lower limit higher than the upper limit results in a dysfunctional quorum scaling. Hence, a strong decrease of fitness can be observed in the left corner of Fig. 4(a).

The final experiment studies the impact of the different configuration spaces on the fitness of the two strategies. Figure 4(b) shows a comparison of the fitness on configuration spaces that are generated by the aforementioned different classes of run time functions (see Table 1). The adaptive ant-inspired strategy performs well on all run time function classes. The reference strategy's fitness drops with an increasing complexity of the configuration space, as the scout units are not able to find good configurations. This shows an important advantage of the ant-inspired strategy: The robustness of the self-organised reconfiguration enables it to deliver nearly constant performance even in dynamic and complex configuration spaces.

4 Sorting Networks of Router Agents

In this section we discuss Emergent Sorting Networks. These were originally described in [2], and we here outline a variant proposed in [20]. The basic components of Emergent Sorting Networks are *router agents*. Each router agent $a \in \mathcal{A}$ has n input buffers $x_1, \ldots, x_n$ and m output buffers $y_1, \ldots, y_m$. A buffer can store exactly one object that can be of one of k different types $t_1, \ldots, t_k$. Router agents are connected to form an (acyclic) network by associating input buffers with output buffers (see Fig. 5). Agents can route objects from one of their input buffers to one of their free output buffers. Consecutively (in random order), the agents apply their local routing rules to move objects. Within each time step every object can be moved at most once. The network is initialised with empty buffers. The agent that represents

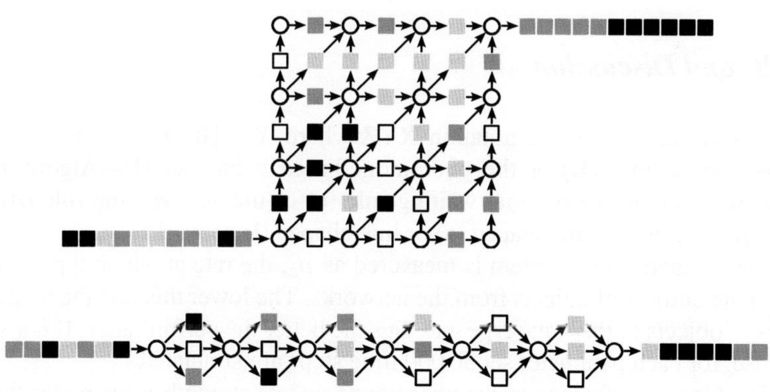

Fig. 5 Square and line shaped sorting networks; *squares* depict buffers; *circles* depict agents; input/output sequences of objects are shown as sequences of *squares*

Algorithm 1 Pheromone-based Behaviour of an Agent $a \in \mathcal{A}$

T is the set of types of the (unmoved) objects in the input buffer of a
if $rand() < |T|/n \wedge \exists$ free y_j **then**
  Choose type $t_s \in T$ according to $\mathbf{p}(t_i) = \frac{\tau_i^a}{\sum_{t_l \in T} \tau_l^a}$ $\forall t_i \in T$
  Move (so far in this step unmoved) object o with type t_s to y_j
  Update pheromone: $\tau_j^a := \tau_j^a + \beta(\mu_j - \tau_j^a)$, with if $(j = s)$ $\mu_j = 1$ else $\mu_j = 0$
end if

the root of the network can access an unlimited queue of objects of random type and injects them into the network.

Original Behaviour. In the agents' routing behaviour as proposed in [2] an agent memorises for each of its output buffer the type of the last object that was moved to this buffer. Upon activation an agents tries to move an object to a buffer to which it has recently moved objects of the same type. If this is not possible it moves with a certain probability a random object to a random output buffer. The probability is determined as the fraction of occupied input buffers of the agent.

Pheromone Based Behaviour. In the following we propose a pheromone based variant of the local routing rules. For each agent $a \in A$ and object type t_i $(i = 1, \ldots, k)$ a pheromone value $\tau_i^a \in (0, 1)$ is introduced. All pheromone values are initially set to $1/k$. An active agent will choose an object based on the pheromone values for the different object types and puts it on a random empty output buffer. The calculation of the probabilities works in the same way as used in the Ant Colony Optimisation (ACO) metaheuristic [8] (see Algorithm 1). The parameter β gives the learning rate, i.e., how strong the pheromones are updated.

New Waiting Rule. In both types of behaviours described so far agents wait to perform an action depending on the number of occupied input buffers. We also test a variant where an agent is only allowed to act if all its input buffers are occupied (this is called *new waiting rule*).

Results and Discussion

In the following we use the notation XYZ where $X \in \{\mathbf{B}, \mathbf{P}\}$ denotes the original agent behaviour (**B**) or the pheromone-based behaviour (**P**—Algorithm 1), $Y \in \{\mathbf{o}, \mathbf{n}\}$ refers to the original waiting rule (**o**) or the new waiting rule (**n**), and $Z \in \{\mathbf{s}, \mathbf{l}\}$ indicates square-shaped (**s**) or line-shaped (**l**) networks.

The performance of a system is measured as p_c, the rate at which type changes occur in the outflow of objects from the networks. The lower this rate the longer the batches of objects of the same type and thus the better the performance. If not stated otherwise, for each parameter set 50 000 time steps are simulated.

In Fig. 6(a) the influence of the pheromone update strength β on p_c for the **Pol** system is depicted. Such parameter test are conducted for all pheromone based sorting networks. The values of β that lead to the best system performance are then

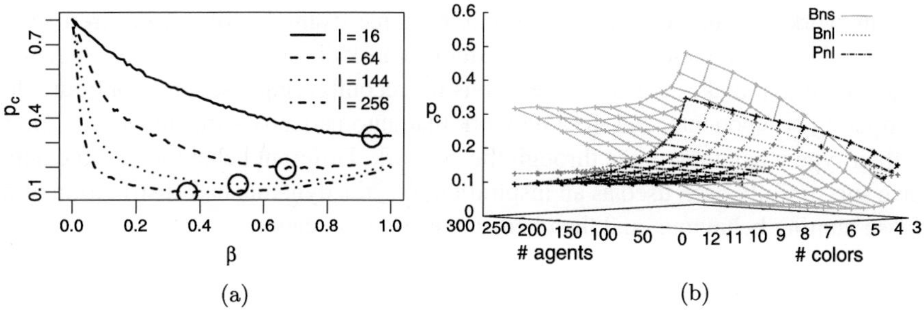

Fig. 6 (**a**) Tuning results for systems for the pheromone based system **Pol**; (**b**) Comparison of systems **Bns**, **Bnl**, and **Pnl**

used for all subsequent experiments. The following conclusions can be made: i) in square-shaped networks an agent should always try to repeat the action of the previous time step ($\beta = 1$), ii) in line-shaped networks the more agents are used, the smaller the value of β should be, and iii) the optimal value for β depends on the applied waiting rule and on the number of object types.

For the original waiting rule the sorting in the line-shaped networks is in general better than the sorting in the square-shaped networks. This trend becomes stronger when more object types are used. Moreover, pheromone-based systems show better performance than the original systems when line-shaped networks with many agents are considered. The opposite is the case for square-shaped networks.

Interestingly, the results for the new waiting rule (systems $*$**n**$*$) are quite different. Here the original system in conjunction with a square-shaped network (**Bns**) works best. When three different types of objects are present, square-shaped networks outperform line-shaped ones. However, when the number of types is increased, the performance of the pheromone-based system on a square-shaped network drops, but the performance of line-shaped networks improves. The results of the three best performing systems **Bns**, **Bnl**, and **Pnl** are given Fig. 6(b). Another interesting observation is that for a small number of object types the system **Bns** outperforms both line-shaped systems. With increased number of the object types the number of agents becomes important: when few agents are used the system **Bnl** is best, whereas for many agents **Pnl** is the best performing system.

Systems that use the new waiting rule (***n***) clearly outperform systems that use the original waiting rule (***o***). The best-performing system with the new waiting rule (**Bns**) performed better than the best-performing system with original rule (**Pol**) in all conducted experiments.

5 Summary

In this article we have shown that the self-organised behaviour of ant colonies can be an inspiration for the design of Organic Computing systems. After an overview on our work in this area we have discussed three examples of ant-inspired self-organised systems in more detail: i) a system for task distribution for mobile agents, ii) a system that makes collective decisions for reconfiguring a set of reconfigurable computing components, and iii) a network that automatically sorts items according to their type when they move through the network. Different behaviours of the ants have successfully been used as an inspiration, i.e., the dynamic task distribution, the house hunting behaviour, and the pheromone communication.

Acknowledgements We are grateful to Cliodhna Quigley for assistance with the manuscript. Arne Brutschy acknowledges support from the Belgian F.R.S.–FNRS. The work was partially supported by the Deutsche Forschungsgemeinschaft (DFG) within SPP 1183.

References

1. Bonabeau, E., Dorigo, M., Theraulaz, G.: Swarm Intelligence: From Natural to Artificial Systems. Oxford University Press, London (1999)
2. Brueckner, S.A.: Return From the Ant—Synthetic Ecosystems for Manufacturing Control. PhD thesis, Humbold University, Berlin (2000)
3. Brutschy, A., Scheidler, A., Merkle, D., Middendorf, M.: Learning from house-hunting ants: Collective decision-making in organic computing systems. In: Proc. ANTS Conference. LNCS, vol. 5217, pp. 96–107 (2008)
4. Beshers, S.N., Fewell, J.H.: Models of division of labor in social insects. Annu. Rev. Entomol. **46**, 413–440 (2001)
5. Diwold, K., Merkle, D., Middendorf, M.: Adapting to dynamic environments: polyethism in response threshold models for social insects. Adv. Complex Syst. **12**(3), 327–346 (2009)
6. Diwold, K., Scheidler, A., Middendorf, M.: The effect of spatial organisation in response threshold models for social insects. In: Proc. European Conf. on Complex Systems (ECCS 2009) (2009)
7. Dorigo, M., Maniezzo, V., Colorni, A.: Ant system: Optimization by a colony of cooperating agents. IEEE Trans. Syst. Man Cybern. **26**(1), 29–41 (1996)
8. Dorigo, M., Stuetzle, T.: Ant Colony Optimization. MIT Press, Cambridge (2004)
9. Hernández, H., Blum, C., Middendorf, M., Ramsch, K., Scheidler, A.: Self-synchronized duty-cycling for mobile sensor networks with energy harvesting capabilities: A swarm intelligence study. In: Proc. IEEE Swarm Intelligence Symposium (SIS 2009), pp. 153–159 (2009)
10. Gautrais, J., Theraulaz, G., Denebourg, J.-L., Anderson, C.: Emergent polyethism as a consequence of increased colony size in insect societies. J. Theor. Biol. **215**(3), 363–373 (2002)
11. Marshall, J.A.R., Dornhaus, A., Franks, N.R., Kovacs, T.: Noise, cost and speed-accuracy trade-off: Decision-making in a decentralized system. J. R. Soc. Interface **3**(7), 243–254 (2006)
12. Merkle, D., Middendorf, M., Scheidler, A.: Decentralized packet clustering in router-based networks. Int. J. Found. Comput. Sci. **16**(2), 321–341 (2005)
13. Merkle, D., Middendorf, M., Scheidler, A.: Self-organized task allocation for computing systems with reconfigurable components. In: Proc. 9th Int. Workshop on Nature Inspired Distributed Computing (NIDISC'06) (2006)
14. Merkle, D., Middendorf, M., Scheidler, A.: Using decentralized clustering for task allocation in networks with reconfigurable helper units. In: Proc. Int. Workshop on Self-Organizing Systems (IWSOS 2006). LNCS, vol. 4124, pp. 137–147 (2006)
15. Merkle, D., Middendorf, M., Scheidler, A.: Swarm controlled emergence - designing an anti-clustering ant system. In: Proc. IEEE Swarm Intelligence Symposium (SIS 2007), pp. 242–249 (2007)

16. Merkle, D., Middendorf, M., Scheidler, A.: Self-organized task allocation for service tasks in computing systems with reconfigurable components. J. Math. Model. Algorithms **7**(2), 237–254 (2008)
17. Merkle, D., Middendorf, M., Scheidler, A.: Organic computing and swarm intelligence. In: Swarm Intelligence, pp. 253–281. Springer, Berlin (2008)
18. Pratt, S.C., Mallon, E.B., Sumpter, J., Franks, N.R.: Quorum sensing, recruitment, and collective decision-making during colony emigration by the ant leptothorax albipennis. Behav. Ecol. Sociobiol. **52**(2), 117–127 (2002)
19. Pratt, S.C., Sumpter, D.J.T., Mallon, E.B., Franks, N.R.: An agent-based model of collective nest choice by the ant temnothorax albipennis. Anim. Behav. **70**(5), 1023–1036 (2005)
20. Scheidler, A., Blum, C., Merkle, D., Middendorf, M.: Emergent sorting in networks of router agents. In: Proc. ANTS Conference. LNCS, vol. 5217, pp. 299–306 (2008)
21. Scheidler, A., Merkle, D., Middendorf, M.: Emergent sorting patterns and individual differences of randomly moving ant like agents. In: Proc. 7th German Workshop on Artificial Life (GWAL-7), pp. 105–115 (2006)
22. Scheidler, A., Merkle, D., Middendorf, M.: Congestion control in ant like moving agent systems. In: Hinchey, M., Pagnoni, A., Rammig, F., Schmeck, H. (eds.) Biologically Inspired Collaborative Computing. IFIP International Federation for Information Processing, vol. 268, pp. 246–256. Springer, Boston (2008)
23. Scheidler, A., Merkle, D., Middendorf, M.: Stability and performance of ant queue inspired task partitioning methods. Theory Biosci. **127**(2), 149–161 (2008)

A. Scheidler · A. Brutschy
IRIDIA - CODE, Université Libre de Bruxelles, Brussels, Belgium

A. Scheidler
e-mail: ascheidler@ulb.ac.be

A. Brutschy
e-mail: arne.brutschy@ulb.ac.be

K. Diwold · M. Middendorf (✉)
Institut für Informatik, Universität Leipzig, Leipzig, Germany
e-mail: middendorf@informatik.uni-leipzig.de

K. Diwold
e-mail: kdiwold@informatik.uni-leipzig.de

D. Merkle
Dept. of Mathematics & Computer Science, University of Southern Denmark, Odense, Denmark
e-mail: daniel@imada.sdu.dk

Chapter 1.7
Organic Computing: Metaphor or Model?

Mathias Gutmann, Benjamin Rathgeber, and Tareq Syed

Abstract Organic Computing is the most recent approach in the field of ambient technologies, aiming at the complete integration of technical systems into the space of human action. As such, the application of this type of technology generates severe technical, societal and political problems. These problems are necessarily connected with a proper understanding of the applied technologies, their potentials as well as their limits. A proper understanding is a necessary prerequisite not only for the "philosophical" evaluation of a technology at hand—it is just as relevant to the immanent development of Organic Computing technologies themselves. In order to reach such an understanding it is useful to scrutinise the metaphorical expressions, which are used by the scientist developing the technologies in question. Accordingly the main aim of this article is the systematic analysis of some central metaphors—particularly "evolution" and "self-organisation". The metaphorical expressions are reconstructed methodologically by applying the culturalist model-procedure, which allows the explication of metaphors in terms of "as if" relations. Based on the results of this reconstruction, the pragmatic and semantic limits and perspectives of Organic Computing can be determined.

Keywords Methodology · Metaphors · Model theory · Evolutionary robotics · Organic computing · Design · Optimisation · Self-x

1 Introduction

Organic Computing (OC) is the last and most advanced offspring of a family of technologies whose origin can be drawn back towards the 60ies of the last century. It is based upon some insights, provided e.g. by evolutionary robotics, which lead to the concept of self-adapting artefacts. Nevertheless, OC proclaims to exceed some immanent limits of evolutionary robotics (ER) by adding self-x-capacities, which enrich the evolutionary paradigm with central aspects of self-organisation. ER and OC differ considerably, particularly concerning the limitations expected; nevertheless, the latter is a *continuation* and advanced *descendant* of the first. Both share some important features concerning the general idea, the research strategy and the technical solutions applied, and both use expressions, which are borrowed from life sciences and in particular from evolutionary biology. Accordingly, it is necessary to

C. Müller-Schloer et al. (eds.), *Organic Computing — A Paradigm Shift for Complex Systems*, 111–125
Autonomic Systems, DOI 10.1007/978-3-0348-0130-0_7,

re-evaluate these common core features in order to provide an adequate understanding of OC as a structurally new technology. This article starts with a methodological reconstruction of the core concept explicated extensively by ER, namely evolution. On the basis of this reconstruction, OC is analysed methodologically considering the concept of self-organisation and self-x-capacities. Finally, a model-theoretical perspective of OC is presented, which avoids the intrinsic shortcomings of ER and articulates some reasonable potentials of OC as a new type of engineering method.

2 Evolutionary Robotics as a Precursor of Organic Computing

In spite of the central role of biological expressions little attention has been paid concerning the methodological status and the explanatory value of the background theory in terms of robotics. The indeterminate status of those expressions transferred to robotics becomes tangible by *Nolfi and Floreano* [25] speculating on the possibility of Lamarckian evolution to be the more adequate type of theory—for *engineering* purposes:

> "One might wonder, whether Lamarckian evolution (...) could be more effective than Darwinian evolution (...). *Ackley and Littmann* (1994) for instance claimed that in artificial evolution, where inherited characters can be easily coded into the genotype, given that the mapping between genotype and phenotype is generally quite simple, there is no reason for not using Lamarckian evolution. Indeed the authors showed that Lamarckian evolution is far more effective than Darwinian evolution in a stationary environment. On the other hand, as shown by *Saski and Tokyo* (1997), Darwinian evolution largely outperforms Lamarckian evolution when the environment is not stationary or when individuals are exposed to different learning experiences." ([25], p. 287)

It should be mentioned, that "optimisation" in engineering and in evolution is a homonymic term—accordingly, the citation above should read, that here is "no *technical*(!) reason for not using Lamarckian evolution". There are indeed severe *biotheoretic* reasons for not using Lamarckian evolution however controversially discussed during the last 30 years. This restriction becomes methodologically relevant, particularly if we insist on the engineering character of both, Organic Computing as well as evolutionary robotics. As such, they represent specific branches of human technology, whereas evolution is first of all a subject matter of scientific theory-construction. The situation becomes worse, when we consider a second theory-type which is of some importance in the context of robotics, that is the theory of self-organisation; in this case it is disputable whether the mechanisms assumed to constitute processes of self-assembly and self-organisation are at all of one and the same type, or whether they are more than pure metaphorical descriptions of processes which could be described properly by avoiding the epistemologically suspicious term "self". According to the stated lack of epistemological reconstruction, it is the aim of this article to identify the methodological status of the central biomorphic expressions, particularly of those connected with the evolutionary paradigm on one hand, the concept of self-organisation on the other hand. This reconstruction only allows us to differentiate between purely metaphorical (and perhaps still heuristically useful) expressions and systematically sound and scientifically relevant models.

3 The Evolutionary and the Engineering Paradigm

Evolution is the central expression in order to describe the structure of the new technology in the field of robotics:

> "The basic idea behind evolutionary robotics goes as follows (...). An initial population of different artificial chromosomes, each encoding the control system (and sometimes the morphology) of a robot, are randomly created and put in the environment. Each robot (physical or simulated) is then let free to act (move, look around, manipulate) according to a genetically specified controller while its performance on various tasks is automatically evaluated. The fittest robots are allowed to preludes (sexually or asexually) by generating copies of their genotypes with the addition of changes introduced by some genetic operators (e.g., mutations, crossover, duplication). This process is repeated for a number of generations until an individual is born which satisfies the performance criterion (fitness function) set by the experimenter." ([25], p. 1)

Very obviously, the description of the evolutionary design of robots is purely (sic) metaphorical, as all the qualifiers and characteristics named do not apply to inanimate objects: the chromosomes are not stainable, they do not build a genotype which becomes expressed via transcription and translation, robots do not act, they do not have sex and are by no definition known born.[1] Our main task now, in order to provide a meaningful reading of this metaphorical (or actually rather allegorical) text, is the systematic reconstruction of those elements of the biological language-game, which are translatable back into the design of technology. That is, we are interested not in the connotation of the "biological" description of technology design, but its possibly reasonable designative function. We will come back to the last statement concerning the role of the experimenter and ask at first, which concept of evolution exactly builds the bio-scientific background.[2] At this starting-point it is only important to notice, that the evolutionary paradigm is defined by some core-features, which differ—some more or less, some fundamentally—from the "engineering" paradigm:

1. In contrast to the engineering paradigm, in the evolutionary paradigm there is no centralised or coordinated planning; the course of evolution is—by definition—not determined by describing a given state of the evolving unit.
2. The transformations the evolving units undergo are not pre-designed. This statement is valid in several respects, from which only some are mentioned here:
   a. The changing of the composition of the unit itself in a given state is determinable but not determined. Thus, "mutations" (in individuals of the unit) are random events though their occurrence is statistically assessable.

[1]The same type of metaphors is at work by dealing with "learning systems", "intelligent agents" or "behaviour" of technical systems [2, 34, 37].

[2]We should keep in mind even at this early state of discussion, that there exists not just one evolutionary theory but a wide and differentiated field of approaches, reaching—when characterised e.g. by the role of reproduction as a mechanism—from Neo-Lamarckian, Neo-Darwinian and Darwinist, towards Non- and even Anti-Darwinist concepts (the latter should not lightly be identified with creationism or Intelligent Design). For further reading see [20] and [21].

b. The result of a changing of the unit depends on a multitude of immanent, internal and external parameters:
   i. Immanent are parameters, which are defined by the specific structure of the individuals the unit is composed of. If we refer, for example, to the species *Pacifastacus leniusculus* (the signal crayfish), the change on the level of the genome of an individual can[3] have consequences that differ from changes in the genome of an individual of the species *Homo sapiens*. The amount as well as the peculiarity of the change will depend on the functional composition of the respective organism.
   ii. Internal parameters refer to the interaction of the individuals the unit is composed of, i.e. a type of change, which unfortunately is often called "behaviour".
   iii. External parameters refer to what is neither covered by i. nor ii.; they might be provided by living as well as non-living aspects of evolving units.
3. The results of the transformation can be described in terms of functional change. In this case an evolving unit at time 2 may be described as presenting better performance in a specific aspect than the same unit at time 1, etc. However, finally a description must be possible, which states that the fitness of the evolving unit was increased.

As a result, evolution seems to provide a type of problem-solving strategies, which allow an indirect optimisation even if the single steps of the transformation, leading from state 1 towards state 2 are not *chosen* by the engineer and insofar are performed by the unit itself. In contrast, the engineering paradigm can be characterised according to *Nolfi and Floreano* ([25], p. 6ff) as follows:

1. Complex units have to be decomposed into smaller, less complex sub-units; this divide et impera-rule holds for all designable units, material as well as immaterial. In the given context of robotic-design, we were dealing with behaviour as well as with tasks.
2. If—according to 1.—robots are supposed to gain environmental information and to act depending on the information, then "perception, data- and information processing, modelling, planning and action" might be a suitable sequence.
3. Following *R. Brook's* approach, the behaviour itself is decomposed into simple, "basic behaviours". By including "self-organisation" processes, the layer-wise construction of the control-system leads to an incremental implementation of both, the behaviour modules as well as the coordination mechanisms.

[3]It is important to realise, that a change on the level of the DNA-sequence does not necessarily lead to phenotypic differences; and vice versa, even one and the same genetic composition may lead to different results (as already emphasised by early embryologists, compare ([22], p. 221). If we take into consideration epigenetic mechanisms, then the relation between genotype and phenotype of a developing unit (which need not be identical with an evolving unit) is represented formally by a many-to-many-relationship. The realisation of this relation in terms of a given organism depends on the description, which rules the biological structuralisation of living entities.

This transformation of the role of the engineer as a central design and control institution leads to a new type of system-building activity, which initiates and uses processes without direct control over the means that are applied or the determination of the subtasks, which have to be realised in order to generate the desired general behaviour of the system:

> "Evolutionary robotics, by relying on an evaluation of the system as a whole and of its global behaviour, releases the designer from the burden of deciding how to break the desired behaviour down into simple basic behaviours." ([25], p. 9)

Thus, by expropriating "natural" processes, quasi-organisms are provided, which out-perform their conventionally engineered antecedents. During this process, the system under construction takes over some parts of the original role of the engineer, and the resulting process is defined as a kind of selection procedure under the rule of a fitness function. Finally, technical systems are generated, which are optimised for several distinct tasks and optimise themselves in the case of a changing environment or a changing fitness function.

4 Methodological Reconstruction I: Is Evolution Design?

The comparison between the engineering and the evolutionary paradigm allows us to determine the exact notion of evolution in terms of evolutionary robotics and insofar of Organic Computing. As the two lists in the previous section show, there are some remarkable differences in definition between "biological" and robotic evolution. Setting aside all the differences in detail, the methodologically most important refer to the role of the "experimenter" on the one hand, and the role of the task, which is defined by the experimenter on the other hand—even if only for the sake of discerning "good" from "bad" behaviour [18]. The experimenter plays a crucial role for the engineering process itself insofar as—at least in principle—it is the experimenter who defines the fitness function:

> "The fitness function is a performance criterion that evaluates the performance of each individual phenotype. Higher fitness values are better. For example, if one wants to optimise a mathematical function, individuals coding for values that generate higher returns will receive higher fitness. Similarly, one may attempt to maximise the difference between the value returned by the function and a target value." ([25], p. 20)

Moreover, the experimenter defines the tasks, for which the robots are supposed to become increasingly better. This description of course is methodologically remarkably similar to the role of "nature" in the evolutionary paradigm, as it is nature in the latter case that defines the relevant aspects of evolutionary fitness. Darwin describes this process as a "wedging" of the organisms by nature itself:

> "One may say there is a force like a hundred thousand wedges trying <to> force every kind of adapted structure into the gaps in the oeconomy of nature, or rather forming gaps by thrusting out weaker ones." ([17], p. 37)

It should be noted, that the metaphors used by Darwin are of teleological and even physico-theological connotations; however they are metaphors, used to describe something which is not to be understood as the *direct* outcome of the doing of an actor, but *as if* it were such an outcome.[4] This brings us to the methodologically leading question, how exactly could *Darwin* know the mechanism of evolution, a process which is not an object of observation[5]—such as the trajectories of planets or the development of an egg into a chicken? We are not allowed for the sake of an answer to refer to reliable knowledge about evolution—because it is exactly this knowledge, which is supposed to be provided.

The only possible and methodologically sound alternative here seems to understand *Darwin's* statement on the wedging as a metaphor for one important aspect of the target-process, evolution, namely the on-going improvement of functional relations to be identified at least between organisms and their environment. In this case, the metaphorical description, which refers to human action and goal-guided activity (such as "wedging" and "adapting",[6]) should be read as a model-operation. The difference can be expressed by stating that "organisms are arranged in their respective environment as if they were wedged by nature into their environment", instead of stating that they actually *are* wedged. The wedging is supposed to result in a specific correspondence between organism and environment; and by explicating this metaphor as a model, we will probably come to the assumption that organisms are understood as if they were functional units which have some specific capacities (according to their respective functional constitution). By referring to living entities as organisms[7] in this ascriptive form,[8] evolution can be conceptualised "as if it were" a process of problem solving, which leads to the increase of efficiency or productivity, i.e. an optimisation-procedure (compare [1]). Accordingly, the term "evolution" has the methodological status of a hypothetical term [3] and belongs to a biological metalanguage, referring to functional structuralisations of living entities and the relations to their respective environment.

So far, our reconstruction refers to some notes taken by the very early *Darwin* of the notebooks; in his later writing, *Darwin* shifts his metaphors significantly. This

[4]Note *Darwin's* familiarity with *Paley's* "Natural Theology": "Although *Paley* is mentioned only once in the Origin, his specter is everywhere apparent in *Darwin's* repeated discussions highlighting the lack of truly intelligent design in nature." ([36], p. 179).

[5]Even in these cases the observation is far from being simple—at least we must be equipped with some correctly working measurement-tools (e.g. for the determination of angles, length and time, compare [13]).

[6]The analysis shows that the predicate is multi-termed, as "y adapts A to B, by using tool x"; so even the physico-theological description itself is gained by an analogy between the creator on the one hand and a craftsman on the other (for further reading see [14], [7], [11]).

[7]The term "organism" then is not synonymous to "living entity"; it rather designates the structuralisation of living entities "as" functional units (see [8] and [9]).

[8]We can discern strong from weak ascription. By (weakly) ascribing intentional descriptions to living entities and their body-parts (such as "A is used to pump blood through the body of x" or "x uses B in order to dig holes") we are able to introduce the term "function" as a basic biological concept, without actually assuming that living entities are in fact setting aims, realising goals or using tools in order to achieve purposes. "Strong" ascription then refers to the application of cognitive expressions to non-human entities such as animals or robots; in this case we assume that they act "as if they were" human actors.

metaphor leads beyond the purely functional metaphors, which are implied by the adaptationist-paradigm (*Gould's and Lewontin's Panglossian* paradigm; see below), as it refers to procedures of breeding. From here to the breeding-metaphor of evolution, which is in fact one of the most important model-types until today, it seems to be a rather short way. But besides the aspects of producing variants and selecting them against some parameters, the main topic in *Darwin's* thinking is the problem of reproduction. And in order to provide a theory of natural reproduction, *Darwin* again refers to human actions; but now, he pays more attention towards breeding and cultivation procedures. It is this type of animal and plant producing procedures he builds his theory of evolution upon; with the consequence that the actor "nature" becomes equipped with some different skills: now it has to define the breeding goals in order to transform evolving units, namely populations.[9] Accordingly, the success of evolutionary transformation is not primarily expressed in terms of functional but of reproductive improvement. And here a very serious methodological problem is generated, by confounding the two descriptions, that of functional improvement on the one hand and that of reproductive success on the other hand. If we stated a direct, causal relation between the efficiency of a function and the producti vity of reproduction, we get caught in the enchanted realm of the Panglossian, adaptationist paradigm; to put with *Gould and Lewontin* in reference to the peculiar spandrels of *San Marco*:

> "Such architectural constraints abound, and we find them easy to understand because we do not impose our biological biases upon them. Every fan-vaulted ceiling must have a series of open spaces along the midline of the vault, where the sides of the fans intersect between the pillars. Since the spaces must exist, they are often used for ingenious ornamental effect. In King's College Chapel in Cambridge, for example, the spaces contain bosses alternately embellished with the Tudor rose and portcullis. In a sense, this design represents an "adaptation", but the architectural constraint is clearly primary. The spaces arise as a necessary by-product of fan vaulting; their appropriate use is a secondary effect. Anyone who tried to argue that the structure exists because the alternation of rose and portcullis makes so much sense in a Tudor chapel would be inviting the same ridicule that *Voltaire* heaped on Dr. *Pangloss*: "Things cannot be other than they are ... Everything is made for the best purpose. Our noses were made to carry spectacles, so we have spectacles. Legs were clearly intended for breeches, and we wear them." Yet evolutionary biologists, in their tendency to focus exclusively on immediate adaptation to local conditions, do tend to ignore architectural constraints and perform just such an inversion of explanation." ([7], p. 148f)

Even though we used "historical" examples to explicate our argument, it sheds some light on the current and on-going discussion within the field of evolutionary theory. The topics of Baldwinian selection, coevolution, cultural evolution or

---

[9] *Darwin* deals very carefully with the term "species", by explicitly holding a nominalist position: "In short: we shall have to treat species in the same manner, as those naturalists treat genera, who admit that genera are merely artificial combinations made for convenience. This may not be a cheering prospect, but we shall at least be freed from the vain search for the undiscovered and undiscoverable essence of the term species." ([4], vol. II, p. 301) In contrast to species, populations as evolving units provide much better possibilities of conceptualising evolutionary transformation. Unfortunately, even populations are introduced by models, and it should be noted, that the intension of "population" depends on the knowledge- and knowhow-basis of the respective models (see e.g. [6, 9, 28, 29, 41]).

epigenetics indicate the tendency of extending the very rigid limits of a strong Darwinist type of evolutionary reconstruction. However, even by integrating functional and structural aspects of evolution, the strict nonteleological framework is indispensable in order to avoid ontological interpretation of design arguments. From this point of view, "evolutionary robotics" presents a logical dilemma, which comes into existence by directly connecting function with reproduction:

1. If we insisted on "robotics" to designate a specific type of technique, that is a kind of human action (however complex it may be), we would have to refer to fitness functions in terms of functional improvement, or at least to functional relations, of an artifact. In that case, there can be no doubt, that the engineer (or the experimenter from the citation above) defines the aim of his artifact—in terms of tasks and the respective fitness function. Consequently, the term "evolution" becomes a metaphor for the fact, that the engineer (or the experimenter) does not necessarily determine the exact way, the task itself is achieved or the tools by which it is achieved.
2. If, on the other hand, we insist, that evolution is not the result of goal-guided action (may it be of human or divine origin), the transformation of a population can be reconstructed ex post festum in terms of a fitness function. In this case, there can be no doubt that evolution is by no means a procedure of problem solving—as there is nobody who defines tasks (a kind of human action, which would overtax even nature itself). Consequently, the term "robotics" becomes a metaphor for the realm, from which the functional models are borrowed, according to which evolution is described as an optimisation process.

"Evolutionary robotics" then has a methodological status, which resembles that of bionics to a certain extent. From this point of view, *Nolfi and Floreano* [25] are absolutely right to state, that "fitness" is used with two meanings, thus, rendering the term "fitness function" to a homonymic expression:

> "This value (which expresses the fitting between organisms and their environment, the authors) is the number of offspring in the case of natural evolution and the fitness value in the case of artificial evolution." ([25], p. 154)

A proper understanding of evolutionary metaphors by transforming them to models rests upon the accurate differentiation between these two types of approaches. By confounding them, we were forced to give a biologically meaningful explanation for the design-principle, which only stands its ground in terms of purposeful human action—e.g. in terms of engineering.

5 Methodological Reconstruction II: Is Evolution Optimisation?

Even if we accept the rejection of evolution as a design process, the methodological alternative remains reasonable, namely to consider evolution as a process of optimisation. This seems to be the more adequate as evolutionists themselves often use optimisation and maximisation routines in order to provide evolutionary reconstructions; technical biology (bionics[10]) provides just as excellent examples for this type of approach as do sociobiology or evolutionary psychology.[11] However, the main task of our methodological reconstruction at this step is the determination of the systematic role of optimisation in connection with the evaluation of evolutionary statements. As we have already ruled out the possibility of evolution as optimisation in terms of design, there remains the possibility of indirect optimisation. In this case, we would prefer those statements on evolutionary transformations that are most parsimonious regarding the transformation steps necessary which lead from a defined earlier state of an evolving unit to a later state of this unit. Parsimony then would be an indirect measure of evolutionary transformation and insofar evolution would have at least some aspects of optimisation. In order to justify this decision, which has the very form of *Ockham's* razor, two argumentative strategies are usually applied:

> "The first is that parsimony is a methodological convention, that compels us to maximise the amount of evolutionary similarity that we can explain as homologous similarity, that is, we want to maximise the similarity that we can attribute to common ancestry. Any character which does not fit a given tree requires us to postulate that the similarity between two sequences shown by that character arose independently in the two sequences—the similarity is due to homoplasy not homology." ([26], p. 190)

In this case, it simply remains open, whether evolution is in fact a process that leads to the "maximisation of similarity"; nevertheless, the reduction of evolutionary tree-reconstruction in terms of purely monophyletic relations, is not a justification of parsimony but its very application. However, no reason is given why we should prefer parsimonious trees and which connection between parsimonious and adequate (not to say true) trees is to be supposed. This unsatisfying reading of parsimony as a convention (which by definition is the result of consensus) leads to a much stronger justification, which now refers to evolution itself:

> "The second view is that parsimony is based on an implicit assumption about evolution, namely that evolutionary change implies that the tree that minimises change is likely to be the best estimate of the actual phylogeny. Under this view, parsimony may be viewed as an approximation to maximum likelihood methods (...), and indeed it was in this context that parsimony methods were first proposed by *Edwards and Cavalli-Sforza*." ([26], p. 191)

This assumption states that evolution actually works by means of parsimony, and the question remains, how we know about this fact. If we referred to actually existing evolutionary trees a vitious circle or a petitio principii were unavoidable. By stating that evolution is—even if only in an indirect sense—connected with optimisation, we would have to refer to a knowledge which can only be provided by not using optimisation procedures; even the outgrouping procedure, which uses taxa that do not belong to the taxa under reconstruction cannot provide a knowledge of the demanded type (see [35], p. 221, for critical reconstruction see [10]). These

[10] See [30] and [24].

[11] Sociobiology and evolutionary psychology are particularly relevant, as they explicitly provide models of behavioural evolution (see e.g., [40] and [38]).

methodological problems can be avoided by assuming, that evolutionary processes can be described as if they were processes of optimisation. In this case, however we had to accept the insight that we know about evolution only insofar, as we are investing some knowhow, which we borrowed from different scientific and technological sources. As a result of our reconstruction we must state that evolution is neither design nor optimisation; nevertheless, it can be understood "as if".

6 Overcoming Evolutionary Robotics: Organic Computing

OC extends ER not only in a methodological but even in a material aspect, by connecting the evolutionary with the self-organisation paradigm [31]. OC then adds the concept of self-organisation as a source of meaningful order and its proponents suppose to iterate the (evolutionary self-) constitution of the controlling system as a property of the system to the system.

However, the theory of self-organisation is well founded in physics and biology. Particularly the theory of dissipative systems showed some potentials for biological conceptualisations connected with the problem of the origin of life and its very early stages.[12] The very idea in using the term self-organisation is to emphasise that there are certain objects, which generate themselves without external aid, intervention or instruction, i.e. they are natural objects in an Aristotelian sense, in contrast to "artificial objects" which are generated in terms of human action or at least in reference towards it.[13] The methodological problems of describing living entities as self-organising systems become tangible, when we try to explicate the term. If a given animal or plant is described as organising itself, e.g. during the process of ontogenesis, the question becomes urgent, what exactly is the bearer of the process of organising. Referring to a well known example it follows:

1. By referring e.g. to a chicken, the "self" is the product of the process and thus cannot act (e.g. organise something) before it came into existence.
2. By referring to the respective egg, the self is generating something, which is not itself (i.e. the chicken) and thus cannot act after coming out of existence.

This well known riddle can be solved easily by referring to two different descriptions, which follow different and mutually irreducible logical grammars. We used the differentiation already for the analysis of the evolutionary paradigm, by referring to living entities on one hand and organisms on the other:

1. In an everyday life description we are stating that an egg becomes transformed into a chicken; by stating that the chicken *develops* from the egg, we are adding

[12]For exemplarily reading concerning the entire paradigm of self-organisation see e.g. [5, 12, 16, 19, 27, 32, 39].

[13]Extending the definition by indirect intervention allows us to include cultivated plants or bred animals as "artificial"; they are not constructed by humans but generated by manipulating living entities in correspondence to human aims (e.g. in order to provide plants or animals with specific, determined and determinable properties) (see [8]).

a specific reflexive connotation simply meaning, that the process takes place—under certain conditions—without human intervention. Neither the chicken nor the egg is the aim of the process, but both phases can be described *as aims*—a description, which transforms our statement on natural processes into a description borrowed from human practise. If we were interested in producing chicken, we were forced to produce eggs and allow their transformation into chicken—and the same is true vice versa, if we were interested in eggs. However, we should keep in mind that there are no aims or means in nature, but that we are describing natural processes as if they were following the logical grammar of human actions.

2. By structuralising living entities as organisms we are providing a scientific description of living entities as if they were functional units. In this description we are using intentional terms in order to introduce "functional relations" such as "the heart exists in order to pump blood", "the startcodon determines the starting point of translation", etc. The intentional expressions provide metaphorical statements, which are referred to explicit causal statements; and these causal statements work without intentional expressions. However, the relation between the intentional description and the causal statement is constituted in terms of "mechanisms" which in fact is expressis verbis a technical relation. The model-procedure, starting with the structuralisation of living entities as functional units and explicating the functional relations in causal terms allows us to scientifically treat living entities as organisms.

We can now explicate the term "self-organisation",[14] and it is useful to start with the second part of the term, which designates a type of human activity such as "A is organising z by means of w". To organise something means to generate a specific kind of order; however to generate order is necessarily connected with purposes as well as the generation of the order is connected with means. In everyday life as well as in sciences we are well acquainted with the generation of orders of different types, generated for different purposes. To give an explicitly "non-biological" example think of *Caesar* organising his troops in front of Alesia in order to allow at the same time the siege of the city and the defence against the *Vercingetorix'* auxiliary troops that surrounded the roman troops. The result of *Caesar* ordering his troops is an organisation of the army, which allows him to achieve the defined purposes. His troops then are "organised in order to", and as such "show a specific kind of order". However, referring to human activity we are dealing with something given, to be ordered in terms of means and ends-in-view. But this reconsideration provides us with the necessary basis to reconstruct the term "self-organisation". By describing living organisms as organised we refer to their functional structuralisation using a causal explication in terms of a mechanism. By adding the term "self" we refer to the functional framework for the causal explication. In terms of an organism we are

[14]For a technical introduction of "self" in terms of recursivity see Chap. 1.1 and [33]; this approach follows the logics of performative (not the resultive) concept, i.e. of "organising", not of "organised", explicated here.

e.g. referring to its metabolic, its neurophysiological or its genetic organisation. The "self" here means the reference to the mechanism, which is under consideration; it provides the invariance of structure despite the facts that each and every element of the mechanism is being replaced during the process itself, i.e. during metabolism, synaptic activity, or homeobox expression.[15] "Self" then expresses neither pure numerical identity (which in fact it does) nor personal identity; it rather determines a semantic identity of a process, described as regenerating a given (e.g. metabolic, physiologic or genetic) order.

7 Self-x Properties and the Order of Descriptions

According to our reconstruction, self-organisation is not just a property of natural units, but a property of something under functional description. Thus, we are not forced to suppose the existence of purposes in nature, we are only dealing with something—in our case living entities—*as if* it were a functional unit:

> "The concept of a thing as in itself a natural purpose is therefore no constitutive concept of Understanding or of Reason, but it can serve as a regulative concept for the reflective Judgement, to guide our investigation about objects of this kind by a distant analogy with our own causality according to purposes generally, and in our meditations upon their ultimate ground." ([14], p. 280)

Thus, there are no purposes in nature, but living entities are described as if they were organised in terms of means and ends-in-view. From this "regulative" point of view, the elements of an—in respect to a description—organised entity served for each other at the same time as means and ends:

> "In such a product of nature every part not only exists by means of the other parts, but is thought as existing for the sake of the others and the whole, that is as an (organic) instrument. Thus, however, it might be an artificial instrument, and so might be represented only as a purpose that is possible in general; but also its parts are all organs reciprocally producing each other. This can never be the case with artificial instruments, but only with nature which supplies all the material for instruments (even for those of art). Only a product of such a kind can be called a natural purpose, and this because it is an organised and self-organising being." ([14], p. 277f)

But *Kant* not only refuses a naturalistic interpretation of ends and means, he also draws an extremely important difference, which is usually neglected in theories of self-organisation. As much as we might describe something as organising "itself" (*organisierend*) in terms of a specific respect, by re-iterating the process under description, we cannot state the same thing as being self-organised (*organisiert*). It is the scientist, who organised something in terms of means and ends, thereby providing the invariances of the process, we refer to as a performance of the described entity—namely its on-going "organising" themselves. As much as we are allowed to

---

[15] It should be noted, that even only the spatial and temporal structuralisation of elements already constitutes order (as the Alesia example showed); consequently we are referring to space as "ordo coexistendi" and to time as "ordo succedendi" (see [15]).

assume the procedural aspects (the organisms' actual organising itself) as a property of the entity under description, we should keep in mind that the statement of something being organised is the result of the activity of the describing scientist. The "self-x properties" organisms then are supposed to show[16], are to be considered as the determination of the invariances of the processes under description. Accordingly, we have to provide two descriptions where the "self" that is supposed to be the subject of organisation, is introduced as the result of the activity of the experimenter. "Self" then refers to the entire system under description—from the point of view of the experimenter.

8 Conclusion: OC as a New Model-Theoretical Perspective

According to our reconstruction OC is not just a material extension of ER, by adding self-organisation in a wide sense to the evolutionary paradigm; it is neither ER nor self-organisation. It might rather be understood as a methodologically extremely creative and promising extension of the tools we can apply to understand the nature of organisms themselves. OC extends the intention of the term "organic" by referring to techniques and technologies, which are only begun to be recognised in biology as a basis for model-theoretical structuralisation of living entities:

> "OC defines an organic computer as a self-organised system that can adapt to a dynamically changing context and achieves the so called self-x-properties as postulated for Autonomic Computing (...): self-configuration, self-optimisation, self-healing, self-explanation, and self-protection. But in spite of being self-organised, an essential feature of organic systems will be their ability to react sensibly to external—in particular human—requirements and to allow for control actions that might be necessary to keep their behaviour within preferred regions of the configuration space—taking into account the effects of emergence." ([31], p. 113)

These "self-x-properties" are methodologically spoken, the result of an ascriptive procedure, which we presented above as the modelling explication of metaphorical expressions. In the case of "self-x" we are confronted with strong ascriptivism because the term "self" refers not primarily towards numerical but towards referential identity. "Self" then designates the invariances of transformational processes, i.e. self is a second order description referring to the relations between given elements as parts of a whole. The self is a performance of an entire system, and not a discrete part. Accordingly, OC aims at the construction of a "controller" which is decentralised and thus a second-order property of the system under (self-)construction:

> "The goal of OC is to build systems that perform their tasks by using (controlled) self-organisation. However, this is independent of using centralised or decentralised observer/controller architectures, since the elements of the system work autonomously and the controller affects some local control parameters only and does not control single elements in detail." ([31], p. 114)

[16] *Timm* [37] names the relevant properties, which show exactly the reference to human activities that provide the logical grammar of the ascriptive structure we proposed.

The example again—as in the case of ER—is supposed to be provided by "nature" in terms of natural systems or organisms. However, nature can be a source of engineering knowhow only after it was brought into a technomorph description. In this case, however, it is not nature, the engineer is following for the design of his autonomous systems: it is but nature described in terms of "as if" relations. And it is only this very form, which allows achieving the meta-norm that is expressed in the motto of "controlling emergence" [23]. This tension cannot be overcome by means of empirical research—it is the result of a research strategy which does not replace the role of the engineer by the indeed wonderful power of evolution. OC rather extends the means exceedingly the engineer has at hand in order to build technical systems that might help to transform our traditional concept of nature, by providing a variety of the most advanced techniques and technologies, which allow the constitution of bio-scientific objects that transcend the limits of classical bionic and technical-biological modelling. Being aware of the fundamental methodological limits of the ascriptive as if-structure, which underlies the modelling of technical systems, Organic Computing provides a promising framework for engineering practise—and avoids at the same time the ontological self-misunderstanding of evolutionary robotics.

References

1. Bonik, K., Gutmann, W.F.: Kritische Evolutionstheorie. Gerstenberg, Hildesheim (1981)
2. Braitenberg, V.: Künstliche Wesen. Vieweg, Braunschweig (1973)
3. Carnap, R.: Einführung in die Philosophie der Naturwissenschaft. Ullstein, Frankfurt (1986)
4. Darwin, C.: Origin of Species, vols. 1, 2. AMS, New York (1897)
5. Eigen, M.: Self-organisation of matter and the evolution of biological macromolecules. Naturwissenschaften **58**, 465–522 (1971)
6. Fisher, R.A.: The Genetical Theory of Natural Selection. New York (1958)
7. Gould, S.J., Lewontin, R.C.: The spandrels of San Marco and the panglossian paradigm: A critique of the adaptationist programme. Proc. R. Soc. Lond. B, Biol. Sci. **205**(1161), 581–598 (1979)
8. Gutmann, M.: Die Evolutionstheorie und Ihr Gegenstand – Beitrag der Methodischen Philosophie zu Einer Konstruktiven Theorie der Evolution. VWB, Berlin (1996)
9. Gutmann, M.: Aspects of crustacean evolution. The relevance of morphology for evolutionary reconstruction. In: Gudo, M., Gutmann, M., Scholz, J. (eds.) Concepts of Engineering and Constructional Morphology: Biomechanical Approaches on Fossil and Recent Organisms, vol. 82, pp. 237–266. Senckenbergiana letheia (2002)
10. Gutmann, M., Janich, P.: Methodologische Grundlagen der Biodiversität. In: Janich, P., Gutmann, M., Prie, K. (eds.) Biodiversität. Wissenschaftliche Grundlagen und gesellschaftliche Relevanz, pp. 281–353. Springer, Berlin (2002)
11. Gutmann, M., Warnecke, W.: Liefert "Intelligent Design" wissenschaftliche Erklärungen? Religion, Staat, Gesellschaft **7**(2) (2006)
12. Jablonka, E., Lamb, M.J.: Evolution in Four Dimensions. MIT Press, Cambridge (2004)
13. Janich, P.: Kleine Philosophie der Naturwissenschaften. Beck, München (1997)
14. Kant, I.: Critique of Judgement. Translated and with Introduction and Notes by Bernard. MacMillan, London (1914)
15. Kant, I.: Kritik der Reinen Vernunft. Reclam, Stuttgart (1966)
16. Kauffman, S.: The Origins of Order. Self-organisation and Selection in Evolution. Oxford Univ. Press, New York (1993)

17. Kohn, D.: The aesthetic construction of Darwin's theory. In: The Elusive Synthesis: Aesthetics and Science, pp. 13–48. Kluwer Academic, Dordrecht (1997)
18. König, L., Mostaghim, S., Schmeck, H.: Decentralized evolution of robotic behaviour using finite state machines. Int. J. Intell. Comput. Cybern. **4**(2), 695–723 (2009)
19. Krohn, W., Küppers, G. (eds.): Emergenz Die Entstehung von Ordnung, Organisation und Bedeutung. Suhrkamp, Frankfurt (1992)
20. Levit, G.S., Meister, K., Hoßfeld, U.: Alternative Evolutionstheorien. In: Krohs, U., Töpfer, G. (eds.) Philosophie der Biologie, pp. 267–286. Suhrkamp, Frankfurt (2005)
21. Levit, G.S., Meister, K., Hossfeld, U.: Alternative evolutionary theories from the historical perspective. J. Bioecon. **10**(1), 71–96 (2010)
22. Maier, W.: Das Säugercranium als natürliche Konstruktion. Mitteilungen des SFB 230 3, 219–226 (1988)
23. Müller-Schloer, C., von der Malsburg, C., Würtz, R.P.: Organic Computing. Informatik Spektrum **27**(4), 332–336 (2004)
24. Nachtigall, W.: Bionik. Springer, Berlin (2002)
25. Nolfi, S., Floreano, D.: Evolutionary Robotics. MIT Press, Cambridge (2000)
26. Page, D.M., Holmes, E.C.: Molecular Evolution. Blackwell, Oxford (1998)
27. Prigogine, I.: Vom Sein zum Werden. Piper, München (1979)
28. Provine, W.B.: The Origines of Theoretical Population Genetics. The University of Chicago Press, Chicago (1992)
29. Provine, W.B.: The R.A. Fisher-Sewall Wright Controversy. In: Sakhar, S. (ed.) The Founders of Evolutionary Genetics, pp. 201–230. Kluwer Academic, Dordrecht (1992)
30. Rechenberg, I.: Evolutionsstrategie Optimierung technischer Systeme Nach Prinzipien der biologischen Evolution. Fromman-Holzboog, Stuttgart (1973)
31. Richter, U., Mnif, M., Branke, J., Müller-Schloer, C., Schmeck, H.: Towards a generic observer/controller architecture for Organic Computing. In: GI Jahrestagung (1), pp. 112–119 (2006)
32. Roth, G.: Selbstorganisation und Selbstrefentialität. Prinzipien der Organisation von Lebewesen. Dialectic **12**, 194–213 (1986)
33. Schmeck, H., Müller-Schloer, C., Çakar, E., Mnif, M., Richter, U.: Adaptivity and self-organization in organic computing systems. ACM Trans. Auton. Adapt. Syst. **5**(3), 1–32 (2010)
34. Schwarzer, J.: Lernverfahren für evolutionär optimierte Künstliche Neuronale Netze auf der Basis Zellulärer Automaten. Logos, Berlin (2003)
35. Sober, E.: Reconstructing the Past. MIT Press, Cambridge (1988)
36. Sulloway, F.J.: Why Darwin rejected intelligent design. J. Biosci. **34**(2), 173–183 (2009)
37. Timm, I.J.: Dynamisches Konfliktmanagement als Verhaltenssteuerung Intelligenter Agenten. Verlagsgesellschaft Berlin, Berlin (2004)
38. Tooby, J., Cosmides, L.: Conceptual foundations of evolutionary psychology. In: Buss, D.M. (ed.) The Handbook of Evolutionary Psychology, pp. 5–67. Wiley, Hoboken (2005)
39. Varela, F.: Autonomy and autopoiesis. In: Roth, G., Schweiger, H. (eds.) Selforganizing Systems, pp. 14–23. Campus, Frankfurt (1981)
40. Wilson, E.O.: Sociobiology. Belknap, Cambridge (2002)
41. Wright, S.: Evolution in mendelian populations. Genetics **16**(1931), 97–159 (1930)

M. Gutmann (✉) · B. Rathgeber · T. Syed
Karlsruhe Institute of Technology (KIT), Institute of Philosophy, Bldg. 20.12, Kaiserstr. 12, 76131 Karlsruhe, Germany
e-mail: mathias.gutmann@kit.edu

17. Kohn, D.: The aesthetic construction of Darwin's theory. In: The Elusive Synthesis: Aesthetics and Science, pp. 13–48. Kluwer Academic, Dordrecht (1997)
18. König, L., Mostaghim, S., Schmeck, H.: Decentralized evolution of robotic behaviour using finite state machines. Int. J. Intell. Comput. Cybern. 2(4), 695–723 (2009)
19. Krohn, W., Küppers, G. (eds.): Emergenz: Die Entstehung von Ordnung, Organisation und Bedeutung. Suhrkamp, Frankfurt (1992)
20. Levit, G.S., Meister, K., Hoßfeld, U.: Alternative Evolutionstheorien. In: Krohs, U., Toepfer, G. (eds.) Philosophie der Biologie, pp. 267–286. Suhrkamp, Frankfurt (2005)
21. Levit, G.S., Meister, K., Hossfeld, U.: Alternative evolutionary theories from the historical perspective. J. Bioecon. 10(1), 71–96 (2008)
22. Maier, W.: Das Säugercranium als natürliche Konstruktion. Mitteilungen des SFB 230 3, 219–226 (1988)
23. Müller-Schloer, C., von der Malsburg, C., Würtz, R.P.: Organic Computing. Informatik Spektrum 27(4), 332–336 (2004)
24. Nachtigall, W.: Bionik. Springer, Berlin (2002)
25. Nolfi, S., Floreano, D.: Evolutionary Robotics. MIT Press, Cambridge (2000)
26. Page, D.M., Holmes, E.C.: Molecular Evolution. Blackwell, Oxford (1998)
27. Prigogine, I.: Vom Sein zum Werden. Piper, München (1979)
28. Provine, W.B.: The Origins of Theoretical Population Genetics. The University of Chicago Press, Chicago (1971)
29. Provine, W.B.: The R.A. Fisher–Sewall Wright Controversy. In: Sarkar, S. (ed.) The Founders of Evolutionary Genetics, pp. 201–229. Kluwer Academic, Dordrecht (1992)
30. Rechenberg, I.: Evolutionsstrategie: Optimierung technischer Systeme nach Prinzipien der biologischen Evolution. Frommann-Holzboog, Stuttgart (1973)
31. Richter, U., Mnif, M., Branke, J., Müller-Schloer, C., Schmeck, H.: Towards a generic observer/controller architecture for Organic Computing. In: GI Jahrestagung (1), pp. 112–119 (2006)
32. Roth, G.: Selbstorganisation und Selbstreferentialität: Prinzipien der Organisation von Lebewesen. Dialektik 12, 194–213 (1986)
33. Schmeck, H., Müller-Schloer, C., Çakar, E., Mnif, M., Richter, U.: Adaptivity and self-organization in organic computing systems. ACM Trans. Auton. Adapt. Syst. 5(3), 1–32 (2010)
34. Schwarzer, J.: Lernverfahren für evolutionär optimierte Künstliche Neuronale Netze auf der Basis Zellulärer Automaten. Logos, Berlin (2003)
35. Sober, E.: Reconstructing the Past. MIT Press, Cambridge (1993)
36. Sullivan, F.R.: Why Darwin rejected intelligent design. J. Biosci. 34(2), 173–183 (2009)
37. [illegible]
38. Tooby, J., Cosmides, L.: Conceptual foundations of evolutionary psychology. In: Buss, D.M. (ed.) The Handbook of Evolutionary Psychology, pp. 5–67. Wiley, Hoboken (2005)
39. Varela, F.: Autonomie und Autopoiese. In: Roth, G., Schwegler, H. (eds.) Self-organizing Systems, pp. 14–23. Campus, Frankfurt (1981)
40. Wilson, E.O.: Sociobiology. Belknap, Cambridge (1975)
41. Wright, S.: Evolution in mendelian populations. Genetics 16(1931), 97–159 (1930)

M. Gutmann (✉) · B. Rathgeber · T. Syed
Karlsruhe Institute of Technology (KIT), Institute of Philosophy, Bldg. 20.12, Kaiserstr. 12, 76131 Karlsruhe, Germany
e-mail: mathias.gutmann@kit.edu

Chapter 2
Methods and Tools

It would have been nice to present a *design methodology* for Organic Computing systems. However, this is, in a way, a contradiction in itself. Normally, the design of a system in the area of engineering means a complete and accurate description and specification of the system, i.e. what it looks like, how it is composed, and how it behaves. In the realm of Organic Computing, the fascinating basic idea is that the system is not and should not be completely "designed". It rather should take shape and adapt its shape as a response to stimuli from its environment. In a way, the major part of the design process of an Organic Computing system takes place dynamically during operation by means of self-organisation based on simple rules. To put it more provocatively: the organic system (re)designs itself based on a few simple rules that guide the self-organisation process towards an intended direction. How these rules should be formulated to achieve the desired system properties and how to safeguard against unwanted exceeding behavioural limits, is not yet generally understood. What research on the design of Organic Computing can offer up to now, however, are methods for *specific domains, tools, guidelines* and—to some degree—*design patterns*. The collection of articles in this chapter gives an overview of different design approaches and—although not providing a general methodology—may offer patterns to be useful for creating Organic Computing systems.

The first article by H. Parzyjegla et al. titled "Model-driven Development of Self-organising Control Applications" (Chap. 2.1) employs a *model driven approach*. Assuming that the programmer of an Organic Computing system should still be able to focus on the functionality of the system and not be burdened with self-organisation aspects it provides a toolbox for developing Organic Computing systems with automatic integration of self-* properties along the tool chain of model transformations. Special attention is devoted to the concept of *self-stabilisation*, which means the provable ability of a system to return to a legal state after a failure.

The article in Chap. 2.2 by H. Seebach et al. on "How to Design and Implement Self-organising Resource-flow Systems" is a guided software development process using the domain of an automated production line as an exemplary application area. Based on an Observer/Controller pattern and constraints it is shown, how even in a self-organising system guarantees can be given concerning the behaviour of the system.

The work presented by D. Kramer et al. (Chap. 2.3) describes "Monitoring and Self-awareness for Heterogeneous Adaptive Computing Systems". It deals with the problem of monitoring at the hardware level and introduces a hierarchical monitoring system tailored to cope with self-organisation. With only little additional hardware effort, the system is able to classify its current state, an important information that may be used by other components for self-optimisation purposes.

"Generic Emergent Computing in Chip Architectures" by M. Reichenbach et al. (Chap. 2.4) focuses on the design of dedicated hardware for high speed processing of high-resolution images. Rather than being a design of an Organic Computing system itself, it describes a parallel system aiming to specifically support Organic Computing algorithms such as the *Marching Pixel* algorithm of Chap. 5.9. It is based on a SIMD (Single Instruction Multiple Data) parallel machine consisting of a two-dimensional array of simple processing elements each designed for fast execution of image processing operations.

In their article "Multi-Objective Intrinsic Evolution of Embedded Systems" (Chap. 2.5) P. Kaufmann and M. Platzner introduce the concept of *Evolvable Hardware* which is defined as a combination of evolutionary algorithms with reconfigurable hardware and run-time optimisation. It presents a toolbox for the evolutionary design of digital circuits. Two applications, one from pattern matching with FPGAs, the other from cache optimisation, demonstrate the usage of the toolbox.

A completely different area and approach is chosen by N. Matsumaru et al. in Chap. 2.6 under the title "Organisation-Oriented Chemical Programming". As well known, chemical reactions can also be seen from an information processing perspective which consequently leads to the perception of a chemical process as being some sort of "computation". The result of that "computation", be it quantitative or qualitative, however, is something that emerges as a global phenomenon. Programming of those chemical systems is therefore radically different from programming of digital computers. The article introduces a theory of chemical organisation, which is used to formulate "programs" for two simple informatics problems. It finally derives some design principles for such chemical systems that although coming from an unfamiliar domain provide some insight into the design of digital Organic Computer systems.

A very vivid example of an—at least compared to the previous article—conventional Organic Computing system is shown by S. Ebers et al. in their contribution "Hovering Data Clouds for Organic Computing" (Chap. 2.7). A *hovering data cloud* can be formed on a mobile ad hoc network, when the mobile nodes that are currently in some geographic vicinity act as temporary hosts for some useful information that may be passed to other nodes as the current nodes leave the geographic area of the cloud. The cloud and the data it contains may be stationary although the carriers of the information may move. This behaviour is achieved with only a small set of rules and operations. Applications for traffic management based on car-to-car communication are presented providing evidence that hovering data clouds can be very useful tools for building Organic Computing systems.

Hans-Ulrich Heiss
School of Electrical Engineering and Computer Science, Technische Universität Berlin, Einsteinufer 17/EN6, 10587 Berlin, Germany
email: heiss@tu-berlin.de

Chapter 2.1
Model-Driven Development of Self-organising Control Applications

Helge Parzyjegla, Arnd Schröter, Enrico Seib, Sebastian Holzapfel, Matthäus Wander, Jan Richling, Arno Wacker, Hans-Ulrich Heiß, Gero Mühl, and Torben Weis

Abstract In this article we present a software development methodology and a supporting algorithm toolbox for Organic Computing applications that were developed in the context of the MODOC project. We focus on control applications for dynamic actuator/sensor networks that have no skilled human administrator and that are subject to frequent random hardware failures. To deal with these characteristics, applications in this setting should be self-organising and self-stabilising. However, realising self-organising and self-stabilising applications is usually a complex task requiring expert knowledge. Thus, we show that the development of such applications can be greatly simplified by shifting complexity into a model-driven tool chain. Developers implement their application in a high-level language based on a role abstraction focusing on application functionality. In order to hide the remaining complexity, we developed a tool chain which is responsible for transforming the high-level code to executable machine code that is automatically enriched by self-x properties such as self-organisation and self-stabilisation, which are not explicitly defined by the application programmer. Thereby, the application benefits from extended fault-tolerance and availability. A major part of the tool chain is a toolbox containing self-x algorithms commonly required for Organic Computing applications. The heart of the toolbox is an algorithm stack realised by our publish/subscribe middleware REBECA that is able to efficiently realise self-organisation even in case of large networks.

Keywords Model-driven development · Publish/subscribe · Self-stabilisation · Self-organisation · Actuator and sensor networks · Composite event detection · Adaptive routing

1 Introduction

Computer networks are becoming more and more omnipresent, but at the same time they are getting too complex and too dynamic to be administrated manually. This is especially true for wireless networks of actuators and sensors (*AS-nets*) which are the basis for most applications in the area of pervasive computing. Here, sensor and actuator devices are continuously shrinking in size, while at the same time also their price declines. However, in absence of a human expert administrator, the system

C. Müller-Schloer et al. (eds.), *Organic Computing — A Paradigm Shift for Complex Systems*, 131–144
Autonomic Systems, DOI 10.1007/978-3-0348-0130-0_8,

must be able to self-organise. Due to scalability reasons, each computing node has usually only a limited local view and, therefore, makes local reconfiguration decisions. Moreover, these cheap pervasive computing devices are more often subject to failures than traditional PCs or even server hardware. Temperature changes, interfering radio signals, and other externally induced forces can cause random hardware failures. These errors can bring the system into an invalid state. Thus, such systems should also be self-stabilising [1], i.e. they should regain a valid state after a bounded time provided that no further errors are induced externally while the system is stabilising. Hence, self-stabilisation and self-organisation are two *self-x properties* [6] that are both required in AS-nets.

In this article, which summarises the main results of our project *Model-Driven Development of Self-Organising Control Applications (MODOC)*, we show how the complexity of developing applications requiring self-x properties can be reduced by shifting complexity into a model-driven tool chain [17] depicted in Fig. 1. With our approach [8, 9], the developer is using a high-level language that is especially tailored towards control applications in AS-nets. To better understand and define the expressiveness of this language, we have devised an automaton that is able to execute all application models, while at the same time guaranteeing self-stabilisation. Simulating this automaton would be very time-consuming, therefore a model transformation tool maps the high-level application model to executable code. Since every bit of RAM, all CPU registers and all messages received by radio are assumed to be subject to errors, the transformation ensures that the generated application code can self-stabilise from all transient errors. Furthermore, we instrument the generated code to back-annotate data into the application model greatly simplifying debugging of the self-organising system at runtime.

The model transformation uses a toolbox of self-x algorithms which are linked with the application code. This toolbox provides communication and coordination features for distributed applications and is self-optimising to provide the best possible QoS. Internally, the toolbox algorithms are based on the publish/subscribe paradigm which provides a loosely coupled content-based communication. With our work we extend the pure publish/subscribe functionality in two ways. First, we introduce advanced programming abstractions, like self-organising role assignment and composite event detection simplifying application development using the high-level language. Additionally, the toolbox contains algorithms that internally adapt the system to maintain a stable and optimal service level. This includes a self-organising clustering as well as self-optimising network algorithms.

The remainder of this article is structured as follows: In Sect. 2 we discuss our development methodology for Organic Computing applications. In Sect. 3 we describe our algorithm toolbox with focus on our self-x publish/subscribe middleware REBECA. The conclusions are presented in Sect. 4.

2 Model-Driven Development

Building self-organising and self-stabilising applications using standard programming languages is very difficult. Especially self-stabilisation is hard to achieve when

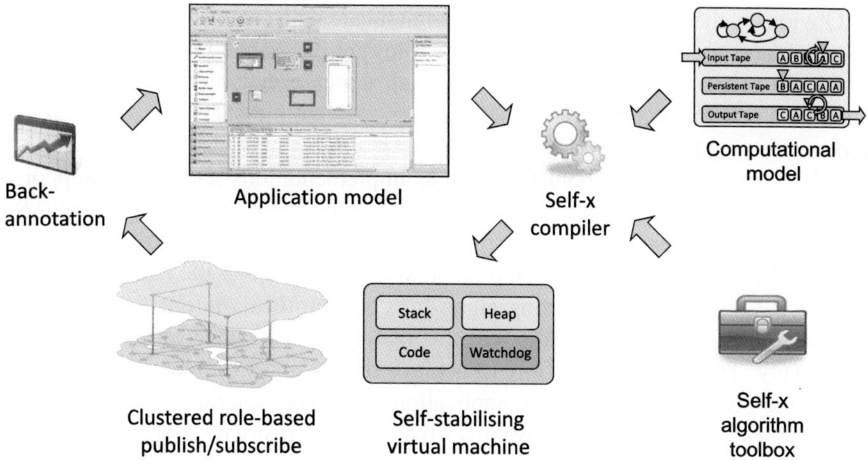

Fig. 1 Software development methodology for Organic Computing applications

the error model allows multi-bit errors in the RAM, CPU registers and messages received by radio. A programmer has to be aware that all information received by radio, stored in the data structures and even the data structures themselves can be wrong or malformed as a result of such bit-errors. Thus, it is very likely that a developer did not consider all possible cases. This means that the application can end up in a state from which it cannot stabilise anymore. Furthermore, testing is very time consuming, because many random bit-error patterns must be tested and for each test the system must demonstrate that it can stabilise again.

For this reason, we have developed a tool chain for model-driven software development as shown in Fig. 1 which takes this burden away from the programmer. In our approach, developers use a high-level programming language specifically tailored towards self-organisation and self-stabilisation. An application in our settings is inherently distributed because it involves sensors and actuators attached to different nodes in a network. However, the developer does not specify which node is responsible for which task. Instead, the application logic is segmented into roles. For example, "sensing a temperature" or "controlling the door of a car port" are roles. The system self-organises by assigning the roles dynamically to those nodes which have the required capabilities, i.e. sensors, actuators, CPU power, memory capacity, network bandwidth etc. Thus, each role is accompanied by a set of requirements and the node knows its capabilities. At runtime each node can detect which roles it is able to perform and a distributed algorithm assigns roles accordingly. Furthermore, the developer can specify whether a role should be performed by all nodes which are capable of executing it or only by one. Usually, sensing is done multiple times to get better results, e.g. all nodes which can measure a temperature do so. Regarding actuators, it depends on the application: while a garage door needs only one controller, turning on the light in the entire house involves all nodes connected to a light.

Obviously, the roles must be able to communicate with each other. The developer does not have to know how and where the roles are reachable. He can simply state that some message is to be sent to one or all nodes executing a certain role. At runtime, a self-organising overlay network ensures that messages are routed to the nodes executing the target role.

Self-stabilisation must hold for all role implementations and for all supporting algorithms and protocols. For example, to achieve self-organisation in an AS-net, we need a network overlay, a routing protocol and a role distribution algorithm. All these parts have to be self-stabilising to ensure that the entire system is self-stabilising. For the supporting algorithms and protocols in our toolbox, self-stabilisation is achieved manually and the algorithms are either verified or tested extensively. The rather high effort to achieve this is not a problem because it is only a one-time effort. In contrast, application developers should not have to struggle with these issues. Therefore, we investigated which restrictions are necessary to our high-level language to guarantee that all programs written in this language are self-stabilising.

2.1 Computational Model

Our computational model is based on a self-stabilising automaton that gives the following guarantee for all algorithms executed on this automaton: If the algorithm is implemented correctly, i.e. it is free of systematic faults, then it stabilises after a constant time from all random transient faults. These faults may alter the state of the automaton in any possible way. Thus, developers do neither have to test nor to verify that their algorithm can stabilise from every possible state because this is already included as an inherent guarantee. Our computational model [18] is based on a modified Turing Machine having three tapes instead of one (Fig. 2). The two additional tapes are dedicated to deal with network communication. A normal Turing Machine has its input on a tape and when it halts, the output is on the same tape. In our settings, each computing node must continuously receive and process messages. Therefore, there is an *input tape* for the incoming message and an *output tape* for the outgoing message. The third tape is called *persistent tape*. Our automaton sleeps until a message arrives on the input tape. Then, it reads this tape and reads/writes

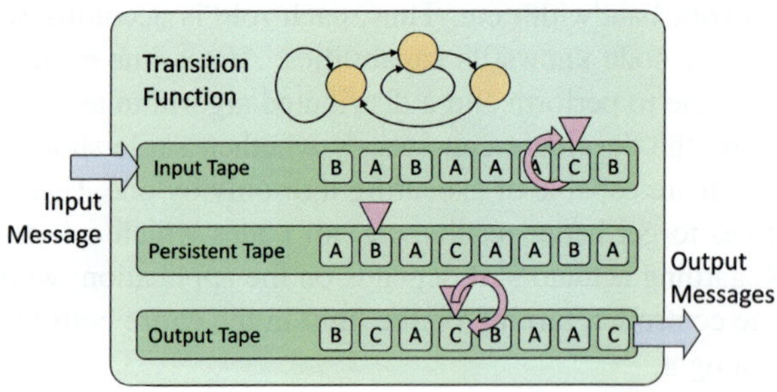

Fig. 2 Self-stabilising automaton

the other two tapes. When the machine halts, it sends the content of the output tape via radio, erases the input and the output tape and returns to its initial state; the persistent tape remains unchanged.

To achieve self-stabilisation, each symbol on the persistent tape as well as the read/write heads of the automaton are ageing and have an *age value*. The automaton has a defined *maximum age value*. If a symbol on the tape becomes older than this maximum, then it is reset to a null-symbol. Should the read/write heads become too old, then the machine resets itself. When a new message arrives, the heads of the automaton are at the initial position, the state machine is on its initial state and the age of the read/write heads is zero. With each step of the machine (i.e. whenever it writes/reads a tape, moves the head or changes states in the state machine), the age of all symbols on the tape and the read/write heads is increased by one. Whenever a symbol is read, the age of the corresponding read/write head is assigned the maximum of its current age and the age of the symbol. When a symbol is written, the age of the new symbol is assigned the age of the read/write head.

Due to these rules of ageing, the machine cannot derive "young" information from "old" information. The only sources of information with age zero are incoming sensor messages. The rationale behind this is that after a transient fault, the sensors always produce new data that are correct and consistent with the real world, whereas all data stored on computing nodes is potentially wrong because of the transient fault. This shows that our computational model is especially geared towards AS-nets, because without sensors there is no way to refresh data in the computing nodes.

The purpose of the computational model is not to serve as a basis for an implementation. Ageing all symbols upon each step is computationally expensive and storing an age for each symbol (i.e. byte or word) wastes memory. The purpose of the model is to identify which programs our high-level language can support. We just have to ensure that all applications developed with our high-level language can be mapped to this computational model. At the same time, all applications that can execute on our automaton should have a representation in the high-level language. Thus, there is an equivalence between the computational model and the expressiveness of our high-level language.

2.2 Model Transformation

Our tool chain translates the model into executable machine code. It first splits the application model into roles and identifies the messages exchanged between roles. Each role is then transformed to source code that has the desired self-stabilisation features. The generated source code behaves as if executed on our computational model, but it runs on standard hardware and is much more efficient than simulating our automaton on standard hardware.

In the following, we want to highlight some solutions which allow the generated source code to greatly outperform a simulated self-stabilising automaton. For data stored on the heap, the generated code tracks the age of entire objects and not the age of single bytes or words. This way, we can save a significant amount of RAM. To avoid increasing the age of each object in every single step, we store time stamps instead of ages and compare the time stamps to a system clock to calculate the age. Periodically we scan the heap for outdated objects or objects which seem to come from the future. Upon a clock screw the heap will be emptied which is equivalent to a machine reset.

The generated code executes a main loop where it waits for new incoming messages. The time allowed for processing an incoming message is limited to t seconds and is observed by a watchdog. This guarantees that the software cannot end up in an infinite loop. After a message has been processed, the stack must be empty again, i.e. the stack pointer (SP) and the program counter (PC) return to their initial positions. When a new message arrives, the generated code does not read a CPU register unless it has been written before. This way, old (and perhaps wrong) values in the CPU registers can do no harm. In addition, the machine stores its current time in one of its registers when it starts processing a new message. If the machine reads an object that is older than the time stamp in this register, then the older time stamp of the object is written in the register. When the machine writes an object, the object time stamp is modified in the same way. Thus, this register corresponds to the age of the read/write head of our automaton.

All data on the stack cannot be older than the time stamp register and data cannot be derived from data that was older than this register because the time stamp register is at least as old as the oldest data it has written. Furthermore, data on the stack can be at most t seconds younger than the time stamp in the register because no data on the stack lives longer than t seconds. This observation relieves us from tracking the time stamp for data on the stack since we have a good bound for the age of stack data. We might assume that data on the stack is t seconds older than it actually is, but we never assume that data is being younger than it is. This is important, because the machine must not derive young from old (and potentially corrupted) information.

The model transformation allows many such optimisations to generate an efficient implementation despite the additional overhead for tracking the age of data. In addition to this, the model transformation uses algorithms from our toolbox. Most notably, it builds on a publish/subscribe infrastructure to send messages to nodes executing certain roles, where roles are dynamically assigned on top of publish/subscribe communication [16].

3 Self-stabilising and Self-organising Algorithm Toolbox

Each distributed application requires a communication infrastructure. In our case, this becomes evident when the high-level description of the application based on abstract roles is transformed into runnable code. At this step, the code has to be enriched by communication capabilities aggregated in a toolbox. Due to the nature of AS-nets where nodes and links are often mobile and unreliable, the communication infrastructure must be robust with respect to failures. Furthermore, the infrastructure must be able to cope with changing and dynamic workloads as they typically occur in AS-nets. Finally, a high-level communication interface significantly reduces the complexity of the code transformation. To meet these requirements, our tool chain employs publish/subscribe as basic communication scheme. In contrast to popular synchronous communication based on request/reply, publish/subscribe is a more flexible, anonymous and loosely-coupled paradigm. This flexibility strongly fits AS-nets and allows to add self-x properties rather easily.

3.1 Self-stabilising and Self-organising Algorithm Stack

Our self-organising and fault-tolerant publish/subscribe infrastructure is based on a self-stabilising algorithm stack as shown in Fig. 3 that consists of three primary layers. Thereby, each layer isolates specific functional aspects, builds upon lower layers, and is proven to be self-stabilising itself. Composing an infrastructure this way allows for comprehensive functionality while guaranteeing that the resulting system is still self-stabilising [3, 10].

The lowest layer of the algorithm stack builds a clustered spanning tree and only presumes a simple radio interface able to receive and broadcast messages. Clustering, as the basis for complex middleware architectures, is achieved by rather simple recursive compositional rules that form hierarchical clusters of nodes and set up many small interacting publish/subscribe networks [11].

The second layer comprises a lightweight publish/subscribe algorithm that provides flexible content-based communication by decoupling publishers and subscribers, i.e. senders and receivers of messages. Thereby, we are able to integrate the necessary routing of messages between clusters into the publish/subscribe substrate itself. This technique reveals several benefits, such as redundant routes and automatic route recovery, which show up as emergent behaviour at no additional cost. As a result of both layers, a clustered network as well as a publish/subscribe based communication infrastructure are established as depicted in Fig. 4. Figure 5 illustrates how a logical path is mapped to a physical path in the next lower layer, where cluster heads act as bridges forwarding application specific subscriptions [16].

Finally, the third layer provides the role abstraction as the foundation for distributed applications in AS-nets. To encounter the challenges derived from self-organisation and self-stabilisation, the role assignment mechanism has to tolerate the addition and removal of devices at runtime and has to be able to recover from any transient fault. Role assignment is realised as an integral part of our light-weight

Fig. 3 Algorithm stack

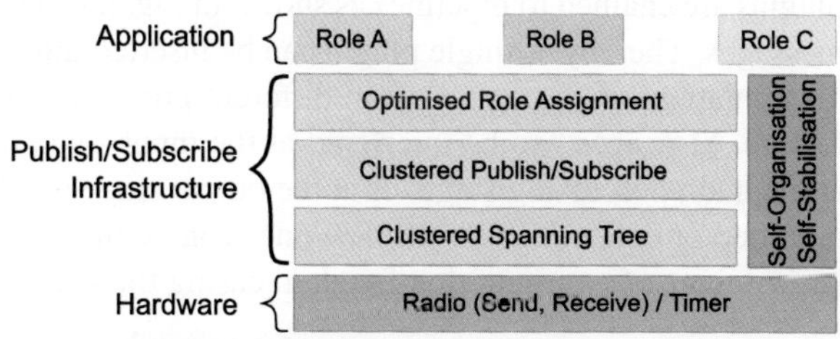

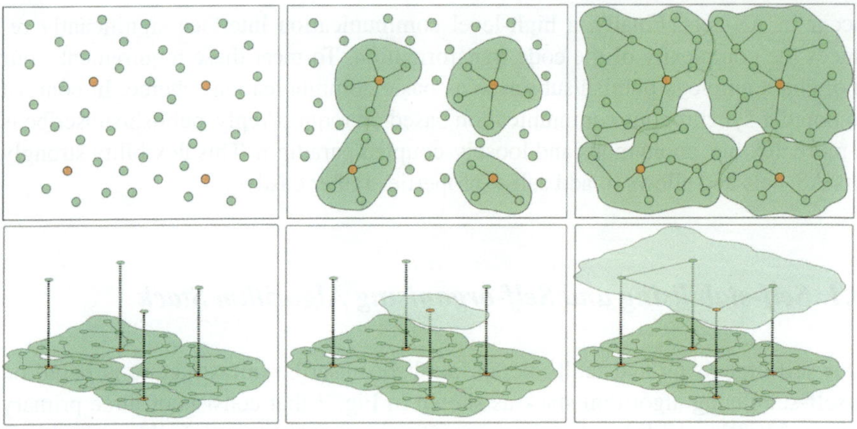

Fig. 4 Clustering schemes

Fig. 5 Path mapping

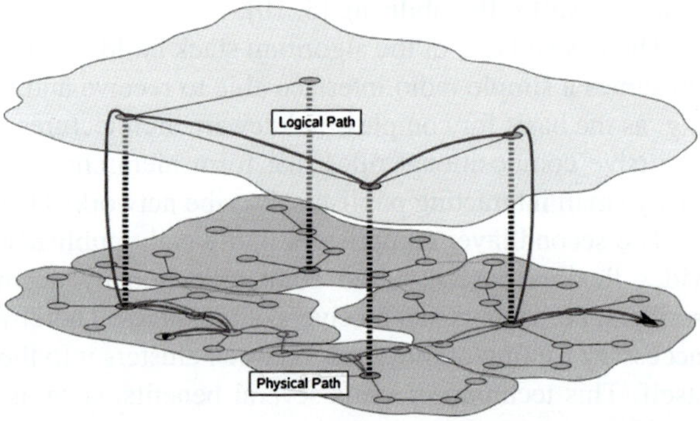

and flexible middleware. The role assignment algorithm assigns the roles such that latency and network traffic are optimised.

The complete algorithm stack has been integrated into our publish/subscribe middleware REBECA [7]. REBECA is a distributed, light-weight publish/subscribe prototype system supporting various matching algorithms, routing schemes, and network protocols. Targeting AS-nets, however, required a complete architectural redesign. The new architecture is primarily built upon the concept of feature composition. A REBECA publish/subscribe broker is merely a container for feature plugins which can be inserted at runtime and provide all functionality. Within a broker, plugins are chained to pipelines as shown in Fig. 6 which are traversed by forwarded messages. Thereby, a single plugin can be inserted into multiple pipelines and, thus, manipulate arbitrary messages at different processing stages to realise a particular feature. Messages can be interfered at the input stage immediately after receiving, at the main processing stage, or at the output stage just before sending; they can be modified or deleted, and even new ones can be inserted. This flexibility provides a large degree of configuration freedom easing the combination of different features.

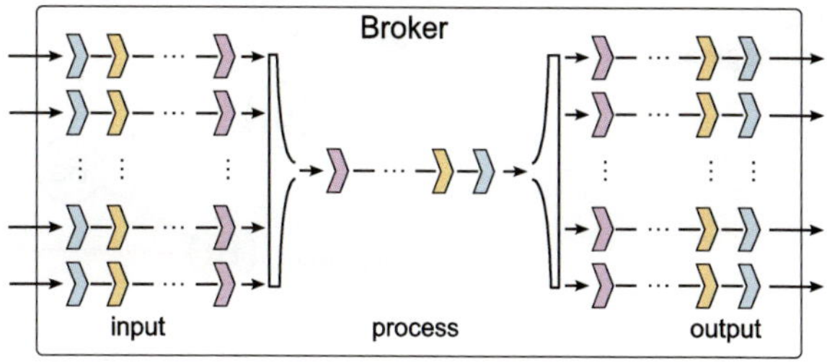

Fig. 6 Pipeline architecture of a broker

3.2 Adaptive and Self-optimising Network Algorithms

Beyond the self-stabilising algorithm stack focusing on fault-tolerance it is also essential to care for aspects regarding the *Quality of Service* (QoS) within the publish/subscribe layer. Since manual administration is neither suited nor generally possible in dynamic environments such as AS-nets, the system must provide means to optimise itself. REBECA contains two adaptive and self-optimising network algorithms that were developed within the scope of the project. Both use local knowledge only to continually push the system towards an efficient operating point. One algorithm adapts the overlay topology while the other fine-tunes the publish/subscribe routing configuration. Due to their orthogonality, both algorithms can be run in parallel.

Adaptive Overlay Topologies Although publish/subscribe infrastructures are used in dynamic environments, the structure of the overlay has been assumed to be static or manually managed in many systems so far. However, in scenarios such as AS-nets where message patterns vary and the network itself may be subject to frequent changes, static topologies cannot keep the system at an efficient working point anymore. Therefore, REBECA contains an algorithm which autonomously adapts the structure of the publish/subscribe overlay network. The algorithm is based on an on-line heuristic which considers communication costs of links, processing costs of nodes, and patterns in the network's message flows [4].

The algorithm works in three phases: In the *evaluation phase*, the heuristic is applied to find inefficient links and potential replacement candidates. In the *consensus phase*, the local knowledge of affected nodes is combined to evaluate the benefit of a reconfiguration. The *reconfiguration phase* finally adapts the topology while ensuring message ordering and avoiding message loss. Figure 7 shows an example. The initial topology connects a publisher of message n to two subscribers with matching filter f via intermediate brokers B_3 and B_4. After two reconfiguration steps, publisher and subscribers are connected directly by replacing the links B_1B_3 and B_4B_5 with the links B_1B_6 and B_5B_6, respectively. Thereby, the intermediate brokers B_3 and B_4 are unloaded which solely act as simple forwarders, otherwise.

Adaptive Routing In publish/subscribe systems, it is usually assumed that the routing algorithm is set only once at design time or system start-up. Since every routing algorithm has its inherent advantages and drawbacks regarding message

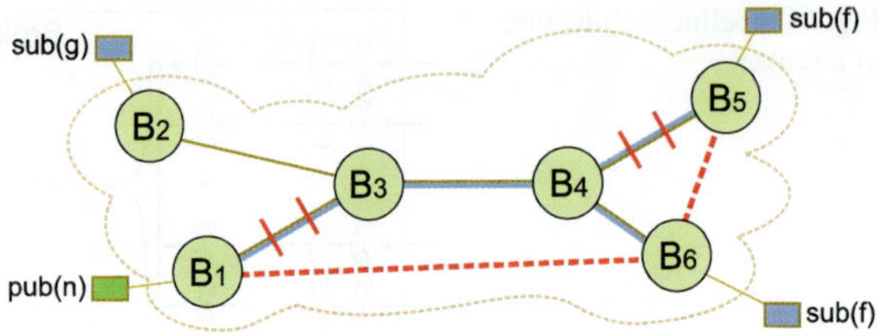

Fig. 7 Overlay reconfiguration

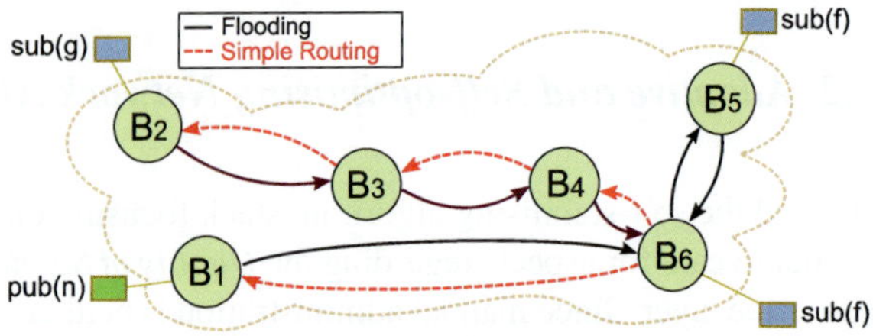

Fig. 8 Hybrid routing configuration

complexity and computational effort, this often leads to inefficient routing configurations when conditions change over time, e.g. in dynamic environments such as AS-nets. Therefore, we introduced a new class of *hybrid routing algorithms* [12, 14] that provide more flexibility. The basic idea stems from the analysis of *hierarchical routing* schemes where notifications are filtered downstream while upstream messages are flooded. Hybrid algorithms extend this idea and allow an edge-wise configuration of the applied routing scheme within the whole network. This way, the advantages of different routing algorithms can be exploited while drawbacks are cancelled out. For example, flooding has no filter forwarding overhead while filtering algorithms reduce the amount of unnecessarily forwarded messages. Whether a combination is beneficial, however, depends on the current relation of publishers to subscribers which dynamically changes over time. By applying a local decision criterion and following a simple coordination protocol, each pair of brokers negotiates about the type of routing algorithm that is used on their common link. Hence, the whole publish/subscribe network becomes adaptive and is able to fine-tune its routing configuration to best suit different conditions in different network parts.

Figure 8 gives an example. Starting point is an overlay configured to completely use notification filtering on all edges for both directions. Assuming that the subscribers at brokers B_5 and B_6 frequently change their interests and update their subscription f, it may be reasonable to switch some links to flooding in order to avoid subscription forwarding overhead. After several reconfiguration steps the depicted configuration is reached. Thereby, the adaptation starts at leaf brokers with the edges B_1B_6, B_2B_3, and B_5B_6 that first switch to flooding. Subsequently, edges B_3B_4, B_4B_6, and B_6B_5 are switched. The remaining edges, however, continue to filter notifications. Thus, there are no messages unnecessarily forwarded to the subscriber at B_2 while, at the same time, the subscription forwarding overhead caused by the subscribers at B_5 and B_6 is eliminated.

It is possible to combine the adaptive routing algorithm with the self-optimising publish/subscribe overlay and run both algorithms in parallel. Figure 9 shows sim-

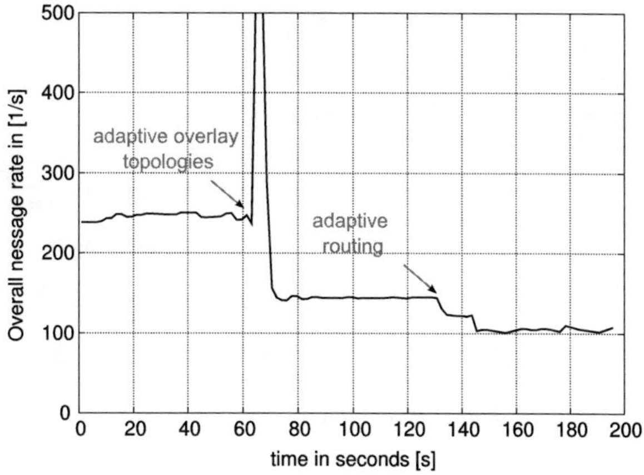

Fig. 9 Cooperative optimisation of both algorithms

ulation results. The first coarse optimisation step is achieved by topology reconfiguration. Thereafter, the adaptive routing algorithm further reduces network traffic by fine-tuning the routing configuration. The short sharp traffic increase after 60 seconds is caused by topology reconfiguration overhead which is usually amortised within seconds. Based on an in-depth analysis including additional publish/subscribe and environment parameters [2, 5, 13], simulation results of both algorithms have been evaluated and compared to offline solutions assuming global information.

3.3 Composite Event Detection

According to our software development methodology as depicted in Fig. 1, we also facilitate applications to monitor composite events that may indicate safety critical situations. Thus, applications might not only be interested in single events, known as *primitive events*, but also in *composite events*, which are spatio-temporal patterns of two or more events being either primitive or composite themselves. By the detection of composite events inside the network, we can effectively reduce message complexity by avoiding every application to subscribe to all events that may form an interesting event pattern. However, placing detectors for composite events such that the required network bandwidth is minimised is a complex on-line optimisation problem. Our approach for adaptive self-organising detector placement decisions is based on a spring relaxation heuristic, where the detectors and subdetectors are modelled as a system of interconnected springs [15]. Starting from an initial detector placement, the algorithm continually improves the placement by applying four basic optimisation steps: *decomposition*, *replication*, *migration*, and *recombination* of event detectors. The application of the steps is triggered by brokers individually based on local knowledge.

An exemplary scenario for an adaptive distributed composite event detection is shown in Fig. 10. We consider an application interested in all warm and moist

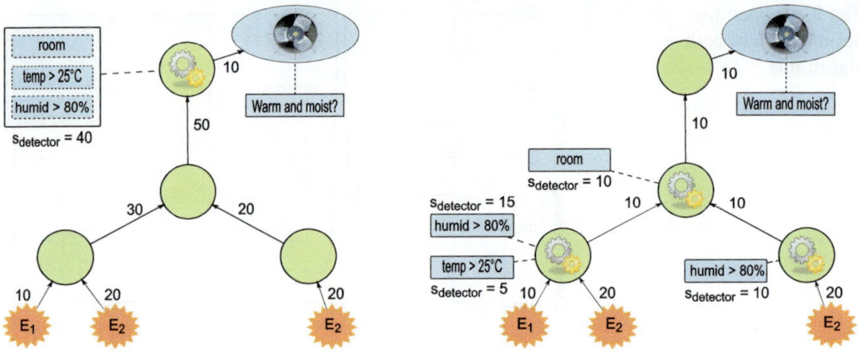

(a) Before relaxation (b) After relaxation

Fig. 10 Optimisation approach based on spring relaxation

rooms. Temperature events are detected by sensor E_1, humidity events by sensor E_2, respectively. The number of events passing each broker is also included in the figures as well as the corresponding selectivity of each detector used for the computation of the individual springs. Figure 10(a) shows the initial state, where the whole composite detector is placed near to the application. Figure 10(b) shows the optimised detector placement after applying the following detector placement steps. First, the whole detector is decomposed into its subdetectors. Second, the subdetectors are migrated as near as possible to the event sources. However, the humidity subdetector has to be replicated before its migration to cover both event streams.

4 Conclusions

Characteristics like self-organisation, self-optimisation and self-stabilisation play an important role in Organic Computing applications. Our application scenario concentrates on AS-nets and, hence, on the area of pervasive computing. We argued that these systems have an inherent requirement for these self-x properties. Using a model-driven software development methodology, we are able to significantly reduce the complexity of building such self-x applications. With our approach, the developer does not have to care about multi-bit errors in RAM, CPU registers or messages received by radio. The tool chain takes care of these complex issues. We have explained how the tool chain maps application models, its control and data structures to a physical machine in such a way that the system can stabilise from the aforementioned errors within a bounded time. We also introduced a new theoretical computational model based on a self-stabilising automaton that acts as a theoretical basis for the development of self-stabilising algorithms. Moreover, based on this automaton, we further developed a more practicable self-stabilising machine with guaranteed self-stabilisation and reasonable performance.

As second major ingredient of our approach, we have presented a toolbox of supporting algorithms and network protocols. These are used by the tool chain to realise common tasks such as role assignment, building a network overlay, and role-based message routing. These tasks have been implemented on top of our publish/subscribe middleware REBECA that has been extended to support fine-grained feature composition by a flexible runtime architecture. REBECA also incorporates mechanisms that renders its broker topology self-stabilising and self-optimising and that self-optimise its routing configuration by applying hybrid routing algorithms. Finally, REBECA supports composite event detection, where the detector placement is continually optimised by a self-organising placement algorithm applying a spring relaxation heuristic.

References

1. Dolev, S.: Self-Stabilization. MIT Press, Cambridge (2000)
2. Jaeger, M.A., Mühl, G.: Stochastic analysis and comparison of self-stabilizing routing algorithms for publish/subscribe systems. In: Proceedings of the 13th IEEE International Symposium on Modeling, Analysis, and Simulation of Computer and Telecommunications Systems (MASCOTS '05), pp. 471–479. IEEE Comput. Soc., Los Alamitos (2005)
3. Jaeger, M.A., Mühl, G., Werner, M., Parzyjegla, H.: Reconfiguring self-stabilizing publish/subscribe systems. In: Proceedings of the 17th IFIP/IEEE International Workshop on Distributed Systems: Operations and Management (DSOM 2006). LNCS, vol. 4269, pp. 233–238. Springer, Berlin (2006)
4. Jaeger, M.A., Parzyjegla, H., Mühl, G., Herrmann, K.: Self-organizing broker topologies for publish/subscribe systems. In: Proceedings of the 2007 ACM Symposium on Applied Computing (SAC'07), pp. 543–550. ACM, New York (2007)
5. Mühl, G., Schröter, A., Parzyjegla, H., Kounev, S., Richling, J.: Stochastic analysis of hierarchical publish/subscribe systems. In: Proceedings of the 15th International Euro-Par Conference on Parallel Processing (Euro-Par '09). LNCS, vol. 5704, pp. 97–109. Springer, Berlin (2009)
6. Mühl, G., Werner, M., Jaeger, M.A., Herrmann, K., Parzyjegla, H.: On the definitions of self-managing and self-organizing systems. In: KiVS 2007 Workshop: Selbstorganisierende, Adaptive, Kontextsensitive verteilte Systeme (SAKS 2007). Springer, Berlin (2007)
7. Parzyjegla, H., Graff, D., Schröter, A., Richling, J., Mühl, G.: Design and implementation of the rebeca publish/subscribe middleware. In: From Active Data Management to Event-Based Systems and More. LNCS, vol. 6462, pp. 124–140. Springer, Berlin (2010)
8. Parzyjegla, H., Jaeger, M.A., Mühl, G., Weis, T.: A model-driven approach to the development of autonomous control applications. In: Proceedings of the 1st Workshop on Model-Driven Software Adaptation (M-ADAPT '07) at ECOOP 2007, vol. 2007-10, pp. 25–27. Technische Universität Berlin, Berlin (2007)
9. Parzyjegla, H., Jaeger, M.A., Mühl, G., Weis, T.: Model-driven development and adaptation of autonomous control applications. IEEE Distrib. Syst. Online **9**(11), 1–9 (2008)
10. Parzyjegla, H., Mühl, G., Jaeger, M.A.: Reconfiguring publish/subscribe overlay topologies. In: Proceedings of the 26th IEEE International Conference on Distributed Computing Systems Workshops (ICDCSW 2006), p. 29. IEEE Comput. Soc., Los Alamitos (2006)
11. Schönherr, J.H., Parzyjegla, H., Mühl, G.: Clustered publish/subscribe in wireless actuator and sensor networks. In: Proceedings of the 6th International Workshop on Middleware for Pervasive and Ad-hoc Computing (MPAC '08), pp. 60–65. ACM, New York (2008)
12. Schröter, A., Graff, D., Mühl, G., Richling, J., Parzyjegla, H.: Self-optimizing hybrid routing in publish/subscribe systems. In: Proceedings of the 20th IFIP/IEEE International Workshop on Distributed Systems: Operations and Management (DSOM 2009). LNCS, vol. 5841, pp. 111–122. Springer, Berlin (2009)

13. Schröter, A., Mühl, G., Kounev, S., Parzyjegla, H., Richling, J.: Stochastic performance analysis and capacity planning of publish/subscribe systems. In: Proceedings of the 4th ACM International Conference on Distributed Event-Based Systems (DEBS '10). ACM, New York (2000)
14. Schröter, A., Mühl, G., Richling, J., Parzyjegla, H.: Adaptive routing in publish/subscribe systems using hybrid routing algorithms. In: Proceedings of the 7th Workshop on Reflective and Adaptive Middleware (ARM '08), pp. 51–52. ACM, New York (2008)
15. Seib, E., Parzyjegla, H., Mühl, G.: Distributed composite event detection in publish/subscribe networks—a case for self-organization. In: Proceedings of the Workshops der wissenschaftlichen Konferenz Kommunikation in verteilten Systemen 2011 (WowKiVS 2011), Electronic Communications of the EASST. European Association of Software Science and Technology (EASST), March 2011
16. Weis, T., Parzyjegla, H., Jaeger, M.A., Mühl, G.: Self-organizing and self-stabilizing role assignment in sensor/actuator networks. In: On the Move to Meaningful Internet Systems 2006: CoopIS, DOA, GADA, and ODBASE. LNCS, vol. 4276, pp. 1807–1824. Springer, Berlin (2006)
17. Weis, T., Ulbrich, A., Geihs, K.: Model metamorphosis. IEEE Softw. **20**(5), 46–51 (2003)
18. Weis, T., Wacker, A.: Self-stabilizing automata. In: Biologically-Inspired Collaborative Computing. IFIP International Federation for Information Processing, vol. 268, pp. 59–69. Springer, Berlin (2008)

H. Parzyjegla · E. Seib · G. Mühl (✉)
University of Rostock, Rostock, Germany
e-mail: gero.muehl@uni-rostock.de

H. Parzyjegla
e-mail: helge.parzyjegla@uni-rostock.de

E. Seib
e-mail: enrico.seib@uni-rostock.de

A. Schröter · J. Richling · H.-U. Heiß
Berlin University of Technology, Berlin, Germany

A. Schröter
e-mail: aschroet@cs.tu-berlin.de

J. Richling
e-mail: richling@cs.tu-berlin.de

H.-U. Heiß
e-mail: heiss@cs.tu-berlin.de

S. Holzapfel · M. Wander · A. Wacker · T. Weis
University of Duisburg-Essen, Duisburg/Essen, Germany

S. Holzapfel
e-mail: holzapfel@vs.uni-due.de

M. Wander
e-mail: wander@vs.uni-due.de

A. Wacker
e-mail: wacker@vs.uni-due.de

T. Weis
e-mail: weis@vs.uni-due.de

Chapter 2.2
How to Design and Implement Self-organising Resource-Flow Systems

Hella Seebach, Florian Nafz, Jan-Philipp Steghöfer, and Wolfgang Reif

Abstract The construction of self-organising systems often leads to very ingenious and specific solutions to a concrete problem. These solutions cannot be easily transferred to other domains or systems. As the development of self-organising systems is a very time consuming and challenging task, instructions, methodologies, and tools to design and construct such systems in a generic and reproducible manner are required. This article presents a software engineering guideline along with a pattern for the class of resource-flow systems and details the steps that are required to implement systems designed according to the pattern. The guideline enables a software engineer to easily and reproducibly construct self-organising resource-flow systems. In addition, the presented concepts and techniques, i.e. the precise definition of the system structure and of behavioural corridors, observation of the corridors at runtime, and the verification of the system components' behaviour allow the engineer to guarantee correct system behaviour despite self-organisation.

Keywords Self-organisation · Software engineering · Multi-agent-systems · Resource-flow systems · Production automation

1 Introduction

In software engineering projects, a decisive virtue of the methods and tools employed is their reusability. The success rate of projects increases if a software engineering process, a modelling tool, a methodology, or an approach has been used before and there are experts in the team that are familiar and feel comfortable with these implements. Unfortunately, Organic Computing (OC) systems and self-organising systems in general do not yet have an established set of tools or even a real foothold within software research. To remedy this situation, we are proposing a software engineering guideline that is based on a pattern for self-organising resource-flow systems. Instances of this important class of systems were used to develop and demonstrate the viability of several techniques that can be used by software engineers within the framework of the guideline. Furthermore, we show how the implementation of such systems is aided by a runtime environment based on multi-agent technology and how a library of plug-in self-organisation mechanisms

C. Müller-Schloer et al. (eds.), *Organic Computing — A Paradigm Shift for Complex Systems*, 145–161
Autonomic Systems, DOI 10.1007/978-3-0348-0130-0_9, © Springer Basel AG 2011

enables an engineer to quickly choose and implement the best option for the specific system under construction.

These tools are applicable in two different situations: (1) a "classically" engineered system is already available; (2) the engineer has to build the system from scratch. Thus, they allow an engineer tasked with the re-design of a system to make it *OC-ready* by mapping the system to the pattern, applying the guideline, and using the runtime environment as an implementation basis. Due to a strict separation of functional and reconfiguration behaviour, the same possibilities are available if the system is newly designed. In both cases, the tools and techniques remain the same.

An important characteristic of self-organising systems is that, although they are able to reduce complexity by their autonomy, their behaviour is often hard to predict. To raise acceptance of self-organisation mechanisms in safety-critical systems, we propose a way to design and construct self-organising resource-flow systems which always behave in a correct manner even though the structure changes by self-organisation. Correct behaviour is ensured by the definition of behavioural corridors (see Chap. 1.5), their observation during runtime and the verification of the system components' functional behaviour. These techniques are made viable by a precise definition of the generic behaviour agents in resource-flow systems exhibit.

The article is structured as follows: First, the system class of self-organising resource-flow systems is defined and specified with the Organic Design Pattern (ODP). Section 3 presents a software engineering guideline that instructs an engineer how to systematically design and construct self-organising resource-flow systems. Then, Sect. 4 shows the implementation of an Observer/Controller architecture for ODP systems and defines the behaviour of the ODP's components. In Sect. 5 different self-organisation mechanisms are discussed and the use of the reference implementation "ORE" for a tool-supported construction of self-organising resource-flow systems is explained. Finally, Sect. 6 discusses future work and concludes the article.

2 Self-organising Resource-Flow Systems

In *self-organising resource-flow systems*, resources are processed according to a task by independent agents. Each agent has a number of capabilities it can apply to the resource. Agents can exchange resources with other agents as, e.g. given by the layout of a shop floor. To fulfil the task for a resource (i.e. to apply the correct capabilities in the correct order), a resource-flow is established that determines how the resource is moved through the system and processed on the way.

In traditional resource-flow systems, the entire processing comes to a standstill when one of the agents involved breaks or cannot apply one of its capabilities anymore. If the system contains redundancy with regard to the capabilities of the agents and flexibility of the resource transportation, a self-organising approach can change the resource-flow and make the system usable again with minimal delay. The agents are capable of locally detecting a failure by constantly monitoring an invariant INV which is usually a combination of several constraints (see Chap. 1.5). If the invariant

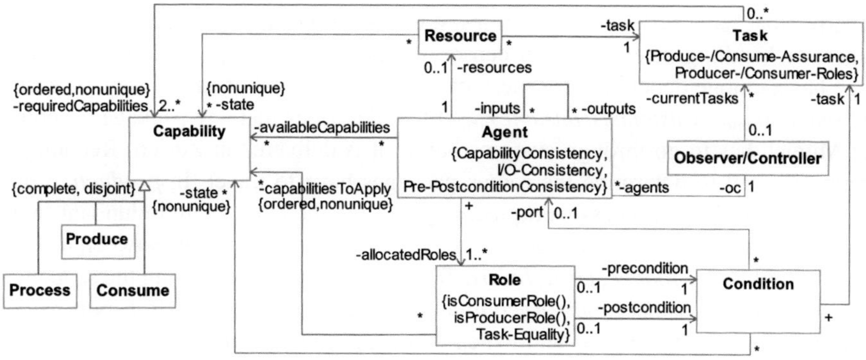

Fig. 1 Organic design pattern for resource-flow systems

does not hold anymore, they initiate a reconfiguration process during which the role that contains their configuration and the information of how to process resources are exchanged. The allocation of roles to agents determines the system setup and consequently the resource-flow. Thus, in self-organising resource-flow systems, self-organisation is a reconfiguration of the resource-flow, accomplished by calculating a new allocation of roles to agents.

The static structure for self-organising resource-flow systems is given by the Organic Design Pattern (ODP) (see Fig. 1). It contains a set of OCL-constraints (Object Constraint Language), e.g. the *CapabilityConsistency* constraint, which can be monitored locally by each agent and is expressed as:

$$self.availableCapabilities \rightarrow includesAll(self.allocatedRoles.capabilitiesToApply)$$

The constraint is defined in the context of the agent (*self*) and asserts that the role allocated to an agent only includes capabilities the agent can actually perform. A description of further constraints, how they can be expressed, and how they are used to specify a correct reconfiguration algorithm can be found in Chap. 1.5.

A new role allocation can be calculated as long as all capabilities required to fulfil the task are still available in the system and the agents that have these capabilities are linked by their potential *inputs* and *outputs* (see Fig. 1). Together with the *capabilities* still available at the agent, a new resource-flow can be computed that is enacted by the agents after they received a new set of *roles* they will have to apply. Both centralised and decentralised reconfiguration approaches are feasible. A brief discussion of different mechanisms is given in Sect. 5.

The nature of roles also enables such a system to be configured for different tasks in parallel. For each task, the role allocation determines a distinct resource-flow. If the system has to be reconfigured, only new roles for those resource-flows that require the agent to apply this capability are needed.

There are a number of examples for actual resource-flow systems that can be described with ODP: in *logistics applications*, the resources are packets that have to be transported from one place to another. The agents are distribution centres that have different capabilities, e.g. accepting packages, delivering packages, or handling customs. Each centre that is able to accept or deliver a package has a unique denotation for these capabilities. The task is then a sequence like $(Produce_{Munich}, Customs, Consume_{Zurich})$ that describes that any package accepted at Munich has to go through customs before it is delivered at Zurich. Reconfiguration is used to determine the routes packages have to travel. In *production automation*, machines process workpieces by altering their state with redundant tools they have at their disposal. Autonomous carts transport workpieces between robots. A task could be $(Produce, Drill, Insert, Tighten, Consume)$, describing the sequence of tool applications necessary. Reconfiguration defines which carts transport workpieces between which robots and which robot applies which tool.

The next section describes a guideline that helps software engineers to design and construct self-organising resource-flow systems that adhere to ODP's structure and thus gain the benefits self-organisation can provide for resource-flow systems.

3 Software Engineering Guideline

The guideline presented here is designed to tailor an existing SE process to be used for self-organising resource-flow systems. It allows to systematically enhance traditional systems with self-x properties or build new self-x systems. An important property of the guideline is that it separates design and construction of productive parts (responsible for the actual resource-flow, e.g. the control of machines or interactions with other physical systems) and self-x parts (responsible for the reconfiguration of the whole system). This is very important when a traditional system already exists and should only be enhanced by adding self-x properties. It also enables domain experts to design the productive part as usual and have an expert design the self-x part or use existing self-x mechanisms.

In this section, the guideline, which is described in detail in [4], and its activities are sketched. Figure 2 depicts a SPEM (Software and Systems Process Engineering Metamodel) model of the guideline. The highlighted activities are novel and will be shortly described in the following while the others are well known classical software

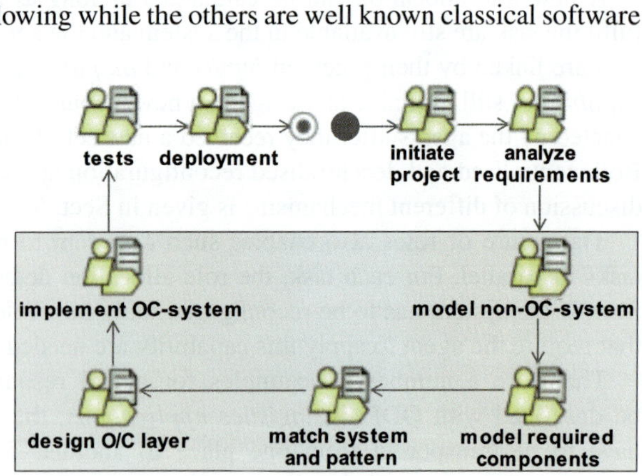

Fig. 2 Software Engineering Guideline in SPEM notation

engineering activities. The novel steps are supported by artefacts and techniques developed in the SAVE ORCA project.[1]

After initiating the project and analysing the requirements for the self-organising resource-flow system, the guideline instructs the engineer to *model the non-OC-system*. Thus, if a productive system already exists, this system must be investigated and the corresponding concepts have to be modelled. If not, the requirements are translated into classical SE models for the concepts needed in the system under construction. In both situations a conceptual model of the "traditional resource-flow system" is the result. This conceptual model is the basis for the activity *model required components*, which includes transforming the concepts into components. Self-organisation regarded as rearrangement of the system's structure needs some degree of freedom, which is added to the components in this task by adding redundant or additional properties. For example, a robot can have several tools multiple times or carts can reach additional robots compared to fixed conveyors. Then the activity *match system and pattern* follows, which encapsulates the mapping of traditional system components to components defined by the pattern presented in Sect. 2. After this, the designer determines the way the system self-organises in the activity *design O/C layer*. The activity also enables the engineer to make statements about the resilience of the system introduced by the self-organising behaviour with regard to the number and kind of failures that can occur. Finally, in *implement OC-system*, the self-organising resource-flow system is implemented in a runtime environment (see Sect. 5). The guideline ends with traditional tests and deployment activities.

To give an impression of what an activity of the guideline includes, the activity *implement OC-system* is described in detail in the following. One important part of the presented approach is the predefined behaviour of the agents in the context of the ODP. This enables a reference implementation (ODP Runtime Environment (ORE), see Sect. 5) where most of the dynamic agent behaviour and other crucial parts of the system are generically implemented. It is generic in a sense that it can be used for all systems modelled with the ODP and created with the help of the guideline, as the agent behaviour and interaction protocols as well as the reconfiguration mechanisms are usable for all instances. Of course, some parts still have to be adapted to the domain and application, as explained in step 12 to 14 of the guideline.

The activity *implement OC-system* describes all steps an engineer has to take to generate code out of the given artefacts (defined in the previous activities and steps 1 to 11) and to adapt the parts of the code which are domain dependent and cannot be automatically generated. In Fig. 3 all the steps and the artefacts needed and produced are depicted. Steps 12 to 14 use input artefacts and transform them into output artefacts as follows:

12. Generate Agent Definitions. The *domain models* (input artefact) and the *instantiation diagram* (input artefact) for classes derived from Agent provide the basis for the automatic generation of *agent definition files* (output artefact) which define the different kinds of agents in the ORE. These files contain clearly defined extension points that have to be manually implemented as follows.

[1] Formal Modelling, Safety Analysis, and Verification of Organic Computing Applications.

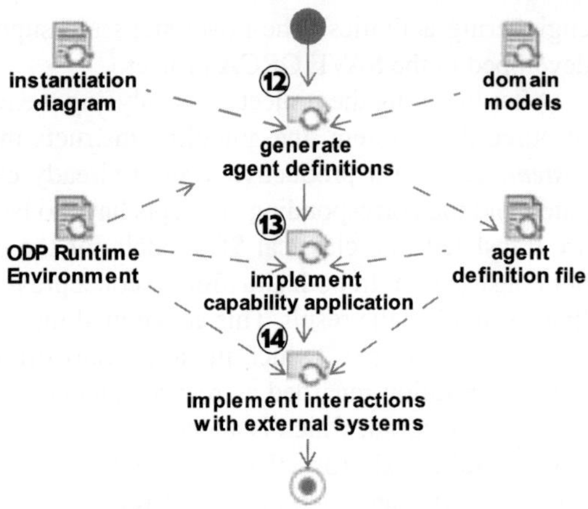

Fig. 3 Last activity of the Software Engineering Guideline: Implement OC-system

13. Implement Capability Application. Each domain processes the resources differently. In a production cell, for example, the robots add parts to a car body, thus changing its status. The engineer has to care about the correct physical change of the resource's status. Therefore, in many examples, additional processing and calculations have to be implemented as well.
14. Implement Interactions with External Systems. If a system has a physical part, e.g. robots, sensors or similar machines, the software agent must be aware of the state of its physical counterpart and able to manipulate it appropriately. As this is coupled to the application of capabilities and the interaction protocols of the agents, these steps are strongly related to the implementation of capability application. At this point it is, e.g. necessary to process sensory feedback, perform path planning and react to the physical environment of the agents.

By incorporating all tasks of the presented guideline into an existing software engineering process the software engineer is able to develop self-organising resource-flow systems systematically and reproducibly. One main benefit of the ODP and the guideline is, as already mentioned, the predefined behaviour of the agents, which is detailed in the next section. This includes functional behaviour as well as reconfiguration behaviour enabled by implementing an Observer/Controller architecture.

4 Functional and Reconfiguration Behaviour

The concept of an Observer/Controller architecture (O/C) can be found in many Organic Computing systems (see, e.g. Chap. 4.7, Chap. 5.1, Chap. 5.7). A generic O/C architecture for OC-systems, which is also the basis for the agent architecture in ODP systems, is given in [2]. The main components of this architecture are the system under observation and control (SuOC), an observer which continuously observes the state of the SuOC, and a controller which takes action if necessary, i.e. if the observer detects a situation that requires intervention by the controller. In the following, it is demonstrated how this abstract architectural description can be implemented in an actual agent system for self-organising resource-flow systems.

4.1 An O/C Architecture with Base Agents and Reconfiguration Agents

There are two driving forces that guide the implementation of an O/C architecture. The first of them is cohesion [6], the grouping of functionality that operates on the same data or serves the same functionality. The second is separation of concerns, the definition of distinct aspects of a system whose functionalities do not overlap.

The abstract description of an observer assumes that it is possible to monitor a system with a separate system and continuously observe all relevant information that is required by the controller. In fact, however, the observer has to perform a more complex task as observation may include interactions with other components, e.g. to find out whether they are still alive. A true separation, even with techniques like aspects, would therefore be hard to realise and somewhat artificial. We thus argue that employing communicational cohesion by pairing the observation part of the O/C with the functional agent which holds all the data and the states that need to be observed (e.g. sensor data in a sensor node) is beneficial in actual implementations. The combination of observer and functional agent is called *base agent*. Thus, separation of concerns is achieved by introducing the base agent and the concept of a *reconfiguration agent* which acts as the controller and is only active when required. The formal counterpart of this separation is explained in Chap. 1.5 and relies on the partitioning of the behaviour into a formal model of an ODP-agent that relates to functionality and one that relates to correct reconfiguration.

Figure 4 shows the implementation of a decentralised Observer/Controller architecture according to above description. The functional part is enhanced by the observer part within the base agent. There are distinct controllers for different tasks

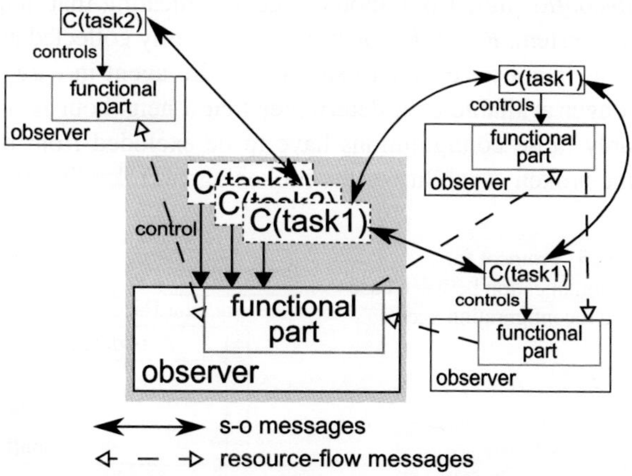

Fig. 4 Proposed implementation of an O/C architecture

that are performed by the base agents. In case a reconfiguration is necessary, the base agent starts a controller for this task, e.g. $C(task1)$. The agent in the centre of Fig. 4 has started several reconfiguration agents to reconfigure tasks 1, 2, and 3 which communicate with the controllers of the other base agents that participate in the reconfiguration. The messages (*s-o messages*) between the controllers are initiated by the respective reconfiguration mechanisms. As soon as the reconfiguration is finished, the controllers terminate and the base agents can resume processing resources according to the task. The way these two agent types interact and their functions are described in the following.

Base Agent

The base agent's main responsibility is to perform a function that helps to solve the problem the system was designed for. This can, e.g. be the control of a physical entity or machine, the application of an algorithm that alters data, or any other processing or control. This basic functionality, detailed in Sect. 4.2 and explained later on, should be modelled independently and separately from the self-x mechanisms and constitutes the *capabilities* of an agent. Several agents collaborate and employ their capabilities to resources to fulfil a common *task*.

The other responsibility of the base agent is the observation of its states and its data and the detection of undesirable behaviour. In resource-flow systems, where an agent is integrated into a resource-flow graph, the base agents also have to check the availability of their neighbours. In case of an internal error or a neighbour that is no longer reachable, the agent has to trigger the self-organisation mechanism. If another agent triggered the mechanism, it is also necessary to react to this event and participate in the reconfiguration.

The responsibilities that come with the observation and with the triggering of a reconfiguration are modelled in a state machine with five parallel states as seen in Fig. 6. The topmost state is the one which determines the reaction of the base agent in case of a failure that results in an invariant violation (INV! = true) or a *ReconfRequest* from another agent, indicating that help is required to reconfigure the system. *ReconfRequest*s are persistently collected as shown in the middle state.

It is not necessary to stop the entire agent in case one of its capabilities is no longer available or it determines that a neighbour is no longer reachable. Instead, only those configurations have to be excluded from processing that make use of the broken capability or the broken agent. Likewise, if another agent requested this

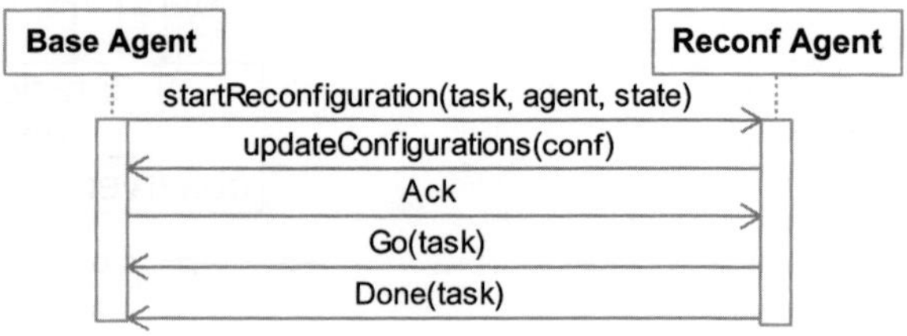

Fig. 5 Protocol for communication between base and reconfiguration agents

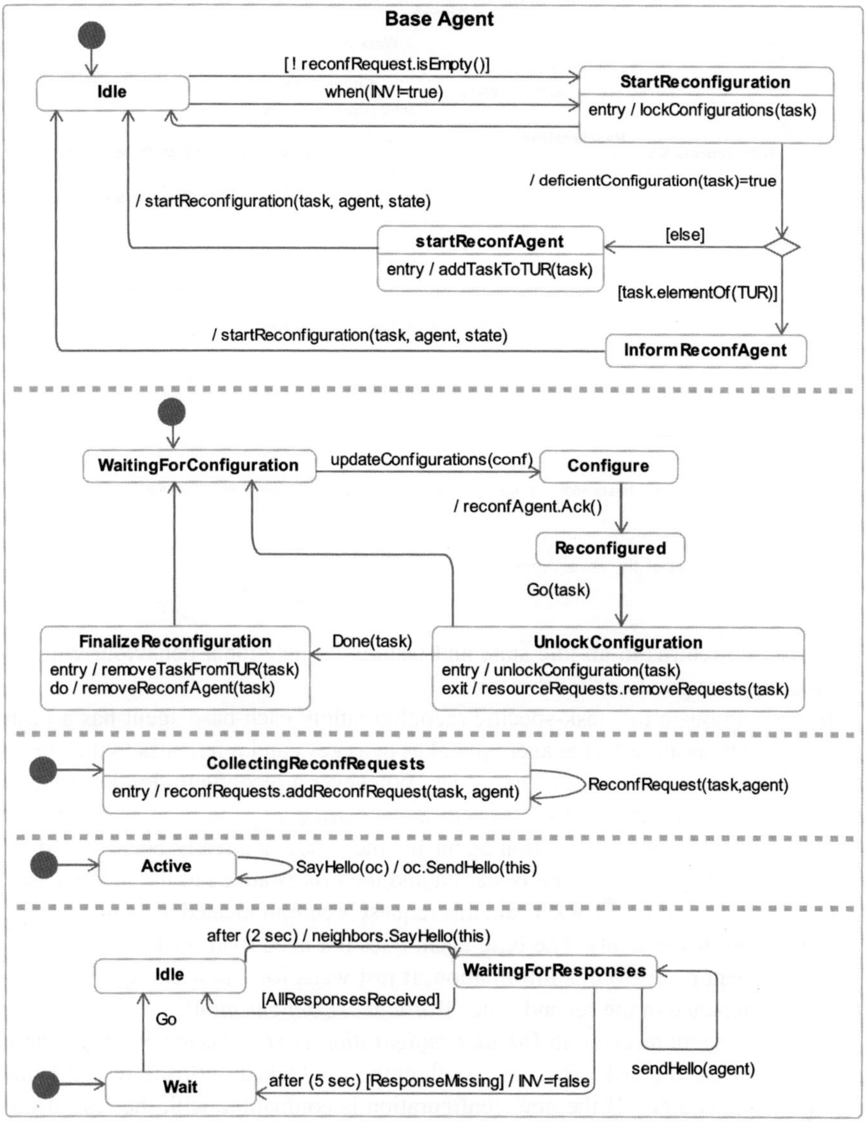

Fig. 6 Observer part of the Base Agent

agent to reconfigure, only the part of the configuration that is of concern for the task that is reconfigured has to be regarded. If the base agent currently performs a role for the task, that should be reconfigured, the base agent needs to be informed about this situation because the result of performing a role in context of a reconfigured task can not be foreseen. Therefore, the base agent always checks the flag *deficient-Configurations(task)* for the task in the role it currently applies, see Fig. 7. If the

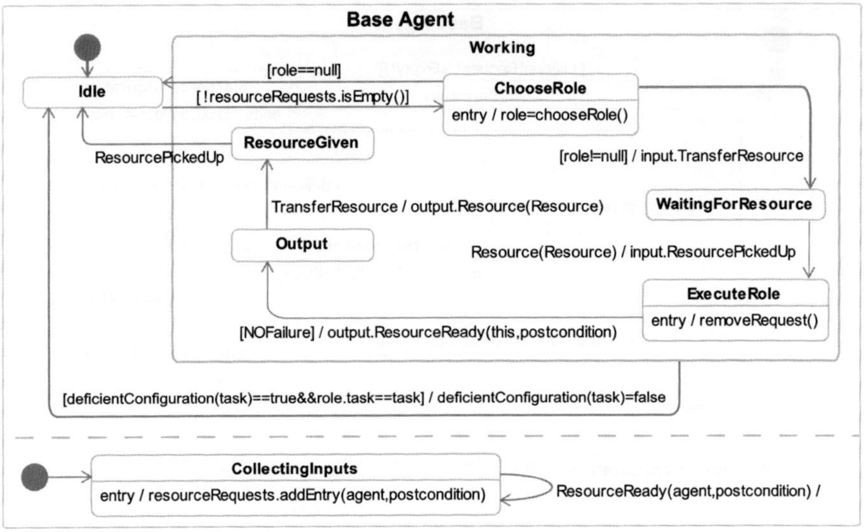

Fig. 7 Functional part of the Base Agent

flag is true, it returns to an idle state and restarts its functional behaviour without considering configurations for this task.

To accommodate this task-specific reconfiguration, each base agent has a hash map that contains all tasks the agent partakes in as keys and references to the corresponding reconfiguration agents as values. Whenever a reconfiguration has to take place to restore the invariant or react to a *ReconfRequest*, the base agent checks the map to see if a reconfiguration agent for the task has already been started. If none is available yet, a new one is started and the task is added to the list of *T*asks *U*nder *R*econfiguration (TUR). Then, this request is communicated to the new or the existing agent, respectively. The base agent does not need to know how the reconfiguration agent reacts to this information. It just waits for a new configuration for the task as depicted in the second state (*WaitingForConfiguration*).

If the base agent receives an *UpdateConfiguration(conf)* message, it sets the new configuration for the task. The new configuration is locked until the reconfiguration agent sends a *Go*. If the new configuration is compatible with the rest of the agent's configuration, the base agent acknowledges the new configuration. In the *Reconfigured* state the agent waits for the *Go(task)* message so that it can unlock the configuration. If the reconfiguration agent sends a *Done(task)* message, the reconfiguration is finished and the task for which the reconfiguration took place is removed from the tasks under reconfiguration (TUR). Otherwise, the agent waits for the next configuration that is calculated.

Apart from the failure of a capability of an agent, failure of an entire agent will also be detected by the neighbours of the failed agent and cause the violation of an invariant. Such failures are recognised by a standard heartbeat mechanism where the agents periodically "ping" their neighbours and if they do not react in time the agent is assumed to be broken. This is modelled in the two states at the bottom of the state machine in Fig. 6.

Reconfiguration Agent

The controller part of the abstract Observer/Controller architecture is implemented in the reconfiguration agent. Each base agent starts a reconfiguration agent whenever an error has been detected or when it has been asked to participate in a reconfiguration triggered by another agent. Different self-organisation mechanisms are implemented with different reconfiguration agents that expose a generic protocol and interface to the base agents. Thus, different mechanisms can be easily exchanged by swapping different reconfiguration agents without any change to the base agent.

The reconfiguration agent communicates with the base agent with the protocol depicted in Fig. 5. A reconfiguration agent receives messages of type *startReconfiguration(task,agent,state)* and queues them internally. This is useful in situations where the reconfiguration agent cannot react immediately to the event because, e.g. it is already in the phase where configurations are being allocated. The event is handled after distributing the configurations to its base agent. By parametrising *startReconfiguration(task,agent,state)* and *Go(task)* with the task, a base agent can selectively reconfigure for specific tasks that are affected by an error.

After the reconfiguration has been started, a new configuration (conf) is calculated and transmitted to the base agent in an *updateConfigurations(conf)* message. The base agent checks the configuration and informs the reconfiguration agent (*Ack*) whether it is compatible with its allocated roles. The reconfiguration agent then sends *Go(task)* to indicate that the base agent can assume its productive state.

After the *Go* message has been sent, the reconfiguration agent checks if there are pending *ReconfRequest*s in its queue. If this is not the case, it will send a *Done(task)* message to the base agent and terminate itself afterwards. Otherwise, it calculates a new configuration and repeats the process. How the reconfiguration agent reacts if there are multiple ReconfRequests waiting depends on the concrete self-organisation mechanism. It can, e.g. discard the older ones and only use the latest one, combine them and handle them at once, or reconfigure for each of them sequentially.

4.2 Functional Behaviour in Self-organising Resource-Flow Systems

After detailing the Observer/Controller part of the agents, an instantiation of the functional part of the base agent is now described for the domain of self-organising resource-flow systems. An agent has a *state* which is instantiated with a tuple of its permissible inputs/outputs, currently allocated roles, and available capabilities. The *configuration* of an agent corresponds to a set of roles (Rs).

The monitoring behaviour of the base agent is extended by the functional part by adding two parallel states which model the characteristics of resource-flow systems: handle the processing of resources and govern the communication between the base agents. In Fig. 7, the state at the bottom depicts that the base agent always collects requests from other base agents that want to give resources to this agent. Whenever the base agent receives a *ResourceReady(agent, postcondition)* event, it puts an entry into the *request* map. This entry contains as key the base agent which sent the event and as values all roles which have a precondition according to the postcondition in the event. If the base agent is reconfigured, this request map is cleared for the according task.

The state at the top represents the agent in running mode. It starts with an *Idle* state which the agent only leaves if there are resources that have to be processed. When the agent is in the *Idle* state it always checks if the *request* map contains an entry. If so, it switches the state to *ChooseRole*. The entry action chooses one of the roles in the map if possible. A fairness function and a deadlock avoidance mechanism [5] determine the choice of a role. To avoid deadlocks, e.g. the agent at the entry of a loop in the resource-flow can decide that no role will be chosen (*role==null*). If that is the case, the agent goes back to the *Idle* state and tries again after a small delay. If a role is chosen (see Fig. 1 for details on the used concepts), the agent asks the agent (*input*) in the *request* map to transfer the resource (*Transfer-Resource*). Then, the agent waits until it receives the resource (*Resource(res)*) and informs the sending agent about picking it up (*ResourcePickedUp*). In the *Execute-Role* state the agent removes the corresponding entry from the *request* map and executes the capability defined in the chosen role. The agent informs the *output* agent defined in the postcondition of the role that the resource is ready and then waits in the *output* state until the *output* agent asks for the transfer of the resource (switch to *ResourceGiven* state). After the agent received the *ResourcePickedUp* event, the agent is in the *Idle* state again and waits for incoming requests. An important part of the resource-handshake between two base agents is that the sending agent has to hold on to the resource until the receiving agent is ready to receive the resource. On the one hand, this behaviour guarantees that an agent is not interrupted by incoming resources and on the other hand this is necessary as an agent does not have the ability to store a resource but can only accept a resource if it is able to process it right away. The main benefit of this design is that the receiving agent has the opportunity to decide which resource it will take next, thus enabling avoidance of deadlocks and starvation as described in [5].

Any failure that occurs in the *Working* state (e.g. the capability that is currently applied breaks) is detected by the observer part of the base agent. It sets *deficient-Configuration(task)* to true. If the flag becomes true for the currently processed task, the agent leaves *Working* and resets *deficientConfiguration(task)*. As the observer part already disabled all roles that belong to the task and the agent is no longer processing resources for it, the task can now be reconfigured while the agent can handle other tasks.

The behaviour of the base agent in self-organising resource-flow systems is now fixed and can be used in different instances. Thus, concrete instances like production automation or logistic scenarios do not have to redefine these concepts and protocols. If they use the unaltered behaviour described here, all verifications and simulations that have confirmed correct functionality on the abstract level are still valid.

The behaviour described above is, as already mentioned, implemented in the reference implementation *ODP Runtime Environment* and thus can be used for the whole system class of self-organising resource-flow systems.

5 ODP Runtime Environment

The *ODP Runtime Environment* (ORE) offers an extensible and convenient platform for the implementation of systems based on the ODP and constructed according to the guideline. It provides the system dynamics and clearly defined extension points for application-specific behaviour as well as a plug-in mechanism for reconfiguration algorithms. A main advantage of using the ORE for the realisation of a system is that the artefacts created during the construction process are the input for a model-to-text-transformation that yields most of the source code required in an implementation. Therefore, the actual implementation effort is minimal and restricted mainly to the actual interactions between the agents and physical machines. The following paragraphs will introduce the architecture of the ODP Runtime Environment, the code transformation and the extension points, as well as different reconfiguration algorithms available for use in an application.

5.1 Architecture and Behaviour

The ORE is based on Jadex [1], a mature multi-agent system that uses the BDI paradigm for the modelling of agents. BDI stands for *Belief*, *Desire*, *Intention* and has originally been used to describe human cognitive processes. Jadex—as a procedural reasoning mechanism—maps this to *believes*, *goals*, and *plans*.

A belief is an arbitrary object that holds any kind of information and is stored in the agent's belief-base, a repository that can be queried and modified by the agent. Goals are the counterpart of BDI's desires and define the states the agent wants to achieve in an abstract way. Most importantly, a goal does not define how it can be achieved but rather only states a desired condition (i.e. retrieve a certain object, maintain a certain state, or perform a certain task). It is possible to define goals that are active all the time, that become active if a condition (no longer) holds, or ones which have to be dispatched manually. Whenever a goal becomes active one or more plans are activated. If there are several plans to achieve a goal, Jadex selects the most suitable one. In case the selected plan does not succeed, the other plans are executed until the goal is fulfilled or no more plans are available. Plans can also be triggered by the arrival of messages from other agents or by internal events. Whenever an agent executes a plan, it makes a commitment to pursue a goal or react to an event with this plan, thus stating its intention. A plan can instantiate goals, send internal and external messages, and has the ability to synchronise with other agents by waiting for replies to messages or the occurrence of internal events.

Jadex also provides a complete communication and discovery infrastructure. Agents are registered at a directory facilitator which can be queried to receive the handles of other agents. Messages can be sent over a standard communication interface and are relayed to their respective recipient(s) by the framework. The reception of a message can be part of a protocol flow or cause an event at the receiver's end.

On top of the Jadex interfaces, an *OcInfrastructure* is provided that represents the Base Agent. This Jadex capability (the unit of reuse in Jadex) defines the generic parts of the base agent. The functional part of the base agent (see Fig. 7) like role selection and execution as well as resource handling and all involved messages are implemented generically. Additionally, the observer part of the base agent (see Fig. 6) is already in place. Monitoring constraints, checking availability of neighbours, and starting of as well as communication with the reconfiguration agents is fully implemented and available to the developer. Generic invariants, like the *CapabilityConsistency* constraint introduced in Sect. 2, are already included and observed.

5.2 Code Transformation and Extension Points

In order to use the predefined functionality, *OcInfrastructure* has to be included in an *agent definition file* (ADF). These files contain descriptions of application-specific beliefs, goals and plans as well as extension points that can be used to implement required behaviour that cannot be defined generically. Most of the information required to create these files is, however, already contained in the artefacts that are created during the software engineering process if the designer sticks to the guideline. Therefore, it is possible to automatically transform the models and thus create an ADF as proposed by step 12 of the guideline that only has to be augmented if specific requirements necessitate it. Furthermore, the transformation also yields skeleton implementations of plans for each of the capabilities defined in the models together with Jadex capability files that are automatically referenced in the ADFs.

Together, these files constitute a runnable implementation of the resource-flow system. The instance diagram created by the designer is transformed into a manager agent which can be started by Jadex and in turn bootstraps the system by starting all the agents defined in the instance. After the transformation, there are three pieces missing: the interactions with the physical machines, application-specific invariants that have to be monitored, and the reconfiguration algorithm. The different choices for the latter are described in the next section, while the former have to be added by augmenting the created files by the implementation of *extension points*.

In an actual resource-flow system, physical machines are represented by software agents. Agents control and structure the work flow on a very high level though and the steps needed to be performed by the machines are not modelled during the design process. Therefore, domain experts will have to design and implement the machinery and the interface between the agents and the machines. The plans, generated for each of the capabilities, contain the appropriate extension points for the application of tools to the resources (step 13 of the guideline). Likewise, the predefined plans that handle the logical resource transmission contain extension points to implement handing over physical resources, according to step 14 of the guideline (see Sect. 3).

Some resource-flow systems also require additional invariants that the agents have to observe during runtime. The generic base agent contains an abstract goal which can be instantiated in an agent implementation. The goal becomes active whenever the application-specific invariants contained in its condition are violated and plans that start the reconfiguration process are automatically executed.

As a positive side effect, the automatic code transformation provides a way to check the consistency of the models. In an iterative process, the designer will refine the models in each iteration and can check the resulting models with the transformation. Also, a runnable system that can be used in simulations to check the viability of the system and evaluate the reconfiguration mechanism is available after each iteration. Testing and simulation are supported by a visual user interface, that shows the resource-flows, allows to add or remove agents, and to modify the capabilities and inputs/outputs available to each agent.

5.3 Plug-in Mechanism for Reconfiguration Algorithms

Just like *OcInfrastructure* represents a generic Base Agent, *OcReconfiguration* represents a generic Reconfiguration Agent. It implements the protocol depicted in Fig. 5 and can thus interact with the generic base agent. As this is the only link between base agents and reconfiguration agents and the base agents are thus agnostic with respect to the actual reconfiguration mechanism, different algorithms can be used with the same base agents, constituting a *plug-in mechanism* for reconfiguration algorithms. At the moment, there are three different algorithms available. Each of them is able to restore the invariant in the system in case it has been violated.

Weak self-organisation with central reconfiguration. The reconfiguration agent that was started after a failure was detected sends the information about its base agent to a central controller. The central controller starts a reconfiguration on all other agents and collects their configuration as well, thus creating a global model of the system. This model can then be used as the input for a solution algorithm, e.g. a constraint solver (see Chap. 1.5) or a genetic algorithm, that finds new configurations for all agents. After the solution has been found, the new roles are distributed to the agents and all of them are restarted.

Strong self-organisation with a wave-like reconfiguration. After a failure was detected, the reconfiguration agent contacts other agents in the vicinity to see if roles can be switched to restore the system. If an agent can provide assistance but in turn has to find a replacement for its own role, it sends a request to its neighbours. This constitutes a fully decentralised mechanism in which reconfigurations propagate through the system in the form of a wave [7].

Strong self-organisation with a coalition-formation algorithm. After a failure was detected, the reconfiguration agent forms a coalition of agents that, together, possess all capabilities required to restore the resource flow. For more information about this algorithm, see Chap. 1.5.

6 Conclusion and Future Work

This article presents a software engineering guideline that simplifies design and construction of self-organising resource-flow systems. The guideline accompanies an engineer towards a robust, flexible, self-organising system. It also enables offline verification of the agents' functional behaviour (see Chap. 1.5) and online adherence to a behavioural corridor, thus guaranteeing correct behaviour. This is achieved by three aspects: (1) a clearly defined system structure along with a predefined behaviour of the agents is provided by the Organic Design Pattern (ODP); (2) the behaviour is implemented as an Observer/Controller architecture to achieve clear separation of concerns; (3) the ODP contains a set of constraints for the domain of resource-flow systems, which specify the behavioural corridor.

The software engineering guideline is supported by the ODP Runtime Environment, an implementation of generic aspects of the ODP. Thus, the implementation can be used for any instance of resource-flow systems modelled with the pattern. In addition, the ORE provides a plug-in mechanism for different reconfiguration mechanisms, enabling reuse of self-organisation mechanisms for this system class.

Future work will include the analysis of other system classes as well as research on the feasibility to adapt the presented techniques to them. For data-flow systems, the structural adaptations of the Organic Design Pattern required were minimal [3].

References

1. Pokahr, A., Braubach, L., Lamersdorf, W.: Jadex: A BDI reasoning engine. In: Multi-Agent Programming, pp. 149–174 (2005)
2. Richter, U., Mnif, M., Branke, J., Müller-Schloer, C., Schmeck, H.: Towards a generic observer/controller architecture for organic computing. In: INFORMATIK 2006 – Informatik für Menschen!, vol. P-93, pp. 112–119 (2006)
3. Seebach, H., Nafz, F., Holtmann, J., Meyer, J., Tichy, M., Reif, W., Schäfer, W.: Designing self-healing in automotive systems. In: Xie, B., Branke, J., Sadjadi, S., Zhang, D., Zhou, X. (eds.) Autonomic and Trusted Computing. Lecture Notes in Computer Science, vol. 6407, pp. 47–61. Springer, Berlin (2010)
4. Seebach, H., Nafz, F., Steghöfer, J.-P., Reif, W.: A software engineering guideline for self-organizing resource-flow systems. In: IEEE International Conference on Self-Adaptive and Self-Organizing System, pp. 194–203. IEEE Comput. Soc., Los Alamitos (2010)
5. Steghöfer, J.-P., Mandrekar, P., Nafz, F., Seebach, H., Reif, W.: On deadlocks and fairness in self-organizing resource-flow systems. In: Müller-Schloer, C., Karl, W., Yehia, S. (eds.) Architecture of Computing Systems – ARCS 2010. Lecture Notes in Computer Science, vol. 5974, pp. 87–100. Springer, Berlin (2010)
6. Stevens, W., Myers, G., Constantine, L.: Structured design. IBM Syst. J. **13**(2), 115–139 (1974)

7. Sudeikat, J., Steghöfer, J.-P., Seebach, H., Reif, W., Renz, W., Preisler, T., Salchow, P.: Design and simulation of a wave-like self-organization strategy for resource-flow systems. In: Proceedings of the 4th Workshop on Multi-Agent Systems and Simulation (2010)

H. Seebach (✉) · F. Nafz · J.-P. Steghöfer · W. Reif
Institute for Software & Systems Engineering, Universität Augsburg, Universitätsstr. 6a, 86159 Augsburg, Germany
e-mail: seebach@informatik.uni-augsburg.de

F. Nafz
e-mail: nafz@informatik.uni-augsburg.de

J.-P. Steghöfer
e-mail: steghoefer@informatik.uni-augsburg.de

W. Reif
e-mail: reif@informatik.uni-augsburg.de

Chapter 2.3
Monitoring and Self-awareness for Heterogeneous, Adaptive Computing Systems

David Kramer, Rainer Buchty, and Wolfgang Karl

Abstract A comprehensive monitoring infrastructure is vital for upcoming heterogeneous, adaptive many-core systems. In order to enable required self-organising capabilities, a monitoring infrastructure has to provide self-awareness. Unfortunately, traditional approaches to monitoring, like hardware performance counters, lack required flexibility and are not suitable for self-organising systems.

We therefore present a flexible, hierarchical monitoring infrastructure for heterogeneous adaptive computing systems being able to provide a detailed and pristine view of the system state. On lower level, an associative counter array performs sustained monitoring of individual components of the system and provides this information to high level instances. These instances analyse and evaluate this information, and finally realise self-awareness. For this purpose, we employ a flexible, rule-based approach for runtime evaluation and classification of the system state. Further system instances, such as the task scheduler, may use the classified state as well as gathered information to realise self-x features, such as self-optimisation.

Keywords Self-awareness · Monitoring · Adaptive computing

1 Introduction and Motivation

Following the current trend, emerging future computing systems will be massively parallel and heterogeneous. For dedicated applications, heterogeneous systems either deliver higher performance, less power consumption, or even both. The increasing amount of available processing cores of emerging architectures allows consolidating many different systems onto a single chip (see the DodOrg architecture in Chap. 4.3). But, as a drawback, this consolidation also increases complexity and makes management of these systems quite challenging. In embedded systems, such heterogeneous many-core systems might, in addition, be able to interact with their environment via various sensors and actuators. This additional feature requires new methods for reacting on external events, making complexity management of heterogeneous systems even more sophisticated.

Organic Computing [11] and self-organising systems can address the growing complexity by employing so-called self-x features, which include self-organisation, self-configuration, self-optimisation, self-healing, and self-protection. These self-x

C. Müller-Schloer et al. (eds.), *Organic Computing — A Paradigm Shift for Complex Systems*, 163–177
Autonomic Systems, DOI 10.1007/978-3-0348-0130-0_10, © Springer Basel AG 2011

features allow self-organising systems to dynamically adapt to changing—external as well as internal—requirements.

Self-awareness is the fundamental basis for any self-x feature: self-awareness gives systems the ability to determine current environmental conditions as well as their own system state. In computing systems, a dedicated holistic monitoring infrastructure forms the basis of self-awareness by performing sustained monitoring of the entire system.

Developing a monitoring infrastructure for heterogeneous, adaptive many-core systems imposes several challenges: firstly, conventional methods for monitoring computing systems only permit concurrently tracking a limited number of events. In addition, it must be determined by the developer at design-time which events should be monitored concurrently. This is in strong contrast to self-organising systems, where a manifold of different events must continuously be monitored in order to provide a pristine system view. Further challenges are sheer amount and plurality of the gathered information. This information must be analysed and evaluated at runtime, requiring intelligent filtering mechanism, and many different events must be correlated in order to determine the circumstances which lead to a bad system state or the cause of performance bottlenecks. These challenges require a coordinated, cooperative, and system-wide monitoring infrastructure.

More challenges arise by creating self-awareness: alike biological systems, computing systems must also be able to classify their current system state as either bad, neutral, or good. This classification is essential if the system must determine the outcome of an optimisation cycle. Traditional approaches are using a fixed set of evaluation rules or so-called objective functions for system-state determination. Conventionally, developers decide at design-time which events to be monitored and how gathered data must be evaluated in order to provide self-awareness. However, due to the complexity and heterogeneity of future self-organising systems, the a-priori creation of such rules becomes increasingly challenging and time-consuming. Therefore, flexible methods for creating such evaluation rules at runtime are required to lower the developer's burden of defining these rules at design-time.

In this article, we therefore present a hierarchical monitoring infrastructure for heterogeneous adaptive many-core systems. On the lower monitoring level, so-called associative counter arrays are used to gather, in principle, every occurring event. Based on gathered data, a flexible, adaptive rule-based approach is employed for realising self-awareness on higher monitoring levels.

The remainder of this article is organised as follows: Section 2 gives an overview of related work. Section 3 introduces our monitoring infrastructure, while Sect. 4 presents the online state-evaluation and classification approach. First results are presented in Sect. 5 and this article concludes in Sect. 6.

2 Related Work

Adaptive, self-organising systems require methods for gathering and processing system events to enable creating a feedback loop. These events can be collected on all system levels, starting with events generated on hardware level ranging up to application-generated events.

Most modern processors offer basic monitoring capabilities by so-called hardware performance counters [13] Here, a certain event is assigned to a hardware counter: every time this event occurs, the counter is incremented. How many events can be concurrently monitored depends on the individual processor architecture, i.e. the number of available counters and how freely they can be assigned to events. Which events are to be monitored is up to the designer, who can use a high-level API such as PAPI [10] for configuring the association between counter and event. According to [14], hardware performance counters suffer from certain basic limitations, such as limited number of registers, sampling delay, or lack of address profiling. These limitations hamper the use of current implementations of hardware performance counters in self-organising systems. To overcome this limitation, we extended the performance-counter approach by introducing uniform event coding and a cache-inspired mechanism, forming the so-called associative counter array (ACA) (see Sect. 3). Due to uniform event coding, the ACA is able to compare arbitrary events, and the cache-inspired mechanism permits monitoring of, in principle, every occurring event.

Current adaptive and self-organising computing systems are using a fixed and tailored monitoring infrastructure for gathering required information about the actual system state. Hence, the methods for analysis of this information are also determined at design time: In [15], Trumler et al. describe the use of self-organising concepts in grid and ubiquitous computing systems. They apply a self-organising mechanism for load balancing in a dynamic environment by migrating services between nodes. Here, the load of the individual nodes is monitored. Different so-called transfer strategies are used to decide whether a service should be migrated or not.

In [16], Zeppenfeld and Herkersdorf also describe the use of decentralised, self-organising concepts for load balancing. But instead of grid and ubiquitous computing systems, they apply these concepts to optimise the task distribution in a symmetrical multi-core processor system. Again, dedicated monitors are employed for observing the actual frequency and utilisation of the individual processors and a dedicated network permits sharing of gathered information. The monitoring instance integrated in each processor are therefore capable of determining, whether their processor has higher or lower load than average. So-called local actuators then use this information for balancing the workload. The area of embedded systems is also addressed in [6]. Kluge et al. are using self-organising methods and the monitoring capabilities of the CAROS operating system for detecting and predicting future performance bottlenecks. As all monitoring information is gathered within the OS, it is a potential single point of failure. In contrast, our approach features many monitoring instances able to communicate with each other in order to increase reliability.

In our approach, we monitor several events emitted by so-called organic processing cells. These events characterise an application's behaviour and are consequently used for system-state evaluation. The concept of using processor events for application characterisation is similar to task-activity vectors (TAV), introduced by Merkel and Bellosa [9]. Here, TAVs are used for steering and optimising task scheduling in order to improve thermal distribution on a chip multi-processor. Due to their uniform structure, TAVs cannot be used for system-state classification and evaluation in heterogeneous many-core systems. Our flexible rule-based approach is able to handle heterogeneity, as it does not rely on a fixed-length vector.

3 Monitoring for Heterogeneous, Adaptive Computing Systems

This section outlines our proposed monitoring infrastructure for heterogeneous, adaptive computing systems, such as the DodOrg architecture (see Chap. 4.3). First, we will introduce the structure of the monitoring infrastructure, before presenting the concept of a uniform event space and the use of so-called associative counter arrays (ACA). The combination of uniform event spaces with ACAs enables capturing every occurring event. In the last part of this section, we will discuss high-level monitoring instances responsible for deriving the system state from a multitude of low-level monitoring data.

3.1 Overall Structure

Our monitoring infrastructure consists of two main parts: low-level monitoring (LLM) and high-level monitoring (HLM). LLM is responsible for local data acquisition, HLM instances for data analysis and finally providing self-awareness. The monitoring infrastructure itself forms the observer component in an observer/controller architecture. Within DodOrg, further system instances, such as the Organic Middleware (see Chap. 4.4) or the Organic Thermal Management (see Chap. 4.3), are using the information provided by our monitoring infrastructure to realise multiple, independent control loops.

Each Organic Processing Cell (OPC) (see Fig. 1) features a single LLM instance residing in the static part of the OPC. Any LLM instance provides an interface to the individual instances of the ACA, which are inherently included in each component, e.g. the artNoC router (see Chap. 4.3) or a general-purpose processor of an OPC.

The gathered information is then sent by the LLM to HLM instances via the regular network infrastructure. These HLM instances then store incoming information in so-called event lists and finally use these for state evaluation and classification. To avoid a performance bottleneck potentially occurring when all monitoring messages from LLM are transmitted to a single, dedicated HLM instance, a system can have multiple HLM instances. In this case, each HLM instance is responsible for a logical sub-system, and a communication protocol is used to overcome the narrow system view and determine a global system state.

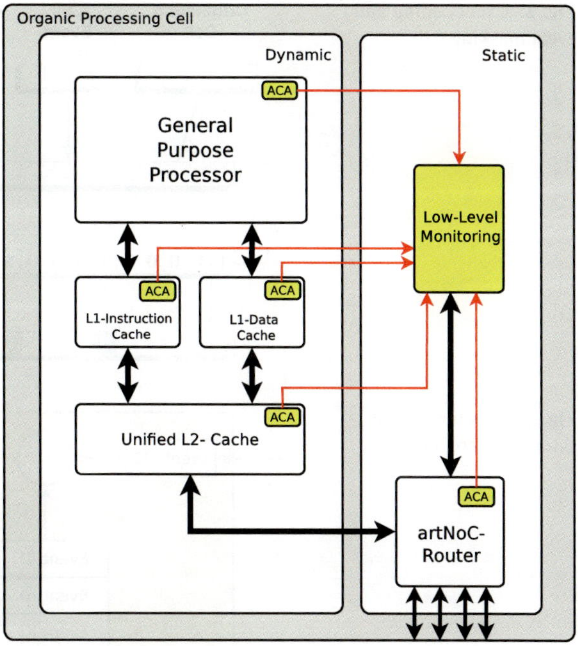

Fig. 1 Distributed and hierarchical monitoring approach

3.2 Event Coding and Event Space

Comparison of arbitrary events is permitted by introducing a biologically inspired uniform event space: in biological systems, hormones, which are used as messengers, are self-defining by their chemical structure. We apply this self-defining scheme to event coding by introducing a unique event ID. This event ID consists of a global part, containing the source of the event, and a local part, denoting the event type itself and associated data, e.g. the memory address (see Fig. 2). The local part in turn uses a format similar to a RISC instruction set. This regular structure allows grouping of certain events. This grouping can be used to create so-called meta events. An easy-to-follow example of a meta event is a general memory access which is formed by memory read and memory write accesses. Grouping is realised using an event mask specifying mandatory and "don't care" fields of an event. This mask is then applied to incoming events by using a simple Boolean *and* operation. The event mask is also used to omit details from the associated data: for example, if memory accesses should be monitored at the granularity of memory pages, the event mask can be used to omit the last twelve bits of the memory address. In this way, all memory accesses to a single page are treated as one event.

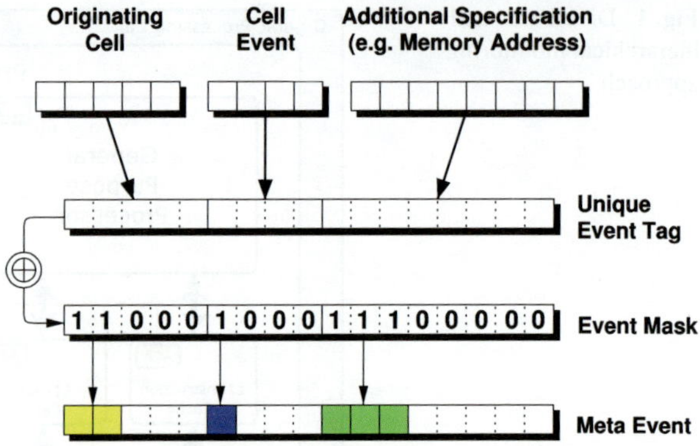

Fig. 2 Event coding and event masking

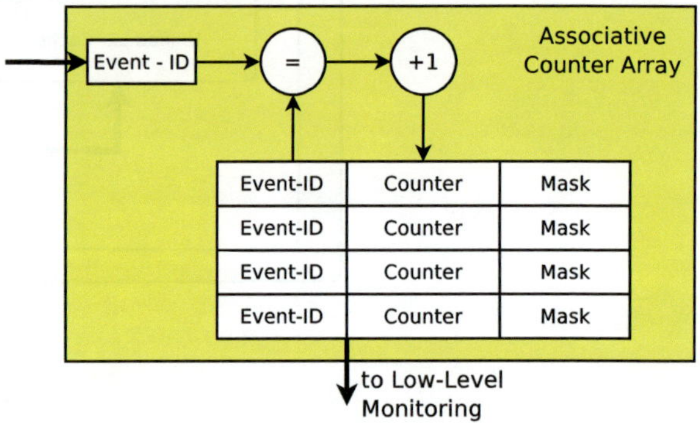

Fig. 3 Structure of the associative counter array

3.3 Associative Counter Array

To permit monitoring of virtually any occurring event, we developed the ACA, combining the hardware performance counter approach with a cache-inspired mechanism. Instead of just monitoring a small fraction of available events, the ACA is able to trigger to all incoming events. The events are generated by the component in which the ACA is integrated. E.g. a cache can provide "cache-hit" and "cache-miss" events, while a general purpose processor can provide events which represent a load/store or floating-point instruction.

The structure of the ACA is similar to a cache, but instead of a data field a counter is employed (see Fig. 3). The tag in caches is replaced by the unique event ID. Alike tag comparison in caches, the ACA compares incoming events with already stored ones. If the event matches a stored one, the associated counter is incremented. Otherwise, the new event will be stored inside the ACA. If a spare counter exists, this counter will be associated with the new event, otherwise one entry of the ACA must be evicted first and the referring counter is set to one and assigned to the new event. Determining which event to be evicted is done with a least-recently used (LRU) replacement strategy. Upon eviction of an event or when the corresponding counter reaches a threshold, a message will be sent to HLM instances. In both cases, the message contains the event ID and the counter value. In case of eviction, the message also contains the event which caused the eviction of the old event in order to permit the HLM instance to react to urgent or critical events. On HLM level, this messages are used for further correlation, e.g. for grouping of events. The ACA also features an interface permitting applications or other system instances to insert their own defined events, or HLM instances to read out the counter value for a given, or all, event at a given time.

3.4 High-Level Monitoring

HLM instances collect information provided by the LLMs and store them inside a so-called event list. The event list is a fixed-length ring buffer with head and tail pointers permitting fast access to both, oldest and newest, events. In addition to received data, a time-stamp is stored. One event list per event ID exists. If the HLM instance receives data from more than one LLM instance, individual event lists per LLM instance and event ID may be used. This permits more sophisticated data analysis at the cost of a higher memory consumption.

The HLM instances also provide a dedicated API for other system instances, such as the task scheduler or power manager, for providing them the current system state or information stored inside the event lists. Other system instances use the current system state for determining the outcome of an optimisation cycle.

4 State Classification and Self-awareness

This section introduces our flexible, rule-based approach for state classification and finally providing self-awareness for heterogeneous, adaptive many-core systems.

The proposed approach shows three major advantages over existing state-classification methods: first, our approach does not rely on pre-defined events. It is therefore applicable to heterogeneous systems where a multitude of different cores may emit a variety of different events. As the approach treats all events equally and does not rely on fixed data structures, even a comparison of different components or sub-systems is possible. Secondly, evaluation rules are derived during runtime and not determined by the developer at design-time. The system dynamically learns how gathered data is to be evaluated in order to determine the current system state during runtime. Thirdly, evaluation rules can be updated at any time including addition of new rules and using new events. Evaluation rules created at design-time do not provide this flexibility.

4.1 Rule Layout and Online Derivation of Evaluation Rules

For each event (or event ID) one rule exists. Such a rule is basically a 5-tuple: the first element denotes the rule type t. Two different types exist, right and left type rules. The difference between these two is the way they are evaluated: for right-type rules, higher values are better, for left-type lower values. The second element, v, depicts whether a rule is valid and can be used for state classification or not; $v = 0$ indicates that the rule is invalid while $v > 0$ indicates validity with the value being the weight of this rule. This weight can be used to determine or adjust the impact of this rule on state classification. An adjustment should be done, e.g. when external requirements are changing. The points p_1, p_2, and p_3 are defining the rules itself and are derived during a dedicated learning phase. We will later on describe, how these points are derived.

$$rule_{ID} = (t, v, p_1, p_2, p_3) \tag{1}$$

During the learning phase, individual HLM instances learn what occurrence ratios for a given event ID are normal and hence are able to determine whether an occurring ratio is above or below average. The learning phase is divided into several intervals of equal length.

$$ratio = \frac{counter\ value}{interval\ length} \tag{2}$$

The individual HLM instances are using the read-out feature of the LLM to retrieve the counter value for a given event ID at the end of each interval. From this counter value and the interval length, a ratio is computed following (2). As an event can occur only once per cycle, the ratio is therefore normalised to the range of 0 and 1. The ratios of individual intervals are used to create a histogram. After a predefined number of entries in the histogram or a predefined number of learning steps, the rule is derived from the histogram. For this purpose, the point p_2 is assigned to the section of the histogram with the median. The point p_2 therefore defines the average case. The two sections of the histogram with the lower and the upper quartile are assigned to the points p_1 and p_2 respectively. Finally, the second element of the rule v is set to 1, indicating that the rule is valid and can be used in the state evaluation and classification process.

Note, the duration of the learning phase for different event ID can differ due to the fact, that not all cores have not necessarily the same events, some events occur in more cores then others.

4.2 State Evaluation and Classification

For determining the state of one event ID, data stored inside the event list is used. HLM instances can compute the current occurrence ratio by applying (3). The counter values of the event are accumulated and normalised by the interval of existing events, i.e. time of the newest (TS_{newest}) entry minus the time of the oldest entry (TS_{oldest}).

$$ratio_{type} = \frac{\sum counter\ values}{TS_{newest} - TS_{oldest}} \tag{3}$$

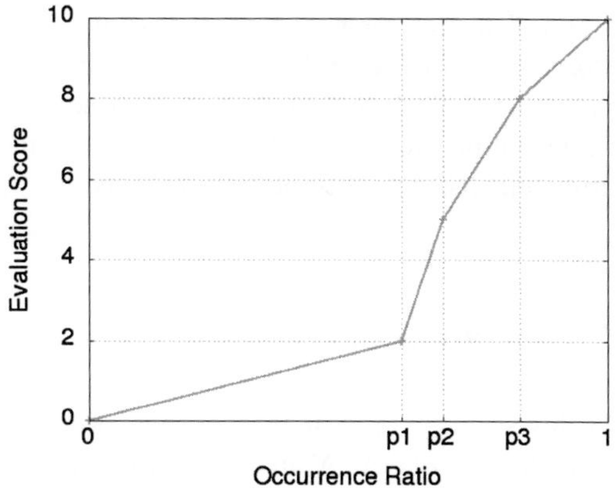

Fig. 4 Evaluation function

The rule plus an evaluation function are then used to translate the ratio into a state. As evaluation function, we employ the function depicted in Fig. 4, which is an approximation to a sigmoid function. The point p_1 is associated with the evaluation score of 2, the point p_2 with score 5, and the point p_3 with score 8. Between those points, a linear interpolation is used.

In order to get the system state for the whole system or a subsystem, the states of the individual events must be aggregated. We employ a simple weighted arithmetic mean for this purpose as shown in (4).

$$classification = \frac{\sum_{event\,ID} \omega_{event\,ID} * State}{\sum_{event\,ID} \omega_{event\,ID}} \tag{4}$$

$\omega_{event\,ID} \in [0, 1]$ denotes the state's weighting. As the weighting of every event ID is set to 1 right after deriving the rule, this simple aggregation scheme treats all events equally. But if necessary, the weighting can be adjusted, e.g. in order to reflect changed external constraints or if the system has changed and an event ID can not occur any longer. In the latter case, v of this event ID is set to zero, therefore excluding the rule from the state-classification process.

4.3 Update of Rules at Runtime

Biological systems are capable of adapting themselves to new situations or changes in environmental conditions. When a changed condition lasts over a longer period of time, biological systems adapt themselves and treat this new condition as normal. Alike biological systems, our rule-based evaluation system has also the ability to adapt to new situations. A new situation in this case could be changed external requirements, e.g. change of power supply, detected by special sensors, or if an application shows a higher performance than average over a longer period of time. In that case, the rules are updated and the new situation defines normality.

For detecting long-term changes of the system state, the system-state history is stored in a ring buffer. When the lower half of the ring buffer differs more than a given threshold from the upper half, a new learning phase will be triggered. For achieving plasticity, newly derived rules are merged with old ones, again using an arithmetic mean for the individual points.

5 Evaluation and Results

In this section, we will first present results of the prototypical hardware implementation of our ACA and then deliver the outcome of our state-evaluation and -classification method.

5.1 Prototypical Hardware Implementation

For evaluation of our low-level monitoring infrastructure, we prototypically implemented the ACA in VHDL and integrated it into a HyperTransport system [1] for monitoring the memory access behaviour delivering the information for memory access optimisations. In this set-up, an AMD Opteron system was connected via HyperTransport with the HTX Board [7]. The HTX Board is equipped with a Xilinx Virtex-4 FX100 FPGA and acts as HT cave. As the interface of the HTX Board to the HT link is realised as a soft-core (HT core) inside the FPGA, monitoring of the HT link itself is permitted. More precisely, the ACA monitored every single memory access from the HTX Board to the systems' main memory and therefore supports memory access optimisations.

Table 1 lists the size of the ACA for various configurations. For this test case, the width of the event IDs was set to 32 bit, as this was the size of a memory address. The used Virtex 4 FPGA has 42176 slices, 84352 flip-flops (FF) and 84352 4-input Look-Up-Tables (LUT). For comparison, the used HT core occupies 4868 slices, 5257 FFs and 7382 LUTs.

Table 1 Size of the associative counter array

No of counter	Counter size (bit)	Used slices	Used FF	Used LUTs
8	16	888	442	1708
8	32	1237	586	2376
16	16	1873	828	3620
16	32	2621	1101	5094
32	16	3160	1599	5520
32	32	4420	2129	8162

5.2 Self-awareness

For demonstrating the proposed state-evaluation and -classification process, we have integrated our approach into a simulation infrastructure modelling a heterogeneous, adaptive many-core such as the DodOrg architecture [2].

While DodOrg is a many-core architecture with potentially more then 100 processing cores, we limit our simulation to a small region comprising 14 general-purpose processors (GPP) and 2 memory cells organised in a 4×4 2D-mesh. The memory cells are located at the inner nodes of the mesh. The small region has one high-level monitoring instance running as a software task on one of the GPPs.

Initial Classification

Our first experiment demonstrates the state-evaluation and -classification process. For this purpose, we use two different sets of applications, which are executed on the available GPPs. The first set of applications consists of four different applications: *susan*, *bitcount*, *qsort* from the MiBench benchmark suite [8], and a discrete wavelet transformation, while the second set consists of only the *bitcount* and the discrete wavelet transformation.

At the beginning of this experiment, three instances of each of the four applications from the first set are executed on twelve GPPs; one GPP is unused in this case. These applications are used for deriving the evaluation rules and definition of the average case.

The initial learning phase for deriving the evaluation rules at the first time is invoked after the first million cycles. It does not start directly with invoking the first applications in order to disregard the starting phase of these applications. Otherwise, the starting phase of these applications would be used for learning, leading to a false definition of "good" even for the starting phase which usually shows worse performance compared to the rest of the application. For learning, a histogram with 1024 sections was used. We read the counter values every 8000 cycles, and the rules are derived when 512 ratios were observed. Learning with many different applications is beneficial, as the derived rules are broader than rules trained with data from only one application. These broader rules result in a more stable system state classification. Within this scenario, the first rule is derived right after 400 000 cycles after initiating the learning phase.

After five million cycles, the second set of applications is scheduled. Here, from each benchmark six individual instances are executed. As depicted in Fig. 5, the evaluation score increases and therefore this set shows a much better overall system performance than the first one. The low evaluation score right after scheduling the second set results from the starting phase of the newly scheduled applications.

An example of a created rule is depicted in Fig. 7. This rule is responsible for evaluating data-cache hit performance. The corresponding histogram used for creation of this rule is depicted in Fig. 6.

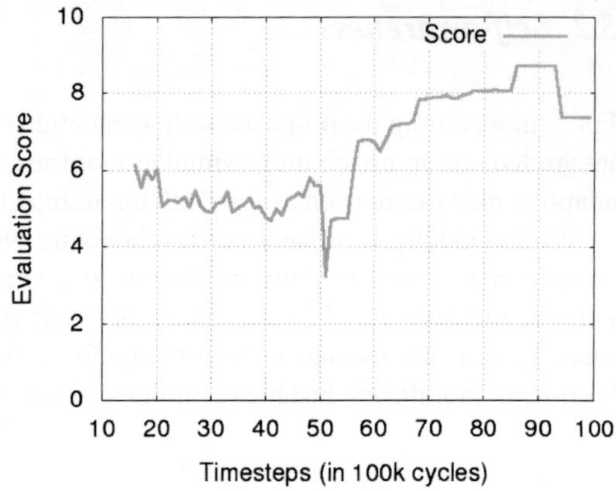

Fig. 5 Initial classification

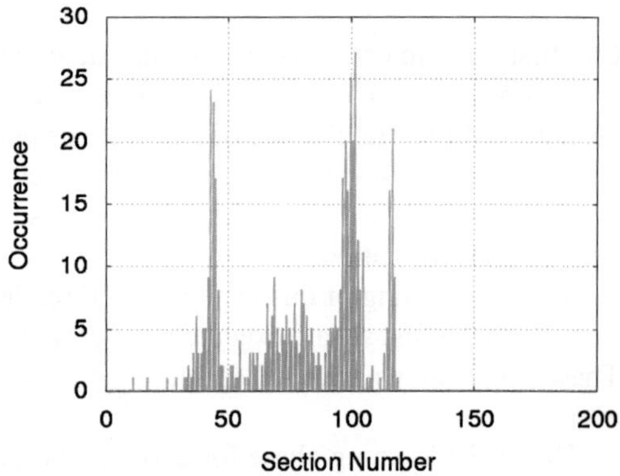

Fig. 6 Histogram used for rule creation

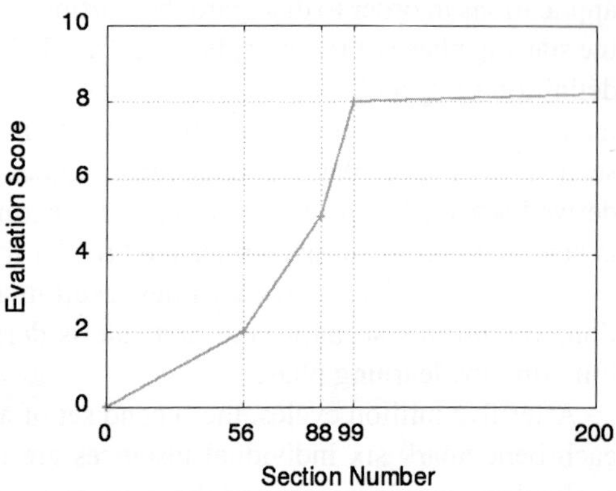

Fig. 7 Resulting rule

Rule-Update at Runtime

Our second experiment demonstrates the adaptation capabilities of the approach. Here, we used the same scenario as described above. For triggering a new learning

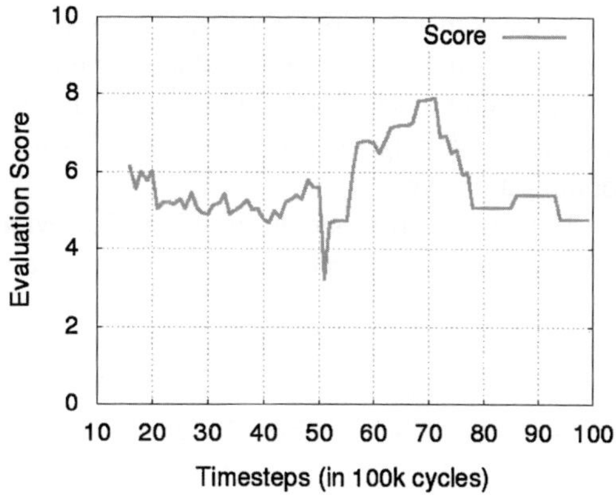

Fig. 8 Update of rules at runtime

phase, we used a ring buffer with 32 entries. The threshold was set to 1.5 being the middle between median and quartiles.

Figure 8 shows the evaluation score for this experiment. Here, the new learning phase was initiated after 6.7 million cycles. The adjustment of the evaluation rule began with the update of the first rule after 7.1 million cycles and ended approximately 600 000 cycles later. Afterwards, our evaluation system treats the new scenario as normal, mimicking plasticity of biological systems.

6 Conclusion and Outlook

In this article, we presented an approach for sustained monitoring of heterogeneous, adaptive computing systems (see also [5]). With respect to future systems, monitoring and evaluation play a vital role in handling system complexity and keeping the system in a sane state [3]. For instance, monitoring is an essential part of guided execution within heterogeneous architectures [4], where the combination of monitoring, evaluation, and eventual user-provided information ensures correct execution of applications [12].

Our hierarchically structured approach comprises a low-level part (LLM), responsible for gathering information about the system, and a high-level part (HLM), responsible for data analysis and finally providing the basic property of self-x features, self-awareness. Within LLM, we employed an associative counter array (ACA) for data acquisition. The ACA introduces flexibility into the domain of hardware counters by decoupling event and counter. A cache-inspired mechanism enables triggering to any occurring event, delivering detailed information about the system. To permit comparison of two arbitrary events, we introduced a uniform event space.

The vast amount of data gathered in heterogeneous, adaptive many-core systems such as the DodOrg architecture is then used for system-state classification by individual HLM instances. These instances feature a flexible, rule-based approach towards evaluating and classifying the current system state. An advantage of this approach is that required evaluation rules are derived during runtime and can be updated at virtually any time. In addition, new events can also be added to the system during runtime. The classified system state is provided to other system instances, such as a task mapper, which are using this state for determining the outcome of an optimisation cycle.

The results from synthesis of the hardware implementation of an associative counter array show that it has low hardware requirements and can be integrated easily into more complex designs like a complete CPU of a NoC router. Using the example of DodOrg, the presented results demonstrate, that the introduced flexible, rule-based approach for state classification is suitable for evaluating the system state of a heterogeneous, adaptive many-core system.

References

1. HyperTransport™I/O Link Specification Revision 3.10 (2008). http://hypertransport.org/docucontrol/HTC20051222-00046-0028.pdf
2. Becker, J., Brändle, K., Brinkschulte, U., Henkel, J., Karl, W., Köster, T., Wenz, M., Wörn, H.: Digital on-demand computing organism for real-time systems. In: Karl, W., Becker, J., Großpietsch, K.-E., Hochberger, C., Maehle, E. (eds.) Workshop Proceedings of the 19th International Conference on Architecture of Computing Systems (ARCS'06). GI-Edition Lecture Notes in Informatics (LNI), vol. P81, pp. 230–245 (2006)
3. Buchty, R., Karl, W.: Design aspects of self-organizing heterogeneous multi-core architectures. In: Information Technology 5/2008 (Issue on Computer Architecture Challenges), pp. 293–299. Oldenbourg Wissenschaftsverlag, October 2008
4. Buchty, R., Kicherer, M., Kramer, D., Karl, W.: An embrace-and-extend approach to managing the complexity of future heterogeneous systems. In: SAMOS '09: Proceedings of the 9th International Workshop on Embedded Computer Systems: Architectures, Modeling, and Simulation, pp. 227–236. Springer, Berlin (2009)
5. Buchty, R., Kramer, D., Karl, W.: An organic computing approach to sustain-ed real-time monitoring. In: Proceedings of WCC2008/BICC (IFIP Vol. 268), pp. 151–162. Springer, Berlin (2008). ISBN 978-0-387-09654-4
6. Kluge, F., Mische, J., Uhrig, S., Ungerer, T.: Building adaptive embedded systems by monitoring and dynamic loading of application module. In: Workshop on Adaptive and Reconfigurable Embedded Systems, St. Louis, MO, USA, April 2008
7. Fröning, H., Nüessle, M., Slogsnat, D., Litz, H., Brüning, U.: The HTX-board: a rapid prototyping station. In: 3rd Annual FPGAWorld Conference (2006)
8. Guthaus, M.R., Ringenberg, J.S., Ernst, D., Austin, T.M., Mudge, T., Brown, R.B.: Mibench: A free, commercially representative embedded benchmark suite. In: Proceedings of the Workload Characterization, 2001. WWC-4. 2001 IEEE International Workshop, pp. 3–14. IEEE Comput. Soc., Washington, DC (2001)
9. Merkel, A., Bellosa, F.: Task activity vectors: a new metric for temperature-aware scheduling. In: Eurosys '08: Proceedings of the 3rd ACM SIGOPS/EuroSys European Conference on Computer Systems 2008, pp. 1–12. ACM, New York (2008)
10. Mucci, P.J., Browne, S., Deane, C., Ho, G.: PAPI: A portable interface to hardware performance counters. In: Proceedings of the Department of Depense HPCMP User Group Conference (1999)
11. Müller-Schloer, C.: Organic computing: on the feasibility of controlled emergence. In: CODES+ISSS '04: Proceedings of the 2nd IEEE/ACM/IFIP International Conference on Hardware/Software Codesign and System Synthesis, pp. 2–5. ACM, New York (2004)

12. Nowak, F., Kicherer, M., Buchty, R., Karl, W.: Delivering guidance information in heterogeneous systems. In: Beigl, M., Cyzorla-Almeida, F.J. (eds.) ARCS 2010 Workshop Proceedings, pp. 95–101. VDE, February 2010
13. Sprunt, B.: Pentium 4 performance-monitoring features. In: IEEE Micro, pp. 72–82 (2002)
14. Sprunt, B.: The basics of performance-monitoring hardware. In: IEEE Micro, pp. 64–71 (2002)
15. Trumler, W., Pietzowski, A., Satzger, B., Ungerer, T.: Adaptive self-optimization in distributed dynamic environments. In: International Conference on Self-Adaptive and Self-Organizing Systems, 320–323 (2007)
16. Zeppenfeld, J., Herkersdorf, A.: Autonomic workload management for multi-core processor systems. In: International Conference on Architecture of Computing Systems, ARCS, Hannover, Germany, pp. 49–60 (2010)

D. Kramer (✉) · R. Buchty · W. Karl
Chair for Computer Architecture and Parallel Processing, Institute of Computer Science and Engineering, Karlsruhe Institute of Technology (KIT), 76128 Karlsruhe, Germany
e-mail: kramer@kit.edu

R. Buchty
e-mail: buchty@kit.edu

W. Karl
e-mail: karl@kit.edu

Chapter 2.4
Generic Emergent Computing in Chip Architectures

Marc Reichenbach, Ralf Seidler, Dietmar Fey, and Benjamin Pfundt

Abstract This article presents some implementations of architectures for emergent algorithms using Marching Pixels (MP). After a short presentation of recent work, we introduce application specific architectures for MP and their hardware implementations. An FPGA solution and an ASIC was made which allows to find the first and second moments of all objects by a specific MP algorithm in less than 10 ms for a megapixel resolution of a given image. In the second part we are presenting a fully programmable SIMD architecture (ParCA) for a fast processing of Cellular Automata (CA). It uses image partitioning with the help of double buffering and has an own assembler which allows an easy implementation of MP-algorithms. FPGA prototyping was accomplished and an ASIC with integrated SRAM and a DRAM Controller for external memory access were layouted. Our results show that easy morphological operations can be done with 390 fps and the implementation of the flooding algorithm is 40 times faster than an Intel Atom processor at 1.6 GHz.

Keywords Cellular array processing · Smart camera chips · Marching pixels

1 Introduction

Image processing is a very important task in industrial applications. Many companies need it for quality control and automation of their production process. Thereby, hard requirements have to be met: Modern applications need high resolution cameras with up to 1.3 megapixels and a high processing speed with up to 500 frames per second [20]. To meet such constraints, parallel processing could be one approach.

In current industrial solutions, standard PC hardware is often used for data evaluation. This results in a high TCO (total cost of ownership) because of the big size and high power consumption. Moreover, acquisition, software and support cause additional costs. In our view, a smart camera with integrated parallel processing is a better solution for the regarded requirements.

In recent works we have shown that emergent image processing algorithms can be implemented utilising Cellular Automata (CA). But since the times of the CAM-8 [10], a parallel and scalable architecture, there has been less work on dedicated hardware for simulating CA. The primary goal in the research on the CAM-8 and its predecessors was to get a hardware accelerator for CA. In the meantime, simulation

C. Müller-Schloer et al. (eds.), *Organic Computing — A Paradigm Shift for Complex Systems*, 179–192
Autonomic Systems, DOI 10.1007/978-3-0348-0130-0_11,

of CA can also be performed satisfactorily on standard computers. Moreover, our research on implementing CA on hardware is driven by using them as a fast and compact parallel computer architecture in an embedded system, exactly for industrial image processing tasks in smart cameras.

In the past, we designed several ASICs with less or only limited programmability which were focused primarily on implementing Marching Pixels (MP) image processing algorithms [15], which are presented in detail in Chap. 5.10.

In order to get more flexibility, we decided to design an application-specific instruction processor array for the execution of MP based emerging computing algorithms. The fine-grain multi-core processor resulting from this research uses a specific instruction set that is not only convenient for the execution of MP algorithms but for arbitrary emergent computing schemes.

In this article, we present the result of that computer architecture design which is called *ParCA*, a fully programmable SIMD architecture. For a high speed processing and high degree of parallelism, image partitioning, and data double buffering are used in that architecture. Building a prototype on FPGA basis was accomplished and some ASICs were designed. The results show a small chip size with low power consumption and a processing which is 40 times faster as standard embedded CPUs.

This article is divided into seven sections. Section 2 describes related work published in the past and shows some advantages and disadvantages in the presented systems. Section 3 deals with some theoretical fundamentals and answers, for example, the question what should be the optimal degree of parallelism for CA. In the next section, an algorithm class is exemplarily presented for which our new architecture was designed. Section 5 presents ParCA in detail including some buffer concepts for an efficient data storing. Section 6 shows performance results achievable with an ASIC of the developed architecture. Finally, conclusion and outlook close the article.

2 Related Work

A hardware realisation of agents in FPGAs, similar to the MP agents, was performed by Halbach and Hoffmann [12, 13] for finding optimal rules for the *Creature Exploration Problem.* This problem refers to so-called creatures that move on a grid according to local rules with the goal to visit all grid points efficiently in an unknown environment. In [21], FPGAs were used for implementing *Ant Colony Optimisation* [4] algorithms. The difference from the work presented here is that MPs are implemented as dynamic elements, i.e. as agents with states and data which are passed from one processor node to the next one whereas, in the cited papers the agents are realised as more or less static data structures whose contents are updated with dedicated hardware.

Published work on pixel processing chips with FPGAs and ASICs is very numerous. Recent work on image processing tasks based on morphological operations realised in FPGAs can be found in [2, 20] and [5]. Application-specific vision chips in CMOS have been designed and demonstrated in analogue, digital and mixed-signal technology. Fatemi and Kleihorst et al. were able to correctly recognise up to 230 faces in real-time with a smart camera [7, 14]. They, therefore, used the parallel processor "Xetal" for low-level image processing and the serial digital signal processor "TriMedia" for high-level recognition tasks. Miao et al. also built a programmable ASIC with pixel-parallel processor elements for low-level operations and line-parallel processors for high-level operations [19]. For an image of size 16×16, simulations showed a maximal clock frequency of 300 MHz and a frame rate of 1000. Using a 180 nm CMOS process, a 16×16 ASIC further was prototyped. It resulted in 0.3 mm^2 chip area and had a power consumption of 8.72 mW.

In the past, some hardware architectures for processing CA were developed. For example, in the 1990s, the architecture CAM-8 [10, 18, 22] was developed as successor of the CAM-6. This architecture proposes only a small degree of parallelism (in most references 8), so all CA cells which are involved in the processing are updated in serial by pipelining. This update process is done with a 16 bit lookup-table. For more complex operations, a sequence of 16 bit functions has to be created and loaded into the lookup-table between two update steps. This is the most criticised point of the architecture. Because of using an older manufacturing technology, the system is very large, shows a high power consumption and is comparatively slow.

A few new architectures using CA were published in recent years. E.g. PHCA [3] was designed as IP (Intellectual Property) core for 1D or 2D array of programmable processor elements (PEs). In order to achieve a small chip size, each PE contains a serial ALU realizing an adder/subtractor. Moreover, a RAM Block of 32 bits was added to store the content of a CA cell. The programmability is just realised by multiplexing data paths between neighbouring ALUs. Some easy data flow transformations were implemented. But more complex algorithms are inappropriate for the PHCA because there is not the possibility to execute jump and loop operations.

In a recent paper [1], an ASIC for image preprocessing is presented, that is partially programmable. It contains a very small application-specific instruction subset of a general-purpose processor. This processor is quite fast, with a processing time of 250 µs of an image with the resolution of 320×240 pixels, but only simple morphological operations (e.g. dilatation, erosion) are possible.

The related works show that either the systems using CAs are not powerful or flexible enough. The purpose of the work presented in this article is to combine those features: designing a flexible and powerful architecture with a high degree of parallelism, coupled with the harsh requirements for embedded systems.

3 Application-Specific Architectures for Marching Pixels Algorithms

Before we present our latest result, the programmable ParCA architecture, we would like to show some results of our preliminary work on application-specific hardware architectures for MP based algorithms.

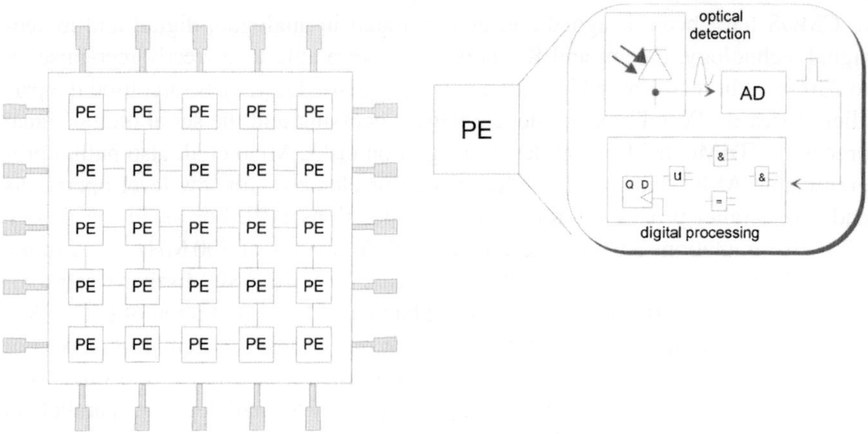

Fig. 1 Architecture of the processor element array and of a single PE

For the application-specific solution, we propose an array of *processor elements* (PEs) as core for a smart CMOS camera chip [8, 9, 16] to find centroids and orientations of objects within an image. Figure 1 shows the principle of the implementation. On the left side, the PE Array, on a chip can be seen. The PEs have a NEWS-neighbourhood, i.e. a direct connection to the four orthogonal neighbours in north, east, west, and south. PEs do not have connectivity across the borders like in a toroidal structure. Outer PEs are used for outside communication instead.

On the right side of Fig. 1, a PE is presented in detail. It has an optical detector which captures a part of the image, e.g. one pixel in the simplest case. The captured image pixel is binarised in the analogue-to-digital converter (AD) and used as input to digital processing. In the digital part, arbitrary digital operations can be used like binary logic, shift operations, multiplexing, or operations of an ALU. Apart from that, it is possible to use flip-flops as local memory for each single PE in which the states of the FSM are stored. State transition is executed in logic in the digital part of every PE.

MPs are realised as data packets that move inside the PE Array. Such a data packet stores the state of an MP and its corresponding local moments' memory values. As local moments, the object centroid and rotation are denoted relatively to the actual pixel. For details concerning the Marching Pixels processing scheme refer to Chap. 5.10. An exemplary execution step works as follows. If a PE does not host a data packet (an agent/MP), it calculates nothing but simply waits in a defined state until an active agent crosses the PE's neighbours. When an agent moves towards the PE, the PE takes the agent by changing its state to the agent's state and copies the agent's memory to its own memory. The previous PE of the agent deletes all information and turns to a passive state in the same step. The new PE of the agent might then, furthermore, calculate some values using the ALU if certain conditions are met. E.g. it updates the values of the local moments and determines if the MP has to turn to state `Final` whether the MP has finished or not.

Table 1 Flooding implementation time requirement

Resolution	k	Time per macro step in μs	$LOE = 32$ in ms	$LOE = 64$ in ms	Worst case in ms
128×128	7	1.51	0.192	0.385	0.771
256×256	8	1.71	0.217	0.436	1.75
1024×1024	10	2.11	0.268	0.538	8.64

The target system of the MP algorithm implementation was an ASIC, but VHDL synthesis was also done for Xilinx FPGAs as rapid prototypes. We used the synthesis results to estimate the required size of the PE Array, which also allows a statement about the pixel resolution and the response time of the detector to find centroids and the orientation. The optical interface and the analogue-to-digital converters have not been designed. For our first implementations, we will attach commercially available sensors to the processing units, consisting of PE Array and observer processor. A bit-serial architecture has been selected for the implementation of the ALU and the data paths between the PEs to transfer the local moments between neighboured PEs. Due to the used serial data paths, one clock cycle is not sufficient to complete one MP step. Therefore, two different notions of the term step are used in the following: *Micro Steps* and *Macro Steps*. A Micro Step equals one clock cycle. A Macro Step refers to a move of a MP from one PE to a neighbour or its decision to be blocked. In other words, a Macro Step corresponds to a state transition in the steering CA. For serial implementations, executing all transformations and operations of one Macro Step described in the algorithms takes several Micro Steps.

The worst case number of required Macro Steps is determined by an object with a border line running exactly along the border line of a whole image of size $2^k \times 2^k$. $4 * 2^k - 1$ Macro Steps are needed for computation where $k \in \mathbb{N}$. By simulations we found out that for a moderate PE clock rate of 100 MHz, one Macro Step for the flooding algorithm, described in Chap. 5.10 takes $(200 * k + 110)$ ns. Table 1 shows the required time of one Macro Step for different image resolutions and the time needed to compute different object sizes (maximum length of objects (LOE) expressed in pixels is 32 and 64) up to the absolute worst case where the size of an object equals the image resolution ($LOE = 2^k$). Even the worst case values are below the required reply time in real time systems of 10 ms.

3.1 Implementation of the Flooding Algorithm on FPGAs and ASICs

We implemented the flooding algorithm on an FPGA. Using FPGA technology has the advantage that a 32 bit *MicroBlaze* soft IP RISC processor from Xilinx or, for Virtex devices, the PowerPC as hard IP could be used as observer processor. For an

Table 2 FPGA utilisation of flooding implementation

Resolution	FPGA type	Slices	LUTs	f_{max} in MHz
16 × 16	Virtex 2 VP30	12,867/13,696	18,786/27,392	262
		93%	68%	
	Spartan3 5000	13,425/33,280	20,084/66,560	155
		40%	30%	
32 × 24	Spartan3 5000	29,684/33,280	44,964/66,560	142
		89%	67%	

image size of up to 32 × 32, the MicroBlaze was used for post-processing determining centroids and orientation of found objects. The PE Array was attached to the MicroBlaze using the FPGA-internal *Fast Simplex Link* bus.

Table 2 shows the results of flooding without observer processor for two Spartan-3 devices. The table shows that images with a size of 16 × 16 can be realised for both device types. For the smaller Virtex-2 device, the number of resources was nearly exhausted. In a Spartan-3 5000, the limit is given for an image with 32 × 24. The achievable clock rates show that it is not a problem to meet the real-time requirements. However, only practically irrelevant small image sizes can be realised. The reason is the limited memory in an FPGA device expressed by the lower per cent values of the LUTs (Look-Up Tables) and the much higher per cent values of the utilised slices. Slices consist of LUTs and flip-flops (FFs). The logic in the LUTs is not the problem, but the required memory in the FFs.

4 The Architecture of ParCA

The following description of the ParCA architecture is top down structured. We start our description with an overview of the main components (Sect. 4.1) of our system architecture. Then we continue with some details on the designed PE architecture (Sect. 4.2), the implemented double buffering mechanism (Sect. 4.3) and the simulation environment (Sect. 4.4) we used for the validation of the architecture.

4.1 System Overview

In Fig. 2, an overview of the ParCA architecture is given. The calculation is carried out by a one dimensional SIMD array of processor elements (*PE Array*) with a special instruction set geared to the execution of emergent algorithms. The *PE Array* is fully programmable and reads the program from a local SRAM memory, generated by the memory compiler *memmaker* [6]. This SRAM is able to hold up to 4096 instructions, filled by a custom assembler developed for fast programming. To reduce

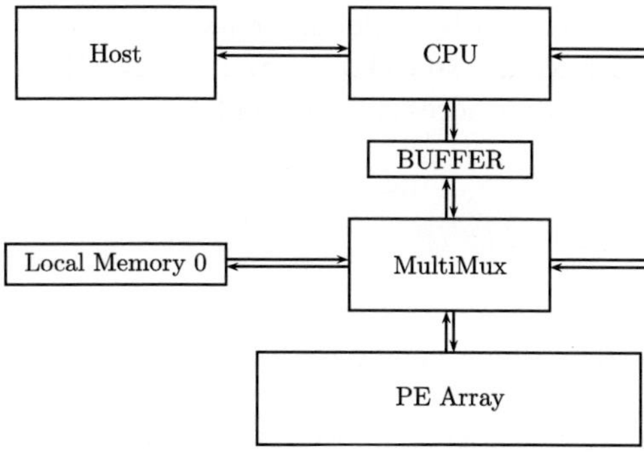

Fig. 2 System overview

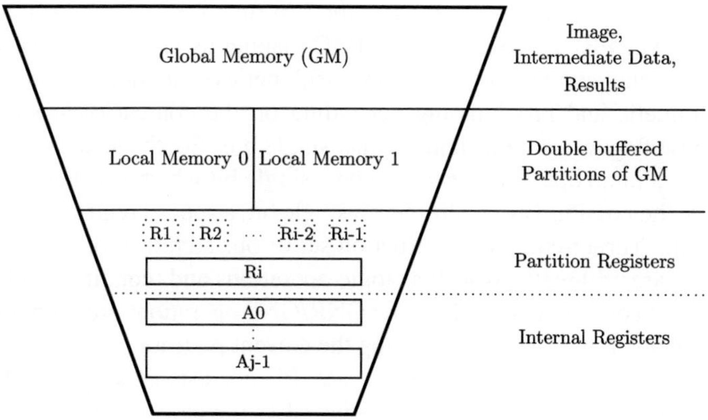

Fig. 3 Memory hierarchy

chip size, the design of each PE is fully serial. That allows to increase the degree of parallelism. The *PE Array* operates on *Local Memory 0* and *Local Memory 1*, which are also implemented in SRAM technology. They contain a rectangular partition of the whole input image. These local memories are filled by the CPU via a buffer and a DMA controller from the *Global Memory* which is designed to be off-chip SDRAM memory. It also contains the program of the CPU and intermediate results. We used a LEON 3 soft IP processor [11] for the realisation of the CPU. For debugging purposes, a PC (*Host*) is connected to the system.

To reduce memory latency, we implemented a memory hierarchy, which is shown in Fig. 3. The global memory (GM) is visible, residing off-chip because of its large size. It is followed by the partitioned on-chip memory in double-buffering manner. Connected to that memory, the local registers follow, where the partition memory is mapped as logical registers into the register set of each PE. To perform fast program control for jump operations, one distinguished register of the partition memory resides in the local register set with the feature to access single bits.

Furthermore, we implemented a word-parallel and bit-serial operating ALU to save chip area. That means, each word is spread over several lines of local memory, but all bits needed by our PE Array are accessible in parallel.

4.2 PE Architecture

Common massively parallel architectures couple PEs in a complete 2D fashion arranged in a von Neumann neighbourhood (NEWS neighbourhood). To realise a small and compact solution, we favoured a simple 1D array of PEs, closely coupled in east and west direction. North and south access is implemented via buffers. Due to the partitioned approach, east and west buffers are required in order to store ghost zone data, too. This is illustrated by Fig. 4, where k PEs are shown, along with the east and west buffers (E, W) and the k buffers for north and south (Ni/Si, $i \in 1, \ldots, k$). The PEs are controlled in SIMD manner using a single control unit.

Since our architecture focuses on simple arithmetic computing cellular automata, integer arithmetic and fast jumping depending on the content of special flags is needed. These flags are not much more than single bits. So, the idea seems common to implement a jump operation operating on a single bit which is called bit-jump. In order to give the possibility of addressing a single bit, a special register is introduced. In Table 3, the Turing-complete instruction set of our architecture is presented. As visible, there are arithmetic as well as logic operations and (conditional)-branching (or jumps), even on single bits (JIS). The GNR/GNP operations are used for moving the PE Array one row further or changing the current partition.

To fulfil above denoted instructions, each PE consists of a register file and a very simple serial ALU. Because in a serial ALU the clock cycles depend on the register width, the register width is set to 16 in the following. The necessary registers are the above mentioned special register (used for bit operations e.g. bit-jumps), a program-counter register, some auxiliary registers and a write-back register. They are implemented as shift-registers. The program counter is present in every PE to manage synchronisation points for different PEs in the case of the execution of nested loops. Since we use SRAM technology, we can read a word in every clock cycle. Writing is more difficult, because read and write operations cannot be done

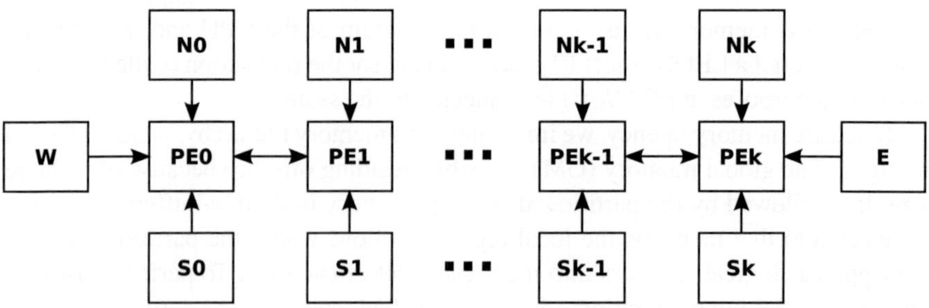

Fig. 4 PE row data path

Table 3 Instruction set for the PEs

Opcode	Name	Cycles	Description
00000	NOP	34	*No Operation*
00001	ADD	34/66	*Addition*
00010	SUB	34/66	*Subtraction*
00011	JMP	18	*Jump*
00100	JEQ	34	*Jump equals*
00101	JGR	34	*Jump greater*
00110	JGE	34	*Jump greater-equals*
00111	JLS	34	*Jump less*
01000	JLE	34	*Jump less-equals*
01001	SLS	51/73	*Shift left*
01010	SLR	51/73	*Shift left-rotate*
01011	SRS	35/67(19/51)	*Shift right*
01100	SRR	35/67(19/51)	*Shift right-rotate*
01101	AND	34/66	*And*
01110	ORS	34/66	*Or*
01111	NOT	34/66	*Not*
10000	ADI	34/66	*Add Immediate*
10001	SBI	34/66	*Sub Immediate*
10010	ANI	34/66	*And Immediate*
10011	ORI	34/66	*Or Immediate*
10100	SET	34/66	*Set Register*
10101	BST	19	*Set Bit*
10110	JIS	18	*Jump if Bit is set*
10111	JNS	18	*Jump if Bit is not set*
11110	GNR	54 + 48	*Get Next Row*
11111	GNP	54 + 48	*Get Next Partition*

at the same clock cycle. Therefore, the special write-back register is introduced to solve this problem.

The ALU is shown in Fig. 5 and includes a Full-Adder (FA) coupled with a carry D-flip-flop (Carry-FF), implementing addition and subtraction. To realise the left-shift operation, a Shift-Register is needed. The Logic device with AND, OR, and NOT as well as a comparator, fed-back with two D-flip-flops for displaying the result of a comparison as greater or less (GR-FF, LE-FF), conclude the ALU.

In the following, we introduce the instruction execution in detail to allow a user to benefit from special operations provided by our architecture. We only need a single CU (control unit) to control all our PEs. In Fig. 6, a state machine of the implemented CU is described. Together with Table 4, we can give the clock cycles for every operation in Table 3. The first value in row cycles of Table 3 shows the normal run time when performing the operation only on auxiliary registers. The second one

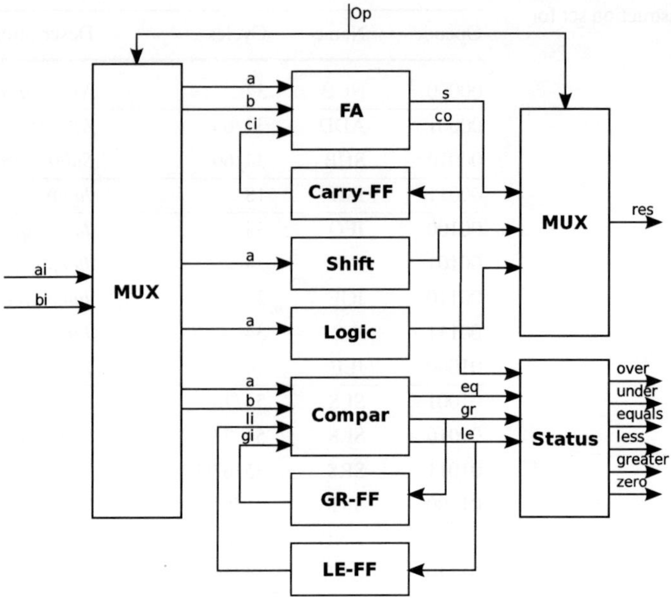

Fig. 5 ALU with all components

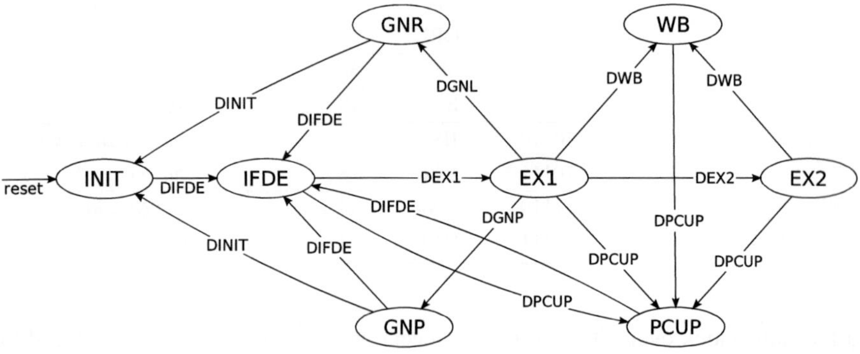

Fig. 6 Control unit state machine

Table 4 Clock cycles per stage

State	INIT	IFDE	EX1	EX2	WB	PCUP	GNR	GNP
Clocks	48	4	16	17	32	14	34	34

shows the run time including write-back into the SRAM. The only exception from that is shown in the shift right operation. If source and destination registers are the same, the faster clock cycle values in parentheses are to be considered.

The INIT state is entered when a new row needs to be processed. Then the buffers and the special registers are filled with new data from the SRAM. Every other operation starts in the IFDE (instruction fetch and decode) state. Where applicable, the next operation is EX1 (execute 1 for all operations) from where the EX2 (execute 2 fore more complex operations e.g. right shift), WB (write-back) or PCUP (PC update) state can be reached. IFDE and PCUP are compulsory for every operation, the execution states are processed for all operations with the exception of bit set and bit jumps. The write-back state is only entered for operations where the destination register is in SRAM memory.

For example, the ADD operation enters IFDE state first (4 clock cycles needed), followed by the EX1 phase (additional 16 cycles), and then, if the destination register is in SRAM, the WB phase is scheduled (32 cycles), causing that the internal PC is updated in the PCUP phase (14 cycles, resulting in either 34 or 66 clock cycles). The bit-jump operations do not need to perform the EX1 state, instead they can rapidly enter PCUP after their IFDE phase, so their execution times are very short. The left shift operations require the EX2 phase. This is necessary since all internal registers are right shift registers. GNR and GNP need 54 and additional 48 clock cycles to perform. The 54 clock cycles result from writing back the special register into the SRAM, involving INIT, EX1 and GNR. The additional 48 clock cycles are needed to perform the INIT state, where the buffers and the special registers are filled with new data from the SRAM.

4.3 Types of Double Buffering

To facilitate fast processing, double buffering and image partitioning are used. This means that data transport and calculation can be done in parallel. While the *PE Array* operates on *Local Memory 0*, new data can be loaded to *Local Memory 1* (see Fig. 2). After finishing the calculation, the system switches a multiplexer, whereby the *PE Array* now works on *Local Memory 1* and *Local Memory 0* can be filled with new data, after reading back the calculation results.

For a reduced size ratio, a further double buffering method, *full buffering*, was examined to find an optimal ratio of size and time [17]. The main idea of *full buffering* is that the *PE Array* only needs three rows to work in a von Neumann neighbourhood. By creating a buffer of four rows, the PE Array can work on three rows and the fourth row can be used to fill the buffer with new data. For a continuous processing the buffer has to be created as ring-buffer in such a way that the successor of the fourth row is the first one. For a parallel read/write access on the buffer a dual port SRAM is common.

The two described methods for double buffering are implemented in the system to choose the best one depending on the application before synthesis ParCA. If chip size is of low concern, then option one should be used to preload as much data as possible. If, especially in embedded systems, a small chip size is necessary, option two should be used. But there are disadvantages: if more than one iteration has to be processed on the input image, data often has to be reloaded from external memory. In option one, the data is stored locally and can be reused without external memory access. In general, option two needs a larger bus bandwidth than option one.

4.4 Simulation Environment

A tool chain was developed for an easy way of simulating a given algorithm on a specific image. It automates the simulation process and returns the resulting image immediately. Therefore, it is possible to test and verify an algorithm or examine the behaviour on a specific input without knowing how the underlying hardware works internally. The first component of the tool chain is a Makefile that invokes all other components. It automatically starts the assembler to translate the given algorithm into the internally used ROM format. The next step is to transform the input image, which has to be in the PGM format (Portable Graymap—an established image exchange format), into the memory layout by means of a transformation program especially written for this purpose. Now the VHDL simulator is started in command line mode where TCL (Tool command language—a powerful script language for development tools) scripts are executed running the actual simulation of our hardware and producing the resultant states of the memory. The last step is to retransform the resulting content of the memory with the mentioned transformation program, that enables the user to compare both, input and output images with a common image viewer. In this way, fast and easy simulations were performed, to evaluate some algorithms before they were tested on the hardware.

5 Results and Layout

The architecture was created as a fully parametrisable low-level-VHDL design. In fact, the amount of registers or the partition's size can be chosen arbitrarily before synthesising the system.

Prototyping was accomplished using a Xilinx Virtex2-FPGA with an embedded PowerPC Core. For a higher degree of parallelism, an ASIC has been built as well. Fig. 7 shows one part of the ASIC which includes the *PE Array* with 32 PEs and the program memory (left cyan blob).

Larger ASICs also were designed by increasing the degree of parallelism to 1024. Therefore, some morphological image processing programs were simulated. The results show a throughput of 390 fps for a one megapixel image at 100 MHz chip clock. The whole chip, including all memory, needs about 22.5 mm^2 space and consumes 4.8 W power.

Furthermore, more complex algorithms like flooding have been implemented. By adjusting the parameters for register width and register count to the flooding algorithm, the chip needs about 24.5 mm^2 and takes 10 ms for one iteration. We implemented the same algorithm on an embedded 1.6 GHz Intel Atom and measured a runtime of about 400 ms.

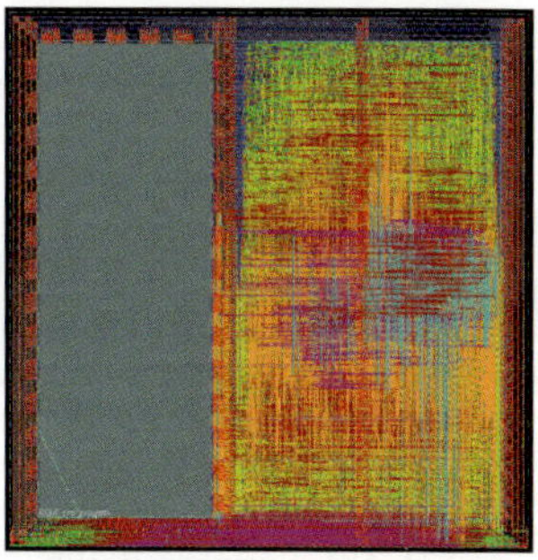

Fig. 7 Layout of the PE Array

6 Conclusion and Outlook

The presented architecture ParCA is one solution for embedded image processing for the use in smart camera chips. It works as a fine grained SIMD architecture with the main application field for cellular automata algorithms. The programmability of the architecture allows the universal use. Some simple image pre-processing algorithms have been implemented with the result, that ParCA is suitable for high speed image processing tasks. More complex image processing tasks were also implemented, where ParCA is always faster than implementations on embedded standard CPUs. Prototyping of the architecture was accomplished and an ASIC layout was taken. We plan to manufacture such a chip to put it into an intelligent camera.

For a smart camera, the development of an SoC is in progress. It includes a DMA Controller, a fast module as image input and an SD Card Reader for a fast swap of the *PE Array* program. Our objective is to manufacture such a SoC and to fit it into a camera in order to get a programmable high speed smart camera.

References

1. Loos, A., Schmidt, M., Fey, D., Gröbel, J.: Dynamically programmable image processor for compact vision systems. In: CIT (2010)
2. Baumann, D., Tinembart, J.: Designing mathematical morphology algorithms on FPGAs: An application to image processing. In: Proceedings of CAIP05, vol. 3691, pp. 562–569. Springer, Berlin (2005)
3. Charbouillot, S., Pérez, A., Fronte, D.: A programmable hardware cellular automaton: example of data flow transformation. VLSI Des. **2008**(2), 1–7 (2008)
4. Dorigo, M.: Optimization, learning and natural algorithms. PhD thesis, Politecnico di Milano, Italy (1992)
5. Draper, B.A., Beveridge, J.R., Bohm, A.P.W., Ross, C., Chawathe, M.: Accelerated image processing on FPGAs. IEEE Trans. Image Process. **12**(12), 1543–1551 (2003)
6. Faraday: MEMAKER Overview Fsd0k_a_sh (2008)
7. Fatemi, H., Kleihorst, R., Corporaal, H.: Real-time face recognition on a smart camera. In: Proceedings of ACIVS 2003 (Advanced Concepts for Intelligent Vision Systems (2003)
8. Fey, D., Komann, M.: Bioinspired architecture approach for a one-billion transistor smart CMOS camera chip. In: Society of Photo-Optical Instrumentation Engineers (SPIE) Conference Series, vol. 6592 (2007)

9. Fey, D., Komann, M., Schurz, F., Loos, A.: An organic computing architecture for visual microprocessors based on Marching Pixels. In: ISCAS, pp. 2686–2689. IEEE Press, Los Alamitos (2007)
10. M.L. for Computer Science. Cam8: a parallel, uniform, scalable architecture for cellular automata experimentation (2001). http://www.ai.mit.edu/projects/im/cam8/
11. Gaisler, A.: GRLIB Ip Core User's Manual (2009)
12. Halbach, M.: Algorithmen und hardwarearchitekturen zur optimierten aufzählung von automaten und deren einsatz bei der simulation künstlicher kreaturen. PhD thesis, Technische Universität Darmstadt (2008)
13. Halbach, M., Hoffmann, R., Both, L.: Optimal 6-state algorithms for the behavior of several moving creatures. In: Cellular Automata. Lecture Notes in Computer Science, vol. 4173, pp. 571–581. Springer, Berlin (2006)
14. Kleihorst, R., Broers, H., Abbo, A., Ebrahimmalek, H., Fatemi, H., Corporaal, H., Jonker, P.: An SIMD-VLIW smart camera architecture for real-time face recognition. In: PRORISC 2003, 14th Workshop on Circuits, Systems and Signal Processing, Proceedings (2003)
15. Komann, M., Fey, D.: Marching pixels—using organic computing principles in embedded parallel hardware. In: PARELEC, pp. 369–373 (2006)
16. Komann, M., Fey, D.: Marching pixels—using organic computing principles in embedded parallel hardware. In: PARELEC '06: Proceedings of the International Symposium on Parallel Computing in Electrical Engineering, pp. 369–373. IEEE Comput. Soc., Washington (2006)
17. Liang, X., Jean, J., Tomko, K.: Data buffering and allocation in mapping generalized template matching on reconfigurable systems. J. Supercomput. **19**(1), 77–91 (2001)
18. Margolus, N.: Cam-8: a computer architecture based on cellular automata. J. Am. Math. Soc. **6**(1), 167–187 (1996)
19. Miao, W., Lin, Q., Zhang, W., Wu, N.-J.: A programmable SIMD vision chip for real-time vision applications. IEEE J. Solid-State Circuits **43**(6), 1470–1479 (2008)
20. Mosqueron, R., Dubois, J., Paindavoine, M.: High-speed smart camera with high resolution. EURASIP J. Embed. Syst. **2007**, 23 (2007)
21. Scheuermann, B., So, K., Guntsch, M., Middendorf, M., Diessel, O., El Gindy, H., Schmeck, H.: FPGA implementation of population-based ant colony optimization. Applied Soft Computing, 303–322 (2004)
22. Weimar, J.: Simulation with Cellular Automata. Logos Verlag, Berlin (1997)

M. Reichenbach (✉) · R. Seidler · D. Fey · B. Pfundt
Chair for Computer Architecture, Department of Computer Science,
Friedrich-Alexander-Universität Erlangen-Nürnberg, Erlangen, Germany
e-mail: marc.reichenbach@informatik.uni-erlangen.de

R. Seidler
e-mail: ralf.seidler@informatik.uni-erlangen.de

D. Fey
e-mail: dietmar.fey@informatik.uni-erlangen.de

B. Pfundt
e-mail: benjamin.pfundt@informatik.uni-erlangen.de

Chapter 2.5
Multi-objective Intrinsic Evolution of Embedded Systems

Paul Kaufmann and Marco Platzner

Abstract The evolvable hardware paradigm facilitates the construction of autonomous systems that can adapt to environmental changes, degrading effects in the computational resources, and varying system requirements. In this article, we first introduce evolvable hardware, then specify the models and algorithms used for designing and optimising hardware functions, present our simulation toolbox, and finally show two application studies from the adaptive pattern matching and processor design domains.

Keywords Evolvable hardware · Evolutionary algorithms · Automatic definition of reusable functions (ADF) · Cartesian genetic programming (CGP) · Multi-objective optimisation (MOEA) · Evolvable caches · Autonomous systems

1 Evolvable Hardware—An Introduction

In the last decades, *natural computing* methods which take problem solving principles from nature have gained popularity. Among others, natural computing includes evolutionary computing. Evolutionary computing covers population-based, stochastic search algorithms inspired by principles from evolution theory. An evolutionary algorithm tries to solve a problem by keeping a set (population) of candidate solutions (individuals) in parallel and improving the quality (fitness) of the individuals over a number of iterations (generations). To form a new generation, genetically-inspired operators such as crossover and mutation are applied to the individuals. A fitness-based selection process steers the population towards better candidates.

Evolvable hardware (EHW) denotes the combination of evolutionary algorithms with reconfigurable hardware technology and run-time optimisation to construct self-adaptive and self-optimising hardware functions (circuits). The term evolvable hardware was coined by de Garis [1] and Higuchi [2] in 1993. The essence of EHW is the usage of optimisation algorithms during the run-time of a system to adapt its functionality to time-variant conditions. EHW is typically applied to two classes of hardware functions. The first class comprises functions with a performance that is dependent on the input data distribution, thus opening up the need for self-adaptation. Functions that should withstand partial defects of the computational resources or react to changes in the resources form the second class of EHW

C. Müller-Schloer et al. (eds.), *Organic Computing — A Paradigm Shift for Complex Systems*, 193–206
Autonomic Systems, DOI 10.1007/978-3-0348-0130-0_12, © Springer Basel AG 2011

compliant functions. Common to both classes is that the optimal solution is either unknown, too cumbersome to compute a-priory, or too complex to be formalised as a consolidated function.

The long-term goal of our work is the development of autonomous embedded systems that implement hardware functions characterised by their functional quality and resource demand [3]. To this, we rely on two concepts to achieve a flexible adaptation: Firstly, an *intrinsic evolutionary search process* adapts the system to slow changes in the environment. Secondly, radical changes in available resources are compensated by replacing the operational circuit with a *pre-evolved alternative* which meets the new resource constraints. To this end, we store at any time an approximated Pareto front of circuit implementations.

In this article, we first discuss models and algorithms for the evolution of hardware functions and then present the MOVES toolbox for development and simulation of EHW. Finally, we focus on two applications of EHW.

2 Models and Algorithms

In this section, we review models and algorithms for evolving hardware. In particular, we discuss cartesian genetic programs (CGP) for representing digital circuits and their extension to automated module creation and multi-objective optimisation. Further, we point to the inherent trade-off between an efficient evolution and the effort required to map evolved circuits to real hardware.

2.1 Cartesian Genetic Programs

A Cartesian Genetic Program (CGP) is a structural hardware model that arranges logic cells in a two-dimensional geometric layout [4]. Formally, a CGP model consists of $n_c \times n_r$ combinational logic blocks, n_i primary inputs, and n_o primary outputs. A logic block has n_n inputs and implements one out of n_f different logic functions of these inputs. While the primary inputs and outputs can connect to any logic block input and output, respectively, the connectivity of the logic block inputs is restricted. The input of a logic block at column c may only connect to the outputs of blocks in columns $c-l, \ldots, c-1$ as well as to the primary inputs. The levels-back parameter l restricts wiring to hardware-friendly local connections. More importantly, as only feed-forward connections are allowed, the creation of combinational feedback loops is avoided. Figure 1a shows an example for a CGP model together with its parameters. The model in this example has five columns, four rows, four primary inputs, and two primary outputs.

Mutation is the commonly used evolutionary operator for CGP. Often formalised as a one-point operator, mutation changes a gene which encodes a node's function and input wiring with a certain probability. The straight-forward implementation of a crossover operator acts on the geometrical structure of the chromosome

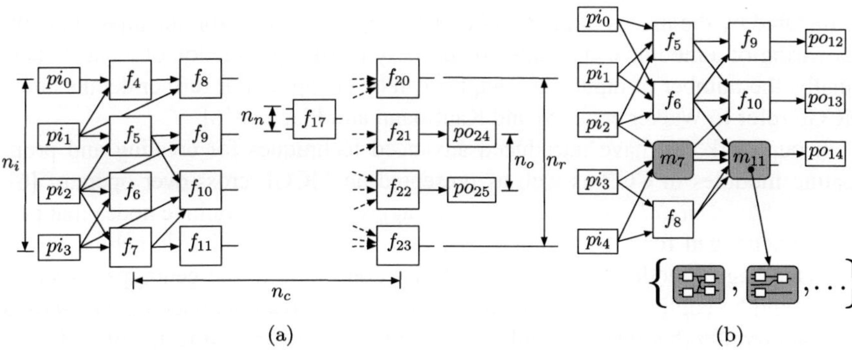

Fig. 1 (**a**) Cartesian genetic programming (CGP) model and (**b**) its extension to the automatic search of reusable sub-functions

exchanging nodes and preserving their wiring. An *n-point crossover* divides the parents' chromosomes into $n + 1$ parts based on the numbering of the nodes. The child's chromosome is then constructed by alternatively selecting partial chromosomes from the parents.

A crucial property of CGP is that nodes which do not contribute to the primary outputs remain in the chromosome and might be propagated through the generations. This property, termed *neutrality*, has been shown to improve significantly the convergence of the search process [4], as it preserves possibly useful sub-functions that can be reconnected to the genotype's active structure by few mutations.

2.2 Modular CGP

In order to improve evolvability of functions with potentially hierarchically-structured solutions, Walker et al. [5] applied the automatic search and definition for reusable sub-functions (ADF) to CGP. The methodology tries to extend a genotype's alphabet by adding more complex nodes to it. The new nodes, denominated as modules, are composed from sub-programs of candidate solutions and are subject to evolutionary selection pressure. The goal behind ADFs is the automatic search of an appropriate functional set for a distinct application. Additionally, ADF aims at the automatic increase of the functional complexity of the representation model. This allows for the evolution to act on a more complex and expressive level, thereby reducing an EA's computational effort.

Figure 1b illustrates a potential modular CGP (MCGP) phenotype structure. Along with the regular population of candidate solutions, MCGP propagates a set of modules through the evolution, dynamically allocating and releasing them by *compress* and *expand* operators. The probability of module creation is lower than that of module expansion, providing pressure towards modules contributing to more successful candidate solutions. When a regular mutation operator hits a module node, it selects randomly from five different actions: The module's internal function can be mutated by the means of the standard CGP mutation operator, the input and output wiring of a module can be altered similarly to wiring mutation of a basic node. Finally, the number of inputs and outputs can be changed. For further details about MCGP refer to Walker et al. [5] and Kaufmann and Platzner [6].

In our work, we have introduced advanced techniques for creating and propagating modules in CGP as well as presented an MCGP cross-over operator [6]. Our *age-based module creation* prefers the aggregation of primitive nodes that persisted unchanged for a large number of generations. The rationale behind this is that such "aged" nodes directly or indirectly contribute to the genotype's success and should be composed into a module. Further, we have implemented *cone-based module creation* that forms modules out of primitive nodes that form cones. Cones are a widely-used concept in circuit synthesis, especially in the area of lookup-table mapping for FPGAs, capsuling functionally related elements as a reusable entity. We have evaluated our novel techniques and compared them to the module creation approach presented in previous work. The results demonstrate the effectiveness of age-based module creation. Cone-based module creation is even more effective but only for regularly structured multiple output circuits such as multipliers. Finally, we have outlined a crossover operator that selects a cone consisting of both primitive nodes as well as modules in a source chromosome and copies this cone into a destination chromosome. Using this novel crossover operator, we are also able to apply multi-objective optimisers to MCGP as crossover allows for intra-population information exchange.

2.3 Multi-objective Optimisation Using CGP

The motivation for multi-objective evolutionary optimisation (MOEA) of hardware lies in the desired ability of an autonomous system to react quickly to changes in the resources by instantiating an appropriately sized solution. This can be achieved by exploiting a modern, Pareto-based MOEA, which optimises concurrently a set of mutually non-dominating solutions. To verify this concept, we have added circuit area and speed as objectives to be optimised [7]. However, multi-objective optimisation of CGP circuits shows a bad convergence behaviour which might be explained by the fact the modern MOEAs are targeted at global optimisation. In contrast, previously known good CGP optimisation algorithms are single-objective local optimisers. To successfully employ a multi-objective heuristic to CGP optimisation, we have extended a state-of-the-art MOEA, SPEA2 [8], to favour a subset of objectives. The resulting algorithm performs similarly to a single-objective optimiser without showing significant negative effects on the secondary objectives. We coined the new algorithm Turtle SPEA2 (TSPEA2) [7, 9]. To improve the run-time on an embedded system, we have implemented the computationally-demanding methods for preserving Pareto-front diversity as an FPGA circuit [10]. Additionally, we have developed a periodisation scheme for single- and multi-objective EAs to combine global and local search techniques [11]. We have shown that periodisation can significantly improve the quality of the evolved Pareto-front approximation.

2.4 Challenges of CGP

The main challenge of the CGP model when applying to real-world applications is scalability. Similar to other EA variants, disproportion in the granularity of the representation model functional blocks and application-inherent functional granularity often results in excessive optimisation effort. As CGP in its original version evolves solutions at the abstraction level of Boolean gates and single wires, it struggles to cope with applications using numerical values and the corresponding mathematical functions. In the last years, a number of approaches have been presented that address this problem. One of them is function-level evolution which uses node functions of coarser granularity and buses instead of single bit wires. We have done preliminary work on the implementation of a coarse-granular CGP using Virtex-5 DSP48E blocks as functional nodes. An algorithmic approach to improve the evolvability of CGP is to enable EAs to create hierarchical solutions by automatically seeking reusable functions [6].

3 Development and Simulation Tools

We have designed the MOVES toolbox [3, 7, 12] for evolutionary design of digital circuits. The toolbox comprises a framework of different digital logic representation models, single and multi-objective optimisers and a set of evolutionary operators. This enables us to easily deal with the setup, control, visualisation, analysis, and distribution of experiments. The overall framework is developed in Java and is, thus, platform-independent. Additionally, for experiments on embedded systems without full-fledged operating systems, some parts such as the CGP representation model and variants of single and multi-objective optimisers are also available in C without using dynamic memory allocation.

The key feature of the MOVES framework is the separation of the different functionalities required for experimenting with hardware evolution. Consequently, the modules of the framework are divided into two major groups, modules that constitute the evolutionary optimisation techniques and modules that serve the experimentation process, including experiment setup and control, statistic analysis, and visualisation.

An excerpt form the MOVES toolbox structure is shown in Fig. 2 and includes the hardware *representation models*, the *evolutionary algorithms*, and the *evolutionary operators*. This separation is suitable for all population-based optimisation techniques that apply evolutionary operators, e.g. genetic programming, genetic algorithms, and evolutionary techniques. Within the representation models, the framework implements the regular CGP, its extension to ADFs—the MCGP, and a coarse-granular CGP model, based on Xilinx Virtex-5 DSP48E blocks as functional elements. The evolutionary algorithms comprise regular GA, $(\mu\{^+_,\}\lambda)$ ES, NSGAII, SPEA2, TSPEA2, μGA, OMOEA, and IBEA2 in the hyper-volume and the ϵ variants. The evolutionary operators can be formalised in a general form, e.g. selection operators, or in a form specific to the representation model, e.g. the fitness

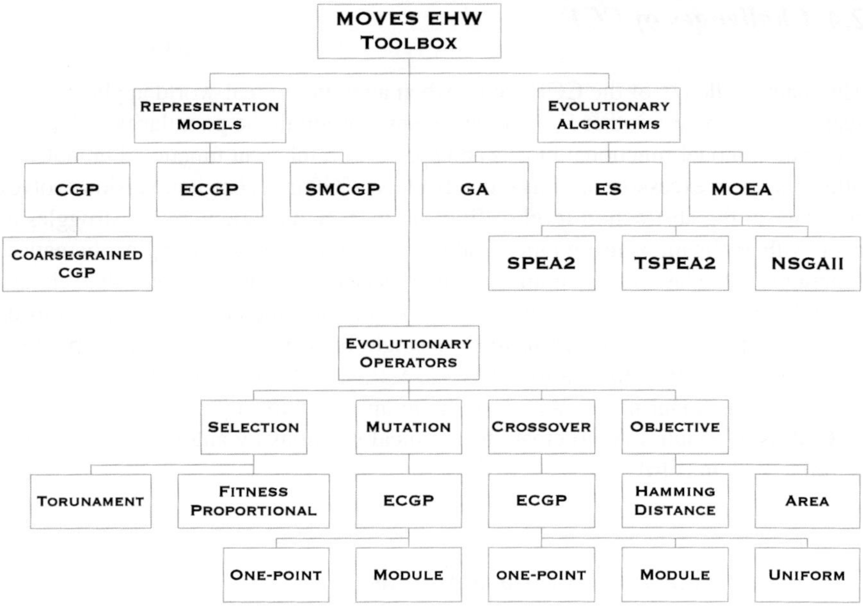

Fig. 2 Excerpt from the MOVES evolvable hardware toolbox structure

evaluation, mutation, and crossover operators. Along with the known mutation and crossover operators, the MOVES framework also implements our novel cone-based and age-based selection and crossover techniques [6].

Within the MOVES framework, an experiment is defined by two human-readable configuration files; one specifies the evolutionary optimiser and another one the experiment setup. The evolutionary optimiser configuration includes the chosen representation model, evolutionary operators, and evolutionary algorithm. The experiment setup configuration comprises termination conditions, visualisation settings, and the logging frequency. The configuration can be provided as regular text files or, alternatively, be entered via the framework's graphical user interface.

Experiments can be run interactively or in batch mode. In the interactive mode, the user can pause, resume, or stop the simulation at any time. Generally, the parameters controlling the experiment setup can be modified during the experiment. For example, the experiment can be executed step-wise or continuously until some of the termination conditions are reached. The user is free to switch between these modes at any time. It is also possible to save the current search state and reload it later on to analyse it. These features are extremely useful to debug and verify new representation models and their corresponding operators, and to tune the parameters of the evolutionary algorithm. The batch mode is used for an unattended simulation. Statistical data can be gathered during the experiment and exported to a text file for later processing.

There are two visualisation tools, the visualisation of the evolved candidate solutions and the visualisation of the evolutionary optimisation process. When visualising a genotype, the user can modify the viewed data, e.g. to manipulate input data or even change the circuit's structure and function. The standard visualisation of the evolutionary optimisation process displays the progress of the best fitness and the population's average fitness over the generations. This is most useful for single-objective optimisers. In experiments with multi-objective optimisers, we can display the progress of all individual fitness values, and the two-dimensional projections of the Pareto-fronts. The visualisations can be done either during an experiment run or offline using previously saved experiment log streams.

For an evolvable hardware experiment, usually dozens of simulation runs are required to evaluate different parameter sets. As evolutionary algorithms are stochastic optimisation methods, several runs with different random number seeds need to be conducted to derive the average behaviour. Such experiments can be tedious to configure and take a very long runtime. The single experiments, however, are independent of each other and amenable to parallel execution. The MOVES framework is able to automatically create a set of experiments where parameters are varied in specified intervals and with defined step sizes, and to execute all simulations as batch jobs on a compute cluster. We employ the grid software Condor that distributes the jobs on the computing nodes in the cluster, monitors the node's activities, and relocates the jobs if it becomes necessary.

4 Applications

To demonstrate our research on evolvable hardware, we select two applications covering the embedded and the high-performance-computing worlds. The first part of this section presents experiments on the classification accuracy of a run-time reconfigurable FPGA pattern matching architecture [13]. At this, the focus is raised towards classification accuracy behaviour and recovery, when dynamically changing the amount of on-chip resources used by the architecture. The second part shows a way to optimise the memory-to-cache address mapping function by the means of EHW. Interposing a small reconfigurable array of look-up tables (LUTs) between the CPU's load/store unit and the cache logic allows to improve the overall execution time of a program.

4.1 Flexible EHW Pattern Matching Architectures

The Functional Unit Row (FUR) architecture for classification tasks was first presented by Glette in [14, 15]. FUR is an architecture tailored to online evolution combined with fast reconfiguration. To facilitate online evolution, the classifier architecture can be reconfigured through partial reconfiguration and the behaviour can be controlled through configuration registers. Figure 3a shows the general organisation of the classifier architecture [16, 17]. For C categories the FUR architecture consists of C Category Detection Modules (CDMs). A majority vote on the outputs

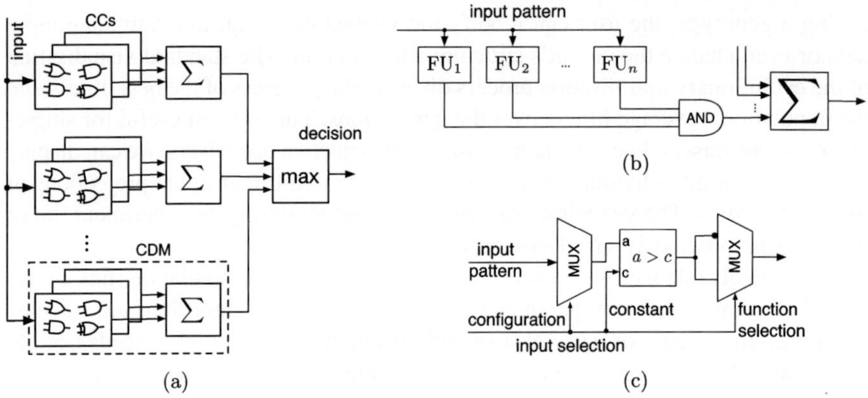

Fig. 3 (**a**) Functional Unit Row (FUR) architecture; (**b**) Category Classifier (CC); (**c**) Functional Unit (FU)

of the CDMs defines the FUR architecture decision. In case of a tie, the CDM with the lower index wins. Each CDM contains M Category Classifiers (CCs), basic pattern matching elements evolved from different randomly initialised configurations and trained to detect the CDM's category. A CDM counts the number of activated CCs for a given input vector, thus the CDM output varies between 0 and M.

In [13], we define a single CC as a row of Functional Units (FU), shown in Fig. 3b. The FU outputs are connected to an AND gate such that in order for a CC to be activated all FU outputs have to be 1. Each FU row is evolved from an initial random bitstream, which ensures a variation in the evolved CCs.

As depicted in Fig. 3c, an FU selects a single value from the input vector and compares it to a constant. While any number and type of functions could be imagined, Fig. 3c illustrates only two functions for clarity. Through experiments, *greater than* and *less than or equal* have shown to work well, and intuitively this allows for discriminating signals by looking at the different amplitudes.

The FUR architecture is parametrised by three values: the number of categories, FU rows in a CDM, and FUs in a FU row. We assume the numbers of categories and FUs in a FU row as constants, reconfiguring the number of FU rows in a CDM. For a sequence $I = \{i_1, i_2, \ldots, i_k\}$, we evolve a FUR architecture having i_j FUs per CDM, then switching to i_{j+1} FUs per CDM and re-evolving the architecture without flushing the configuration evolved so far.

For our investigations, we rely on the UCI machine learning repository [18] and, specifically, on the Thyroid benchmark. The Thyroid data set splits into three groups with cardinalities of 6.666, 166 and 368 samples. To evolve a FUR classifier, we employ a $1+4$ ES scheme. With a mutation operator changing three genes in every FU row, the configuration for a complete FUR architecture is evolved during a single ES run. In preparation for the experiments on the reconfigurable FUR architecture, we investigate the FUR's general performance and over-fitting effects by evaluating it on a set of useful FU rows per CDM and FUs per FU row configurations. To this, we evaluate the FUR-architecture on combinations of 2 to 20 FUs per FU row

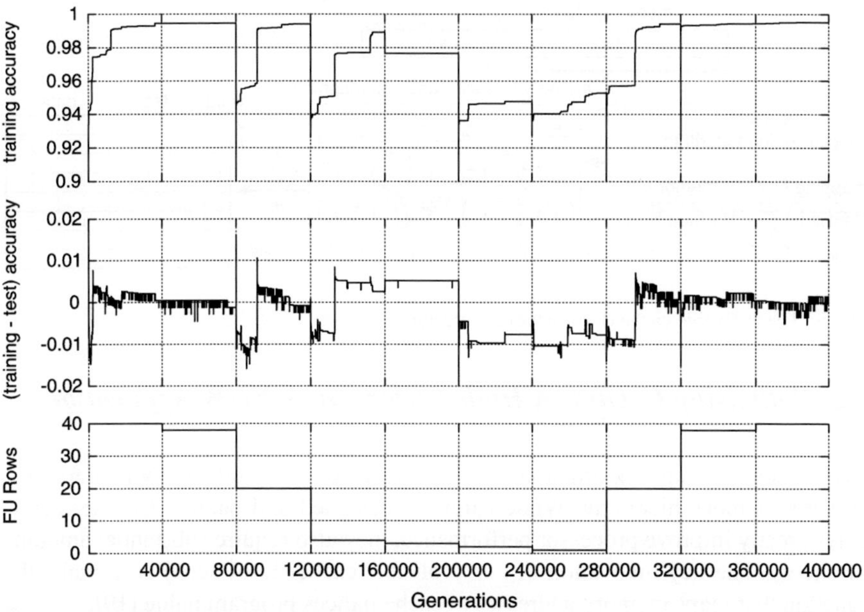

Fig. 4 Reconfigurable Thyroid benchmark: Changing the classifier's resources (number of FU rows) during the optimisation run. The *first diagram* shows the training accuracy, the *second diagram* the difference between the training and test accuracy and the *last diagrams* shows the changes in the FU rows

and 2 to 80 FU rows. In these experiments, the FUR architecture shows error rates which are very close to the error rates of the best known conventional classification algorithms [13].

Figure 4 visualises the FUR's classification behaviour under changes in the available resources while being under optimisation. We execute a single experiment and configure a FUR architecture with 4 FUs per FU row and change the number of FUs every 40 000 generations. We split the data set into disjoint training and test sets and start the training of the FUR classifier with 40 FU rows. Then, we gradually change the number of employed FU rows to 38, 20, 4, 3, 2, 1, 20, 30, 40 executing altogether 400 000 generations. We observe the following: the training accuracy drops significantly for almost any positive and negative change in the number of FU rows and recovers subsequently. The recovery rate of the test accuracy depends on the amount of FU rows. While for periods with few FU rows the recovery rate is slow, for periods with 20 and more FU rows the evolutionary process manages to recover the test accuracy much faster. Additionally, for configurations with very few FU rows, the test accuracy begins to deteriorate. This can be observed in Fig. 4 at generations 120 000 to 280 000.

In summary, as long as the FUR configuration contains enough FU rows, the FUR's test accuracy behaviour is stable during reconfigurations. Additionally, more FU rows leverage faster convergence.

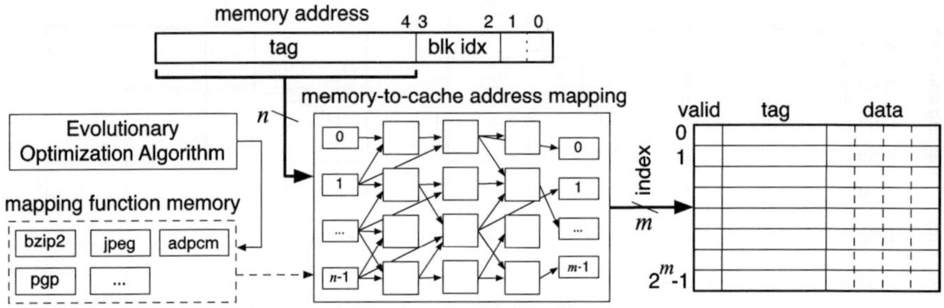

Fig. 5 The evolvable cache (EvoCache) architecture

4.2 Optimising Caches: A High-Performance EHW Application

Cache memories are important and well-investigated elements of any modern processor's memory hierarchy. While carefully designed and balanced cache hierarchies greatly improve processor performance, they also require substantial amounts of energy. The key innovation of our evolvable cache (EvoCache) is to make the function that maps memory addresses to cache indices programmable [19].

The EvoCache approach is presented in Fig. 5. A very small reconfigurable logic fabric implements a hashing function that maps a part of a memory address to a cache line index. The hashing function is optimised to achieve a low overall execution time for a specific application. The algorithmic methods for optimisation originate in the EHW domain. Our architecture provides a mapping function memory that can store several configurations for the reconfigurable logic fabric, which allows for quickly switching to different memory-to-cache address mappings. To prevent aliasing, i.e. storing several potentially dirty copies of the same physical address at different indices in the cache, we flush the cache when a new mapping is activated.

Including EvoCaches into a processor architecture will increase the logic area, the hit time, and the overall number of memory cells for the cache. The increase in logic area is due to the reconfigurable fabric itself which is assumed to be small as the fabric comprises only a handful of look-up tables (LUTs). Additionally, we require a mapping function memory to store the configurations for the logic fabric. The size of a configuration is architecture dependent. The architecture used for this case study comes with a configuration size of 151 bytes. The increase of the cache size is due to the fact that the flexibility in the hashing function requires us to store the full address excluding block and byte offsets as tags in the cache. The additional overhead depends on the actual cache configuration. For example, a conventional 4-way set associative cache of 16 KByte data with block size of two words for a byte-addressable architecture with 32 bit addresses comes with an overhead of 25.56%, where the overhead includes for each cache block the valid bit and the tag. Switching to an EvoCache of same data size and organisation increases the overhead to 34.88%. We think this overhead is bearable since today most processor designs are not restricted by silicon area but by performance and performance per energy.

The increase in hit time is more critical. The additional delay depends strongly on the depth of the LUT network. This depth can be restricted in the optimisation process to satisfy timing constraints. Moreover, for many embedded processors with clock frequencies well below one GHz, the pressure on the timing is moderate. High-performance processors, on the other hand, have several levels of cache where only the first level is optimised for hit time. Here, the EvoCache approach can still be applied to higher level caches.

For the experiments, we leverage our MOVES toolbox [12], which allows us to generate a set of jobs for fitness evaluation and distribute them on a compute cluster. For system simulation we rely on SimpleScalar [20] as it is easily extensible and it models a variant of the widely-used MIPS instruction set architecture. SimpleScalar allows us to establish a fine-grained energy metric, based on the memory and cache access patterns. To this, our energy estimation model splits the energy demand into a static and a dynamic part. We derive the static and access energies from the CACTI cache model, the standard V58C2256 DDR SDRAM module, and a 200 MHz ARM at 90 nm, respectively.

To evaluate the EvoCache concept, we have configured a processor and its memory hierarchy in a configuration similar to those of current ARM processors [21]. The configuration includes a split first level cache and a unified second level cache. The L1 caches are 2-way associative with a hit latency of one cycle, 64 sets and a block size of 16 bytes. The L2 cache has an associativity of four ways with a hit latency of 6 cycles, 128 sets and a block size of 32 bytes. The memory bus between the L2 cache and the external memory is 8 bytes wide. The external memory shows an access time of 18 cycles and a 2-cycle delay for consecutive data transfers in burst mode. Hence, the miss penalty for the L2 cache amounts to 24 cycles. Using this configuration, a conventional cache system for a byte-addressable architecture with 32 bit addresses has a 22 bit tag and a 6 bit index for the L1 caches and a 20 bit tag and 7 bit index for the L2 cache, respectively. For an EvoCache, the original tags and indices merge into a single tag of 28 and 27 bits for the L1 and L2 caches, respectively. We have evolved mapping functions for two optimisation scenarios. In the first optimisation scenario, only the first level caches (LI:I and L1:D) are EvoCaches with evolved mapping functions while in the second scenario all three caches receive evolved mapping functions. We simulate the execution of two benchmarks, `bzip2` (version 1.0.4) and `jpeg` (version 6a). For each combination of benchmark and optimisation scenario, we first evolve a mapping function on a training data set. This optimisation step has been repeated for 16 times to compute the average behaviour. Then we evaluate EvoCaches using the best evolved `bzip2` and `jpeg` circuits on a large, diverse set of test data disjoint to the training data, This time, we additionally log the miss rates and the estimated energy consumptions.

Figure 6 summarises EvoCaches generalisation results. The following observations can be made here: Compared to a conventional cache of equal size, EvoCaches generalise well and deliver up to 10.98% execution time improvement and up to 10.70% reduction in energy for the `bzip2` benchmark. The `jpeg` benchmark gains even higher improvements with up to 14.21% in the execution time and up

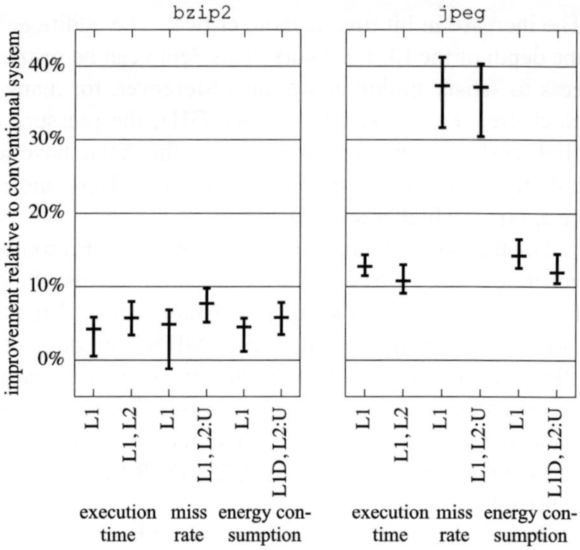

Fig. 6 EvoCache generalisation performance for `bzip2` and `jpeg`

to 16.43% in the energy consumption. Interestingly, `bzip2` profits from optimisation of L1:I, L1:D, and L2:U caches, while `jpeg` suffers from the optimisation of the second level cache. Noteworthy for the `jpeg` benchmark is the disproportion in the miss-rate gain. While the execution time improves by roughly 11% to 12%, the miss-rate gains about 35% improvement. This demonstrates that for multiple levels of cache (or sophisticated processor architectures) the total miss rate is not necessarily a suitable metric for quantitatively determining a performance improvement.

5 Conclusion

This article gives a short introduction to evolvable hardware, presents an overview over our models and tools for evolutionary digital circuit design and concludes with two case studies. In the first case study, we leverage the FUR classifier architecture for creating evolvable hardware systems that can cope with fluctuating resources. We demonstrate that the FUR's generalisation performance is robust to changes in the available resources as long as a certain amount of FU rows is present in the system. Furthermore, the FUR's capability to recover from a change in the available resources benefits from additional FU rows. In the second case study, we present the EvoCache concept which relies on two main ideas. First, the memory-to-cache-mapping function is implemented by a small reconfigurable logic fabric. Second, the function is optimised by an evolutionary algorithm with the goal to achieve a minimal overall execution time with respect to a specific application. We evaluate the concept on the `bzip2` and `jpeg` benchmarks. Compared to conventional caches, we observe runtime improvements of up to 10.98% for `bzip2` and up to 14.31%

for `jpeg` and energy reductions of up to 10.70% for `bzip2` and up to 16.43% for `jpeg`.

References

1. de Garis, H.: Evolvable hardware: genetic programming of a Darwin machine. In: Intl. Conf. on Artificial Neural Nets and Genetic Algorithms, pp. 441–449. Springer, Berlin (1993)
2. Higuchi, T., Niwa, T., Tanaka, T., Iba, H., de Garis, H., Furuya, T.: Evolving hardware with genetic learning: a first step towards building a Darwin machine. In: From Animals to Animats, pp. 417–424. MIT Press, New York (1993)
3. Kaufmann, P., Platzner, M.: Toward self-adaptive embedded systems: multi-objective hardware evolution. In: Architecture of Computing Systems (ARCS). LNCS, vol. 4415, pp. 199–208. Springer, Berlin (2007)
4. Miller, J., Thomson, P.: Cartesian genetic programming. In: European Conf. on Genetic Programming (EuroGP), pp. 121–132. Springer, Berlin (2000)
5. Walker, J.A., Miller, J.F.: Evolution and acquisition of modules in Cartesian genetic programming. In: European Conf. on Genetic Programming (EuroGP), May. LNCS, vol. 3003, pp. 187–197. Springer, Berlin (2004)
6. Kaufmann, P., Platzner, M.: Advanced techniques for the creation and propagation of modules in Cartesian genetic programming. In: Genetic and Evolutionary Computation (GECCO), pp. 1219–1226. ACM, New York (2008)
7. Kaufmann, P., Platzner, M.: Multi-objective intrinsic hardware evolution. In: Intl. Conf. Military Applications of Programmable Logic Devices (MAPLD) (2006)
8. Zitzler, E., Laumanns, M., Thiele, L.: SPEA2: Improving the Strength Pareto Evolutionary Algorithm. Technical Report 103, ETH Zurich (2001)
9. Knieper, T., Defo, B., Kaufmann, P., Platzner, M.: On robust evolution of digital hardware. In: Biologically Inspired Collaborative Computing (BICC). IFIP International Federation for Information Processing, vol. 268, pp. 2313. Springer, Berlin (2008)
10. Schumacher, T., Meiche, R., Kaufmann, P., Lübbers, E., Plessl, C., Platzner, M.: A hardware accelerator for k-th nearest neighbor thinning. In: Proc. Intl. Conf. on Engineering of Reconfigurable Systems & Algorithms (ERSA), pp. 245–251. CSREA Press (2008)
11. Kaufmann, P., Knieper, T., Platzner, M.: A novel hybrid evolutionary strategy and its periodization with multi-objective genetic optimizers. In: IEEE World Congress on Computational Intelligence (WCCI), Congress on Evolutionary Computation (CEC), pp. 541–548. IEEE, New York (2010)
12. Kaufmann, P., Platzner, M.: MOVES: a modular framework for hardware evolution. In: Adaptive Hardware and Systems (AHS), pp. 447–454. IEEE, New York (2007)
13. Knieper, T., Kaufmann, P., Glette, K., Platzner, M., Torresen, J.: Coping with resource fluctuations: the run-time reconfigurable functional unit row classifier architecture. In: Intl. Conf. on Evolvable Systems (ICES). LNCS, vol. 6274, pp. 250–261. Springer, Berlin (2010)
14. Glette, K., Torresen, J., Yasunaga, M.: An online EHW pattern recognition system applied to face image recognition. In: Applications of Evolutionary Computing (EvoWorkshops). LNCS, vol. 4448, pp. 271–280. Springer, Berlin (2007)
15. Torresen, J., Senland, G., Glette, K.: Partial reconfiguration applied in an on-line evolvable pattern recognition system. In: NORCHIP 2008, pp. 61–64. IEEE, New York (2008)
16. Glette, K., Gruber, T., Kaufmann, P., Torresen, J., Sick, B., Platzner, M.: Comparing evolvable hardware to conventional classifiers for electromyographic prosthetic hand control. In: Adaptive Hardware and Systems (AHS), pp. 32–39. IEEE, New York (2008)
17. Glette, K., Torresen, J., Kaufmann, P., Platzner, M.: A comparison of evolvable hardware architectures for classification tasks. In: Intl. Conf. on Evolvable Systems (ICES). LNCS, vol. 5216, pp. 22–33. Springer, Berlin (2008)

18. Asuncion, A., Newman, D.J.: UCI Machine Learning Repository. University of California, Irvine, School of Information and Computer Sciences (2007)
19. Kaufmann, P., Plessl, C., Platzner, M.: EvoCaches: Application-specific adaptation of cache mappings. In: Adaptive Hardware and Systems (AHS), pp. 11–18. IEEE, New York (2009)
20. Austin, T., Larson, E., Ernst, D.: SimpleScalar: An infrastructure for computer system modeling. IEEE Comput., **35**(2), 59–67 (2002)
21. ARM: ARM10E Processor Family. http://www.arm.com/products/CPUs/families/ARM10EFamily.html (2010)

P. Kaufmann (✉) · M. Platzner
Department of Computer Science, University of Paderborn, Warburger Str. 100,
33098 Paderborn, Germany
e-mail: paul.kaufmann@upb.de

M. Platzner
e-mail: platzner@upb.de

Chapter 2.6
Organisation-Oriented Chemical Programming

Naoki Matsumaru, Peter Kreyssig, and Peter Dittrich

Abstract *Chemical information processing* possesses a variety of valuable properties, such as robustness, concurrency, fault-tolerance and evolvability. However, it is difficult to predict and program a chemical system because the computation emerges as a global phenomenon from microscopic reactions. For programming chemical systems a theoretical method to cope with that *emergent behaviour* is desirable. Here we will review *design principles* for chemical programs. We focus on programs that should compute a qualitative and not a quantitative result. The design principles are based on chemical organisation theory, which defines a chemical organisation as a closed and self-maintaining set of molecular species. The fundamental assumption of so-called *organisation-oriented programming* is that computation should be understood as a movement between chemical organisations. In this case we expect that the resulting system is more robust and fine-tuning of the kinetic laws will be less important. As examples for the usage of this design method we show a logic gate and a solution to the maximal independent set problem implemented as artificial chemistries.

Keywords Self-organisation · Maximal independent set problem · Chemical computing · Unconventional computing · Artificial chemistry · Sensor networks · Emergent control

1 Introduction

By employing a large number of simple components interacting with each other in an orchestrated way, biological systems invented a variety of information processing mechanisms, which are robust, self-organising, adaptable, decentralised, asynchronous, fault-tolerant and evolvable. This principle of biological information processing has been exploited to cope with the fast-growing complexity of technical information processing systems [28, 35, 36]. Since all known life forms process information using chemical processes [20], the chemical reaction metaphor has been proposed as a source of inspiration [2, 10]. Using chemical reactions for formal computations has initially been suggested by Banâtre and Métayer [2]. In their GAMMA system [3], a chemical reaction is defined as a rewriting operation on a multiset, mimicking a well-stirred reaction vessel. In order to capture the spatial

C. Müller-Schloer et al. (eds.), *Organic Computing — A Paradigm Shift for Complex Systems*, 207–220
Autonomic Systems, DOI 10.1007/978-3-0348-0130-0_13, © Springer Basel AG 2011

context of chemical systems, chemical rewriting systems have been extended to the chemical abstract machine (CHAM) [6]; P-Systems [31] stressing the importance of membranes; and MGS [14] allowing arbitrary topologies [4]. However, there is still a lack of knowledge of how to actually design chemical computing algorithms.

The difficulty of engaging the chemical reaction process for computing is that the solution appears as an emergent global behaviour based on the manifold local interactions [5]. An emergent behaviour of biological systems is simulated by combining simple biochemical signalling pathways [7]. As recently demonstrated by Tsuda et al. [34] with a slime mould in the genus of Physarum, the discrepancy between local and global behaviour may also be problematic in a practical case. A satisfying theory of emergence is lacking [27].

This article reviews the contributions towards the establishment of a theoretical analysis of the emergent behaviour in chemical computing. We suggest chemical organisation theory [11, 29, 33] as a tool helping to construct (program) and analyse (describe and understand) chemical computing systems.

Inspired by Fontana and Buss [13], we define a (chemical) organisation as a set of molecular species that is (algebraically) closed and (stoichiometrically) self-maintaining [11]. It is important to note that when we talk about organisations, we abstract details like concentration levels or the spatial distribution of a chemical species. On this relatively high level of abstraction, a system state is characterised by the molecular species present only and we can describe the dynamics of a system more qualitatively, namely, as a movement between sets of species instead of a movement in a more complex state space [33].

In Sect. 2 we describe the concepts from chemical organisation theory needed here. As an example of chemical computing, an XOR logic gate is implemented using an (artificial) chemical reaction network with the help of chemical organisation theory in Sect. 3.1. A second example is given by a chemical algorithm for the solution of the maximal independent set problem in Sect. 3.2. In Sect. 4 we review three different ways of designing chemical programs guided by chemical organisation theory. Finally, in Sect. 5 we discuss the potential of the theory as a theoretical base for emergence analysis.

2 Chemical Reaction Networks, Chemical Organisation Theory, and Movement between Organisations

By a *chemical reaction network* we mean a pair $\langle \mathcal{M}, \mathcal{R} \rangle$ where $\mathcal{M}$ is a set and $\mathcal{R}$ is a subset of $\mathcal{P}_{mult}(\mathcal{M}) \times \mathcal{P}_{mult}(\mathcal{M})$. Here $\mathcal{P}_{mult}(\mathcal{M})$ is the set of multisets over $\mathcal{M}$. We call the elements of $\mathcal{M}$ *molecular species* and the elements of $\mathcal{R}$ *reactions* resembling the notions of chemistry. Reaction networks can also be described by Petri nets or multiset rewriting. We summarise the needed concepts of *chemical organisation theory* [11].

We fix a reaction network $\langle \mathcal{M}, \mathcal{R} \rangle$. By *applying* a reaction $(l, r) \in \mathcal{R}$ to a multiset over $\mathcal{M}$ we mean replacing the subset l by the subset r. We assume that the multiset is always large enough.

For $(l,r) \in \mathcal{R}$ we also write $l \longrightarrow r$ or

$$\sum_{m \in \mathcal{M}} l_m m \longrightarrow \sum_{m \in \mathcal{M}} r_m m$$

where we denote by $l_m, r_m \in \mathbb{N}_0$ the multiplicity of m in l, r respectively. This resembles notation from chemistry. Furthermore the *support* and the *product* of (l,r) are

$$\mathrm{supp}(l,r) := \{m \in \mathcal{M} \mid l_m > 0\}, \quad \mathrm{prod}(l,r) := \{m \in \mathcal{M} \mid r_m > 0\}.$$

Let $\mathcal{A}$ be a subset of $\mathcal{M}$. We define $\mathcal{R}_A$ by setting

$$\mathcal{R}_{\mathcal{A}} := \{(l,r) \in R \mid \mathrm{supp}(l,r) \subseteq \mathcal{A}\}.$$

The *stoichiometric matrix* $\mathbf{M}_{\mathcal{A}} \in \mathbb{R}^{|\mathcal{A}| \times |\mathcal{R}_{\mathcal{A}}|}$ for $\mathcal{A}$ is given by

$$(\mathbf{M}_{\mathcal{A}})_{a,(l,r)} = r_a - l_a, \quad a \in \mathcal{A}, (l,r) \in \mathcal{R}_{\mathcal{A}}.$$

Definition 1 A subset $\mathcal{A}$ of $\mathcal{M}$ is *closed* if for all reactions $(l,r) \in \mathcal{R}_A$ we have $\mathrm{prod}(l,r) \subseteq \mathcal{A}$, i.e. if $(\mathcal{A}, \mathcal{R}_{\mathcal{A}})$ is a reaction network.

$\mathcal{A}$ being closed means that by applying reactions from $\mathcal{R}_{\mathcal{A}}$ to multisets over $\mathcal{A}$ we do not get molecules outside $\mathcal{A}$.

Definition 2 A subset $\mathcal{A}$ of $\mathcal{M}$ is *semi-self-maintaining* if for every $a \in \mathcal{A}$ and $(l,r) \in \mathcal{R}_{\mathcal{A}}$ with $l_a - r_a > 0$ there is an $(l',r') \in \mathcal{R}_{\mathcal{A}}$ with $r'_a - l'_a > 0$.

$\mathcal{A}$ being semi-self-maintaining means that if a reaction application destroys a species, there is also a reaction producing this species.

Definition 3 A subset $\mathcal{A}$ of $\mathcal{M}$ is *self-maintaining* if there is a vector $v \in \mathbb{R}^{|\mathcal{R}_A|}$ with strictly positive entries such that $\mathbf{M}_{\mathcal{A}} v \in \mathbb{R}^{|\mathcal{A}|}$ has only non-negative entries.

$\mathcal{A}$ being self-maintaining means that applying reactions from $\mathcal{R}_{\mathcal{A}}$ at certain rates to a multiset over $\mathcal{M}$ does not reduce the number of molecules of any species of $\mathcal{A}$.

Definition 4 A subset of $\mathcal{M}$ is a *chemical (semi-)organisation* if it is closed and (semi-)self-maintaining.

The organisational analysis decomposes the reaction network given into overlapping sub-networks or organisations. We visualise the set of all organisations by a Hasse diagram in which the ordering is given by the subset relation, e.g., Fig. 1(A).

What we mean by *movement between chemical organisations* is described in the following. For the dynamics of the reaction networks, we assume the standard *law of mass action*, remarking that other kinetics can be used equally well. After assigning constant reaction rates to each reaction, we can derive a system of ordinary

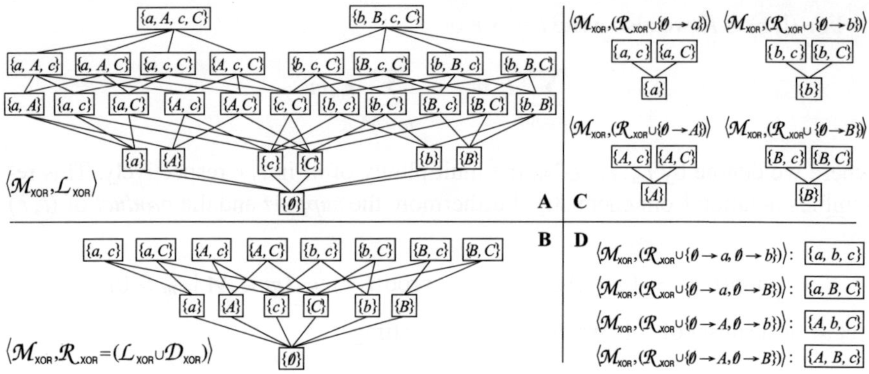

Fig. 1 Hierarchy of organisations for the chemical reaction network implementing an XOR logic gate. (**A**) The network consists only of the logical reactions $\mathcal{L}_{\mathrm{XOR}}$. (**B**) Destructive reactions $\mathcal{D}_{\mathrm{XOR}}$ are added to exclude contradictions. The resulting reaction network $\langle \mathcal{M}_{\mathrm{XOR}}, \mathcal{R}_{\mathrm{XOR}} \rangle$ implements the XOR logic gate without any input specified. (**C**) One input is defined by adding one influx reaction. (**D**) Adding the second input. The hierarchy of organisations collapses from (**A**) to (**D**), with the desired output as the only organisation left in (**D**). (Appeared in [25])

differential equations to describe the change in concentration of the species. The solution of the system of equations yields a trajectory in the space of species concentrations. For every point on this trajectory we can find a set of present species, i.e., the species with non-zero concentration, called the *abstraction* of this point. Following the trajectory we get a sequence of abstractions, i.e., subsets of the set of species. This also yields a "movement" between chemical organisations. The choice of the reaction rate constants only influences the quantitative and not the qualitative aspects of the movement.

The fundamental assumption of the so-called *organisation-oriented programming* is that computation should be understood as such a movement between chemical organisations. The underlying hypothesis is that when a computation can be explained like this, it is more robust and fine-tuning of the kinetics is less important, since the organisations only depend on stoichiometric information. For example, classical DNA Computing [1] can also be understood in terms of chemical organisation theory. For each solution there is at least one organisation and the experimental steps assure that the system will end up in such an organisation.

3 Examples

3.1 A Chemical XOR—*Reaction Network, Organisations, and Dynamics*

To demonstrate how chemical organisation theory can be used for chemical computing, an (artificial) chemical reaction network is designed to implement an XOR logic gate. We follow the general recipe described by Matsumaru and Dittrich [23].

The XOR logic gate is defined as a set of three boolean variables $\{\mathsf{c}, \mathsf{a}, \mathsf{b}\}$ and a set of one boolean function $\{F_{\mathsf{c}}\}$ where the function is $\mathsf{c} = F_{\mathsf{c}}(\mathsf{a}, \mathsf{b}) := \mathsf{a} \oplus \mathsf{b}$. An algebraic chemistry $\langle \mathcal{M}_{\text{XOR}}, \mathcal{R}_{\text{XOR}} \rangle$ is generated to implement the logic gate. Since there are $N = 3$ boolean variables, the set of molecular species consists of six molecular species:

$$\mathcal{M}_{\text{XOR}} = \{a, A, b, B, c, C\} \tag{1}$$

where the lower and upper case version of the variable name are assigned to the boolean variable of that name. For example, molecular species a represents boolean variable $\mathsf{a} = 0$ and A stands for $\mathsf{a} = 1$. The set of reaction rules $\mathcal{R}_{\text{XOR}}$ is decomposed into two parts:

$$\mathcal{R}_{\text{XOR}} = \mathcal{L}_{\text{XOR}} \cup \mathcal{D}_{\text{XOR}} \tag{2}$$

where $\mathcal{L}_{\text{XOR}}$ is a set of reactions for the logical operation and $\mathcal{D}_{\text{XOR}}$ is a set of destructive reactions:

$$\mathcal{L}_{\text{XOR}} = \mathcal{L}^{\mathsf{c}}_{\text{XOR}} = \{a + b \rightarrow c,\ a + B \rightarrow C,\ A + b \rightarrow C,\ A + B \rightarrow c\}, \tag{3}$$

$$\mathcal{D}_{\text{XOR}} = \{a + A \rightarrow \emptyset, b + B \rightarrow \emptyset, c + C \rightarrow \emptyset\}. \tag{4}$$

Now we set the input variables of the boolean network a and b to initiate the computational process. For the initialisation, an inflow reaction is added to the reaction network. We start with providing one input only, leaving the other input variable undefined. Figure 1(C) shows the results for the four resulting algebraic chemistries. We can see that providing one input signal has further reduced the behavioural freedom of the reaction system. Only three combinations of molecular species are left, which may be encountered in the reaction vessel as a stationary state. Furthermore we can see that—in this special case—the output is not determined from a stoichiometric point of view since, in all four Hasse diagrams, sets containing c and C are found to be closed and self-maintaining.

When we finally provide both inputs, the Hasse diagram of organisations collapses so that only one organisation remains for every input condition, see Fig. 1(D). This implies that, no matter how we chose the dynamics, no other molecular species than those of the organisation can be sustained in the reaction vessel regardless of the initial state. We can see that the remaining organisation contains the desired output molecular species c or C, respectively.

To validate the results from applying organisation theory to the XOR reaction network, stochastic simulations are performed using the simulator packages *MGS* [14] and *Copasi* [32].

Figure 2 shows a typical simulation run. The influx is defined as an irreversible constant flux with kinetic parameter set to 1. For all other reactions we chose irreversible mass action kinetics. The parameters for the destructive reactions $\mathcal{D}_{\text{XOR}}$ are set to $k = 0.1$ and those for the logical reactions $\mathcal{L}_{\text{XOR}}$ are set to $k = 0.001$. At several simulation times the input is changed in order to observe the switching of the XOR gate. Initially, there exist no molecular particles in the reactor and two influxes of a and b are present. This corresponds to the case in which both the input variables

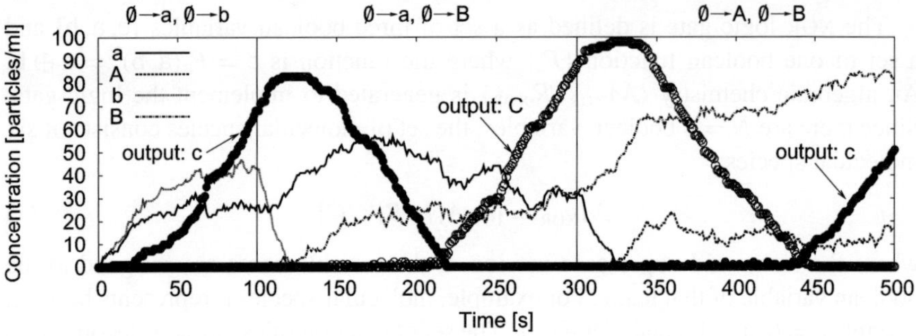

Fig. 2 Dynamic behaviour of the chemical reaction network implementing an XOR logic gate. The time course of all six molecular species is shown. Irreversible mass action kinetics are assumed for all reactions. Reaction rates are set to $k = 0.001$ for logical reactions. Reaction rates of destruction reactions are set to $k = 0.1$. For all irreversible constant influxes (e.g. $\emptyset \rightarrow A$), the rates are set to $k = 1$. The reaction system is stochastically simulated with the biochemical network simulator *Copasi* using a compartment size of 1 ml. See text for details. (Appeared in [25])

a and b are set to 0. Since molecular species c is generated, the output is computed to $\mathsf{c} = 0$.

At simulation time 100 s the content of input variable b is switched to 1 by replacing the influx of molecular species b with the influx $\emptyset \rightarrow B$. The molecular particles b and c, whose concentrations are still high from the previous computation, deteriorate and finally vanish. The desired output C does not appear until the time point of approximately 200 s. Then, instead of a, the molecular species A is applied as an input starting from simulation time 300 s. The remaining molecules of species a and C from the previous computation decay first and the desired answer c appears in the end.

3.2 Maximal Independent Set Problem—A Chemical Algorithm and a Small Example

General Algorithm

The *maximal independent set (MIS) problem* is formally defined as follows. Let an *undirected graph* $G = (V, E)$ be defined by a set of N vertices $V = \{v_1, \ldots, v_N\}$ and a set of edges E. When two vertices v_p and v_q are connected, the pair of the vertices is in the set of edges, i.e. $(v_p, v_q) \in E$. Note that the order of the pair is insignificant, that is $(v_p, v_q) = (v_q, v_p)$. A set of vertices $I \subset V$ is *independent* if no two vertices in the set are adjacent, i.e. $\forall v_p, v_q \in I : (v_p, v_q) \notin E$. An independent set is *maximal* if no vertex can be added to the set while keeping the property of independence. Including another vertex in a maximal independent set would violate the independence property.

A reaction network $\langle \mathcal{M}, \mathcal{R} \rangle$ is designed as follows. For each vertex v_j, we assign two molecular species s_j^0 and s_j^1 representing the membership of the vertex in the MIS. A high concentration of species s_j^1, higher than a threshold chosen to be smaller than any positive coordinate of any fixed point, means that the vertex v_j is included in the MIS. A high concentration of species s_j^0 expresses that the vertex v_j is not included in the MIS. Thus, the set of molecular species $\mathcal{M}$ contains $2N$ molecular species:

$$\mathcal{M} = \{s_j^0, s_j^1 \mid j = 1, \dots, N\}. \tag{5}$$

The set of reaction rules $\mathcal{R}$ is constructed by assembling reactions for each vertex:

$$\mathcal{R} = \bigcup_{i=1}^{N} \mathcal{R}^i = \bigcup_{i=1}^{N} (\mathcal{V}^i \cup \mathcal{N}^i \cup \mathcal{D}^i). \tag{6}$$

A reaction rule to *produce species* s_i^1 is the first:

$$\mathcal{V}^i = \left\{ \overbrace{s_j^0 + s_k^0 + \cdots + s_l^0}^{n_i} \rightarrow n_i s_i^1 \right\} \tag{7}$$

where n_i is the number of vertices connected to vertex v_i and $v_j, v_k, \dots, v_l$ are its neighbouring vertices, that is $(v_i, v_j), (v_i, v_k), \dots, (v_i, v_l) \in E$. This reaction is interpreted as follows. When no neighbouring vertex is included in the MIS, the target vertex v_i should be included in the set.

The negation of this predicate is considered by a set of n_i reactions:

$$\mathcal{N}^i = \{s_j^1 \rightarrow s_i^0 \mid (v_i, v_j) \in E\}. \tag{8}$$

This is the second type of reactions, which produce species s_i^0 from any species corresponding to the neighbouring vertices with superscript 1. This rule can be interpreted as follows. If there exists at least one neighbouring vertex included in the MIS, then the target vertex v_i should be excluded from the maximal independent set (otherwise the definition of the MIS would be violated).

The last component of set $\mathcal{R}^i$ is a *destructive reaction*. Since the membership of the MIS is a binary state, the state becomes undefined when neither or both of the species are present. In order to avoid the latter case, the two opposite molecular species are defined to vanish upon collision:

$$\mathcal{D}^i = \{s_i^0 + s_i^1 \rightarrow \emptyset\}. \tag{9}$$

Note that the reaction network is defined such that molecules react only if they are located on the same vertex or are neighbours. Thus the resulting (artificial) chemical system can be interpreted as a spatially distributed compartmentalised reaction system, where a compartment j holds only the two chemical species representing a vertex v_j, namely s_j^0 and s_j^1 and where the topological structure of the compartments is equivalent to the undirected graph.

Small Example

Provided that an undirected graph $G = (V, E)$ consists of three vertices and those vertices are connected linearly as shown in Fig. 3(A), i.e.,

$$G = (\{v_1, v_2, v_3\}, \{(v_1, v_2), (v_2, v_3)\}). \tag{10}$$

Following the recipe, a reaction network $\langle \mathcal{M}, \mathcal{R} \rangle$ is constructed. The set of molecular species $\mathcal{M}$ consists of six species because the graph contains $N = 3$ vertices:

$$\mathcal{M} = \{s_1^0, s_1^1, s_2^0, s_2^1, s_3^0, s_3^1\}. \tag{11}$$

According to the previous section the set of reactions $\mathcal{R}$ results in:

$$\begin{aligned} \mathcal{R} = \{ & s_2^0 \to s_1^1, s_2^1 \to s_1^0, s_1^0 + s_3^0 \to 2s_2^1, \\ & s_1^1 \to s_2^0, s_3^1 \to s_2^0, s_2^0 \to s_3^1, s_2^1 \to s_3^0, \\ & s_1^0 + s_1^1 \to \emptyset, s_2^0 + s_2^1 \to \emptyset, s_3^0 + s_3^1 \to \emptyset\}. \end{aligned}$$

The reaction network is analysed for its hierarchical chemical organisational structure within the reaction network. In our example, the reaction network $\langle \mathcal{M}, \mathcal{R} \rangle$ possesses five organisations:

$$O = \{\emptyset, \{s_1^0\}, \{s_3^0\}, \{s_1^0, s_2^1, s_3^0\}, \{s_1^1, s_2^0, s_3^1\}\}. \tag{12}$$

Figure 3(B) visualises these organisations as a Hasse diagram. We note that the organisations do not form a lattice, because there is not a unique largest organisations. The two largest organisations represent the two desired solutions to the MIS problem, namely "010" and "101". This explains that in a dynamical reaction system implementing the designed reaction network, the species combinations representing desired solutions are more likely to stay in the dynamical system and the other solutions consisting of species that are not an organisation cannot stably exist [11].

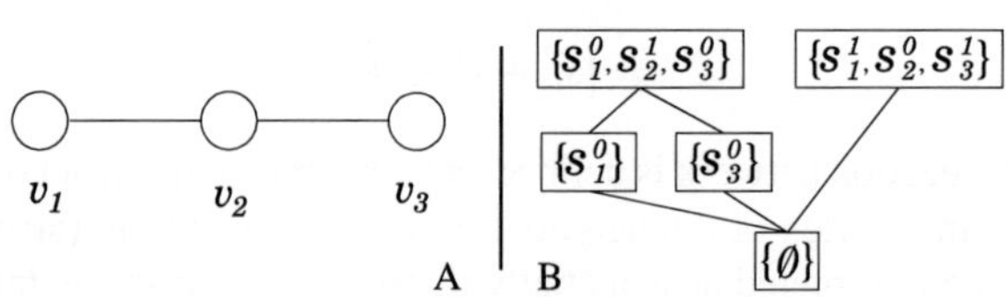

Fig. 3 Analysis of a chemical program with organisation theory. (**A**) Graph structure and (**B**) hierarchy of organisations within the chemical reaction network for the maximal independent set problem for the linear 3-vertex graph. (Appeared in [26])

4 Design Principles

4.1 Design Principles Derived from Heuristics

When following an organisation-oriented approach, we first concentrate on the reaction network neglecting kinetic laws. The reaction network is designed with respect to its organisational structure, considering the following principles **P1–P6**. Then, in the second step, the kinetics including kinetic parameters is specified for fine tuning of the computation as stated in **P7**. The kinetic laws determine the dynamics between and inside organisations.

P1: There should be one organisation for each output behaviour class. Assume that computation appears as a movement between organisations and the output behaviour can be categorised in different discrete behaviour classes. That is, species combinations in the computational reaction system uniquely identify the output behaviour. The reaction network should be designed so that there should exist at least one organisation corresponding to each output behaviour class categorised. For instance there are two organisations in the MIS problem example corresponding to the results, cf., the two biggest organisations in Fig. 3(B).

P2: The set of molecular species (and the organisation) representing a result should be in the closure of the species representing the initial input. This principle assures that there is a reaction path from the initial input configuration to the desired output species. Otherwise, the desired output will not appear as a result of the computation. The chemical XOR for example are designed following this principle. Furthermore, it is expected that the desired output set is contained in a self-maintaining set within that closure. The self-maintenance property of the set of molecular species indicates theoretical possibilities to sustain all the species in the dynamical reaction systems, so the desired output species may be sustained in the reaction system until the outcomes of the computation is observed. The ideal case is that the desired output is represented by a largest self-maintaining set within that closure. In case that there exists a larger self-maintaining set than the desired output set, the dynamics may settle above the desired one. This argument leads to the next principle.

P3: The set of molecular species representing an input should generate the organisation representing the desired output. This principle will be fulfilled on the following two conditions. The desired output is contained within the closure of the input (**P2** is fulfilled) and the largest self-maintaining set contained in the closure corresponds to the desired output. For the chemical XOR examples this was also the case. When inflow reactions are added, the closure of the input species turns out to be also self-maintaining. The largest self-maintaining set within a closure is not always unique in general. But this is not required and it can even be beneficial.

P4: Eliminate organisations not representing a desired output. Since each organisation potentially includes fixed points, the reaction system's dynamics may converge to one of the organisations. Hence, it makes sense to eliminate organisations not representing an output in order to avoid false computational outputs. This can be achieved by destroying either its closure property or its self-maintenance.

P5: An output organisation should have no organisation below.
The dynamics of the reaction system that moves from one organisation O_1 to another O_2 below (i.e. $O_2 \subseteq O_1$) is called a downward movement. This dynamical move can be theoretically prevented by the self-maintenance property with the right kinetics. Practically speaking, this move may occur spontaneously due to, e.g. stochastic effects, because the self-maintenance property only ensures the possibilities to sustain all species. Following this principle, a downward movement can be restricted.

P6: Assure, if possible, stoichiometrically the stability of an output organisation.
Instead of eliminating organisations below the desired output as in the previous principle **P5**, the downward movement can be ruled out by a purely stoichiometric argument. It may be possible to design the reaction network such that the organisation representing the desired output is stable for any kinetic law. As a simple example consider the system $\mathcal{R} = \{a \to b, b \to a\}$, which has the two organisations $\{\emptyset\}$ and $\{a, b\}$. Due to mass-conservation, the system can never move spontaneously from the organisation with two species to the empty one.

P7: Use kinetic laws for fine tuning.
The kinetic laws determine the system's behaviour within an organisation and the transition dynamics between organisations. One of rationales for the right kinetics is to assure that the dynamical reaction systems are stable in the output organisations, restricting mainly the downward movement. The existence of such laws is ensured by chemical organisation theory to a certain extent. With classical dynamical systems theory it is even possible to rigorously derive dynamical stability from network structure at least in some cases [9, 12]. Another point of consideration is a trade-off between stability and the speed of computation since chemical reaction systems may compute by moving amongst organisations.

4.2 Design by Evolution

In the previous section we described heuristics for a constructive programming strategy. This is similar to conventional programming, for which the target problem should be divided into sub-problems or at least the programmer should be aware of the structure of the problem intuitively or logically. While that constructive approach for designing the chemical reaction network in vivo has also been pursued [16], our focus here is to design chemical systems in an *autonomous* manner by *evolution*. The main difference is that the programmer is now freed of the structural analysis of the problems. In other words, it is not necessarily clear to the programmer how the target problem is solved.

Comparing these two approaches of constructing and evolving, the engineered system tends to be simpler and more effective because irrational components are usually omitted. These simplicity and effectiveness lead to a stiffness such that the system is intolerant to modifications. Tiny modifications cause unpredictable and often negative effects on the system such as function failure. It may also be the case that major system changes cause no effects at all. On the other hand, evolved systems embrace more components than necessary. This extra complexity may lead to distinctive characteristics such as *robustness* and *adaptability*. The evolved systems may be robust against a functional failure of components, for example, by assigning an identical sub-function to different components in order to sustain the functionality as a whole. As an example we mention the comparison between the two approaches for a reaction network realising a flip-flop [22].

Making use of the advantages of evolvability is believed to distinguish the chemical computing systems from conventional computers because conventional computing systems exhibit severe difficulties on adaptability etc. due to the intolerance to modifications.

4.3 Design by Exploration

We discuss an approach for *autonomous design*, cooperating with principles of *exploration* [24]. This explorative approach is fundamentally different because programming is not associated with modification of chemical computing systems. Instead, systems are explored and searched for interesting behaviours. The basic idea is that an autonomous system is used, as a preliminary step, to explore the behaviour of the chemical reaction system. Then a specific aspect of the system's behaviour will be utilised for a particular computational purpose. When dealing with real chemical systems, this approach has an apparent advantage because modifying the reaction network is very much restricted.

The scouting algorithm developed by Pfaffmann and Zauner [30] is an evolutionary experimentation method for autonomous experimentation. Experiments are dynamically scheduled to explore the systems' behaviour such that maximal information gain at each step is achieved. In accordance with communication theory, information is quantified as the surprise value of arriving data [8]. Motivations of this algorithm are to obtain experimental data sufficient to build quantitative system-level models as intended in Systems Biology [18]. For this purpose computational techniques have not only to discover regularities in existing data, but rather the experimental procedure itself has to be embedded in a closed-loop discovery process [17, 21].

Kulkarni and Simon [19] developed a program that attempts to generate experiments in which unexplained phenomena are enhanced. Notably, the program does not start out with a pre-set goal as is common in optimisation experiments but decides on its objectives dynamically. This work demonstrates that an algorithm can successfully navigate an immense search space by emulating the interplay of adjusting hypotheses and modifying experiments, which is characteristic of human experimenters [15].

5 Conclusion

In this article we reviewed a theoretical analysis method that helps to discover and implement computing capabilities in (artificial) chemical reaction networks. Given a list of molecular species and a list of reaction rules, the reaction network is decomposed into a hierarchy of closed and self-maintaining sub-networks called organisations. We have shown that this structure helps to assess the emergent dynamical behaviour of the chemical reaction network under study. In particular, we argue that the computation can be understood as a movement between organisations.

When the approach is applied to a reaction network implementing an XOR logic gate, the hierarchy of organisations helps to predict its emergent dynamical behaviour. Defining different inputs leads to different organisations corresponding to the various states of the gate. Even though a few of the logic gates are connected, the hierarchy of organisations is helpful for analysing the emergent dynamical behaviour. The second example shows how to implement a distributed and robust chemical algorithm to solve the MIS problem. The analysis via organisations gives an insight to the possible dynamical behaviour.

When designing a system with numerous small, extensively interacting components, its global behaviour cannot be easily predicted from the known local interactions. A general theory of emergence is desirable not only for analytical purposes but also for engineering such systems. We mentioned three different approaches for the design of artificial chemical reaction networks. Firstly, the design principles **P1–P7** are heuristic guidelines for the manual construction of chemical algorithms. Secondly and thirdly, the two approaches—evolution and exploration—share the common assumption that the effects of changing a reaction system are hard to predict in advance. The evolutionary design approach modifies reaction networks and evaluates the changes afterwards. A prediction process is not involved in this way of programming. For the explorative approach, the intention to alter reaction systems is even disregarded. When dealing with natural systems, however, this assumption is believed to be appropriate since interactions established within the systems are complex and tangled. That way, there is the possibility to utilise the abundant complexity embraced within natural systems.

The results reviewed in this article suggest that the theory of chemical organisations is a promising candidate to contribute to a general theoretical framework to master self-organisation in complex chemical-like information systems.

References

1. Adleman, L.M.: Molecular computation of solutions to combinatorical problems. Science **266**, 1021 (1994)
2. Banâtre, J.-P., Métayer, D.L.: A new computational model and its discipline of programming. Tech. Rep. RR-0566, INRIA (1986)
3. Banâtre, J.-P., Métayer, D.L.: The GAMMA model and its discipline of programming. Sci. Comput. Program. **15**(1), 55–77 (1990)

4. Banâtre, J.-P., Fradet, P., Radenac, Y.: Principles of chemical programming. In: Abdennadher, S., Ringeissen, C. (eds.) RULE'04 Fifth International Workshop on Rule-Based Programming, pp. 98–108. Tech. Rep. AIB-2004-04, Dept. of Comp. Sci., RWTH Aachen, Germany (2004)
5. Banzhaf, W., Dittrich, P., Rauhe, H.: Emergent computation by catalytic reactions. Nanotechnology **7**(1), 307–314 (1996)
6. Berry, G., Boudol, G.: The chemical abstract machine. Theor. Comput. Sci. **96**(1), 217–248 (1992)
7. Bhalla, U.S., Iyengar, R.: Emergent properties of networks of biological signalling pathways. Science **283**, 381–387 (1999)
8. Cherry, C.: On Human Communication: A Review, a Survey, and a Criticism, 2nd edn. Chap. 5. MIT Press, Cambridge (1966)
9. Clarke, B.L.: Stability of complex reaction networks. Adv. Chem. Phys. **42**, 1–213 (1980)
10. Dittrich, P.: The bio-chemical information processing metaphor as a programming paradigm for organic computing. In: Brinkschulte, U., Becker, J., Hochberger, C., Martinetz, T., Müller-Schloer, C., Schmeck, H., Ungerer, T., Würtz, R. (eds.) ARCS '05—18th International Conference on Architecture of Computing Systems 2005, pp. 95–99. VDE Verlag, Berlin (2005)
11. Dittrich, P., Speroni di Fenizio, P.: Chemical organisation theory. Bull. Math. Biol. **69**(4), 1199–1231 (2007)
12. Feinberg, M., Horn, F.J.M.: Dynamics of open chemical systems and the algebraic structure of the underlying reaction network. Chem. Eng. Sci. **29**(3), 775–787 (1974)
13. Fontana, W., Buss, L.W.: "The arrival of the fittest": Toward a theory of biological organization. Bull. Math. Biol. **56**, 1–64 (1994)
14. Giavitto, J.-L., Michel, O.: MGS: a rule-based programming language for complex objects and collections. In: van den Brand, M., Verma, R. (eds.) Electr. Notes in Theor. Comput. Sci., vol. 59. Elsevier, Amsterdam (2001)
15. Gooding, D.: Experiment and the Making of Meaning. Kluwer Academic, Dordrecht (1990)
16. Guido, N.J., Wang, X., Adalsteinsson, D., McMillen, D., Hasty, J., Cantor, C.R., Elston, T.C., Collins, J.J.: A bottom-up approach to gene regulation. Nature **439**(7078), 856–860 (2006)
17. King, R.D., Whelan, K.E., Jones, F.M., Reiser, P.G.K., Bryant, C.H., Muggleton, S.H., Kell, D.B., Oliver, S.G.: Functional genomic hypothesis generation and experimentation by a robot scientist. Nature **427**, 247–252 (2004)
18. Kitano, H.: Systems biology: A brief overview. Science **295**, 1662–1664 (2002)
19. Kulkarni, D., Simon, H.A.: Experimentation in machine discovery. In: Shrager, J., Langley, P. (eds.) Computational Models of Scientific Discovery and Theory Formation, pp. 255–273. Morgan Kaufmann, San Mateo (1990)
20. Küppers, B.-O.: Information and the Origin of Life. MIT Press, Cambridge (1990)
21. Langley, P., Simon, H.A., Bradshaw, G.L., Zytkow, J.M.: Scientific Discovery: Computational Exploration of the Creative Processes. MIT Press, Cambridge (1987)
22. Lenser, T., Matsumaru, N., Hinze, T., Dittrich, P.: Tracking the evolution of chemical computing networks. In: Bullock, S., Noble, J., Watson, R.A., Bedau, M.A. (eds.) Proceedings of the Eleventh International Conference on Artificial Life, pp. 343–350. MIT Press, Cambridge (2008)
23. Matsumaru, N., Dittrich, P.: Organization-oriented chemical programming for the organic design of distributed computing systems. In: 1st International Conference on Bio Inspired Models of Network, Information and Computing Systems (BIONETICS), Cavalese, Italy, 11–13 December 2006. ACM International Conference Proceeding, vol. 275. IEEE, New York (2006)
24. Matsumaru, N., Centler, F., Zauner, K.-P., Dittrich, P.: Self-adaptive scouting - autonomous experimentation for systems biology. In: Raidl, G.R., Cagnoni, S., Branke, J., Corne, D.W., Drechsler, R., Jin, Y., Johnson, C., Machado, P., Marchiori, E., Rothlauf, F., Smith, G.D., Squillero, G. (eds.) Applications of Evolutionary Computing, EvoWorkshops2004: EvoBIO, EvoCOMNET, EvoHOT, EvoIASP, EvoMUSART, EvoSTOC, Coimbra, Portugal, 5–7 Apr. 2004. LNCS, vol. 3005, pp. 52–61. Springer, Berlin (2004)
25. Matsumaru, N., Centler, F., Speroni di Fenizio, P., Dittrich, P.: Chemical organization theory as a theoretical base for chemical computing. Int. J. Unconv. Comput. **3**(4), 285–309 (2007)

26. Matsumaru, N., Lenser, T., Hinze, T., Dittrich, P.: Toward organization-oriented chemical programming: A case study with the maximal independent set problem. In: Dressler, F., Carreras, I. (eds.) Advances in Biologically Inspired Information Systems. Studies in Computational Intelligence, vol. 69, pp. 147–163. Springer, Berlin (2007)
27. Müller-Schloer, C.: Organic computing: On the feasibility of controlled emergence. In: Proceedings of the 2nd IEEE/ACM/IFIP International Conference on Hardware/Software Codesign and System Synthesis, CODES+ISSS2004, pp. 2–5. ACM Press, New York (2004). ISBN 1-58113-937-3. doi:10.1145/1016720.1016724
28. Müller-Schloer, C., von der Malsburg, C., Würtz, R.P.: Aktuelles Schlagwort: Organic Computing. Inform. Spektr. **27**(4), 332–336 (2004)
29. Peter, S., Dittrich, P.: On the relation between organizations and limit sets in chemical reaction systems (in print). Adv. Complex Syst. **14**(1), 77–96 (2011)
30. Pfaffmann, J.O., Zauner, K.-P.: Scouting context-sensitive components. In: Keymeulen, D., Stoica, A., Lohn, J., Zebulum, R.S. (eds.) The Third NASA/DoD Workshop on Evolvable Hardware, Long Beach, California, 12–14 July 2001, pp. 14–20. Jet Propulsion Laboratory, California Institute of Technology, IEEE Comput. Soc., Los Alamitos (2001)
31. Păun, G.: Computing with membranes. J. Comput. Syst. Sci. **61**(1), 108–143 (2000)
32. Sahle, S., Gauges, R., Pahle, J., Simus, N., Kummer, U., Hoops, S., Lee, C., Singhal, M., Xu, L., Mendes, P.: Simulation of biochemical networks using copasi—a complex pathway simulator. In: Proceedings of the Winter Simulation Conference, 2006. WSC 06, pp. 1698–1706 (2006)
33. Speroni di Fenizio, P., Dittrich, P.: Artificial chemistry's global dynamics. movement in the lattice of organisation. J. Three Dimensional Images **16**(4), 160–163 (2002)
34. Tsuda, S., Aono, M., Gunji, Y.-P.: Robust and emergent *physarum* logical-computing. Biosystems **73**(1), 45–55 (2004)
35. von der Malsburg, C.: The challenge of organic computing. Memorandum, Comp. Sci. Dept. (1999)
36. Würtz, R.P.: Organic computing for face and object recognition. In: Dadam, P., Reichert, M. (eds.) Informatik 2004, vol. 2, pp. 636–640. Gesellschaft für Informatik (2004)

N. Matsumaru
Gifu University Hospital, Gifu University, Gifu City 501-1193, Japan
e-mail: n_matsu@gifu-u.ac.jp

P. Kreyssig · P. Dittrich (✉)
Bio Systems Analysis Group, Jena Centre for Bioinformatics and Department of Mathematics and Computer Science, Friedrich Schiller University Jena, 07743 Jena, Germany
e-mail: peter.dittrich@uni-jena.de

P. Kreyssig
e-mail: peter.kreyssig@uni-jena.de

Chapter 2.7
Hovering Data Clouds for Organic Computing

**Sebastian Ebers, Sándor P. Fekete, Stefan Fischer, Horst Hellbrück,
Björn Hendriks, and Axel Wegener**

Abstract As part of our project *AutoNomos*, we have investigated traffic information and management systems that motivate the usage of new methods and tools inspired by Organic Computing paradigms. Current traffic monitoring and management approaches with stationary infrastructure lack flexibility with respect to system deployment and have difficulties with detecting unpredictable events (e.g., accidents). One goal of *AutoNomos* is the development of a distributed and self-organising traffic information and management system without a centralised infrastructure. Our system relies on a GPS-based navigation system and a wireless radio interface; vehicles can gather information about the current position on the road network and form a vehicular ad-hoc network (VANET) to share information about traffic phenomena. In this article, we introduce *Hovering Data Clouds* (*HDCs*) as a tool to collect, aggregate and disseminate application-specific data. HDCs evolve in a self-organising manner at locations of relevant data in the system. Although their data is hosted on the nodes, HDCs exist independent of the individual carriers. While HDCs float between physical carriers, their corresponding HDC messages are disseminated in the network by a new effective transport protocol named *AutoCast*, designed according to Organic Computing paradigms. Finally, we demonstrate that HDCs detect traffic phenomena reliably and propagate them robustly within the network.

Keywords Organic computing · Self-organising systems · Wireless ad-hoc networks · Hovering data clouds

1 Introduction

During the last decades, computing systems have evolved enormously with respect to their processing power, storage and memory. At the same time, the size of these systems continuously decreases resulting in powerful mobile systems. Additional wireless interfaces enable ad-hoc networks consisting of a notable number of these mobile systems. However, with the transition from wired to wireless technologies in computer networks, the complexity of systems increases significantly. Distributed applications built on top of wireless networks must be well designed to work in this difficult and changing environment. Especially ad-hoc wireless communication

C. Müller-Schloer et al. (eds.), *Organic Computing — A Paradigm Shift for Complex Systems*, 221–234
Autonomic Systems, DOI 10.1007/978-3-0348-0130-0_14,

between devices with wireless transceivers based on multi-hop topologies is an open research field.

Protocols designed in the past based on simulations solely using simple transmission models and graph-theoretic assumptions fail to work reliably in real field tests. Therefore, systems that flexibly adapt to their environment in order to exhibit robust behaviour are being studied in a number of research programs, among them the Autonomic Networking Initiative and the Organic Computing Initiative. Organic Computing systems are expected to exhibit "life-like" properties, including self-adaptation and self-organisation in the sense of being adaptive to changing conditions. In this article we explore Organic Computing approaches for system design, including protocol design in wireless multi-hop networks.

As a motivating example, we consider traffic monitoring, in particular traffic jams on a highway. Each traffic jam itself has life-like properties and consists of several parts, i.e., front, back and middle region. Although this article focuses on traffic monitoring, there are a number of other scenarios [3] that can be dealt with by our approach. The important aspects are *self-organisation* and *mobility of data*, which evolves when phenomena like a traffic jam occur. *Hovering Data Clouds* are a new concept for creating applications for ad-hoc networks. We show that HDCs are optimised to handle complex and organic phenomena.

The rest of this article is organised as follows: Section 2 discusses work related to our approach. In Sect. 3, the concept of *Hovering Data Clouds* is introduced as a self-organising entity that is able to migrate seamlessly from one physical device to another. Two major building blocks of HDCs (data aggregation and data dissemination with *AutoCast*) are discussed in Sect. 4 and Sect. 5. Section 6 presents evaluation results of HDC implementation and applications based on HDCs. The article concludes in Sect. 7 with a summary and future work.

2 Related Work

Research in the field of artificial intelligence builds and investigates self-learning systems that are able to adapt to changes in the environment. These self-learning systems (e.g., neural networks) have intrinsic emergent properties. Particularly in automation, the challenge is to predict the properties of these intelligent systems. Supervised learning is introduced to ensure the correct behaviour of a system in real applications, as the behaviour of artificial-intelligent systems is difficult to predict. These approaches of controlled learning are part of the practical solutions in Organic Computing [9]. A system architecture as introduced in [8] called "Observer/Controller architecture" provides feedback that leads to emergent properties, but also controls the subsystems. The Observer/Controller architecture can be implemented on different layers of a system and helps to develop well-defined hierarchies in a complex system.

Another field of related work is *Content-Centric Networking (CCN)* [5]. In CCN, information is not bound to a specific physical resource either. A consumer "asks" the net for specific content and every entity that is aware of the request can deliver the content if the data is available. However, CCN focuses on content delivery, which originates by certain resources, e.g., an on-line magazine. The content is no longer bound to a dedicated network node; however, it always originates at a fixed resource. In contrast to HDCs, CCN does not foresee data creation and data aggregation by network participants. In our work, we examine practical solutions. We are motivated by the notion of emergence by De Wolf [1], where control is decentralised and local mechanisms influence the global behaviour. In particular, we do not introduce a hierarchy or structures like topologies or clusters or an Observer/Controller architecture at a macroscopic level.

To the best of our knowledge, this is a novel approach for designing applications and protocols for wireless ad-hoc networks. Furthermore, the application of the presented organic principles is not limited to the traffic domain.

3 Concept

In this section, we introduce *Hovering Data Clouds* (*HDC*) as a tool to detect, monitor and follow a phenomenon efficiently. We describe the requirements and architecture of our approach, providing an easy-to-use, well-structured programming interface.

Our goal is the detection and monitoring of short- as well as long-term phenomena, which may even evolve over time. This is evaluated in Sect. 6 on the basis of journey-time predictions that need to be continuously updated due to differences in traffic volume. To reach scalability, flexibility and responsiveness, we forego a centralised infrastructure and employ distributed algorithms executed on mobile nodes that are able to communicate via wireless ad-hoc networks. Because we do not want to depend on the availability of a certain number of mobile nodes, our approach has to cope with sparse and dense networks. Additionally, participating nodes may completely change over time.

In most cases, a phenomenon worthy of being monitored can be linked to some characteristic data. Without a centralised infrastructure, nodes apply local rules on this characteristic and local data. To gain robustness (e.g. to cope with measurement or observation errors), local data gathered by a node needs to be correlated to data collected by other, in most cases neighbouring, nodes. To detect a phenomenon, characteristic data has to be detected and aggregated by multiple distributed nodes. Thus, nodes exchange messages via the described wireless ad-hoc network. In addition, the message exchange will keep data available near the location of the phenomenon over a longer period of time. Thus, the system has to adapt to the dynamics of the network topology. Even if nodes that detected and monitored a phenomenon are completely exchanged, the data needs to stay in place. To achieve this goal, data "hovers" over a certain area and is decoupled from physical nodes. Thereby, data migrates seamlessly from node to node, using the wireless ad-hoc network.

Unfortunately, communication in wireless networks is not reliable. Systems need to cope with message and data loss. Thus, we do not assign roles to nodes, because roles need to be maintained continuously, which is too error-prone under these conditions. Without roles we achieve a stateless system in which nodes act autonomously and have equal rights to contribute data to the system. Besides improving robustness, this enhances scalability of the system, because each node is only responsible for its own actions and data.

To fulfil these requirements, we propose the concept of *Hovering Data Clouds* (*HDC*), which comprise a set of rules and a local storage containing data about a phenomenon. HDCs are not bound to certain nodes, they use them as temporary hosts and exploit their hosts' communication, processing and data storage capabilities. Data that is temporarily stored by a node on behalf of a HDC defines the node's perspective of this HDC and thereby the monitored phenomenon. HDC rules are applied to new data in the same way that results from local sensors or other nodes. This may constitute new results that are integrated into the local data stored on behalf of the HDC. If integration takes place, it corresponds to an update of the node's perspective of the HDC. A corresponding notification of update of local data is sent to other nodes, which may in turn result in an update of their local data. Due to different sensor values and received messages, data stored on the various hosting nodes may differ. However, as all nodes apply the same rules to correlate and integrate data into their local storage, an application programmer using our approach defines correlation and integration rules that make the data converge to ensure that nodes develop a common understanding of the monitored phenomenon. Because updates are local actions, there is no need for coordination and we can omit the assignment and maintenance of rules.

If the system handles more complex phenomena that comprise of multiple and possibly spatial distributed phenomena, the simple correlation and aggregation of local data is not sufficient. To achieve this goal, aggregation of HDCs is required that provide data about phenomena that can be semantically linked to a superior one (cf. Sect. 4). This results in a shift from data to information. To manage and provide this information, we employ a new structure called *Organic Information Complex* (*OIC*).

Aggregation imposes another advantage: it abstracts from details of the monitored phenomenon, which results in data reduction. For scalability reasons, the key is to minimise data to be disseminated.

Within *Transport Layer*, we suggest to use *AutoCast*, which we introduce in more detail in Sect. 5. *AutoCast* disseminates *data units* via a wireless ad-hoc network and provides abstraction from communication tasks to upper layers.

Components located in the *HDC Layer* are responsible for storing and handling data and information, respectively. The *HDC Repository* contains all currently known HDCs and therefore represents the knowledge of all detected and currently monitored phenomena. The *Correlation Handling* component correlates incoming data to HDCs available in the local HDC repository. If correlation to an existing HDC fails, a new HDC is generated based on provided data. This data may result from local data mining (1), the transport layer (2), i.e., from other nodes, as well as from the upper layer (3) that forms the aggregation of semantically linked HDCs. If correlation results in an update of one or multiple HDCs, these updates are processed by the *Information Extraction* unit. The extraction of valuable information

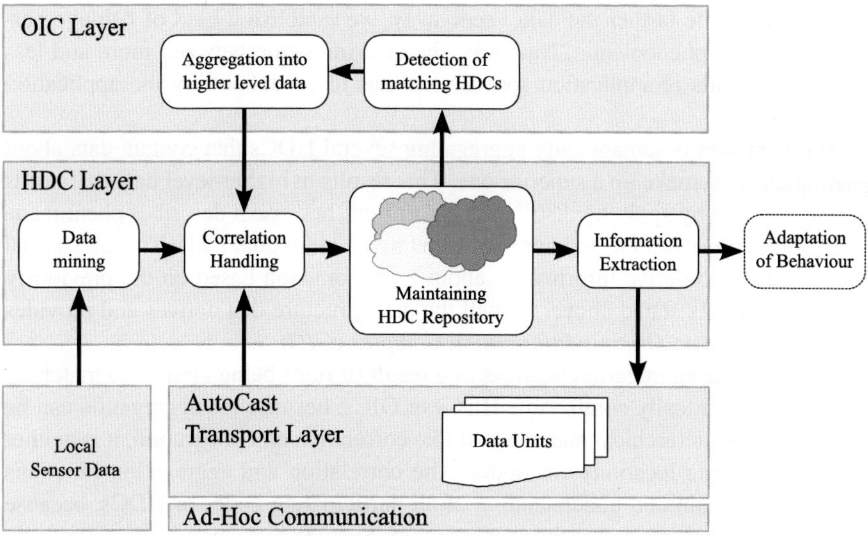

Fig. 1 Software architecture running on every node

results in messages sent via the transport layer. Additionally, the behaviour of the hosting node can be adapted due to the new findings. Note that this feature is optional and if a node does not offer an interface to adapt its behaviour, it will be omitted without having an impact on the rest of the functionality.

The *OIC layer* examines the HDC repository for candidates that can be semantically aggregated to an OIC, which is an application-specific procedure. Because rules applied in this step produce deterministic results in terms of generating the same OIC when using the same HDCs as input, there is no need to store OICs in a repository separate from HDCs. The *Aggregation* unit performs semantic aggregation and feeds the result back to the HDC layer's correlation handling component.

4 Data Aggregation

As described in the previous section, HDCs can detect, monitor and follow a phenomenon efficiently. HDCs arise when multiple (in most cases neighbouring) nodes aggregate their local sensor values that are characteristic of the phenomenon. They are updated by integrating new findings that result from the underlying sensor values.

In terms of scalability, it is necessary to reduce the amount of data to be transmitted to a minimum. The information about a nearby phenomenon needs to be precise and up to date, which means that data needs to be rich in detail and update processes need to be fast. However, if the phenomenon is far away, one can afford to get semantically summarised data to prevent high delays. By constantly decreasing the level of detail the farther the data veers away, we establish a kind of fish-eye perspective of the phenomenon. Naturally, the discrimination between more and less important details is application specific and has to be defined by the application developer.

The next step is semantically aggregating several HDCs that contain data about phenomena that make up a superior one. This results in higher-level data that needs to be computed by applying distributed algorithms, because it describes phenomena that could neither be detected nor monitored by an individual node. The actual goal of this step is to derive information about a phenomenon based on the previously gathered data. As stated above, we call the data structure that derives and provides this information an *Organic Information Complex* (*OIC*).

Note that the aggregation happens as a result of rules being applied to matching data, i.e., semantically combinable HDCs or OICs, because the aggregation can be continued in a hierarchical manner. Just like correlation and integration, it is neither bound to specific locations nor nodes. The correlation and aggregation functions converge to a common understanding of all participating nodes or HDCs, because the same rules are applied to the same data (cf. Sect. 3).

The characteristic data that needs to be gathered, correlated and aggregated is obviously application specific and requires domain-specific knowledge. Thus, an application developer using our concept has to take care of the correlation and aggregation functions to converge and the evaluation of data provided by local sensors.

To abstract from the communication and sensor hardware, the lower layers facilitate the following *Application Programming Interfaces* (*APIs*):

API provided by every sensor:

1. *registerListener(Listener)*—registers a component that will be informed of changes of the sensor's values.
2. *getValue()*—returns the sensor's value.

API provided by the transport layer:

1. *registerListener(Listener)*—registers a component that will receive every incoming data unit.
2. *getDataUnits()*—returns all currently stored data units.
3. *disseminate(DataUnit)*—disseminates a data unit via the ad-hoc network.

As an example, consider the applicability of our approach in the medical domain to discover an epidemic by aggregating data about many people suffering from the same disease.

Another example from a different domain is shown in Fig. 2. It depicts three different stages of data aggregation on a highway section. The topmost picture shows two *Traffic Jam Back* and two *Traffic Jam Front* HDCs. Simplified, the back of a traffic jam is detected, when the breaking events and the resulting velocities of neighbouring vehicles get aggregated. We described the algorithms employed to compute these HDCs in detail in [3].

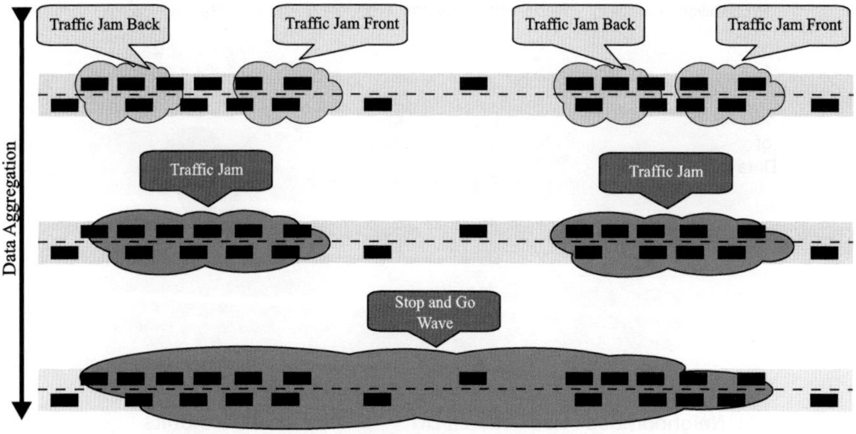

Fig. 2 Data aggregation to recognise a traffic jam

The aggregation of a *Traffic Jam Back* and a *Traffic Jam Front* HDC results in an OIC describing a traffic jam as a whole. As can be seen in the middle part of Fig. 2, the four initial HDCs get aggregated to two OICs. Because the correlation and aggregation mechanisms also work for OICs as well, we can continue the aggregation process in a hierarchical manner. The result is the detection of a *Stop-and-Go Wave*, as depicted in the lower part of Fig. 2. For these aggregations, the location of the monitored phenomena needs to be known, e.g., by using GPS-based navigation systems.

5 Data Dissemination—AutoCast

In order to disseminate data within a wireless ad-hoc network, we developed the Organic Computing protocol *AutoCast* [4, 10]. In the following we focus on the organic aspects of the *AutoCast* approach and refer to the described articles for further details.

As the protocol is designed independent of data types used in the application, we use the term *data unit* for the rest of this section if we refer to data that needs to be disseminated in the network. For a simple protocol design, we restrict ourselves to a single message format that contains *protocol information*: *neighbourhood information*, *hashes of data units*, and *data units* illustrated in Fig. 3. To reflect a node's state of information, hash values for data units are computed and a list of hash values is added to each transmitted message. Those hash values uniquely identify data units from the sender by adding only a fractional amount of overhead. The forwarding strategy is based on the locally observed neighbourhood.

We distinguish between the following actions that are depicted in Fig. 3. *Beacons* are sent periodically to track the neighbourhood. A data unit, initially sent by a node entitled "source of data" in the figure, is forwarded by *probabilistic flooding*, resulting in a wave moving through the network partition. Each node decides

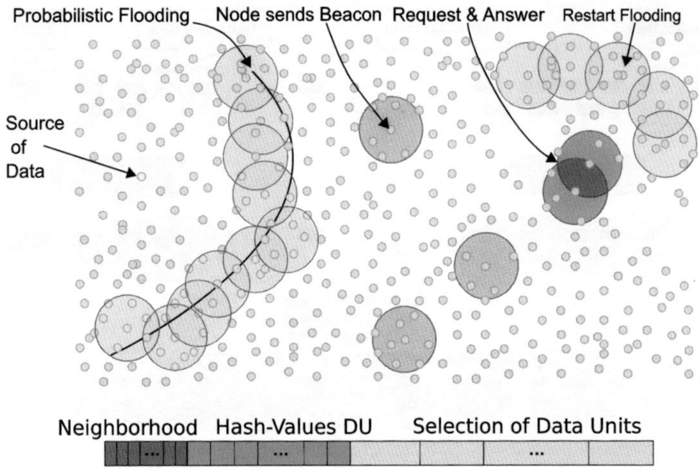

Fig. 3 Illustration of AutoCast actions and data format

independently whether it participates in the flooding process considering the neighbourhood and a randomised criterion, so that on the average two nodes will become active. We describe the criterion in the following. When network partitions merge, as is the case in the upper right of Fig. 3, *beacons* are treated as (implicit) *requests* for data units by receiving neighbours. As a consequence, receiving neighbouring nodes *answer* with missing data units. Thus, the process of *probabilistic flooding* is restarted. All these actions are executed independently by each node.

We distinguish the following dissemination rules and their delays:

1. Nodes answer to requests for data units after 1δ, so that currently missing data units can be added to subsequent flooding messages.
2. Nodes flood data units with a delay of 2δ, so that requests from other nodes do not interrupt this action too much.
3. Nodes request missing data units after 3δ.

To avoid jamming the wireless channel, which we experienced during design of the protocol, the strategy "Listen before Talk" is used. Therefore, actions get postponed by the described delays every time a new message is received.

Particularly in highly mobile and sparse scenarios links between nodes break frequently, because the distance between nodes increases faster than the communication range, and thereby the network splits into several partitions. In order to cope with partitions, *AutoCast* implements a recovery strategy. Nodes piggyback data units. When "new" nodes from other partitions enter their communication range, nodes restart the dissemination again.

The following situations can occur at a receiving node:

- When a node detects that the received message contains more hashes of data units than it has stored, it can *request* the missing data units, which are identified by the additional hashes.

- When a received message does not contain hashes of data units, a node can *answer* to such an (implicit) request message with a message including those missing *requested* data units.

The periodic beacon messages maintain neighbourhoods and also act as implicit requests, if one or more receiving nodes have stored data units that are unknown to the sender of the beacon message.

Thus, each node processes all *AutoCast* messages for information and data units. Every node is particularly attentive to beacon messages sent by other nodes. If a beacon message is received by a node, the node is inserted in the local neighbourhood table. The next expected beacon from this new neighbouring node is scheduled due to the beacon interval given in the beacon. A node is removed from the neighbourhood table if its scheduled beacon did not arrive in time.

If a node decides to answer, request or flood data units, the message also includes beacon data. Thus, a periodic beacon without data units is only required if nodes perform no other action for a duration of the beacon interval.

Because every node is aware of the number of its direct neighbours, it adapts its data dissemination strategy to this number. Thus, it is not important which node sends out the data, as long as nodes that receive a new data unit forward it. We balance the protocol, so that on the average two nodes forward new data units by means described in [4].

AutoCast exploits local neighbourhood information as a metric for the data dissemination strategy of a node. Neighbourhoods change continuously due to movement of the nodes. By choosing the optimal size for a beacon interval as in [4], the deviation between real and observed neighbourhoods can be bounded by a maximum. Due to this fast and reliable neighbourhood detection, the dissemination strategy adapts seamlessly to local network densities.

6 Evaluation

In this section, we evaluate the performance of the data aggregation presented in Sect. 4 and the *AutoCast* protocol presented in Sect. 5. For this evaluation, we used a combination of the network simulator *ns-2* [2] and the traffic simulator *SUMO* [6], a microscopic traffic simulator that uses the car-following model from Krauß [7] that provides realistic mobility traces for this scenario. By coupling these simulators using our *Traffic Control Interface* (*TraCI*) [11], the nodes in the network simulator are linked to vehicles in the traffic simulator. The movements of a vehicle result in the corresponding movement of the associated node in the network simulator that hosts our software architecture. Thus, we can evaluate the accuracy of our journey-time predictions by comparing them to the actual journey-times reported by the simulator. Because in addition, a node can influence the driving behaviour of an associated vehicle, the optional *Adaption* component (cf. Sect. 3) is available. Hence we can evaluate the result of dynamic re-routing based on the computed predictions. The vehicles are aware of their individual position and routes, which is achieved in real-world scenarios by using a GPS-based navigation system.

6.1 Data Aggregation

Now we turn to evaluating the application of our approach using the scenario briefly described in Sect. 4. Once again, we focus on a traffic-related scenario, even though our approach is applicable in many different fields. We show that precise journey-time predictions can be generated by aggregating information about congestions and correlated travel times on individual sections in a road network. In order to simulate different fractions of vehicles that are equipped with our system, we vary the *penetration rate* between 0% and 100%.

As described above, the correlation of individual travel times results in HDCs. The data of these HDCs comprises the following fields:

- identifiers of the intersections
- mean travel times
- number of vehicles with correlated travel times
- time of the last update
- time of the last message sent to indicate the mean travel time

The scenario used for this evaluation is depicted in Fig. 4. It consists of a 78.5 km-long circuit that comprises two pairs of lanes leading in opposite directions. On this circuit, the two intersections *A* and *B* are linked by a route section that is 7.5 km long. In addition, a 15 km-long bypass exists that links both intersections. Because speed limits on the circuit and the bypass are similar, the travel time when using the bypass is usually longer. Because the vehicles try to circulate as fast as possible, the bypass is normally not used.

At the beginning of the simulation that lasts 120 minutes, 1500 vehicles are placed on each pair of lanes on the circuit. After 65 minutes, a vehicle is stopped shortly before it enters an intersection, simulating an accident and blocking one of the two lanes. This results in a congestion that increases the travel time between *A* and *B*.

As parameter for our evaluation, we use the mean travel times between *A* and *B* that are reported once per minute by SUMO.

The result of our evaluation is shown in Fig. 5. On the x-axis, the simulation time and on the y-axis the mean travel times are depicted for various penetration rates, illustrated by individual curves in the diagram. Prior to the simulated accident, the mean travel time is about 5 minutes, which means that a vehicle's velocity is about 90 km/h. After the accident, the mean travel time increases according to the increasing length of the traffic jam.

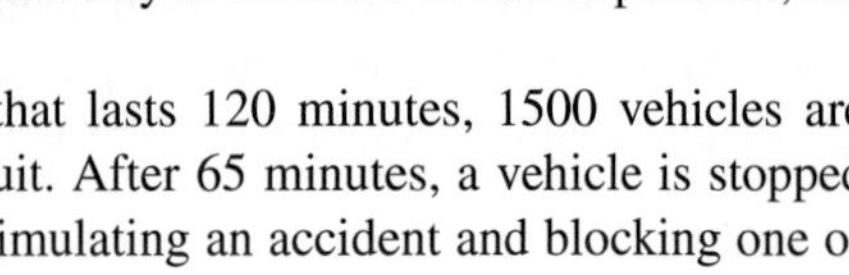

Fig. 4 Road traffic scenario for evaluation of adaptive rerouteing based on HDCs

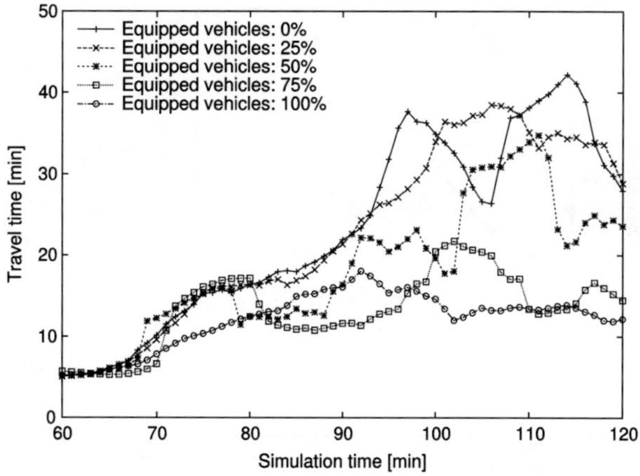

Fig. 5 Mean travel time from *A* to *B* (cf. Fig. 4) with varying penetration rates

With 0% equipped vehicles and thereby without journey-time predictions, the mean travel time continuously increases for about 30 minutes after the accident. After the 95th minute, the mean travel time oscillates between 26 and 42 minutes, which is due to the inhomogeneity of the traffic flow.

With increasing equipment rate, the maximum mean travel time continuously decreases, as cars decide to take the detour. With an equipment rate of 100%, the maximum mean travel time does not exceed 18 minutes.

6.2 AutoCast

In [4] we provide results for a wide range of scenarios. Here we select the most dynamic and challenging scenario, where vehicles as network nodes drive on a 10 km section of a highway with two lanes in each direction. They form a so-called vehicular ad-hoc network (VANET). The VANET scenario is challenging, because vehicles move with approximately 25 m/s (≈90 km/h), tend to form clusters, and create many network partitions. We use the combination of the two simulators SUMO and ns-2 as in Sect. 6.1. To vary the network density, we equip different fractions of vehicles with a wireless interface and the particular protocol, so the average neighbourhood size increases from 0.75 neighbours (5% equipped vehicles) to 15 neighbours when all vehicles are equipped. We produce data units each second with a lifetime of 50 s at a vehicle that is closest to the 5 km mark of the road. The generated data unit has a size of 200 bytes and a dissemination area that covers the complete 10 km road section.

The communication parameters in *ns-2* are set according to the standard 802.11 implementation with a communication range of 250 m.

The results in Fig. 6 show *AutoCast's* performance in comparison to the references flooding and theoretical optimum. In all subfigures, the x-axis depicts the penetration rate and the y-axis shows one performance indicator of the protocol.

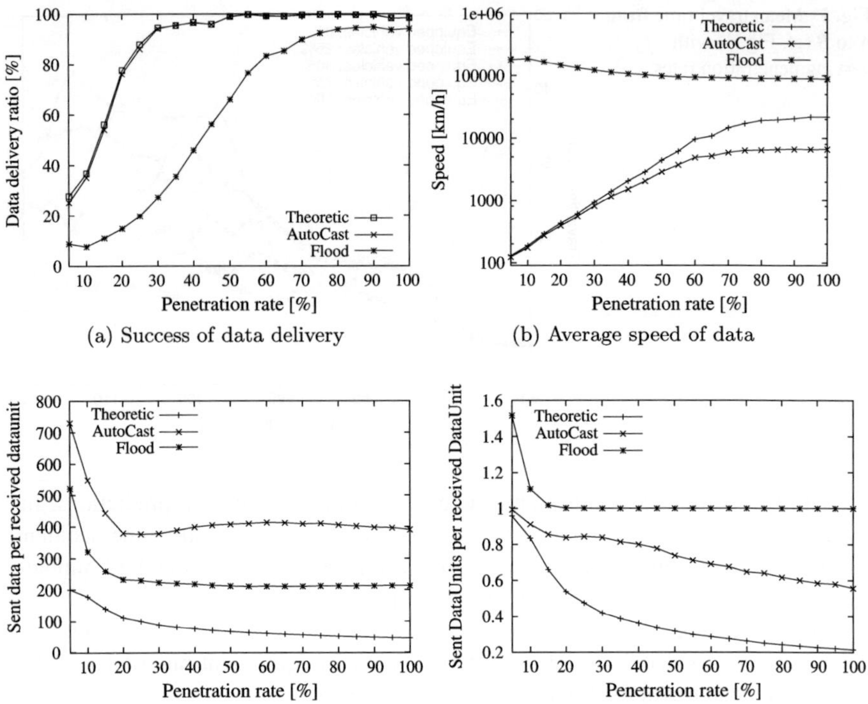

(a) Success of data delivery (b) Average speed of data

(c) Sent bytes per delivered data unit (d) Relation between number of sent and delivered data units

Fig. 6 *AutoCast* performance evaluation for VANET scenario

The most important performance metric for data dissemination is the success of data delivery, as depicted in Fig. 6a. *AutoCast* reaches nearly the same number of nodes as the theoretical optimum. In contrast, flooding performs badly, because a recovery mechanism is missing and consequently the flooding process stops at the border of network partitions.

The speed of data delivery is shown in Fig. 6b. For a low penetration rate, data speed for *AutoCast* and the theoretical optimum is limited to the vehicles' speed on the road, since data travels piggybacked with the vehicles most of the times. With increasing penetration rate and thereby increasing network density, data speed rises as more data is forwarded by wireless communication. The "Listen before Talk" principle of *AutoCast* limits the speed of data for high penetration rates to around 6600 km/h (which is fast enough for any application that we imagine so far), while the theoretical optimum reaches around 21000 km/h. Surprisingly, flooding performs best with respect to the speed of data. Flooding delivers data immediately or never, because each node receiving a data unit rebroadcasts it exactly once. To put this result into the right perspective, it should be kept in mind that flooding delivers only a fraction of data units, in particular when the network is sparse, as illustrated in Fig. 6a.

Figure 6c and Fig. 6d show the average amount of data that needs to be sent in order to deliver one data unit to one receiving node. For flooding, every node forwards a data unit once; thus, the corresponding lines converge to the size of a data unit (200 bytes, cf. Fig. 6c) and one sent data unit per received data unit (cf. Fig. 6d). The increase for lower penetration rates results from the very first broadcast of a data unit. *AutoCast* performs worse than the theoretical optimum; but since we set the average number of forwarders to two, *AutoCast* is designed to send at least twice the number of data units compared to the theoretical optimum. In the case of network partitions, answer messages increase the number of data units. Recall that answer messages re-initiate a selective forwarding of data units that are piggybacked between network partitions.

7 Conclusion and Future Work

In this article we motivated the need for a new tool to collect, aggregate and disseminate application specific data in a self-organising manner. We introduced *Hovering Data Clouds* as a generic tool for Organic Computing systems, based on distributed algorithms, local knowledge and rules. HDCs present self-organising entities in an ad-hoc network. *AutoCast* has demonstrated to be an efficient data distribution protocol for ad-hoc networks following the Organic Computing paradigm and well suited to distribute HDC messages.

In the near future we will publish *Hovering Data Clouds* as a software design pattern for Organic Computing systems. We are about to complete the implementation of the *AutoNomos* system according to the architecture depicted in Fig. 1.

References

1. De Wolf, T., Holvoet, T.: Emergence versus self-organisation: Different concepts but promising when combined. Eng. Self-Organising Syst. **3464**, 1–15 (2005)
2. Fall, K., Varadhan, K.: The Ns Manual. The VINT Project—a collaboration between researchers (2010)
3. Fekete, S.P., Schmidt, C., Wegener, A., Hellbrück, H., Fischer, S.: Empowered by wireless communication: Distributed methods for self-organizing traffic collectives. ACM Trans. Auton. Adapt. Syst. **5**(3), 439–462 (2010)
4. Hellbrück, H., Wegener, A., Fischer, S.: Autocast: A general-purpose data dissemination protocol and its application in vehicular networks. Ad Hoc Sens. Wirel. Netw. J. 91–22 (2008)
5. Jacobson, V., Smetters, D.K., Thornton, J.D., Plass, M.F., Briggs, N.H., Braynard, R.L.: Networking named content. In: CoNEXT '09: Proceedings of the 5th International Conference on Emerging Networking Experiments and Technologies, pp. 1–12. ACM, New York (2009)
6. Krajzewicz, D., Bonert, M., Wagner, P.: The open source traffic simulation package SUMO. In: RoboCup 2006 Infrastructure Simulation Competition, Bremen, Germany (2006)
7. Krauß, S.: Microscopic modeling of traffic flow: Investigation of collision free vehicle dynamics. PhD thesis, Universität zu Köln (1998)
8. Müller-Schloer, C., Sick, B.: Emergence in Organic Computing Systems: Discussion of a Controversial Concept. In: Yang, L., Jin, H., Ma, J., Ungerer, T. (eds.) Autonomic and Trusted Computing. Lecture Notes in Computer Science, vol. 4158, pp. 1–16. Springer, Berlin (2006)

9. Schmeck, H.: Organic computing-vision and challenge for system design. In: Proceedings of the International Conference on Parallel Computing in Electrical Engineering, p. 3. IEEE Comput. Soc., Washington (2004)
10. Wegener, A., Hellbrück, H., Fischer, S., Schmidt, C., Fekete, S.: Autocast: An adaptive data dissemination protocol for traffic information systems. In: Proceedings of the 66th IEEE Vehicular Technology Conference Fall 2007 (VTC2007-Fall), Baltimore, USA, October, pp. 1947–1951 (2007)
11. Wegener, A., Piórkowski, M., Raya, M., Hellbrück, H., Fischer, S., Hubaux, J.-P.: Traci: An interface for coupling road traffic and network simulators. In: Proceedings of 11th Communications and Networking Simulation Symposium (CNS'08), Ottawa, Canada, pp. 155–163 (2008)

S. Ebers (✉) · S. Fischer · A. Wegener
Institute of Telematics, University of Lübeck, Lübeck, Germany
e-mail: ebers@itm.uni-luebeck.de

S. Fischer
e-mail: fischer@itm.uni-luebeck.de

A. Wegener
e-mail: wegener@itm.uni-luebeck.de

S.P. Fekete · B. Hendriks
Institute of Operating Systems and Computer Networks, TU Braunschweig, Braunschweig, Germany

S.P. Fekete
e-mail: s.fekete@tu-bs.de

B. Hendriks
e-mail: b.hendriks@tu-bs.de

H. Hellbrück
Department of Electrical Engineering and Computer Science, Lübeck University of Applied Sciences, Lübeck, Germany
e-mail: hellbrueck@fh-luebeck.de

Chapter 3
Learning

There is no Organic Computing (OC) without learning. With appropriate machine learning techniques we equip OC systems with various kinds of self-$*$ properties that are needed for their interaction with humans or other OC systems in dynamic environments. Moreover, these environments are typically noisy. This term summarises various influences such as measurement errors, incomplete observations, transmission errors, etc. that must be treated like random influences. In the field of OC, a key issue is to increase the degree of autonomy in learning. This chapter shows how this challenge can be met, e.g., by self-organised gathering of sample data, knowledge exchange between OC systems, mutual imitation of OC systems, or by effecting a good compromise between exploitation and exploration phases of an OC system.

The first article, *Aspects of Learning in OC Systems* (Chap. 3.1, by Cakar et al.), focuses on a problem, which is common to many applications in the field of OC: The self-adaptation of systems in noisy, dynamic environments. The solution advances eXtended Classifier Systems (XCS), a variant of Learning Classifier Systems (LCS). Learning in these classifiers is based on evolutionary techniques and reinforcement learning. To improve the online learning speed of XCS, the authors propose a rule combining mechanism. To facilitate the search for an optimum in the dynamically changing fitness landscape they suggest a population-based optimisation technique, the Role Based Imitation algorithm.

The second article, *Combining Software and Hardware LCS for Lightweight On-chip Learning* (Chap. 3.2, by Bernauer et al.), also employs XCS, but it addresses the specific challenges of hardware implementations. Such implementations are needed for task allocation or system-on-chip parametrisation at runtime, for instance. The authors combine the advantages of a software-based XCS with a hardware-based variant, the learning classifier tables (LCT): First, the XCS has to learn a set of rules for a given problem at design time. Then, the learned rules are translated into a form that is suitable for LCT. With these rules, the LCS, which is not able to generate new classifiers, continues to learn at runtime.

The third article, *Collaborative Learning by Knowledge Exchange* (Chap. 3.3, by Fisch et al.), deals with a kind of "collective intelligence" of distributed systems: Components of such a system learn classification rules in a partially autonomous way and exchange these rules with other components to improve the behaviour of the overall system. Like in Chap. 3.1, the environment is assumed to be noisy and dynamic. The techniques proposed in this article are based on probabilistic considerations. The authors present a new classifier paradigm, techniques to detect the need for self-adaptation (new rules are needed or obsolete rules may be discarded), and techniques for rule exchange, numerical interestingness assessment, and rule fusion.

The fourth article, *Framework for Controlled Self-optimisation in Modular System Architectures* (Chap. 3.4, by Brockmann et al.), also tackles the problem of self-optimisation in dynamic and noisy (e.g., due to disturbances and anomalies) environments. The authors describe a novel framework that addresses various challenges in a holistic way: The stability-plasticity trade-off, the exploration-exploitation problem, stability and safety constraints, the need for human-interpretable rules, for incremental learning, etc. The functional core behaviour in this framework is realised by so-called basic control units, which implement certain neuro-fuzzy techniques (zero order Takagi-Sugeno systems).

The fifth article, *Increasing Learning Speed by Imitation in Multi-robot Societies* (Chap. 3.5, by Jungmann et al.), also deals—such as Chap. 3.3—with the problem of learning from others. Here, robots learn from other robots by observing and imitating their behaviour. The main challenges are that there are no fixed demonstrator-imitator relationships and (typically) only one example of the type of behaviour from which the imitator has to learn. Thus, the imitator tries to model sequences of the demonstrator's state transitions with probabilistic methods. The article shows how a robot decides whom and when it will imitate, how it interprets observed behaviour, and how it fuses the knowledge gained from these observations with its own.

The sixth article, *Learning to Look at Humans* (Chap. 3.6, by Walther and Würtz), solves a computer vision problem with OC techniques: The learning of a generalisable model of the visual appearance of humans from video data. The article shows how OC principles can be integrated into a posture estimation cycle. The main goals are to provide robust techniques, i.e., techniques that are largely independent from different individuals, backgrounds, and attires. For that purpose the authors adopt a large variety of computer vision and machine learning techniques and enhance them by OC techniques such as (controlled) self-organisation in order to generate all required model information autonomously, for instance.

Altogether, the articles in this chapter cover a wide range of machine learning techniques. These techniques are used at design time or at runtime and realised either in hardware or (mostly) in software to equip OC systems with "intelligent" behaviour in order to operate in noisy, dynamic environments.

Bernhard Sick
Computationally Intelligent Systems Lab, University of Applied Sciences Deggendorf, Deggendorf, Germany
email: bernhard.sick@fh-deggendorf.de

Chapter 3.1
Aspects of Learning in OC Systems

Emre Cakar, Nugroho Fredivianus, Jörg Hähner, Jürgen Branke, Christian Müller-Schloer, and Hartmut Schmeck

Abstract Organic Computing (OC) aims at handling the growing complexity in technical systems by endowing them with *life-like* capabilities such as self-organisation, self-configuration, and self-adaptation. OC systems with these capabilities can tolerate disturbances and continue working properly while adapting their behaviour to the changes in their environment. In this context, the two-layer Observer/Controller architecture has been developed to determine the optimum set of parameters for an OC system that operates in a continuously changing environment. Layer 1 of this architecture, which is implemented using an eXtended Classifier System (XCS), allows for quick response to changes if a situation appears, which is *close* to a previously encountered situation. Thus, Layer 1 acts as a kind of memory. Layer 2 is triggered if the new situation is not covered by the population of the XCS on Layer 1. In that case, different parameter sets are evaluated using an optimisation algorithm on a simulation model of the real system. After that, the best parameter set found is given to the XCS on Layer 1 for further evaluation in the real world.

The contribution of this article is two-fold: Firstly, we present a rule combining mechanism for XCS that infers maximally general rules from the existing population to increase the on-line learning speed on Layer 1. Secondly, we present a new population-based optimisation algorithm for Layer 2, which can be used to find high quality solutions for OC systems that operate in continuously changing environments. Furthermore, we provide experimental results for both mechanisms and show that the proposed techniques improve both the learning rate and the solution quality.

Keywords Learning · Extended classifier system · Population-based optimisation · Two-layer Observer/Controller architecture · Self-adaptation

1 Introduction

Technical systems that are able to adapt their behaviour according to the changes in the environmental conditions have gained increasing attention in the last few years. In OC [7] we deal with technical systems that consist of a large number of elements (agents) interacting with each other in order to accomplish a given goal. One of the important characteristics of OC systems is that the agents search for the optimum behaviour on the given fitness landscape and at the same time affect it so that the

C. Müller-Schloer et al. (eds.), *Organic Computing — A Paradigm Shift for Complex Systems*, 237–251
Autonomic Systems, DOI 10.1007/978-3-0348-0130-0_15,

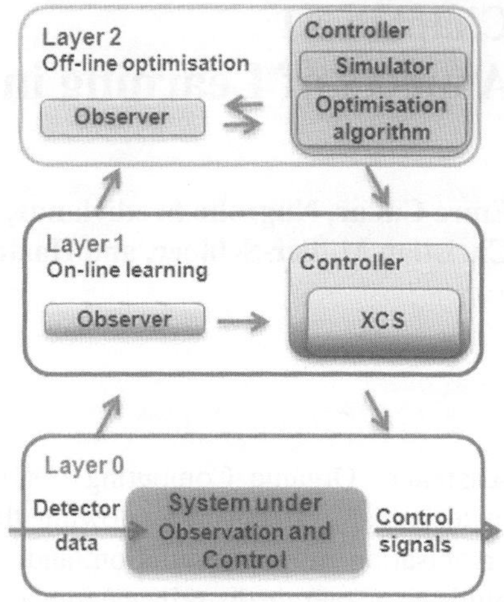

Fig. 1 The two-layer Observer/Controller architecture

form of the fitness landscape changes as a result of the agent behaviour. This kind of fitness landscapes are called *self-referential*, meaning that the optimum in the particular fitness landscape is not fixed (as is the case in static fitness landscapes), rather it *moves* as a function of the agent behaviour. This fact presents additional challenges requiring:

1. A fast and resource-efficient on-line learning mechanism, which facilitates a quick response to the situations that are close to previously encountered *known* situations and
2. An effective off-line optimisation mechanism to evaluate different (possibly sub-optimal) solutions for new previously *unknown* situations.

In this context, we developed the two-layer Observer/Controller architecture (see Chap. 4.1) for OC systems (see Fig. 1) to intelligently explore the fitness landscape and to react reasonably on changes in a short period of time.

On the lowest layer (Layer 0) of the proposed architecture is the productive system (System under Observation and Control—SuOC). The SuOC may be any kind of decentralised and parameterisable system, which consists of a set of elements possessing certain observable attributes. On-line parameter selection for the SuOC is implemented on Layer 1 using an eXtended Classifier System (XCS) [15]. Here, an observer component determines the current situation in the SuOC and the controller selects an adequate action for the corresponding situation using an XCS facilitating a quick response to the encountered situation (see requirement 1 above). Layer 2 is triggered each time the observed situation on Layer 0 is not covered by the population of the XCS on Layer 1. In this case, an optimised classifier covering the corresponding situation is created by using a model-based optimisation on Layer 2. Here, an optimisation algorithm evaluates different solutions for the observed situation using a simulation model of the SuOC. At the end of the optimisation a new classifier is created, which maps the observed situation to the best solution found by the optimisation algorithm facilitating a fast evolution of different solutions for a given situation (see requirement 2 above). In this article, we investigate different learning and optimisation techniques to improve the efficiency of Layer 1 and Layer 2 to find high-quality solutions for OC systems that operate in dynamic and noisy environments.

This article is organised as follows: Section 2 summarises related work. Section 3 presents a novel discovery component for the XCS to increase the on-line learning speed on Layer 1. Section 4 presents a new population-based optimisation algorithm for Layer 2, which can be used to find high quality solutions for OC systems that operate in dynamic environments. Finally, Sect. 5 concludes and summarises this article.

2 State of the Art

In recent years, many efforts have been devoted to developing algorithms for optimisation of numerical, real-valued problems, where the goal is to find the best possible set of parameters (i.e., solution) according to some given criteria typically expressed as mathematical functions. In this context, each algorithm proposes a different technique for the examination of new trial solutions (exploration) and for using the available information about the fitness landscape to get closer to a (possibly local) optimum (exploitation). In the following, we present some well-known optimisation algorithms, which we have used for comparison purposes in our investigations.

Differential Evolution (DE) is a population-based optimisation algorithm, which has been proposed by Storn and Price [12]. The main idea behind DE is that it generates new parameter vectors by adding a weighted difference vector between two population members to the parameter vector of a third one using the following formula:

$$X^i = X^i_{r1} + F(X^i_{r2} - X^i_{r3}) \tag{1}$$

where F stands for the scaling factor and X^i_{r1}, X^i_{r2} and X^i_{r3} are the parameter values of individuals $r1$, $r2$ and $r3$ in i^{th} dimension. DE has shown very good results in several real-world applications [13].

Particle Swarm Optimisation (PSO) is another population-based optimisation algorithm developed by Kennedy and Eberhart [4] inspired by the behaviour of bird flocking, where individuals spread in the environment to look for food and move around independently. Each particle knows its best solution and the best global solution found so far and is propelled towards these positions at each optimisation step. Here, a particle determines its velocity vector using the following formula:

$$V^i_{k+1} = \omega * V^i_k + \varphi_1 * U_1[0,1) * (P_{id} - X_{id}) + \varphi_2 * U_2[0,1) * (P_{gd} - X_{id}) \tag{2}$$

where V^i_k stands for the current velocity of the particle, ω for the inertia weight, X_{id} for the current position of the particle, P_{id} for the personal best position of the particle, and P_{gd} for the best global position found so far. φ_1 and φ_2 are the acceleration coefficients while $U_1[0,1)$ and $U_2[0,1)$ are uniformly distributed random numbers generated between [0, 1). PSO is also used successfully in different real-world applications [8].

DE and PSO are both population-based optimisation algorithms. In order to provide a more comprehensive investigation, we also deal with Simulated Annealing (SA) [6], which is a trajectory-based optimisation algorithm. SA is based on an analogy taken from thermodynamics and uses a temperature-based approach to escape local optima accepting new solutions if they are better than the current one or in the other case if the following condition is met:

$$random[0,1) < e^{\frac{eval(v_n)-eval(v_c)}{T}} \tag{3}$$

In (3), T stands for the temperature while v_n and v_c are the current and the new trial solutions, respectively. T is decreased over time, and the probability to accept bad solutions decreases so that the final stages of SA merely resemble an ordinary hill-climbing algorithm.

The algorithms presented above possess no strict distinction between exploration and exploitation so that both of them take place at the same time. Since we deal with self-referential fitness landscapes in OC, there is a need of an optimisation algorithm that facilitates an effective optimisation scheme, where previously found good solutions are kept while other parts of the fitness landscape are further explored, simultaneously. This requires an intelligent distinction between exploring and exploiting individuals. In Sect. 4 of this article, we propose a new optimisation algorithm (Role Based Imitation algorithm—RBI) that utilises a role assignment strategy to determine the exploring and exploiting individuals to effectively explore the fitness landscape and to find the optimum.

Beside the optimisation heuristics, which are important in the context of Layer 2 of the two-layer Observer/Controller architecture (see Fig. 1), we also deal with machine learning using an eXtended Classifier System (XCS) [15] on Layer 1. In this context, XCS is utilised to recognise the characteristics of the problem and to gain the ability to respond properly to changes in the environment.

XCS is a Learning Classifier System (LCS) variant which implements the reinforcement learning framework and the Q-Learning evaluation method [14]. It reacts to each environmental state and learns by evaluating the action using the given reward. The learning result (i.e., knowledge) is kept in a rule base called population, in form of rules called *classifiers*. Mainly, a classifier consists of a condition C containing ternary alphabets 0, 1, and wildcard, an action A and a reward prediction P (commonly written as "*C:A→P*"). E.g., a classifier "*01:X→100*" means if $C_0 = 0$ and $C_1 = 1$, then propose action X predicting the reward of 100.

XCS aims at having highly accurate classifiers and a small population size (i.e., the number of classifiers in the population) by creating maximally general classifiers. In Sect. 3 of this article, we present a Rule Combining mechanism (XCS-RC) which replaces the discovery component of standard XCS. XCS-RC *infers* maximally general and highly-accurate classifiers using an inductive reasoning method to provide a small population size and to increase the on-line learning speed on Layer 1.

3 Online Learning Using XCS

In learning, OC systems focus on the ability to cope with new and unknown environments. An agent, starting with an empty knowledge base, is given the task to model a problem by collecting information step by step. It observes the current condition and reacts accordingly. Then, it learns using the reward given by the environment. After some time, the system is expected to be capable of modelling the environment correctly and to respond to a given state appropriately.

One of the key factors for successful learning is that the agent guarantees minimum loss of the gathered knowledge, thereby providing an efficient resource usage. Commonly, there is a trade-off between maintaining all the learned rules and minimising the used resources. In case of on-line learning, the difficulty increases even more. The learning process must not include too many trial-and-error cycles because it is potentially harmful to the system. This means, the learning speed is also important.

To achieve the goals mentioned above, a proper classifier generalisation mechanism is required. For this purpose, the standard XCS uses wildcards "(#)" to enable double representation of a state, which can be either one or zero. Hence, the task is to define how to place the wildcards properly.

XCS employs a discovery component consisting of *covering*, *subsumption*, *deletion*, and *genetic operators*. Covering creates classifiers and generalises the conditions by placing wildcards instead of taking value from the given input. The generalisation occurs after a probability parameter $P_{\#}$ is satisfied by a random value. Subsumption deletes less general classifiers to strengthen the more general ones, deletion occurs when the number of classifiers in the population exceeds a threshold, and genetic operators are used to get better classifiers from the population using evolutionary algorithms.

However, the technique has some drawbacks. Firstly, the effect of the randomly placed wildcards in covering is brought through all the XCS learning cycles. Secondly, it is difficult to evaluate the generalising capability of the system (cf. [1]). Therefore, a Rule Combining mechanism (XCS-RC) is proposed here, which implements an inductive reasoning method to generalise classifiers while keeping the accuracy high.

3.1 XCS with Rule Combining (XCS-RC)

The rule combining technique is derived from the concept of making conclusions solely based on the already learned knowledge (i.e., population members) [5]. It generalises classifiers by combining a pair of existing classifiers and replacing them by the result of the combining process. E.g., the rules "*0101:X→100*" and "*0011:X→100*" can be combined into the classifier "*0##1:X→100*" meaning if $C_0 = 0$ and $C_3 = 1$, then propose action X predicting the reward of 100, regardless of the values of C_1 and C_2.

[P] before combining

Index	Cond.	Act.	Pred.
1	0001	X	100
2	0111	X	100
3	0101	X	95
4	0011	Y	10

[P] after combining

Index	Cond.	Act.	Pred.
1*	0##1	X	100
2*	0011	Y	10

(a) Classifiers (1) and (2) are combined while (3) is subsumed, no disproving classifiers currently exist

[P]

Index	Cond.	Act.	Pred.
1	0001	X	100
2	0111	X	100
3	0011	X	10
4	0011	Y	10

(b) Not combinable due to a disproval (3)

Fig. 2 Rule combining

Properly combined rules lead to more general classifiers with high accuracy and less resource usage. A pair of classifiers must have an equal action and a small prediction difference to be combined. Another prerequisite is that, within the population, there are currently no rules disproving the conclusion.

A disproving occurs, if a classifier matching the general condition exists, which has the same action, but whose prediction is outside the tolerable value. E.g., if the value of tolerable prediction is ± 5, then the classifiers "*0001:X→100*" and "*0111:X→100*" cannot be combined into "*0##1:X→100*" if a disproving classifier "*0011:X→10*" exists (see Fig. 2(b)). Contrary to that, a classifier "*0101:X→95*" does not disprove the conclusion, as depicted in Fig. 2(a).

By combining pairs of classifiers, wildcards are inserted based on the gathered information without any random processes. Besides, the subsumption technique, which aims at reducing the number of classifiers, is also implemented after combining. Less general classifiers advising the same action can be subsumed into a newly created classifier, if they satisfy the prediction tolerance constraint (see Fig. 2(a)). This reduces the number of rules without losing any learned knowledge while decreasing the resource usage. Finally, the learning methods are compared as follows.

3.2 Comparison of XCS and XCS-RC

The following single-step and multi-step problems are used to test and compare the generalising capabilities of XCS-RC and XCS: the multiplexer scenario for the single-step problem, and Markovian environments of Woods and Maze for the multi-step tasks (as in [15]). The same parameter settings for XCS and XCS-RC are used, and both are initialised with an empty knowledge base.

In the 6-bit multiplexer, the first two bits are the address, pointing out one of the four value bits. For example, a binary string of "100101" has "10" as the address and "0101" as value bits. The value bits are indexed from left to right and the one pointed out by the address bit is the correct "answer" (in this example is "0"). A series of randomly created binary strings are given to both learning methods, correct answers earn a reward of 1000, and incorrect ones will be awarded by zero. The comparison uses 6-bit, 11-bit, and 20-bit multiplexers having two, three, and four address bits respectively.

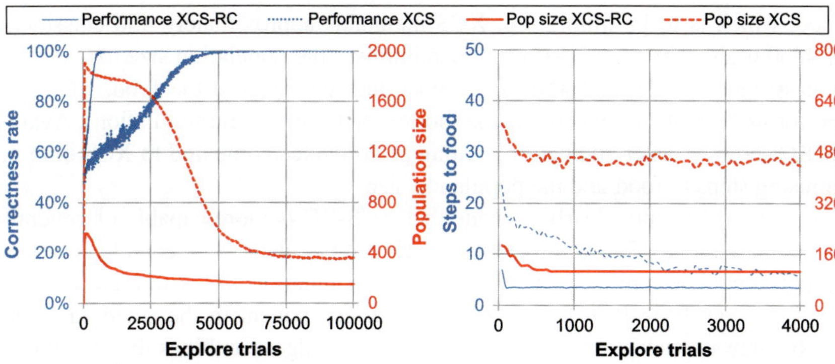

(a) 20-bit multiplexer (b) Maze4 environment

Fig. 3 Comparison of the learning results

For the multi-step problems, the Woods and Maze environments are used to compare the generalisation techniques of XCS and XCS-RC. At the beginning of every trial, an animat is placed randomly within a cellular automaton, where each cell may contain food, rock, or is empty. The animat observes its Moore neighbourhood and chooses the direction to move, either horizontally, vertically, or diagonally. Moving to a cell containing rock is not possible. If the animat tries to do that, its position will remain unchanged.

The objective is to minimise the number of steps to get to the nearest cell containing food. Afterwards, rewards are distributed to all the steps taken by the animat from the starting cell and the next trial begins. The simulations are run in several environment sets (i.e., Woods1, Woods2, Maze4, Maze5 and Maze6).

The challenge for XCS and XCS-RC is to quickly maximise their performance and minimise the population size. The XCS-RC uses neither "normal" subsumptions nor genetic operators from the standard XCS. For every trial, the simulations switch between exploration and exploitation. To highlight the learning effectiveness, only the results from the exploit mode are recorded. Data are taken every 50 exploit trials and average values over 20 simulations are displayed to compare the generalisation techniques of XCS and XCS-RC. Due to space restrictions, only results with the 20-bit multiplexer and the Maze-4 environment are presented here (see [5] for further details).

In Fig. 3(a), the horizontal axis denotes the number of the explore trials and the black lines indicate the correctness rate of the learning progress. A correctness rate of 100% means that all of the last 50 given problems have been answered correctly. The grey lines depict the number of classifiers in the population, which is to be minimised. XCS-RC requires 12 000 trials to achieve a correctness rate of 99.8%, while XCS needs more than 61 000 trials for the equal level of results. Here, XCS-RC shows a higher capability of producing more general classifiers with high accuracies compared to XCS.

Figure 3(b) gives the simulation results in facing multi-step problems using the Maze4 environment for XCS and XCS-RC. The black lines are the number of steps to food, which is to be minimised. XCS maintains approximately 450 classifiers after 4000 trials, while XCS-RC quickly minimises the population size to less than 120 classifiers after around 450 trials. Regarding the steps to the food, XCS-RC outperforms XCS and is able to find the shortest path early in the simulations. Again, XCS-RC shows a faster and better learning performance compared to XCS both in minimising steps to food and the population size.

The provided results clearly indicate that XCS-RC is more capable of generalising compared to XCS. Several goals are achieved at once: (1) General classifiers with high accuracies are created and (2) a small population size with high learning rate is achieved. Instead of using the discovery component in the standard XCS, XCS-RC draws conclusions from the learned knowledge, combines the classifiers, and later ends up with better and quicker learning performance together with efficient usage of resources.

4 Optimisation

As mentioned in Sect. 1, we perform an off-line optimisation on Layer 2 of the two-layer O/C architecture in order to create adequate classifiers for previously unseen conditions. In this context, we have developed a new population-based optimisation algorithm (*Role Based Imitation algorithm*—RBI) that utilises a role assignment strategy for exploring and exploiting agents to facilitate an intelligent search for a system optimum in self-referential fitness landscapes. In the following, we present RBI in more detail.

4.1 The Role-Based Imitation Algorithm (RBI)

One of the major issues in optimisation is the well-known exploration/exploitation dilemma [11] with respect to determining an adequate balance between exploring and exploiting individuals to effectively find the optimum on a given fitness landscape. In this context, RBI proposes a strict distinction between exploring and exploiting agents providing an effective optimisation scheme, where previously found good solutions are kept while other parts of the fitness landscape are further explored. Every element of the population is viewed to be an agent exploring the n-dimensional search space for optimal solutions, i.e. every agent is associated with a solution. At any point in time the assignment of a role ("explorer" or "exploiter") is adapted with respect to (1) the current degree of convergence of a (sub-)population and (2) the relative quality of the agent's solution (see Table 1).

At each optimisation step, an agent determines (1) a random subset of parameters (dimensions) to be optimised and (2) a random comparison set of agents to gather information about their solutions and the corresponding fitness values. Afterwards, the agent calculates the standard deviation of parameter values for each considered

Table 1 The RBI-scheme defining different roles for the agents

	High-convergence (case 1)	Low-convergence (case 2)
Agent with a low fitness	Random exploration	Exploitation (imitation)
Agent with a high fitness	Do nothing	Inductive exploration

Procedure 1 The random exploration procedure

1: $probability = U[0, 1)$;
2: $x = Pi_k * (1 - probability) + Pi_l * probability$;
3: $choice = random(true, false)$
4: **if** $choice == true$ **then**
5: $\quad Pi = x + |x - max(Pi_k, Pi_l)| * U[0, 2)$
6: **else**
7: $\quad Pi = x - |x - min(Pi_k, Pi_l)| * U[0, 2)$
8: **end if**

dimension based on its comparison set and determines whether these parameter values in particular dimensions are sufficiently converged or not. For each considered dimension "i", we distinguish between two cases:

Case 1: High-convergence in the ith dimension
In this case, the agents within the comparison set have similar parameter values for the given dimension. Here, the agent does not change its parameter value and maintains a (possibly) good solution if its fitness value is larger than the average fitness value of the agents in its comparison set (see "Do Nothing" in Table 1). Otherwise, it takes on the role of an explorer (see "Random exploration' in Table 1). Since the agents in the comparison set have similar parameter values in the considered dimension, there is no meaningful information available according to which the agent can change its parameter value. Thus, the agent **randomly** chooses two additional agents (e.g., $agent_k$ and $agent_l$) from the population that were originally **not** in its comparison set and changes its parameter value using Procedure 1.

According to Procedure 1, the agent calculates a random value (variable x) between Pi_k and Pi_l, which are the parameter values of the randomly selected agents $agent_k$ and $agent_l$ in i^{th} dimension, respectively (line 2). Afterwards, the agent determines where to explore using the variable "choice", which is either *true* or *false* with a probability of 50% (line 3). If the variable "choice" is *true*, the agent explores around $max(Pi_k, Pi_l)$, which gives the larger parameter value of Pi_k and Pi_l. Here, the agent calculates an offset using the difference between x and $max(Pi_k, Pi_l)$ and multiplies it with a uniformly distributed random number between 0 and 2 (line 5). The parameter value of the current agent in the ith dimension is determined by simply adding this offset to the variable x (line 5). Line 7 defines the exploration around $min(Pi_k, Pi_l)$, correspondingly.

Case 2: Low-convergence in the ith dimension
In this case, the parameter values of agents in the comparison set have not converged and the agent takes on the role of an explorer or an exploiter according to its fitness

value (see "inductive exploration" and "exploitation" in Table 1). Here, the agent divides other agents in its comparison set into two groups: (1) Agents that have a higher fitness value (the set φ) and (2) agents that have a lower fitness value (the set λ) than itself. Then, the agent checks whether it is *successful* or not in comparison to the more successful agents from the set φ. An agent (e.g., A_1) is successful, if its fitness value is *close enough* to the fitness values of agents that have a higher fitness value than itself, i.e., if the following condition holds:

$$\mathit{fitness}(A_1) > \mathit{meanF}(\varphi) - \mathit{stdDevF}(\varphi) * \beta \quad (4)$$

where *meanF*() and *stdDevF*() are functions to calculate the arithmetic mean and standard deviation of fitness values of agents from the set φ, respectively. The control parameter β is used to determine the closeness between the fitness value of A_1 and $\mathit{meanF}(\varphi)$. According to the condition above, A_1 chooses one of the two possible actions: (1) Inductive exploration and (2) exploitation.

Case 2a: Inductive exploration (anti-imitation)

If condition 4 holds (i.e., agent A_1 is successful), A_1 takes on the role of an explorer and changes the value of its parameter in the considered dimension "away from" those agents that have a lower fitness than itself using the set λ. The underlying assumption is that the parameter values of agents from the set λ are not optimal, and a "repulsion" from these values **might** lead to a better fitness value. Since it is not guaranteed that this kind of repulsion always provides a better fitness, we call this step "inductive exploration". The inductive exploration is implemented using Procedure 2.

Case 2b: Exploitation (imitation)

If condition 4 does not hold (i.e., agent A_1 is not successful), A_1 takes the role of an exploiter and imitates the successful agents (the agents in φ) in its comparison set according to Procedure 3.

In the next section, we present experimental results regarding the optimisation with RBI in self-referential fitness landscapes.

Procedure 2 The inductive exploration procedure	**Procedure 3** The exploitation procedure
$\mathit{offset} = \lvert P_i - \mathit{meanP}_i(\lambda)\rvert * U[0,1)$	$\mathit{offset} = \lvert P_i - \mathit{meanP}_i(\varphi)\rvert * U[0,1)$
if $P_i > \mathit{meanP}(\lambda)$ **then**	**if** $P_i > \mathit{meanP}(\varphi)$ **then**
$P_i = P_i + \mathit{offset}$	$P_i = P_i - \mathit{offset}$
else	**else**
$P_i = P_i - \mathit{offset}$	$P_i = P_i + \mathit{offset}$
end if	**end if**

Fig. 4 The inductive exploration and exploitation procedures. $\mathit{meanP}_i(\lambda)$ and $\mathit{meanP}_i(\varphi)$ are the averaged parameter values of agents from the sets φ and λ in the ith dimension, respectively. U[0,1) is a uniformly distributed random number generated between 0 and 1

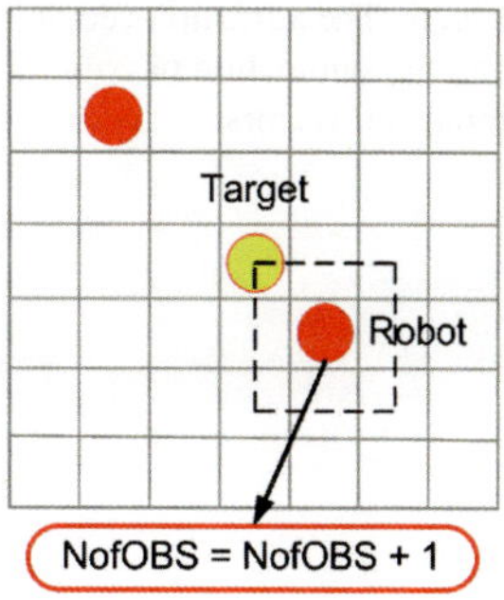

Fig. 5 The observation horizon of a robot

4.2 Optimisation in Dynamic Fitness Landscapes

We compared RBI to the well-known optimisation algorithms Differential Evolution (DE) [12], Particle Swarm Optimisation (PSO) [4] and Simulated Annealing (SA) [6] in static and self-referential fitness landscapes. For the comparison in static fitness landscapes, we used different benchmark functions from the literature, where each algorithm tries to find the global minimum in the given fitness landscape [3].

Since in OC we deal with self-referential fitness landscapes, we focus in this article on the investigation of RBI in a test scenario from the predator/prey domain [2], where the optimal behaviour of a predator depends on the behaviour of other predators in the system. This situation corresponds to a self-referential fitness landscape. In the following, we present our predator/prey scenario in more detail.

We use the simulation toolkit *RePast* [9] to implement our scenario. A single time step in a RePast simulation is called a *tick*. In our experiments, we also use the notion of ticks in producing experimental results. We use the terms "robots" for the predators and "target" for the prey. Each robot has an internal variable "number of observations" (abbr. NofOBS), which is incremented at each tick if the target is within the *observation horizon* of the robot (see Fig. 5). A robot's local objective is to *maximise* the value of its NofOBS, whereas the target evades the robots in order to stay unobserved as often as possible. The moving behaviour of the robots is determined according to an attraction/repulsion model together with a hard constraint, which prevents a robot to move into the observation horizon of another robot (see [2] for more details). Thus, in our scenario the target moves continuously and cannot be captured.

Figure 6 shows that a robot (R) is attracted by the target and repelled by other robots ($R1$ and $R2$). The sizes of the attraction and repulsion vectors are determined by two attributes: (1) the distance between the robots and (2) an optimisation parameter P used to adapt the repulsion forces. Each robot R calculates the sum of these vectors as:

$$\vec{V}_R = P_R \times \sum_{i=1}^{n} \vec{V}_{Ri\_R} + \vec{V}_{R\_Target} \tag{5}$$

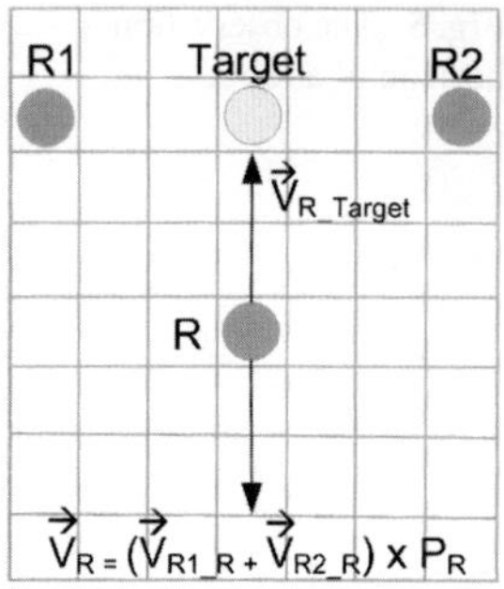

Fig. 6 The attraction vector and the summation of repulsion vectors

where n is the number of robots in the system other than R, and P_R is the optimisation parameter that is used by R to adapt the size of repulsion vectors from other robots.

The behaviour of the target is determined similarly to the behaviour of a robot; it is repelled by the robots and also from the edges of the grid environment. Furthermore, the target is two times faster than each robot in the system.

According to the experimental setup presented above, the fitness landscape of the defined problem (i.e., the optimisation of repulsion parameters P) is self-referential, since we can only call a specific P_R value of a robot (R) "optimal" according to the current distribution of all P values in the system. Since this distribution changes over time, the optimal value for P_R also changes. We provide experimental results in the remainder of this section based on this test scenario.

Parameter Settings and Experimental Results

We compared RBI to DE, PSO, and SA using the experimental setup presented above. In our experiments, we investigated two different scenarios using 25 and 50 robots. A large number of robots result in a more complex scenario, since there are more elements which co-determine the form of the corresponding fitness landscape. Thus, an optimisation algorithm can benefit from the large population size only if it can cope with this complexity.

The search range for P values is set to $[-10, 10]$, and they are initialised randomly between $[-10, -5]$. We used the total number of observations (i.e., $\sum NofOBS$) after 50,000 ticks to measure the system performance. Each robot determines its fitness value **every 100 ticks** by calculating the increase in its NofOBS between the **current** (t_i) and the **previous** optimisation step ($t_i - 100$). Since the robots optimise their behaviour every 100 ticks and we measure the system performance after 50,000 ticks, the number of function evaluations for a single robot is limited to 500 (50,000/100).

In the less complex scenario with 25 robots, the P values obtained by both DE and RBI converge to the same range of values between 1.3 and 1.7, but DE requires approximately 15,000 ticks and RBI only 4,000 ticks for the convergence. Thus, RBI's overall performance is better than DE's in this scenario. The P values with

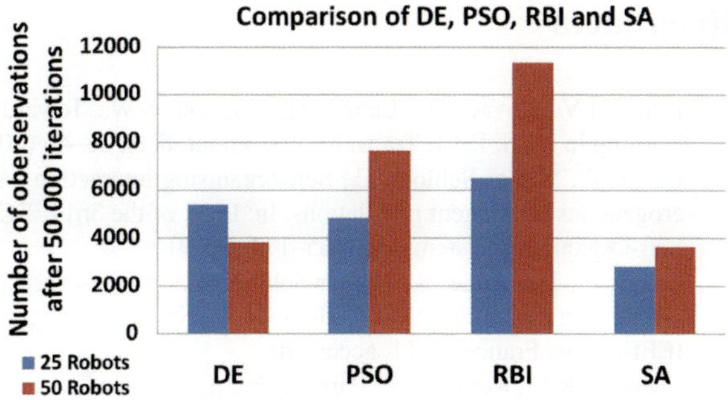

Fig. 7 The total number of observations with different number of robots using DE, PSO, SA and RBI

PSO converge to a range of values between 2.4 and 2.8 after approximately 3,000 ticks. PSO has a higher convergence speed than DE and RBI, but finds only a suboptimal solution, while SA does not converge at all so that the parameter values remain distributed in the search range $[-10, 10]$ over the whole simulation time providing the worst system performance.

In the more complex scenario with 50 robots, we observed that the P values with RBI converge to a range of values between 0.4 and 0.8 after approximately 5,000 ticks, while the P values with PSO converge to a range of values between 0.8 and 1.2 after approximately 4,500 ticks. Since we have a large number of robots in the same environment, small repulsion forces lead to a better system performance. Thus, PSO again finds a suboptimal solution in this scenario. We have observed that DE cannot cope with the increasing number of robots so that the P values remain distributed in the search space $[-10, 10]$ without any convergence as this is the case with SA. Hence, both algorithms provide a worse system performance compared to RBI and PSO. Overall, only RBI and PSO can cope with the increasing complexity in this scenario, while RBI outperforms all its competitors [3].

5 Conclusion

In this article, we investigated different techniques to improve the efficiency of both Layer 1 and Layer 2 of the two-layer Observer/Controller architecture. Firstly, we presented a rule combining mechanism (XCS-RC), which replaces the discovery component of the standard XCS in order to acquire maximally general classifiers and to increase the learning speed. Secondly, we presented a new population-based optimisation algorithm (RBI) that utilises a role assignment strategy to determine the exploring and exploiting agents in order to increase the solution quality in dynamic fitness landscapes. In both cases, we provided experimental results and showed that the presented techniques are effective in finding high-quality results in a short period of time.

Our future work includes the investigation and validation of XCS-RC and RBI in the context of Organic Traffic Control (see Chap. 5.1 and [10]).

References

1. Butz, M.V., Kovacs, T., Lanzi, P.L., Wilson, S.W.: Toward a theory of generalization and learning in XCS. IEEE Trans. Evol. Comput. **8**(1), 28–46 (2004)
2. Cakar, E., Müller-Schloer, C.: Self-organising interaction patterns of homogeneous and heterogeneous multiagent populations. In: Proc. of the 3rd IEEE Int. Conf. on Self-Adaptive and Self-Organizing Systems, pp. 165–174 (2009)
3. Cakar, E., Tomforde, S., Müller-Schloer, C.: A role-based imitation algorithm for the optimisation in dynamic fitness landscapes. In: Swarm Intelligence Symposium, 2011. SIS 2011. IEEE, Paris, France (2011, accepted)
4. Eberhart, R.C., Kennedy, J.: Particle swarm optimization. In: Proc. of the 1995 IEEE Int. Conf. on Neural Networks, pp. 1942–1948 (1995)
5. Fredivianus, N., Prothmann, H., Schmeck, H.: XCS revisited: A novel discovery component for the eXtended Classifier System. In: Proceedings of the 8th Int. Conf. on Simulated Evolution And Learning. LNCS, vol. 6457, pp. 289–298. Springer, Berlin (2010)
6. Kirkpatrick, S., Gelatt, C.D., Jr., Vecchi, M.P.: Optimization by simulated annealing. Science **220**, 671–680 (1983)
7. Müller-Schloer, C.: Organic computing—on the feasibility of controlled emergence. In: CODES+ISSS '04: Proceedings of the International Conference on Hardware/Software Codesign and System Synthesis, pp. 2–5. IEEE Comput. Soc., Washington, DC (2004)
8. Nenortaite, J., Butleris, R.: Application of particle swarm optimization algorithm to decision making model incorporating cluster analysis. In: Human System Interactions, 2008 Conf., pp. 88–93 (2008)
9. North, M.J., Howe, T.R., Collier, N.T., Vos, J.R.: The repast symphony development environment. In: Proceedings of the Agent 2005 Conference on Generative Social Processes, Models and Mechanisms (2005)
10. Prothmann, H., Rochner, F., Tomforde, S., Branke, J., Müller-Schloer, C., Schmeck, H.: Organic control of traffic lights. In: Proceedings of the 5th International Conference on Autonomic and Trusted Computing (ATC-08). LNCS, vol. 5060, pp. 219–233. Springer, Berlin (2008)
11. Rejeb, L., Guessoum, Z., M'Hallah, R.: The exploration-exploitation dilemma for adaptive agents. In: Proceedings of the Fifth European Workshop on Adaptive Agents and Multi-Agent Systems (2005)
12. Storn, R., Price, K.: Differential evolution—a simple and efficient heuristic for global optimization over continuous spaces. J. Glob. Optim. **11**(4), 341–359 (1997)
13. Ursem, R.K., Vadstrup, P.: Parameter identification of induction motors using differential evolution. In: The 2003 Congress on Evolutionary Computation, 2003. CEC '03, vol. 2, pp. 790–796 (2003)
14. Watkins, C.: Learning from delayed reward. Ph.D. thesis (1989)
15. Wilson, S.W.: Generalization in the XCS classifier system. In: Genetic Programming 1998: Proceedings of the Third Annual Conference, pp. 665–674 (1998)

E. Cakar (✉) · J. Hähner · C. Müller-Schloer
Institute of Systems Engineering, Leibniz Universität Hannover, Appelstr. 4,
30167 Hannover, Germany
e-mail: cakar@sra.uni-hannover.de

J. Hähner
e-mail: haehner@sra.uni-hannover.de

C. Müller-Schloer
e-mail: cms@sra.uni-hannover.de

N. Fredivianus · H. Schmeck
Institute AIFB, Karlsruhe Institute of Technology (KIT), Bldg. 05.20, 76128 Karlsruhe, Germany

N. Fredivianus
e-mail: nugroho.fredivianus@kit.edu

H. Schmeck
e-mail: hartmut.schmeck@kit.edu

J. Branke
Warwick Business School, University of Warwick, Coventry CV4 7AL, UK
e-mail: Juergen.Branke@wbs.ac.uk

Chapter 3.2
Combining Software and Hardware LCS for Lightweight On-chip Learning

Andreas Bernauer, Johannes Zeppenfeld, Oliver Bringmann, Andreas Herkersdorf, and Wolfgang Rosenstiel

Abstract In this article we present a novel two-stage method to realise a lightweight but very capable hardware implementation of a Learning Classifier System for on-chip learning. Learning Classifier Systems (LCS) allow taking good run-time decisions, but current hardware implementations are either large or have limited learning capabilities.

In this work, we combine the capabilities of a software-based LCS, the XCS, with a lightweight hardware implementation, the LCT, retaining the benefits of both. We compare our method with other LCS implementations using the multiplexer problem and evaluate it with two chip-related problems, run-time task allocation and SoC component parameterisation. In all three problem sets, we find that the learning and self-adaptation capabilities are comparable to a full-fledged system, but with the added benefits of a lightweight hardware implementation, namely small area size and quick response time. Given our work, autonomous chips based on Learning Classifier Systems become feasible.

Keywords Autonomic System-on-Chip (ASoC) · System-on-Chip (SoC) · Learning classifier system (LCS) · XCS · Learning

1 Introduction

As the number of functions integrated in a single chip increases, the complexity of a chip grows significantly. Furthermore, increasing transistor variability [4, 6], process variation [1], and degradation effects [18] make it increasingly difficult to ensure the reliability of the chip [16]. The International Technology Roadmap for Semiconductors (ITRS) [13] estimates that until 2015, up to 70% of a chip's design must be reused to keep up with the increasing complexity.

Autonomic System-on-Chip (ASoC) ([15], see also Chap. 4.7) add a logical, autonomic layer to contemporary SoCs that helps the designer to manage the com-

---

plexity and reliability issues: decisions that are hard to take at design time because many parameters are uncertain, can be taken at run time by the autonomic layer. Learning Classifier Systems (LCS) have been shown to be able to take the right run-time decisions ([2, 3], see also Chap. 5.3) and even adapt to events that due to the chip complexity have not been foreseen at design time. LCS use a genetic algorithm and reinforcement learning to evolve a set of rules, the interaction of which propose a preferably optimal action to any situation the chip may encounter. Although LCS allow very capable systems for autonomous run-time decisions and self-adaptation, current hardware implementations either require large portions of the chip [5], increasing total chip costs, or have limited learning capabilities [24].

In this article, we present a novel two-stage method to realise an on-chip Learning Classifier System (LCS) that is small, takes the good run-time decisions, and can adapt to unexpected events. In the first stage at design time, we learn a rule set in software using a particular LCS, the XCS [23]. In the second stage, we use the rule set to initialise the lightweight LCS hardware implementation Learning Classifier Table (LCT) [24]. The idea is that the XCS learns just enough rules so that the LCT can adapt to the actual manifestation and conditions of a particular chip and even to unexpected events, albeit in a limited way.

We first compare our method to other LCS implementations using the multiplexer problem, a traditional testbed for LCS [23], and then apply it to two chip-related problems, namely task-allocation and SoC component parameterisation. We show that the LCT can adequately learn and still react to unexpected events. To the best of our knowledge, this is the first study of a lightweight but still capable hardware implementation of an LCS. We think that our work makes using LCS to control chips conceivable.

This work is structured as follows. Section 2 gives an overview of related work. Section 3 introduces the XCS and the hardware implementation LCT. Section 4 describes our proposed method. Section 5 presents the three benchmarks multiplexer, task-allocation and SoC component parameterisation that we use to assess our method. Section 6 shows the results of our assessment and Sect. 7 concludes this article.

2 Related Work

Learning Classifier Systems were originally introduced in [12]. The XCS was first presented in [21] and later refined in [23]. The XCS has been used in a large range of learning and classification problems, including controlling a robotic mouse [10], a system-on-chip (SoC) [3], the lights of a traffic junction [17], and for finding suitable partitions in hardware-software co-design [11]. A first hardware implementation of an XCS has been presented in [5], named XCS_i, which uses fixed-point arithmetic. The implementation shows good learning rates of the XCS_i, but is quite large. In [24], the authors present an optimised hardware implementation of an LCS, called the Learning Classifier Table (LCT), which is small but has no mechanism to create new classifiers. Using a hand-crafted initial rule set, the authors show that the LCT can adjust the frequency of a SoC according to a given objective function.

The most popular machine learning algorithms for which hardware implementations exist are neural networks [9, 19] and, more recently, support vector machines [14]. Along with the fact that for these systems, "the actual rules implemented [are] not apparent" [19], their implementations are about five times as large as the LCT [14].

3 XCS and LCT

We briefly describe the XCS and LCT and refer to [7, 22–24] for further details.

The XCS learns a minimal set of *classifiers* (or *rules*) the interaction of which, in the ideal case, provide an optimal response (called *action*) for a given situation. The learning is based on a genetic algorithm and reinforcement learning. Each classifier (or rule) consists of a condition, an action, a reward prediction, the reward prediction accuracy, and some other house keeping values. The condition is a string of bits ('0', '1', and the don't-care symbol '#'). At each learning step, the XCS matches the input signal with the condition of each classifier and notes the actions and accuracy-weighted reward predictions that each classifier proposes. The XCS then selects an action to apply: in the exploit mode, it chooses the action that promises the highest reward, while in the explore mode, it chooses a random action to find new alternatives. After the action has been applied, the XCS receives a reward depending on the new state and updates its reward predictions and classifier set accordingly. After some number of iterations, the genetic algorithm repeatedly creates new, possibly better suited rules.

The LCT consists of a memory, which holds a fixed number of classifiers, and hardware-based mechanisms for action lookup and fitness update. There is no mechanism to generate new classifiers. The classifiers in the LCT consist only of a condition, an action and a fitness, similar to the fitness in the strength-based ZCS [20]. To realise the don't-care bits, the LCT first logically ANDs the monitor signal with a mask before comparing it with the bit value. The LCT selects the action of a matching classifier randomly according to the classifier's relative fitness (roulette-wheel selection) using weighted reservoir sampling to ensure a fixed lookup time. After receiving the reward for a previously applied action, the LCT distributes the reward r to the classifiers of the action set and updates the fitness f according to $f \leftarrow \beta r + (1-\beta) f$ with the learning rate $0 \leq \beta \leq 1$.

4 Methodology

One major trade-off of hardware-based machine learning lies between the learning capabilities of the implementation and the allotted hardware resources: the system is either very capable but requires a lot of resources or it requires little resources but is less capable. We address this problem with the following two-stage approach:

1. At design time, the software-based XCS learns a (preferably optimal) set of rules to solve a given problem.
2. We translate the XCS rules with our `xcs2lct` tool into a form that is suitable for the LCT. Initialised with these rules, the hardware-based LCT continues to learn at run time.

With this setup, we can use all the resources that are available to a capable software implementation (the XCS) and use the acquired knowledge in a lightweight hardware implementation (the LCT). The idea is that the XCS learns a rule set that allows the LCT to adapt to the actual manifestation and conditions of a particular chip and even to an unexpected event, despite its limited learning capabilities.

As the chip area that is necessary to store the classifiers in memory constitutes the largest part of the LCT, we would like to minimise the number of necessary classifiers to keep the chip area requirement small. We therefore consider translating both all XCS rules to corresponding LCT rules (all-XCS translation) and only the top performing rules (top-XCS translation). The `xcs2lct` translates the rules according to the following algorithm, which ensures that the XCS and the LCT classifiers match the same input values:

```
foreach b ← xcs-rule[i] do
    if b == '#' then lct-rule[i].(mask,bit) ← ('0', '0');
    else             lct-rule[i].(mask,bit) ← ('1', b);
```

To compare our method with the base performance of the LCT, we also consider two more ways to generate LCT rules, full-constant and full-reverse. Both translations provide all possible LCT rules, that is, a complete condition-action table[1] as there is no known method to generate an appropriate rule table for the LCT. The full-constant translation initialises the rule fitness to half the maximum reward (500) and, as it is independent of the XCS rules, represents the bottom line of LCT's own learning capabilities. The full-reverse translation sets the rule fitness to the highest predicted reward of all matching XCS rules, or zero, if no XCS rule matches, and represents the combined learning capability of the LCT and the XCS.

The original action selection strategy for the LCT is roulette-wheel, which selects actions randomly according to the relative predicted reward of the matching classifiers, similar to the explore mode of the XCS. Additionally, we also consider the winner-takes-all strategy, which selects the action whose matching classifiers predict the highest reward, similar to the exploit mode of the XCS. However, unlike in the XCS, in the LCT the accuracy of the prediction does not influence the action selection.

While the XCS is usually configured to alternate between the explore and exploit mode, in our experiments the LCT uses only one of either strategies. We leave the analysis of alternating strategies in the LCT as future work.

[1] Of course, the memory requirements of the classifiers generated with full-* grow exponentially with the problem size. We use them only for comparison.

5 Experimental Setup

We use three problem types to assess our method: multiplexer [21], task allocation [2], and SoC component parameterisation. Additionally, we define an unexpected event for each problem type to explore LCT's learning ability. As the XCS has already been shown to be able to solve these problem types and adapt to unexpected chip events [2], in this work we concentrate on the performance of the LCT.

The *multiplexer* problem is a typical LCS benchmark [23]. The n-multiplexer-problem is defined over binary strings of length $n = k + 2^k$. The first k bits index a bit in the remaining bits. The correct action for the LCS is the value of the indexed bit. For example, in the 6-multiplexer problem, $m_6(011101) = 0$ and $m_6(100100) = 1$. We define the *inversed multiplexer* as the unexpected event for the multiplexer, that is, the LCS is supposed to return the inversed value of the indexed bit. For example, in the inversed 6-multiplexer problem, $\overline{m}_6(011101) = 1 - m_6(011101) = 1$. We use the same XCS parameters as the full-fledged FPGA implementation of XCS presented in [5] to have comparable results: $\alpha = 0.1$, $\beta = 0.2$, $\delta = 0.1$, $\varepsilon_0 = 10$ (which is 1% of the maximum reward), $\nu = 5$, $\theta_{\text{GA}} = 25$, $\chi_{\text{GA}} = 0.8$, $\mu_{\text{GA}} = 0.04$, $P_{\#} = 0.3$; GA subsumption is on with $\theta_{\text{GAsub}} = 20$, while action set subsumption is off. We do not use generalisation or niche mutation. The reported results are averages over 20 runs.

The *task-allocation* problem has been first introduced in [2] and is motivated by the advent of multi-core systems, where tasks can be run on several cores simultaneously to increase overall reliability. In the (L,i)-task-allocation problem, the LCS must allocate i available tasks on $L \geq i$ cores, some of which are known to be occupied and thus not available. The system input is a binary string of length L, where each bit represents the occupation of a particular core. There is one action for each possible allocation plus a special action that indicates that no allocation is possible (e.g., when all cores are already occupied), totalling $\binom{L}{i} + 1$ possible actions. An action is valid and returns the maximum reward if the corresponding allocation only allocates available cores; otherwise the reward is zero. The unexpected event for the task-allocation problem is the unmonitored failure of a core: although reported as available, the core cannot be occupied, and an allocation of that core returns zero reward. For the task-allocation problem, we use the XCS parameters from [2] to have comparable results, which differ from the multiplexer settings only in the following parameters: $\alpha = 1$, $\theta_{\text{GA}} = 250$, $\chi_{\text{GA}} = 0.1$, $\mu_{\text{GA}} = 0.1$, $P_{\#} = 0.4$; GA subsumption is off. The reported results are averages over 5 runs, due to the longer simulation time for the many problem instances.

The *SoC component parameterisation* problem demonstrates the ability of LCS to dynamically parameterise a system-on-chip at run time, similar to [3]. The system consists of a processing core that is subject to random load fluctuations. As the load changes, the LCS is responsible for setting the operating frequency of the core as low as possible (i.e., maintaining as high a utilisation as possible), while ensuring that the core can keep up with the workload. The monitor input consists of the core's current frequency as well as its utilisation. There are five possible actions: four actions to increase or decrease the core's operating frequency by 10 or 20 MHz over a range from 50 to 200 MHz, and one action to keep the core's frequency unchanged. The reward for each action is calculated by comparing the value of a system-wide objective function before and after the action is applied. The objective function indicates how far from the designer-specified optimum of high utilisation and low error rate the system is currently operating and is defined as $f_{obj} = (100\% - \text{utilisation}) + \text{error\_rate}$, where a low value indicates that the system is operating near its optimum. A base reward of half the maximum reward (500) is given when the objective function returns the same value before and after the action is carried out. This is the lowest possible reward without actively worsening the system's operating state. The unexpected event for the component parameterisation problem is a manufacturing defect that causes critical timing errors for operating frequencies in excess of 100 MHz. As a result, increasing the frequency above 100 MHz causes the core to cease functioning, resulting in wasted cycles for error correction and providing lower rewards to the LCS. With timing errors, the LCT must therefore learn to cap the frequency at 100 MHz, even when the workload would warrant higher operating frequencies. We use the same XCS parameters as for the task-allocation problem, except for $\alpha = 0.8$ and $P_{\#} = 0.1$. The reported results are averages over 100 runs.

We use the implementation of the XCS in the programming language C as described in [8] as the software version of XCS. We use a SystemC-based simulation model of the LCT hardware implementation described in [24], with the additional winner-takes-all strategy described in Sect. 4.

We compare the performance of the LCT that has been instructed using our method with the base performance of the LCT, the performance of the full-fledged hardware implementation of the XCS presented by [5], the performance of the XCS reported in [2], and the performance of the software version of the XCS. We also check whether the LCT retains the capability of LCS to adapt to unexpected events.

6 Results

In this section, we present the results on the three problem types multiplexer, task-allocation, and SoC component parameterisation mentioned previously.

6.1 Multiplexer

Figure 1 shows the correctness rate (x-axis) and population size (y-axis) for the 6-, 11-, 20-, and 37-multiplexer problem for all eight possible combinations of translations and action selection strategies for the LCT. Note that the x-axis starts at 70% correctness rate and that the scales of the y-axes differ. The top-XCS translation uses only classifiers that predict the maximum reward with perfect accuracy. As we aim for a small but correct LCS, in each graph lower right is better. The figures show that

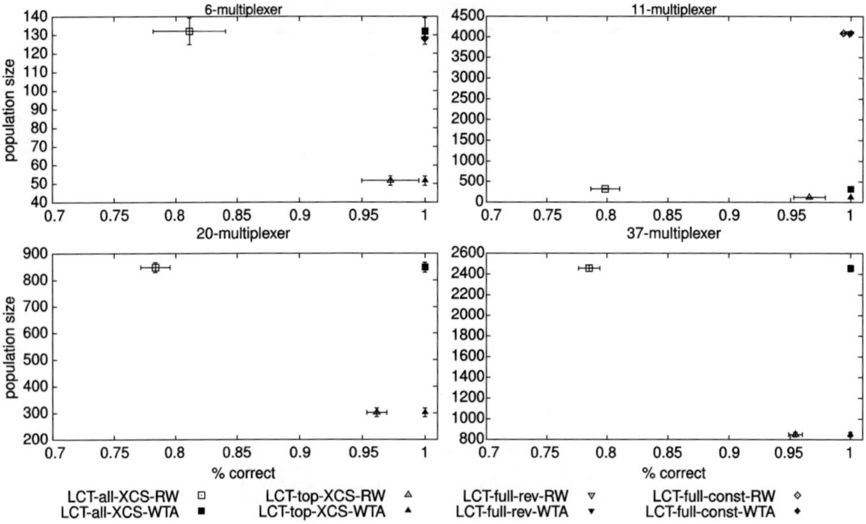

Fig. 1 Performance in the multiplexer problem. *Clockwise from upper left:* 6-, 11-, 20-, and 37-multiplexer. Within each graph, lower right is better. Note that the y-axes differ in scale. *Error bars* are standard deviations σ in the respective dimension

in the new winner-takes-all (WTA) strategy (solid symbols), the LCT solves the multiplexer problem perfectly, while in the original roulette-wheel (RW) strategy (empty symbols), it solves only between 80% and 97% of the problem instances. With the winner-takes-all strategy, the LCT shows the same results as the full-fledged XCS implementation presented in [5]. The figure also shows that the population size of the all-XCS translation (square symbol) is about three times the population size of the top-XCS translation (upwards triangle symbol) for all multiplexer problems. As the population sizes for the full-* translations rise exponentially, we excluded them from the 20- and 37-multiplexer problem.

All LCT configurations were able to perfectly adapt to the unexpected event of the inversed multiplexer problem (not depicted), given a rule base that the XCS has learned for the (regular, non-inversed) multiplexer problem. However, the LCT can only adapt to the inversed multiplexer problem, if the XCS was able to solve the multiplexer problem sufficiently well (e.g., because XCS' learning process was terminated prematurely). Otherwise, even if the XCS shows a correctness rate of 100%, not all LCT configurations can adapt to the inversed multiplexer. Figure 2 illustrates the case for $\overline{m}_{11}$. While the configurations all-XCS and full-const solve 80–100% of the inversed multiplexer problem, the top-XCS and full-rev solve no more than 30%. The correctness rate did not change further until 1 million steps. We assume that the prematurely terminated XCS contains too many high-rewarding rules that are falsely marked as accurate because they were trained on only few problem instances, disturbing the results of the top-XCS and full-rev translations.

From the results in the multiplexer problem, we conclude that with the all-XCS translation the LCT shows both a high correctness rate and retains the capability to

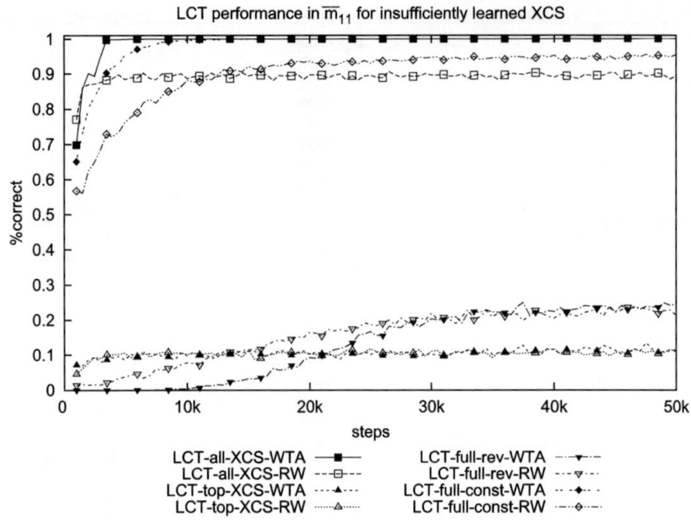

Fig. 2 LCT performance in the inversed multiplexer problem $\overline{m}_{11}$ using rules from an insufficiently learned XCS. $\sigma < 0.005$ if rate $> 80\%$; $\sigma < 0.1$ if rate $< 30\%$

adapt to unexpected events. When using the full-const translation, we find similar results. Combining XCS' knowledge and LCT's own learning capabilities in the full-rev translation leads to an LCT whose capability to adapt to unforeseen events is very sensitive to the quality of the XCS rules. Similar is true when using only the top performing XCS rules with the top-XCS translation. As for more real-world problem types the XCS cannot always learn perfectly, we will concentrate on the all-XCS translation in the following experiments.

6.2 Task Allocation

Figure 3 shows the rate R_{LCT} of valid task allocations of the LCT for the (L,i)-task-allocation-problems, $1 \leq i < L \leq 10$, and R_{XCS} for comparison. The x-axis shows the problem instances and the y-axis shows run-time R_{LCT} and design-time R_{XCS}. From the figure we note that the LCT uses rule bases for which the XCS correctly allocates more than 90% of the problem instances for $L < 9$ and more than 80% for $9 \leq L \leq 10$, comparable to what has been reported in [2]. We find that the LCT using the winner-takes-all strategy (WTA) has very similar rates to the XCS, with a larger difference only for $L = 10$. Using the roulette-wheel strategy (RW), the LCT finds valid allocations considerably less often; in particular for $1 < i < L-1$, R_{LCT} drops as low as 22%. The reduced performance in the $(10,5)$ and $(10,6)$ problem instances concurs with the findings in [2] that these two problem instances are the most difficult for the XCS.

To test LCT's ability to adapt to unexpected events, we initialise the LCT with the all-XCS-translated XCS rules and let the cores fail randomly every 5 000 steps.

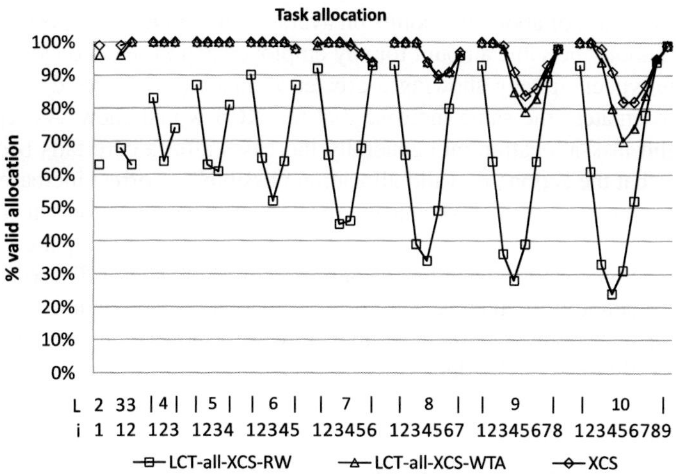

Fig. 3 Rate R_{LCT} of valid task allocations in the LCT and R_{XCS} for comparison. $\sigma < 11\%$ or better for any setting

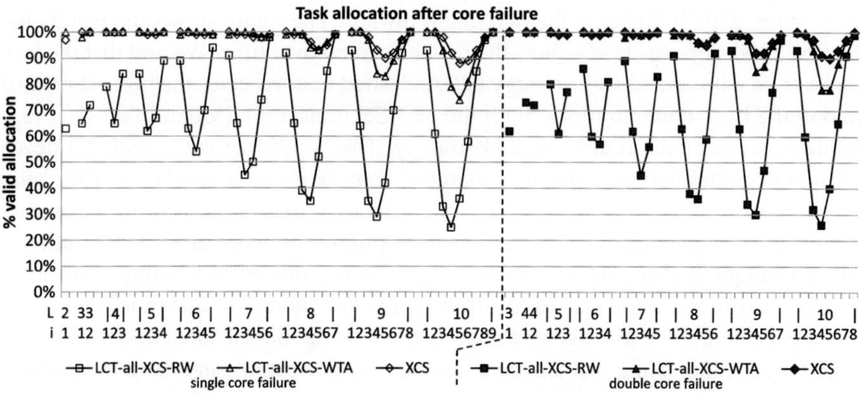

Fig. 4 R_{LCT} after one or two randomly chosen cores have failed, and R_{XCS} for comparison. After one core has failed, $\sigma < 7\%$; after two cores have failed, $\sigma < 5\%$

Note that there is no further rule sharing between the XCS and the LCT besides the initialisation of the LCT; we depict the XCS solely for comparison purposes.

Figure 4 shows R_{LCT} and R_{XCS} after the first (left half) and the second (right half) randomly chosen cores have failed. Note that the diagram shows fewer problem instances for the second core failure, as not every instance allows the failure of two cores (e.g., when allocating three tasks out on four cores, the failure of two cores turns the problem unsolvable). We find that the rate of valid task allocations of the LCT increases slightly, on average by about 1%-point (maximum 10%-points) after the first core has failed and an additional 1%-point (maximum 11% -points) after the second core has failed. Compared to the rates before any core has failed, we find an increase of about 2%-points on average (maximum 17%-points). The increase is of about the same amount for any employed action selection strategy, with the roulette-wheel strategy showing a greater variance (not depicted). The results show approximately the same increase that the XCS would show. As reported in [2], the valid task-allocation rate generally increases after a core fails because the probability that the action "no valid allocation possible" is correct increases.

Summarising, we find that when using the winner-takes-all action selection strategy, the LCT shows rates of valid task allocations which are comparable to what we find in the XCS and to what has been reported in [2]. The LCT also retains the capability to adapt to the unexpected failure of two cores, as previously shown for the XCS in [2]. The roulette-wheel strategy, however, shows high rates of valid task allocations only for some border cases.

6.3 Component Parameterisation

Figure 5 shows the reward returned to the LCS in the SoC component parameterisation problem before (left) and after (right) the unexpected event of malfunctioning in the core, with 1000 being the maximum reward. The figure shows the results for the first 3000 steps to clearly show the reward's trend over time. We find that the less explorative winner-takes-all strategy (WTA, dashed line) receives the highest reward among the LCT configurations, with the all-XCS translation (square) being on top. While on average the roulette-wheel strategy (RW, solid line with symbols) never actively degrades performance, it is unable to achieve even the level of performance that a static, non-learning winner-takes-all strategy (cross on dashed line) achieves given the XCS-generated rule set as a starting point. The more explorative roulette-wheel strategy is also unable to show a significantly improved learning behaviour, clearly making the winner-takes-all strategy a better choice for this problem.

As expected, the initial average reward when using the full-const translation (triangle) is 500, indicating that an equal number of rules benefit and harm the system.

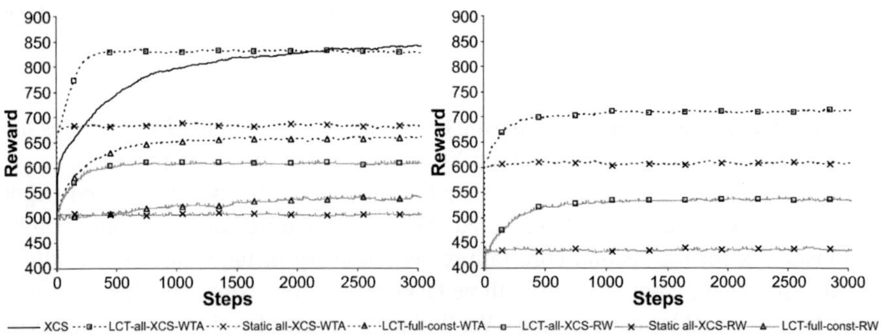

Fig. 5 Reward averaged over 100 runs for component parameterisation with fully functional (*left*) and defective (*right*) component. After stabilisation, $\sigma < 20$. Learning rate of XCS used to generate LCTs' initial rule set included for comparison

Even though the winner-takes-all strategy is quickly able to achieve higher rewards, it is not able to achieve the same level of reward as a system initialised with a design-time generated rule set (all-XCS, square). The roulette-wheel strategy is only able to attain a very slight improvement to its average reward.

Comparing the final reward of the design-time XCS (solid line with no symbols) with the initial rewards of the run-time LCT using all-XCS translation shows a surprising discrepancy. Although the LCT uses the rules learned by the design-time XCS, we find a severe drop in the initial reward (from ~840 to ~650). We presume that this is because the LCT does not incorporate the complete functionality of the XCS. For example, the LCT cannot sufficiently represent XCS rules with high accuracy but low prediction, as the LCT does not store accuracy. Thus, the LCT must initially re-learn portions of the design space. Fortunately, the LCT is able to perform this initial re-learning fairly quickly within the first 500 steps.

Figure 5 shows the results of the component parameterisation problem with the unexpected event as explained in Sect. 5. The results are very similar to those of the non-defective system, except that the average reward achieved by the system is somewhat lower than before. In fact, the starting rewards of less than 500 for the roulette-wheel strategy (solid line) indicate that, initially, a majority of actions are taken that disadvantage the system. As before, the learning capabilities of the LCT quickly achieve an increase in the average reward. However, the fact that any frequency above 100 MHz results in timing errors prevents the system from adapting to heavier load scenarios, forcing the system to operate at a lower degree of optimality and generally reducing the achievable maximum rewards.

In summary, we find that the LCT using the winner-takes-all action selection strategy and the all-XCS translation is capable to solve the SoC component parameterisation problem, even in the event of a unexpected manufacturing defect.

7 Conclusions

In this article, we have presented a two-stage method that combines the capability of the software-based XCS with the area efficiency of the LCS hardware implementation LCT. In the first stage at design time, the XCS initially learns a set of classifiers based on a software simulation of a given problem. In the second stage, we translate the classifiers into rules that are suitable for the LCT and apply the LCT to the same problem at run time.

We showed that with our newly introduced winner-takes-all action selection strategy for the LCT, the LCT can solve the multiplexer, the task-allocation and the SoC component parameterisation problem, if we initialise it with all rules that the XCS has learned (all-XCS). In addition, the LCT retains the capability to adapt to the unexpected events of the problems, which includes the unexpected failure of two cores and the manufacturing defect of a core. We also found that the performance of the LCT is less sensitive to the performance of the XCS when using the all-XCS translation.

In summary, the results show that our proposed method allows a small and lightweight yet very capable hardware implementation of an LCS, with which the autonomic control of chips using LCS becomes feasible.

In future work, we will investigate alternating between roulette-wheel and winner-takes-all action selection for quicker adaptation to unexpected events in the LCT. We will also examine ways to reflect XCS' knowledge of reward prediction accuracy in the reward of the generated LCT rules, avoiding the initial drop in returned reward, and we will look for a trade-off between the good performance of all-XCS and the smaller classifier set of top-XCS.

Acknowledgements This work has been partially sponsored by Deutsche Forschungsgemeinschaft under the grants 'HE 4584/3-3' and 'RO 1030/14-3'.

References

1. Agarwal, A., Zolotov, V., Blaauw, D.T.: Statistical clock skew analysis considering intra-die process variations. IEEE CAD **23**(8), 1231–1242 (2004)
2. Bernauer, A., Bringmann, O., Rosenstiel, W.: Generic self-adaptation to reduce design effort for system-on-chip. In: IEEE SASO, pp. 126–135 (2009)
3. Bernauer, A., Fritz, D., Rosenstiel, W.: Evaluation of the learning classifier system XCS for SoC run-time control. In: LNI, vol. 134, pp. 761–768. Springer, Berlin (2008)
4. Bernstein, K., Frank, D., Gattiker, A., Haensch, W., Ji, B., Nassif, S., Nowak, E., Pearson, D., Rohrer, N.: High-performance CMOS variability in the 65-nm regime and beyond. IBM J. Res. Dev. **50**(4/5), 433 (2006)
5. Bolchini, C., Ferrandi, P., Lanzi, P.L., Salice, F.: Evolving classifiers on field programmable gate arrays: Migrating XCS to FPGAS. J. Syst. Archit. **52**(8–9), 516–533 (2006)
6. Borkar, S.: Thousand core chips: a technology perspective. In: DAC, pp. 746–749. ACM, New York (2007)
7. Butz, M., Wilson, S.W.: An algorithmic description of XCS. In: Lanzi, P.L., Stolzmann, W., Wilson, S.W. (eds.) IWLCS '00. Lecture Notes in Artificial Intelligence, vol. 2321, pp. 253–272. London Math. Soc., London (2001)
8. Butz, M.V., Goldberg, D.E., Tharakunnel, K.: Analysis and improvement of fitness exploitation in XCS: bounding models, tournament selection, and bilateral accuracy. Evol. Comput. **11**(3), 239–277 (2003)
9. Dias, F.M., Antunes, A., Mota, A.M.: Artificial neural networks: a review of commercial hardware. Eng. Appl. Artif. Intell. **17**(8), 945–952 (2004)
10. Dorigo, M.: ALECSYS and the AutonoMouse: Learning to control a real robot by distributed classifier systems. Mach. Learn. **19**(3), 209–240 (1995)
11. Ferrandi, F., Lanzi, P.L., Sciuto, D.: Mining interesting patterns from hardware-software codesign data with the learning classifier system XCS. Evol. Comput. **2**, 8–12 (2003)
12. Holland, J.H.: Adaptation. In: Rosen, R., Snell, F.M. (eds.) Progress in Theoretical Biology, pp. 263–293. Academic Press, New York (1976)
13. International Roadmap Committee. International Technology Roadmap for Semiconductors (2008). http://www.itrs.net/reports.html
14. Irick, K., DeBole, M., Narayanan, V., Gayasen, A.: A hardware efficient support vector machine architecture for FPGA. In: FCCM '08, pp. 304–305. IEEE Comput. Soc., Washington (2008)
15. Lipsa, G., Herkersdorf, A., Rosenstiel, W., Bringmann, O., Stechele, W.: Towards a framework and a design methodology for autonomic soc. In: ICAC (2005)

16. Narayanan, V., Xie, Y.: Reliability concerns in embedded system designs. Computer **39**(1), 118–120 (2006)
17. Prothmann, H., Rochner, F., Tomforde, S., Branke, J., Müller-Schloer, C., Schmeck, H.: Organic control of traffic lights. LNC **5060**, 219–233 (2008)
18. Schlunder, C., Brederlow, R., Ankele, B., Lill, A., Goser, K., Thewes, R., Technol, I., Munich, G.: On the degradation of p-MOSFETs in analog and RF circuits under inhomogeneous negative bias temperature stress. In: IEEE IRPS, pp. 5–10 (2003)
19. Widrow, B., Rumelhart, D.E., Lehr, M.A.: Neural networks: applications in industry, business and science. Commun. ACM **37**(3), 93–105 (1994)
20. Wilson, S.W.: Classifier systems and the animat problem. Mach. Learn. **2**(3), 199–228 (1987)
21. Wilson, S.W.: ZCS: A zeroth level classifier system. Evol. Comput. **2**(1), 1–18 (1994)
22. Wilson, S.W.: Classifier fitness based on accuracy. Evol. Comput. **3**(2), 149–175 (1995)
23. Wilson, S.W.: Generalization in the XCS classifier system. In: Koza, J.R., Banzhaf, W., et al.(eds.) Genetic Programming Conference, University of Wisconsin, Madison, Wisconsin, USA, pp. 665–674. Morgan Kaufmann, San Mateo (1998)
24. Zeppenfeld, J., Bouajila, A., Stechele, W., Herkersdorf, A.: Learning classifier tables for autonomic systems on chip. In: Hegering, H.-G., Lehmann, A., Ohlbach, H.J., Scheideler, C. (eds.) GI Jahrestagung (2). LNI, vol. 134, pp. 771–778 (2008)

A. Bernauer (✉) · W. Rosenstiel
University of Tübingen, 72076 Tübingen, Germany
e-mail: bernauer@informatik.uni-tuebingen.de

J. Zeppenfeld · A. Herkersdorf
Technische Universität München, 80290 München, Germany

J. Zeppenfeld
e-mail: zeppenfe@tum.de

O. Bringmann
Forschungszentrum Informatik, 76131 Karlsruhe, Germany
e-mail: bringman@fzi.de

Chapter 3.3
Collaborative Learning by Knowledge Exchange

Dominik Fisch, Edgar Kalkowski, and Bernhard Sick

Abstract Humans act efficiently in a dynamic environment by learning from each other. Thus, it would be highly desirable to enable intelligent distributed systems, e.g., multi-agent systems, smart sensor networks, or teams of robots, to behave in a way which follows that biological archetype. The constituents of a such a distributed system may learn in a collaborative way by communicating locally learned classification rules, for instance. This article first gives an overview of the techniques that we have developed for knowledge exchange. Then, their application is demonstrated in a realistic scenario, collaborative detection of attacks to a computer network.

Keywords Knowledge exchange · Collaborative learning · Classification

1 Introduction

Humans learn from other humans and we claim that intelligent technical systems operating in a dynamic environment—such as collaborating robots, smart sensors of a sensor network, or software agents in the Internet—should also learn from others. Humans do not only learn from other humans by communicating recently made observations but also by exchanging *knowledge* (e.g., rules) they have gained from that observations. In a dynamic environment, where new knowledge emerges or old knowledge becomes obsolete, intelligent systems—simply referred to as (organic) agents in the following—must adapt on-line to their local environment by means of appropriate machine learning techniques. If they exchange learned knowledge, they will efficiently be enabled to cope with particular situations they were not confronted with before. As an example, consider an agent that must detect network-based attacks to a computer connected to the Internet. If this agent learns to recognise new attack types or specific attack situations, for instance, it may share this knowledge with other intrusion detection agents in order to improve the detection capabilities of the overall distributed intrusion detection system of which it is a constituent (cf. [5, 14]). The overall agent system learns in a collaborative way.

In this article, *knowledge* refers to *classification rules* that represent uncertain knowledge. We consider applications where rules that are locally acquired by the agents are *potentially useful* for other agents. That is, if those agents observe the same situation (i.e., data originating from the same processes in their environment),

C. Müller-Schloer et al. (eds.), *Organic Computing — A Paradigm Shift for Complex Systems*, 267–280
Autonomic Systems, DOI 10.1007/978-3-0348-0130-0_17,

the rules can be applied successfully. References to related work on knowledge exchange can be found in [7]. It can be stated that most related work focuses on "crisp" rules which makes knowledge exchange a much easier task.

The remainder of the article is structured as follows: In Sect. 2 we give an overview of the techniques that have been developed to realise collaborative learning by knowledge exchange. Section 3 demonstrates the application of these techniques in a realistic scenario, namely collaborative attack detection as outlined above. The major findings are summarised in Sect. 4.

2 Overview of Methodological Foundations

In our work, we are focusing on classification problems and, thus, the kind of knowledge we want to exchange between organic agents are classification rules. If we think about how knowledge exchange can be realised, we are facing a large variety of research issues. The main questions are:

1. How can knowledge be represented or, in other words, which classifier paradigms can be used and how can these be trained from sample data (either off-line or on-line)?
2. How can a need to acquire new knowledge (novelty) or a possibility to discard outdated knowledge (obsoleteness) be detected by an organic agent?
3. How can knowledge (in form of rules) be extracted from one classifier and how can it be integrated in another classifier (e.g., fused with already existing knowledge)?
4. How can various knowledge properties be assessed numerically (i.e., importance, uniqueness, comprehensibility) and how can a knowledge exchange process be improved by using this kind of meta-knowledge?
5. How can solutions for the mentioned problems be combined in an organic agent for knowledge exchange?

Other important questions that will not be addressed here are, for instance: How can knowledge be distributed in an agent system (e.g., broadcasting or distribution on demand)? How do agents assess the competence of others? How can the knowledge exchange process itself be assessed numerically?

2.1 Layered Architecture of an Organic Agent

An agent for knowledge exchange has a three-layered architecture [2]:

- The main component of the **reaction layer** is the active classifier. Data observed by sensors—so-called samples—are classified and responses are triggered depending on the classification result. Then, the environment is controlled by means of actuators.

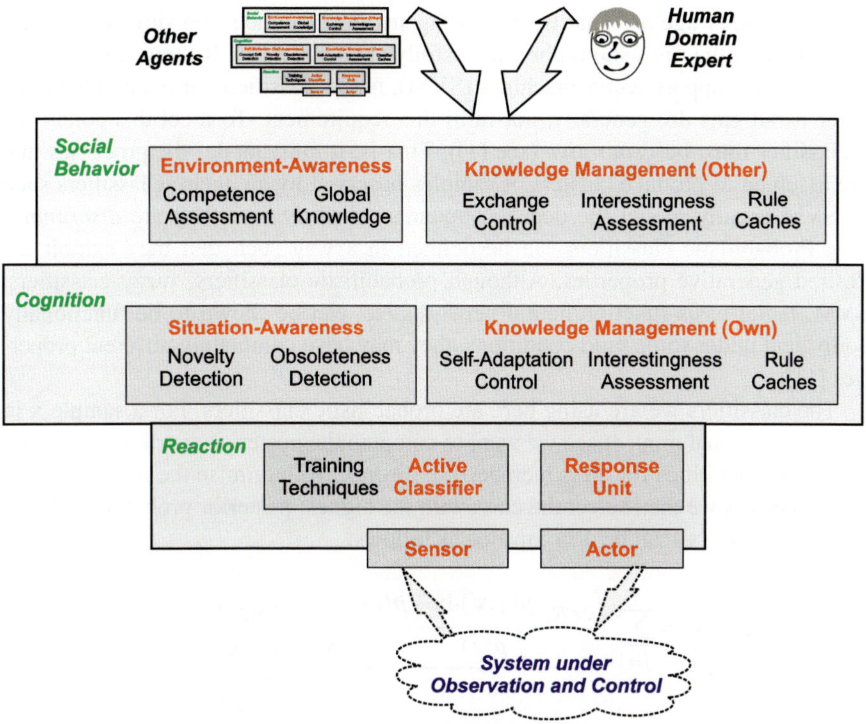

Fig. 1 Layered architecture of an organic agent

- The **cognition layer** realises a kind of *situation awareness*: The agent is aware of what it does know and what it does not know. It is able to detect the need for new knowledge and to detect when existing knowledge becomes obsolete due to changes in the dynamic environment. Additionally, it is able to react accordingly by either learning new rules on its own or by taking over rules received from other agents (which are intermediately stored in rule caches). The agent also assesses new rules with respect to their interestingness.
- The **social behaviour layer** is concerned with the interaction with other agents to exchange knowledge and with the communication with human experts to obtain rule conclusions, for instance. This requires a certain *environment awareness*.

A number of such agents (see Fig. 1) collaborates in a multi-agent framework.

2.2 Knowledge Representation and Off-line-Training

An important pre-condition for knowledge exchange is that the classification rules must be represented in a way that can be termed to be "objective" in the following sense: Rules that can successfully be applied by one agent must be potentially useful for other agents. That is, if these agents observe the same situation in their environment, the rules should be successfully applicable, too. It turns out that rules contained in support vector machines (SVM), fuzzy classifiers, or many other classifier paradigms do, generally, not fulfil this requirement. To meet this condition, a classifier must be *generative* (see [1]). That is, it must model the processes that are assumed to produce clusters of samples observed by an agent. Classifiers such as SVM usually model the decision boundary directly, i.e., they are *discriminative*. Probabilistic classifiers can be defined in a way such that they exhibit the desired generative properties. Although probabilistic classifiers, fuzzy classifiers, SVM, radial-basis-function neural networks etc. can be shown to be functionally equivalent under some mild conditions, they may have completely different properties [11].

The classifiers we are using here are probabilistic classifiers. For a sample $\mathbf{x}$ in an I-dimensional input space we want to compute the posterior distribution $p(c|\mathbf{x})$, i.e., the probabilities for class membership given $\mathbf{x}$. To minimise the risk for classification errors we then select the class with the highest posterior probability. As set out in [12], $p(c|\mathbf{x})$ can be decomposed as follows:

$$p(c|\mathbf{x}) = \sum_{j=1}^{J} \underbrace{\frac{\int_{\mathbf{x}' \in \mathcal{R}_c} p(j|\mathbf{x}')\mathrm{d}\mathbf{x}' \cdot p(c)}{p(j)}}_{p(c|j)} \cdot \underbrace{\frac{p(\mathbf{x}|j)p(j)}{\sum_{j'=1}^{J} p(\mathbf{x}|j')p(j')}}_{p(j|\mathbf{x})} . \tag{1}$$

In this classification approach based on a so-called *mixture density model* $p(\mathbf{x})$,

- the conditional densities $p(\mathbf{x}|j)$ ($j \in \{1, \dots, J\}$) are the *components* (or *rules*) of the mixture model,
- $p(j)$ is a multinomial distribution with parameters π_j (the *mixing coefficients* or rule "weights"),
- $\mathcal{R}_c$ is the (not necessarily connected) region of the input space associated with class c, and
- the $p(c|j)$ are multinomial conditional distributions with parameters $\xi_{j,c}$.

That is, we have a classifier (rule set) consisting of J rules, where each rule j is described by a distribution $p(j|\mathbf{x})$ (which we call *rule premise*) and a distribution $p(c|j)$ (which we call *rule conclusion*). For a particular sample $\mathbf{x}_n$ in a data set with N samples (i.e., $n \in \{1, \dots, N\}$), the values $p(j|\mathbf{x}_n)$ are called *responsibilities* $\gamma_{n,j}$ (i.e., of the component for the sample).

Which kind of density functions can we use for the components? Basically, we may have categorical as well as numerical input dimensions. In the former case, multinomial distributions can be used whereas Gaussian distributions can be used in the latter. For many practical applications, the use of Gaussian models can be motivated by the generalised *central limit theorem* which roughly states that the sum of independent samples from any distribution with finite mean and variance converges to a normal distribution as the sample size goes to infinity (cf., e.g., [3]). Altogether, components (rules) are based on hybrid models that combine the two kinds of distributions. Details concerning different modelling possibilities can be found in [10].

How can parameters of the classifier be determined in an off-line approach from sample data? Basically, rule premises can be trained in an unsupervised way from sample data, while class labels are needed for a supervised training of the rule conclusions.

For the unsupervised part it would be possible to use standard maximum likelihood (ML) techniques such as expectation maximisation (EM) for mixture models. We decided to use another technique: Variational Bayesian inference (VI) realises the Bayesian idea of regarding the distribution parameters as random variables whose distributions must be found (*second-order probabilities*). To understand this idea, consider tossing a coin in order to estimate the probability for throwing "heads". Throwing a coin ten times with five "heads" and five "tails" and applying a frequentest parameter estimation approach (EM) would result in an 50% estimate for the probability of throwing "heads". We get the same result if the coin is thrown 1 000 times with 500 "heads" and 500 "tails", but in this case we would be much more certain that the coin is an "ideal" one. It would be advantageous to get a kind of "certainty" estimate together with the parameter estimate. A Bayesian approach offers this possibility [1]. A prior distribution is introduced for the parameter μ that has to be estimated. This prior distribution is combined with the likelihood function for the set of observations $\mathbf{X}$ which results in a posterior distribution of the parameter:

$$\textit{posterior} \propto \textit{likelihood} \cdot \textit{prior} \qquad p(\mu|\mathbf{X}) \propto p(\mathbf{X}|\mu) \cdot p(\mu). \tag{2}$$

In the coin tossing example, the likelihood function is based on a binomial distribution and the prior over μ must be a beta distribution. In our classifier described above, we work with multinomial/Dirichlet distributions and Gaussian/Gaussian-Wishart distributions, respectively (see [10, 12] for details). A Bayesian parameter estimation approach inherently offers the possibility for on-line training. Also, the uncertainty in parameter estimates is expressed by the posterior distribution over the parameters with a variance which monotonically decreases with the number of observations. A concrete estimate for a parameter can be obtained by computing the expectation of the hyper-distribution, for instance.

Now, we still need to determine the parameters $\xi_{j,c}$ of the rule conclusions $p(c|j)$. These can be obtained in a second, supervised step using labelled samples as shown in [12]. But, this step can also be realised in a different way if labelled data were not available: After the unsupervised step, the components may be labelled by a human domain expert.

2.3 Novelty and Obsoleteness Detection and Reaction

The detection of a need for new knowledge (new rules) must be realised without feedback from the environment, e.g., information about the "success" of the agent's

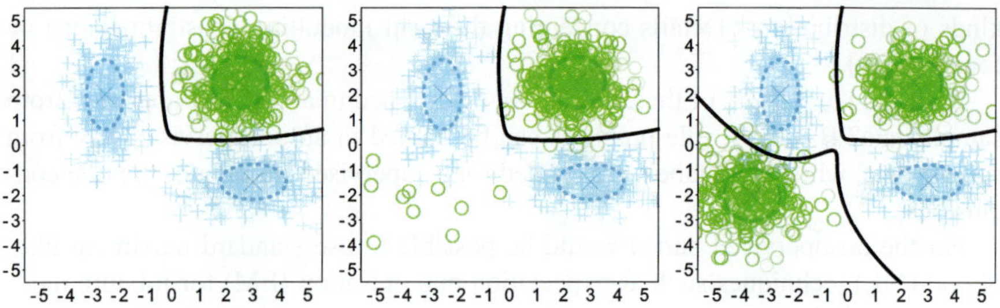

(a) A two-class classifier with three rules that model three processes producing data.

(b) A new process emerges and produces data that must be classified ("novelty").

(c) The new situation is recognised and the decision boundary is modified by creating a new rule.

Fig. 2 Example motivating the need for novelty detection and reaction

actions. Assume that a new process "emerges", i.e., it starts producing samples that the agent observes in the input space of its classifier (see Fig. 2; a detailed explication can be found in [7]). Such a situation must be detected by the agent and it must react accordingly by adding a new rule to the classifier. Without any feedback from the environment, the samples must be analysed in order to decide whether the currently available rule set is "sufficient" for the observed situation. Single samples, however, may be outliers from a process for which an appropriate rule already exists. Only the assessment of a certain number of samples allows for a reliable novelty detection. The mechanism for novelty detection, which is related to a divergence measure for probability densities, basically determines how well a sequence of observed samples "fits" the $p(\mathbf{x}|j)$ contained in the rule premises. For a detailed description we refer to Chap. 1.3 of this book.

Whenever novelty is stated, the rule set must be adapted accordingly by adding new rules. Basically, we may use the VI technique on a sliding window of recent samples to find rule premises. To avoid changes of the already existing ones, the parameters of existing components are fixed and only those of new components and all the mixture coefficients are adapted. A rule conclusion, i.e., an estimate of the parameters of the distribution $p(c|j)$ for a new component j can then be obtained in various ways:

1. Application experts can be asked to label a set of recently observed samples (e.g., measured within a sliding window). These labels are then used to determine values for the parameters $\xi_{j,c}$.
2. Application experts can be asked to label a new rule j, i.e., to assign it uniquely to one of the classes.
3. In the case of rule exchange, certain rules, and in particular their conclusions, may be taken over from other agents.
4. Rule conclusions may also be given implicitly if, for instance, certain regions of the input space are known to be assigned to certain classes. This includes the case that an agent stores discarded rules in a rule cache for later use.

In any case, the output distributions of all other rules must be adapted as well (details can be found in [6]).

Obsoleteness detection is basically very similar to novelty detection. The main difference is that individual measures are needed for every rule in the classifier. Whenever obsoleteness is stated, the corresponding component is deleted and the parameters π_j and $\xi_{j,c}$ of the remaining components are determined anew in a way which is very similar to the "novelty reaction" case.

2.4 Knowledge Extraction and Integration or Fusion

Due to the specific classifier paradigm chosen here, the extraction of rules from a classifier becomes a trivial task. Rules could, for instance, be communicated "on demand", i.e., whenever another agent asks for certain knowledge. Here, we have chosen a rather simple communication mechanism: Rules are broadcasted to other agents in a multi-agent system whenever an agent acquires a new rule with the techniques described above.

An agent that receives a rule either integrates the new rule immediately into the active classifier, discards it immediately, or stores it in a rule cache and observes its behaviour to decide later about its possible application. This behaviour is realised by means of a measure which we call *usefulness* [6]. Basically, a rule may receive positive usefulness ratings (rewards) with a mechanism that is very similar to the novelty/obsoleteness detection mechanism described above. Every time a cached rule is "covered" by a sample, its usefulness increases. Time, on the other hand, has a decreasing influence on the usefulness value. If the cached rule is covered by a sufficient number of observations within a certain time frame, rewards outweigh the temporal decay and its usefulness increases. When the usefulness reaches a user-defined threshold, the rule is integrated into the active classifier. The mixing coefficients must be adapted in this case. If the rule is not integrated within a certain time interval, it is considered useless and discarded (i.e., deleted from the rule cache).

If a rule is very similar to an existing one—similarity is again measured with a divergence measure for probability densities—it may also be fused with that rule. In this case, we exploit the second-order probabilities obtained with VI by multiplying the corresponding posterior distributions of parameters:

$$\textit{fused} \propto \textit{posterior1} \cdot \textit{posterior2} \qquad p(\mu|\mathbf{X}_1 \cup \mathbf{X}_2) \propto p(\mu|\mathbf{X}_1) \cdot p(\mu|\mathbf{X}_2). \tag{3}$$

Again, the computation of expectations leads to the desired estimates.

2.5 Interestingness Assessment

With the term *interestingness* we summarise a set of properties of rules that may be used to improve the knowledge exchange process. The term originates from the field of data mining where interesting knowledge must be extracted from often huge amounts of data. There, "interesting" means "valid, novel, potentially useful, and ultimately understandable" [4]. For our probabilistic classifier we focus on some other attributes of knowledge that can be measured numerically. They are objective on the one hand and related to attributes such as novelty, usefulness, and understandability on the other:

1. *Informativeness:* A rule is considered as being very informative if it describes a really distinct kind of process "generating" data.
2. *Uniqueness:* This measure describes how unambiguous the represented knowledge is, i.e., to which degree samples belonging to different classes are covered by that rule.
3. *Importance:* The importance of a rule measures the relative weight of the corresponding component within the classifier.
4. *Discrimination:* The discrimination measure evaluates the influence of a certain rule on the decision boundary—and, thus, on the classification performance—of the overall classifier.
5. *Representativity:* The performance of a generative classifier highly depends on how well it "fits" the data. This kind of fitness, which we call representativity, is determined for the continuous dimensions of the classifier.
6. *Comprehensibility:* With certain constraints concerning distribution parameters, human-readable rules may be extracted from our probabilistic classifier. Comprehensibility measures how well the rules can be interpreted by a human domain expert.

In [9] we set out how these attributes can be measured numerically using information taken from parameters of our probabilistic classifier. In [6] we show how such measures can be applied to improve a knowledge exchange process. Measures such as informativeness and importance, for instance, may be used to initialise the usefulness assessment or to set the reward.

Various interestingness ratings may also be sent together with rules to other agents. That is, we not only exchange knowledge but also *experience* (a kind of meta-knowledge).

3 Experiments

In several publications we described results of various simulation experiments where we investigated the new techniques for novelty and obsoleteness detection and reaction with artificial or real data and the knowledge exchange process with artificial data, respectively (see, e.g. [2, 6–8, 10]). Here, we describe the application of the techniques for knowledge exchange in the field of distributed intrusion detection to demonstrate the potential benefits of collaborating agents.

Figure 3 shows the setting of the experiment and how collaboration is realised. We simulate three agents that are placed within the same computer network (illustrated by the cloud) and consider the following attack scenario: In the first step, an

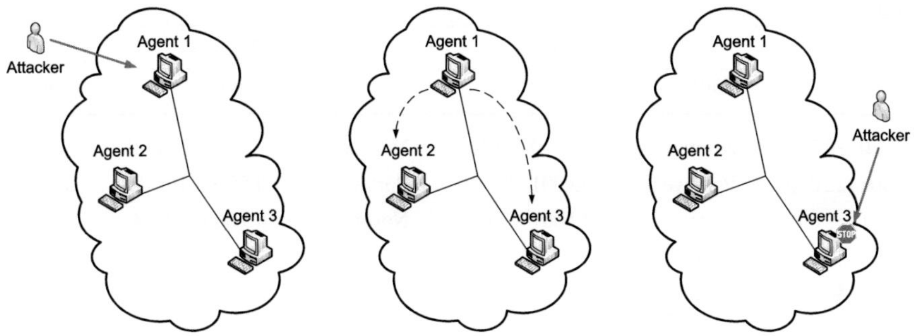

(a) Unknown attack started against agent 1.

(b) Agent 1 broadcasts its new knowledge.

(c) Agent 3 immediately detects the attack.

Fig. 3 Collaborative intrusion detection agents

attacker launches an attack against agent 1. This attack is unknown to all agents. Since agent 1 is situation-aware (with novelty detection and reaction capabilities) it is able to learn new detection rules for this attack. Then, it broadcasts this new knowledge to the remaining agents. Eventually, agent 3 is the target of the same kind of attack. However, as it already possesses the required knowledge it is able to detect this attack from the beginning and appropriate countermeasures can be applied immediately.

To construct the scenario for this experiment, we use parts of the well-known KDD-Cup '99 network intrusion data set that contains nearly 5 million connection records (each with 34 continuous and 7 categorical attributes) of captured network traffic (i.e., both legitimate background traffic and attacks) [13]. From these 41 attributes we select the—in our opinion—two most generic and basic ones, namely the number of transmitted data bytes from source to destination (*src_bytes*) and the number of transmitted data bytes from destination to source (*dst_bytes*), both of which being integers with a large value range such that they can be treated like continuous variables.

For every agent we create a data set that consists of 50 000 samples, here referred to as connection records (both legitimate background traffic, i.e., *normal* traffic and attack traffic). We use two attack types, namely *Back* and *Portsweep*. Back is a denial-of-service (DoS) attack that is launched against Apache web servers. An attacker sends requests to the web server that contain up to thousands of leading slashes. With the server's attempt to process these requests, it will significantly slow down and will then be unable to process any further requests. Portsweep is a relatively simple tool for probing networks. It conducts network scans and is able to search for ports and hosts listening on a specific network.

At every time step, all agents get a new connection record from their data set as input to be classified. The structure of the data sets (i.e., the attack schedule) for this experiment is set out in Table 1. The entries for Back and Portsweep represent a mixture of records of the corresponding attack and background traffic (ratio 1:3). It can be seen that the Back attack is first launched against agent 3 at time 15 001 and

Table 1 Attack schedule

Agent 1		Agent 2		Agent 3	
Time	Traffic	Time	Traffic	Time	Traffic
0–10 000	Normal	0–30 000	Normal	0–15 000	Normal
10 001–22 000	Portsweep	35 001–47 000	Portsweep	15 001–19 803	Back
22 001–32 000	Normal	47 001–50 000	Normal	19 804–50 000	Normal
32 001–36 000	Back				
36 001–50 000	Normal				

later against agent 1 at time 32 001. Thus, agent 1 could potentially benefit from the knowledge learned by agent 3. The same holds for the Portsweep attack and agents 1 and 2.

All agents are equipped with novelty detection (parametrised with $\alpha = 0.5$, $\eta = 0.1$, cf. Algorithm 1 in Chap. 1.3) and novelty reaction capabilities. Their initial active classifiers are trained with 2 000 background traffic records, i.e., when the experiment starts they can only classify normal traffic and no attacks. If novelty is detected (threshold $\tau = 0.2$), the agent waits for another 100 connection records (to ensure an sufficient amount of novel samples) until it performs novelty reaction based on a sliding window of the last 100 records that yielded a decrease of the novelty status. The class assignments of new rules are provided by a simulated human domain expert (i.e., they are calculated based on the true sample labels). If a new rule is assigned to an attack, it is immediately sent to the remaining agents where it is placed in a rule cache for further evaluation. We define a usefulness status s_{use} for every rule in the cache which is initialised with $s_{\text{use}} = 0$. Every record that is not covered by the existing rules but by a cached rule increases the corresponding status s_{use} by 0.1. Time has a decreasing influence on s_{use}, i.e., the usefulness status of all cached rules is reduced by 0.01 every time step as long as it is above its initial value. If the usefulness status of a rule exceeds 1.0, it is removed from the cache and integrated into the active classifier. The mixture coefficient of the rule is estimated based on a sliding window of the most recent 50 records.

The trajectory of the novelty status of agent 1 is depicted in Fig. 4. This agent detects the start of the Portsweep attack around time 10 500 and learns a new rule which is sent immediately to the remaining two agents. The Back attack starting at time 32 001 does not lead to the creation of a new rule as agent 1 integrates the rule learned by agent 3 before it detects novelty.

Agent 2 does not detect novelty at all because it integrates the Portsweep rule sent by agent 1 before the novelty detection threshold is reached. The trajectory of the usefulness of the received Portsweep rule of agent 2 is illustrated in Fig. 5.

Figure 6 shows the novelty status of agent 3. Around time 15 500 the start of the Back attack is successfully detected and a corresponding rule is learned and broadcasted.

The classification results of the three agents in this experiment are set out in Tables 2–4. Rows and columns filled with zero values only indicate that this particular

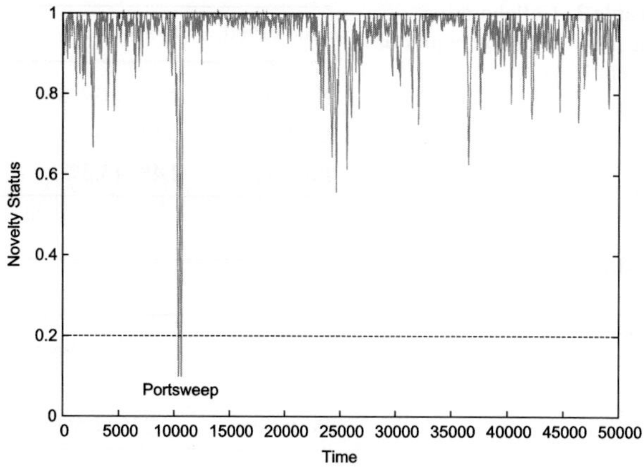

Fig. 4 Novelty status of agent 1

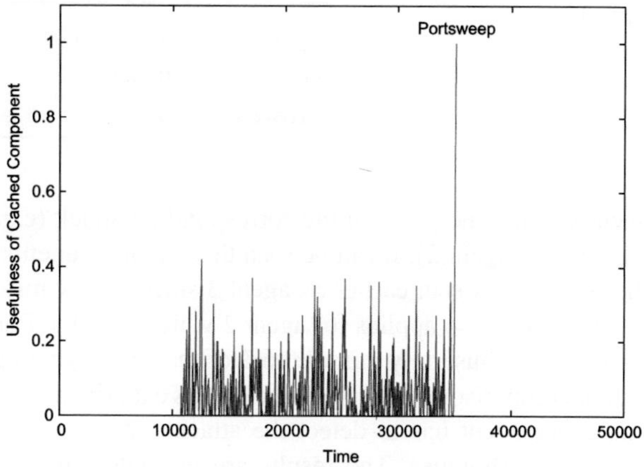

Fig. 5 Usefulness of cached Portsweep rule of agent 2

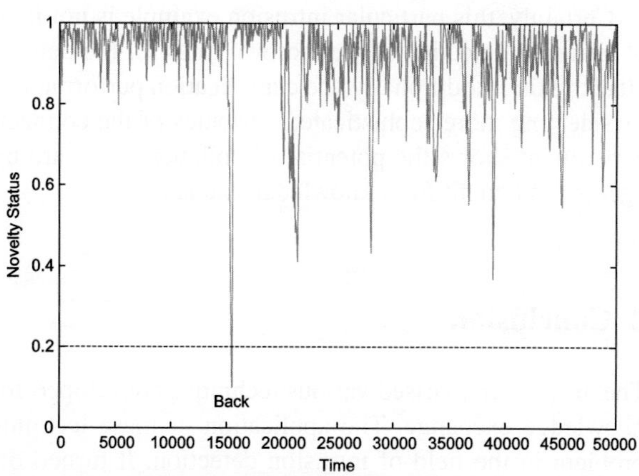

Fig. 6 Novelty status of agent 3

Table 2 Collaborating agents: Confusion matrix of agent 1

True \ Pred.	Normal	Back	Portsweep
Normal	44 513 (96.8%)	33 (3.3%)	179 (6.0%)
Back	0 (0.0%)	967 (96.7%)	0 (0.0%)
Portsweep	1 487 (3.2%)	0 (0.0%)	2 821 (94.0%)

Table 3 Collaborating agents: Confusion matrix of agent 2

True \ Pred.	Normal	Back	Portsweep
Normal	46 274 (98.5%)	0	19 (0.6%)
Back	0	0	0
Portsweep	726 (1.5%)	0	2 981 (99.4%)

Table 4 Collaborating agents: Confusion matrix of agent 3

True \ Pred.	Normal	Back	Portsweep
Normal	48 797 (100%)	153 (12.7%)	0
Back	0 (0.0%)	1 050 (87.3%)	0
Portsweep	0	0	0

agent was not the target of the corresponding attack (e.g., there was no Back attack launched at agent 2). It can be seen that agent 1 benefits from the knowledge about the Back attack sent earlier by agent 3 since it only misclassifies 3.3% of the Back records. This also applies for agent 2 which uses the Portsweep knowledge sent by agent 1 and, thus, only misses 0.6% of the corresponding attack records.

For comparison we also conducted this experiment without knowledge exchange, i.e., every agent has to detect the attacks on its own using the proposed novelty detection techniques. The results are presented in Tables 5–7. This time, agents 1 and 2 exhibit significantly higher classification error rates for the Back and the Portsweep attacks.

Certainly, this particular intrusion example is not innovative from the viewpoint of intrusion detection as well-known rules for the detection of Back and Portsweep attacks exist. Additionally, the classification performance could be further improved by selecting more sophisticated attributes of the connection records. However, this experiment shows the potential of collaborative learning and it demonstrates how agents can benefit from knowledge exchange.

4 Conclusion

The article summarised various techniques developed for collaborative learning by knowledge exchange. The application of these techniques has been shown for a problem in the field of intrusion detection. It turned out that several of the tech-

Table 5 Non-collaborating agents: Confusion matrix of agent 1

True \ Pred.	Normal	Back	Portsweep
Normal	44 497 (96.7%)	123 (12.3%)	179 (6.0%)
Back	0 (0.0%)	877 (87.7%)	0 (0.0%)
Portsweep	1 503 (3.3%)%	0 (0.0%)	2 821 (94.0%)

Table 6 Non-collaborating agents: Confusion matrix of agent 2

True \ Pred.	Normal	Back	Portsweep
Normal	45 974 (97.8%)	0	482 (16.0%)
Back	0	0	0
Portsweep	1 026 (2.2%)	0	2 518 (84.0%)

Table 7 Non-collaborating agents: Confusion matrix of agent 3

True \ Pred.	Normal	Back	Portsweep
Normal	48 797 (100%)	153 (12.7%)	0
Back	0 (0.0%)	1 050 (87.3%)	0
Portsweep	0	0	0

niques are not only interesting from the viewpoint of knowledge exchange: Novelty detection techniques may be used for emergence detection and measurement in organic systems, interestingness measures are needed in a broader data mining context, knowledge fusion techniques may be used in the field of distributed data mining, etc. Thus, the results of our work will influence several research fields. However, many interesting research issues in the field of knowledge exchange still need to be addressed, e.g., the active rating of other agents' expertise, knowledge brokerage, trust mechanisms, etc. Also, new application scenarios will be addressed in the future, e.g., collaborative danger warning in vehicles.

Acknowledgement This work was supported by the German Research Foundation (DFG) under the grants SI 674/3-1, SI 674/3-2, and SI 674/3-3 within the Priority Programme Organic Computing.

References

1. Bishop, C.M.: Pattern Recognition and Machine Learning. Springer, New York (2006)
2. Buchtala, O., Sick, B.: Functional knowledge exchange within an intelligent distributed system. In: Lukowicz, P., Thiele, L., Tröster, G. (eds.) Architecture of Computing Systems – ARCS 2007. LNCS, vol. 4415, pp. 126–141. Springer, Berlin (2007)
3. Duda, R.O., Hart, P.E., Stork, D.G.: Pattern Classification. Wiley, Chichester (2001)
4. Fayyad, U.M., Piatetsky-Shapiro, G., Smyth, P.: Knowledge discovery and data mining: Towards a unifying framework. In: Proc. of the Second Int. Conf. on Knowledge Discovery and Data Mining (KDD 1996), Portland, OR, pp. 82–88 (1996)

5. Fisch, D., Hofmann, A., Hornik, V., Dedinski, I., Sick, B.: A framework for large-scale simulation of collaborative intrusion detection. In: Proc. of the 2008 IEEE Conf. on Soft Computing in Industrial Applications (SMCia/08), Muroran, Japan, pp. 125–130 (2008)
6. Fisch, D., Jänicke, M., Kalkowski, E., Sick, B.: Learning by teaching versus learning by doing: Knowledge exchange in organic agent systems. In: Proc. of the IEEE Symp. on Intelligent Agents (IA 2009), Nashville, TN, pp. 31–38 (2009)
7. Fisch, D., Jänicke, M., Kalkowski, E., Sick, B.: Techniques for knowledge acquisition in dynamically changing environments. ACM Trans. Auton. Adapt. Syst. (2011, accepted)
8. Fisch, D., Jänicke, M., Sick, B., Müller-Schloer, C.: Quantitative emergence - a refined approach based on divergence measures. In: Proc. of the Fourth IEEE Int. Conf. on Self-Adaptive and Self-Organizing Systems (SASO 2010), Budapest, Hungary, pp. 94–103 (2010)
9. Fisch, D., Kalkowski, E., Sick, B.: In your interest: Objective interestingness measures for a generative classifier. In: Proc. of the 3rd Int. Conf. on Agents and Artificial Intelligence (ICAART) Rome, Italy, pp. 414–423 (2010)
10. Fisch, D., Kastl, F., Sick, B.: Novelty-aware attack recognition – intrusion detection with organic computing techniques. In: Proc. of the 3rd IFIP Conf. on Biologically-Inspired Collaborative Computing (BICC 2010) at the World Computer Congress (WCC 2010), Brisbane, Australia, pp. 242–253 (2010)
11. Fisch, D., Kühbeck, B., Sick, B., Ovaska, S.: So near and yet so far: New insight into the relationship between some well-known classifier paradigms. Inf. Sci. **180**(18), 3381–3401 (2010)
12. Fisch, D., Sick, B.: Training of radial basis function classifiers with resilient propagation and variational Bayesian inference. In: Proc. of the Int. Joint Conf. on Neural Networks (IJCNN 2009), Atlanta, GA, pp. 838–847 (2009)
13. Hettich, S., Bay, S.D.: The UCI KDD archive (1999)
14. Hofmann, A., Sick, B.: On-line intrusion alert aggregation with generative data stream modeling. IEEE Trans. Dependable Secure Comput. **8**(2), 282–294 (2010)

D. Fisch (✉) · E. Kalkowski · B. Sick
Computationally Intelligent Systems Lab, University of Applied Sciences Deggendorf, Deggendorf, Germany
e-mail: dominik.fisch@fh-deggendorf.de

E. Kalkowski
e-mail: kalkowsk@fim.uni-passau.de

B. Sick
e-mail: bernhard.sick@fh-deggendorf.de

Chapter 3.4
A Framework for Controlled Self-optimisation in Modular System Architectures

Werner Brockmann, Nils Rosemann, and Erik Maehle

Abstract Organic Computing tackles design issues of future technical systems by equipping them with self-x properties. A key self-x feature is self-optimisation, i.e. the system's ability to adapt its dynamic behaviour to its current environment and requirements. In this article, it is shown how self-optimisation can be realised in a safe and goal-directed way, but also why it has to be enhanced and embedded into a suitable, modular system architecture. Then, a suitable framework for controlled self-optimisation is developed, which enables the system designer to give a priori guarantees of important dynamic system properties, and which ensures the system's ability to cope dynamically with anomalies. The key features are online machine learning, which is complemented by incremental, local regularisation in a local Observer/Controller architecture as well as expressing anomalies by health signals, which are exploited to guide the learning process dynamically in order to achieve fast, but safe learning.

Keywords Organic computing · Self-optimisation · Framework · Online learning · Incremental regularisation

1 Introduction

1.1 Background

Future technical systems will get more and more integrated into natural and human environments, e.g. robotics or driver assistance systems. Because of dependability issues, many of such upcoming areas raise unforeseeable problems for the safety of the control systems like faults, anomalies and disturbances, which drastically increase engineering efforts or are even intractable by static control system approaches, which are fixed at design time. As an example, classic control approaches do not scale well as it takes a lot of time to achieve suitable formal models. Organic Computing opens up many opportunities here due to introducing a flexible run-time behaviour by self-x properties. In this work, we focus on self-optimisation as a method to effectively reduce the engineering effort and to enable a flexible adaptation to (changing) environments.

C. Müller-Schloer et al. (eds.), *Organic Computing — A Paradigm Shift for Complex Systems*, 281–294
Autonomic Systems, DOI 10.1007/978-3-0348-0130-0_18, © Springer Basel AG 2011

This problem can formally be stated as follows. Let $x(t)$ denote the relevant system state (e.g. given by sensor readings) of the self-optimising system at time t, and let $x_d(t)$ denote the target value of x at time t. Of course, the goal is to have $x(t) = x_d(t)\,\forall t$. To achieve this, the self-optimising system, whose control function is denoted as f, can influence the system state by some actuator, which is set by the output variable $u(t)$. Of course, the control function must be implemented by a learning function approximator. So its output is determined by a certain number of internal parameters θ which may change over time by learning. Thus,

$$u(t) = f(x(t), x_d(t), \theta(t)). \tag{1}$$

Self-optimisation now means to change the internal parameters $\theta(t)$ by machine learning principles in such a way that the performance of the system increases for re-occurring situations. This is a non-trivial task as $x(t)$ might be subject to anomalies and because learning can only be based on prior observations of the system state.

If this kind of learning can be done safely, it is a further and highly desirable property for such a self-optimising system to be able to learn from scratch, making it a so called self-tuning system, as engineering efforts can be significantly reduced further. Hence, very effective and fast, but safe learning methods are required, which must run on-line and be real-time capable. This then allows also to quickly re-learn optimal behaviour at run-time if the underlying physical system, its environment or requirements change. As it is a non-trivial task to still meet safety requirements, the resulting properties a self-optimising system should show are worked out in detail in the following section.

1.2 Desired Properties of Safe Self-optimisation

In self-optimising systems, one has to deal with a potentially chaotic feedback loop, because the system behaviour determines the learning stimuli, which in turn determine the future system behaviour, which again determines the future learning stimuli, and so on. Hence, small deviations may cause an arbitrary effect. Thus, one has to ensure safety constraints. This is the most crucial point. Actually, it covers two aspects. On the one hand, it has to be guaranteed that the self-optimising process avoids safety critical system states, especially while it is still learning. This is hard to meet if the system operates in an uncertain environment and might be subject to faults and anomalies. On the other hand, it has to be ensured that the process of self-optimisation will ultimately converge to an optimal functional behaviour. This has to be done in an autonomous way, i.e. without an external teacher or reference. In this context, a self-optimising system usually has to be able to approximate any desired functional behaviour (plasticity) while it has to be robust to all kinds of disturbances and anomalies at the same time (stability). In order to address the different challenges of self-optimisation safely, the underlying methods should also show the following more specific properties.

1) They need to learn incrementally. This means that learning stimuli are generated one after the other and processed in an incremental way, i.e. step by step without storing them all. This keeps the methods real-time capable and able to compensate time-variant process behaviour on-line.

2) They need to work without a formal model of the underlying physical system. This is required to achieve the overall goal of reduced engineering effort, as it also becomes potentially much easier to compensate time-variant effects and to address disturbances and anomalies.

3) They need to be able to learn out of the normal system operation, i.e. without the need to make specific excitations to generate learning input by trial and error. Essentially, in a system which is supposed to work productively, especially in a human environment, learning has to occur in a strictly guided way. Only if all possible actions can be deemed safe a priori, then methods based on guided trial and error can be exploited.

4) The knowledge representation should be in an interpretable form. Compared to other general parametric representations, e.g. neural networks, rule based systems like fuzzy systems have the advantage of being understandable and interpretable by a system designer, e.g. for safety issues. On the one hand, this allows to incorporate a priori knowledge into the system. On the other hand, the learnt knowledge can always be analysed, which is crucial, e.g. to determine liabilities in the case of an accident.

5) In addition, the requirements imposed upon a general self-optimising system might change over the system's lifetime. For example, during learning from scratch or a re-learning phase it is more important to achieve stable and safe behaviour. But as soon as stability is reached, the focus is switched towards optimality. Thus, self-optimisation should be enhanced by approaches to observe and control the learning process dynamically as such, i.e. to achieve *controlled* self-optimisation. This is especially important if more than one self-optimising module is present within one architecture, as this may cause dynamic interferences between the interconnected learning processes.

6) Finally, self-optimisation might run into convergence and stability issues in case of anomalies and uncertainties. Hence specific means are required to control the learning process dynamically such that anomalies or uncertain information do not corrupt learnt knowledge on the one hand, and that learning adapts quickly to new situations caused by anomalies on the other hand.

Controlled self-optimisation thus requires a specific framework to tackle learning from scratch and re-learning as well as to fight anomalies and disturbances at the same time in a safe and goal-directed way.

2 State of the Art

The multitude of methods proposed for online learning systems has shown different trade-offs between the requirements stated above. Especially the extreme cases of learning based only on a reward function, i.e. reinforcement learning, and learning based on a model where only parameters are learnt, i.e. adaptive control, are not always the best solutions for practical problems.

The requirement of being model-free is put to an extreme by reinforcement learning. The influence of the system designer is usually limited to providing the reward function, i.e. an evaluation of the performance, and setting up a suitable learning structure. Meanwhile, reinforcement learning has evolved to high data efficiency, i.e. high speed learning. Non-linear controllers can be learnt in a comparatively short training period [9, 15]. A key ingredient for this success are policy gradient methods [16]. They allow to accelerate the learning process by choosing an appropriate policy representation and, thus, to incorporate prior domain knowledge [13]. Reinforcement learning algorithms are unable to avoid unsafe states because they require full system exploration, i.e. specific operations to generate learning input aside the normal system operation. Risk sensitive approaches [8] aim at minimising the danger caused by that, but require additional engineering efforts, which counteracts the basic idea of reinforcement learning. So basically, reinforcement learning tries to find practically usable trade-offs between absence of a model, learning speed and prior knowledge. It is still a problem that the actual process of learning is hard to predict and to guide.

In contrast, adaptive control usually assumes that the system at hand belongs to a certain class of systems which can be formally described. Based on this formalism and the design goal, e.g. good tracking performance, powerful adaptive controllers can be derived, and even general function approximators like neural networks can be used [7, 10, 14]. This strategy allows to prove stability and convergence of the closed loop control as well as of learning in advance. Thus, high-speed yet safe learning can be achieved. Adaptive control approaches can be used for MIMO control problems [23], but they do not scale as well as policy gradient learning methods concerning engineering effort. The major drawback of adaptive control methods is the basic assumption of the system class. Without strong a priori knowledge, the system designer can hardly determine this class. In addition, many design choices have to be made for which no engineering rules can be applied. These methods further are not applicable in case of system changes due to anomalies and faults. All in all, these approaches usually focus on formal safety rather than on reduced engineering effort and system flexibility in a changing environment.

Other online learning approaches fall between these two borderline cases, e.g. neural networks, probability-based learning systems and learning fuzzy systems. In any case, only one or very few of the requirements summarised above are met by a specific approach. Especially dealing explicitly with uncertainties and anomalies in an effective, goal-directly and safe way is not addressed at all.

In order to allow for self-optimisation in complex and uncertain scenarios, a framework has to integrate the specific properties of self-optimisation into a safety concept. Especially, this means to protect the overall system from safety issues due to incomplete learning and to protect learning itself from corruption by anomalies. Thus the Organic Robot Control Architecture (ORCA) [4, 12] was developed for robotic applications. See also Chap. 4.5 of this compendium. One of its key features is to explicitly handle anomalies dynamically. This is done by representing the degree of anomaly of any relevant signal or system entity by specific attributes. These attributes capture the notion of normality, validity or trustworthiness, which roughly corresponds to the notion of *health* of a biological system. Hence, these additional attributes are called *health signals* [11].

A health signal has a scalar value from the interval[1] [0; 1]. If the health signal is zero, then the corresponding signal should have no influence on further calculations, because it corresponds to no trustworthiness. Instead, a safe and robust fall-back strategy should be used then. In contrast to that, the effect of a signal should be maximal if its health signal is equal to one, because it relates to absolute trustworthiness, i.e. there is no doubt about the correctness of the respective signal value. Due to the generality of this representation, different kinds of anomalies can be represented in a uniform way [2, 11].

Within ORCA, there are basically two kinds of modules, namely *Basic Control Units* (BCUs) and *Organic Control Units* (OCUs). The BCUs implement the functional core behaviour, i.e. the system is fully functional (in principle) by using only the BCUs. They are complemented by OCUs which observe the online behaviour of the BCUs, i.e. their state of operation, their performance, and their interplay, in order to detect anomalies and in order to optimise the BCUs. Detected anomalies are then represented by health signals. This is the basis of the framework for controlled self-optimisation, which exploits the health signal mechanism in two ways: first, in order to protect learnt knowledge from corruption, and second, as information about the validity of the learnt knowledge, which can be used to ensure safety in the remainder of the system architecture.

3 Framework for Controlled Self-optimisation

3.1 Overview

The starting point for self-optimisation methods, which show the properties stated above, is Directed Self-Learning (DSL) [1]. Its basic schematic architecture is shown in the left of Fig. 1. It is composed of a control loop consisting of a controller and the underlying system (which includes the physical system and the rest of the system as well, i.e. other modules). In our case, a learning BCU is implemented by a so-called neuro-fuzzy element—a universal function approximator, which represents its knowledge locally in the form of rules. While the neuro-fuzzy element (BCU) controls the process, i.e. during normal operation, the law of adaptation, which is implemented within an associated OCU, monitors the behaviour of the control loop and generates incremental learning stimuli to optimise the neuro-fuzzy element if the control behaviour is not optimal. The law of adaptation is given by the system designer and describes indirectly how to approach optimal control behaviour by incremental learning stimuli.

[1] Within this work, let $[\cdot, \cdot]$ denote an interval of integral numbers and $[\cdot; \cdot]$ an interval of real numbers.

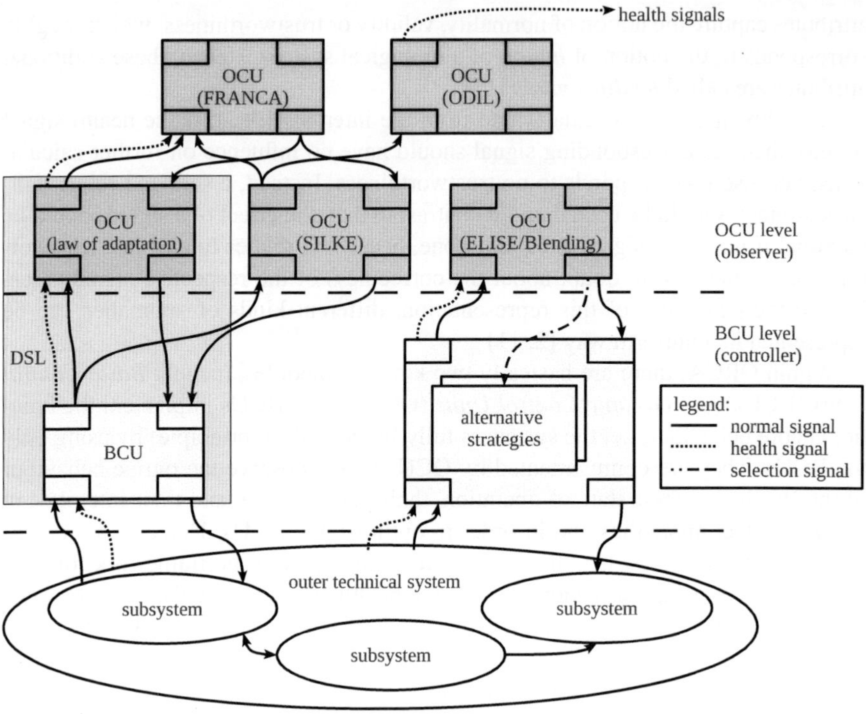

Fig. 1 Schematic representation of the basic and enhanced self-optimisation architecture

This DSL learning process occurs completely online, i.e. while the system is under normal operation. Methodologically, it already fulfils the first four properties from the last section: It allows to self-optimise the system behaviour in an incremental way, it does not need a formal model, it learns in a guided, trial-and-error free way, and it represents the learnt knowledge in the form of rules. But by embedding DSL into an architecture like ORCA, it can be significantly improved concerning anomaly handling, concerning stability-plasticity issues and concerning interactions among multiple self-optimising subsystems by specific extensions.

The key extension is the so called SILKE approach (System to Immunise Learning Knowledge-based Elements) [3, 17–19, 22]. Its core idea is to regularise the ill-posed problem of incremental learning based on meta-level characteristics like smoothness of the learnt functional behaviour. It therefore generates additional correcting changes to the learnt parameters immediately at each learning step. This way, the learning process as such can be guided online towards solutions which fulfil the desired meta-level property. Hence, the SILKE approach can influence the stability-plasticity tradeoff in a positive way. Besides this, a mechanism similar to the SILKE approach is used in the ODIL approach (Online Diagnosis for Incremental Learning) [5]. Here, local violations of the desired meta-level property are detected and propagated in the form of health signals. These health signals thus represent weaknesses and anomalies of the learning process.

Controlled self-optimisation by the SILKE approach is a general technique, but within a complex architecture like ORCA the specific way of how to apply it depends on the overall system state, e.g. on uncertainties and on the overall progress of self-optimisation. Hence, the so called FRANCA approach (Fast Rate Adaptation for Neuro-(fuzzy) Control Applications) [6] provides a generic OCU-level meta-control by automatically controlling the dynamic interplay of self-optimisation and the SILKE approach by tuning the learning rate λ as well as the adjustment rate α of the SILKE approach. This way, it can actively influence the stability-plasticity trade-off depending, e.g. on the appropriateness of learnt knowledge and the presence of uncertainties and anomalies.

But the presence of *critical* anomalies requires a more determined proceeding. For example, self-optimised behaviour might take dangerous actions given uncertain, i.e. potentially wrong inputs. Thus, the so called health signal based strategy blending scheme was developed as an additional means which can switch to a robust fallback strategy in the case of too strong anomalies and uncertainties or too weak learning [20].

A separate class of anomalies arises from multiple interacting learning subsystems. Here, two situations have to be distinguished. On the one hand, multiple self-optimising systems can be active alternatively, i.e. only one is active at a given point in time but interacts with the other one due to sequential activation. On the other hand, multiple self-optimising systems can be active in parallel, i.e. they interact due to physical coupling. Within the framework for controlled self-optimisation, the former is addressed by the so called ELISE approach (Exploiting Learning stimuli in Interrupted SElf-optimisation) [21]. Here, potentially harmful alternations between the different self-optimising systems are detected and counteracted. A specific emphasis is put on generating useful learning stimuli in the switching region, which is crucial for the dynamic interplay. The second case is addressed by a specific variant of the FRANCA and the SILKE approach which allows to prioritise the different input dimensions and to focus the guidance effect of the SILKE approach on the prioritised dimensions. By this technique, overall stability can be reached much easier and faster. In the following, features of the general framework are presented in more detail for single self-optimising systems.

3.2 Directed Self-learning

Based on the problem statement from the introduction and given that the designer knows the approximate time delay τ which occurs between changing an actuator value $u(t)$ and the reaction $x(t)$ of the process to this change, one can look at $x_d(t-\tau)$ in order to compare it to $x(t)$. If $x(t) = x_d(t-\tau)$, then $u(t-\tau)$ was a good choice. On the other hand, if $x(t) \neq x_d(t-\tau)$, then $u(t-\tau)$ was either too low or too high. Whether $u(t-\tau)$ was too low or too high is determined by the law of adaptation. In our case, it is formulated by a function $H(x(t), x_d(t-\tau))$. This function H is given in advance by the designer, encapsulates the desired system behaviour, and determines learning stimuli with respect to the known monotonicity properties of the process.

So at time step t, a learning stimulus Δf is generated for the input vector $x(t-\tau)$ by

$$\Delta f(t) = \lambda \cdot H(x(t), x_d(t-\tau)), \tag{2}$$

with $\lambda \in [0; 1]$ as the learning rate.

This means that the next time the process is in the same state as at time $t-\tau$, the controller (BCU) will produce the output $u(t-\tau) + \Delta f(t)$. Obviously, for this to take place, the learning stimulus $\Delta f(t)$ has to change the internal parameters θ of the controller. Here, it is essential to use a *local* function approximator, so that the learning stimulus, which is only locally valid, only changes the output of f in a small region close to $x(t-\tau)$.

From a more global point of view, it is assumed that there exists a parameter vector θ^* which yields $x(t) \approx x_d(t-\tau) \forall x, x_d$, i.e., such that $f(x, \theta^*)$ is globally optimal. It is then possible to define a global error E between the current controller and the optimal controller:

$$E(\theta, \Omega) = \int_{\Omega} |f(a, \theta) - f(a, \theta^*)| da, \tag{3}$$

with Ω as the whole input space. The question is now, whether each learning step reduces this global error. By construction, given that the law of adaptation generates a correct estimate of the sign of change, and that the learning rate is appropriate, the error will be reduced for $x(t)$. But each learning step changes only a small number of parameters (local learning), so the global error cannot be reduced to 0 in one step. And the *true* error between $f(\theta)$ and $f(\theta^*)$ is not exactly known, as H is just a local estimate which is also only given incrementally.

For clarity, it is helpful to split the error into a sum of three terms, $E(\theta(t), \Omega) = \hat{E}(\theta(t), B(x(t), \epsilon)) + \tilde{E}(\theta(t), B(x(t), \epsilon)) + E(\theta(t), \Omega \backslash B(x(t), \epsilon))$. Here, $B(a, \epsilon)$ denotes a ball of radius ϵ in the input space around some point a. Basically, this symbolically splits the global error into the error in the vicinity of $x(t)$ and the error "far" away from $x(t)$ (i.e., outside the ball B around $x(t)$). And the error in the vicinity of $x(t)$ is further decomposed into the error $\hat{E}$ which is the error between the current controller and the controller defined by the learning stimulus, and the error $\tilde{E}$, which is implicitly given by a potentially wrong estimate, i.e., a wrong learning stimulus.

The idea of controlled self-optimisation can now be stated based on these terms: The only *observable* error term is $\hat{E}$, so only this can be minimised directly by changing the parameters θ. But if the estimate of the law of adaptation H or the learning rate λ are wrong, this does not minimise $\tilde{E}$. Besides that, even when using a local function approximator, each parameter has a certain region of influence, which might be higher than the validity region ϵ of the current learning stimulus. Thus, $E(\theta(t), \Omega \backslash B(x(t), \epsilon))$ might even become larger if one just minimises $\hat{E}(\theta(t), B(x(t), \epsilon))$.

Because of that, additional means are required to ensure the success of learning. The idea is to introduce another correction term $\Delta g(x(t), \theta(t))$ to make the current local rule base comply with an a priori known meta-level property in a larger region, which is the essence of the SILKE approach. This way, the parameters θ are changed based on both the learning stimulus Δf and this new correction term Δg. The actual influence of each of these terms is determined by individual factors, namely the learning rate λ and the adjustment rate α, which will be formally introduced in Sect. 3.4. Changing λ can be used to balance the error between $\hat{E}$ and $\tilde{E}$, and changing α can be used to incorporate the knowledge stored in the parameters of the region around $x(t)$, this way balancing between $\hat{E}$ and $E(\theta(t), \Omega \backslash B(x(t), \epsilon))$. This means that the effect of a local learning stimulus is related to the regional context such that desired meta-level properties are fulfilled in a better way. For a more thorough analysis of the SILKE approach, the used function approximator and the learning algorithm are outlined in detail first.

3.3 Neuro-fuzzy Elements

The neuro-fuzzy elements used within this framework of controlled self-optimisation are zero order Takagi-Sugeno fuzzy systems with triangular membership functions, sum-prod-inference and a normalised rule base. For details, see [18]. Here, it is important that each rule of such an n-dimensional neuro-fuzzy element can be identified by an n-dimensional index J and that each rule has a rule conclusion $c(J) \in \mathbb{R}$. Here, incremental learning is performed by instantaneous gradient descent learning, which changes these rule conclusions. The conclusions of a neuro-fuzzy element at rule index J after a single learning step are then given by

$$c_{t^+}(J) = c_{t^-}(J) + \frac{\lambda \cdot \mu(J, x)}{\sum_K \mu(K, x)^2}(y - V(t^-, x)),$$

with μ denoting the activation degree of the respective rule, $V(t^-)$ as the output of the whole neuro-fuzzy system before learning and $y \in \mathbb{R}$ as the target value for learning, i.e. $y - V(t^-, x)$ is the error Δf, which is to be corrected by learning. The learning rate λ determines the degree to which the learning stimulus (x, y) is obeyed.

3.4 DSL and the SILKE Approach

In order to learn as fast as possible, the idea is to follow the (potentially biased or anomalous) teacher H, but to use an appropriate SILKE template afterwards to obtain a correction towards the unknown optimal rule conclusions by taking their context into account [18].

To do this, additional assumptions about this unknown optimal rule conclusions are needed. The SILKE approach proposes to use local meta-level characteristics like local smoothness, curvature or linearity. In most practical scenarios, the system designer knows such characteristics in advance as they reflect similar properties of the controlled process. In order to express such local meta-level characteristics, one has to look at the local neighbourhood of rule conclusions.

It is now possible to describe the so called SILKE approach in more detail. Each learning step is followed by another correction step, which is not based on the learning stimulus, but rather on the local neighbourhood of the learning stimulus within the neuro-fuzzy element. This additional modification of the neuro-fuzzy element is based on the so called *adjustment rate* $\alpha \in [0; 1]$, which determines the influence of the SILKE approach. A value of $\alpha = 0$ means no modification, while value of $\alpha = 1$ means a full modification in compliance with the meta-level property.

The description of the SILKE approach is based on $U = [-1, 1]^n$ denoting the nominal area of effect of the SILKE approach, and $U'(J)$ denoting the subset of this nominal area which is valid at the rule index vector J. The core of the SILKE approach is the so called mask $m : \mathbb{N}^n \mapsto \mathbb{R}$, which has support only over U, so $m(u) = 0$ for $u \notin U$ and which represents the desired meta-level property to be enforced by the SILKE approach (see below). The so called SILKE *template* $T_m(J)$ convolves this mask with the local neighbourhood of rule conclusions:

$$T_m(J) = c_a(J) - \sum_{u \in U'(J)} c_a(J-u) \cdot m(u).$$

The template value $T_m(J)$ is then used to correct the rule conclusions according to

$$c_{t^+,s}(J) = (1 - \alpha \cdot \mu(J,x)) \cdot c(J) + \alpha \cdot \mu(J,x) \cdot (c(J) - T_m(J)).$$

Based on the desired meta-level property, one has to define a suitable mask m in a way which ensures convergence of the SILKE approach towards local compliance to this meta-level property. For this design problem, a formal framework has been derived, which allows to create templates which are convergent by construction [19]. Thus, the designer can give a priori guarantees that the dynamic process of self-optimisation can be forced to show the desired local meta-level property. Fundamentally, this design framework is based on solving a matrix equation

$$(M - I)\tilde{N}\hat{N}\hat{c} + \tilde{m}\hat{c} = 0, \tag{4}$$

while maintaining that a specific eigenvalue constraint is met. For further explanations and a definition of the symbols within (4), see [18].

Now, given a suitable template, the above mentioned task to balance the different local errors given above can be tackled by dynamically adapting the learning rate λ and the adjustment rate α to the current interplay of learning and control performance. This is where the FRANCA approach comes into play. It features heuristic rules to dynamically balance the stability-plasticity trade-off of the self-optimising system. Hence, in the presence of anomalies or uncertainties, stability can be increased, yielding higher dynamic robustness [6].

3.5 *Self-optimisation and Uncertainties*

As stated before, for severe anomalies, additional means have to be taken by suitable, architectural approaches which go beyond general approaches for self-optimisation. This is where *health signal based strategy blending* (HS-based blending) comes into play. As before, two modules are involved, a BCU and an OCU. And like before, the task of the OCU is to generate learning stimuli for the BCU for self-optimisation. This basic structure is extended in two ways (see Fig. 1, right): It is assumed that the BCU gets a health signal from ORCA, i.e., from some other module(s). And within the BCU, *two* alternative strategies are implemented together with a blending scheme.

The two strategies are a safe emergency or fallback strategy, which is not allowed to be changed, and a flexible strategy, which is optimised online by learning stimuli from the OCU in order to improve performance. The safe strategy should be designed in a way that it yields safe behaviour even if one or more of the input signals can not be trusted. To the contrary, the self-optimising strategy should yield (nearly) optimal behaviour under the prerequisite that all input signals can be fully trusted.

The blending scheme has two tasks. First, it has to map the health signal to activation degrees, $\alpha_i \in [0; 1], i \in \{1, 2\}$, for the two strategies within the BCU. For a better interpretability and understandability, it is useful that $\alpha_1 + \alpha_2 = 1$. These activation degrees are used to blend between the strategies, as the output u of the BCU is the weighted sum of the internal outputs u_1, u_2 of the two strategies. As the seconds task, the blending scheme has to reduce the *learning rate* for the learning strategy whenever it is not fully active. Here, it is reasonable to learn only if the self-optimising controller is fully active.

The HS-based blending scheme ensures safety of a self-optimising BCU mainly for the case of external anomalies. But it is also possible that learning itself causes an anomalous operation, which might be the case for a sudden change of the underlying system or for a suboptimal configuration of learning. Here, the ODIL approach comes into play. Based on the fact that the optimal BCU behaviour is not explicitly known (else no learning would be necessary), it monitors whether or not the currently learnt BCU behaviour complies to a desired meta-level property, similar to the SILKE approach. But instead of a direct feedback on the learnt knowledge, detected violations are transformed into health signals in order to take countermeasures at some other place within the system architecture. In the simplest case, the health signal of a rule base at a given input vector x is the weighted sum

$$h(x) = \sum_J \mu(J, x) \left(1 - \bar{N}(T_m(J))\right), \tag{5}$$

where $\bar{N}$ denotes a suitable normalisation function like the one defined in [5]. It fits into the ORCA architecture and supports the framework of controlled self-optimisation in a natural way.

4 Discussion

A key aspect of Organic Computing systems is their ability to adapt to changing conditions, environments and requirements. In order to meet rigid safety constraints they have to take the underlying system dynamics as well as disturbances and anomalies into account. Hence, self-optimisation has to cope with the stability-plasticity dilemma, the exploration-exploitation dilemma, dynamically changing uncertainties, incremental data, safety constraints and interpretability issues. The framework presented above tackles all these challenges at once in an integrated way. Thereby it goes beyond other state-of-the-art-approaches, which tackle only one or two of them. It makes anomalies from different sources explicit and thus allows to address them in a goal-directed manner in order to achieve fast and flexible, but stable and safe self-optimisation.

The framework consists of a set of methods, which were developed with a focus on robotic systems and which tackle self-optimisation from the controller's point of view, i.e. online learning function approximation. Single methods, like the SILKE approach, or combinations of them allow to ensure safety of the online learning system. They thus also support re-learning as well as learning from scratch. As the individual methods fit naturally and intuitively together, the framework ultimately allows to relieve the system designer from performing tedious design tasks and engineering work when engineering dependable self-optimising systems. No other approach is known so far, which would tackle safe online learning in such a broad manner.

5 Conclusion and Outlook

Being able to handle the complexity of future technical systems like robotics will only be possible if the engineering problem can be eased, and a self-optimising system shows a lot of properties which ensure safe and dependable operation. The framework for controlled self-optimisation, which has been presented in this article, allows to give a priori guarantees of critical system properties at design time. In addition, anomalies and dynamically changing system properties can be compensated. It is based on a suitable, modular system architecture and enhances the principal learning architecture by algorithms which can inherently deal with health signals at their inputs, at their learning stimuli, and which can also express their own health.

Within this article, it was argued how online function approximation can be done in a safe and fast way. This holds, e.g. for learning optimal behaviour as fast as possible during the initial learning phase, i.e. learning from scratch, as well as during any phases of re-learning, and of mission success even during phases of anomalies and uncertainties. Thus, the presented framework for controlled self-optimisation within a modular system architecture can be also used to achieve self-tuning and self-healing.

Many challenges presented in this article are of general nature, i.e. they also apply to other forms of adaptation. Hence, the approaches shown in this article are of general applicability either indirectly, e.g. by transferring the concept of meta-level guidance to self-organisation, or directly, e.g. by using the SILKE approach even for open-loop function approximation. Here, first investigations have shown promising results. Future work will deal with a more thorough analysis to formally prove convergence on a more general level.

References

1. Brockmann, W.: Online machine learning for adaptive control. In: Proc. IEEE Int. Work. Emerging Technologies and Factory Automation, pp. 190–195 (1992)
2. Brockmann, W., Buschermöhle, A., Hülsmann, J.: A generic concept to increase the robustness of embedded systems by trust management. In: Proc. IEEE Conf. Systems, Man, and Cybernetics, SMC, pp. 2037–2044 (2010)
3. Brockmann, W., Horst, A.: Stabilizing the convergence of online-learning in neuro-fuzzy systems by an immune system-inspired approach. In: Proc. IEEE Int. Conf. Fuzzy Systems, FUZZ-IEEE, pp. 351–356 (2007)
4. Brockmann, W., Maehle, E., Mosch, F.: Organic fault-tolerant control architecture for robotic applications. In: IARP/IEEE-RAS/EURON Workshop on Dependable Robots in Human Environments (2005)
5. Brockmann, W., Rosemann, N.: Instantaneous anomaly detection in online learning fuzzy systems. In: Hoffmann, F., Cordón, O., Angelov, P., Klawonn, F. (eds.) 3rd Int. Workshop on Genetic and Evolving Fuzzy Systems, pp. 23–28. IEEE Press, Piscataway (2008)
6. Brockmann, W., Rosemann, N., Lintze, C.: Dynamic rate adaptation in self-adapting real-time control systems. In: Lohweg, V., Niggemann, O. (eds.) Proc. Workshop Machine Learning in Real-Time Applications—MLRTA 09. Lemgo Series on Industrial Information Technology, vol. 3 (2009). ISSN 1869-2087
7. Farrell, J., Polycarpou, M.: Approximation Based Control: Unifying Neural, Fuzzy and Traditional Adaptive Approximation Approaches. Wiley-Interscience, Hoboken (2006)
8. Geibel, P., Wysotzki, F.: Risk-sensitive reinforcement learning applied to control under constraints. J. Artif. Intell. Res. **24**, 81–108 (2005)
9. Hafner, R., Riedmiller, M.: Neural reinforcement learning controllers for a real robot application. In: Proc. IEEE Int. Conf. Robotics and Automation, ICRA, pp. 2098–2103 (2007)
10. Hsu, C.F.: Self-organizing adaptive fuzzy neural control for a class of nonlinear systems. IEEE Trans. Neural Netw. **18**(4), 1232–1241 (2007)
11. Kleinlützum, K., Brockmann, W., Rosemann, N.: Modellierung von Anomalien in einer modularen Roboter-Steuerung. In: Berns, K., Luksch, T. (eds.) Autonome Mobile Systeme 2007, pp. 89–95. Springer, Berlin (2007)
12. Mösch, F., Litza, M., Auf, E.S., Jakimovski, B., Maehle, E., Brockmann, W.: Organic fault-tolerant controller for the walking robot OSCAR. In: Proc. Work. Dependability and Fault Tolerance, ARCS. VDE Verlag GmbH (2007)
13. Peters, J., Schaal, S.: Natural actor-critic. Neurocomputing **71**(7–9), 1180–1190 (2008)
14. Polycarpou, M.: Stable adaptive neural control scheme for nonlinear systems. IEEE Trans. Autom. Control **41**(3), 447–451 (1996)
15. Riedmiller, M., Montemerlo, M., Dahlkamp, H.: Learning to drive a real car in 20 minutes. In: Proc. Frontiers in the Convergence of Bioscience and Information Technologies, FBIT, pp. 645–650 (2007)
16. Riedmiller, M., Peters, J., Schaal, S.: Evaluation of policy gradient methods and variants on the cart-pole benchmark. In: Proc. IEEE Int. Symp. Approximate Dynamic Programming and Reinforcement Learning, ADPRL, pp. 254–261 (2007)
17. Rosemann, N., Brockmann, W.: Concept for controlled self-optimization in online learning neuro-fuzzy systems. In: Hertzberg, J., Beetz, M., Englert, R. (eds.) KI 2007: Advances in Artificial Intelligence. LNAI, vol. 4667, pp. 498–501. Springer, Berlin (2007)

18. Rosemann, N., Brockmann, W.: Incremental regularization to compensate biased teachers in incremental learning. In: Proc.World Congress Computational Intelligence, pp. 1963–1970. IEEE Press, Piscataway (2010)
19. Rosemann, N., Brockmann, W., Neumann, B.: Enforcing local properties in online learning first order ts-fuzzy systems by incremental regularization. In: Proc. Int. Fuzzy Systems Assoc. World Congress / 2009 European Soc. for Fuzzy Logic and Technology Conf.—IFSA/EUSFLAT, pp. 466–471 (2009)
20. Rosemann, N., Buschermöhle, A., Brockmann, W.: Beschleunigung der Selbstoptimierung durch Selbstsimulation. In: Hoffmann, E., Hüllermeier, E. (eds.) Proc. Workshop Computational Intelligence, pp. 114–128. KIT Scientific, Karlsruhe (2009)
21. Rosemann, N., Hülsmann, J., Brockmann, W.: Disrupted Learning—Lernen Bei Harten Zustands - Oder Strukturwechseln. In: Proc. Workshop Computational Intelligence, pp. 105–117. Universitätsverlag Karlsruhe, Karlsruhe (2008)
22. Rosemann, N., Neumann, B., Brockmann, W.: Formale Eigenschaften des Silke-Ansatzes zur Kontrolle selbstoptimierender Systeme. In: Hegering, H.-G., Lehmann, A. (eds.) Proc. Informatik 2008 Beherrschbare Systeme—dank Informatik. LNI, vol. 134, pp. 755–762. Köllen, Bonn (2008)
23. Zhang, T., Ge, S.: Adaptive neural control of MIMO nonlinear state time-varying delay systems with unknown dead-zones and gain signs. Automatica **43**(6), 1021–1033 (2007)

W. Brockmann · N. Rosemann (✉)
University of Osnabrück, Osnabrück, Germany
e-mail: nils.rosemann@uni-osnabrueck.de

W. Brockmann
e-mail: werner.brockmann@uni-osnabrueck.de

E. Maehle
University of Lübeck, Lübeck, Germany
e-mail: maehle@iti.uni-luebeck.de

Chapter 3.5
Increasing Learning Speed by Imitation in Multi-robot Societies

Alexander Jungmann, Bernd Kleinjohann, and Willi Richert

Abstract The paradigm of imitation provides a powerful means for increasing the overall learning speed in a group of robots. While separately exploring the environment in order to learn how to behave with respect to a pre-defined goal, a robot gathers experience based on its own actions and interactions with the surroundings, respectively. By accumulating additional experience via observing the behaviour of other robots, the learning process can be significantly improved in terms of speed and quality. Within this article we present an approach, that enables robots in a multi-robot society to imitate any other available robot without imposing unnecessary restrictions regarding the robots' design. Therefore, it benefits not only from its own actions, but also from actions that an observed robot performs. In order to realise the imitation paradigm, we solve three main challenges, namely enabling a robot to decide whom and when to imitate, to interpret and thereby understand the behaviour of an observed robot, and to integrate the experience gathered by observation into its individual learning process.

Keywords Autonomous systems · Societies of robots · Imitation · Reinforcement learning · Markov decision process · Affordances

1 Introduction

The benefits of imitation in terms of drastically reducing the exploration space has been acknowledged by different robotics researchers in the previous decade. The reason seems to be obvious: imitation is one of the most powerful learning techniques applied in nature [2, 4, 5]. Although learning by imitation is used in several applications like learning to swing a tennis bracket or to play the drums [9], the demonstrator-imitator relationship is almost always determined in advance. Approaches based on imitating skill sequences constitute an exception. Referring to pre-determined demonstrator-imitator relationships, a task that has to be learned by imitation is repeated several times (often by humans), so that an observing robot can understand the observed task, derive a generalised representation and integrate it in its own task repertoire to gain the ability of utilising it afterwards. When imitation has to take place in a group of robots, or, to be more general, in a multi-robot society, such fixed conditions do not need to hold any more.

C. Müller-Schloer et al. (eds.), *Organic Computing — A Paradigm Shift for Complex Systems*, 295–307
Autonomic Systems, DOI 10.1007/978-3-0348-0130-0_19,

Imitation is very important when robots in groups should benefit from each others' learning efforts. Typically, the imitation process should not interrupt the observed robot, so that the observing robot often has only one example of a just observed and interesting behaviour to learn from. Although the example usually does not provide enough information in order to learn a generalised version of the observed task, it can help the observing robot to reduce the exploration space for learning and thus helps to increase the overall learning speed.

Our approach overcomes the restrictions invoked by a pre-defined imitator-demonstrator relationship, and implements a solution for increasing the overall learning speed in multi-robot societies, while minimising the negative interference between imitators and demonstrators. When a robot (imitator) tries to understand the observed behaviour of another robot (demonstrator), it more precisely tries to understand the observed behaviour *episode*, consisting of the recorded perception and the demonstrator's state of "well-being" visible to all other robots. The imitator first translates the observations into its own perception to see what it would perceive in place of the demonstrator. It then scans the subjective perception and initiates its low-level skills to give votes about how well each skill could have achieved the perception changes. Together with the likeliness of the demonstrator's state space, those votes are then used to find the most likely sequence of actions corresponding to the observations.

With the presented approach comprising the strategy and low-level skill layers, an observing robot can benefit from the imitation process by observing new state sequences for which it could spend more exploration efforts and by gathering additional episodes in order to improve the speed and the quality of the overall learning process. Furthermore, it can observe new behaviours for already known state transitions and incorporate transition data of other robots by merging it into its own strategy.

2 Related Work

Most approaches regarding imitation of robotic behaviour are based on Hidden Markov Models (HMM) and use the Viterbi algorithm to synthesise behaviour thereof. For example, Billard et al. [3] use the Viterbi algorithm to let the upper part of a robot replay a limited set of arm movements that move coloured objects. In their work the demonstrator-imitator roles are known and fixed. Also, the start and end points of the behaviour to imitate is known to the robot. Billard et al. split the imitation task into the observation and imitation processes, having the goal to minimise the discrepancy between the demonstrated and imitated data sets. In their approach, the robot is only able to learn low-level behaviour and this can only be done from scratch. In contrast to Billard, we do not aim to imitate for the sake of copying another robot's low-level behaviour, but to gather new inspiration for the imitating robot to improve its learning efforts. For this purpose, all levels of abstraction have to be included, instead of only considering the low-level behaviour.

Inamura et al. [10, 11] come closest to our approach with their *Mimesis Loop* approach, which is able to symbolise observed low-level behaviour traces. This is used as top-down teaching from the user's side in combination with the bottom-up learning from the robot's side. As this is useful to decrease the programming effort, it is an exclusive solution, not allowing to be used with other learning techniques like Reinforcement Learning. Also, their approach is not able to use the already existing abstract states of the imitator in the recognition process. Once a robot has extracted enough information to construct a HMM based on the recognised low-level behaviours, it is fixed to that HMM—no exploratory actions on the abstract states are possible any more. Furthermore, the segmentation process that splits the continuous movement trajectories into basic movements uses a fixed scheme. With this, it is not possible to allow for ambiguities in the recognition phase.

In our approach, we assume that the robot has already decent self-learning capabilities. Imitation is used to guide the robot to the "salient" points in exploration space. With more experience, the robot will collect better skills and a more realistic strategy representation. This in turn will enable it to extract more knowledge from its observation efforts.

3 ESLAS—An Imitation Supporting Architecture

To make robots capable of integrating the imitation paradigm into their overall learning process, we designed the ESLAS architecture along the *Triple Tower Architecture* developed by Nilsson [13]. Nilsson distinguishes towers for perception, modelling and action. Our approach focuses on the layered modelling tower, which is responsible for the control of the robot's behaviour, thus the overall learning and additional imitating process as well. Within the ESLAS architecture, the modelling tower is divided into three different levels of abstraction: the motivation layer, the strategy layer and the skill layer. They are briefly described in the following passages. For a more detailed description we refer to Chap. 5.9.

3.1 Motivation Layer

For the evaluation of the robot's overall state, we use biologically inspired evaluation methods. We specify all high-level goals in the form of a drive system having one drive in the interval [0, 1] for each goal. The robot's overall motivation is then to minimise every drive's value. For that reason, it has to develop and adapt its strategy and skill set. The vector of the current drive state to the point of origin can thus be seen as the current motivation. Its change is used as a reward for the strategy briefly described in the next section. Regarding the imitation process, the drive value is furthermore used to express the robot's overall well-being to the other robots. It will guide the observing robots to imitate only obviously beneficial behaviour.

3.2 Strategy Layer

In order to satisfy the motivation layer the robot has to learn a proper strategy. As this would not be feasible with the original state space we use an abstraction method to generalise actual state realisations into abstract regions. The environmental model is updated during runtime and therefore changes quite often in the beginning. A model-free Reinforcement Learning method, like Q-learning [16] is not practicable, because all experience gets lost each time the underlying model changes. Instead, we use model-based Reinforcement Learning with prioritised sweeping to derive a strategy based on a Semi-Markov Decision Process (SMDP). Because the regions are merged and split at run-time, we use Value Iteration to determine the best policy.

At the beginning, all states belong to only one region. Over the course of interaction with the environment the model is updated by several revisors which are invoked recurrently to maintain a mapping of observations in the perception space $\mathbb{R}^d$ to states in the abstracted region space $\mathbb{S}$ by means of nearest neighbour, where d is the number of dimensions of the perception space. The revisors use heuristics to split or merge regions so that the model and underlying statistics reflect the world experience.

3.3 Skill Layer

The skill layer provides a generalised learning method for learning reactive low level skills. Each skill is accompanied by a goal function, that the skill layer tries to learn: $a_f : \mathbb{R}^d \to \mathbb{R}^+$. The function takes the state corresponding to the actual perception as an input and returns an error value that the skill has to minimise.

The skill layer starts with an exploration phase, where it executes random actions by applying random values for the output vector. It records action/effect-traces to get information about how the goal function behaves in certain situations. It then creates likely hypotheses about dependencies between action and goal function changes. It uses these hypotheses in the execution phase to execute actions that minimise the goal function which in turn optimises the skill. For our strategy-learning algorithm, we assume that all skills have finished building hypotheses and are ready for execution.

There are two major benefits with the skill layer over using atomic actions. First, it gives the possibility to create quite complex actions, just by specifying an error function. Second, with the error function, a skill can recognise itself from a trace of observations.

4 Enabling Robots to Learn by Imitation

Based on the previously described ESLAS architecture we can now adopt the paradigm of learning by imitation. In our approach we solve three main challenges that arise when enabling robots in a multi-robot society to learn from each other without interfering and negatively influencing each other:

1. Whom and when should a robot imitate within a multi-robot society?
2. How does a robot interpret and thereby understand the behaviour of another robot, in order to decide whether to imitate it or not?
3. How should an imitating robot integrate the demonstrator's behaviour in its own behaviour repertoire in order to benefit from it, namely to increase its overall learning speed?

4.1 Deciding Whom and When to Imitate

In order to overcome the restriction of pre-determined demonstrator-imitator relationships in a multi-robot society, a robot has to decide autonomously *whom* and *when* to imitate. Especially in heterogeneous robot societies, this turns out to be a real problem, since not all members necessarily provide the same set of capabilities. Therefore the robots need the ability to measure the behavioural difference between each other. To solve this problem, our approach constructs and maintains for each robot a Bayesian network of *affordances* detected in the observation, a so-called affordance network (Fig. 1). This is done at runtime.

The term affordance has been introduced by the psychologist Gibson [7], who observed that our human perception of the world depends on our interactions with it. In his notation, an affordance is a property, that describes a possible interaction with an object. According to this notation, affordance networks encode dependencies between the interaction possibilities, environmental objects offer to the different robots. Using the concept of affordances, a robot is able to reason about behavioural differences between robots without having to take into account their diverse hardware and software conditions. With a metric on those networks, a robot is able to calculate the difference between its own capabilities and those of the other robots.

The observing robot can then decide whether it should imitate the second robot by observation or ignore it. In contrast to randomly choosing an available robot without knowing if imitating the robot's observed behaviour will be meaningful for the individual learning process at all, a robot now has a mechanism to decide whether to continue or to cancel observing a demonstrator. Applying this concept

Fig. 1 Example for an affordance network

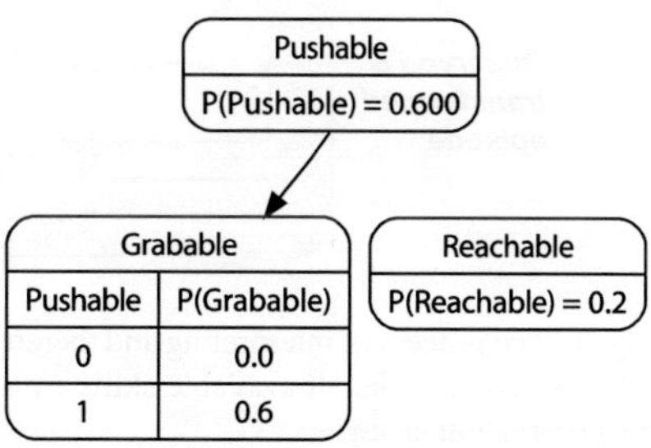

of *guided imitation* brings two advantages. First, an imitation process can be prematurely abandoned and unnecessary computations minimised. Second, the quality of the additional episodes increases and thus the quality and speed of the individual learning process increase as well. In our approach we assume the set of affordances to be pre-defined and constant. However, this is not a restriction to our algorithm as learning affordances at run-time [12] does not pose a problem at all.

The complexity of the behavioural difference measurement depends on the specified explorable affordances of the objects available in the environment. By increasing the quantity and diversity of the affordances that a robot is able to explore, it can compare its own behaviour with other robots in a more fine grained way. Explored affordances can thus be seen as samples of the behaviour function of the robot. While exploring the environment, the robots gather samples. Those samples can be used to approximate the robots' behavioural possibilities. A more detailed description can be found in [8].

4.2 Interpreting Observed Behaviour

After deciding whom and when to imitate, the challenge of how to interpret a behaviour while observing a demonstrator arises. In Fig. 2 an example is shown in which an imitating robot tries to understand the observed behaviour episode of another robot (demonstrator). The observed episode consists of the recorded perception and the demonstrator's visible overall drive state. The imitator first translates the observations into its own perception to see what it would perceive, if itself had been in the demonstrator's situation. It then scans the subjective perception change in time and allows its low-level skills to give so-called *votes* about how well each

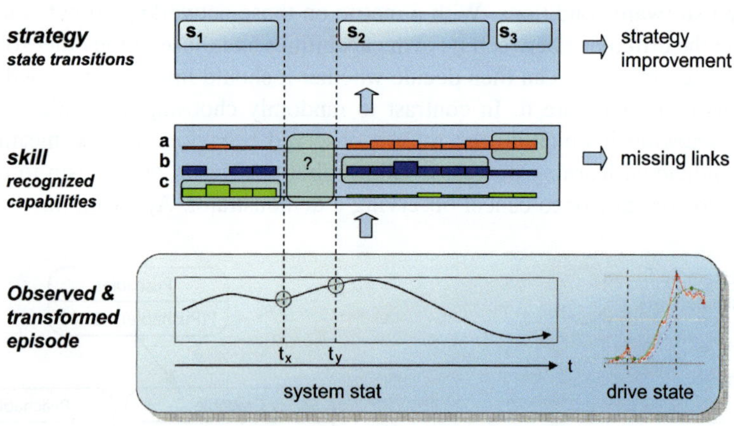

Fig. 2 The process of interpreting and thereby understanding a demonstrator's performance within an observed episode: all available skills emit votes regarding how well they could have achieved the observation changes

of them could have achieved the perception changes. These votes represent the error functions for which appropriate skills have been learned (cf. Sect. 3.3). Those votes are then used together with the likeliness of the demonstrator's state space to find the most likely path corresponding to the observations. Our implemented algorithm is inspired by the Viterbi algorithm [15] which is often used to imitate using Hidden Markov Models (HMM). In general, the Viterbi algorithm tries to find the most likely hidden state sequence $s_1^T = \langle s_1, s_2, \ldots, s_T \rangle$ (*Viterbi path*), that explains the observation sequence o_1^T. This can be done by maximising the following constraints:

$$s_1^{T*} = \arg\max_{s_1^T} P(s_1^T \mid o_1^T) \tag{1}$$

Using Bellman's dynamic programming algorithm [1] the Viterbi algorithm determines the maximum efficiency in time $O(Tn)$ where n is the number of non-zero transition probabilities. It recursively calculates the probability

$$V(s,t) = \max_{s_1^{t-1}} P(o_1^t, s_1^{t-1}, s_t = s) \tag{2}$$

that s is the hidden state at time t given the observations o_1^t for all $s \in S$, with S being the observer's state space:

$$V(s,t) = P(o_t \mid s_t = s) \max_{s'} P(s_t = s \mid s_{t-1} = s') V(s', t-1) \tag{3}$$

V is initialised with $V(s, 1) = \max_{s_1} P(o_1 \mid s_1 = s) P(s_1 = s)$ for all $s \in S$. The most likely path can now be extracted using

$$\varphi(s,t) = \arg\max_{s'} P(s_t = s \mid s_{t-1} = s') V(s', t-1), \tag{4}$$

which determines the best predecessor of state s at time t.

The imitation approaches usually found in literature calculate the Viterbi path to find the state sequence the imitator should realise in order to exactly copy the observed behaviour. This is done using the state space (assumed to be fix) of the inferred HMM, which is assumed to reflect the demonstrator's state space. In contrast to those methods it is important to see that we use a method similar to the calculated Viterbi path to explain the observations recorded from the demonstrator with the imitator's already existing state and action space. Thereby, the imitator tries to understand the demonstrator with the knowledge it already has in terms of its own state space (cf. Sect. 3.2) and behaviour repertoire (cf. Sect. 3.3).

If the observations provide enough information to infer the corresponding state, $P(o_t \mid s_t)$ could be straightforwardly calculated out of the state representation chosen for the specific domain. However, this is seldom the case in realistic applications. In order to be able to use the Viterbi algorithm for inference on the imitator's self-learned knowledge, the robot has to infer the probable state transitions and to guess which of its behaviours could have realised those observed state transition.

The calculation of $P(s_t = s \mid s_{t-1} = s')$ (see (3)) is more involved. If one would just take the transition probability of its greedy action in s_{t-1} the robot would not get new insights about other and maybe better state transition behaviours in that specific state. Instead, it should guess from the observations which of the behaviours in its own behaviour repertoire would best match the recorded observations.

Let us now consider state transition $\langle s_{t_a}, s_{t_b} \rangle$, where $s_{t_a} \neq s_{t_b}$, and that all observations o_t are already translated into a first-person model using geometrical transformations. Firstly, for every recorded observation step $\langle o_{t-1}, o_t \rangle$ ($t \in [t_a, t_b]$) all the behaviours are asked to give a vote $P_b(o_t \mid o_{t-1})$ representing the ability of behaviour b to be able to realise that step. These are determined by means of the corresponding error function with which the behaviours have been learned (Sect. 3.3). These votes are then divided by the time span of the full state transition:

$$P_b(s_{t_b} \mid s_{t_a}) = \frac{\sum_{t=t_a}^{t_b} P_b(o_t \mid o_{t-1}, s_{t_a})}{t_b - t_a} \tag{5}$$

At every state transition, one can now determine the most likely transition action $b_{ml} = \arg\max_b P_b(s_{t_b} \mid s_{t_a})$. It can be used to retrieve the transition probability in the observer's SMDP that would most probably correspond to the observation of the demonstrator:

$$P(s_{t_b} \mid s_{t_a}) = P(s_{t_b} \mid s_{t_a}, b_{ml})$$

Thereby, we get the following recursive solution

$$\begin{aligned} V(s,t) = {} & \max_b P_b(o_t \mid s_t = s, o_{t-1}) \\ & \times \max_{s'} P(s_t = s \mid s_{t-1} = s', a_{t-1} = b_{t-1}) \\ & \times V(s', t-1) \end{aligned} \tag{6}$$

in which $P(s_t = s \mid s_{t-1} = s', a = b_{t-1}) = T(s', a, s)$ are the transition probabilities learned in the strategy layer. In case of a missing link, i.e. if a transition or skill likelihood probability drops below the user defined threshold θ, the algorithm has to account for additional bootstrapping regarding the state values: After having found again understandable observation changes regarding the skill votes, it reassigns the state values in the same way it is done at the beginning of the algorithm (see (1)–(4)).

4.3 Incorporating the Extracted Knowledge

In order to benefit from imitation the imitating robot first observes the observed robot's state $o \in \mathbb{R}^d$ over a certain period of time, where d is the number of dimensions of the state space. An observation is a trace of observed raw states with length n:

$$O_{t_1}^{t_n} = (o_{t_1}, o_{t_2}, \ldots, o_{t_n}) \tag{7}$$

This observation sequence is projected by a mapping function $\varphi : \mathbb{R}^d \rightarrow \mathbb{R}^d$ from the demonstrator's point of view to the point of view of the imitator. In the projected state $\hat{o}$, the imitator seems to be in the position of the demonstrator himself. $\hat{O}$ is calculated for every observation:

$$\hat{O}_{t_1}^{t_n} = (\hat{o}_{t_1}, \hat{o}_{t_2}, \ldots, \hat{o}_{t_n}) \tag{8}$$

For the projected trace, the observer has to find out the missing information in order to be able to integrate this observation into its own strategy. To create a trace of interactions it has to approximate action a, duration d, and reward r for the salient observation episodes. Action a is determined by the recognition process described in the previous section. It uses the robots own skills to get the most likely corresponding action. The duration d can be retrieved by comparing timestamps for every observation. Reward r is also not directly observable, but can be approximated by the means of the demonstrator's drive state. With a, d and r determined, the robot is then able to create a trace of interactions:

$$K_{t_1}^{t_n} = (I_{t_1}, I_{t_2}, \ldots, I_{t_n}) \tag{9}$$

$K_{t_1}^{t_n}$ is imported into the strategy of a robot just like his own experiences I. However, it has to be implemented as a new episode of experience—not connected to the robot's own stream of experience. Otherwise the state transitions would be inconsistent. The current state s_{cur} of the robot is backed up and inserted again after the imitation, so that the robot can continue with s_{cur}. The whole process of learning observed behaviour is transparent to the underlying strategy. It does not know whether its input comes from an observation or from its own perception.

5 Results by Simulation

Our approach is evaluated by simulation using the PlayerStage/Gazebo framework [6] simulating a robot similar to the well-known Pioneer2DX. The dynamics are simulated using the *Open Dynamics Engine (ODE)* [14].

The simulation scenario (Fig. 3) consists of three bases to which pucks, that are dispersed in the field, have to be transported. The bases have different colours: red,

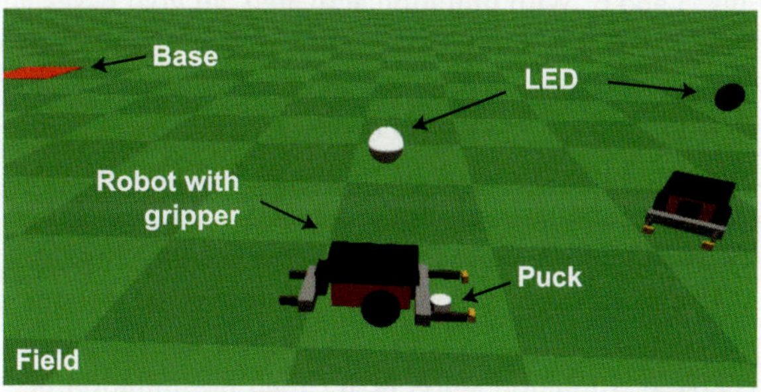

Fig. 3 The robots have grippers to grab a puck and LEDs for showing their drive state

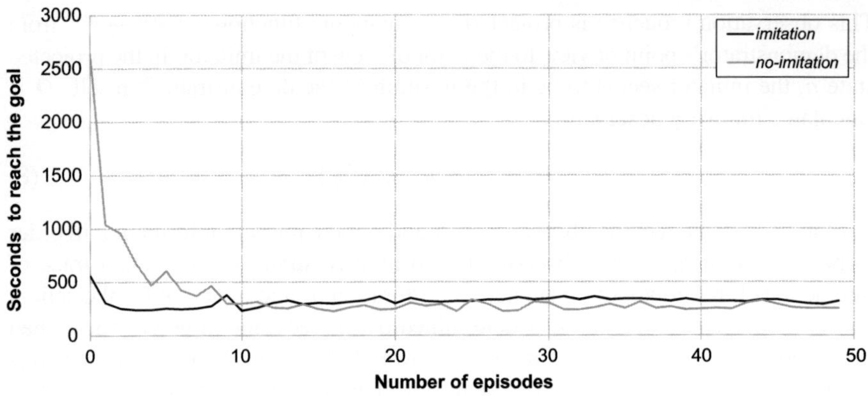

Fig. 4 The average time needed for reaching a goal over 50 runs

yellow, and black. The robots have some predefined actions, provided by the skill layer described in Sect. 3.3: approaching the puck, as well as approaching the different bases. A positive reward of 10 points is given for collecting the puck. Once the puck reaches one of the bases, the robot receives another certain amount of positive reward. For the yellow and red base the reward is 20 points. The black base is *much* farther away and thus more difficult and unlikely to reach. For transporting a puck to this base a robot gets 10000 points in order to take the considerable expenditure for reaching the black base into account. Furthermore, the raw state is a three-dimensional vector containing the distance of the robot to the closest puck, the distance of the closest puck to the closest goal, and the ID of the closest goal. We perform all experiments in two phases: with imitation activated and with imitation deactivated for comparison.

Figure 4 depicts the time in seconds, that a robot needs to catch one puck and bring it to a base over 50 consecutive episodes. The values for all charts are averaged over more than 500 robots. As one can see, in the experiment with imitation activated, the goal is reached much faster than in the one with imitation deactivated at the beginning. The time values meet each other after eight episodes and stay nearly the same, with a little advantage for no-imitation. The reason for no-imitation being a little bit faster in the end does not mean it is better. It is because those robots visit the black base less frequently although it has the highest reward. This is shown in Figs. 5 and 6. With imitation activated, all goal bases get nearly the same amount of pucks at the beginning. After some exploration, more and more robots find out that its most beneficial to only use the black base.

Without imitation the robots explore and learn to prefer the black base, too. However, the number of robots knowing that fact is much lower than with imitation. Therefore, the average time to reach a goal is higher for imitation than without, because the distance to the black base is much higher.

The reward per second (Fig. 7) is a good overall indicator of how successful a strategy is. It takes into consideration the reward, as well as the time needed to receive it. Imitation starts better than no-imitation because of the little amount of

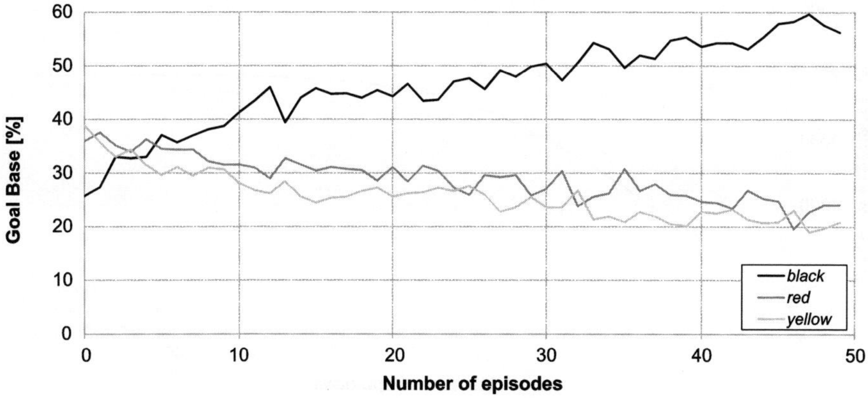

Fig. 5 The percentage of pucks brought to the respective base with imitation *activated*

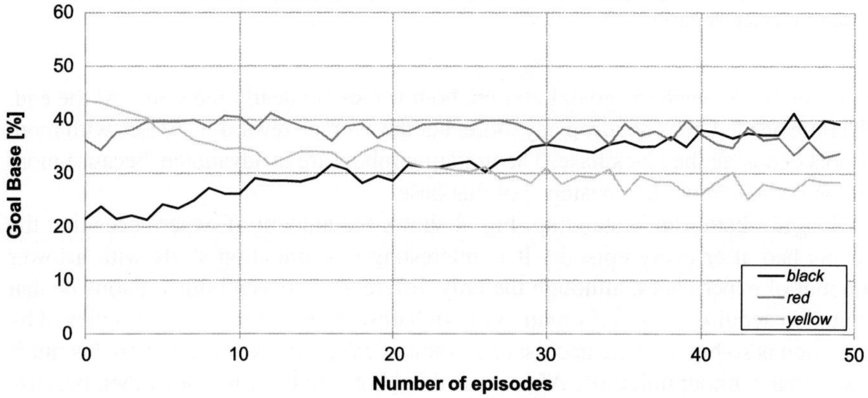

Fig. 6 The percentage of pucks brought to the respective base with imitation *deactivated*

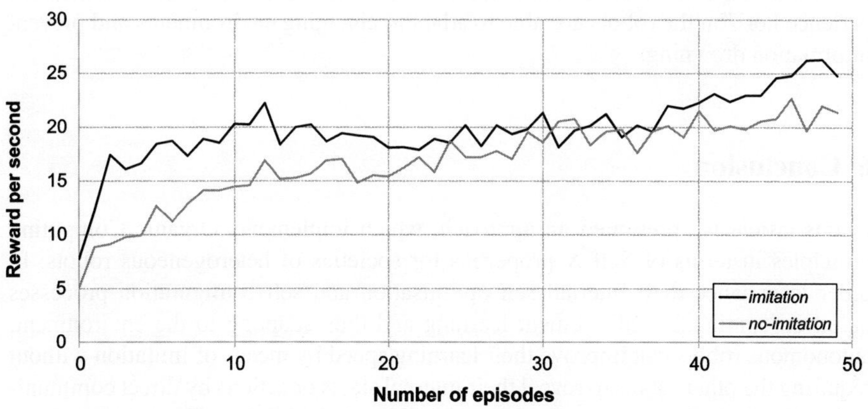

Fig. 7 Reward per second

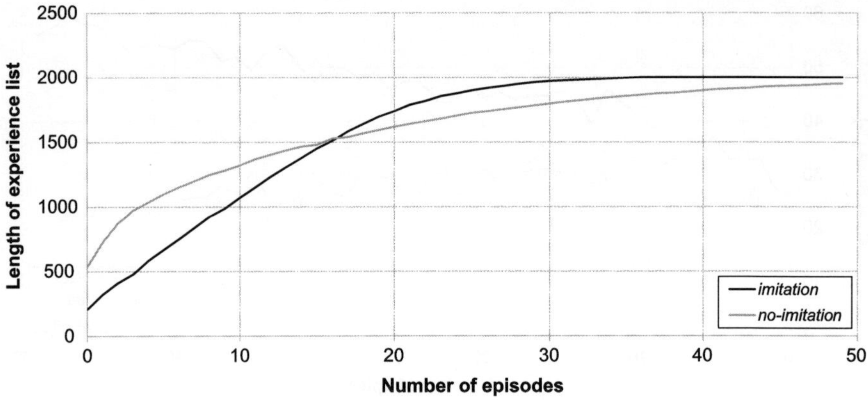

Fig. 8 Length of the experience list. First imitation uses less memory, because it needs less time to reach a goal. Later no-imitation is nearly equally fast, but the imitation still observes other robots. Therefore the amount of experience increases faster, until the limitation of 2000, where the oldest observations are dropped

time needed to reach the goal. Later on, both values are nearly the same. At the end, the time needed does not drop any more, but the average reward increases with more robots choosing the black base. The imitating robots are in advantage, because more of them know about the existence of that base.

To get a better understanding, Fig. 8 shows the amount of experiences that the robots had after every episode. It is interesting that imitation starts with a lower amount of experiences, although the only difference between both versions is, that imitation acquires more information through observation. The speedup achieved by imitation is so big, that the necessary amount of experiences to reach a goal is much lower than without imitation. After some time, both charts cross each other, because the time needed to reach a goal becomes nearly the same for both, so imitation acquires more experiences by observation. The number of experiences is bounded by 2000, where old observations are dropped as new ones arrive. Using this sliding experience horizon the robots are able to adapt to changing environments and prevent information drowning.

6 Conclusion

In this article we presented an approach, which implements Organic Computing principles in terms of Self-X properties for societies of heterogeneous robots. In order to enhance their internal self-optimisation and self-configuration processes based on continual reinforcement learning and thus adapting to the environment, autonomous robots can improve their learning speed by means of imitation without requiring the other robots to reveal their internal states or actions by direct communication. The imitating robots only use externally perceivable data. Thus we overcome the restriction of a pre-determined demonstrator-imitator relationship. Furthermore robots do not have to be explicitly prepared beforehand any more in order to gain the ability of imitating other robots or to be imitated by other robots in a society of heterogeneous robots. Using the presented approach robots can now improve their strategies observing any robot that is around only requiring that it expresses its overall state in terms of how well it reaches its goals.

References

1. Bellman, R.E.: Dynamic Programming. Dover, New York (2003)
2. Billard, A.: Learning motor skills by imitation: a biologically inspired robotic model. Cybern. Syst. **32**(1–2), 155–193 (2000)
3. Billard, A., Epars, Y., Calinon, S., Cheng, G., Schaal, S.: Discovering optimal imitation strategies. Robot. Auton. Syst. **47**(2–3), 69–77 (2004)
4. Borenstein, E., Ruppin, E.: Enhancing autonomous agents evolution with learning by imitation. In: Second International Symposium on Imitation in Animals and Artifacts (2003)
5. Demiris, J., Hayes, G.R.: Imitation as a Dual-Route Process Featuring Predictive and Learning Components: A Biologically Plausible Computational Model, pp. 327–361. MIT Press, Cambridge (2002)
6. Gerkey, B.P., Vaughan, R.T., Howard, A.: The player/stage project: Tools for multi-robot and distributed sensor systems. In: Proceedings of the International Conference on Advanced Robotics (ICAR), pp. 317–323 (2003)
7. Gibson, J.J.: The theory of affordances. In: Perceiving, Acting, and Knowing: Toward and Ecological Psychology, pp. 62–82. Erlbaum, Hillsdale (1977)
8. Golombek, R., Richert, W., Kleinjohann, B., Adelt, P.: Measurement of robot similarity to determine the best demonstrator for imitation in a group of heterogeneous robots. In: Biologically-Inspired Collaborative Computing vol. 268, pp. 105–114. Springer, Boston (2008)
9. Ijspeert, A.J., Nakanishi, J., Schaal, S.: Movement imitation with nonlinear dynamical systems in humanoid robots. In: Proceedings of the IEEE International Conference on Robotics and Automation (ICRA), vol. 2, pp. 1398–1403 (2002)
10. Inamura, T., Nakamura, Y., Ezaki, H., Toshima, I.: Imitation and primitive symbol acquisition of humanoids by the integrated mimesis loop. In: Proceedings of the IEEE International Conference on Robotics and Automation (ICRA), vol. 4 (2001)
11. Inamura, T., Toshima, I., Nakamura, Y., Saitama, J.: Acquiring motion elements for bidirectional computation of motion recognition and generation. In: Experimental Robotics VIII (2003)
12. Lopes, M., Melo, F.S., Montesano, L.: Affordance-based imitation learning in robots. In: IEEE/RSJ International Conference on Intelligent Robots and Systems (IROS), pp. 1015–1021 (2007)
13. Nilsson, N.J.: Artificial Intelligence: A New Synthesis. Morgan Kaufmann, San Mateo (1998)
14. Smith, R.: Website of ode (open dynamics engine) (2008). http://www.ode.org/
15. Viterbi, A.: Error bounds for convolutional codes and an asymptotically optimum decoding algorithm. IEEE Trans. Inf. Theory **13**(2), 260–269 (1967)
16. Watkins, C.J.C.H., DayanBellman, R.E.: Dynamic Programming. Dover, New York (2003), P.: Q-learning. Mach. Learn. **8**, 279–292 (1992)

A. Jungmann (✉) · B. Kleinjohann · W. Richert
C-LAB, University of Paderborn, Paderborn, Germany
e-mail: global@c-lab.de

Chapter 3.6
Learning to Look at Humans

Thomas Walther and Rolf P. Würtz

Abstract The problem of learning a generalisable model of the visual appearance of humans from video data is of major importance for computing systems interacting naturally with their users and other humans populating their environment. We propose a step towards automatic behaviour understanding by integrating principles of Organic Computing into the posture estimation cycle, thereby relegating the need for human intervention while simultaneously raising the level of system autonomy. The system extracts coherent motion from moving upper bodies and autonomously decides about limbs and their possible spatial relationships. The models from many videos are integrated into meta-models, which show good generalisation to different individuals, backgrounds, and attire. These models even allow robust interpretation of single video frames, where all temporal continuity is missing.

Keywords Image understanding · Autonomous learning · Organic computing · Pose estimation · Articulated model

1 Introduction

Any interaction of a computing system with the external world is plagued by the imperfection of the available sensory data. This problem is particularly hard in the case of visual data, but applies to any sensors which are required to keep the system informed about the world and thus is the basis of any intelligent behaviour. This imperfection can only be overcome by building robust and general internal models of the world, which can partly override the data and lead to robust behaviour. This remains true if the best available methods for input processing are used.

Humans are extremely relevant tenants of the environment of Organic Computing systems. They are required as guides for behaviour, they must not be hurt, and they should be able to interact with a computing system in a natural way. Therefore, the analysis of human body poses by mere observation is a topic of growing interest in computer vision—with application potential ranging from surveillance over man-machine communication to motion picture animation. Yet, the artificial pose estimation (PE) approaches developed over the last two decades are nowhere close to matching human visual capabilities. This may be due to fundamentally

C. Müller-Schloer et al. (eds.), *Organic Computing — A Paradigm Shift for Complex Systems*, 309–322
Autonomic Systems, DOI 10.1007/978-3-0348-0130-0_20,

different working principles of artificial and biological vision systems. In the following, we aim at levelling these differences by Organic Computing (OC) [39] concepts. In particular, we propose a PE system that acquires knowledge in a completely unsupervised manner directly from video input; this knowledge is then generalised to novel situations, mimicking human skills in non-trivial and continuous learning [25]. We combine state-of-the-art computer vision techniques, biological principles like (controlled) self-organisation and machine learning to autonomously assemble a higher-level, upper-body *meta-model* of captured human beings. The learned model is shown to generalise well to different movements, attire, and individuals during subsequent matching attempts in complex environments.

2 Learning Upper Body Models

From a technical point of view, a human body is an articulated ensemble of non-rigid objects (clothed limbs) performing relatively smooth and coherent movements over time. Articulation between body parts is enforced by skeletal joint constraints. Thus, a basic model describing human upper body characteristics should capture shape/appearance of the single limbs and encode kinematic relations between them.

Semi- and unsupervised extraction of body parts from given video footage has been addressed in the literature before, yet either a lack of system autonomy (e.g., [14]) or unwieldy system complexity (e.g., [27, 41]) render these approaches impractical for our purpose. We here extend the simple strategy by [14] to a fully autonomous method that extracts body parts from fronto-parallel video sequences without human intervention.

In the first instance, the input stream is parsed for the frame f_{max} displaying maximum motion. This frame is likely to allow for precise foreground detection by a frame differencing mechanism. A *multi-frame difference image* (inspired by [13]) is learned from f_{max} and used as a precise *foreground map* to 'seed', together with its complementary background map, a binary *graph-cut* [5] procedure that cleanly separates the moving foreground entity from bland background structures. Figure 1a depicts results of the graph cut stage for a typical input sequence.

(a) Graph cut results (b) Motion trajectories

Fig. 1 Background segmentation from motion and motion trajectories form the basis for the model extraction

We then use the extracted area as a mask to evenly distribute an intensity patch feature population on the retrieved foreground entity. This allows to track motion of the captured subject through all frames $f \neq f_{\max}$. The required tracking scheme has been adopted from [33], using the GPU-based implementation of [31], in order to track large sets of features with near-real-time performance. Figure 1b shows some exemplary feature trajectories resulting from the tracking process.

To eventually identify the sought-after body parts, coherently moving features have to be grouped together, based on the similarity of their motion trajectories. We use *spectral clustering* [34] for that purpose, which appeals by its ease of implementation and the fact that it has already proven useful in motion trajectory delineation, see, e.g., [2, 26, 38, 40].

Following [34], the basic operation of spectral clustering is as follows. A data collection **E** of N arbitrary data objects $e_1, \ldots, e_N$ and some similarity function w_{ij} defining their pairwise similarity yields a similarity matrix **W**. Each data element e_i is assigned a degree d_i, its total similarity with all data elements in **E**, and the degree matrix **D** is the diagonal matrix with entries d_i. By analysing the spectrum and eigenvectors of the Laplacian matrix $\mathbf{L} = \mathbf{D} - \mathbf{W}$ or of some normalised form of **L**, it is possible to detect clusters in the original data **E**. For details and theory of spectral clustering see, e.g., [1, 3, 34].

The recursive subdivision in the spectral clustering scheme of [29] allows for automatic estimation of the number N_L of clusters by *self-tuning* introduced in [42]. Intermediate clusters are assigned a Θ-score, which is calculated from the eigenvectors and measures the quality of the proposed subdivision. If this value is low, the subdivision of the cluster stops, in an intermediate interval the proposed subdivision can be improved by removing outliers, and clusters with high Θ-scores are further subdivided recursively. The two thresholds on the Θ-score allows for a completely data-driven decision about the final number of clusters. This automatically yields the correct body topology in all of our input sequences. Possible excess clusters induced by soft cloth motion are eliminated using appropriate merging strategies. Exemplarily, Fig. 2a shows found feature groups, which correspond nicely to the underlying limbs; for completeness, Fig. 2b demonstrates the fair segmentation performance also in the presence of a cluttered background.

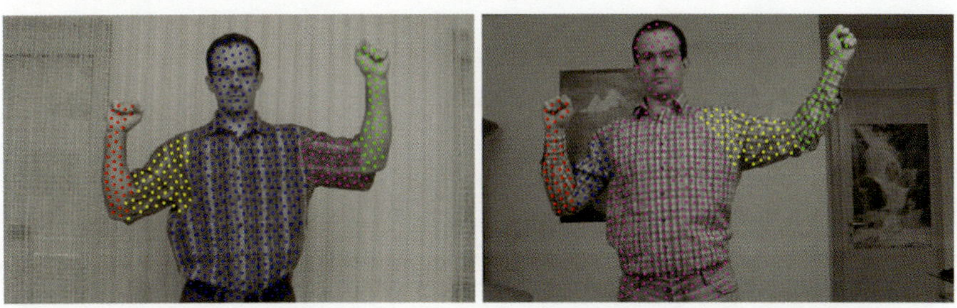

(a) Simple background (b) Cluttered background

Fig. 2 Feature segmentation; features with identical colour are part of the same group

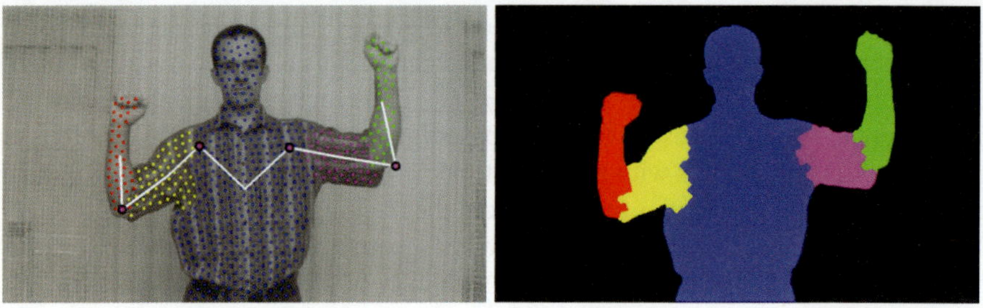

(a) Kinematic skeleton (b) Limb segmentation

Fig. 3 Feature segmentation; features/pixels with identical colour are part of the same group

Given the limb clusters, we use the method from [14] to construct a *skeleton*. Physical plausibility of a joint between any two clusters is used to guide probabilistic construction and pruning of a spanning tree connecting all identified body parts. Tree branches surviving the pruning process then represent the most likely kinematic configuration of the articulated foreground entity. Figure 3a shows a kinematic skeleton derived from a typical input sequence.

Eventually, full limb templates are cut out for the reference frame. While this step footed on graph-cut-based multi-class segmentation in earlier work [36], the current methodology is much more efficient and shows comparable performance w. r. t. model quality: define the distance $D_i(\mathbf{x})$ between a foreground pixel $\mathbf{x}$ and body part i as the smallest Euclidean distance between $\mathbf{x}$ and any element in feature cluster i. Given that there exist N_L body parts, $\mathbf{x}$ is assigned to limb $\arg\min_l D_l(\mathbf{x})$, where $l \in [0 \ldots N_L - 1]$. Results of this assignment procedure are depicted in Fig. 3b.

With these prerequisites, all elements of a baseline upper human body model are available; it remains to cast them into a concise formulation. We favour *pictorial structure* (PS) [11] models for that purpose. PS models are straightforward to construct and can be matched efficiently to 2D image content. Our current body representations contain the learned limb templates (by means of shape and colour) and employ elastic kinematic constraints to connect single body parts. Joint angle limits are also inferred to prevent retrieval of awkward postures during subsequent model matching.

3 Meta-model Construction

Clearly, the simple models learned above are highly scenario specific and will hardly perform well in unconstrained posture retrieval. Yet, they prove useful for construction of abstract patterns that capture characteristic traits of the human upper body while remaining generic enough to represent a wide range of different subjects. Generating these meta-models is straightforward: let $\mathcal{M} = [M_0, \ldots, M_{N_M-1}]$ be a set of PS body models learned from N_M input sequences using the above techniques. The meta-model is initialised by copying limb templates, kinematic constraints and joint

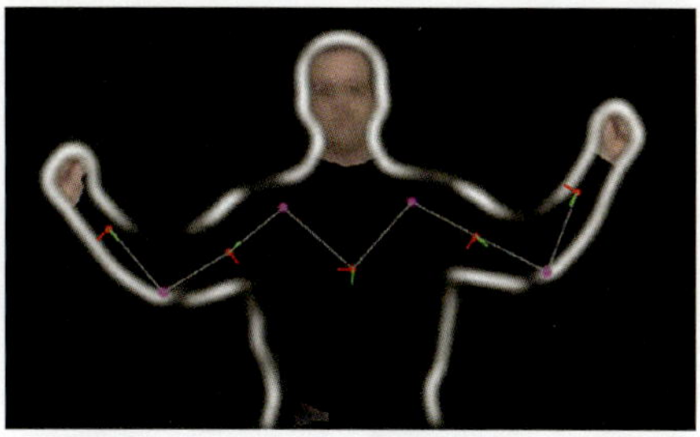

Fig. 4 Extracted meta-model

angles from M_0. Model $\mathcal{M}_1$ is then matched and aligned with the current meta-representation, such that orientation of the limb coordinate systems and limb and joint enumeration coincide. Coarse model matching is performed using registration routines from [10], whereas precise alignment of the single limbs is guided by iterative closest point procedures [4] (accelerated according to [28]). Superposition of the registered limb shapes yields new, prototypical outlines of all *meta-limbs*. During superposition, modified Gaussian voting (remotely inspired by the approach presented in [18]) is employed to extract smooth limb shapes and counter the influence of cloth deformation.

However, to construct colour prototypes of the meta-limbs, more care has to be taken: generalisation over strong variation in cloth colour is required, yet *persistent* colour areas that remain similar in all input sequences (like head and hand colour patches) have to be represented reliably. Thus, colour information from any limb of M_1 is integrated into the meta-colour prototype if and only if colours in both representations are similar; colour similarity is measured by histogram-based correlation in HSV (hue, saturation, value) colour space. Repeating the above process for $M_2 \ldots M_{N_M-1}$ continuously refines the shape and colour meta-prototypes; the resulting meta-model is depicted in Fig. 4.

Compared to the prototyping approaches used for shape and colour features, skeleton prototyping is straightforward. Whereas the overall *meta-skeleton* is necessarily identical to the skeleton of M_1 w. r. t. connectivity, relative locations of the meta-joints are found by averaging the joint locations from all $M_i \in \mathcal{M}$. Similarly, the distribution of relative meta-joint angles is learned by aggregating information from all input models.

4 Matching Considerations

To match the fully evolved meta-model to novel input images, we employ a pictorial structure matching scheme similar to [10]. Due to the tree-like structure of the learned models, this dynamic programming approach allows to speed up model matching significantly while guaranteeing globally optimal results. We enhance their approach in several aspects: firstly, we allow the matching algorithm to not only find the location (shift, rotation, scale) of each meta-limb, but also to infer if a body part is flipped or not. This enables the system to cope with *kinematic flips*

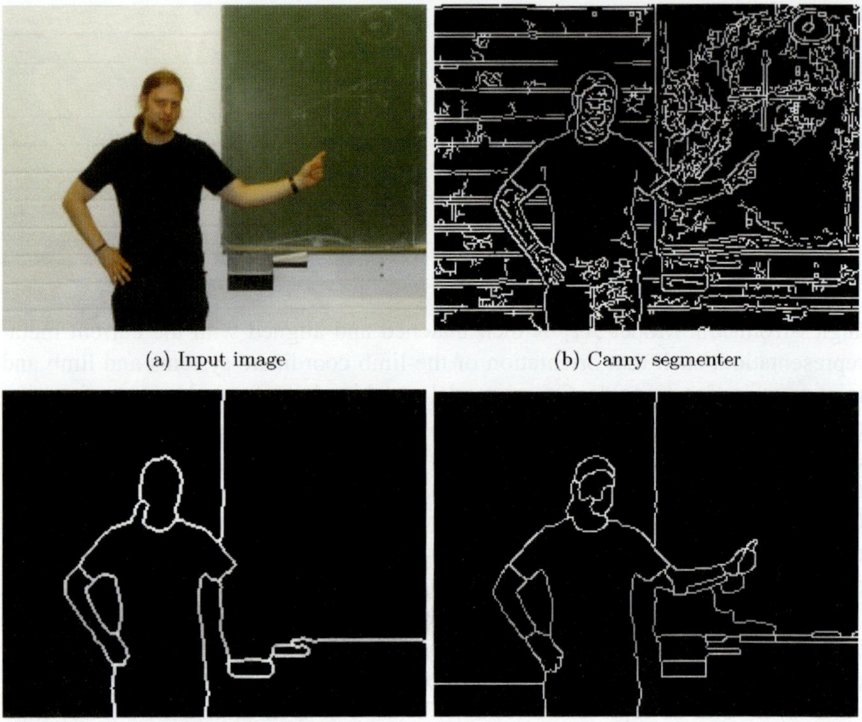

(a) Input image (b) Canny segmenter

(c) JSEG (d) EDISON

Fig. 5 Different image segmentation schemes

(terminology chosen in allusion to [32]). Such flips occur due to the 3D nature of the captured scenario and have to be taken into account to allow analysis of a broader range of body postures. It is assumed that each body part can only be flipped orthogonally to the image plane (around each limb's major principal axis); the limits and angular statistics of each joint attached to the flipped body part are updated automatically.

Secondly, matching reliability is increased by refining the shape matching cost function constructed in [36]: matching is now computed using the *oriented chamfer* distance (cf. [30]) between the meta-limb shapes and a line segmentation of the given query image. The stand-alone JSEG ([8]) algorithm utilised in [36] to generate this line segmentation has been replaced by the EDISON [6] image segmentation scheme that is fully integrated into our system. Quality of line images generated by EDISON perceptually compares to or even outperforms the JSEG output (cf. Figs. 5c and 5d) and is by far superior to simple Canny edge segmentation (cf. Fig. 5b). Note that we outsource oriented chamfer calculations to the GPU (using a CUDA [24]-based implementation), to compensate for the increased computational effort inherent in this more powerful approach. To save computation time, the above

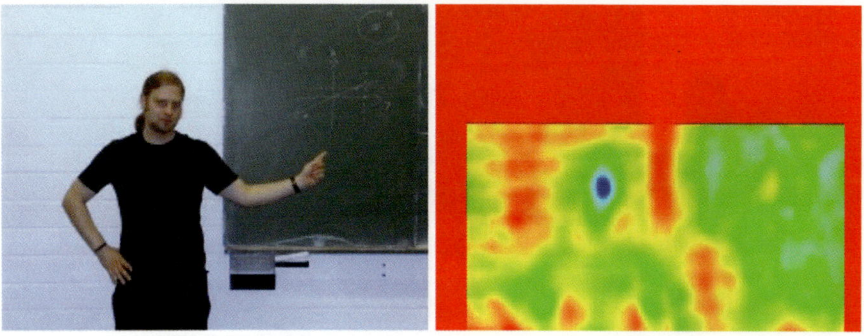

(a) Input image (b) Response map

Fig. 6 Gabor-based torso detection, minima (*blue*) in the response map correspond to optimal barycentre locations of the body trunk, *opaque red* areas indicate Gabor jets falling outside the image plane

fuzzy meta-shapes are thinned (thinning algorithm from [9]) prior to being used for oriented chamfer matching.

Adding up to the above, it is straightforward to exploit the persistent colour feature stored in each meta-limb for a per-limb *colour cue map*: for that, we first transform the RGB (red-green-blue) representation of the query image to HSV colour space. Let then $\mathbf{W}(\mathbf{x})$ define a window of 7×7 pixels centred at a given position $\mathbf{x}$ in the HSV representation of the query image. Assume that an HS-histogram can be derived (during a batch-processing step not described here due to spatial constraints) from the meta-limbs' persistent colour regions. A similar histogram is deemed available for the window patch. Note that we again drop the 'value' component during histogram construction to achieve better invariance to illumination variation. The map value at $\mathbf{x}$ is then calculated as the correlation of the two HS-histograms. Loosely following [10], the final colour cue map is used (after re-scaling) to define an additional *colour matching cost* that backs up the shape cue described above and renders overall matching behaviour more robust.

To capture texture properties, Gabor features [7], which are both biologically motivated and well-known in face recognition [17], are used to learn prototypes for the single limbs using a batch procedure: a Gabor grid graph is generated for all body parts in M_0 and adopted to the limbs' shapes. Each graph node corresponds to a Gabor jet [17] (8 orientations, 5 scales) extracted at the node's position. Running over all following frames of all input sequences, the reliability of each putative graph node is assessed by calculating the temporal variance of the attached jets; nodes corresponding to jets with high variance are deemed unreliable and become deleted. The employed Gabor features are not rotation invariant and are highly susceptible to changes in image texture, thus, fast rotating body parts like forearms or upper arms, which display vividly varying surface texture, do not form valid Gabor prototypes. Contrarily, the torso remains upright in all input sequences and face texture stays similar; as a consequence, a Gabor prototype organises in the head region. The evolved prototype allows to scan novel images for face-like structures; as head and

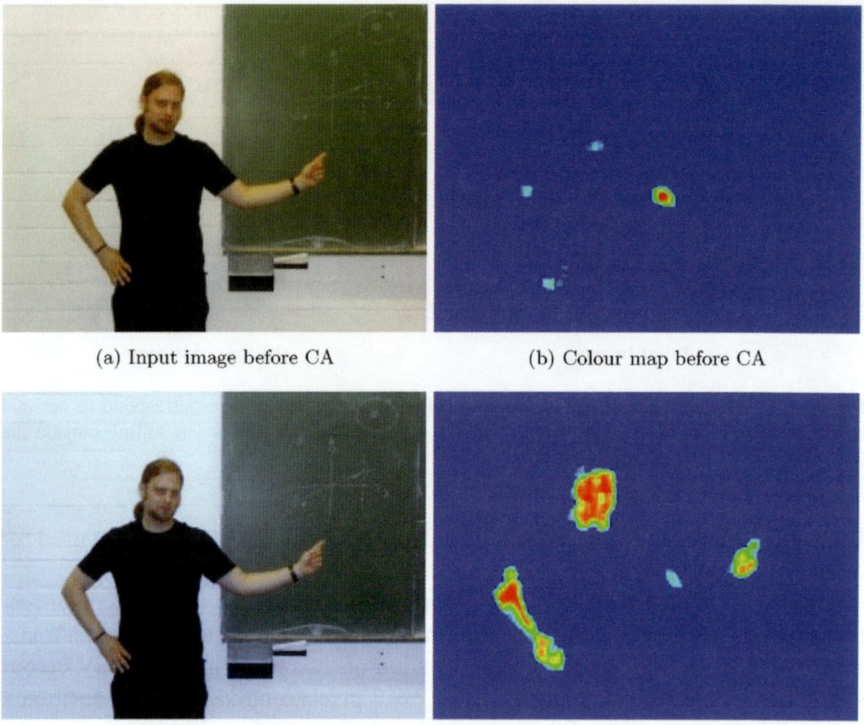

(a) Input image before CA

(b) Colour map before CA

(c) Input image after CA

(d) Colour map after CA

Fig. 7 Colour maps without and with chromatic adaptation (CA), note the significant improvement in colour map quality

body trunk are rigidly coupled in our approach, full torso detection becomes trivial as shown in Fig. 6.

However, illumination conditions in novel input images might deviate significantly from the illumination observed during training, thereby adversely impacting the colour mapping process described above. To counter this issue, we employ automatic *chromatic adaptation* (inspired by [20]): at first, the torso detector is used to find the face region of the captured subject. Then, the colours of the input image are modified in the Lab colour space unless the face colour distribution in the input image and the learned face colour distribution of the meta-torso coincide. The modified image allows for more precise detection of persistent colour and significantly increases the robustness of the matching process, as demonstrated in Fig. 7.

5 Experimental Results

The following matching results on complex image data show the performance of the learned meta-model in the retrieval of human body postures. It is evident that not only pure 2D motion patterns can be handled; the system is also able to cope with foreshortening effects and 'out-of-plane' poses, as long as influence of 3D motion components does not become dominant. Figure 9e demonstrates this feature nicely. Background clutter is tolerated to a significant degree, as seen, e.g., in Figs. 8h or 8c. We include some outdoor examples (Figs. 8g, 9e), to prove that drastic illumination changes do not hamper pose retrieval. Even quite awkward body postures are analysed correctly, even if not seen during the learning phase (see, e.g., Fig. 8d). Given that our upper body detector is tuned to upright humans, it is interesting to see that the posture in Fig. 8b is mostly correctly retrieved, notwithstanding the strong unnatural bending of the spine. Variations in subject scale and illumination conditions are compensated, as seen in Figs. 9e and 9f. Eventually, Fig. 9b demonstrates fairly stable system performance in the presence of camouflage, like gloves.

6 Conclusion and Further Work

The system proposed in this work complies with Organic Computing directives in that all required model information is generated autonomously; achieved generalisation performance is good, as demonstrated experimentally. Notwithstanding these encouraging results, stepping from pure theory into practical applications is required: for instance, we intend to use our system to render a humanoid robotic device capable of understanding and mimicking human upper body motion. Besides that, lifting the presented approach from two to three dimensions is a future research topic, requiring use of stereoscopic input data and learning of fully-fledged 3D body models.

2D approaches for human posture identification are not a new idea: [23] employs cue combination similar to ours to make limb matching more robust; however, the model used there is trained manually, requiring human supervision, which is not acceptable in the current OC context. [12] presents an interesting, learning-based approach for posture estimation employing pictorial structures. Yet, PS model initialisation as well as body pre-detection are still relying on human hand-crafting and domain knowledge. Furthermore, spatio-temporal constraints are exploited to make posture recognition more reliable, restricting the fully trained system to operate on video sequences. [19] strives to solve the pose estimation problem on single 2D input images; their technique shows impressive capabilities, yet relies on higher level domain knowledge provided by human supervisors. On the contrary, our system performs quite well in gathering similar knowledge (e.g., colour cues or kinematic constraints) on its own to improve the model matching. Kumar [15] learns body models (of humans and animals) with occlusions in a fully autonomous way from given video input. Their approach could well be used for limb segmentation in our framework. Nevertheless, the role of non-rigidity is not thoroughly accessed in Kumar's work and no explicit skeleton is extracted. In [16], a pictorial structure upper body model is learned from input data, yet, the given input is already hand-labelled (contradicting OC ideas) and the learned PS model's rectangular shapes inevitably

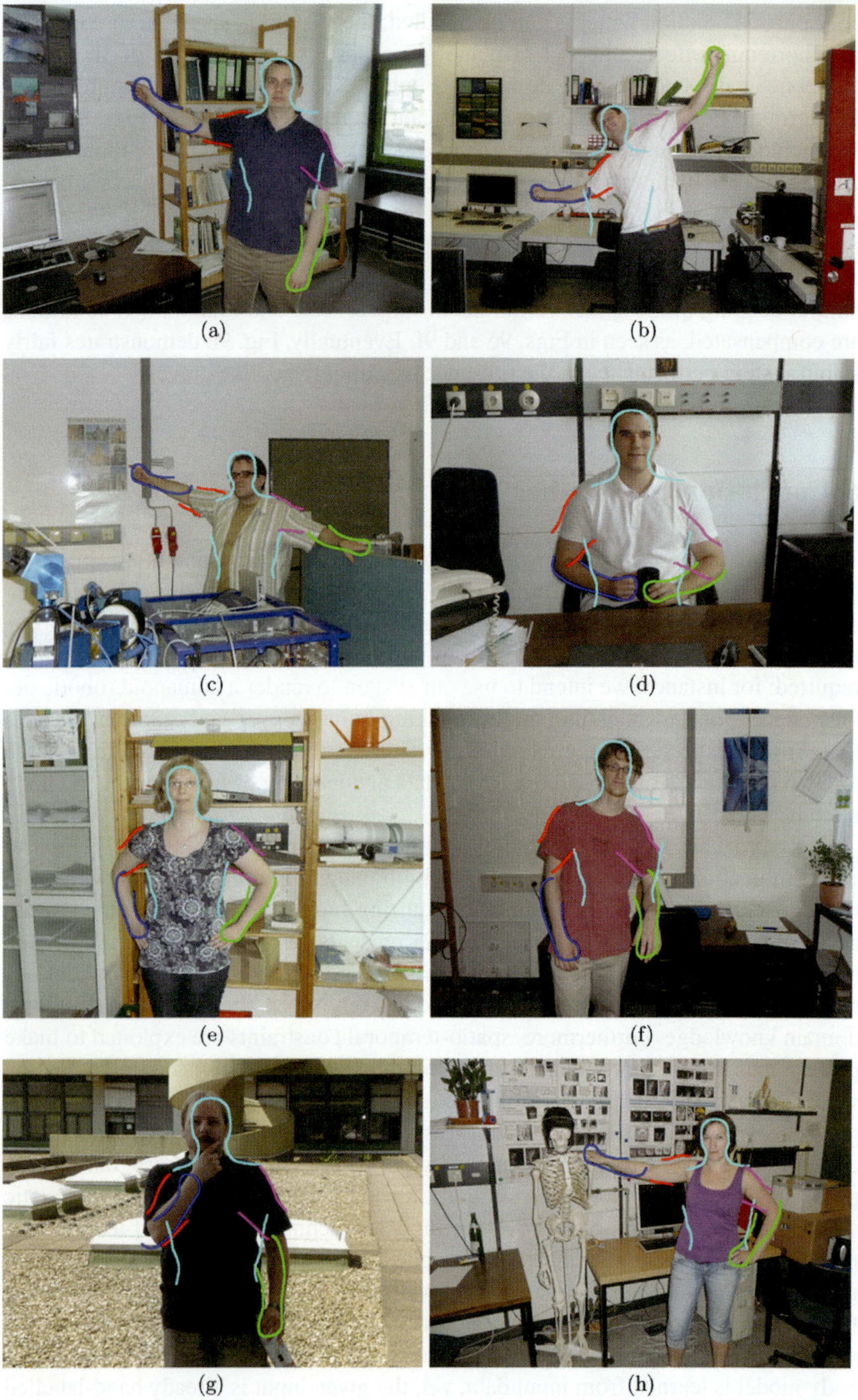

Fig. 8 Experimental evaluation

Fig. 9 Experimental evaluation

display less detail than our meta-limbs. The interesting approach of [22] also proposes to rely on pictorial structures for extraction of moving people from video streams; however, their solution falls short of OC requirements for the same reasons as [12].

As no model can capture all eventualities, the generalisation capabilities are of utmost importance. We have shown here that the learned model can be successfully applied to the interpretation of even static images of human postures. Mature autonomous learning systems must carry this a step further in that the dimensions of generalisation are not an empirical issue but can be controlled by the designer of the system. In that way, the supervision required for controlled self-organisation can be lifted onto a higher level. In [21] we have described a learning system, which can learn invariant face recognition purely on the basis of example images, with minimal supervision providing the information about which images belong to the same personal identity.

Beside the good results for looking at humans, we believe that the strategy followed here can serve as a guideline for many other Organic Computing systems as well. Stability of sensor interpretation can be greatly enhanced by using the best available processing methods for many different cues and integrate them into an autonomous learning system that builds up models of the external world on the basis of examples encountered earlier. This also brings the systems a good step closer to learning like in the natural antetype.

Acknowledgements The authors gratefully acknowledge funding from the DFG in the priority program "Organic Computing" (MA 697/5-1, WU 314/5-2, WU 314/5-3). Partial results have been published in [35–37]. We thank our colleagues at the Institut für Neuroinformatik for posing for the testing data.

References

1. Alavi, E.Y., Chartrand, G., Oellermann, O.R., Schwenk, A.J. (eds.): Graph Theory, Combinatorics and Applications, vol. 2, pp. 871–898. Wiley, New York (1991)
2. Atev, S., Masoud, O., Papanikolopoulos, N.: Learning traffic patterns at intersections by spectral clustering of motion trajectories. In: Proc. Intl. Conf. on Intelligent Robots and Systems, pp. 4851–4856 (2006)
3. Auffarth, B.: Spectral graph clustering. Course report, Universitat de Barcelona, Barcelona, January 2007
4. Besl, P.J., McKay, N.D.: A method for registration of 3-D shapes. IEEE Trans. Pattern Anal. Mach. Intell. **14**(2), 239–256 (1992)
5. Boykov, Y.Y., Jolly, M.-P.: Interactive graph cuts for optimal boundary & region segmentation of objects in N-D images. In: Proc. ICCV, Vancouver, Canada, vol. 1, pp. 105–112 (2001)
6. Christoudias, C., Georgescu, B., Meer, P.: Synergism in low-level vision. In: Proc. ICPR, Quebec City, Canada, vol. 4, pp. 150–155 (2002)
7. Daugman, J.G.: Complete discrete 2-d Gabor transforms by neural networks for image analysis and compression. IEEE Trans. Acoust. Speech Signal Process. **36**(7), 1169–1179 (1988)
8. Deng, Y., Manjunath, B.: Unsupervised segmentation of color-texture regions in images and video. IEEE Trans. Pattern Anal. Mach. Intell. **23**(8), 800–810 (2001)
9. Eriksen, R.D.: Image processing library 98 (2006). http://www.mip.sdu.dk/ipl98/

10. Felzenszwalb, P.F., Huttenlocher, D.P.: Efficient matching of pictorial structures. In: Proc. CVPR, vol. 2, pp. 66–73 (2000)
11. Felzenszwalb, P.F., Huttenlocher, D.P.: Pictorial structures for object recognition. Int. J. Comput. Vis. **61**(1), 55–79 (2005)
12. Ferrari, V., Marin-Jimenez, M., Zisserman, A.: Progressive search space reduction for human pose estimation. In: Proc. CVPR, pp. 976–983 (2008)
13. Kameda, Y., Minoh, M.: A human motion estimation method using 3-successive video frames. In: International Conference on Virtual Systems and Multimedia, Gifu, Japan (1996)
14. Krahnstoever, N., Yeasin, M., Sharma, R.: Automatic acquisition and initialization of articulated models. Mach. Vis. Appl. **14**(4), 218–228 (2003)
15. Kumar, M.P., Torr, P., Zisserman, A.: Learning layered motion segmentation of video. Int. J. Comput. Vis. **76**(3), 301–319 (2008)
16. Kumar, M.P., Torr, P.H.S., Zisserman, A.: Efficient discriminative learning of parts-based models. In: Proc. ICCV (2009)
17. Lades, M., Vorbrüggen, J.C., Buhmann, J., Lange, J., von der Malsburg, C., Würtz, R.P., Konen, W.: Distortion invariant object recognition in the dynamic link architecture. IEEE Trans. Comput. **42**(3), 300–311 (1993)
18. Lee, Y.J., Grauman, K.: Shape discovery from unlabelled image collections. In: IEEE Conference on Computer Vision and Pattern Recognition, pp. 2254–2261. IEEE Press, New York (2009)
19. Marcin, E., Vittorio, F.: Better appearance models for pictorial structures. In: Proc. BMVC, September 2009
20. Montojo, J.: Face-based chromatic adaptation for tagged photo collections (2009)
21. Müller, M.K., Würtz, R.P.: Learning from examples to generalize over pose and illumination. In: Alippi, C., Polycarpou, M., Panayiotou, C., Ellinas, G. (eds.) Artificial Neural Networks—ICANN 2009. LNCS, vol. 5769, pp. 643–652. Springer, Berlin (2009)
22. Niebles, J.C., Han, B., Ferencz, A., Fei-Fei, L.: Extracting moving people from Internet videos. In: Proc. ECCV, pp. 527–540. Springer, Berlin (2008)
23. Noriega, P., Bernier, O.: Multicues 2D articulated pose tracking using particle filtering and belief propagation on factor graphs. In: Proc. ICPR, pp. 57–60 (2007)
24. NVIDIA. NVIDIA CUDA Compute Unified Device Architecture—Programming Guide. NVIDIA (2007)
25. Poggio, T., Bizzi, E.: Generalization in vision and motor control. Nature **431**, 768–774 (2004)
26. Porikli, F.: Trajectory distance metric using hidden Markov model based representation. Technical report, Mitsubishi Electric Research Labs (2004)
27. Ross, D.A., Tarlow, D., Zemel, R.S.: Learning articulated structure and motion. Int. J. Comput. Vis. **88**(2), 214–237 (2010)
28. Rusinkiewicz, S., Levoy, M.: Efficient variants of the ICP algorithm. In: Proc. Third Intl. Conf. on 3D Digital Imaging and Modelling, pp. 145–152 (2001)
29. Shi, J., Malik, J.: Normalized cuts and image segmentation. IEEE Trans. Pattern Anal. Mach. Intell. **22**, 888–905 (2000)
30. Shotton, J., Blake, A., Cipolla, R.: Efficiently combining contour and texture cues for object recognition. In: British Machine Vision Conference (2008)
31. Sinha, S.N., Frahm, J.-M., Pollefeys, M., Genc, Y.: Gpu-based video feature tracking and matching. Technical report 06-012, Department of Computer Science, UNC Chapel Hill (2006)
32. Sminchisescu, C., Triggs, B.: Estimating articulated human motion with covariance scaled sampling. Int. J. Robot. Res. **22**, 371–391 (2003)
33. Tomasi, C., Kanade, T.: Detection and tracking of point features. Technical Report CMU-CS-91-132, Carnegie Mellon University (1991)
34. von Luxburg, U.: A tutorial on spectral clustering. Stat. Comput. **17**, 395–416 (2007)
35. Walther, T., Würtz, R.P.: Learning to look at humans—what are the parts of a moving body. In: Perales, F.J., Fisher, R.B. (eds.) Proc. Fifth Conference on Articulated Motion and Deformable Objects. LNCS, vol. 5098, pp. 22–31. Springer, Berlin (2008)

36. Walther, T., Würtz, R.P.: Unsupervised learning of human body parts from video footage. In: Proceedings of ICCV Workshops, Kyoto, pp. 336–343. IEEE Comput. Soc., Los Alamitos (2009)
37. Walther, T., Würtz, R.P.: Learning generic human body models. In: Perales, F., Fisher, R. (eds.) Proc. Sixth Conference on Articulated Motion and Deformable Objects. LNCS, vol. 6169, pp. 98–107. Springer, Berlin (2010)
38. Wang, H., Culverhouse, P.F.: Robust motion segmentation by spectral clustering. In: Proc. British Machine Vision Conference, Norwich, UK, pp. 639–648 (2003)
39. Würtz, R.P. (ed.): Organic Computing. Springer, Berlin (2008)
40. Yan, J., Pollefeys, M.: Automatic kinematic chain building from feature trajectories of articulated objects. In: Proc. of CVPR, pp. 712–719 (2006)
41. Yan, J., Pollefeys, M.: A factorization-based approach for articulated nonrigid shape, motion and kinematic chain recovery from video. IEEE Trans. Pattern Anal. Mach. Intell. **30**(5), 865–877 (2008)
42. Zelnik-Manor, L., Perona, P.: Self-tuning spectral clustering. In: Advances in Neural Information Processing Systems. NIPS, vol. 17 (2004)

T. Walther (✉) · R.P. Würtz
Institut für Neuroinformatik, Ruhr-Universität Bochum, Bochum, Germany
e-mail: Thomas.Walther@ini.rub.de

R.P. Würtz
e-mail: Rolf.Wuertz@ini.rub.de

Chapter 4
Architectures

Organic Computing (OC) systems are distributed systems that include OC techniques to ease system management and adaptivity. Self-X techniques of Autonomic and Organic Computing are applied to observe and control the managed system. Such self-X techniques feature self-configuration, self-optimisation, self-healing, and self-protection, but can also expand towards self-adaptivity, self-explanation, self-awareness and several more. Typically, an observer (system monitoring and analysis) and a controller (planning and execution) form a control loop with the managed component or system.

The system architecture of an OC system can more or less integrate the Observer/Controller loop into the components of the distributed system. The control instance can either be fully integrated leading to a self-organising system, where only the local rules implement the self-x features, or can be distributed and node-based. Each node features a so-called Organic Manager and the community of the Organic Managers organise mutual monitoring, control and distribution of load. Furthermore, it can be distributed with Organic Manager nodes that are separated from the managed functional nodes. Another possible implementation can be hierarchically distributed, which allows Observer/Controller loops within the managed components. Even centralised implementations are possible: The Organic Managers can consist of multiple levels of Observer/Controller loops.

The articles in this chapter demonstrate several approaches to OC systems. The different approaches are motivated by the application domains and demonstrate the variety of OC systems.

Chapter 4.1 presents the generic Observer/Controller design pattern as an architectural blueprint for self-organising OC systems. It explains in detail a multi-level Observer/Controller system architecture to manage a *System under Observation and Control*. Three design variants of OC systems—central, distributed, and multi-level—are discussed and demonstrated in different applications.

The further articles after the general Chap. 4.1 demonstrate different approaches to OC systems. Chapter 4.2 focuses on the *Organic Computing Middleware for Ubiquitous Environments* that features an Organic Manager integrated into each node of a distributed system. The Organic Manager of a node monitors its own node and neighbouring nodes by information exchange piggybacked on application messages. A planner-based control loop determines actions in case of failures of neighbouring nodes or node overloads.

Chapter 4.3 introduces the DodOrg (Self-adaptive Organic Many-Core) architecture that is inspired by biological concepts and distinguishes several levels similar to brain, organs, and cells. It is based on Organic Processing Cells implemented by reconfigurable hardware, features an Organic middleware at organ level, and application at brain level. Each level contains own Observer/Controller control loops and is enhanced by a thermal management. The full system should be real-time capable.

The Organic Middleware of DodOrg is explained in more detail in Chap. 4.4. It is based on an artificial hormone system for task allocation. An "Eager" hormone determines how suitable a cell is to execute a specific task, a "Suppressor" hormone inhibits task execution on a cell, and an "Accelerator" hormone favours task execution on a cell. Accelerators are used to cluster cooperating tasks in neighbouring cells thus forming an organ.

Chapter 4.5 introduces the Organic Robot Control Architecture ORCA. It features a functionally centred approach where separate Organic Control Units monitor so-called Basic Control Units that implement functions of a robotic system. The Basic Control Units can be distributed. ORCA defines a variant of an Observer/Controller system architecture. The distributed system consists of separated function units and Observer/Controller units.

Chapter 4.6 introduces the EPOC architecture—enabling system evolution under hard constraints. It aims at bridging the gap between unpredictable evolutionary behaviour and predictability of system properties that are subject to hard system constraints, i.e. hard guarantees on the real-time behaviour. A layered contracting architecture allows to separate applications from the underlying platform and to evolve service guarantees.

Finally, Chap. 4.7 introduces an Autonomic/Organic system-on-chip platform for SoC design, called ASoC (Autonomic System-on-Chip). The ASoC is split into two logical layers: The functional layer contains the functional elements, i.e., processing elements, memories, buses or various hardware units; the autonomic layer consists of interconnected autonomic elements that monitor the functional elements, evaluate the monitored data and stir actions on the associated functional element if required.

Theo Ungerer
University of Augsburg, Germany
email: ungerer@informatik.uni-augsburg.de

Chapter 4.1
Observation and Control of Organic Systems

Sven Tomforde, Holger Prothmann, Jürgen Branke, Jörg Hähner, Moez Mnif, Christian Müller-Schloer, Urban Richter, and Hartmut Schmeck

Abstract Organic Computing (OC) assumes that current trends and recent developments in computing, like growing interconnectedness and increasing computational power, pose new challenges to designers and users. In order to tackle the upcoming demands, OC has the vision to make systems more life-like (organic) by endowing them with abilities such as self-organisation, self-configuration, self-repair, or adaptation. Distributing computational intelligence by introducing concepts like self-organisation relieves the designer from exactly specifying the low-level system behaviour in all possible situations. In addition, the user has the possibility to define a few high-level goals, rather than having to manipulate many low-level parameters.

This article presents the generic Observer/Controller design pattern that serves as an architectural blueprint for self-organised OC systems. The major components of the Observer/Controller architecture and their responsibilities are introduced. Besides the general design, we discuss several distribution variants of the architecture. Furthermore, a survey presents research projects that use the Observer/Controller paradigm to solve technical problems from various domains.

Keywords Organic computing · System architecture · Design of adaptive systems · On-line learning · Observer/controller architecture · Learning classifier system · Evolutionary algorithm

1 Introduction

During the last decades, an impressive progress in computing technology has led to an exponential increase in available computing power and a decrease of chip size to a miniature format. The combination of both trends allows for an ubiquitous presence of small and interconnected devices that simplify today's life. The advantages of these new achievements are manifold as systems become more flexible and multifunctional and thereby almost indispensable in daily life. But these benefits are accompanied by several drawbacks. Some years ago, malfunctions of technical devices due to mutual influences, unpredictable environmental conditions, or decreasing manageability caused by vast configuration spaces have not been considered as a practical problem. In order to cope with these aspects of complexity and the corresponding challenges in system design, Organic Computing (OC) proposes to

C. Müller-Schloer et al. (eds.), *Organic Computing — A Paradigm Shift for Complex Systems*, 325–338
Autonomic Systems, DOI 10.1007/978-3-0348-0130-0_21, © Springer Basel AG 2011

distribute the computational intelligence among large populations of smaller entities and to provide appropriate degrees of freedom for self-organised behaviour [21].

The most important aspect of OC systems in this context is the adaptability to changes in the environmental conditions, in particular with respect to human needs. Hence, an "organic" system should be aware of its own capabilities—typically referred to as *self-x properties*. Originally defined by the Autonomic Computing community [9] in the context of self-managed IT-infrastructure, these properties include characteristics like self-configuration, self-healing, self-optimisation, and self-protection. Freedom in terms of self-organisation and self-x properties provides the basis for the anticipated adaptiveness and allows for reducing the complexity of system management.

The new requirements are supported by a paradigm shift in system design—dynamic, adaptive, and self-managing systems are needed instead of monolithic and static systems [15]. This has impact on the work of a system designer, because the existing techniques and models cannot be reused. Thus, the question arises, how such self-organised and highly interconnected OC systems can be designed, and how they can be made reliable and usable. Obviously, the designer is neither able to foresee all possible system configurations, nor to prescribe proper behaviours for all cases. In addition, we need to relieve the user from controlling all system parameters in detail—in contrast, we need mechanisms that allow to influence the system on a higher level by defining strategies or setting goals.

The paradigm shift in system development is accompanied by the need of adequate methods, techniques, and system architectures to be able to design and control complex systems, since there is no existing generalised approach. Therefore, a regulatory feedback mechanism—the *generic Observer/Controller architecture*—has been proposed [3, 20]. The architecture constitutes a generalised way to achieve controlled self-organisation in technical systems. This article summarises this generic Observer/Controller design pattern. Besides an explanation of the concept and its main components, different design variants are introduced. Furthermore, the article gives an overview of applications that are built upon the presented design pattern or are at least inspired by it.

This article is organised as follows: Sect. 2 describes the Observer/Controller design paradigm for OC systems. This design paradigm formulates a generic approach which has to be customised to the specific application. Section 3 describes possible design variants and outlines their particular advantages and disadvantages. Although the architecture is a recent theoretic concept, there are already some systems that are based on the general design. Some of these systems are classified and outlined in Sect. 4. Finally, Sect. 5 summarises the presented design and outlines further application domains and extensions of the proposed architectural concept.

2 Generic Observer/Controller Architecture

OC systems are characterised by the need of an adequate response to environmental or internal changes. Typically, this response results in an adaptive behaviour and incorporates further aspects like robustness and flexibility. In order to allow for such

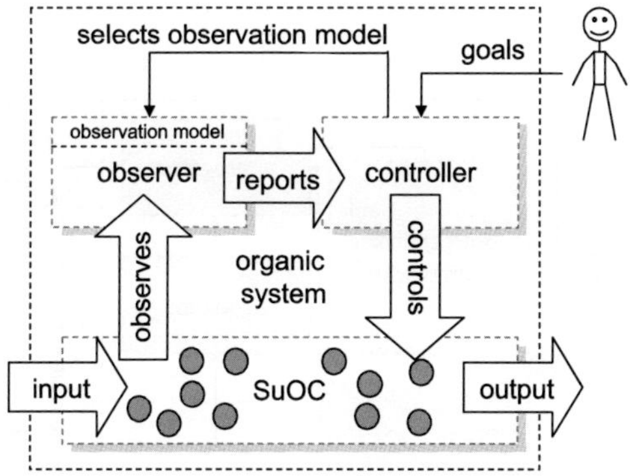

Fig. 1 Observer/Controller architecture

an adaptation process, the system's design has to provide a regulatory feedback mechanism that is capable of monitoring, analysing, and reacting to changing conditions. Therefore, we propose a generic Observer/Controller architecture, which constitutes a generalised way to achieve controlled self-organisation in technical systems [3, 20]. As depicted in Fig. 1, this regulatory feedback mechanism contains three major components:

System under Observation and Control (SuOC) The SuOC is the "productive" part of the system that serves a specific purpose. It is functional without observer and controller and it will remain operable if higher layers fail (i.e. Observer/Controller components).

Observer The SuOC's state and dynamics are monitored by the observer in order to give an appropriate description of the current situation for the whole system at each point of time.

Controller Based on the observer's aggregated information, the controller influences the SuOC with respect to the goals given by the user.

Figure 2 illustrates the architectural design in more detail and describes the workflow within the whole system. The regulatory feedback mechanism of the generic Observer/Controller architecture has some similarities with other architectural concepts like IBM's MAPE cycle [9]. Since this article does not discuss related architectural concepts, the reader is referred to *Richter* [19] for a comparative survey in the context of the generic Observer/Controller architecture. The remainder of this section focuses on describing the depicted key components of the architecture and their tasks.

2.1 System Under Observation and Control

The lowest layer of the architecture encapsulates the productive part of the system. This productive system can serve various purposes, examples range from urban traffic light control [17] over elevator control [18] to data communication systems [24] (see Sect. 4 for a survey). Higher layers of the architecture monitor and adjust (if

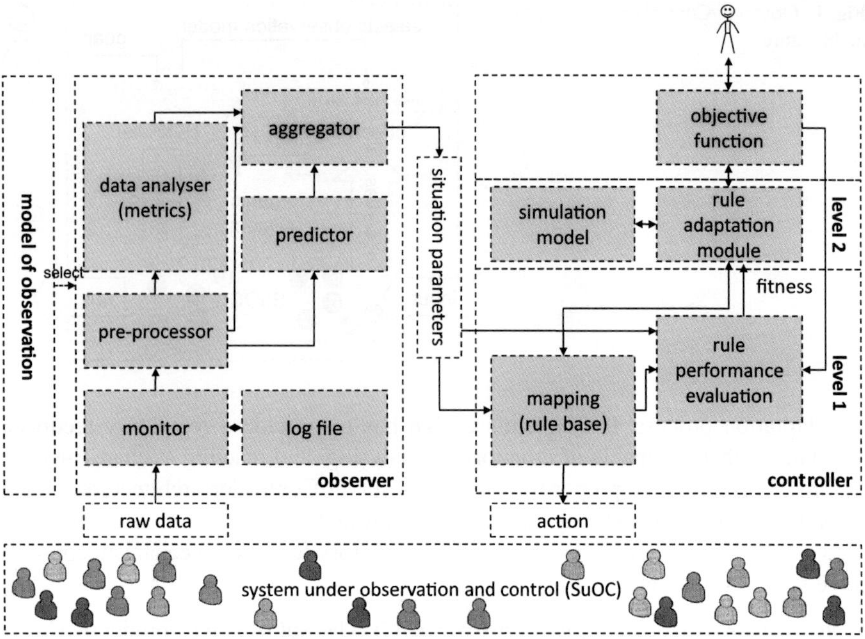

Fig. 2 Generic Observer/Controller architecture

necessary) the parameter configurations of the productive system in discrete time intervals.

OC postulates the distribution of computational intelligence among large populations of smaller entities—thus, the SuOC in Fig. 2 might refer to single systems or groups of autonomous systems. In both cases, the SuOC needs to fulfil some basic, application-specific requirements:

- The SuOC's behaviour and its environmental conditions have to be observable.
- The performance of the SuOC according to some goal given by the designer or user has to be measurable.
- The SuOC has to possess a set of variable parameters that can be dynamically adapted at run-time and that have certain impact on the performance of the system.

2.2 Observer

It is the observer's task to measure and quantify the current state of the SuOC and to predict its future development. The observation process consists of the steps monitoring, pre-processing, data analysis, prediction, and aggregation. An observation model customises the observer by selecting the observable attributes of the SuOC and by determining appropriate analysis and prediction methods.

Monitor and log file The monitor samples attributes of the SuOC according to a given sampling rate. All measured data is stored in a log file for every loop of observing/controlling the SuOC. The stored time series form the basis for the pre-processing, analysis, and prediction steps.

Pre-processor Typical tasks performed during pre-processing include the smoothing and filtering of stored time series and the extraction of derived attributes. The pre-processed data is used in the data analysis and prediction steps.

Data analyser The data analyser provides a system-wide description of the SuOC's current state. The implemented analysis techniques largely depend on the observed system and the purpose of the Observer/Controller. Cluster computation, emergence detection (according to the definition in [14]), or further mathematical and statistical methods can be applied to extract information from the pre-processed data.

Predictor While the data analyser is dedicated to the current system state, the predictor's task is to forecast future developments. This enables the controller to base its control decisions not only on historic and current data, but also on predicted developments. Prediction techniques are again specific to the organic system's domain.

Aggregator The results of the pre-processor, data analyser, and predictor are handed on to the aggregator where they are combined to situation parameters which are transmitted to the controller.

Depending on the requirements of the controller, the obtained situation parameters can be adapted using the *model of observation*. The model specifies which properties of the SuOC are observed and which sampling rate is used for observation (this selection is obviously limited by the available detectors and their capabilities). Furthermore, the applied analysis and prediction methods are selected. By influencing the model of observation, the controller can focus in detail on critical aspects within the SuOC even when its resources (like communication bandwidth, processing power, or energy) are limited. Further details on the observer with a special focus on determining emergent behaviour have been discussed by *Mnif* [13].

2.3 Controller

Based on the received situation parameters, the controller influences the SuOC to achieve the objectives specified by the user. The controller is internally composed of two layers: *Layer 1* is dedicated to the on-line learning of appropriate actions for the received situation parameters. It consists of a mapping component that assigns possible actions to known situations. Furthermore, the performance of stored situation-action mappings that have been applied in the SuOC is evaluated by a performance evaluation component. *Layer 2* aims at improving the mapping of Layer 1 by providing off-line optimisation capabilities. It is composed of an adaptation module and a simulation model. Using optimisation or machine learning algorithms, the adaptation module creates additional mappings and can rely on the simulation model in the process. The components of both layers are discussed in the following:

Mapping The mapping component is responsible for an immediate reaction to received situation parameters. It stores previously learnt situation-action mappings (e.g. in form of Learning Classifier System rules [4]) which determine the reaction to known situations. When a situation is unknown, an action might be deduced from stored mappings for similar situations.

Performance evaluation The performance evaluation component calculates quality updates for the situation-action mappings based on their performance observed in the SuOC. When the execution of an action changes the state of the SuOC, this change is subsequently reflected in the situation parameters derived by the observer. The controller's performance evaluation component evaluates the change with respect to the objective function and updates previously applied situation-action mappings through the adaptation module. Machine learning techniques like Reinforcement Learning [22] can be applied here. The update of existing mappings resembles an on-line learning based on observations in the SuOC.

Adaptation module The adaptation module's main task is to create new situation-action mappings and to delete mappings of insufficient quality. For the exploration of additional mappings, various optimisation techniques (including, e.g., Evolutionary Algorithms [7]) can be applied. Optimisations are supplemented by a simulation model that allows for the safe evaluation of candidate solutions. The adaptation module resembles an off-line optimisation that is performed in parallel to the on-line learning mentioned above.

Simulation module The simulation module supports the exploration of new situation-action mappings. It allows to quickly and safely estimate the effect of a new action or mapping before its application in the SuOC.

Objective function The user-defined objective function guides the learning and optimisation processes on both layers of the controller. It is used for fitness evaluation in the performance and adaptation modules.

The combination of on-line learning (Layer 1) and off-line optimisation (Layer 2) in the controller leads to a *2-layered learning* concept that is inspired by the *Anytime Learning* approach that has been proposed by *Greffenstette and Ramsey* [8]. The most prominent examples of 2-layered learning in OC systems are traffic control [17] and data communication networks [24]. A detailed investigation of the controller with a special focus on learning aspects has been performed by *Richter* [19].

3 Design Variants of the Observer/Controller Architecture

The generic Observer/Controller architecture needs to be customised to different scenarios by adapting the various components of the observer and the controller. As stated in [5] and depicted in Fig. 3, this customisation of design variants ranges from fully central to fully distributed architectures. The former case describes a single Observer/Controller that regulates various components of the SuOC and that directly intervenes into all of these entities (see Fig. 3a). In contrast, the latter example defines one Observer/Controller for each component of a technical system (see Fig. 3b). These two variants—the fully central and the fully distributed

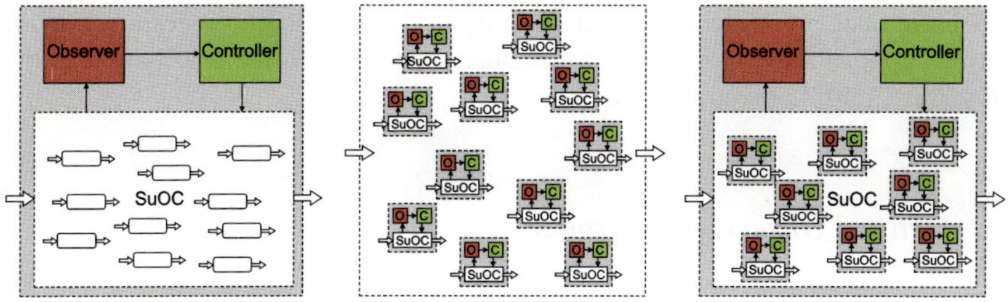

(a) Central (b) Distributed (c) Multi-level

Fig. 3 Distribution possibilities of the generic Observer/Controller architecture

architecture—define the two extreme points in the design space. Nevertheless, there are also many other distribution possibilities like a multi-level architecture (see Fig. 3c).

Based on these various possibilities to realise and customise the generic Observer/Controller architecture, the designer of a technical system has to decide about the most promising approach for his/her context. In the course of this decision process, the need for different design variants can be classified according to increasing size, complexity, and heterogeneity of the contained subsystems. The simplest case is an isolated system with a clearly defined purpose and a restricted configuration space where no distribution is needed. In contrast, larger and more complex systems are characterised by a drastic increase of situation and configuration spaces that cannot be handled by one single Observer/Controller component. With growing complexity, a hierarchical or multi-levelled decomposition of the control problem becomes more recommendable. A common analogy for multi-levelled systems is the organisational structure of large companies: The management serves as highest instance that defines abstract global goals or strategies. Lower layers of the hierarchy convert these abstract goals for their area of responsibility into more specific goals—hence, high level administration units are not involved in low level decisions.

OC introduces the variability of systems as a measurement for the quantification of complexity during the design process for technical systems. The term *variability* is defined as the number of possible configurations of a SuOC [6] (see Chap. 1.1). Obviously, the variability tends to increase with the complexity of the SuOC. However, introducing hierarchical and multi-levelled Observer/Controller structures is a powerful instrument to reduce the externally visible variability and, therefore, hide the complexity of a system.

4 Application Survey

The generic architectural design (as presented in Sect. 2) and its distribution variants (as presented in Sect. 3) describe an architectural blueprint. Several projects from varying application domains have developed their own systems that are built upon this basic concept or are at least inspired by it. This section provides a survey of various applications. The survey is structured according to the distribution of the Observer/Controller components.

4.1 Central Observer/Controller

When a technical system cannot be subdivided into subsystems or when its subsystems are highly interrelated, a centralised Observer/Controller architecture (like in Fig. 3a) is often the method of choice for monitoring and controlling the system's behaviour. Applications that have been successfully implemented with a centralised Observer/Controller include, but are not limited to, elevators, cleaning robots in buildings, and machine management in off-highway machines.

Elevator Control

An *organic elevator control* system [18] serves as one example for a centralised variant of the generic Observer/Controller architecture. In a group of elevators, emergent effects can appear that have to be detected and controlled to avoid undesired processes. Typically, elevators stop at the nearest hall call in their current running direction and only change their direction after serving all requests for the current one. In buildings with several parallel elevators working according to this simple concept, a synchronisation effect can be observed when all elevators move up and down as a parallel wave. This so-called *bunching* effect results in increasing waiting times for passengers and has been proven as inefficient [1]. From the viewpoint of OC, the bunching effect represents an interesting emergent behaviour in a technical scenario which has to be detected and avoided. Therefore, the observer contains corresponding emergence detectors and the controller can intervene to discontinue the synchronisation by manipulating the behaviour of individual elevator cabins. As one example, the elevator does not notice calls for the next hall and passes them without stopping. Considering the distribution variants given in Sect. 3, the organic elevator control system consists of several autonomous elevators and one centralised Observer/Controller component.

Organic Computing in Off-highway Machines

The project *Organic Computing in Off-highway Machines* (OCOM) focuses on machine management in off-highway machines like tractors or wheel loaders [28]. An off-highway machine consists of several subsystems (like the traction drive, the power take-off, the hydraulic system, and varying auxiliary components) that are closely interrelated. An adaptive machine-wide management of these subsystems is crucial for the machine's efficient operation (e.g. with respect to a minimal fuel consumption).

To adapt the subsystems to the machine's various operation scenarios and to improve their coordination, OCOM relies on an Observer/Controller architecture that builds upon the generic concept presented in this article. The SuOC is formed by the machines' subsystems, while the Observer/Controller is responsible for a reliable, adaptive, and robust machine management. The major contribution of the project is the modular development and implementation of the generic design pattern.

To support the basic investigations, a simulation model for OC-based off-highway machines has been developed. The model allows to investigate the quantification and control of emergent global effects which can occur in a collection of subsystems. Based on these simulative studies, comprehensive measurements are conducted on a real off-highway machine (a *Fendt Vario 412* tractor) rigged with additional sensor technology. Future work will refine the architectural design and generalise the achieved results to improve the control strategies of similar machines. Further details on OC in off-highway machines are given in Chap. 6.1.8.

Cleaning Robots

Self-organised *cleaning robots* have been investigated as another application scenario for the generic architecture [19]. These cleaning robots follow a local strategy to search in their local neighbourhood for dirty places, clean these places, and self-optimise their search behaviour by learning. The group of robots is applied to a dirty environment and tries to maximise the cleaning performance. Robots are able to indirectly communicate with each other by placing "pheromones" at places they have already cleaned—these pheromones are observed by other robots. The strategy of the robot is based on avoiding double-cleaning of areas. Based on this simple scenario, two different variants of the generic architecture have been investigated: a centralised and a distributed variant. In the former scenario, one centralised component is responsible for generating and exchanging the robots' behaviour strategies [16], while the latter scenario focuses on distributed learning of autonomous robots without the need of centralised control [12].

4.2 Distributed Observer/Controller Components

When a technical system consists of several subsystems, it can be advisable to observe and control each subsystem with a separate Observer/Controller component (see Fig. 3b). The predominant application area of distributed OC systems are applications where the subsystems are loosely coupled, where the subsystems are locally distributed (like in communication or road networks), or where subsystems are controlled by different authorities (like in the context of service-oriented architectures).

Service-oriented architectures (SOAs) are typical distributed applications with manifold interactions among various components. Since conventional management systems are restricted in their capabilities to deal with sophisticated interactions between components, management responsibility at run-time is distributed to each component. To this end, design principles of OC are leveraged to design Organic Service-oriented Architectures (OSOAs) [23]. In detail, each SOA entity is equipped with an Observer/Controller component to achieve controlled self-organisation locally [11]. The observer monitors its respective SOA component and determines its current operational state. Based on this information, the controller adapts the behaviour of the underlying SOA component according to given objectives specified in service-level agreements (SLA). In addition, service components coordinate their run-time behaviour via automatically negotiated SLAs, which ensures a compliance of the entire SOA-based system with given business objectives [10]. As result, the increasing complexity of SOAs is handled in a completely distributed manner without the need of centralised or hierarchical components.

Organic Network Control

The *Organic Network Control* (ONC) system has been developed to dynamically adapt parameter configurations of data communication protocols to changing environmental conditions [24]. Typically, such parameters are buffer sizes, delay values, interval lengths, or counters. In traditional networking, these values are pre-configured using a static setup that works well on average. In contrast, the goal of the ONC system is to find the best parameter setting for each occurring situation. As one example, a broadcast protocol for mobile ad-hoc networks is adapted according to the movements of other nodes within the communication distance of the wireless communication infrastructure [25]. In principle, protocols from all layers of the protocol stack can be adapted, if they can be observed, modified, and evaluated using locally available information. Currently, research has focused on exemplary representatives from the domains of peer-to-peer networks, wireless sensor networks, and mobile ad-hoc networks. The ONC system is based on an adapted variant of the generic Observer/Controller architecture with a special focus on the 2-layered learning concept. The SuOC is one instance of a network protocol that can be observed locally and that is characterised by a set of variable parameters that can be altered at run-time. This altering process is performed by an Observer/Controller component situated at Layer 1 that works on a set of rules with a fixed set of actions. Layer 2 is responsible for evolving new rules (the mapping between situations and protocol parameter configurations) in a simulation-based environment. Since each participant in the data communication network is equipped with an Observer/Controller component and no further hierarchical elements are contained in the resulting system, ONC can be considered as one example for a completely distributed system. Further details on Organic Network Control are given in Chap. 6.1.11.

Organic Traffic Control

Signalised intersections in urban road networks are in the focus of the project *Organic Traffic Control* [17]. The project investigates an integrated traffic control system for urban environments and focuses in particular on the traffic-responsive control of traffic lights, on the self-organised coordination of signals, and on route guidance mechanisms.

The generic Observer/Controller architecture is applied at a signalised intersection where it evaluates and optimises the signalisation on-line. The observer monitors the local traffic demands and evaluates the active signal plan's performance (e.g. with respect to the vehicular delay), while the controller optimises the signalisation by learning new signal plans. Internally, the controller combines on-line reinforcement learning on Layer 1 and simulation-based optimisation on Layer 2, thereby being a successful example of 2-layered learning in a technical system.

Collaboration mechanisms that allow for the traffic-responsive coordination of intersections are a second focus of the project. The self-organised mechanisms establish progressive signal systems (also called *green waves*) in a road network. To assess the possibilities and limitations of decentralised traffic control systems, a completely decentralised mechanism (like in Fig. 3b) and a hierarchical variant (like in Fig. 3c) have been developed and compared [26].

The project's current focus is an integrated traffic management system that provides route guidance information to the drivers. Route recommendations are based on traffic demands measured at the network's intersections and on a local communication between the intersection controllers. Recent results of organic traffic control are summarised in Chap. 5.1.

4.3 Multi-levelled Observer/Controller Components

Multi-levelled OC systems consist of distributed Observer/Controller components that are hierarchically organised. Observer/Controller components on a higher level influence organic subsystems on lower levels (see Fig. 3c). The predominant application area are technical compounds with several subsystems that are sufficiently complex to require their own Observer/Controller. To achieve system-wide objectives, higher level Observer/Controller components influence groups of locally or logically related subsystems based on aggregated data gathered from lower levels. Examples include traffic networks with regional traffic light control (see Chap. 5.1) or smart-home environments with several controlled appliances.

MeRegioMobil

To adapt the consumers' energy demand to the power generation in the grid, the project *MeRegioMobil* investigates a smart-home environment equipped with several household appliances and an electric vehicle [2]. Within the smart-home environment, the vehicle's charging periods and the operation of various appliances (like the washing machine or the freezer) are automatically rescheduled. The rescheduling is based on price signals that reflect a load-prediction of the energy grid, but considers user constraints (like a deadline for finishing a washing programme).

Rescheduling is performed by a multi-levelled Observer/ Controller framework: Each appliance is equipped with a local Observer/Controller component that observes the appliance's current state and can turn the appliance on or off according to current conditions. The local Observer/Controller components communicate their data (e.g. power consumption profiles) to a higher-level smart-home management device that centrally derives timing strategies for the smart-home. Chapter 6.1.9 presents more details about the smart home environment and its multi-levelled Observer/Controller architecture.

5 Conclusion

Driven by the initial motivation and corresponding vision of OC, this article discussed a generalised design pattern for OC systems. Based on the insight that the emerging ubiquitous systems consisting of flexible networks of smart objects will no longer be manageable with current methods of systems engineering, a paradigm shift in system development is needed towards a generalised way to design OC-like systems. Therefore, the generic Observer/Controller architecture has been presented that incorporates several necessary mechanisms for OC systems like adaptivity, robustness, self-improvement in terms of learning, and flexibility in case of changing goals. Based upon the general design pattern, we introduced different distribution variants in order to cope with varying types of applications ranging from isolated to highly interconnected and complex systems.

To demonstrate the status of system design based on the principles of OC, this article presented current research projects that are built upon the proposed design pattern. Future work related to architectures of OC systems will proceed along two main aspects: the further refinement of the design pattern and the development of market-ready OC systems. The refinement of the design pattern and its contained components will consider characteristics like robustness and flexibility as well as an adaptation of the observation model at run-time due to some stimuli. Furthermore, the design pattern will find its way into market-ready systems. According to the initial vision for OC as postulated in 2003, we are moving towards OC-like applications ranging from smart production (like factories or warehouses) over networked systems (like data communication, smart grids, or urban traffic systems) to support systems for the daily life (like household robots or assisted living systems) [27]. Within the last six years of research on OC, a big step forward has been taken for some of these ideas (like urban traffic control systems), but others remain open for ongoing work.

References

1. Al-Sharif, L.R.: Bunching in lifts: Why does bunching in lifts increase waiting time? Elevator World **11**, 75–77 (1996)

2. Becker, B., Allerding, F., Reiner, U., Kahl, M., Richter, U., Pathmaperuma, D., Schmeck, H., Leibfried, T.: Decentralized energy-management to control smart-home architectures. In: Müller-Schloer, C., Karl, W., Yehia, S. (eds.) Architecture of Computing Systems—ARCS 2010. LNCS, vol. 5974, pp. 150–161. Springer, Berlin (2010)
3. Branke, J., Mnif, M., Müller-Schloer, C., Prothmann, H., Richter, U., Rochner, F., Schmeck, H.: Organic Computing—Addressing complexity by controlled self-organization. In: Margaria, T., Philippou, A., Steffen, B. (eds.) Proc. 2nd Int. Symp. on Leveraging Applications of Formal Methods, Verification and Validation (ISoLA 2006), pp. 185–191. IEEE Press, New York (2006)
4. Butz, M.V.: Rule-based Evolutionary Online Learning Systems—A Principled Approach to Lcs Analysis and Design. Springer, Berlin (2005)
5. Cakar, E., Hähner, J., Müller-Schloer, C.: Investigation of generic observer/controller architectures in a traffic scenario. In: Hegering, H.-G., Lehmann, A., Ohlbach, H.J., Scheideler, C. (eds.) INFORMATIK 2008: Beherrschbare Systeme – dank Informatik. LNI, vol. 134, pp. 733–738. Köllen Verlag, Bonn (2008)
6. Cakar, E., Mnif, M., Müller-Schloer, C., Richter, U., Schmeck, H.: Towards a quantitative notion of self-organisation. In: Proc. IEEE Congress on Evolutionary Computation (CEC 2007), pp. 4222–4229. IEEE Press, New York (2007)
7. Eiben, A.E., Smith, J.E.: Introduction to Evolutionary Computing, 2nd edn. Springer, Berlin (2007)
8. Grefenstette, J.J., Ramsey, C.L.: An approach to anytime learning. In: Proc. 9th Int. Workshop on Machine Learning, pp. 189–195 (1992)
9. Kephart, J.O., Chess, D.M.: The vision of autonomic computing. IEEE Comput. **36**(1), 41–50 (2003)
10. Liu, L., Schmeck, H.: Enabling self-organising service level management with automated negotiation. In: IEEE/WIC/ACM Int. Conf. on Web Intelligence and Intelligent Agent Technology, pp. 42–45 (2010)
11. Liu, L., Thanheiser, S., Schmeck, H.: A reference architecture for self-organising service-oriented computing. In: Brinkschulte, U., Ungerer, T., Hochberger, C., Spallek, R.G. (eds.) Architecture of Computing Systems—ARCS 2008. LNCS, vol. 5974, pp. 205–219. Springer, Berlin (2008)
12. Lode, C., Richter, U., Schmeck, H.: Adaption of XCS to multi-learner predator/prey scenarios. In: Proc. 12th Annual Conf. on Genetic and Evolutionary Computation (GECCO 2010), pp. 1015–1022. ACM, New York (2010)
13. Mnif, M.: Quantitative Emergenz: Eine Quantifizierungsmethodik für die Entstehung von Ordnung in selbstorganisierenden technischen Systemen. PhD thesis, Leibniz Universität Hannover, Institute for Systems Engineering, System and Computer Architecture Group (2010)
14. Mnif, M., Müller-Schloer, C.: Quantitative emergence. In: Proc. 2006 IEEE Mountain Workshop on Adaptive and Learning Systems (SMCals 2006), pp. 78–84. IEEE Press, New York (2006)
15. Müller-Schloer, C.: Organic Computing: On the feasibility of controlled emergence. In: Proc. 2nd Int. Conf. on Hardware/Software Codesign and System Synthesis, pp. 2–5 (2004)
16. Pathmaperuma, D.: Lernende und selbstorganisierende Putzroboter. Master's thesis, Institut AIFB, Univ. Karlsruhe (TH) (2008)
17. Prothmann, H., Branke, J., Schmeck, H., Tomforde, S., Rochner, F., Hähner, J., Müller-Schloer, C.: Organic traffic light control for urban road networks. Int. J. Auton. Adapt. Commun. Syst. **2**(3), 203–225 (2009)
18. Ribock, O., Richter, U., Schmeck, H.: Using organic computing to control bunching effects. In: Brinkschulte, U., Ungerer, T., Hochberger, C., Spallek, R.G. (eds.) Architecture of Computing Systems—ARCS 2008. LNCS, vol. 4934, pp. 232–244. Springer, Berlin (2008)
19. Richter, U.: Controlled self-organisation using learning classifier systems. PhD thesis, Universität Karlsruhe (TH) (2009)
20. Richter, U., Mnif, M., Branke, J., Müller-Schloer, C., Schmeck, H.: Towards a generic observer/controller architecture for organic computing. In: Hochberger, C., Liskowsky, R. (eds.) Informatik 2006 – Informatik für Menschen, pp. 112–119. Köllen Verlag, Bonn (2006)

21. Schmeck, H.: Organic Computing—A new vision for distributed embedded systems. In: Proc. 8th IEEE Int. Symp. on Object-Oriented Real-Time Distributed Computing (ISORC'05), pp. 201–203 (2005)
22. Sutton, R.S., Barto, A.G.: Reinforcement Learning—An Introduction. MIT Press, Cambridge (1998)
23. Thanheiser, S., Liu, L., Schmeck, H.: Towards collaborative coping with IT complexity by combining SOA and organic computing. Int. Trans. Syst. Sci. Appl. **5**(2), 190–197 (2009)
24. Tomforde, S., Cakar, E., Hähner, J.: Dynamic control of network protocols—a new vision for future self-organised networks. In: Filipe, J., Cetto, J.A., Ferrier, J.-L. (eds.) Proc. 6th Int. Conf. on Informatics in Control, Automation, and Robotics (ICINCO'09), pp. 285–290. INSTICC (2009)
25. Tomforde, S., Hurling, B., Hähner, J.: Dynamic control of mobile ad-hoc networks—Network protocol parameter adaptation using organic network control. In: Filipe, J., Cetto, J.A., Ferrier, J.-L. (eds.) Proc. 7th Int. Conf. on Informatics in Control, Automation, and Robotics (ICINCO'10), pp. 28–35. INSTICC (2010)
26. Tomforde, S., Prothmann, H., Branke, J., Hähner, J., Müller-Schloer, C., Schmeck, H.: Possibilities and limitations of decentralised traffic control systems. In: WCCI 2010 IEEE World Congress on Computational Intelligence, pp. 3298–3306. IEEE Press, New York (2010)
27. VDE/ITG/GI: Positionspapier Organic Computing (2003)
28. Wünsche, M., Mostaghim, S., Schmeck, H., Kautzmann, T., Geimer, M.: Organic computing in off-highway machines. In: 2nd Int. Workshop on Self-Organizing Architectures (SOAR 2010), pp. 51–58. ACM, New York (2010)

S. Tomforde (✉) · J. Hähner · M. Mnif · C. Müller-Schloer
Institute of Systems Engineering, Leibniz Universität Hannover, Appelstr. 4,
30167 Hannover, Germany
e-mail: tomforde@sra.uni-hannover.de

J. Hähner
e-mail: haehner@sra.uni-hannover.de

M. Mnif
e-mail: mnif@sra.uni-hannover.de

C. Müller-Schloer
e-mail: cms@sra.uni-hannover.de

H. Prothmann · U. Richter · H. Schmeck
Institute AIFB, Karlsruhe Institute of Technology (KIT), Bldg. 05.20, 76128 Karlsruhe, Germany

H. Prothmann
e-mail: holger.prothmann@kit.edu

U. Richter
e-mail: urban.richter@kit.edu

H. Schmeck
e-mail: hartmut.schmeck@kit.edu

J. Branke
Warwick Business School, University of Warwick, Coventry, CV4 7AL, UK
e-mail: juergen.branke@wbs.ac.uk

Chapter 4.2
Organic Computing Middleware for Ubiquitous Environments

Michael Roth, Julia Schmitt, Rolf Kiefhaber, Florian Kluge, and Theo Ungerer

Abstract The complexity of computer systems has been increasing during the past years. To control this complexity organic computing introduces the self-x features. The Organic Computing Middleware for Ubiquitous Environments eases to manage distributed computing systems by using self-configuration, self-optimisation, self-healing and self-protection. To provide these self-x features the latest version of our middleware uses an Observer/Controller architecture with an automated planner. Planning is time consuming so we introduced additionally reflexes for faster reactions. The reflexes are learned from previous plans and can be distributed to resource restricted nodes.

Keywords Organic planner · Organic middleware · Self-x techniques · Self-x features · Two level reflexes · Ubiquitous middleware

1 Introduction

The complexity of computer systems is steadily increasing. There is a growing variety of mobile computing devices, which are highly connective and can be used for different tasks. Cellular phones have Internet access, play music and can act as navigation systems. Such highly dynamical devices can connect to other devices in the vicinity to fulfil their tasks. So these devices must adapt unsupervised to the new environment. The concepts of Organic Computing enable devices to do so and handle failures without user interaction by introducing self-x features [4].

The Organic Computing Middleware for Ubiquitous Environments (OCμ) is designed to facilitate a device-independent application of Organic Computing methods in ubiquitous environments. In these environments we expect a heterogeneous collection of devices with diverse capabilities regarding computing power, memory space, and energy supply. To ensure that many devices are capable of running OCμ we used Java and implemented an application case study called Smart-Doorplates [1] which resembles a Smart Office environment.

To implement the Smart Doorplate scenario we developed an initial version of OCμ and investigated techniques for self-configuration [20], self-optimisation [21], self-healing [14], and self-protection [10, 11]. The self-x properties are implemented as separated services and can be used as needed. This approach has the advantage

C. Müller-Schloer et al. (eds.), *Organic Computing — A Paradigm Shift for Complex Systems*, 339–351
Autonomic Systems, DOI 10.1007/978-3-0348-0130-0_22, © Springer Basel AG 2011

that self-x services can be easily replaced, but the disadvantage that no cooperation between the self-x properties is possible, and that the self-x services might even interfere with each other. To resolve this problem, we devise a new version of OCμ middleware and replaced the separated self-x services by an automated planner which combines functionality of three self-x services into one component. Thus, a loose cooperation between self-x is achieved, whereas interferences are precluded.

Because planning can be time consuming we introduce reflexes. The Reflex Manager stores system states and plans from previous system interactions. This means the Reflex Manager takes the current system state as input and outputs a sequence of actions to transform the system into a valid state. The output of the automated planner can be used to generate a reflex to allow a faster reaction if the same situation reoccurs. In this way the system will become more efficient by using its experience. On restricted nodes, where no planning is possible, the Reflex Manager will get reflexes generated from other nodes so we can provide self-x features.

This article is organised as follows: In Sect. 2 we discuss related work. Section 3 describes our work on the initial OCμ architecture. In Sect. 4 we present the new Organic Manager with the automated planner. Section 5 concludes this chapter with a summary and a brief outlook on further research of OCμ.

2 Related Work

The Tuples On The Air (TOTA) middleware [7] uses what the authors call *spatial computing*. In spatial computing each node is placed in a virtual n-dimensional space, assigned an area of this virtual space, and logically connected to its neighbours. TOTA combines tuple spaces and event-based approaches. Information is sent in tuples to the neighbours of a node. Each node changes this tuples using a change function which is sent with the tuple and forwards this tuple to the node's neighbours. So each node receives the tuple. The tuples can change with each node. This allows other nodes to read the tuples and use this information. Each TOTA node must know how to read tuples. If a node fails the tuples will be changed in a different way and the network can react to the node failure.

CARISMA [8], a service-oriented real-time organic middleware architecture, is developed towards real time embedded systems. Each service uses local information to decide whether a job can be done and which quality can be achieved. Based on these estimates the middleware decides which service shall execute the job by using an auction system. Each service optimises its own performance and so the whole system becomes optimised. The CARISMA middleware provides the communication infrastructure for the services and the functionality for the optimisation.

The artificial hormone system AHS (see Chap. 4.4) uses hormones for task allocation. It is inspired by the hormone system of higher mammals. The AHS middleware concentrates on mapping tasks to processing elements. Hormones are realised as messages. With this completely decentralised approach it is possible to implement self-configuration, self-optimisation, and self-healing. Additionally AHS guarantees an upper bound for the self-configuration.

The Organic Robot Control Architecture ORCA (see Chap. 4.5) is based on a hierarchical system architecture including several control units. It aims to improve the system behaviour and adapt to malfunctions without a formal model. ORCA uses supervised learning to control the self-organisation of the system.

The HYDRA project [5] develops a middleware for networked embedded systems that allows developers to create intelligent applications for heterogeneous physical devices. The project objective is to develop a middleware based on a service-oriented architecture for ubiquitous environments but did not offer any self-x features.

Our goal is to design a middleware for embedded systems but also for networked personal computers. The middleware does not need to provide real-time capabilities like CARISMA and AHS. Another point is that we target the middleware to provide self-x features. Services shall only provide their actual functionality, but should not care about self-x properties. In TOTA many additional messages are created. We also minimise communication by using piggy-backing. This is especially important for energy-constrained and mobile devices, which are usually connected by some radio technique. Radio communication is very energy inefficient.

3 Initial OCμ Architecture

The basic architecture of OCμ is similar to other state-of-the-art ubiquitous middleware systems. Figure 1 gives an overview of the initial OCμ [1, 18] architecture that enhanced the basic middleware by an Organic Manager and Monitoring Queues. On the left of the figure is the basic middleware system. It is comprised of three layers. The Transport Connector layer is responsible for the delivery of the messages to other nodes on different communication infrastructures. The Event Dispatcher layer

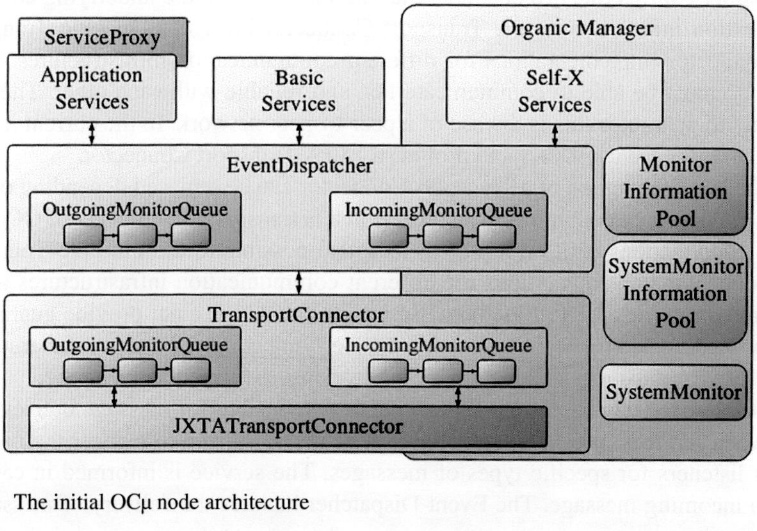

Fig. 1 The initial OCμ node architecture

is capable of finding the recipient of a message that was sent by another service either locally or from another node. The applications as well as some basic services of the middleware reside on the top layer. Basic Services are necessary for the operation of the middleware and must run on every node. These are e.g. the configuration service and the discovery service. On the right side is the Organic Manager with the self-x-services and the Information Pools. These Information Pools store the monitored data. The self-x-services use this information to provide the self-x features.

There are three additional parts in the middleware that differentiate OCμ from other middleware systems: first, a typed messaging that adds further freedom in terms of message delivery and service requests (Sect. 3.2); second, a monitoring on the lower levels of the middleware (Sect. 3.3); and third, the Organic Manager with the self-x services (Sect. 3.4).

Applications are separated into services that can be distributed on the nodes of a network. Future ubiquitous systems will be comprised of many computing nodes, so applications should be composed of interactive components instead of being a monolithic block of software. There are two advantages to this approach: The services of an application can be reused much easier than parts of a code, and an application can be distributed over the network to increase the computational performance of the application. In the Smart-Doorplates scenario [1] for example location tracking and location prediction are some of the implemented application services.

3.1 Middleware Components

We will give a short overview of the components in OCμ. All components are still used with the updated Organic Manager. We changed the implementation of some parts during our work but maintained their functionality.

Transport Connector: To decouple the middleware from the underlying communication infrastructure, the Transport Connector layer uses specific Transport Connector implementations for different communication infrastructures. Each node must be able to communicate fast and reliable with each other. The best way to guarantee this is the use of a peer-to-peer network. In the current implementation, we use JXTA and provide a JXTA Transport Connector.
The implementation of a Transport Connector can be replaced depending on the given communication infrastructure, which is transparent to the rest of OCμ and the applications built on top of it. It is also possible to use multiple Transport Connector implementations for different communication infrastructures at the same time (i.e., UDP, Bluetooth, Serial Line). OCμ can not provide guaranties for message delivery but relies on the underlying communication infrastructure. A service must resend a message if no response arrives in a given time.

Event Dispatcher: The Event Dispatcher is responsible for message delivery between services. It enables services to send messages and to register themselves as listeners for specific types of messages. The service is informed in case of an incoming message. The Event Dispatcher handles the delivery of messages.

It knows whether a message can be delivered locally or whether the message must be sent to a remote node. This information is collected during runtime explicitly by using the discovery service if the location of the receiving service is unknown, or implicitly by collecting the information from service advertisements. Each service has to register at the Event Dispatcher and must provide a service advertisement that holds basic information describing the service. The Event Dispatcher propagates these service advertisements to other nodes. Service advertisements have a limited lifetime. After this lifetime the advertisement becomes invalid. To get a new service advertisement the discovery service must be used.

Services: A service needs to implement a special interface to participate in OCμ and to receive messages delivered by the Event Dispatcher. We differentiate two kind of interfaces for services. A *simple service* and a *relocatable service*. Each interface has the full functionality needed to communicate within the middleware. Beyond the functionality of a *simple service*, a *relocatable service* can be transferred to another node, whereas a *simple service* is bound to the node it was started on. The binding on a special node is important for services that need a special hardware or software environment, e.g., fixed sensors or databases. Each service provides information about the system requirements to enable OCμ to determine if a node is capable of running a service. The system requirements can also be used to ensure that a *relocatable service* will only be relocated to nodes with special hardware if necessary.

Service Proxy: The service proxy is used to forward messages for a service that was recently relocated to another node. During the relocation of a service the service proxy stores incoming messages and forwards these as soon as the service is started at the new node. The service sends his answer directly back and the receiving node can then update its entry for this service. The service proxy has a limited lifetime and automatically terminates after that time. A proxy's lifetime depends on the lifetime of the service advertisement of the corresponding service. So a service proxy will work until every service advertisement with the old location of the service in the middleware becomes invalid. If the service proxy cannot be reached, the sending node must discover the service anyway. Thus, the overhead of updating all nodes can be deferred to the point where it cannot be avoided any more, but it is always less than a broadcast to all nodes. When a service is stopped for relocation the internal states will be saved on a self-healing distributed data store [19]. After relocation the service can resume the work at the point it was relocated. This mechanism guarantees the availability of services during and even after a relocation.

Organic Manager: OCμ implements an Observer/Controller architecture where the Organic Manager including the self-x services takes the part of the controller. The monitors which are introduced in Sect. 3.3 are the observers. All relevant information collected by the local monitors is analysed within the monitors and stored at the Organic Manager in the Information Pools, i.e. System Monitor Information Pool, Monitor Information Pool. They provide the collected information to other services and the self-x services employing a publish/subscriber mechanism.

3.2 Messaging

In OCμ services use asynchronous messages to communicate. To decide which service is the receiver of a message OCμ uses message types. Each service can register on message types and will receive messages that are directed to this type or any subtype. So for example the location tracking service can register for the type *ocm.locationTracking* and will receive messages with this type and also of the type *ocm.locationTracking.update* but not messages of the type *ocm*. Each subtype can consist of several subtypes. There is no limit on the depth.

Services can send unicast and broadcast messages. If no message type is given all services an the target nodes will receive this message. By combining this two options there are four kinds of messages.

Unicast Message with Type: The message is sent to all services which are registered on the message type on one node. A service uses this kind to send a request or response directly to another service.

Broadcast Message with Type: The message is sent to all services who are registered on the message type in the entire middleware. This kind is used to send messages to a given kind of service on all nodes.

Unicast Message without Type: This message is sent to all services on one node. It is used to get an alive response from each service on a given node.

Broadcast Message without Type: This message is sent to all services on each node. It can be used to get an alive response from each service in the middleware.

3.3 Monitoring

To get information about services, resources and the node itself, monitoring is a vital point in OCμ. To avoid the single point of failure and the communication overhead of a centralised monitoring, information is collected locally on each node. To obtain information as detailed as possible, it is necessary to use more than just one monitoring point and also to use specialised monitors for different tasks. OCμ not only uses a single monitor on each layer (Application, Transport Connector, and Event Dispatcher) but also Monitor Queues for a fine-grained monitoring. In OCμ separate Monitor Queues exist for each direction. The advantage of the Monitor Queues is that monitors can be added and removed as needed and that the monitors can be kept simple and fast.

Monitoring on the Transport Layer: The monitors at the Transport Layer are responsible for monitoring transport-dependent information. Incoming and outgoing messages can be used to monitor the latency between peers and the amount of data exchanged between services. This information can be used to decide if it is better to run a service locally or at a remote node.

Monitoring at the Event Dispatcher: As the Event Dispatcher delivers incoming and outgoing messages, we use it to monitor service dependant information concerning the message exchange. If a service requests information from another service, the dispatcher can measure the local and remote response time. It can also collect *alive information* about services by monitoring the messages to assure that a service is still available. This prevents the overhead of *alive messages* by just looking at the ongoing communication. If a service does not send messages the Event Dispatcher can not conclude on its own whether the service has failed or simply has completed its task and does no longer communicate with this node. For this reason we implemented the Failure Detector [15] as part of the self-healing service described in the next section.

The System Monitor: The System Monitor is used to gather information about the physical hardware a node is running on. Information about e.g. memory, processing power, and communication capabilities might be vital for the Organic Manager to reason about the distribution of the services.

3.4 Self-X Services

The self-x features are implemented by separated self-x services. Each service uses the information stored in the Information Pools. If additional information is needed each service gathers it on its own. There is no other interaction between the self-x services.

Self-Configuration Service [18, 20]: Each service must provide information on the required resources. The self-configuration service calculates a quality value for each service. This value states how reliable the service can be executed on the local node considering the resource requirements. Then an auction is triggered and the node with the best quality is selected to execute the service. This auction uses a distributed algorithm so there is no single point of failure. The self-configuration service on the winning node updates the quality for the other services to consider the resources which are now bound by the newly executed service.

Self-Optimisation Service [18, 21]: The Self-Optimisation Service uses the node information stored in the Monitor Information Pool to compare the load of the local node with other nodes. If the load is higher than a calculated threshold, a service is relocated. Using the information about required resources provided by each service the threshold can be calculated for each service. So it is possible to relocate a service with large demands on resources to another node where more resources are available.

Self-Healing Service [13, 16]: For the incorporation of self-healing capabilities into distributed systems the detection of failures is a crucial part. We implemented a Failure Detector [14, 15] that uses an adaptive algorithm. It has been designed for flexible and generic usability as a basis to realise self-healing of distributed systems. Additionally we developed a self-healing distributed data store [19] that allows to recover from service or node failures.

Self-Protection Service [9, 11]: In our research we transferred techniques of a biological immune system to a computer based environment in order to design a self-protecting service which protects OCμ from malicious events. Antibodies are distributed in the system and detect malicious messages. To refine this method we devised probodies [10] to recognise known messages. The combination of these two types enables us to discover and discard malicious messages.

3.5 Shortcomings

The approach of using separate self-x services has some disadvantages. Through the separation of the self-x services there is no information exchange or cooperation. So each self-x service has to collect his own data and works without knowledge of the other algorithms. Two services can influence the system in opposite ways. The self-healing service can start more instances of a service on the local node when a failed node is detected. Meanwhile the self-optimisation has detected that the local node is overloaded and is relocating services to other nodes.

Moreover, the planning introduced into the self-healing service is time and resource consuming. Therefore, we investigate a two-level approach with additional reflexes for fast reacting, respectively resource restricted nodes.

4 The Refined Architecture

To address the identified problems we introduce an automated planner as proposed by Satzer et al. [13, 16] for the self-healing service. Figure 2 shows the new architecture of OCμ. The basic middleware functionality on the left of the new architecture

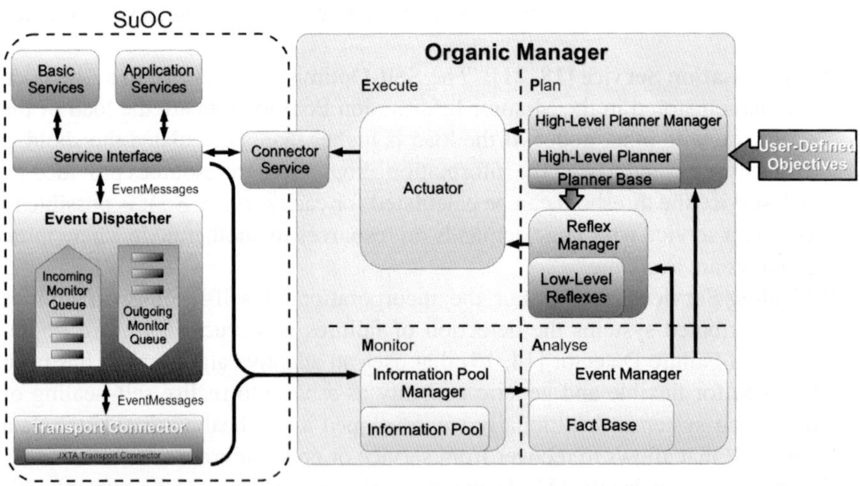

Fig. 2 The refined OCμ node architecture with MAPE cycle

remains nearly unchanged. There are no longer Monitor Queues in the Transport Connector layer. We discovered that all information previously monitored in the Transport Connector can be monitored in the Event Dispatcher. So we decided to move all monitors into the Event Dispatcher.

The focus on our work is the new Organic Manager which implements the self-x properties. The Organic Manager is based on the concepts of the observer/controller architecture [12] and organised like the MAPE cycle introduced by Kephart and Chess [6].

The first stage monitors the system. The second stage analyses this monitored data and transforms it into facts. If a malfunction is detected the third stage uses these facts and creates a plan to bring the system into a valid state again. The last stage executes the plan. So the first two stages act as Observer and the other two stages as Controller for our system.

4.1 Monitor

The Monitoring stage consists of the message monitors in the Event Dispatcher, the Information Pool Manager and the Information Pool. Information about the local node is gathered by the system observer service which polls the information about CPU, memory usage and the number of running services in periodic intervals. The system observer service is a basic service and runs on each node.

Monitor Queues in the Event Dispatcher: Each node has system observer monitors for incoming and outgoing messages. On each outgoing message the system information is sent via piggy-back. The incoming monitor on the receiving node gathers this information and passes it to the Information Pool Manager. We also implemented information forwarding so that a node can forward information about other nodes. Thus we accomplish that information about a node with only few communication partners is distributed.

Information Pool Manager: The Information Pool Manager stores the collected data in the Information Pool. The Manager is responsible for deciding whether the received information is relevant. So it can happen that forwarded node information is discarded, e.g. outdated information will not be stored. The Information Pool Manager also triggers the Event Dispatcher in the Analyse stage if any node information has changed.

Information Pool: The Information Pool stores the information collected by the monitors. It does not interact with any class on its own accord. Instead, all interaction must be triggered through the Information Pool Manager.

4.2 Analyse

The Analyse stage is triggered by the Information Pool Manager if any information in the Information Pool has changed. The main objective of this stage is to transform the raw information into more descriptive *facts* that can be processed by a planner. For example the information *remaining battery capacity 3%* is transformed to the fact *BATTERY_LOW*. The conversion rules from raw data to facts are statistically defined.

Event Manager: The Event Manager analyses the data in the Information Pool and creates facts. The Event Manager uses data mining methods to aggregate the raw information. Among these are cluster analysis and time series analysis.

Fact Base: The Fact Base stores the facts. It is filled by the Event Manager. In case the automated planner is triggered the Fact Base is passed to the High Level Planner Manager which converts the facts into a format the planner can process.

4.3 Plan

If a fact has changed, the High-Level Planner Manager and the Reflex Manager are triggered. The planner has information about the desired system state, called goals, and a list of actions that can be taken. The actions have preconditions and effects. Some actions like start, stop, and relocation of a service are provided on start up by the middleware itself. Goals and further actions can be added by any service. For the communication with the planner we use the Planning Domain Definition Language (PDDL [2]). It has been developed as a generic way to define planning problems in order to compare different planning algorithms. In parallel to the planner the Reflex Manager tries to find a stored reflex reaction for the current problem.

User Defined Objectives: A user can state additional goals for the planner, e.g. how often a service runs in the network. These objectives are stored in the Planner Base.

Planner Base: The Planner Base stores information about the goals and actions for the planner. This information is passed to the planner along with the facts from the Analyse stage by the High-Level Planner Manager.

High-Level Planner Manager: The High-Level Planner Manager converts the facts from the Event Manager to PDDL and passes these converted facts to the planner. Because of this abstraction every planner which can handle PDDL can be used. The automated planner tries to find the appropriate actions to direct the system from the actual state to a target state. This plan is passed back to the Planner Manager. If a valid set of actions is found the actions are sent to the actuator.

High-Level Planner: We use JavaFF as High-Level Planner, which is based on the *Fast Forward* planning system [3]. It is a planner which is written in Java and can plan with Boolean and numerical values. By changing the actions and the facts accordingly we tested both types of planning. The planner is exchangeable as long as it can process PDDL.

Reflex Manager: The Reflex Manager stores plans and reuses them if the same or a similar problem reoccurs. When only using Boolean variables in the plan the

Reflex Manager behaves like a cache. Each system state can simply be compared to previous states.
If instead numerical values in plans are used we realised that a metric is required to compare the initial system state with the stored states. In this scenario it is also possible that the Reflex Manager returns a plan which has solved a similar problem in the past but is not suited for the current problem. So two contradicting plans are sent to the Actuator and it has to solve the conflict. Each node starts with an empty Reflex Manager. Resource restricted nodes not running a planner must rely on more powerful nodes to generate reflexes for them. Without interaction with other nodes no additional services can be started. As soon as new service were started there is an interaction necessary which will also provide reflexes.

4.4 Execute

The Execute stage executes the plan either locally or by delegating single steps of the plan to other nodes.

Actuator: The Actuator is responsible for executing the plans. If an action cannot be executed on the own node the actuator sends a message to another node and the action is executed there. If the Reflex Manager has sent a plan for a similar situation and the planner has devised another plan the Actuator will receive a plan from the Reflex Manager and later a plan from the High-Level Planner Manager. Then the Actuator must stop the execution of the actual plan from the Reflex Manager and undo the contradicting steps of the plan. After that the execution of the correct plan from the Planner Manager starts.

Connector Service: As only services can send messages we implemented the Connector Service which allows the Actuator to send and receive messages.

5 Summary and Outlook

The new OCμ architecture provides self-organisation, self-configuration and self-healing features with the help of an automated planner. Our middleware uses the MAPE cycle to implement an observer/controller architecture and employs threads to enable services to run in parallel on multi-core processors. The Monitors collect information of the system. This information is analysed in the Analyse stage and transformed into facts. The Planner Manager uses these facts to determine if the system is in a desired state. If this is not the case the automated planner devises a list of actions to guide the system in a valid state. The Reflex Manager is used to avoid the time consuming planning on reoccurring problems. The Executor is responsible for executing the plans generated by the automated planner. Services can provide goals and actions for the automated planner. OCμ uses PDDL to represent goals and actions. So any automated planner that processes PDDL can be used. With the given interfaces OCμ services can be created simply and fast without knowledge of the whole system.

Self-protection techniques are fundamentally different and do not fit into the planner scheme at the moment, but the use of the automated planner for self-protection will be researched in the future. We will improve the Analyse stage to use distributed data mining techniques and adapt the conversion rules during run-time. So the system will be able to react on changes and provide accurate facts for the planer. We will investigate several automated planners for the Planning Stage. A representation for problems and metric must be found to enable the Reflex Manager to compare them. The information in the Monitor Stage can be outdated or incomplete. Thus the Analyse Stage must be capable of dealing with these uncertainties.

We further plan to port OCμ to an FPGA-based many-core simulator platform (developed at the department) and investigate the usage of OCμ's self-x-techniques for configuration, optimisation and healing in case of core failures.

OCμ is also used in the DFG project OC-TRUST [17] as middleware (see Chap. 6.1.6). Since services can be relocated without the interaction of the user the system behaviour is unpredictable. OC TRUST investigates the influence of trust on organic systems. It aims to measure the reliability of services and investigates how a relocation to another node will influence the reliability and the measurement. Furthermore the OC-TRUST investigates the use of organic systems in computing and energy grid scenarios.

References

1. Bagci, F., Kluge, F., Satzger, B., Pietzowski, A., Trumler, W., Ungerer, T.: Experiences with a smart office project. In: Research in Mobile Intelligence, pp. 294–319. Wiley, New York (2010)
2. Ghallab, M., Aeronautiques, C., Isi, C.K., Penberthy, S., Smith, D.E., Sun, Y., Weld, D.: PDDL-The Planning Domain Definition Language. Technical report, Yale Center for Computational Vision and Control (1998)
3. Hoffmann, J., Nebel, B.: The FF planning system: fast plan generation through heuristic search. J. Artif. Intell. Res. **14**(1), 253–302 (2001)
4. Horn, P.: Autonomic computing: IBM's perspective on the state of information technology. IBM Corp. **15**, 1–39 (2001)
5. Jahn, M., Pramudianto, F., Al-Akkad, A.: Hydra middleware for developing pervasive systems: a case study in the ehealth domain. In: 1st International Workshop on Distributed Computing in Ambient Environments (DiComAe 2009), pp. 13–21 (2009)
6. Kephart, J.O., Chess, D.M.: The vision of autonomic computing. IEEE Comput. **36**(1), 41–50 (2003)
7. Mamei, M., Zambonelli, F.: Spatial computing: The TOTA approach. In: Self-star Properties in Complex Information Systems, pp. 307–324 (2005)
8. Nickschas, M., Brinkschulte, U.: CARISMA—a service-oriented, real-time organic middleware architecture. J. Softw. **4**(7), 654–663 (2009)
9. Pietzowski, A.: Selbstschutz in Organic- und Ubiquitous-Middleware-Systemen unter Verwendung von Computer-Immunologie. PhD thesis, Universität Augsburg, Germany (2009)

10. Pietzowski, A., Satzger, B., Trumler, W., Ungerer, T.: Using positive and negative from immunology for detection of anomalies in a self-protecting middleware. In: INFORMATIK 2006—Informatik für Menschen, vol. P-93, pp. 161–168 (2006)
11. Pietzowski, A., Trumler, W., Ungerer, T.: An artificial immune system and its integration into an organic middleware for self-protection. In: The 8th Annual Conference on Genetic and Evolutionary Computation, pp. 129–130 (2006)
12. Richter, U., Mnif, M., Branke, J., Müller-Schloer, C., Schmeck, H.: Towards a generic observer/controller architecture for organic computing. GI-Jahrestagung **1**, 112–119 (2006)
13. Satzger, B.: Self-healing distributed systems. PhD thesis, Universität Augsburg, Germany (2008)
14. Satzger, B., Pietzowski, A., Trumler, W., Ungerer, T.: Variations and evaluations of an adaptive accrual failure detector to enable self-healing properties in distributed systems. In: Architecture of Computing Systems (ARCS 2007), 171–184 (2007)
15. Satzger, B., Pietzowski, A., Trumler, W., Ungerer, T.: A lazy monitoring approach for heartbeat-style failure detectors. In: The Third International Conference on Availability, Reliability and Security (ARES 2008), pp. 404–409 (2008)
16. Satzger, B., Pietzowski, A., Trumler, W., Ungerer, T.: Using automated planning for trusted self-organising organic computing systems. In: The 5th International Conference on Autonomic and Trusted Computing (ATC 2008), pp. 60–72 (2008)
17. Steghöfer, J.P., Kiefhaber, R., Leichtenstern, K., Bernard, Y., Klejnowski, L., Reif, W., Ungerer, T., André, E., Hähner, J., Müller-Schloer, C.: Trustworthy organic computing systems: challenges and perspectives. In: The 7th International Conference on Autonomic and Trusted Computing (ATC 2010), pp. 62–76 (2010)
18. Trumler, W.: Organic Ubiquitous Middleware. PhD thesis, Universität Augsburg, Germany (2006)
19. Trumler, W., Ehrig, J., Pietzowski, A., Satzger, B., Ungerer, T.: A distributed self-healing data store. In: The 4th International Conference on Autonomic and Trusted Computing (ATC 2007), pp. 458–467 (2007)
20. Trumler, W., Klaus, R., Ungerer, T.: Self-configuration via cooperative social behavior. In: The 3rd International Conference on Autonomic and Trusted Computing (ATC 2006), pp. 90–99 (2006)
21. Trumler, W., Pietzowski, A., Satzger, B., Ungerer, T.: Adaptive self-optimization in distributed dynamic environments. In: First International Conference on Self-Adaptive and Self-Organizing Systems (SASO 2007), pp. 320–323 (2007)

M. Roth (✉) · J. Schmitt · R. Kiefhaber · F. Kluge · T. Ungerer
Department of Computer Science, Augsburg University, Augsburg, Germany
e-mail: roth@informatik.uni-augsburg.de

J. Schmitt
e-mail: schmitt@informatik.uni-augsburg.de

R. Kiefhaber
e-mail: kiefhaber@informatik.uni-augsburg.de

F. Kluge
e-mail: kluge@informatik.uni-augsburg.de

T. Ungerer
e-mail: ungerer@informatik.uni-augsburg.de

Chapter 4.3
DodOrg—A Self-adaptive Organic Many-core Architecture

Thomas Ebi, David Kramer, Christian Schuck, Alexander von Renteln, Jürgen Becker, Uwe Brinkschulte, Jörg Henkel, and Wolfgang Karl

Abstract The Digital on-Demand Computing Organism (DodOrg) project addresses a new computer architecture principle. The complexity of today's computing systems and the dynamics which arise from the interaction of various subsystems with each other as well as with the environment, restrain an efficient optimisation and maintenance of such systems.

DodOrg is a novel, biologically inspired, heterogeneous, and adaptive computer architecture, that features self-x properties in order to ease management and optimisation. Multiple interleaved control loops, that span all system layers, are employed for realisation of these self-x properties, such as self-optimisation. A dedicated monitoring infrastructure provides the basis for these control loops and realises the basic property of self-awareness. The modular architecture of Organic Processing Cells (OPC) provide a flexible hardware infrastructure with reconfiguration capabilities, that are essential for realisation of adaptive systems. Distributed task mapping is achieved through an Artificial Hormone System (AHS) within the organic middleware. The organic thermal management increases the system's mean time to failure by balancing the thermal distribution between OPCs.

Keywords Self-adaptive computing · Self-x · Distributed middleware · Monitoring · Thermal management · Reconfigurable computing

1 Introduction

DodOrg is inspired by biological concepts, mainly focusing on self-x mechanisms that range from self-awareness and self-organisation in general to self-optimisation, self-protection, and self-healing in particular. The architecture concept covers the complete computing system. It consists of three levels, 'brain', 'organ' and 'cell', as shown on the left of Fig. 1. Beginning with the brain level, a modular software architecture for embedded scenarios provides the application input for the DodOrg organism. The application is partitioned into a set of self-contained tasks, whereas different realisation alternatives are available for each task. A set of attributes describes the properties of each realisation alternative and the relations between the tasks.

At the organ level, a distributed middleware layer is responsible for mapping the application tasks down to the cell-based organic hardware. During the mapping

C. Müller-Schloer et al. (eds.), *Organic Computing — A Paradigm Shift for Complex Systems*, 353–368
Autonomic Systems, DOI 10.1007/978-3-0348-0130-0_23, © Springer Basel AG 2011

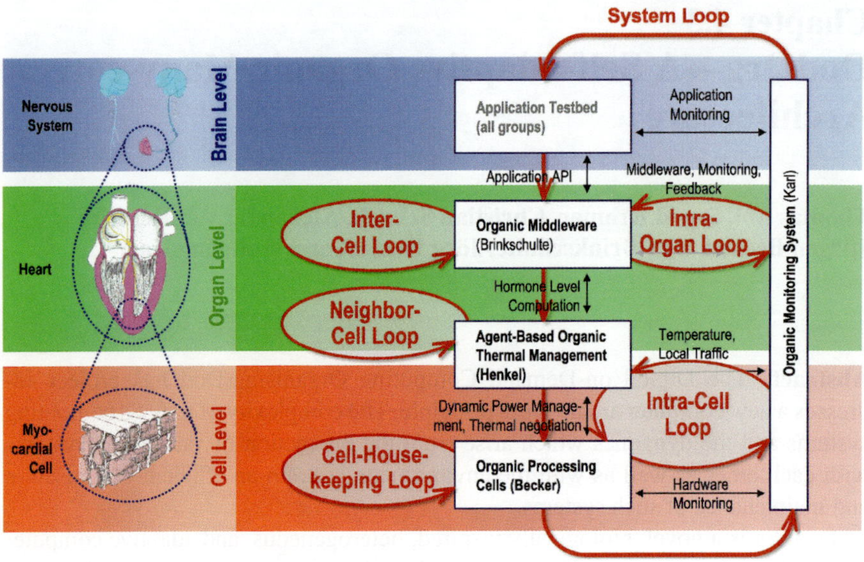

Fig. 1 Project structure

process, which is controlled by a novel artificial hormone system, related tasks are grouped together, establishing so-called *virtual organs* and thereby minimising the communication overhead. Due to the completely decentralised approach of the organic middleware, a high dependability is achieved. In close cooperation with the organic middleware, an agent-based thermal management layer coordinates the brain level and cell level for avoiding of peak temperatures on the chip. This prevents early ageing effects in future sub-micro architectures.

At the cell level, a novel adaptive and dynamically reconfigurable hardware architecture is capable of implementing self-x features in an efficient way. It consists of an array of adaptable organic processing cells, which communicate through a specially tailored network-on-chip. As a prerequisite for enabling self-x features, all system levels are supplied with a comprehensive monitoring architecture.

In biology, multiple interacting control loops can be observed for achieving the self-x properties and especially for realising a robust system behaviour. Consequently, exploiting the information delivered by the monitoring layer, different closed control loops on all system layers control the global behaviour of the DodOrg organism (see Fig. 1).

On the lowest level, the organic processing cells (OPCs) autonomously perform certain housekeeping tasks, e.g. providing stable inter-cell communication by employing cell-level control loops based on local, low-level monitoring data, provided by the individual cell's observer unit.

The functionality of an OPC is rated according to several parameters, which in turn are weighted independently and according to input requirements, leading to the computation of hormones. These hormones are then used by the middleware guiding task-to-OPC assignment and triggering migration of tasks from one OPC to another. For these processes, local monitoring is crucial for collecting local OPC status data regarding communication, load, and derived suitability for given tasks. Derived suitability is an actual verification of the assumed match of an individual OPC to given tasks.

Thermal management takes place between neighbouring cells, forming a second control loop on an inter-cell level, but restricted to neighbouring cells. This loop ensures the avoidance of hot-spots by influencing the task assignment process with respect to thermal considerations, using low-level monitoring information (load, temperature) from the neighbouring cell's observer units.

The middleware-driven task assignment, in turn, is a further inter-cell control loop. Unlike thermal management, however, it spans the global range of all cells. It is responsible for organ forming and reforming, based on received and sent hormones that indicate the current conditions and functional abilities of the OPCs. These hormone-based control loops allow robust reactions to environmental changes and failures. A newly introduced immune component enhances this with respect to malicious malfunctions and attacks, comparable to dealing with infections and malicious diseases in biology. Further, system control is achieved by additional monitoring layers, spanning an entire organ up to the entire system, forming a topmost system-wide control loop fostering overall system robustness. Our comprehensive approach to organic computing represents the first step towards more adaptive, more power-efficient and more flexible future embedded systems.

The remainder of this article is organised as follows: Section 2 gives an overview of our hardware infrastructure, especially the Organic Processing Cells architecture and the communication infrastructure. Section 3 outlines the employed monitoring infrastructure, while Sect. 4 introduces the organic middleware. The organic thermal management is presented in Sect. 5. Finally, Sect. 6 concludes this article.

2 Organic Hardware

The organic self-x hardware architecture developed within the scope of the DodOrg project represents a new kind of adaptive hardware platform [11]. Its architecture, as well as the hardware data path and interconnection structures can be continuously adapted and self-customised during runtime. Therefore, it is possible to efficiently support the required flexibility, e.g., to react to different load conditions or to compensate possible hardware faults. Following a modular concept, the basic entities of the DodOrg hardware are the *Organic Processing Cells* (OPCs). The entire hardware is composed of an OPC-mesh, which is connected through the *artNoC* network. Controlled by the middleware and the power-management, this node-based approach enables a seamless shifting of responsibilities and tasks among the OPCs.

The OPC internal structure, shown in Fig. 2, consists of two main parts. The first part, on the right, contains the cell-specific functionality. This part realises the OPC data path and therefore defines its characteristic. The cell mix is determined during design time and can be tailored according to a given application domain.

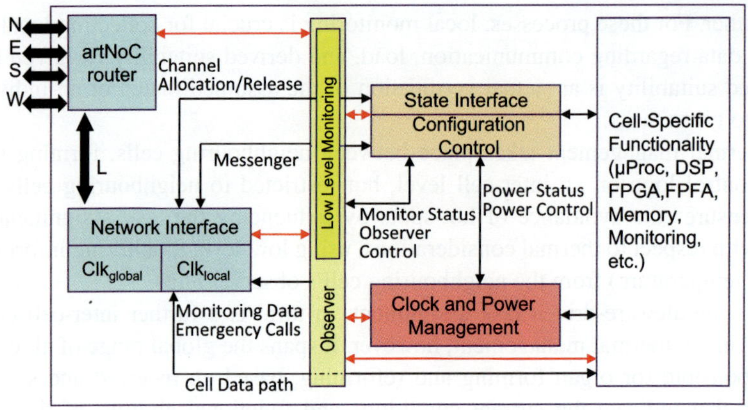

Fig. 2 Organic Processing Cell architecture

The second part on the left is the same for all OPCs. It contains the communication infrastructure and hardware units to support the self-x features of the software layers. In addition, the OPCs are equipped with own local intelligence and hardware monitoring, in order to realise low-level cell control loops. This ensures that basic services, e.g., a reliable communication among OPCs, can be guaranteed.

2.1 Communication Infrastructure

A reliable and efficient communication infrastructure is essential for each distributed computing system. The communication infrastructure of the DodOrg hardware consists of three main components: artNoC-router, network interface, and chip-to-chip-communication.

According to the Organic Computing principles, the organisation and control of the artNoC (adaptive real-time Network-on-Chip) [14] is completely decentralised. Decisions about the packet-routing and the packet-transmission schedule are based on the local view of each single router node only. Three different traffic classes are supported, which have been derived from the communication requirements of the DodOrg system: *best effort* (BE), *real-time* (RT) and *broadcast* (BC). BE traffic, as well as RT traffic are unicast traffic from OPC to OPC. BE traffic dynamically uses the available network bandwidth. In addition, guaranties for communication latency and bandwidth can be given for RT traffic, e.g., in case of control packets. Both guarantees are the result of local interaction of artNoC routers without a global knowledge of time or a global schedule. Further, a special feature of the RT traffic is the one-bit *feedback channel*, that is routed in the opposite direction from receiver OPC to sender OPC. It allows a fast handshake, which is essential for the implementation of the end-to-end protocol in the network interface. The realised BC traffic class is based on a dynamic *spanning-tree* approach, which allows an efficient one-to-many communication by reducing the packet count significantly. This is an essential feature of the DodOrg system, because the hormone-based middleware and power-management software layers broadcast hormones on a regular basis. With respect to fault tolerance, a maximum connectivity in case of faulty links was achieved by a flooding phase that establishes the dynamic spanning tree for each BC packet from scratch. For unicast traffic, different adaptive routing techniques were implemented and evaluated. The most advanced fully adaptive fault-tolerant routing algorithm uses a state-transmission protocol in order to create local fault regions and a detour path around the faulty region. This includes link monitoring units for lifelock and deadlock detection. In case of a blocked packet, the broadcast channel is used as a *recovery channel*, because it is deadlock free and guarantees maximum connectivity.

The network interface is the link between the OPC cell units and the artNoC. It realises the decoupling of the artNoC clock domain from the local OPC-specific clock domain. Further, to support the shifting of software, as well as the reconfiguration of hardware tasks from one cell to another cell, it provides special *connection tables*. Each task in the system has an unique task ID. The connection table maps the task ID to the physical artNoC address where the task is currently executed. This allows software and especially hardware tasks to refer to the task ID as the communication target without knowing the physical OPC address. Connection table entries are dynamically updated either from the middleware, from the OPC configuration control or directly from the network interface. The latter possibility is used to establish bidirectional communication paths between two tasks or cell units on the fly. Also, other end-to-end communication services are implemented in the network interface, such as packet retransmission or packet acknowledge.

Through chip-to-chip communication, different DodOrg chips can be connected in order to develop one bigger DodOrg organism. Thus, the seamless extension of the on-chip communication services across the borders of a single chip enables a dynamic OPC resource pool and can be seen as a physical growth or cutback of the DodOrg organism. In order to reduce the pin count, a high-speed serial connection was realised. The ports of the artNoC routers at the edge of the OPC grid are connected to an off-chip interface (OCI). The OCI translates the parallel artNoC data and handshake signals into a serial data stream, which is sent over the serial link. On the other chip, the corresponding OCI converts the serial input stream back to the artNoC protocol. To achieve a good serial-link utilisation, an *adaptive control flow protocol* was realised. Depending on the number of virtual channels that are transmitted and the flit buffer status in the receiving OCI, an improvement of serial-link efficiency of up to 81% can be achieved. Further, a hot plug and unplug of DodOrg chips is supported by each OCI through a link negotiation-protocol.

2.2 Power Management

Two features of the DodOrg hardware architecture provide the basis for the Organic Thermal Management software layer (see Sect. 5). On the one hand, different hardware data paths are available, which can be used to execute different realisation

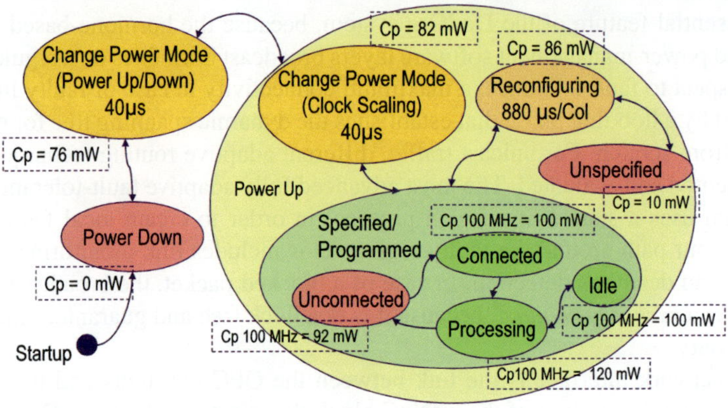

Fig. 3 OPC hardware life cycle model

alternatives of a tasks. Each alternative has specific power and performance characteristics. On the other hand, each OPC data path is driven by a dedicated clock domain. Through the power management module within each OPC, the Thermal Management can adapt the clock frequency according to the current power budget [12].

Based on the FPGA hardware prototype, an abstract hardware model was developed as an input for the Thermal Management energy trading. A graphic representation of the model is shown in Fig. 3. It describes the states during the *OPC processing* and *development phases*. To reach or change a processing state, the OPC enters an intermediate development state. For each state, the power consumption was determined. In addition, the processing times for each development state are given. With respect to power consumption, because the DodOrg hardware consists of several OPCs, the power consumption of the DodOrg hardware is the sum of those individual OPC models.

2.3 Low-Level Monitoring

Several fixed hardware monitoring units are incorporated at cell level. On the one hand, they serve as feedback for the cell housekeeping control loops. For example, counters for the artNoC link utilisation are used to detect the root of a deadlocked packet, or the neighbour OPC status is used to identify detour routes around faulty routers. On the other hand, a unified interface for all cell level monitoring data is available, which can be accessed by the monitoring layer (see Sect. 3) over the artNoC.

Further, data-path-specific monitoring probes are available, e.g., an idle counter for a µC-data path. Within FPGA OPCs, even monitoring with dynamic monitoring probes is supported. In [13], we used a dynamic ring-oscillator monitoring probe, in order to determine the speed grade of a complete FPGA data path area. Therefore,

Table 1 Hardware resource usage based on the number of virtual communication channels (VCs) and artNoC features: RT—real-time, FB—feedback channel, XY—xy routing, WF—west-first routing, BC—broadcast channel, FA—full adaptive routing, RC—packet recovery; artNoC buffer-depth = 2 flits; link-width = 8 Bits

	4 VCs			3 VCs			2 VCs		
	Slices	FFs	LUTs	Slices	FFs	LUTs	Slices	FFs	LUTs
artNoC									
XY	1092	559	2092	834	422	1553	510	286	983
XY+RT+FB	1501	683	3006	1048	504	2043	685	335	1318
WF+RT+FB	1737	739	3489	1241	565	2414	759	363	1441
WF+RT+FB+BC	2361	752	3055	1568	591	2168	1297	447	1924
FA+RT+FB+BC+RC	3417	1145	5401	2426	438	3536	1669	598	2665
Network Interface	610	395	1131	705	571	1205	638	547	1088
Off-Chip Interface	389	268	737	307	216	585	232	162	442
Temperature Monitor	73	63	140	73	63	140	73	63	140
Virtual-ICAP Interface	480	280	867	480	280	867	480	280	867

it is possible to adapt the maximum clock frequency according to dynamic effects, such as changing ambient temperature, supply-voltage variations or local-device-specific speed variations. Each OPC is also equipped with a temperature sensor as an input for the Thermal Management layer (see Sect. 5).

2.4 *Hardware Prototype*

The DodOrg hardware as described in the previous sections was implemented as an FPGA-based prototype. Various design time parameters on HDL-level exist, that allow to generate a customised hardware. As an example, Table 1 shows the resource usage based on the number of virtual channels and the artNoC features. The amount of resources increases almost linear with the number of virtual channel. The runtime flexibility of the OPC hardware data path is realised through dynamic and partial reconfiguration of the OPC-data-path area. In order to enable a decentralised access to the FPGA configuration memory from each OPC, the *Virtual-ICAP Interface* was developed [11]. By default, a MicroBlaze soft-core processor is used as a μC data path. A dynamic OPC resource pool and further flexibility is achieved through the chip-to-chip communication approach, which allows to connect several DodOrg chips together on demand.

3 Organic Monitoring

Self-awareness is the basic self-x property and fundamental for achieving further self-x features, such as self-adaptation, self-healing or self-optimisation. Self-awareness is the ability to gain and interpret data for determination of the actual system's state. Therefore, a dedicated monitoring infrastructure is essential, that continuously observes the entire system in order to get information about the system state at every time.

Within DodOrg, we employ a hierarchical monitoring infrastructure that spans all system layers and comprises low-level monitoring (LLM) and high-level monitoring (HLM). The low-level monitoring components gather a huge amount of data that has to be continuously analysed and interpreted by HLM instances with respect to external and internal requirements. Individual HLM instances also closely interact with all DodOrg system layers, providing required data for steering processes conducted by the organic middleware and the organic thermal management. In the context of DodOrg, each Organ has its own HLM instance. As many Organs can be executed concurrently, HLM instances can communicate with each other for, e.g., exchanging information.

For realisation of the LLM components, we extended the well-known hardware-counter approach [7, 15]. Usually, as in modern processors, hardware-counters have a fixed association between counter and event. This means that only a small number of events can be counted at a given time. When other events occur at the same time, they are simply ignored and not counted. Which events are to be counted must be determined by the system developer at design time. Within DodOrg, we extended the hardware-counter approach with a cache-inspired principle. This uncouples the association between counter and event and permits the monitoring of any occurring event. We introduced a hormone-inspired event coding, which allows the proper comparison of arbitrary events.

To achieve self-awareness, HLM instances can analyse the information gathered by the LLM and therefore determine the current system state. Traditional approaches use a fixed set of evaluation rules, determined by the developer at design time.

To free the developers from the tedious task of defining the evaluation rules at design time, we employ a flexible, adaptive, rule-based approach. Within this approach, a dedicated learning phase is used for deriving the evaluation rules at run-time. These simple evaluation rules are used to determine the current state of a single event. In order to generate a system state for the entire (sub-)system, an aggregation method is required that combines the individual states of the events. For this purpose, we use a weighted arithmetic mean, which has the advantage that it has a low computational complexity and, as the weights for certain events can be adjusted, can reflect external requirements. The main advantage of our rule-based approach is that rules can be updated if necessary and new rules can be added at any time.

Using this approach, HLM instances can continuously evaluate the state of the system and provide this state and also the gathered information to the organic middleware and the organic thermal management via a dedicated API. These instances can use this classified state and the provided data for determining the outcome of an optimisation cycle and for further problem analysis. For detailed information about the employed monitoring infrastructure, the inclined reader is referred to Chap. 2.3.

4 Organic Middleware

Middleware is a common layer in distributed systems. It manages the cooperation of tasks on processing elements and hides the distribution and communication between processing elements from the application. It is responsible for seamless task interaction on distributed hardware. As shown in Fig. 4, all tasks are interconnected by the middleware layer and are able to operate beyond processing-element boundaries as if they would reside on a single hardware platform.

After the system's start-up, the middleware determines an initial task mapping that takes into account the capabilities (e.g. computational power, memory, etc.) and the state (e.g. operation temperature, energy level, etc.) of the OPCs. The middleware reacts to changing conditions by changing the task allocation and thereby optimising the system's tasks mapping. In case of malfunctions or breakdowns of processing elements, the middleware remaps missing tasks, hence realising self-healing.

Our organic middleware, the Artificial Hormone System (AHS), is highly inspired by the hormone system of higher animals. In order to avoid a single point of failure, the AHS is completely decentralised and provides self-x properties. It pays attention to related tasks, which often have a high communication rate, and clusters them closely together, thus forming "virtual organs". Besides that, it is real-time capable and causes only a negligible communication overhead. Tight upper bounds for the real-time behaviour of the AHS as well as for the data load induced by the artificial hormones can be guaranteed [3, 4].

In the following, we give a short description of how the AHS works. For task allocation, only three types of hormones are used: *Eager value*, *Suppressor* and *Accelerator*. The *Eager value* hormone determines the suitability of an OPC for executing a task. *Suppressors* weaken the possibility of an execution, while *Accelerators* favour the execution of a task on an OPC.

Figure 5 sketches the basic control loop used for assigning a task T_i to an OPC PE_y. This closed control loop is executed for every task on every OPC. Based on the level of the three hormone types, it determines whether a task T_i is executed on an OPC. Details of the functioning of the AHS, the hormone cycle, timing constraints and further theoretical analysis can be found in [4] and Chap. 4.4. The reliability of the AHS with improved timing constraints and further test series that illustrate the timing behaviour of self-configuration and self-healing were presented in [17].

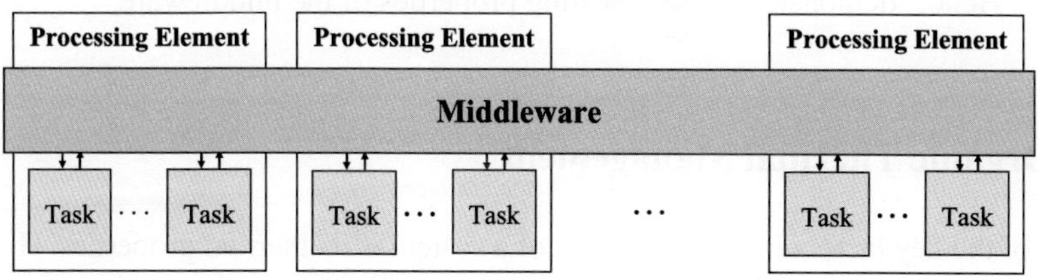

Fig. 4 Middleware in a distributed system

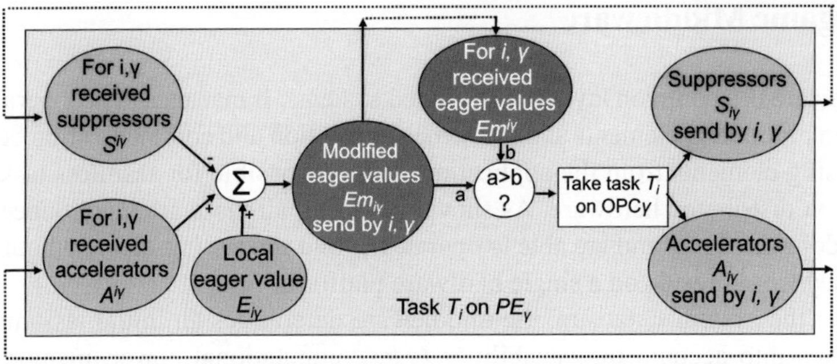

Fig. 5 Hormone cycle

In [6], we investigated the quality of task mappings from the AHS on a grid of heterogeneous processing elements. An algorithm was introduced to measure the quality of such task mappings. We showed that the AHS's mapping was up to nearly 20% better than mapping results of a simple load balancing. With the ability to quantify the quality, further improvements and fine-grained adjustments can be realised.

Stability and robustness of the AHS were examined in [5]. With no central element, the stability of the AHS cannot be controlled by a single processing element, instead the hormone values have to be chosen carefully to guarantee system stability. With theoretical analysis of the AHS, we were able to introduce upper and lower bounds that have to be met to guarantee system stability. We specified ranges in which the AHS will work stably. Additionally, we analysed the limits of single PEs as a step towards being able to set a good system size. With this work, it is now easily possible to analyse whether the system will be stable or not.

In [18], we presented an implementation of the AHS, written in pure ANSI C. With this ANSI C implementation, it is now possible to run the AHS on a variety of devices, from basic microcontrollers to PCs. We showed first real-world scenarios to demonstrate the capability and show a smoothly working organic system. Network load of the AHS was analysed and compared to the network traffic of the application running on top of the organic middleware. We could show that the AHS uses significantly less network capacity than the application, which was not disturbed or constricted by the AHS in any way. Another test showed an error scenario in which a processing element went dead and the system had to recover from that loss. The measured data showed that the AHS managed to remap the tasks and restore their states. Hence, demonstrating self-healing properties of the middleware.

5 Organic Thermal Management

One of the key factors in the robustness of a system is its thermal properties. Higher temperatures correspond to an increase in ageing effects (e.g., due to elector migration), which reduces the system's mean time to failure (MTTF) considerably. These effects are becoming more pronounced, because the technology size decreases, resulting in an increase of power density on the chip. Thus, in order to create a robust system, thermal properties must be considered.

The organic hardware (see Sect. 2) of OPCs supplies us with a power-management infrastructure that provides basic mechanisms for controlling the power output of each OPC. It allows the Organic Thermal Management to set the maximum frequency at which each OPC may run and, in case of multiple voltage domains (e.g., created by the multi-chip extensions: Sect. 2), also the maximum allowed voltage.

Traditionally, the focus of low-power techniques has been on minimising either energy or power. These approaches, however, fail to consider heat-specific factors regarding the spatial on-chip power distribution and the resulting heat transfer across the multi-core chip. Better results have been achieved by emerging methods, which consider these effects. These methods typically rely on a *Central Control Instance* (CI), which reacts to *thermal hotspots* through a near-uniform distribution of power over the entire chip by means of runtime task migration [8, 19]. Along with *DVFS*, these approaches can keep the generated heat through a core's power consumption below a design-time threshold temperature [19]. The use of *central control instances*, however, will soon lead to scalability issues, as multi-core architectures are beginning to incorporate hundreds or thousands of cores [2]. In particular, the increasing amount of sensor data transmitting thermal values and the computational overhead for managing *DTM* in hundreds of cores cannot be handled efficiently with *central control instances*; instead these will have to yield to distributed management techniques [1, 9, 10].

Thermal Management in DodOrg is realised as a fully distributed agent-based system [9]. Agent-based systems are one of the fundamental approaches for implementing organic self-x properties. Agents are able to exploit their locality in order to reduce the problem space and exhibit emergent behaviour through negotiation and cooperation with other agents. We consider an agent as a situated, autonomous, and flexible computational entity that acts on the behalf of an OPC. They are situated at each OPC and are responsible for controlling its power management, e.g. determining at which frequency it is allowed to run.

The agent negotiation techniques used for thermal management in DodOrg are based on the economic models used in agent-based computational economics (ACE) [16]. These are loosely based on the distribution of supply and demand of energy on the chip. Initially, a predefined number of energy units are distributed evenly throughout the system among all agents. Each of these energy units is assigned to one of the following two categories: *used* units are currently used by the OPC for running tasks, i.e., the number of used units determine the OPC's frequency and voltage through the power management module presented in Sect. 2, and *free* units that are energy units available on an OPC that are not currently used by the OPC to run tasks, because the required frequency and voltage set by used units already meet task requirements. Together, the two types of units are used by an OPC's agent to calculate *buy* and *sell* values used for trading energy units based on (1). Here the weights $w_{i,j}$ are determined by using the OPC power model presented in Sect. 2

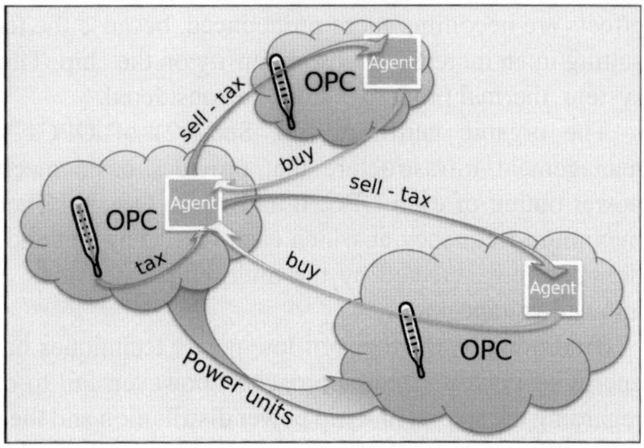

Fig. 6 Energy budget trading among agents

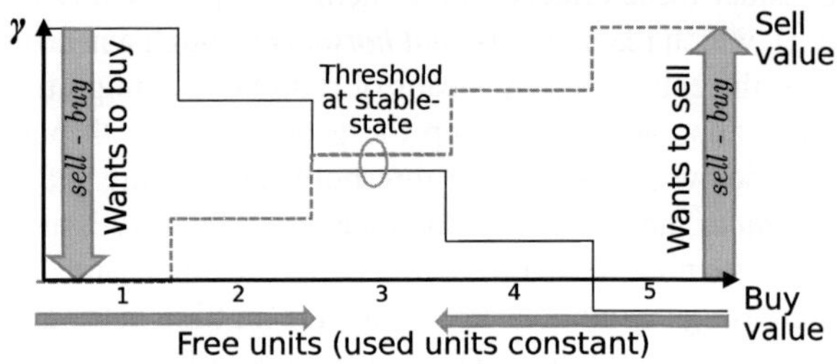

Fig. 7 Overview of *buy* and *sell* values assuming the same number of *used* energy units

and may vary for different OPC types, while γ is a normalising parameter used to offset buy and sell values:

$$\begin{aligned} \text{sell-value agent } n: \quad & sell_n = w_{u,s} \cdot used_n + w_{f,s} \cdot free_n \\ \text{buy-value agent } n: \quad & buy_n = w_{u,b} \cdot used_n - w_{f,b} \cdot free_n + \gamma \end{aligned} \tag{1}$$

Buy and sell values vary dynamically at runtime, depending on the current supply/demand of energy units calculated by each agent. An overview of how these values relate and the effect of γ is shown in Fig. 7. The underlying main concept is that it becomes increasingly difficult to obtain energy units if there are already many energy units that are being used in a local area (high demand), even though there is no defined upper bound, because agents at the edge of a local area will trade with immediately neighbouring agents outside this local area. Through their interface with the organic monitoring presented in Sect. 3, agents are able to obtain temperature values from the on-chip thermal sensors and incorporate these values in the negotiation process by adding a "tax" to the buy values. This has the effect that increasing temperatures make it more and more difficult for an agent to obtain additional energy units, so it will conversely start losing the ones it has.

If the difference between buy and sell values ($sell_i - buy_i$) of an agent i is exceeded in one of its neighbours by more than a given threshold, the agent relinquishes an energy unit to the neighbour with the maximum ($sell_n - buy_n$) value. This difference, in effect, expresses the agent's "desire" to obtain or give up energy units, as shown in Fig. 7. If the agent has free energy units, giving up energy units is simply a matter of decrementing its number of free energy units. However, if there are no free energy units, the local agent must give up one of its used power units. This means that the agent must also reduce its power consumption through DVFS.

Apart from deciding what energy units are traded and when, the difference between the buy and sell values also gives an indication of how well an OPC is suited for additional tasks from an energy/thermal point of view. As such, they relay this information to the Organic Middleware by emitting suppressors based on these values. In our economic approach, these suppressors are proportional to the difference between an OPC's sell and buy value.

By limiting trading to immediate neighbours, the communication overhead of DTM when compared to a central approach is reduced by a factor of 11.9 in a 96-core architecture [9]. Our agents can be implemented in either hardware or software, with each implementation resulting in different overheads. In particular, hardware agents require additional chip area (143 slices in our FPGA prototype), while software agents require execution time on an OPC, which, although minimal, takes away from task execution. Our approach was simulated using power traces obtained from the MiBench benchmark suite, running multiple applications simultaneously. Our approach implements a dynamic thermal threshold, because task migration is dependent on the suppressors sent to the Organic Middleware and other factors that influence an OPC's eager value (see Sect. 4), and not the specific OPC temperature. This results in slightly higher peak temperatures than pure DTM approaches, but at the same time reduces overall energy consumption. For instance, the state-of-the-art PRTM approach [19] reaches a peak temperature that is 4.6% below the one achieved by our approach, which is able to reduce peak temperatures by 11% compared to not using thermal management. However, the DodOrg Organic Thermal Management at the same time reduces application execution time and energy consumption by 44% each in comparison with PRTM [19].

Current work aims at improving the trading of energy units. While the approach so far distributes power consumption pro-actively across the chip, the trading itself is done reactively. This improved approach uses the state classification of the Organic Monitoring (see Sect. 3) system in order to determine the effectiveness of past trading decisions and to find out in which system states trading will likely be needed in the near future. This evaluation is done using economic reinforcement learning, where the state classification serves as a fitness function. Additionally, because the DodOrg system allows the connecting of multiple physically separated chips (Sect. 2), it is made up of independent thermal domains. In this case, the economic learning provides the means for differentiating between inter-domain and intra-domain trading.

6 Conclusion

In this article, we described the architecture and the interaction of the layers of the DodOrg system, a biology-inspired computing architecture. The brain level provides the application scenario for DodOrg system. This scenario consists of a set of tasks. At the organ level, the organic middleware uses an artificial hormone system for mapping these tasks to available hardware resources. For optimisation, the organic middleware groups related tasks together, forming so-called virtual organs. At cell level, an adaptive hardware architecture supports the self-x features in an efficient way. In between the organ and cell level, the organic thermal management controls the power distribution of the cells, while influencing task mapping decisions of the organ level. Self-awareness, the basic property for self-x features, is provided by a dedicated organic monitoring infrastructure, which performs a sustained real-time monitoring of all components and system layers.

An intrinsic feature of many biological systems is their capability of self-healing, self-adapting, self-configuring, etc., or self-x features for short. In contrast, today's computing systems hardly feature any of these characteristics, even though their complexity has risen to an amount where this very complexity is increasingly hampering maintenance tasks such as optimisation, protection, and repair.

Our DodOrg architecture presents a complete organic system from top to bottom. Its distributed nature allows it to scale to large system sizes. It provides the necessary mechanisms for system maintenance: through the system's self-organisation capabilities, the configuration costs remain low while self-awareness allows us to detect changing system properties. This further enables the DodOrg architecture to self-optimise for changing system states. Self-healing properties allow us to recover from failures and malfunctions. Interleaved control loops provide robustness, while stability is guaranteed by the analytical bounds of the artificial hormone system.

The achievements are promising and form a good foundation in the research area of self-adaptive computing architectures. Many results of DodOrg motivate further research.

References

1. Al Faruque, M., Jahn, J., Ebi, T., Henkel, J.: Runtime thermal management using software agents for multi- and many-core architectures. IEEE Des. Test Comput. **27**(6), 58–68 (2010)
2. Borkar, S.: Thousand core chips: a technology perspective. In: Proceedings of the 44th Annual Design Automation Conference (DAC), pp. 746–749 (2007)
3. Brinkschulte, U., Pacher, M., von Renteln, A.: Towards an artificial hormone system for self-organizing real-time task allocation. In: 5th IFIP Workshop on Software Technologies for Future Embedded & Ubiquitous Systems (SEUS) (2007)
4. Brinkschulte, U., Pacher, M., von Renteln, A.: An artificial hormone system for self-organizing real-time task allocation. In: Organic Computing. Springer, Berlin (2008)
5. Brinkschulte, U., von Renteln, A.: Analyzing the behaviour of an artificial hormone system for task allocation. In: ATC, Brisbane, QLD, Australia (2009)
6. Brinkschulte, U., von Renteln, A., Pacher, M.: Measuring the quality of an artificial hormone system based task mapping. In: ACM Autonomics, Turin, Italy (2008)
7. Buchty, R., Kramer, D., Karl, W.: An Organic Computing Approach to Sustained Real-time Monitoring. In: Proceedings of WCC2008/BICC. IFIP, vol. 268, pp. 151–162. Springer, Berlin (2008)
8. Coskun, A.K., Rosing, T.S., Gross, K.C.: Temperature management in multiprocessor socs using online learning. In: Proceedings of the 45th Annual Design Automation Conference (DAC), pp. 890–893 (2008)

9. Ebi, T., Al Faruque, M.A., Henkel, J.: TAPE: thermal-aware agent-based power economy for multi/many-core architectures. In: International Conference on Computer-Aided Design (ICCAD), pp. 302–309 (2009). Best Paper Award
10. Ebi, T., Al Faruque, M.A., Henkel, J.: NeuroNoC: neural network inspired runtime adaptation for an on-chip communication architecture. In: Proceedings of the Eighth IEEE/ACM/IFIP International Conference on Hardware/Software Codesign and System Synthesis. CODES/ISSS'10, pp. 223–230. ACM, New York (2010)
11. Schuck, C., Haetzer, B., Becker, J.: An interface for a decentralized 2D reconfiguration on Xilinx Virtex-FPGAs for organic computing. Int. J. Reconfigurable Comput. **2009**, 1–12 (2009)
12. Schuck, C., Haetzer, B., Becker, J.: Dynamic online reconfiguration of digital clock managers on Xilinx Virtex-II/Virtex II-pro FPGAs: A case study of distributed power management. In: ReCoSoC2010 (2010)
13. Schuck, C., Haetzer, B., Becker, J.: Reconfiguration techniques for self-x power and performance management on Xilinx Virtex-II/Virtex-II-Pro FPGAs. Int. J. Reconfigurable Comput. **2010**, 1–17 (2010)
14. Schuck, C., Lamparth, S., Becker, J.: artNoC—A novel multi-functional router architecture for organic computing. In: FPL, pp. 371–376 (2007)
15. Sprunt, B.: The basics of performance-monitoring hardware. In: IEEE Micro, pp. 64–71 (2002)
16. Tesfatsion, L.: Agent-based computational economics: Growing economies from the bottom up. Artif. Life **8**(1), 55–82 (2002)
17. von Renteln, A., Brinkschulte, U.: Reliability of an artificial hormone system with self-x properties. In: Parallel and Distributed Computing and Systems, Cambridge, Massachusetts, USA, 2007
18. von Renteln, A., Brinkschulte, U.: Implementing and evaluating the AHS organic middleware—a first approach. In: ISORC, Carmona, Spain, 2010
19. Yeo, I., Liu, C.C., Kim, E.J.: Predictive dynamic thermal management for multicore systems. In: Proceedings of the 45th Annual Design Automation Conference (DAC), pp. 734–739 (2008)

T. Ebi · D. Kramer (✉) · J. Henkel · W. Karl
Institute for Computer Science and Engineering, Karlsruhe Institute of Technology,
Karlsruhe, Germany
e-mail: kramer@kit.edu

T. Ebi
e-mail: thomas.ebi@kit.edu

J. Henkel
e-mail: henkel@kit.edu

W. Karl
e-mail: karl@kit.edu

C. Schuck · J. Becker
Institute for Information Processing Technologies, Karlsruhe Institute of Technology,
Karlsruhe, Germany

C. Schuck
e-mail: schuck@kit.edu

J. Becker
e-mail: becker@kit.edu

A. von Renteln · U. Brinkschulte
Institute of Computer Science, University of Frankfurt, Frankfurt, Germany

A. von Renteln
e-mail: renteln@es.cs.uni-frankfurt.de

U. Brinkschulte
e-mail: brinks@es.cs.uni-frankfurt.de

Chapter 4.4
The Artificial Hormone System—An Organic Middleware for Self-organising Real-Time Task Allocation

Alexander von Renteln, Uwe Brinkschulte, and Mathias Pacher

Abstract This article presents an artificial hormone system for a completely decentralised realisation of self-organising task allocation. We show tight upper bounds for the real-time behaviour of self-configuration and self-healing. We also present stability criteria and a AHS implementation coded in pure ANSI C together with some real-world scenario test series and results.

Keywords Organic Computing · Organic middleware · Task mapping

1 Introduction

Today's computational systems are growing increasingly complex. They are built from large numbers of heterogeneous processing elements with highly dynamic interaction. Middleware is a common layer in such distributed systems, which manages the cooperation of tasks on the processing elements and hides the distribution to the application. It is responsible for seamless task interaction on distributed hardware. As shown in Fig. 1, all tasks are interconnected by the middleware layer and are able to operate beyond processing element boundaries like if they would reside on a single hardware platform.

Self-organisation techniques are necessary to handle the complexity of today's and even more tomorrow's distributed systems. Such a system should be able to find a suitable initial configuration by itself, to adapt or optimise itself to changing environmental and internal conditions, to heal itself in case of system failures or to protect itself against attacks. These so-called self-X features are essential for the idea of Organic Computing. Middleware is a good place to implement such self-X features. By autonomously choosing an initial task allocation, which means finding the best initial processing element for each task, middleware can configure the distributed system. By changing the task allocation, middleware can optimise the system in case of changing environmental conditions and heal it in case of processing element or task failures. Especially for self-healing, it is important that these organic features are decentralised to avoid single points of failure.

This work presents an artificial hormone system for task allocation to heterogeneous processing elements. The proposed approach has the following properties:

C. Müller-Schloer et al. (eds.), *Organic Computing — A Paradigm Shift for Complex Systems*, 369–384
Autonomic Systems, DOI 10.1007/978-3-0348-0130-0_24, © Springer Basel AG 2011

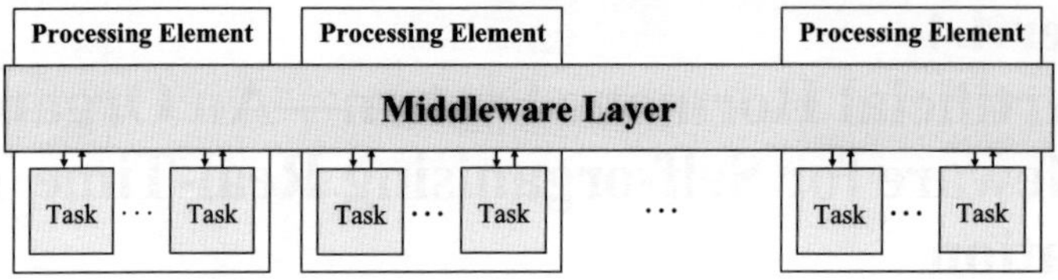

Fig. 1 Middleware in a distributed system

- It is **completely decentralised**. There are no central decision making instances to determine the task allocation. Each processing element (PE) in the heterogeneous distributed system decides according to simple local rules and information received from other processing elements which tasks to take.
- It is **self-organising**. There is no external organisation instance which influences the task allocation. This is done by the interaction of the PEs only.
- It is **self-configuring**. The presented approach determines an initial task allocation, which takes into account the capabilities (e.g. computational power, memory, etc.) and the state (e.g. operation temperature, energy level, etc.) of the heterogeneous PEs.

  The artificial hormone system is also able to respect related tasks (which often have a high communication rate) in order to cluster them close together, thus forming "organs".
- It is **self-optimising**. The task allocation autonomously adapts to changing environmental conditions and states of the PEs (e.g. decreasing energy level, increasing temperature) during operation. Self-optimisation also includes the assignment of newly arriving tasks to PEs.
- It is **self-healing**. Due to the lack of central instances and due to the capability of self-optimisation the presented approach automatically compensates the effects of failing tasks or PEs by reordering the task allocation.
- It is **real-time** capable. There are tight upper time bounds for self-configuration and self-optimisation. This bounds are partially valid for self-healing, too.
- It produces **limited communication overhead**, which is reasonable for embedded applications.

The term "artificial hormone system" has been chosen, because our approach was highly inspired by the hormone system of higher animals. There are several comparable properties between the hormone system in biology and our technical system:

- In biology, chemical signals called messengers or hormones are unspecifically spread to certain regions of the body or the whole body to cause some effects. The messengers (or hormones) of our artificial hormone system are also not addressed to a specific processing element (PE); rather they are spread in the neighbourhood of a processing element or over the whole processor grid.
- The reaction of a cell to a hormone depends on the cell itself. In the same way, the reaction of a PE to a messenger in our system depends only on the specification of the PE itself (see properties mentioned above).

- A PE is able to react to received messages in different ways: It starts, stops, continues or quits the execution of a task. In reaction to this, the PE itself is also able to spread messengers over the system establishing a closed control loop which stabilises the system. Such loops can also be found in nature: By the hormones T3 and T4 of the thyroid a closed loop controlling the temperature of the body is realised.
- Like in the biological hormone system, these closed loops are completely decentralised. As for cells, removing PEs from the loop does not harm the system unless there are enough PEs left to execute tasks and send or receive messengers.
- The hormones of higher animals are reduced by their metabolism, so they are not effective after some time (if no new ones are produced). In our implementation of the artificial hormone system, the effectiveness of the messengers is bounded by time stamps. If not renewed, the messengers of our system expire too.

Note that our "artificial hormone system" is not a copy of the biological hormone system. It has been rather inspired by nature and its strategies. In biology, hormones are chemical objects transmitted via chemical processes and reactions. In our approach, the messengers are bits and bytes transferred via communication links. However, the effects and principles are similar. This is why we have called our messengers hormones as well.

The remainder of this article is organised as follows: In the following Sect. 2 the AHS will be presented in detail. In Sect. 3 we will discuss how to guarantee stability. Section 4 introduces our AHS framework written only in ANSI C. Some results of test series are presented in Sect. 5. Section 6 contains related work. Section 7 concludes this article with a conclusion and an outlook of future work.

2 The Basic Principle of the Artificial Hormone System

For the task allocation of the AHS, only three types of hormones are used:

Eager value: This hormone determines, how suited a PE is to execute a task.
Suppressor: It represses the execution of a task on a PE. Suppressors can be used to limit task execution and to indicate a degrading PE state.
Accelerator: In contrast to the suppressor, the accelerator hormone favours the execution of a task on a PE. They can be used to cluster related or cooperating tasks in the neighbourhood (thus forming organs) or to indicate an improved PE state.

Figure 2 sketches the basic control loop used to assign a task T_i to a processing element. The notation scheme is as follows: $H^{i\gamma}$ means a hormone for task T_i executed on PE_γ and $H_{i\gamma}$ means a hormone from task T_i executed on PE_γ. Hereby, Latin letters are task indices and Greek letters are processing element indices. The closed control loop is executed for every task on every processing element. It determines based on the level of the three hormone types, if a task T_i is executed on a processing element PE_γ or not.

The local static eager value $E_{i\gamma}$ indicates how well task T_i executes on PE_γ. From this value, all suppressors $S^{i\gamma}$ received for task T_i on PE_γ are subtracted and

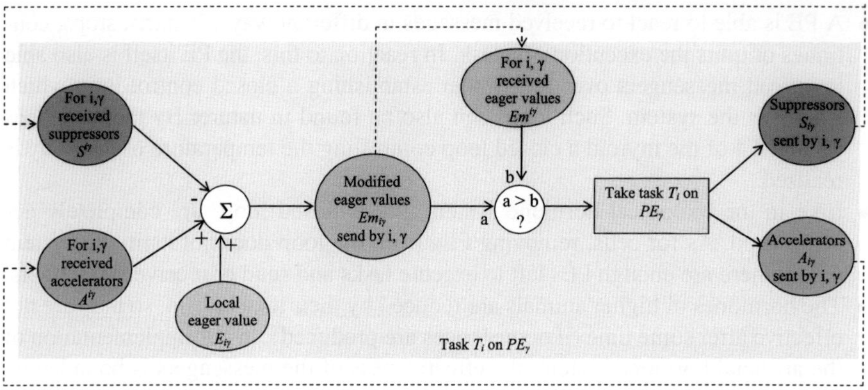

Fig. 2 Hormone based control loop

all accelerators $A^{i\gamma}$ received for task T_i on PE_γ are added. The result of this calculation is a modified eager value $Em_{i\gamma}$ for task T_i on PE_γ. The modified eager value is sent to all other PEs in the system and compared to the modified eager values $Em^{i\gamma}$ received from all other PEs for this task. Is $Em_{i\gamma}$ greater than all received eager values $Em^{i\gamma}$, task T_i will be taken by PE_γ (in case of equality a second criterion, e.g. the position of a PE in the grid, is used to get an unambiguous decision). Now, task T_i on PE_γ sends suppressors $S_{i\gamma}$ and accelerators $A_{i\gamma}$ to the others. This procedure is repeated periodically—each cycle is called a *hormone cycle*. At this point, we emphasise that the initial strength of the hormone values is set by the applicants who want to influence the task allocation. The organic middleware evaluates the hormones to allocate the different tasks, but it does not set their initial strength.

2.1 Different Kinds of Hormones

After the introduction of the hormone cycle of the AHS, we will now show the different kinds of hormones and what they are used for.

We begin explaining the eager values:

Local eager value $E_{i\gamma}$: This value states the initial suitability of PE_γ for task T_i. It ensures that the task allocation is oriented on the capabilities of the PEs.

Modified eager value $E_{i\gamma}^{i\Omega}$: This value is calculated by adding the received accelerators for task T_i on PE_γ and subtracting the received suppressors for task T_i on PE_γ from the local eager value $E_{i\gamma}$. After the calculation, it is sent out to task T_i on all the other PEs.

The following suppressors are used for the artificial hormone system:

Acquisition suppressor $Sa_{i\gamma}^{i\Omega}$: This suppressor is sent to task T_i on all the other PEs in the system, as soon as PE_γ has taken task T_i. Hence, this suppressor determines how often task T_i will be allocated in the overall system. A very strong acquisition suppressor enforces that task T_i is taken only once, while a weaker suppressor enables a multiple allocation of this task.

Load suppressor $Sl_{i\gamma}^{M\gamma}$: This suppressor is sent out only locally to that PE_γ which has taken task T_i. It affects not only task T_i, but all tasks on this PE. Hence, the load suppressor determines how many tasks can be taken by a PE. A very strong load suppressor makes sure, that a PE can only take one task, while a weaker suppressor allows multiple tasks to be allocated on this PE.

Monitoring suppressor $Sm_{M\gamma}^{M\gamma}$: Like the load suppressor, this suppressor is also only sent out locally to a PE by local monitoring and affects all tasks on this PE. With this suppressor, the common state of a PE influences task allocation. The lower e.g. the energy level or the higher the temperature of a PE, the stronger gets this suppressor.

The following kinds of accelerators are used for the artificial hormone system:

Organ accelerator $Ao_{i\gamma}^{V_i\Phi_\gamma}$: This accelerator is sent out to all tasks V_i related to task T_i on the PEs Φ_γ neighboured to PE_γ, if PE_γ has taken task T_i. Hence, this accelerator attracts tasks related to task T_i to settle on the same or neighboured PEs. The stronger the accelerator, the stronger the attraction. The basic idea behind this is that related tasks work on common problems and have to communicate frequently. Therefore, short communication distances are crucial. Related tasks form a kind of virtual organ which works on a bigger problem.

Stay accelerator $As_{i\gamma}^{i\gamma}$: As soon as PE_γ has taken a task T_i, this assignment is initially fixed. It will not be rechecked any longer. This leads to a stable task allocation in the context of self-configuration. But to allow self-optimisation, the possibility of changes in the task allocation is necessary. Therefore, a task assigned to a PE can offer itself periodically for reallocation. To achieve this, the task suspends the transmission of its acquisition suppressor $Sa_{i\gamma}^{i\Omega}$ and starts sending its modified eager value $E_{i\gamma}^{i\Omega}$ again. This enables other PEs to take this task, if they are more suitable in the meanwhile. But as a task migration introduces costs, these additional expenses also have to be considered. The stay accelerator expresses these costs by favouring the stay of task T_i on PE_γ. It is sent from task T_i on PE_γ to itself (i, γ). The stronger the stay accelerator, the better another PE must be suited for task T_i to be able to take it from PE_γ.

Monitoring accelerator $Am_{M\gamma}^{M\gamma}$: This accelerator is sent locally to a PE by local monitoring and affects all tasks on the PE. It is the opponent of the monitoring suppressor. Hence, the local monitoring can strengthen a PE if it is currently very powerful, e.g. due to a high energy level (solar cell in plain sun).

The described approach is completely decentralised, each PE is responsible for its own tasks, the communication to other PEs is realised by a unified hormone concept. Furthermore, it implements the described self-X properties:

- The approach is **self-organising**, because no external influence controls the task allocation.
- It is **self-configuring**, an initial task allocation is found by exchanging hormones. The self-configuration is finished as soon as all modified eager values become

zero meaning no more tasks wants to be taken. This is done by sending suppressors. Of course, these suppressors have to be chosen strong enough to inhibit an infinite task assignment (the suppressors must be stronger then the accelerators), otherwise the system would become unstable.

- The **self-optimisation** is done by offering tasks. The point of time for such an offer is determined by the task respectively the PE itself. It can be done periodically or at a point in time where the task or the PE is idle. Furthermore, an offered task continues its operation on the old PE as long as it is not taken by a new PE.
- The approach is **self-healing**, in case of a task or PE failure all related hormones are no longer sent, especially the acquisition suppressors. This initiates an automatic reassignment of the task to the same PE (if it is still active) or another PE. The only additional requirement is a hormone $H_{i\gamma}^{j\delta}$ sent from task T_i on PE_γ to task T_j on PE_δ has an expiration time. If task T_j on PE_δ receives no new hormone value within this expiration time, the old value is discarded. This enables the detection of missing hormones after the expiration time.

2.2 Constraints of the Artificial Hormone System

In the following, we will shortly present the results of our theoretical analysis of timing behaviour and constraints of the AHS. A detailed discussion of the real-time behaviour, especially of upper time bounds for self-configuration, self-optimisation and self-healing as well as an analysis of the communication overhead can be found in [3].

To guarantee a correct behaviour of the AHS, the following equation must be true:

$$t_C \geq 2t_{DS} + 2t_K$$
$$\text{with } t_{DS} \to 0 : t_C \geq 2t_K$$

With t_C being the time one hormone cycle takes, t_{DS} being the time between the decision making and the sending out of hormones, and t_K being the maximal communication distance.

As an upper bound for the timing behaviour of the self-configuration, we found the following Worst Case Timing Behaviour (WCTB):

$$\text{WCTB (standard)} = 2m - 1 \text{ cycles}$$
$$\text{WCTB (refined)} = m_{max} + v_{max} \text{ cycles}$$

With m being the number of tasks to be allocated by the AHS, m_{max} being the largest number of tasks a PE is applying for, and v_{max} being the largest number of related tasks of a task. Due to the first formula, with longer communication distances the cycle time will of cause increase.

As an upper bound for self-healing, we came up with the following equation:

$$\text{WCEB} = e_\gamma + \max_{T_i \in E_\gamma} \{v_i\} + a \text{ cycles}$$

With e_γ being the number of tasks that were lost due to the failure, $\max_{T_i \in E_\gamma} \{v_i\}$ being the greatest number of tasks related to tasks that were lost, and a being the number of additional cycles until former suppressors have faced away.

3 Stability Analysis of the AHS

A variety of parameters and possible configurations can be seen that can influence the outcome and the stability of the task mapping. First, there are the hormones themselves: the eager values, the accelerators, and the suppressors. Apart from the already mentioned hormones above, there can be additional local types of accelerators and suppressors for system monitoring, load indication, etc. (see Sect. 2.1). Second, there are the degrees of relationship of the tasks. Of course, the whole scenario also depends on the number of (different) tasks and the number of processing elements. We define stability as follows: the AHS is stable iff only a finite number of tasks are mapped. To be able to determine bounds we start with a very strict configuration with only a few variable parameters and will drop these limitations further and further to get to a generalised approach. In the following, we will only present the results—detailed derivations can be found in [4].

3.1 AHS Stability Without Accelerators

With a scenario where the AHS uses only suppressors and eager values but no accelerators, we can show the following stability criteria and give an upper bound for the task allocation:

If $sup_{\min} > 0$ and $sup_{\max} > 0$:

$$\left\lceil \frac{ev_{\max} - lsup_{\max}}{sup_{\max}} \right\rceil \leq n_i \leq \left\lceil \frac{ev_{\max} - lsup_{\min}}{sup_{\min}} \right\rceil$$

With n_i being the number of tasks mapped on PE_i, $ev_{\max}$ being the highest eager value, and $(l)sup_{\min\,|\,\max}$ being the lowest|highest (local) suppressor. (If either side of the inequation below 0, it is to be set to 0.) Holding these conditions, the AHS will work stable and the number of tasks will be bounded by the given limits.

3.2 AHS Stability with Equal Suppressors, Accelerators and Eager Values

When the AHS only uses equal hormone levels for eager values, equal levels for the suppressors, and equal levels for the accelerators, the system has the following stability criteria:

If $sup > (v-1)*acc$:

$$n_i = \left\lceil \frac{ev}{sup - (v-1)*acc} \right\rceil$$

With v being the number of related tasks, and acc being the accelerator value—and the rest as in Sect. 3.1 above. The AHS will work in a stable way as long as the constraint will be met. If not, the AHS will try to allocate related tasks infinitively and therefore be unstable.

3.3 AHS Stability with Varying Hormones

We can draw the following stability criteria for the AHS using different hormone levels for eager values, accelerators, and suppressors:

If $sup_{\min} > (v-1)*acc_{\max}$:

$$\left\lceil \frac{ev_{\max}}{sup_{\max} - (v-1)*acc_{\min}} \right\rceil \le n_i \le \left\lceil \frac{ev_{\max}}{sup_{\min} - (v-1)*acc_{\max}} \right\rceil$$

3.4 AHS Stability with Additional Local Suppressors and Accelerators

With additional local suppressors and accelerators, the following stability criteria can be made:

If $sup_{\min} > (v-1)*acc_{\max}$:

$$\left\lceil \frac{ev_{\max} + lacc_{\min} - lsup_{\max}}{sup_{\max} - (v-1)*acc_{\min}} \right\rceil \le n_i \le \left\lceil \frac{ev_{\max} + lacc_{\max} - lsup_{\min}}{sup_{\min} - (v-1)*acc_{\max}} \right\rceil$$

We took the same notations as in the last subsection. The load suppressor is a local suppressor, its strength grows with the number of tasks which have been allocated by a PE. With enough processing elements, the upper bound will not change, as the load suppressor only applies locally and there are still processing elements left without load. Therefore, with enough processing elements the load suppressor can therefore be neglected in regards to the upper bound.

4 AHS Implementation

The AHS is completely coded in pure ANSI C in a mostly platform independent way. Figure 3 shows the structure of the AHS with the different modules. Only two modules need to be changed when the environment changes—the bottom two modules: AHS Basic OS Support and AHS Basic Communication Support. An easy porting to different kind of target systems like PCs or microcontrollers is possible in this way. Furthermore, the AHS is kept very compact: less than 8000 Total Physical Lines of Code (SLOC) are used. E.g. using Visual C++ as a compiler, the resulting AHS object file to be linked to the application occupies only 22 kBytes of code.

Each node has to execute an instance of the AHS to run the middleware on a distributed system. The implementation of the AHS which is running on every node is structured as follows:

Each AHS instance has to perform several actions. For one it has to run the **control loop** which we described in Sect. 2. It starts with the calculation and sending of the eager values. Then, the hormone values of the other nodes in the system are received. Finally, a decision on local task assignment is made. Then the cycle starts over again. The cycle period can be adjusted to the scenario in which the AHS shall be used (e.g. a cycle-time of 300 ms for a autonomous guided vehicle application, see next section). Also adjustable is the number of cycles for a hormone to "fade out" i.e. to lose it's significance. This property is used by the AHS to be able to realise self-healing. Additionally, fine grained adjustments can be set which change the method the AHS will reevaluate the mapping of the tasks in several ways. AHS cycle period, number of fade out cycles, and fine grained adjustments are set either by an organic monitoring or the application.

Furthermore, the AHS on a node has to **communicate** with the other nodes i.e. sending and receiving messages. The send operation can be a broadcast of the hor-

Fig. 3 AHS structure

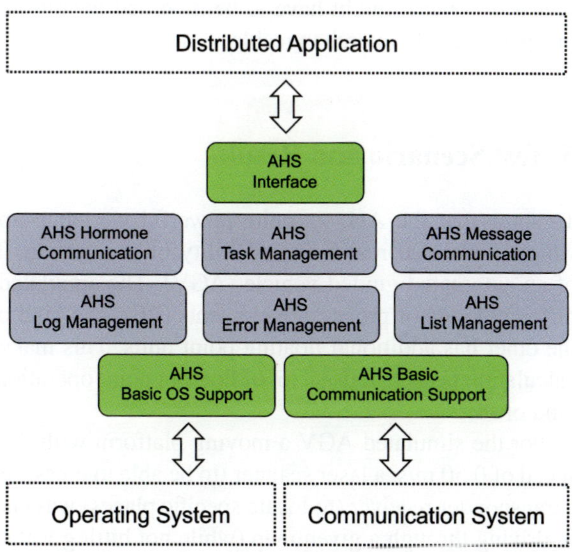

Fig. 4 Variables and their ranges

data type	value range	size
hormone values	0 - 255	8 bit
processor id	0 - 255	8 bit
task id	0 - 255	8 bit
relationship	0 - 10	4 bit
task priority	0 - 255	8 bit

mone, a send only to the neighbours, and of course also the sending to single communication partners. Tasks can run several times on one node hence it is necessary to address the instance of the task when sending respectively receiving. Messages to be sent are hormones as well as application specific data. The hormones types are: eager value, task suppressor, load suppressor, organ accelerator, and offer accelerator. A hormone message consists of the hormone type, the hormone value, the sender and recipient address (network id, processor id, task id), and task specific values (i.e. relationship degree). As there are no real-valued variables, ranges and granularity were chosen as shown in the table in Fig. 4.

The communication part is built in a way that allows the developer to easily change protocols and methods of sending. For our implementation we chose UDP as it is very simple to use and offers broadcast and multicast as well as unicast. Although the AHS normally should have a safe and reliable way to communicate (i.e. each hormone should arrive) the UDP approach worked fine and single lost hormone messages did not corrupt the AHS. We added some redundancy in the communication protocol, to make the AHS more reliable using error prone wireless networks. We repeat each packet twice—one time directly after the first sending of the packet and another time a bit later. Both delay times can be configured. As well, additional repetitions can be configured. This helps to encounter the most common problems in wireless communication: receiving a faulty network packet and the loss of a bundle of packages due to electromagnetic disturbances. This method worked fine in our scenarios—but it might have to be modified for different network scenarios hence this part is easily interchangeable.

5 Test Scenario and Results

To show that the AHS is quite powerful we came up with an example scenario which demanded real-time capability of the system. The scenario is a simulator of an automated guided vehicle (AGV). The simulated AGV is controlled by two different kinds of processing elements (PE): one kind of PE only has integer units, the other has additional floating point units. This makes the latter better suited for calculation tasks that use a lot of floating point operations as they run faster on these kind of PEs.

For the simulated AGV a moving platform with the size of 0.50×0.75 m, top speed of 0.50 m/s, a laser scanner (to be able to measure distances of obstacles), and a transponder receiver (to locate specific places) was chosen. The whole assignment of driving through a given map (while not hitting walls and obstacles) was divided

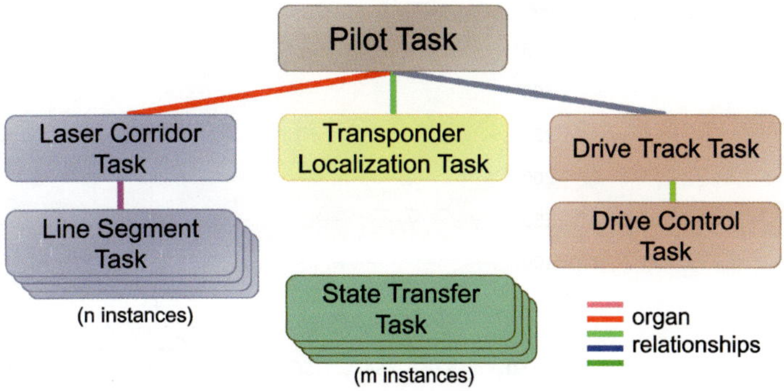

Fig. 5 Tasks and task dependencies of the AGV

Fig. 6 Network load

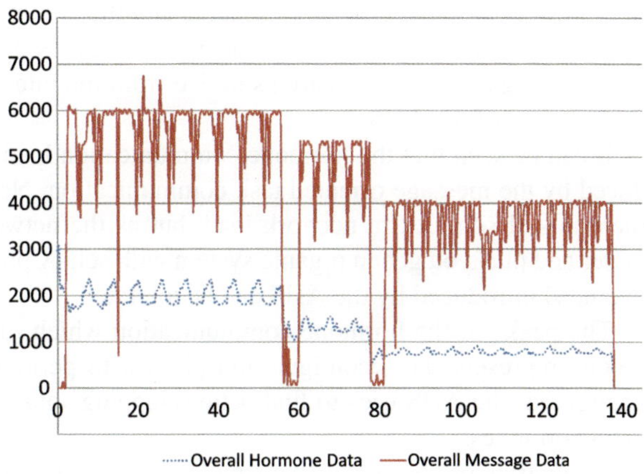

in the following small tasks: *Laser Corridor Task*, *Line Segment Task*, *Pilot Task*, *Drive Track Task*, *Drive Control Task*, *State Transfer Task*, and *Transponder Localisation Task*. Figure 5 illustrates the tasks which are used for the AGV and their dependencies. For the test series we chose a processing power of four PEs consisting of three simple PEs and one floating point unit PE. This scenario is not very challenging for the AHS which is far mightier as we discussed in earlier publications. Nevertheless mapping the 11 tasks on four PEs and providing the communication for the running task shows a real-world application and is therefore well suited. With the speed of the AGV a hormone cycle of 300 ms was chosen to be able to quickly react to changes in the environment.

Figure 6 shows the network activity in the simulated AGV scenario. The blue line displays the sum of all hormones which are sent by the AHS. At the beginning, when the mapping of the tasks is done, the hormone load is about 3.2k messages per second. The upper line represents the messages between the different interacting tasks running on the organic middleware. As the tasks are not mapped yet the network load of the tasks communication is zero. After the primary mapping of all tasks the network load of the AHS drops to around 1.75k messages per second. After

Fig. 7 Self-healing in detail

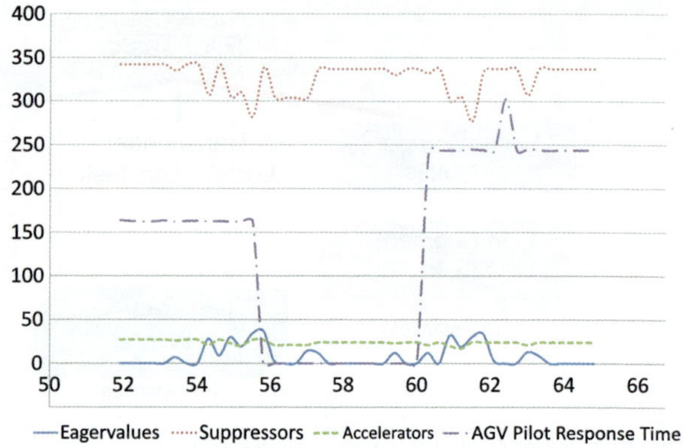

the starting phase the tasks are executed and the communication between the tasks starts. The communication load runs up to 6.8k messages per second and the load is quite irregular. The fluctuations derive from the interactions of the tasks and are independent of the AHS.

It can be seen that the hormone communication load is lower than the load produced by the message oriented task communication. Nevertheless, due to the AHS there is still some extra network load, but as the network capacity is dimensions larger and plus you get an organic system with self-X properties the communication overhead introduced by the AHS can be tolerated.

The peaks of the hormone communication which run up to 2.2k messages per second represent the reconfiguration phases. In periodic intervals (period can be configured) the AHS tries to find better mapping solutions and therefore sends out extra hormones.

At the time of 55 seconds we simulated a failure of one PE. The hormone communication load drops to 1k messages per second. At the same time the communication load of the tasks drops to nearly zero. This happens because the drive control task realises that the real-time demands cannot be met. The AGV stops and waits until the system has reconfigured itself. The self-healing can be seen at the time of 57 seconds when the hormone communication spikes to 1.5k messages per second. At the 60 second mark the AHS has reconfigured, the drive control task starts the AGV and the system is working again. The overall communication dropped because only messages between different PEs are measured. At around 80 seconds after the start we simulate a second failure and again the AHS is able to recover and prove it's self-healing ability.

The reconfiguration can be seen in Fig. 7 in more detail. Eagervalues, Suppressors, and Accelerators have the same colours as before. Additionally, the AGV Pilot Response Time is plotted into the graph with a purple line. The response time was multiplied by 10 for visibility reasons. So at the 52 second mark the AGV Pilot response time is 17 ms. At that time the suppressors are nearly at a level of 350, the accelerators at around 30 and the eager values at 0. When the failure occurs the accelerators increase as the suppressors stop being submitted and fade out. Soon afterwards the real-time time limit is reached and the pilot task stops the AGV. It takes

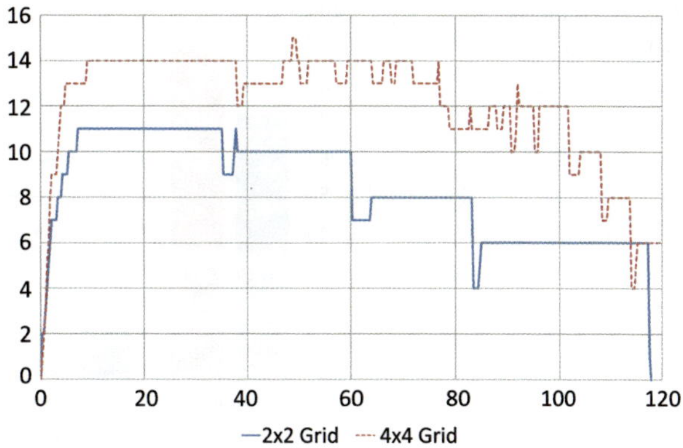

Fig. 8 Timeline of task distribution

about 2 seconds (6 hormone cycles) for the AHS to do the self-healing process and remap the tasks from the broken PE onto a working PE. The pilot task starts the AGV as soon as all the dependent tasks start working and start providing data. During the break-down the AGV stopped for 4.2 seconds. This might seem long but compared to the damages the system took it is quite amazing how fast the whole system could recover. Thanks to the AHS the system self-healed and continued working.

In the final part of this section, we want to show the impact of the numbers of PEs. We therefore made two PE grid configurations: a 2 × 2 PE grid and a 4 × 4 PE grid. In each test we continuously reduced the numbers of PEs by killing PEs—on at a time. In Fig. 8 the timeline of the task distribution can be seen. The blue line represents the 2 × 2 PE grid where PEs were killed at the timestamps: 35.4, 60.3, 83.4. Each time after the loss of a PE the AHS had to self-heal by remapping the tasks which had just died with the PE. After a short settling phase the AHS mapped the tasks again. The same result can be seen with the line which represents the 4 × 4 PE grid. In comparison to the 2 × 2 PE grid, where the AHS mapped 11 tasks, the AHS now mapped 14 tasks namely 3 more State Transfer Tasks. These tasks were additionally mapped because of the excessive supply of PEs. This is also the reason why it takes the AHS with the configuration of a 4 × 4 PE grid longer to startup. Figure 9 shows the time it took the AHS to self-configure with the two configurations and the minimal, average, and maximal time it took the AHS to self-heal from a PE loss. The blue bars represent measurements from the 2 × 2 PE grid and the bars the 4 × 4 PE grid—same colours as used above. As mentioned above, the self-configuration of the 4 × 4 PE grid takes a little bit longer due to the additional tasks. The 4 × 4 PE grid can react faster to PE loss as most of the time an idling or almost idling PE is available—this can be seen at the bars for minimal and average recovery time. But with bad timings the time can also increase—as can be seen at the bars of the maximal recovery time. This increase is however seldom and is limited to a worst-case time of the self-configuration time (when all tasks have to reconfigure). In general a larger system will always be able to react faster to changes and break-downs with self-optimisation and self-healing.

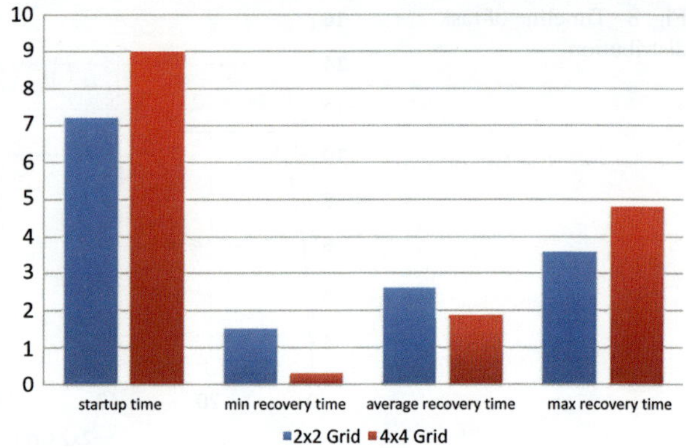

Fig. 9 Timings of the self-healing

6 Related Work

There are several approaches for task allocation in middleware. In [2], the authors present a scheduling algorithm distributing tasks onto a grid. It is implemented in the Xavantes Grid Middleware and arranges the tasks in groups. This approach is completely different from ours because it uses central elements for the grouping: The Group Manager (GM), a Process Manager (PM) and the Activity Managers (AM). Here, the GM is a single point of failure because, if it fails there is no possibility to get group information from this group anymore. In our approach there is no central task distribution instance and therefore no single point of failure can occur.

Another approach is presented in [9]. The authors present two algorithms for task scheduling. The first algorithm, Fast Critical Path (FCP) ensure keeping of time constraints. The second one, Fast Load Balancing (FLB) schedules the tasks so that every processor will be used. Using this strategy—especially the last one—it is not guaranteed that related tasks are scheduled nearby each other. In contrast to our approach, these algorithms do not include the failing of processing elements.

In [8], a decentralised dynamic load balancing approach is presented. Tasks are considered as particles which are influenced by forces like e.g. a load balancing force (results from the load potential) and a communication force (based on the communication intensities between the tasks). In this approach, the tasks are distributed according to the resultant of the different types of forces. A main difference to our approach is that we are able to provide time bounds for the self-configuration. Besides our approach covers self-healing which is absolutely not considered by this decentralised dynamic load balancing.

[10] presents a load balancing scheme for task allocation based on local workpiles (of PEs) storing the tasks to be executed. The authors propose to execute a load balancing algorithm between two PEs to balance their workload. The algorithm is executed with a probability inversely proportional to the length of the workpile of a PE. Although this approach is distributed it does not consider aspects like self-healing and real-time constraints.

Other approaches of load balancing are presented in [1, 5–7, 11]. None of them cover the whole spectrum of self-x-properties, task clustering, and real-time conditions like our approach.

7 Conclusion

This article presented the AHS—an artificial hormone system to assign tasks to processing elements within a processor grid. The assignment is completely decentralised and holds self-X-features. The basic principle of the AHS with all its hormones and the hormone cycle was explained. Upper bounds for the self-configuration and self-healing and a stability analysis of the AHS were shown.

After the theoretical part, we presented our implementation of the AHS. This implementation was coded in pure ANSI C to be able to run it on a variety of platforms from microcontrollers to PCs. Test series showed the self-configuration and self-healing in a real-world scenario.

The achieved results show that the AHS is quite powerful despite it's simplicity. The AHS can be used to construct a robust and stable system. The research builds a good foundation and motivates further research (e.g. the use of an Artificial Immune System).

The AHS was also investigate in the scope of a practical example, the DoDORG project, which deals with a grid of processing elements to be organised by an organic middleware using the artificial hormone system (see Chap. 4.4).

References

1. Becker, W.: Dynamische adaptive Lastbalancierung für große, heterogen konkurrierende Anwendungen. Dissertation, Universität Stuttgart, Fakultät Informatik, Dezember 1995
2. Bittencourt, L.F., Madeira, E.R.M., Cicerre, F.R.L., Buzato, L.E.: A path clustering heuristic for scheduling task graphs onto a grid. In: 3rd International Workshop on Middleware for Grid Computing (MGC05), Grenoble, France (2005)
3. Brinkschulte, U., Pacher, M., von Renteln, A.: An artificial hormone system for self-organizing real-time task allocation in organic middleware. In: Würtz, R.P. (ed.) Organic Computing, pp. 261–284. Springer, Berlin (2008)
4. Brinkschulte, U., von Renteln, A.: Analyzing the behaviour of an artificial hormone system for task allocation. In: ATC, Brisbane, QLD, Australia (2009)
5. Decker, T., Diekmann, R., Lüling, R., Monien, B.: Universelles dynamisches task-mapping. In: Konferenzband des PARS'95 Workshops in Stuttgart, PARS-Mitteilung 14, pp. 122–131 (1995)
6. Finke, J., Passino, K.M., Sparks, A.: Cooperative control via task load balancing for networked uninhabited autonomous vehicles. In: 42nd IEEE Conference on Decision and Control, 2003. Proceedings, vol. 1, pp. 31–36 (2003)
7. Finke, J., Passino, K.M., Sparks, A.: Stable task load balancing strategies for cooperative control of networked autonomous air vehicles. In: IEEE Transactions on Control Systems Technology, vol. 14, pp. 789–803 (2006)
8. Heiss, H.-U., Schmitz, M.: Decentralized dynamic load balancing: The particles approach. In: Proc. 8th Int. Symp. on Computer and Information Sciences, Istanbul, Turkey (1993)
9. Radulescu, A., van Gemund, A.J.C.: Fast and effective task scheduling in heterogeneous systems. In: IEEE Computer—9th Heterogeneous Computing Workshop, Cancun, Mexico (2000)
10. Rudolph, L., Slivkin-Allalouf, M., Upfal, E.: A simple load balancing scheme for task allocation in parallel machines. In: ACM Symposium on Parallel Algorithms and Architectures, pp. 237–245 (1991)

11. Xu, C., Lau, F.: Decentralized remapping of data parallel computations with the generalized dimension exchange method. In: Proceedings of Scalable High-Performance Computing Conference, pp. 414–421 (1994)

A. von Renteln (✉) · U. Brinkschulte · M. Pacher
Institute for Computer Science, University of Frankfurt, Frankfurt, Germany
e-mail: renteln@es.cs.uni-frankfurt.de

U. Brinkschulte
e-mail: brinks@es.cs.uni-frankfurt.de

M. Pacher
e-mail: pacher@es.cs.uni-frankfurt.de

Chapter 4.5
ORCA: An Organic Robot Control Architecture

Werner Brockmann, Erik Maehle, Karl-Erwin Grosspietsch, Nils Rosemann, and Bojan Jakimovski

Abstract Mastering complexity is one of the greatest challenges for future dependable information processing systems. Traditional fault tolerance techniques relying on explicit fault models seem to be not sufficient to meet this challenge. During their evolution living organisms have, however, developed very effective and efficient mechanisms like the autonomic nervous system or the immune system to make them adaptive and self-organising. Thus, they are able to cope with anomalies, faults or new unforeseen situations in a safe way. Inspired by these organic principles the control architecture ORCA (Organic Robot Control Architecture) was developed. Its aim is to transfer self-x properties from organic to robotic systems. It is described in this article with a specific focus on the way ORCA deals with dynamically changing uncertainties and anomalies.

Keywords Organic computing · Control architecture · Self-organisation · Self-optimisation · Self-healing

1 Background

In the field of embedded systems, the trend towards more and more complexity persists. Being able to handle this complexity is going to open up many opportunities, e.g. for robotic or automation systems. It is a characteristic of these application areas that the system designers are faced with strong non-linearities, time variance and a tight integration into natural and safety critical environments. But for future embedded systems, the system complexity will increase, e.g. the number of sensors and actuators and thus also the probability of disturbances and faults.

Because a perfect model of the environment, of the outer physical system and of all other agents around is always hard to get, there is a lot of uncertainty in the design process. The embedded system further relies on sensory information in order to fulfil its mission. But basically all sensors are subject to noise and outliers, have limited resolution and can fail. So, any kind of further processing of sensory data is also subject to these uncertainties.

A key point is here that data sources and hence their signals do not have the same degree of uncertainty all the time, but the uncertainties vary over time depending on the environment, faults or simply wear-out. Designing the further signal processing

C. Müller-Schloer et al. (eds.), *Organic Computing — A Paradigm Shift for Complex Systems*, 385–398
Autonomic Systems, DOI 10.1007/978-3-0348-0130-0_25, © Springer Basel AG 2011

in a way which can cope with, e.g. the maximal amount of noise, will result in a too conservative behaviour. But ignoring the uncertainties obviously threatens safety, which is of course not acceptable for most systems, especially such which operate in human environments.

Although the range of possible dynamic interactions of such a system, e.g. a mobile robot, with its environment is hardly predictable, safety has to be ensured all the time. This results in a design problem. Finding a suitable control strategy is hence a great challenge, even for the case of error-free hardware. The key point is to design the whole system architecture in a way which requires neither complete formal models of the underlying physical system, nor of its fault states, nor of its environment. Building such models at design time is usually very time consuming, error prone, and even impossible for dynamically changing, e.g. natural environments. First there are faults or errors (in the classical meaning), i.e., severe damage of hardware, but also software faults due to undiscovered flaws which stem from design errors or gaps and misparametrisations due to the engineering complexity. Suboptimal parameter settings may lead to an undesired or even unsafe system operation. Additionally, sensor readings are subject to noise which introduces ambiguity and uncertainty into the system. And finally, the interaction with the environment may not be the way it is expected to be, i.e. the actuation itself is uncertain or even unsafe.

The approach taken by Organic Computing is to solve this problem at run-time by increasing the system's flexibility and adaptability, i.e. by proper self-x features like self-organisation, self-optimisation and self-healing. In this article, self-organisation denotes the ability to organise the interplay of function modules by local rules in order to achieve an improved global functionality. Self-optimisation denotes the ability of a function module or a set of function modules to systematically improve their performance while in operation and without an external teacher, which is fundamentally linked to autonomous online machine learning. Finally, self-healing concerns the ability of a system to take countermeasures, e.g. change the module interplay structure or re-learn internal behaviour, in order to fulfil and improve the system's mission even in the case of faults, anomalies or severe changes.

This was the starting point of the ORCA project. This article reviews the main architectural results and focuses on dealing with anomalies by the so called health signals and their processing within the *Organic Robot Control Architecture*, called ORCA. Within this compendium, Chap. 5.7 describes the application and experimental evaluation of ORCA on the six legged walking robot OSCAR. In addition, Chap. 3.4 presents a framework for controlled self-optimisation within ORCA.

2 Organic Robot Control Architecture

The main purpose of the Organic Robot Control Architecture (ORCA), whose principal architecture is sketched in Fig. 1, is to ease the engineering process for complex control systems, which encompasses the behaviour for the *normal* cases as well as for cases where *faults*, other anomalies or unforeseen situations are likely

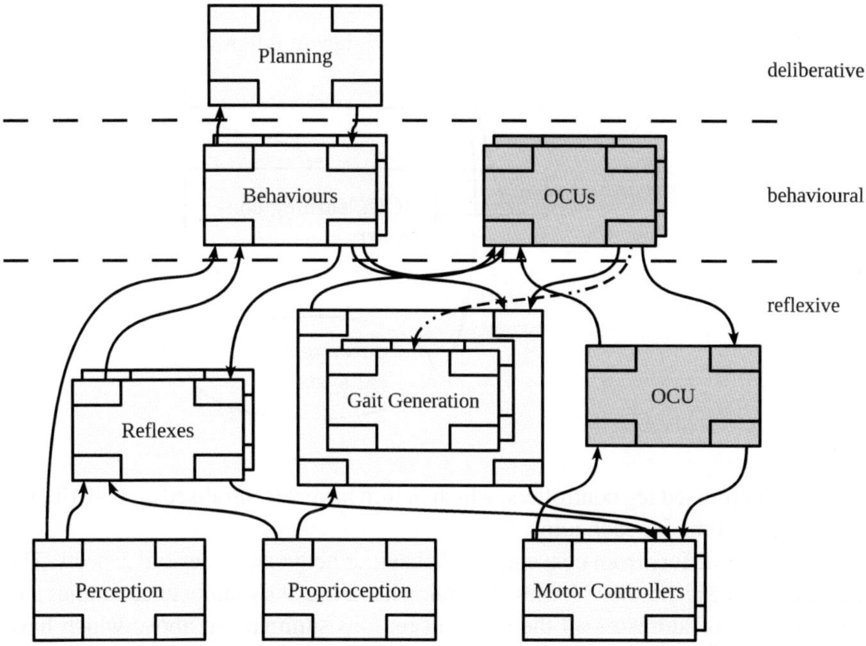

Fig. 1 Schematic representation of the Organic Robot Control Architecture ORCA

to occur. It is a generic architecture which integrates organic principles, like self-organisation, self-optimisation and self-healing.

Therefore, ORCA features a functionality centred approach, similar to behaviour networks [1]. The whole control architecture is hence split into separate interacting modules. These modules cooperate in a data driven way, similar to the Subsumption Architecture [5] and the Modular Controller Architecture [19]. These modules are arranged in a hierarchical fashion into three layers [6], the reflexive, the behavioural and the deliberative layer. Depending on the perspective, one can also use the categories hardware layer, functional layer and decisional layer, see Fig. 1. Hence there is an upstream flow of (enriched sensory) data from the sensors to the planning level, and a downstream flow of (expanded) control data from there to the actuators with shortcuts at every level.

Each module implements basically two functions. The first one is supposed to update the module's internal state based on new sensory data. As soon as every module has taken part in updating the sensory data flow, the second one is performed. Here, the architecturally highest modules, e.g. planning algorithms, starts to update the control data flow, resulting in an update of each module concerning this new overall control state.

The modular structure of ORCA favours decentralisation whenever it is feasible, just like behaviour based approaches given in [1, 19]. For example, the gait pattern of our experimental testbed OSCAR (see Chap. 5.7 and [16]) organises itself by dif-

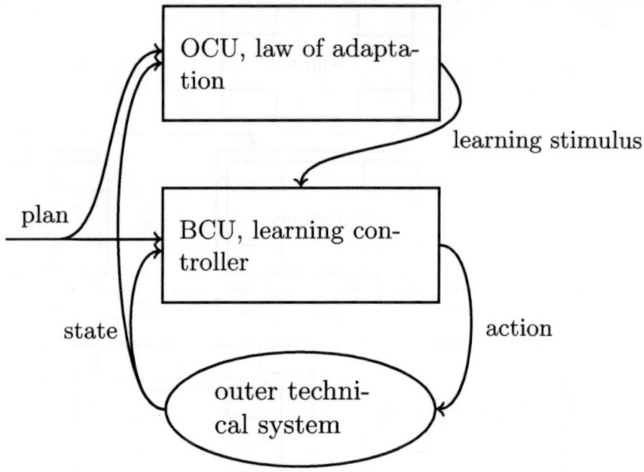

Fig. 2 Schematic representation of the basic self-optimising control loop

ferent, decentralised leg controllers, which in turn have decentralised and distributed control algorithms for each joint.

But ORCA differs from conventional behaviour networks in several aspects. The main difference is that ORCA uses an explicit handling of dynamic uncertainties and anomalies. This addresses all the different reasons summarised above which have an (unwanted) influence on the system, i.e., which should be treated as abnormal. All these kinds of uncertainties are handled by so called *health signals* [12]. They allow to handle anomalies dynamically at run-time as locally as possible and this way to achieve a self-organising and self-healing architecture.

The second difference concerns the system modules themselves. In ORCA, the conventional functional behaviour is implemented by so called *basic control units* (BCU). This means that, in principle, the system, e.g. a robot, can fulfil its tasks or missions just by the BCUs alone. The BCUs are complemented by a second type of module. These modules observe the performance and the dynamic interaction of a single BCU or a group of them. Based on these observations, they can activate or deactivate certain modules, can change module parameters or can change module interaction patterns. In many cases, they also generate, propagate or respond to health signals. Thus, they are responsible for the self-x properties of the ORCA architecture, and are hence called *organic control units* (OCU). It should be emphasised here that an OCU can not only monitor BCUs but also other OCUs. This OCU/BCU architecture can be considered as a variant of the Observer/Controller architecture [17, 18].

The OCUs perform a third task as well. In many crucial aspects, one can only approximately describe the desired functional behaviour of a BCU at design time, e.g. by some heuristic (fuzzy) rules. One can then employ an OCU to monitor and optimise the closed-loop performance of the BCU by adapting its behaviour during the (normal) system operation at run-time by incremental *online* learning.

This self-adaptation mechanism is realised by a local architecture as shown in Fig. 2. It implements another (higher level) feedback loop for OCU/BCU self-optimisation which effectively eases the engineering process if a safe operation can be assured despite the danger of building a chaotic (feedback) loop by online learning. On the one hand, it allows learning from scratch, i.e., to start the whole system with very little a priori knowledge, followed by a phase of self-tuning, thus relieving the design engineer from detailed parameter setting. On the other hand, the system is very flexible and can thus re-learn an optimal behaviour if the underlying physical system, the environment, or the mission changes. But within ORCA, the goal is not just easier engineering for the normal behaviour, but especially the flexible reaction of the system to all kinds of anomalies and system changes. Hence, the OCUs are extended by methods to control the process of self-optimisation as such. For details on controlling the learning process as such and the guidance of its dynamics by *controlled self-optimisation* see Chap. 3.4 of this compendium. Architecturally, the handling of anomalies and uncertainties is performed by health signals, which are addressed by the next section.

3 Health Signal Principles

3.1 Health Signals

Data from a broken sensor are completely uncertain and should be discarded, whereas data from an undisturbed sensor can be fully trusted. The same holds for information processed out of such sensor data. The approach taken within ORCA to tackle such dynamic uncertainties is to attach an additional attribute to each potential source of uncertainty at any point in time, which is called a *health signal*. Its task is to dynamically provide a measure for the estimated uncertainties of the data source.

As a formal approach, we propose to assign a real number to each sensor reading at every time step and to compare the uncertainties by comparing these assigned real numbers. Thus, a health signal for a given data source $a(t)$ with uncertainty is denoted by $h_a(t) \in \mathbb{R}$. As a data source can not be arbitrarily certain or uncertain, all health signals are normalised to the range [0; 1]. Hence, a health signal of 0 corresponds to a completely uncertain, i.e., unhealthy state of the data source, while a health signal of 1 corresponds to a fully trustworthy state, i.e., the data source is working as intended. In between, the health signal values can be used for comparing the uncertainty of one entity at two distinct time steps.

This explicit representation of the uncertainty of a data source naturally extends from sensor readings to other system entities. Thus, the output of a processing module is, of course, uncertain, depending on its inputs, but in addition depending on its internal states or the trustworthiness of the implemented algorithm. It is important to notice here that learnt knowledge in a self-optimising BCU is of course also subject to uncertainties, e.g. when learning from scratch or during re-learning. Health signals are thus a general mechanism to represent uncertainties, anomalies, vagueness, ambiguity as well as faults and errors. But in order to apply this representation throughout the whole architecture, it first has to be clarified how health signals have to be generated consistently, i.e. one needs specific approaches to estimate the uncertainty of a system entity dynamically. Second, often multiple health signals have to be fused into a single one depending on the way their corresponding signals are processed. This means that one needs approaches to automatically estimate the uncertainty of a system entity which depends on several uncertain data sources. And third, health signals have to be incorporated into the behaviour of the system, i.e. one needs (general) algorithms which can be used in BCUs to directly react on health signals.

3.2 Health Signal Generation

In general, all mechanisms for generating health signals work by implicitly or explicitly defining what is *normal*, and then by detecting deviations from this notion of normality. This detection is crucial for the health signal metaphor and has to address any source of uncertainty. Thus many different specific ways to determine health signals are needed, depending on the kind of data source at hand and depending on the knowledge about the source. Most of the following mechanisms can be employed here to define normality flexibly out of the system behaviour itself, i.e. without detailed pre-engineering.

Working range:
: Use an interval to define an expected or allowed working range for the data source at hand, e.g. a sensor, which is normal.

Noise level:
: Use a filter and compare the filtered value of a signal to its unfiltered value. Strong deviations indicate a change in the noise level [2].

Situational accuracy:
: Use a priori knowledge about the variation of the accuracy of a data source depending on the operating condition or system state, e.g. a sensor along its input range. Examples are a distance dependent resolution of a laser range finder or its sensitivity depending on the reflexivity of the environment, and glaring of a camera, or odometric uncertainty depending on the driving speed.

Reference behaviour:
: Memorise the characteristical temporal behaviour of a set of signals for specific situations, e.g. after taking a specific action or when the system is following a known trajectory. Then compare the temporal system behaviour if a typical action or a previous trajectory re-occurs to this reference behaviour [12].

A priori models:
: Use (rough) a priori models about physical or general expected characteristics, e.g. a rise of the torque given a rise of the electric current in a motor.

Predictive models:
: Use a function approximator and have it learn how to predict the next system state based on the current system state and the action taken, i.e. learn a *self-model*. Then, always compare the predicted state with the observed state. This

approach is more flexible than a priori models because of its adaptability. But additional means should be applied which allow the function approximator to *know* about the trustworthiness of its predictions and to incorporate it into learning (and re-learning). Especially, if the prediction of the approximator for a given situation has been correct many times, then a sudden temporal decrease in prediction correctness indicates an anomaly, whereas a steady decrease should be handled by re-learning.

Adaptive filters:

Use an *adaptive filter* [7, 8] to represent the current dynamic behaviour of a certain signal, e.g. a sensor or a servo controller. This way, the dynamics are captured by the filter coefficients. Then, compare the current filter coefficient vector to a set of previously learnt filter coefficient vectors which are known to be normal, or use characteristic patterns of the dynamic evolution of the filter coefficients or the filter output, respectively. Adaptive filters can quickly detect anomalies with a very low computational demand [15].

Mutual information:

Use the statistical notion of mutual information, i.e.

$$I(X,Y) = \sum_{x \in X} \sum_{y \in Y} p(x,y) \log \left(\frac{p(x,y)}{p_1(x) p_2(y)} \right),$$

to determine how much information two signals share. Here, X and Y are the (random) system signals at hand, $p(x,y)$ is their joint probability distribution and $p_1(x)$ and $p_2(y)$ are the marginal distributions of X and Y, respectively. Memorise the value of the mutual information for the normal case. Then, compare the current value, i.e. while in operation, to the memorised one. A deviation indicates the appearance of a new causal effect, i.e. an anomaly [14].

Self-/Non-self discrimination:

Similar to the immune system of organic systems, use a multitude of so called *detectors*. Each individual detector responds to a specific constellation of observable features of the system, e.g. certain sensor values or certain dynamic properties. Then, divide the multitude of detectors into the so called self set, containing all detectors responding to the normal operation, and the non-self set, containing all detectors responding to abnormal conditions. An example is the so called RADE (Robot Anomaly Detection Engine) approach. It uses fuzzy rules as detectors and rule weights to adjust the self and the non-self sets dynamically [9].

Behavioural self-tests:

At a higher level, trigger a so called self-test for a behaviour module which is to be estimated concerning its *health status*. For this to be possible, a global situation has to be brought about for which the dynamic performance of the behaviour module at hand is known and the system is allowed to make some test operations. Thus, one can compare the current performance of the behaviour module with the known one to estimate its health status.

Learning dynamics:

Monitor a self-optimising system module. It consists of two intertwined closed-loop interactions, one between the BCU and the outer technical system, and one between the OCU and the BCU (see Fig. 2). Thus, the learning dynamics between the OCU and the BCU will reflect characteristic properties of the dynamics between the BCU and the outer technical system. Thus, anomalies in any of the two interaction loops can be detected by observing the learning dynamics, e.g. the strength of the learning stimuli, the performance of the learning system or its health state reflecting the current degree of learning. A further specific uncertainty feature is generated by the ODIL approach (Online Diagnosis of Incremental Learning) [3]. It captures violations of certain expected meta-level characteristics within the learnt rule base in an incremental and local way.

The challenge is of course to find the methods which are most suitable for the system at hand. This is strongly influenced by the knowledge about the system, by the expected dynamics, by the criticality of specific anomalies for the success of the system's mission and of course by the available computational resources.

Often, a small deviation from normality is not critical, so it still gets mapped to a health signal of 1. And a strong deviation, i.e., beyond a certain threshold, always gets mapped to a health signal of 0. In between, it is a simple, but practical trade-off to assume a linear dependency of the health signal on the strength of the deviation from normality. This trapezoid mapping is thus a general mechanism for health signal generation.

3.3 Health Signal Fusion

After having generated all relevant health signals, they obviously have to be exploited. Here, one has to distinguish system modules (BCU, OCU) which can directly handle health signals and modules which can not. The latter modules have to work with the potentially unhealthy input values as they are to generate their output values. The key point is then to automatically estimate the health signal of their output values or of the system module itself based on the health of the input signals. This is called *health signal fusion* [12].

In the simplest case, a functional module has one input and one output. Then, the output can not be more certain, i.e. healthier, than the input. But it can be less certain, depending on the actual operation of the module. As an extreme case, the health signal of the output might even be 0 if the health signal of the input signal is less than 1. To a certain degree, this can be extended to more than one input signal. If the algorithm of a module with more than one input critically depends on the normality of all inputs, than the health signal of the output can not be higher than the smallest input health signal. Formally, this is expressed by so called *t-norms* (triangular norms) known from fuzzy set theory. Examples for t-norms are the minimum and the algebraic product. They perform a pessimistic estimation in cases where input signals contain non-redundant information.

But not all functional modules are of this type. For example, a module might depend on the temperature which is estimated by two independent temperature sensors. Intuitively, if both sensor readings have a health signal of less than 1, but agree concerning the temperature value, then the health signal of the module's output can be higher than even the highest one of the two input health signals. Thus, for the case of redundant uncertain information, *compensatory* operators or *t-conorms* can be used.

These operators are part of the general class of fusion operators which can be expressed by the *discrete Choquet integral*, which is widely used in multi-criteria decision making [13]. The Choquet integral fuses n health signals to an overall health signal H which is given by

$$H = \int_0^1 \nu(\{s|h(s) \geq x\})dx. \tag{1}$$

Here, $h(s)$ is the health signal belonging to the input number s. And the set S contains the indices of all n input signals. In addition, $\nu : \mathbb{P}(S) \mapsto [0; 1]$ is a fuzzy measure or capacity and assigns a weight to each subset of S, i.e. each member of the power set $\mathbb{P}(S)$.

As an example, consider the six legged walking robot OSCAR. Assume that after some preprocessing, each leg $s = 1, \ldots, 6$ has been assigned a health signal $h(s)$. The question is how to determine the health signal H of the whole robot. As some redundancy is present, the robot can keep on walking with four legs, if the legs are distributed properly, e.g. by the SIRR approach (Swarm Intelligence for Robot Reconfiguration, [10]) and can still stand with three legs. Hence, a fusion by t-norms and t-conorms is not adequate. Instead, a sample fusion by the Choquet integral is given by (1) with

$$\nu(A) = \begin{cases} 1 & \text{card}(A) \geq 4 \\ 0.5 & \text{card}(A) = 3 \\ 0 & \text{else} \end{cases}.$$

Here, card(A) denotes the number of elements in the set A, i.e. the cardinality. The modelling in this example only takes into account that at least four legs are favoured, no matter how they are distributed on the robot's body. A more advanced modelling could take this into account by providing an improved function ν. It should look at specific indices rather then only the cardinality of the subset which is to be measured. But even in this simple example, the presented health signal fusion can take the intricate redundancy of OSCAR into account, which is hardly possible with less expressive fusion operators.

3.4 Health Signal Processing

ORCA is a data driven architecture, i.e. it can be understood as a function which calculates the output (the motor commands) based on the input (the whole sensory data) and depending on the "flow" of the sensory data through the architecture. In this context, the health signals are generated at the beginning, i.e. along with the sensory input data. During the course of further processing, some modules do not take the health signals into account, so their output health signal is calculated outside of them by health signal fusion. But naturally, some modules take the health signals into account in order to gain the intended benefits. Thus, one needs general algorithms to process health signals and which hence can be applied generically in many functional modules within ORCA.

Based on the semantics of health signals, an input signal with a health signal of 0 is completely unhealthy and uncertain, and must not be used within the calculations. Instead, some fall-back mechanism has to be used which can produce a safe result even without using the uncertain input value. For health signals larger than 0, the influence of the corresponding input value should increase gradually.

A simple thresholding mechanism has been demonstrated in [12]. Here, a four legged walking robot uses health signals generated by a reference behaviour for the electric currents within the servo motors of each joint. These joint health signals are then fused by a t-norm to get a health signal for the whole leg, and finally for the whole robot. Now if this health signal drops below a certain threshold, a reflex is triggered which causes the robot to increase its step height and hence to overcome small obstacles.

A more gradual health signal processing is given by health signal based blending (HS-based blending). In a sample investigation, an automation system for a pick-and-place task is periodically faced with a change of the underlying physical system. It hence has to update its self-model. Depending on the uncertainty about this self-model, the system gradually blends between a fixed and safe, but slow fall-back BCU and a self-optimising BCU. This blending between alternatives is controlled by the health signal of the self-model such that the safe BCU is used if the health signal is 0 and the faster, but more critical BCU if it is 1. In between, both output signals are weighted accordingly and summed. The approach also demonstrates the interplay of health signals and learning systems. It features a reduction of the *learning rate* of the self-optimising BCU depending on the health signal, i.e. to learn more carefully or even stop learning in uncertain situations.

The same mechanism can be used at a higher architectural level by switching or blending between different behaviours. For example, on the walking robot OSCAR, one can use an OCU to decide about the range sensor BCU which is to be used by a BCU for the wall following behaviour. If the health signal of, e.g. the camera drops to a low value, the OCU can re-organise the BCU interplay so that another data source is used, e.g. a set of infra-red sensors.

Another strategy in this context is to use self-organisation, i.e. to re-organise the BCU interplay based on local rules which depend on health signals. As an example, OSCAR can detect that one of its legs is completely broken. The broken leg hampers the whole movement. OSCAR features the ability to trigger a leg amputation in such a case [11]. That means that the broken leg is removed while OSCAR is still in operation, without external assistance. The neighbouring legs can respond to such a broken leg by locally re-organising their movement parameters. This response can be seen as if the legs formed a swarm and self-organise according to swarm mechanisms. This approach is hence termed Swarm Intelligence for Robot Reconfiguration (SIRR) [10]. As a result, a new gait pattern emerges which uses only five legs.

At the highest architectural level, namely the deliberative level, another important response type arises from appropriate planning algorithms. So far, it has only been conceptualised for mobile robots. Based on a map of the environment, an advanced robot planning algorithm will be used to explicitly incorporate the health status of the OSCAR robot. For example, one approach is to virtually block all regions in the map which the robot cannot move through with its current number or power of legs depending on the health signals. Then the planning algorithm tries to find a route to the goal which is not blocked by real or virtual obstacles. The important characteristic is here that health signals are used no longer as attributes, but as normal signals which influence the system operation directly. Hence decisions are made not only based on signals, but also explicitly on health signals.

4 Discussion

To demonstrate the feasibility of the Organic Robot Control Architecture, the six legged walking robot OSCAR was chosen because it shows many practically relevant aspects. The main aspect is, of course, that designing a control system for such a robot is a complex task, as finding formal models for arbitrary environments and all possible fault states is hardly possible. But nonetheless, there is some intrinsic redundancy present due to the number of legs and due to different sensors to detect the environmental characteristics.

As shown in Chap. 5.7 within this compendium and in several experiments on OSCAR [9–11], organic principles can be exploited to design a robust control architecture for such a robot by the ORCA architecture. For this implementation, no formal models had to be found or used. Instead, the robot features a self-organising walking pattern, a self-organising behaviour selection, a self-optimising sensor selection and a high robustness against different kinds of anomalies due to self-healing. This self-healing is an issue of the whole architecture, even though it employs the locality principle. According to this principle, all anomalies are treated as locally as possible. Thus, weak anomalies can often be compensated by local adaptive filters or by local self-optimisation. In other cases, local, event based countermeasures, i.e. reflexes, are appropriate. The higher levels hence only have to deal with global issues, which, e.g. threaten the success of the whole mission.

But besides these robotic results, strong indications have been found that the organic principles have a much wider applicability than only robotic systems. Especially the combination of controlled self-optimisation with the architectural issues of health signals seem to be a promising approach for similar problems in a broader class of automation systems. The basic structural problems, i.e. designing complex control systems for uncertain environments, are basically the same as in robotics. Thus, many aspects of the Organic Robot Control Architecture might be transferred into the domain of general automation systems to form an Organic *Robust* Control Architecture.

It should be noted here that some architectural elements of ORCA are, of course, an implementation of the Observer/Controller architecture (O/C). The BCUs are then the system under observation and control (SuOC). In a certain sense, the OCUs of ORCA are observer and controller at the same time. In ORCA, there are thus many SuOCs with the OCUs as their associated observers and controllers. Level 2 Observer/Controllers are given by an OCU monitoring another OCU. From a theoretical point of view, ORCA is thus as generic as the O/C architecture. It is characterised by a separate OCU architecture which complements the BCUs, which incorporate the basic functionality of the system. The OCUs hence form their own subarchitecture.

Here again, the flow of health signals, which accompany the normal data flow as attributes, can be viewed as forming a separate health signal (generating and processing) architecture. Especially this view on health signals has triggered a new research line called *trust management* which generalises the idea of explicit uncertainty handling and is outlined in the outlook chapter of this compendium, in Chap. 6.1.5. The interrelation of health signals and online learning for self-optimisation and self-adaptation in uncertain environments could only be sketched in this article. For more details on this topic see [4] and Chap. 3.4 in this compendium.

5 Conclusion and Outlook

As a summary of this architectural overview, ORCA is an approach to tackle the design problem of complex control systems by Organic Computing principles. More specifically, self-organisation, self-optimisation and self-healing are achieved by using a 3-layered, decentralised, hierarchical and modular system architecture with two types of modules, namely BCUs and OCUs. Complementing BCUs with OCUs makes the ORCA architecture more flexible than the subsumption architecture [5] and avoids its engineering problems. It also goes beyond behaviour networks and other modular control architectures by combining it with self-optimisation and the health signal mechanism. Hence, a flexible, but dependable operation is achieved by local (reactive) countermeasures, i.e. without explicit fault monitoring and models.

On the one hand, it turned out that the health signal mechanism is more generally applicable to address several types of uncertainties in technical systems. The research will hence proceed to increase their trustworthiness of the operation in uncertain environments. Learning in such environments is a challenge on the other hand because of the stability-plasticity dilemma and because it introduces another source of uncertainty. Ongoing research thus addresses methodological extensions to incorporate uncertainties explicitly into online machine learning principles.

References

1. Albiez, J., Luksch, T., Berns, K., Dillmann, R.: An activation-based behavior control architecture for walking machines. Int. J. Robot. Res. **22**(3–4), 203–211 (2003)
2. Brockmann, W., Buschermöhle, A., Hülsmann, J.: A generic concept to increase the robustness of embedded systems by trust management. In: Proc. IEEE Conf. Systems, Man, and Cybernetics, SMC, pp. 2037–2044 (2010)
3. Brockmann, W., Rosemann, N.: Instantaneous anomaly detection in online learning fuzzy systems. In: Hoffmann, F., Cordón, O., Angelov, P., Klawonn, F. (eds.) 3rd Int. Workshop on Genetic and Evolving Fuzzy Systems, pp. 23–28. IEEE Press, Piscataway (2008)
4. Brockmann, W., Rosemann, N., Lintze, C.: Dynamic rate adaptation in self-adapting real-time control systems. In: Lohweg, V., Niggemann, O. (eds.) Proc. Workshop Machine Learning in Real-Time Applications. Lemgo Series on Industrial Information Technology, vol. 3 (2009)
5. Brooks, R.: A robust layered control system for a mobile robot. IEEE J. Robot. Autom. **2**(1), 14–23 (1986)
6. Gat, E.: On Three-Layer Architectures. In: Kortenkamp, D., Bonasso, R.P., Murphy, R. (eds.) Artificial Intelligence and Mobile Robots, pp. 195–210. MIT Press, Cambridge (1998)
7. Großpietsch, K.-E., Silayeva, T.A.: Organic computing—a new paradigm for achieving self-organized dependable behaviour of complex IT-systems. In: Hoyer, C., Chroust, G. (eds.) Proc. IDIMT 2006 Conf., pp. 127–138. Trauner, Linz (2006)
8. Großpietsch, K.-E., Silayeva, T.A.: Fault monitoring for hybrid systems by means of adaptive filters. In: Doucek, P., Chroust, G., Oskrdal, V. (eds.) Proc. IDIMT 2010 Conf., pp. 177–185. Trauner, Linz (2010)
9. Jakimovski, B., Maehle, E.: Artificial immune system based robot anomaly detection engine for fault tolerant robots. In: Proc. Int. Conf. Autonomic and Trusted Computing, ATC, pp. 177–190. Springer, Berlin (2008)
10. Jakimovski, B., Meyer, B., Maehle, E.: Swarm intelligence for self-reconfiguring walking robot. In: Swarm Intelligence Symposium, SIS. IEEE Press, St. Louis (2008)
11. Jakimovski, B., Meyer, B., Maehle, E.: Self-reconfiguring hexapod robot oscar using organically inspired approaches and innovative robot leg amputation mechanism. In: Int. Conf. Automation, Robotics and Control Systems, ARCS (2009)
12. Kleinlützum, K., Brockmann, W., Rosemann, N.: Modellierung von Anomalien in einer modularen Roboter-Steuerung. In: Berns, K., Luksch, T. (eds.) Autonome Mobile Systeme 2007, pp. 89–95. Springer, Berlin (2007)
13. Klement, E.P., Mesiar, R., Pap, E.: A universal integral as common frame for choquet and sugeno integral. IEEE Trans. Fuzzy Syst. **18**(1), 178–187 (2010)
14. Larionova, S., Jakimovski, B., El Sayed Auf, A., Litza, M., Mösch, F., Maehle, E., Brockmann, W.: Toward a fault tolerant mobile robot: mutual information for monitoring of the robot health status. In: Int. Workshop on Technical Challenges for Dependable Robots in Human Environments, IARP, EURON. IEEE/RAS (2007)
15. Mladenov, M., Mock, M., Großpietsch, K.-E.: Fault monitoring and correction in a walking robot using lms filters. In: Kucera, M., Roth, R., Conti, M. (eds.) Proc. Workshop Int. Solutions in Embedded Systems, pp. 95–104 (2008)
16. Mösch, F., Litza, M., Auf, E.S., Jakimovski, B., Maehle, E., Brockmann, W.: Organic fault-tolerant controller for the walking robot OSCAR. In: Proc. Work. Dependability and Fault Tolerance, ARCS. VDE Verlag GmbH (2007)
17. Müller-Schloer, C., von der Malsburg, C., Würtz, R.P.: Organic computing. Inform.-Spektrum **27**(4), 332–336 (2004)
18. Richter, U., Mnif, M., Branke, J., Müller-Schloer, C., Schmeck, H.: Towards a generic observer/controller architecture for Organic Computing. In: Hochberger, C., Liskowsky, R. (eds.) INFORMATIK 2006—Informatik für Menschen!, pp. 112–119. Köllen, Bonn (2006)
19. Scholl, K.-U., Kepplin, V., Albiez, J., Dillmann, R.: Developing robot prototypes with an expandable modular controller architecture. In: Proc. Int. Conf. Intelligent Autonomous Systems, pp. 67–74 (2000)

W. Brockmann · N. Rosemann (✉)
University of Osnabrück, Osnabrück, Germany
e-mail: nils.rosemann@uni-osnabrueck.de

W. Brockmann
e-mail: werner.brockmann@uni-osnabrueck.de

E. Maehle · K.-E. Grosspietsch · B. Jakimovski
University of Lübeck, Lübeck, Germany

E. Maehle
e-mail: maehle@iti.uni-luebeck.de

K.-E. Grosspietsch
e-mail: karl-erwin.grosspietsch@online.de

B. Jakimovski
e-mail: bojan@iti.uni-luebeck.de

Chapter 4.6
The EPOC Architecture—Enabling Evolution Under Hard Constraints

Moritz Neukirchner, Steffen Stein, and Rolf Ernst

Abstract Organic Computing Systems adapt to changing requirements, environment conditions or failing components. As these external influences are hard to predict, the evolution these systems undergo throughout their lifetime becomes as unpredictable. This however stands in contrast to hard constraints, that the system may have to satisfy e.g. due to safety requirements.

The EPOC architecture aims at bridging the gap between unpredictable evolutionary behaviour and predictability of system properties that are subject to hard system constraints. We show how the principles of a layered contracting architecture allows evolving service guarantees, how separation between an admission control scheme and an execution environment allows safe evaluation of possible evolution steps and how Observer/Controller loops can be used to allow long-term optimisation as well as quick reflexes.

Keywords Real-time · Runtime environment · Contracting · Middleware · Performance analysis · Scheduling analysis · Embedded systems

1 Introduction

Organic Computing has the vision of developing computer system with more "life-like" properties than today's computers [27]. Organic Computing systems shall for example adapt to changes in the environment or be capable of self-healing, i.e. autonomously recover from fault conditions. In general, adding such properties to a given system requires the system to change in functionality and/or structure over time. In some applications, it is acceptable that Organic Computing systems learn in their environment, i.e. they may tentatively perform some change in function or structure to then evaluate whether this has been of advantage for the system.

Systems that have to accommodate applications that require hard guarantees on non-functional properties such as timing, cannot simply apply configuration changes and monitor the resulting behaviour. Instead the system behaviour after the configuration change has to be conservatively estimated and verified beforehand. Only then it can be safely applied without the risk of any constraint violations.

The EPOC project has focused on exactly these kind of systems and has developed a runtime environment implementing a sophisticated admission control

C. Müller-Schloer et al. (eds.), *Organic Computing — A Paradigm Shift for Complex Systems*, 399–412
Autonomic Systems, DOI 10.1007/978-3-0348-0130-0_26,

scheme that enables evolution and adaptation also of systems that are subject to hard constraints. This section describes the architecture of this runtime environment and motivates the design choices taken. Furthermore, it gives insight into the employed admission control scheme and extensions for self-optimisation and self-healing capabilities within the EPOC architecture.

The remainder of this section is structured as follows. First we present the general architectural approach to the design of the runtime environment, which follows three design principles. Then we go into detail on how each of these principles has been incorporated into the runtime environment's architecture to finally conclude the section.

2 Architectural Approach

The EPOC Architecture aims at allowing evolving applications on a shared platform, while providing hard guarantees on real-time behaviour. Thus, it has to bridge the gap between conservative design strategies that allow verification of hard constraints, and novel Organic Computing approaches that allow for self-adapting software components.

The EPOC Architecture follows three major design principles. It is based on a layered software architecture to separate applications from the underlying platform. Admission of the applications to the platform is controlled through contracting interfaces. This principle of layered contracting architectures is mainly known from design processes of systems that require updates throughout their lifetime. The second principle is the separation of the admission control—or *Model Domain*, and of the execution of the actual applications within the *Execution Domain*. Separating both domains allows to verify system configurations model-based prior to their application in the system. Once accepted the hard guarantees of an application can efficiently be enforced within the Execution Domain. The last design principle is that of Observer/Controller loops to allow self-adaptation of the system to changing environment or application requirements. Observer/Controller loops are used in the Model as well as the Execution Domain to allow long-term planning as well as quick reactions.

In the following three sections we will focus on each design principle in more detail and show how it is being applied within the EPOC Architecture.

3 Layered Contracting Architecture

In this section, we will elaborate on the principles of the architecture principle of layered contracting.

Layered software architectures are common in current systems. They abstract implementation details of lower levels to high level applications and thus allow for high portability by means of clearly defined interfaces between the layers. This

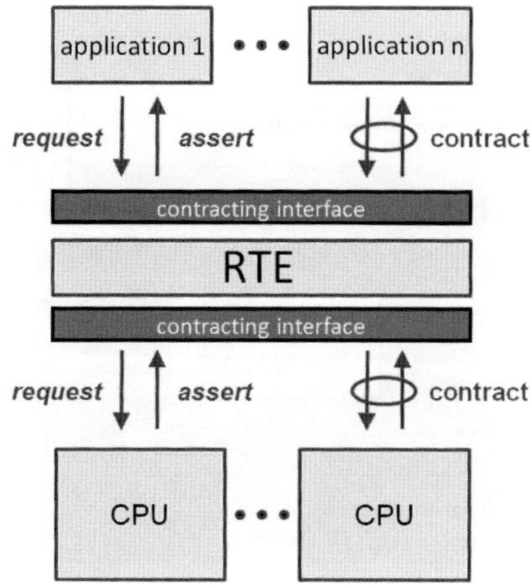

Fig. 1 Architectural approach

also allows replacement of low level components without adaptations of higher levels. This modularity is especially important for self-adapting applications, as often found in Organic Computing systems. If applications or system components modify their own behaviour due to environment influences, the layered approach allows to limit the influence of this change on other parts of the system.

Keeping the interfaces between the layers static throughout the systems lifetime strongly limits the capability of single components to self-adapt. The EPOC architecture uses a *layered contracting architecture* (Fig. 1) to bridge the gap between clearly defined interfaces and component adaption. In this architectural principle a runtime environment (RTE) resides between user applications and platform and controls resource accesses. The RTE and the user applications negotiate the required and provided services in form of *contracts*. Before an application executes on the platform it has to provide a description of its own structure and behaviour (e.g. task graph, communication dependencies) and to request services from the RTE. In our case of real-time applications these requests may include timing constraints such as e.g. worst-case response times or path latencies. The RTE then decides based upon platform capabilities and contracts of other applications whether these services can be granted/guaranteed to the requesting application. If so, it asserts that these services can be provided with the current configuration and thus closes the contract.

In order to provide hard service guarantees every change in configuration has to follow the contracting procedure. The RTE becomes a contract broker that manages service guarantees and thus allows safe evolution of the overall system.

4 Domain Separation

The EPOC architecture aims at providing hard guarantees, while allowing evolution of the system. This is addressed through the contracting architecture introduced in the previous section. This raises the question how these hard guarantees can be obtained.

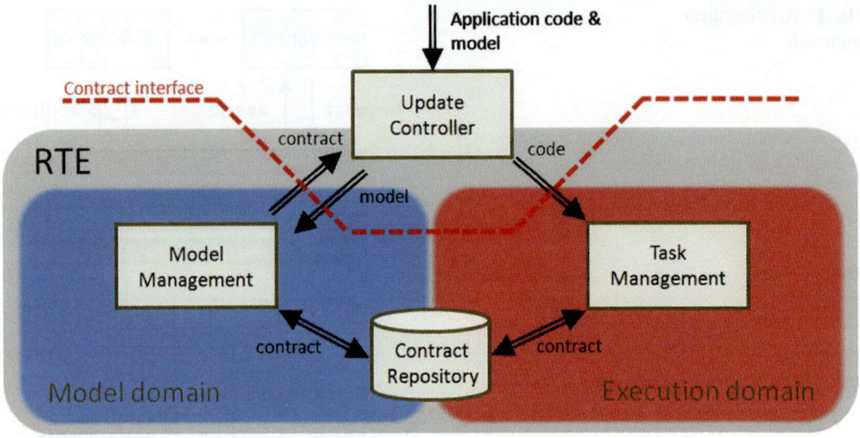

Fig. 2 Framework architecture

Relying on monitoring data of running applications is not an option, as this data may not reflect the worst-case behaviour and thus is not suitable to provide hard guarantees. Instead the RTE of the EPOC architecture relies on a model-based performance verification algorithm [31]. This verification is executed within the RTE upon admission control, to determine feasibility of a configuration change. The overall architecture of this runtime environment is depicted in Fig. 2.

The architecture is divided into a *Model Domain* and an *Execution Domain*. While the former is responsible for contract verification and negotiation, the latter enforces parameter settings and specifications contained in the contracts. New or changed application are inserted into the system via the *Update Controller*. This service request consists of a specification of the application and its constraints (e.g. task graph, scheduling parameters, timing properties and constraints) and the program code of the actual application. First the model is inserted into the *Model Management component* where the effects of the configuration change on the overall system behaviour are analysed via the verification algorithm. If the change is acceptable, a contract containing the application specification and constraints is stored in the contract repository and the Update Controller is notified. The Update Controller then transfers the application code to the *Task Management* component. Task management then reads the contract data from the contract repository and sets all execution parameters according to the model and starts execution of the new configuration. During execution it can monitor whether all applications adhere to their own contract description.

This strict decoupling of model and execution domain has the advantage that system configurations can be evaluated in the model domain without the necessity of loading the actual application onto the embedded system. This limits the temporary impact on required resources, such as memory, on the system to a minimum, in case an application or update is rejected. Furthermore the separation enforces that only verified configurations are executed. As the configuration data is read from the contracts, configuration parameters are forced to conform to the model specification.

Both domains in combination ensure safe evolution of the system, as new configurations can be verified and evaluated w.r.t. to their performance while running applications can be monitored and controlled according to their model.

4.1 Model Domain

The Model Domain performs the contract negotiation based on a performance analysis algorithm. Figure 3 shows the detailed view of the model domain, which further refines the Model Management component from Fig. 2.

The *Contract Negotiation* component receives contract requests from the Update Controller through the Contract Interface. The request contains the model of the requesting application. Contract Negotiation inserts the model into the model-based performance analysis of the *Model Analysis* component. The analysis then determines a conservative estimation of the timing behaviour after the configuration change and reports relevant timing metrics back to Contract Negotiation. Based on these timing metrics and the constraints of all applications, Contract Negotiation determines feasibility of the service request. A closed contract is stored in the Contract Repository and reported to the Update Controller.

This feasibility evaluator can be complemented by a model-based optimisation algorithm (*Model Optimisation*). This enables the EPOC RTE to evaluate the performance of possible configuration changes autonomously. This can be used in an Observer/Controller loop across Contract Negotiation, Model Analysis and Model Optimisation for self-optimisation. This OC loop will be discussed further in Sect. 5.1.

If the admission control process itself is not timing critical, all Model Domain components can be executed on lowest priorities. This way the feasibility evaluation

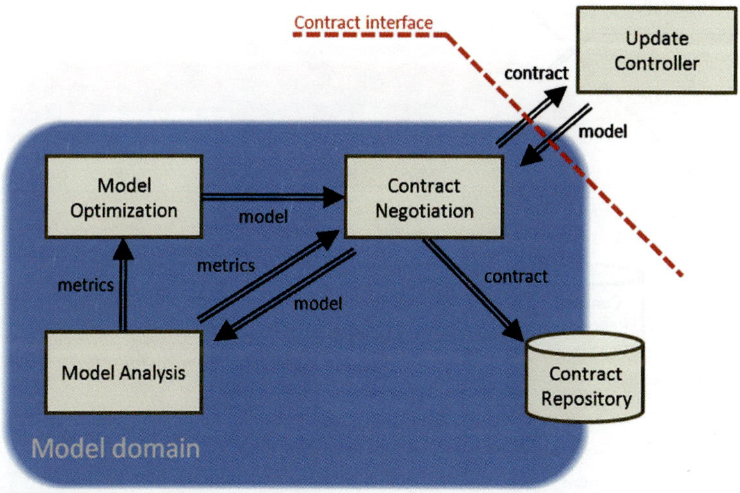

Fig. 3 Detailed view of the model domain

and self-optimisation processes will not interfere with running user applications and will thus have no influence on their real-time behaviour. This allows to use complex verification and optimisation algorithms, while providing tight hard real-time constraints in user applications.

4.2 Execution Domain

This section gives a more detailed view on the Execution domain as introduced in Fig. 2. The Execution Domain provides an execution environment for the software to be run on the EPOC system. Within this domain, the configuration parameters previously evaluated in the model domain are used to configure the execution environment of each single task.

The overall structure of this domain is refined in Fig. 4. The executable code of the user tasks is inserted into the execution domain using a specific network protocol implemented by the contracting interface which is initiated by the update controller. The received tasks and associated data are managed in dedicated memory areas which we call *task slots*. This organisation fosters observability of the tasks' behaviour as will be discussed in Sect. 5.2, where the Observer/Controller loop closed by the Contract Supervision component is described in greater detail.

Once the tasks have been placed in vacant slots, the contract enforcement component reads the contract data from the repository in order to configure the execution parameters of the specific task. This includes provisioning of operating system resources, such as execution stacks and priorities, as well as establishment of communication channels to associated tasks. This procedure ensures that tasks only execute

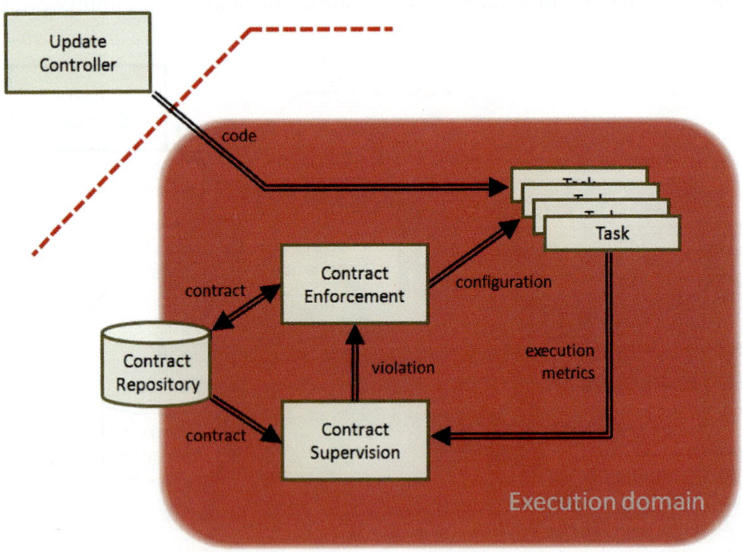

Fig. 4 Detailed view of the execution domain

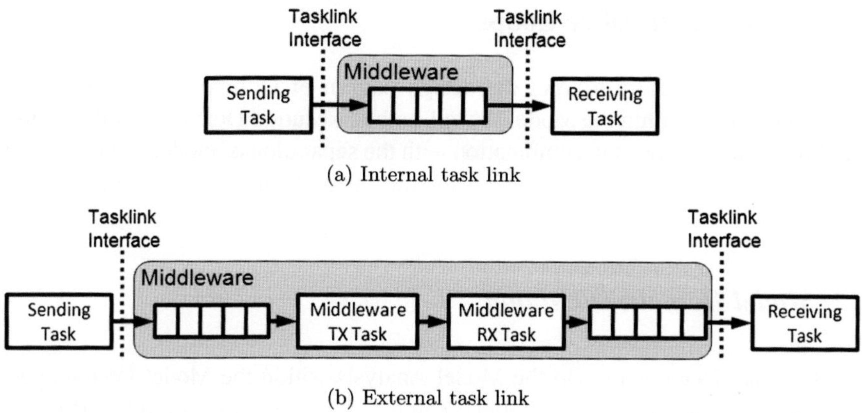

Fig. 5 Task link implementation (sender/receiver)

in a configuration setting that has previously been verified by the admission control scheme implemented in the model domain.

As a second design aspect of the Execution Domain (not shown in Fig. 4) tasks may only communicate using a middleware, which is also configured using contract data. This ensures, that tasks cannot arbitrarily send and receive data without prior analysis of the effects in the model domain.

This middleware couples two tasks by means of a queue as provisioned by the micro kernel (Fig. 5a). In case the tasks run on different processors, this queue is augmented by network communication using communication stacks. The necessary send and receive tasks are part of the Execution Domain (Fig. 5b). This design pattern can be implemented at very low overhead and imposes only few requirements on the operating system, i.e. support for queues and communication stacks. It does, however, allow tight observation and control of the communication behaviour of the tasks.

The presented middleware design leaves the track of applying concepts from general purpose computing also to embedded systems [1, 8, 9, 12, 28, 30, 32, 36, 37] to avoid the overhead incurred [29], and instead offers communication services similar to the AUTOSAR specification [4], which has been designed for use by design time tools, but enables efficient implementation of communication relationships. This is enabled by the contracting principle, which limits dynamism in the system to few, well-defined points in time.

Although the admission control scheme achieved by combining Model and Execution Domain is very powerful, the overhead during execution remains minimal, as the contracting protocol only needs to be employed during an update. During normal operation of the system, tasks are run on a very efficient operating system that only needs to provide the basic services common in embedded systems, such as locking.

5 Observer/Controller Loops

The EPOC architecture follows the design pattern of Observer/Controller loops (O/C-loops) [24] within the Model as well as the Execution Domain. In this section we show how O/C-loops in combination with the separation of model and execution allow for long term planning as well as monitoring and quick reaction to failures.

5.1 Model Domain O/C-Loop

The Contract Negotiation and the Model Analysis within the Model Domain perform the admission control as outlined before. This admission control scheme allows to protect the system from infeasible configuration changes.

This self-protection can be extended by the introduction of an optimisation component, as shown in Fig. 3, to form an O/C-loop. The loop is formed by the analysis component as the observer, providing complex metrics about the expected timing behaviour of the system and the optimisation component as the controller, which may change the model, so that timing properties change in a desired way.

To follow the distributed nature of the targeted embedded systems, both the model analysis as well as optimisation are implemented distributedly. Thus the components form a collaborating swarm of O/C-loop entities, each controlling only a part of the system model.

Thus, the key components in this O/C-loop are the model analysis and the optimisation component. In the remainder of this section we give a short introduction into the algorithms employed in the EPOC framework.

Observer—Model Analysis

The model analysis component in the EPOC framework is based on the basic theory of *Compositional Performance Analysis* [17, 18, 26, 33], which is concerned with the analysis of worst-case timing behaviour of a given real-time system. As a model, it relies on task graphs of the applications running on the system, which are annotated with worst-case execution times and activation patterns at the system inputs. The first can be derived by e.g. static code analysis [35] or simulation, whereas the latter is oftentimes given by the system setup, e.g. sensors delivering data at a given rate. In compositional performance analysis, activation patterns are described by η-functions, which define the minimum (η^-) and maximum number of events that may arrive in a given time window Δt. Usually, these functions are parametrised by a periodicity $\mathcal{P}$ and a jitter $\mathcal{J}$ by which events may deviate from the periodic behaviour [14, 26].

From this data, performance analysis derives system timing properties. This includes worst-case response times of tasks or task chains ("end-to-end latency"), as well as the output jitter of tasks. These can be compared to given constraints on latency or input jitter of subsequent system components to determine feasibility of a system setup.

In case a system is deemed infeasible, this is reported to the controller, in this case the optimisation component, to find a configuration under which all system constraints are satisfied—thus rendering it feasible.

Controller—Model Optimisation

The model optimisation component of the EPOC framework aims at exploring the system configuration space within the model. Timing metrics obtained from the model analysis component guide this exploration process. This way, system configurations can be optimised w.r.t. real-time properties.

Optimisation techniques for real-time systems are well-known from the fields of scheduling analysis [2, 3, 5, 10, 11, 19, 21–23, 25] and performance analysis [13, 16]. In the scope of the EPOC project we focus on assignment of execution priorities under constraints on end-to-end path latency. This shall be used to transition system setups, that are deemed infeasible by the admission control process, into the feasible region. While some of the above approaches are able to solve this problem effectively, the realisation within a distributed RTE becomes difficult due to distribution of the underlying performance analysis and due to resource limitations.

The optimisation component used in the EPOC RTE uses a lightweight distributed heuristic algorithm in an iterative process. Based on the metrics of worst-case task response times, worst-case path latencies and the associated path latency constraint, scheduling parameters of each individual task are calculated. Then, the modified model is analysed again through the observer component. If the modified model still does not satisfy all system constraints, additional optimisation steps are performed.

While this algorithm is capable of finding a feasible configuration for an infeasible system setup, the architecture is general enough to employ algorithms for more complex target metrics, as meta-charac teristics as robustness against design variations or faults [15].

5.2 Execution Domain O/C-Loop

The O/C-loop of the Execution Domain is formed by the Contract Supervision Component and the Contract Enforcement Component (Fig. 4).

In the EPOC framework, this loop is used to enable quick reactions ("reflexes") to cope with misbehaving tasks of applications. The main goal of the O/C-loop is to ensure isolation and fault containment. In the scope of this project, this means that tasks must not exceed their timing budget (even in case of a fault), as this might also cause other applications to violate their timing requirements contrary to the given contract guarantees.

A secondary goal addressed in the implementation of the EPOC RTE is isolation of tasks from each other as well as from the runtime-environment. Here, we focus on malicious or erroneous memory access patterns which, if not handled, may cause applications or even the runtime environment to fail at arbitrary points in time due to memory corruption.

In this section, we outline the observation techniques enabling monitoring of timing behaviour of tasks as well as our approach to the detection of memory access errors. The latter requires a memory protection unit (MPU) within the processor, which is becoming increasingly common also in the embedded context.

Monitoring Timing Aspects

As described in Sect. 3 contract information provides a complete specification of the allowed timing behaviour, including activation periods, jitter, as well as best-case and worst-case execution times.

The framework presented in this article allows to easily check and to enforce adherence of software modules to their own temporal specification. Time-triggered activation of tasks and message transmission on communication channels is readily enforced through the task management component, as the tasks must rely on operating system primitives to achieve periodic activation. These are already under the control of the runtime environment and thus do not need to be monitored. Similarly, the transmission time on communication resources is directly enforced through the middleware. Communication channels are set up with communication buffers according to the contract description. Only data within these communication buffers is transmitted. Consequently a task cannot transmit longer messages than specified.

Execution times of tasks, the activation period and jitter of tasks and communication, that are activated through external events can easily be monitored by means of watchdog timers [34], timed monitoring tasks [7], or heartbeat monitors [6]. The micro kernel used in our implementation [20] automatically monitors execution time of scheduled tasks providing an easy-to-use interface to gather this data.

Thus, all relevant timing metrics can be monitored using state-of-the-art methods to achieve a sufficient amount of isolation between the tasks to ensure adherence to the contracts at all times.

Monitoring Memory Access Patterns

To also achieve protection against failures due to corrupted memory, we also implemented a scheme capable of detecting accesses to memory that has not been allocated to the currently running task. This is achieved by the use of a memory protection unit (MPU), and is enabled by the special memory organisation in the Execution Domain.

The memory organisation is depicted in Fig. 6. The lower memory areas are reserved for the code of the runtime environment and space for shared and global

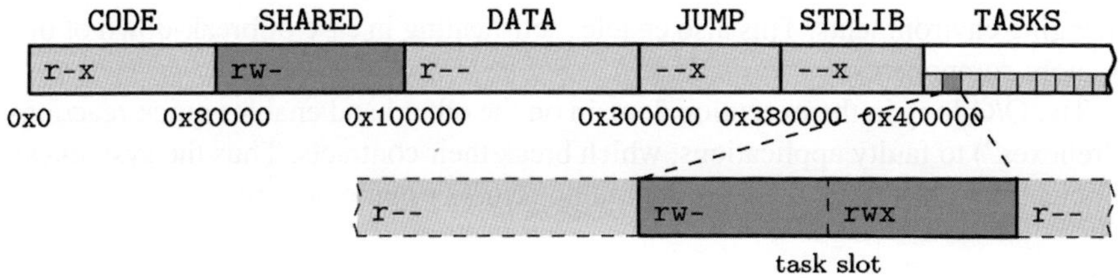

Fig. 6 Memory alignment for memory protection

variables of the RTE (CODE, SHARED, DATA). This is followed by memory areas for code that serves as entry points to the RTE services for application tasks (JUMP,STDLIB). These memory regions are followed by the segments for the application tasks—the *task slots*. They are divided into two distinct memory areas for the task's stack (left half) and the remaining data (code, global variables, queue memory—right half).

Using a memory protection unit, these memory regions can be assigned an access permission scheme depending on the currently running task. The figure shows the access configuration for the task of an application.

As can be seen from Fig. 6, each task of a user application only has write-access to a limited amount of the memory (dark shaded area). Thus, the memory protection scheme efficiently protects correct tasks from faulty ones, as these cannot corrupt their memory. Allocating the stack at the beginning of a task slot has the advantage that stack overflows, that are otherwise hard to find, can be detected by the memory protection scheme. As stacks on our platform grow downwards and adjacent task slots are not writable by the given task, a growth of the stack over the stack boundary immediately causes a memory access violation.

Note that the memory address 0x0 is not writable in the task context. If memory is zeroed at startup many illegal accesses due to uninitialised pointers can be identified. This allows efficient detection and thus handling of erroneous access patterns during run time.

Using these two advanced detection schemes, potential violations of contracts can efficiently be detected by the runtime environment. Together with a fitting policy, this ensures that even in the presence of faulty tasks all contracts (except for the one of the failing application) are satisfied at all times. Other control policies, allowing critical applications to break their contracts, while accepting that less critical applications do not get the service specified in their contracts are also possible, but have not yet been exploited in this project.

5.3 Long-Term Evolution and Quick Reflexes

The two O/C-loops presented in the previous paragraphs allow the resulting system to exhibit two key properties of Organic Computing systems. The O/C-loop in the model domain allows for long-term self-optimisation and thus self-adaptation to changing environments. This also enables self-healing in case of break-down of one or more components.

The O/C-loop in the execution domain on the other hand enables quick reactions ("reflexes") to faulty applications, which break their contracts. Thus the system can efficiently protect itself from programming errors or malicious software.

6 Conclusion

In this article, we have shown how the EPOC architecture allows evolution of Organic Computing systems while providing hard guarantees to user applications. The three main architectural characteristics are a layered contracting architecture, the separation of the model domain and the execution domain and the use of Observer/Controller loops within both domains.

The layered contracting architecture provides an interface to user applications to request services from the runtime environment which has the exclusive right to grant access to system resources. This allows to request configuration changes, while at the same time retaining control over the platform. The separation of model and execution domain ensures that the model-based admission control scheme can run as background service. This enables the system to execute complex verification and optimisation algorithms while providing real-time services to user applications. Finally, the use of Observer/Controller loops in the model domain and execution domain allow to perform high-level optimisation and long-term planning in the model, but also to quickly react to faulty behaviour within the execution domain.

References

1. Abdelzaher, T., Dawson, S., Feng, W.-C., Jahanian, F., Johnson, S., Mehra, A., Mitton, T., Shaikh, A., Shin, K., Wang, Z., Zou, H., Bjorkland, M., Marron, P.: Armada middleware and communication services. Real-Time Syst. **16**, 127–153 (1999)
2. Andersson, B.: Global static-priority preemptive multiprocessor scheduling with utilization bound 38. In: Principles of Distributed Systems. Springer, Berlin (2008)
3. Audsley, N.C., Burns, A., Richardson, M.F., Wellings, A.J.: Hard real-time scheduling: The deadline-monotonic approach. In: Proc. IEEE Workshop on Real-Time Operating Systems and Software (1991)
4. AUTOSAR—Automotive Open System Architecture. http://www.autosar.org/
5. Bertogna, M., Cirinei, M., Lipari, G.: New schedulability tests for real-time task sets scheduled by deadline monotonic on multiprocessors. In: Principles of Distributed Systems. Springer, Berlin (2006)
6. Chen, X., Feng, J., Hiller, M., Lauer, V.: Application of software watchdog as a dependability software service for automotive safety relevant systems. In: 37th Annual IEEE/IFIP Int'l. Conf. on Dependable Systems and Networks, 2007. DSN '07 (2007)
7. Chodrow, S.E., Jahanian, F., Donner, M.: Run-time monitoring of real-time systems. In: Proc. of Real-Time Systems Symp. (RTSS) (1991)
8. Chung, P.E., Huang, Y., Yajnik, S., Liang, D., Shih, J.C., Wang, C., Wang, Y.: Dcom and corba side by side, step by step, and layer by layer (1997)

9. Cucinotta, T., Palopoli, L.: QoS Control for Pipelines of Tasks using Multiple Resources. IEEE Trans. Comput. **59**, 416–430 (2010)
10. Davis, R., Burns, A.: Optimal priority assignment for aperiodic tasks with firm deadlines in fixed priority pre-emptive systems. Inf. Process. Lett. **53**, 249–254 (1995)
11. Davis, R.I., Burns, A.: Priority assignment for global fixed priority pre-emptive scheduling in multiprocessor real-time systems. In: 30th IEEE Real-Time Systems Symp. (RTSS) (2009)
12. Gill, C.D., Levine, D.L., Schmidt, D.C.: The design and performance of a real-time corba scheduling service. Real-Time Syst. **20**, 117–154 (2001)
13. Glaß, M., Lukasiewycz, M., Teich, J., Bordoloi, U., Chakraborty, S.: Designing heterogeneous ECU networks via compact architecture encoding and hybrid timing analysis. In: Proc. of the 2009 Design Automation Conference (DAC) (2009)
14. Gresser, K.: Echtzeitnachweis ereignisgesteuerter Realzeitsysteme. PhD thesis, Technische Universität München, Fakultät für Elektrotechnik und Informatik (1993)
15. Hamann, A.: Iterative design space exploration and robustness optimization for embedded systems. PhD thesis, Technische Universität Braunschweig (2008)
16. Hamann, A., Jersak, M., Richter, K., Ernst, R.: A framework for modular analysis and exploration of heterogeneous embedded systems. Real-Time Syst. **33**, 101–137 (2006)
17. Henia, R., Hamann, A., Jersak, M., Racu, R., Richter, K., Ernst, R.: System level performance analysis - the symta/s approach. IEE Proc., Comput. Digit. Tech. **152**, 148–166 (2005)
18. Jersak, M.: Compositional scheduling analysis using standard event models. PhD thesis, Technische Universität Braunschweig (2005)
19. Jonsson, J., Shin, K.G.: Robust adaptive metrics for deadline assignment in distributed hard real-time systems. Real-Time Syst. **23**, 239–271 (2002)
20. Labrosse, J.J.: Microc/os-ii—The Real-Time Kernel, 2nd edn. (2002)
21. Lehoczky, J.P., Ramos-Thuel, S.: An optimal algorithm for scheduling soft-aperiodic tasks in fixed-priority preemptive systems. In: Real-Time Systems Symp. 1992, pp. 110–123 (1992)
22. Leung, J.Y.T., Whitehead, J.: On the complexity of fixed-priority scheduling of periodic, real-time tasks. Perform. Eval. **2**, 237–250 (1982)
23. Liu, C.L., Layland, J.W.: Scheduling algorithms for multiprogramming in a hard-real-time environment. J. ACM **20**, 46–61 (1973)
24. Müller-Schloer, C.: Organic computing—on the feasibility of controlled emergence. In: IEEE/ACM/IFIP Int'l. Conf. on Hardware/Software Codesing and System Synthesis (CODES + ISSS 2004) (2004)
25. Natale, M.D., Stankovic, J.A.: Dynamic end-to-end guarantees in distributed real time systems. In: Real-Time Systems Symp. (1994)
26. Richter, K.: Compositional scheduling analysis using standard event models. PhD thesis, Technical University of Braunschweig, Department of Electrical Engineering and Information Technology (2004)
27. Schmeck, H.: Organic computing—a new vision for distributed embedded systems. In: Eighth IEEE International Symposium on Object-Oriented Real-Time Distributed Computing, 2005. ISORC 2005, pp. 201–203 (2005)
28. Schmidt, D.C.: Middleware for real-time and embedded systems. Commun. ACM **45**(6), 43–48 (2002)
29. Schmidt, D.C., Deshpande, M., O'Ryan, C.: Operating system performance in support of real-time middleware. In: Proc. of the 7th IEEE Workshop (2002)
30. Schmidt, D.C., O'Ryan, C.: Patterns and performance of distributed real-time and embedded publisher/subscriber architectures. J. Syst. Softw. **66**, 213–223 (2003)
31. Stein, S., Hamann, A., Ernst, R.: Real-time property verification in organic computing systems. In: Second Int'l. Symp. on Leveraging Applications of Formal Methods, Verification and Validation (2006)
32. Thomas, A.: Enterprise javabeans technology. Technical report, Sun Microsystems, Inc. (1998)
33. Tindell, K., Clark, J.: Holistic schedulability analysis for distributed hard real-time systems. Microprocess. Microprogram. **40**(2 3), 117–134 (1994). Parallel Processing in Embedded Real-time Systems

34. Tindell, K., Kopetz, H., Wolf, F., Ernst, R.: Safe automotive software development. In: Design, Automation and Test in Europe Conference and Exhibition (2003)
35. Wilhelm, R., Engblom, J., Ermedahl, A., Holsti, N., Thesing, S., Whalley, D., Bernat, G., Ferdinand, C., Heckmann, R., Mitra, T., Mueller, F., Puaut, I., Puschner, P., Staschulat, J., Stenström, P.: The worst-case execution-time problem—overview of methods and survey of tools. ACM Trans. Embed. Comput. Syst. **7**(3), 1–53 (2008)
36. Wollrath, A., Riggs, R., Waldo, J.: A distributed object model for the javatm system. In: Proc. of the 2nd USENIX Conf. on Object-Oriented Technologies (COOTS) (1996)
37. Yau, S., Karim, F., Wang, Y., Wang, B., Gupta, S.: Reconfigurable context-sensitive middleware for pervasive computing. IEEE Pervasive Comput. **1**, 33–40 (2002)

M. Neukirchner (✉) · S. Stein · R. Ernst
Institut für Datentechnik und Kommunikationsnetze, Technische Universität Braunschweig, Braunschweig, Germany
e-mail: neukirchner@ida.ing.tu-bs.de

S. Stein
e-mail: stein@ida.ing.tu-bs.de

R. Ernst
e-mail: ernst@ida.ing.tu-bs.de

Chapter 4.7
Autonomic System on Chip Platform

Abdelmajid Bouajila, Johannes Zeppenfeld, Walter Stechele, Andreas Bernauer, Oliver Bringmann, Wolfgang Rosenstiel, and Andreas Herkersdorf

Abstract CMOS technology downscaling has enabled the integration of more complex algorithms in System-On-Chips (SoCs). Nevertheless, design complexity and deep-submicron related reliability problems are hindering CMOS evolution. Organic Computing is a new research direction addressing these challenges by embedding life-like principles in SoCs. In this article, we present an autonomic SoC (ASoC) platform. It extends the functional SoC components with autonomic elements in order to build a distributed autonomic SoC. We also present how different autonomic components, including a learning classifier based decision system, adapt to changing environments and globally optimise SoC parameters. A simulation based evaluation is then presented.

Keywords Autonomic SoC · Organic computing · Self-x properties · Learning · Architectures

1 Introduction

The evolution of integrated circuits, usually referred to as "Moores Law", is characterised by the doubling of number of transistors per CMOS chip every two to three years. The good news behind this evolution is that progress in CMOS micro- and nanotechnology has over decades been exploited to build more and more powerful generations of microprocessors, memory and ASIC devices for a wide range of applications. Today we are able to manufacture SoC solutions that have a capacity of a few billion transistors. These enormous chip capacities allow integrating an ever-increasing number of heterogeneous functions onto a single chip carrier. The economic drivers behind denser functional chip integration are system cost and power reduction as well as support for new applications and services.

Structured SoC design [1] is the de facto industry standard architecture platform to develop complex application or customer specific ICs (integrated circuits). The objective in platform-based SoC design is to assemble an as large as possible portion of the target system from already existing function blocks (components) which are maintained in so called IP (Intellectual Property) Core Libraries. Only the application- or customer-specific functions, which do not exist in libraries, have to be developed in the conventional standard cell design method. The platform-based

C. Müller-Schloer et al. (eds.), *Organic Computing — A Paradigm Shift for Complex Systems*, 413–425
Autonomic Systems, DOI 10.1007/978-3-0348-0130-0_27,

SoC approach maximises design reuse to significantly shorten the overall system design cycle, reduces development expenses and increases the probability for a "right-first-time" success (in contrast to "first-time-right" after several attempts).

The downside of CMOS capacity scaling is that already today and even more in the future, we have an inherent complexity problem when designing systems of such size. The complexity problem becomes apparent when we set the microprocessor complexity growth in relation to the designer productivity growth. We observe an increasing productivity gap between available capacity and attainable productivity, resulting in increasing development or engineering costs and/or increasing development time cycles. Thus, the capacity problem, which dominated IC progress over decades, has turned into a complexity problem.

As a consequence of increasing chip capacities, the physical MOSFET transistor parameters, i.e. channel length Lmin, transistor width, gate oxide thickness and interconnect width had to shrink by a factor of $\sqrt{2}$ between consecutive CMOS generations. However, scaling down the physical transistor parameters increased the vulnerability against parasitic and stochastic effects. The margin between nominal device operation and device failure due to supply noise, thermal variations, alpha particle radiation and manufacturing process variations are shrinking. As a consequence, manufacturing defects or temporary device failures during operation will further increase when CMOS technology approaches its physical limits in the sub 10 nm Lmin range. Bottom line, this means that besides the inherent complexity problem we also have to deal with an increasing reliability problem in CMOS SoC design. Dealing with these problems requires new design methods and tools to guarantee reliable and robust IC systems in spite of unsafe and faulty functions at the lowest process levels [2, 3].

Another key problem of current SoC designs is their relative rigidness or inflexibility with respect to dealing with failures or changing workload conditions. A physical defect somewhere in the system typically affects the operation and performance of the entire chip. That is true for both hardware and software. Fault tolerant properties have explicitly to be designed into a SoC but are not an integral aspect of the SoC method or IP cores. Thus, the SoC in general has no possibility to replace a malfunctioning element at run-time or to work around a failure. In the scope of increasing SoC complexities and decreasing CMOS reliabilities, SoC drop-outs or performance degradation will become more frequent and yield in a quality problem. Degrading performance, which leads to the breaking of hard real-time constraints, has disastrous implications for safety critical applications, e.g. ABS in a car. For solving these problems we have to revise the whole process of designing SoCs.

In conclusion, the classical, conservative worst-case design approach in IC system design and manufacturing will be no longer feasible. New conceptual approaches to design methods, SoC architectures and EDA design tools are necessary.

In 2001, IBM declared the hardware/software complexity of IT (information technology) systems as the grand challenge for continued progress in this industry [4]. Autonomic or Organic Computing in future IT systems is inspired by the way how nature deals with complexity. For example: The human organism has an autonomic nervous system that fulfils life critical control and supervision functions (like heart beat control, breathing, temperature control through skin functions) without the active and conscious awareness of the human being. Th goal of Autonomic Computing is to integrate such self-x functions (where x stands for -control, -management, -organisation, -healing) into IT equipment such that this equipment attains a higher level of autonomous operation.

Most existing approaches and projects on Organic or Autonomic Computing are targeting at box level systems and designate the operating system and higher layer middleware software to provide the self-x capabilities which are characteristic for autonomous system behaviour. Our focus is on the chip level hardware layer of SoCs, which form the underpinning base technology for all IT infrastructure and computing equipment. Due to CMOS technology evolution, future SoCs will witness a continued exponential increase in transistor capacity resulting in a complexity and reliability problem. Therefore, we propose to rededicate a fraction of the abundant transistor capacity of future SoCs to implement Organic Computing principles for the sake of higher fault tolerance, performance, power efficiency, easier system diagnosis and the capability to autonomously adapt to changing environmental conditions, be it either externally imposed workloads or temperature variations. Such a conceptual shift in the approach to IC design requires a fresh and holistic view on the implications for SoC architecture platforms. In Sect. 2, we present our Autonomic SoC architecture platform. In Sect. 3, we present architectural building blocks of our ASoC platform. A test scenario will be presented in Sect. 4 and conclusion in Sect. 5.

2 Autonomic SoC Architecture

Figure 1(a) shows the proposed ASoC architecture platform. The SoC is split into two logical layers: the functional layer contains the IP components or Functional Elements (FEs). FEs are either general purpose CPUs, memories, on-chip buses, special purpose processing units (PUs) or system and network interfaces as in a conventional, non-autonomic design. The autonomic layer consists of Autonomic Elements (AEs) and an interconnect structure among the AEs. The interconnect structure within the autonomic layer—AE Interconnect—have not to be identical to the on-chip interconnect at the functional layer. Note, although organic enablement of next generation standard IC and ASIC devices represents a major conceptual shift in IC design, the proposed ASoC platform represents a *natural evolution* of today's SoCs. We regard this property as instrumental for gaining industry acceptance into this approach. Major investments have been made into the establishment of IP component libraries at the functional layer. The capability to reuse existing cores as they are and to augment them with corresponding AEs preserves this investment and enables a gradual evolution path towards increasing organic content in SoCs. Furthermore, this strategy allows subsequent IP core generations to dissolve the logical separation between FE and AE IP components and merge both parts, either partially or entirely, when indicated.

Each AE shall contain a monitor (M), an evaluator (E) and an actuator (A). The monitor (also referred to as observer in literature [5]) senses signal or state information from the associated FE. The evaluator merges and processes the locally obtained

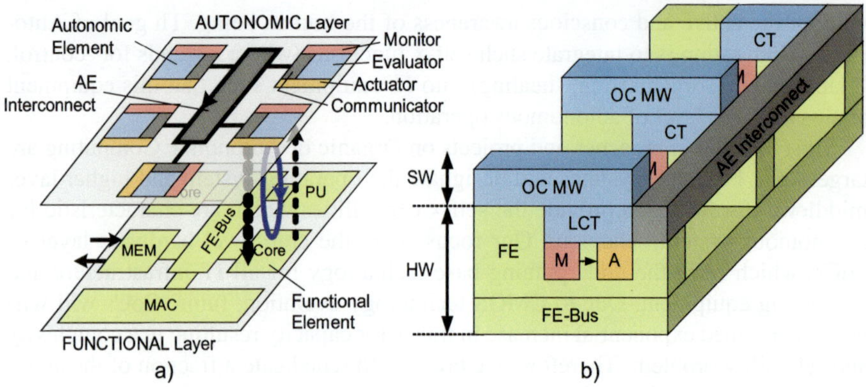

Fig. 1 ASoC Architecture platform. Part (**a**) is a logical 2-layer presentation. The functional layer contains functional components. The upper autonomic layer contains the Autonomic Elements (AE) which supervise the proper and optimal operation of the functional layer. Part (**b**) is a HW/SW function stacking: Monitor (M), Actuator (A), Learning classifier table (LCT), Organic Computing Middleware (OC MW). OC MW of hardware accelerators runs on remote cores

information with the state information obtained from other AEs (coming through the AE interconnect) and/or memorised local knowledge. The actuator executes a possibly necessary action on the local FE. The combined evaluator and actuator can also be considered as a controller. Therefore, each FE and its corresponding AE (monitor + controller) is an *Observer/Controller architecture* [5]. In Fig. 1(a), we see a *control loop* formed by a core (FE) and its AE (Observer/Controller). Hence, our two-layer Autonomic SoC architecture platform can be viewed as *distributed (decentralised) Observer/Controller architecture* [5] which allow communication between different Observers/Controllers.

Let's consider an AE which monitors the reliability and workload of the corresponding FE. This AE will then *decide* which actions the corresponding FE should take in order to optimise its state. AE actions can be split into two classes. The first class are actions which have to be applied rapidly and therefore does not involve the evaluator in the action selection process or give the evaluator very short time for it. For instance in a processor, not initiating a recovery mechanism before a data is sent to an output peripheral means that these wrong data will spread on neighbouring IPs which means that recovery can be much more difficult than correcting the error locally. We presented this fast reaction by a direct arrow going from the monitor to the actuator in Fig. 1(b). Analogy in the human nervous system are the reflexes. The second class of actions are not time critical such as migrating one task from one core to another. This can happen if a core reliability error rate crosses a critical threshold although some undertaken actions (frequency decrease, etc.). In addition to this, it is impossible to simulate all possible operation scenarios of an integrated circuit, the SoC should learn to live under unexpected working environments. Based on this, we have to embed a sophisticated decision system on the AE evaluators. The decision system must be able to learn and adapt itself at run time, while, at the same time, it should be possible to train and instruct it within acceptable time frames at design time. However, it is difficult to find a single decision system which both obeys the restrictions of a hardware implementation and allows quick training and instruction, as the hardware restrictions usually result in a reduced learning rate. Nevertheless, we were able to find a possible solution for the problem of instructing and implementing a hardware decision system: instead of using a single decision system for both run time and design time, we use two much related decision systems. At design time, we use the learning classifier system XCS [6] to learn a rule table, which, at run time, our decision system uses to realise the organic properties of ASoC. The hardware rule table (Learning Classifier Table—LCT) contains condition/action pairs which recommend an appropriate action for a given chip condition. In addition to the evaluator's hardware part (LCT), a software part (Organic Middleware—OC MW) runs complex parts of the learning classifiers such as genetic operators (see Fig. 1(b)). The OC MW runs on the corresponding FE if it is a core otherwise it runs on a remote core. We should mention that as both the XCS and the LCT use the same condition/action pairs, we can use the power of a full-fledged XCS implementation at design time to learn the condition-action pairs that our run-time decision system uses to realise self-correction and self-optimisation in ASoC. More in-depth details of the architecture and implementation of the decision system in ASoC will be detailed under Sect. 3.2.

3 Autonomic SoC Architectural Building Blocks

We present in Sect. 3.1 how to build a self-correcting processor pipeline whose monitors trigger directly an action without the need for an evaluator's decision system. We then present in Sect. 3.2 how the AE evaluator takes more sophisticated decisions—through its learning classifier table—in order to tackle reliability decrease/environment changes. The autonomic interconnect which shares data among the AEs will be presented in Sect. 3.3. All these building blocks are essential to build an autonomic SoC.

3.1 Autonomic Processor Core

As processor cores are a main building block in SoCs, we present how to extend them in order to integrate self-x properties. For the self-healing property, the challenge is towards the growing problem of soft and timing errors. Integrating self-healing mechanisms in software leads to delays in error detection and can lead to many undetected errors (for instance if a critical register got an error). In Fig. 1, an autonomic element monitors the state of the IP and sends actions through its actuator. We still do not have off-the-shelf IPs with monitor/actuator ports that can plug to an autonomic element. Therefore we have to modify the functional IP with monitors towards soft and timing errors and add a recovery mechanism for error correction. One efficient error monitor towards soft- and timing error is the shadow register

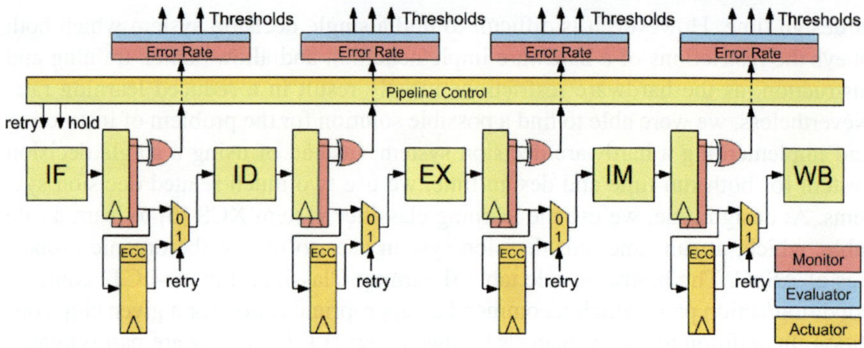

Fig. 2 ASoC Architectural building block: Self-correcting processor pipeline towards soft- and timing errors

technique [7, 8]. Detecting timing and transient errors in a processor pipeline implies the insertion of shadow registers and comparators to each register stage within the data path [8]. However, this simple and effective approach has a performance pitfall: When an error is detected by the comparator, the pipeline input register for the logic stage in which the error occurred has already been overwritten by the next instruction. A re-computation of the previous stage is not possible, requiring correction schemes to flush the pipeline and restart the errant instruction. To overcome this problem, we proposed a new self-correcting CPU scheme that extends the shadow-register detection technique to also correct detected errors (see [9]). For correction, we introduce a customised micro-rollback technique which does not require a complete pipeline flush but incurs a constant pipeline stall penalty of only two cycles. This is accomplished by saving the most recent value of the inter-stage registers in "history registers" (see Fig. 2). With our rollback distance of one clock cycle, a single (history) word register is sufficient in lieu of a costly multi-word FIFO [10]. This makes retry operations feasible without the need to flush the pipeline. It is worth mentioning that a single clock cycle error detection latency is critical for avoiding the complex undoing of write operations to the memory system (in case the error occurred in the memory access stage). An efficient processor pipeline error correcting scheme must therefore react quickly (within one clock cycle) to detected errors. Thus, whenever the monitor detects an error, the actuator initiates a micro-rollback (see signals "hold" and "retry" of the pipeline control in Fig. 2). The evaluator does not intervene in this error recovery mechanism. This is because there is no time left for making complex decisions and that the decision is already known: *whenever a soft- or timing error is detected, do a micro-rollback*. We can see an analogy between this fast reaction and the human nervous system, spine-triggered reflexes (example *if hand touches hot metal, remove it*). The self-correcting pipeline reliability status (see error rate in Fig. 2) is forwarded to the AE evaluator which determines if local correction is enough, if the processor frequency should be decreased or if the processor should be switched off. Such decisions require a sophisticated algorithm and are taken at a higher level than the systematic correction using micro-rollback.

Higher level decisions involve the FE, monitor, evaluator and actuator, as described in the following sections.

We added this autonomic extensions in a LEON3 processor pipeline [11]. Tests were conducted in a Virtex4 FPGA under error injection and the obtained self-correcting processor pipeline detected and corrected them. FPGA slices overhead were of 14% of the Leon3 core.

3.2 AE Evaluator Architecture

The AE evaluators play a central role as the decision makers of the autonomic layer. They receive status information signals from the AE monitors, and use this information to determine appropriate actions to be taken by the AE actuators. As mentioned before, we decided on using reinforcement learning techniques in the form of learning classifier systems (LCS) to provide this evaluation functionality. Whereas a full-fledged extended classifier system (XCS, [12]) is used for design time evaluation, at run time we employ a simplified variant of LCS, which we will refer to as a learning classifier table (LCT). While the simplifications associated with LCT are necessary to allow for an efficient hardware implementation, we will show that they do not hinder the LCT's learning capabilities.

Figure 3 shows the classical structure of an XCS as proposed by Wilson [12] in comparison to the LCT structure in Fig. 4. The environment, monitors and actuators of the XCS correspond to the autonomic system's functional elements, monitors and actuators, respectively. Due to the potentially high rate at which the monitor provides data, an aggregator is inserted to generate a condensed monitor signal. This signal then passes to the learning classifier, where an appropriate action is determined and passed to the actuator. Both the XCS and LCT contain a list of rules that make up the classifier population, where each rule consists of a condition (C), an action (A), and a fitness (F) value. A typical rule may for example state that *if the utilisation is high and the number of errors is low* (condition), then *to increase the frequency* (action). In LCS, all rules with a condition matching the monitor signals are combined into what is referred to as the match set, each rule of which will generally propose a different action. The preferred action is selected from the match set based on the fitness value of the proposing rule. Whereas the fitness value of XCS reflects the accuracy (ϵ) of the rule's reward prediction (ρ), for simplicity the rules in our LCT will directly use the predicted reward as their fitness, similar to XCS' predecessor ZCS [13]. This makes calculating a prediction array based on the matching rules unnecessary.

Numerous methods have been suggested in the literature for selecting a single rule or action from the match set based on its predicted reward. These include always selecting the action with the highest reward prediction, random selection with individual probabilities weighted by their prediction, or completely random selection [12]. Although selecting the rule with the highest prediction may appear to best utilise the momentary knowledge of the classifier, it does not allow for learning, since learning depends on the trial of numerous options to determine the rule

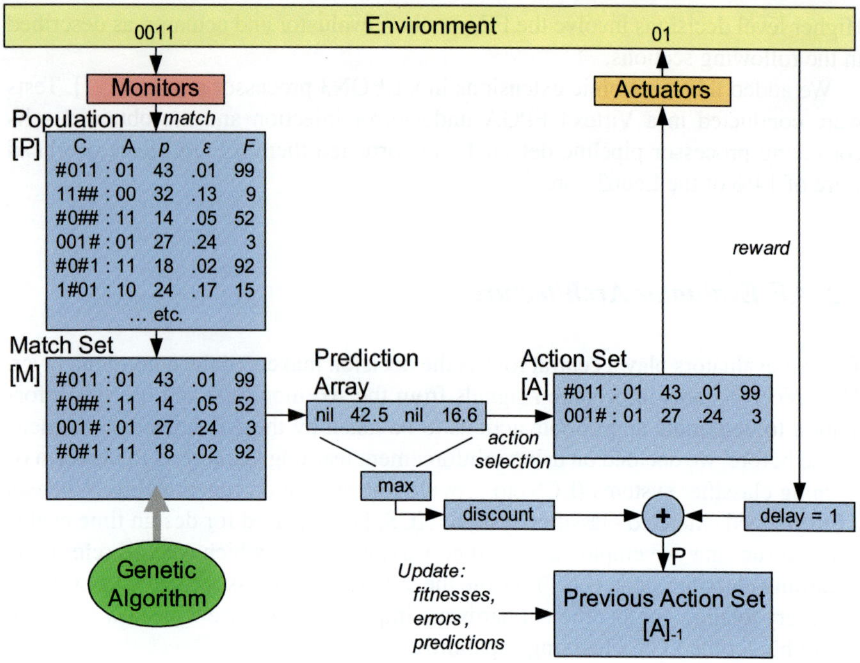

Fig. 3 Classical structure of an XCS as proposed by Wilson [12]

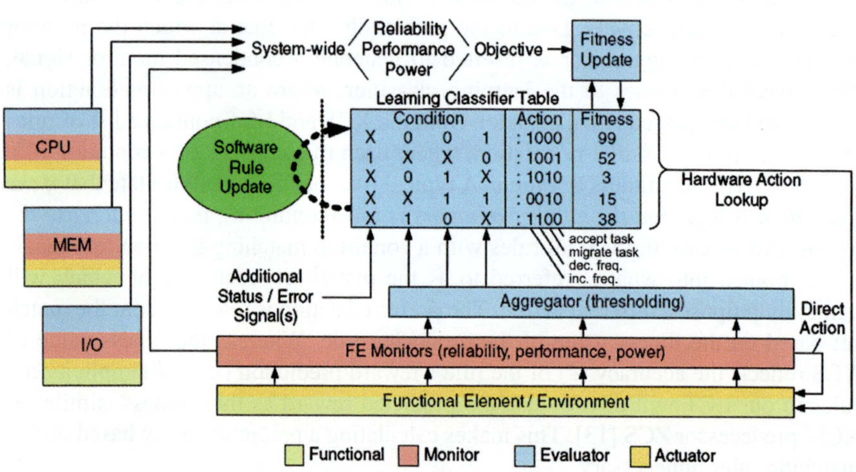

Fig. 4 Autonomic Evaluator based on Learning Classifier Table, LCT

which performs best in a given situation. The opposite extreme, completely random selection, disregards predictions entirely, instead randomly selects any one of the matching rules' actions. In order to balance the exploitation of existing knowledge stored in the classifiers and the exploration of new or previously untested rules,

Fig. 5 Pseudo code for efficient randomised rule selection in hardware

```
∑F = 0
LOOP over all rules
  IF rule matches
    ∑F += rule fitness
    IF Random % ∑F < rule fitness
      act = rule action
  END IF
END LOOP
Perform act
```

we employ a randomised selection scheme, where the selection probability of each rule is weighted by its prediction (commonly known as roulette-wheel selection). Rules with a high prediction have a larger chance of selection, but do not entirely bar the selection of even those rules with a very low prediction. The pseudo code presented in Fig. 5 shows the algorithm used for efficient randomised rule selection in hardware [14], the novelty of which lies in the fact that it includes both the match set "creation" and roulette wheel selection of an appropriate rule from within the classifier population, without the need to first generate an explicit match set by copying matching rules to an intermediary memory (see Fig. 3 where there is a match set whereas in Fig. 4 there is no need for additional match set memory). The algorithm is based on single-pass weighted reservoir sampling [15, 16], with the special case of a reservoir size of one, since only a single rule will be selected for execution. The core concept of the algorithm is to keep track of the total prediction (fitness) sum of any already matched rules, and to weight the fitness of further matching rules against the fitness sum seen so far. This basically translates into a probability for selecting a new matching rule (over the previously selected rule) of the new rule's fitness against the sum of the fitness of all matching rules seen so far, or $P(n) = Fn/\sum Fn$, where Fn is the new rule's fitness and $\sum Fn$ is the sum of the fitness of all matching rules seen so far. This results in a probability of $P(i) = Fi/\sum FN$ for the selection of any rule i, where $\sum FN$ is the sum of the fitness of all matching rules. Besides taking into account the dynamically changing reward prediction during rule selection, genetic operators used to modify the classifier population over time are another important part of learning within an LCS/XCS. Although genetic operators such as cross over and mutation can be readily implemented in hardware, choosing appropriate rules to be crossed or replaced (selection) can be difficult. More importantly, a hardware system must avoid generating rules with unpredictable results at run time, which is difficult to guarantee across essentially random genetic operators. Our current LCT implementation therefore limits its learning capabilities to run-time prediction updates, while allowing for rule updates by mapping the LCT's classifier population into the system's address space. Synthesis results of the LCT in a Xilinx Virtex4 FPGA reveal less than 1.4% overhead compared to a standard LEON3 processor core. Note that this figure only reflects the rule selection algorithm and does not include the fitness update logic, which we do not however expect to cause a significant increase in the numbers presented here.

3.3 Autonomic Element Interconnect

The autonomic element interconnect provides the link necessary for different AEs to communicate with each other. As mentioned in the sections before, the AE evaluator will accept as its input a status vector composed of both local monitor signals and global information regarding the status of other AEs or the system as a whole. Therefore, each AE must send messages to inform other AEs with the states of their corresponding FEs. The question of sharing the same medium between the functional interconnect and the autonomic interconnect or assigning a separate medium arises. It is well-known that SoC designers make a large effort to estimate their functional interconnect performance and to check that timing and priorities are respected in all design steps. Suggesting a new Autonomic SoC architecture that shares the functional interconnect means that all the design/verification steps of the functional SoC have to be redone. Therefore, it becomes clear that separating the autonomic and functional design will decrease drastically the design complexity of our autonomic SoC. The second question when designing the AE interconnect regards the required performance. An AE evaluator requires monitor information from other AEs in around 1 ms time intervals (parameter determined by the AE evaluator learning classifier). This means that our AE interconnect should be able to transport at least 1 message/ms for each AE. The AEs also send requests (e.g. for task migration). In this case the evaluator of the AE sender should wait for all responses of the AEs on the interconnect and after that send an acknowledge to the station which will receive a task. These messages should also complete in a 1 ms time slot *as the time between two actions is set by the AE evaluator to 1 ms*.

These requirements should guide us when making the choice of the interconnect structure. Standard buses like AMBA AHB would be completely over-dimensioned (bandwidth peak up to 5.31 Gbps at 166 MHz). An alternative is to use a serial ring which our estimations proved that it is enough for our application. In order to avoid repetition, we present in the following the performance of our designed serial ring both analytically and from FPGA measurements and which prove that our choice is correct.

An interface unit (AE communicator) connects each AE to the ring (see Fig. 1 and Fig. 6). It has a serial input from the previous station in the ring and a serial output to the next station. The AE communicator is able to receive, insert and remove data from the ring. Data from the AE is transmitted on the ring when a request to

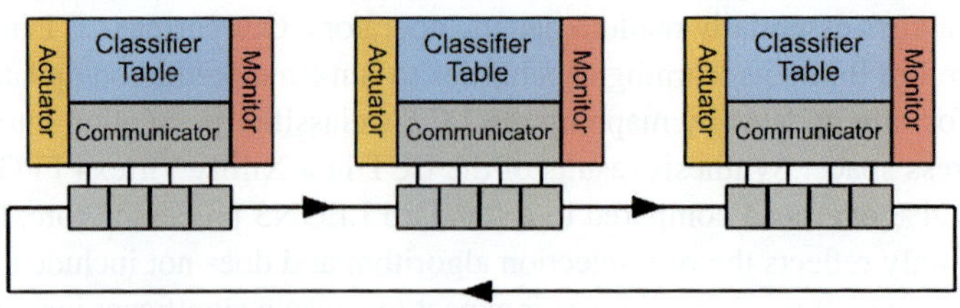

Fig. 6 Serial ring structure for the AE interconnect

send is applied and a free token (permission to send) has been received. Data from other AEs addressed to the AE station is directly copied by the AE communicator to the attached AE. The data to be exchanged can be a data frame or a token. The frame type field is used to distinguish between the token and data frame. Besides this, the frame contains the destination and source address of the message. The frame payload field has a variable size adjustable during run-time. The payload field is followed by a CRC checksum of the message and a 4-bit frame control field to carry acknowledgement information from the frame receiver to its sender. If we set the payload to 32 bit and if there are 8 AEs in the serial ring. The maximum round trip latency (when all stations are requesting to send) equals 8*(Token_Size + Frame_Size) in clock cycles. Calculating the token and frame size from the above fields description and with a clock frequency of 50 MHz, we get a worst round trip of 15.04 μs (752 clock cycles). This translates to 66 messages/ms for each AE. This is much more than the design requirement of being able to send 1 message/ms for each AE. FPGA synthesis in a Virtex4 FPGA gives a resource utilisation of 333 slices per AE communicator.

4 ASoC Evaluation

In the following, ASoC is evaluated using a multi-core case study—a video frame processing application distributed on two cores. The simulation model used here consists of two processor cores, a shared bus and a shared memory. An Input/Output interface provides a constant workload to the system which is evenly shared between the two cores. When one core is switched off (basically mimicking a core failure), the second adapts according to a pre-defined objective in the LCT, i.e. to keep utilisation close to fifty percent by varying the frequency.

Simulation results of such a system are presented in Fig. 7. Initially, the LCT requires too much time to properly raise the frequency when the second core fails, resulting in a loss of frames. Towards the end of the simulation, the LCT has learned to maintain a slightly higher average frequency so that the processor is prepared for the periodic doubling in workload. This basically eliminates the number of dropped frames occurring after the first processor switches off. However, since learning requires continual exploration of new territory (in this case system parameter configurations), the LCT continues to adapt the system in order to find even better operating states. This results in occasional dropped frames even towards the end of the simulation. We consider this to be acceptable—in order to learn it is necessary to be allowed to make a few minor mistakes. A more complete evaluation of ASoC concepts based on a network processing application running in an FPGA hardware demonstrator can be found in Chap. 5.3.

5 Conclusion

We have tackled the complexity and reliability problems of current SoC designs and implementations using organic principles. Current and future SoCs integrate an

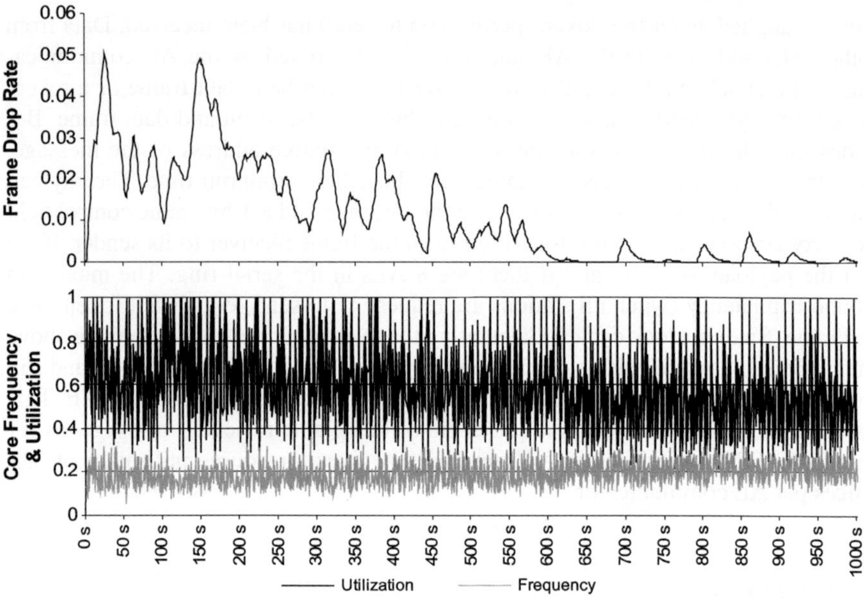

Fig. 7 Multi-core case study, video processing. *Below*: component behaviour, core frequency/utilisation. *Above*: system behaviour, frame drop rate

increasing number of functionalities, making it more and more difficult for today's chip designers to analyse the interaction of different chip components and to choose optimal design parameters for chips, as the production of chips faces increasing variation, the chips operate in changing environments, and they work under varying loads. Besides, current tools to design SoCs do not sufficiently allow modelling and analysing faults and fault tolerant behaviour, which is necessary to address the increased wear out of chips with feature sizes approaching physical limits. Addressing the issues of current and future SoC designs and implementations, we have investigated a flexible hardware architecture that integrates self-x properties and learning capabilities so that it can react to failures and optimise the system under changing environments and varying workloads at run time. We have shown that the learning capabilities of the hardware will lighten the chip designer's burden to specify every single aspect of the design and to find the optimal parameter settings, because the hardware can self-correct and self-optimise at run-time and take decisions that formerly had to be taken at design time. To ease the transition to our new architecture and to preserve the huge investments that have already been made into various IP libraries, we have to reuse as many existing functional elements (FEs) as possible, leaving the functional layer of current SoCs—almost—untouched. To this functional layer, we added a (logical) autonomic layer consisting of autonomic elements (AEs), which take care of the run-time decisions to achieve self-correction and self-optimisation and to realise the organic concepts. The ultimate objective is that IP vendors integrate in their functional IPs (FEs) interfaces to the autonomic elements (AEs). In that case the ASoC design flow will be exactly the same as the IP-based design used today. A central feature of the ASoC architecture is the hardware decision system that is necessary to accomplish self-correction and self-optimisation at run time. We have used an extended classifier system (XCS) for design time evaluation, whereas in hardware we have implemented a simplified variant of LCS in order to keep overheads low. Complex functions of LCS such as genetic operators are planned to be implemented in an organic middleware.

References

1. Sangiovanni-Vincentelli, A.L., Carloni, L.P., De Bernardinis, F., Sgroi, M.: Benefits and challenges for platform-based design. In: DAC 2004, pp. 409–414 (2004)
2. De Micheli, G.: Designing robust systems with uncertain information. In: Asia and South Pacific Design Automation Conference (ASPDAC 03) (2003)
3. Borkar, S.Y.: Designing reliable systems from unreliable components: the challenges of transistor variability and degradation. IEEE MICRO **25**(6), 10–16 (2005)
4. Horn, P.: Autonomic Computing: IBM's Perspective on the State of Information Technology. IBM Corporation, Oct. 2001. http://www.research.ibm.com/autonomic/manifesto
5. Schöler, T., Müller-Schloer, C.: An observer/controller architecture for adaptive reconfigurable stacks. In: ARCS 2005, pp. 139–153 (2005)
6. Wilson, S.W.: Generalisation in the XCS classifier system. In: Koza, J.R., et al. (ed.) Genetic Programming 1998: Proceedings of the Third Annual Conference, pp. 665–674. Morgan Kaufmann, San Francisco (1998)
7. Nicolaidis, M.: Time redundancy based soft-error tolerance to rescue nanometer technologies. In: 17th IEEE VLSI Test Symposium (1999)
8. Ernst, D., et al.: Razor: a low-power pipeline based on circuit- level timing speculation. In: The 36th Annual International Symposium on Microarchitecture (Micro-36), December 2003
9. Bouajila, A., et al.: Organic computing at the system on chip level. In: Proc. International Conference on Very Large Scale Integration of System-on-Chip (VLSI-SoC06), Nice, France, Oct., pp. 338–341. (2006)
10. Tamir, Y., Tremblay, M.: High-performance fault-tolerant VLSI systems using micro-rollback. IEEE Transactions on Computers **39**(4) (1990)
11. Aeroflex Gaisler, http://www.gaisler.com/
12. Wilson, S.W.: Classifier fitness based on accuracy. Evol. Comput. **3**(2), 149–175 (1995)
13. Wilson, S.W.: ZCS: a zeroth level classifier system. Evol. Comput. **1**(2), 1–18 (1994)
14. Zeppenfeld, J., Bouajila, A., Stechele, W., Herkersdorf, A.: Learning classifier tables for autonomic systems on chip. In: Lecture Notes in Informatics, vol. 134, pp. 771–778. Springer, Berlin (2008)
15. Efraimidis, P.S., Spirakis, P.G.: Weighted random sampling with a reservoir. Inf. Process. Lett. **97**(5), 181–185 (2006)
16. Gregable, Reservoir Sampling, http://gregable.com/2007/10/reservoir-sampling.html

A. Bouajila (✉) · J. Zeppenfeld · W. Stechele · A. Herkersdorf
Institute for Integrated Systems, Technische Universität München, München, Germany
e-mail: a.bouajila@tum.de

A. Bernauer · W. Rosenstiel
Wilhelm-Schickard Institut für Informatik, Universität Tübingen, Tübingen, Germany

O. Bringmann
Microelectronic System Design, FZI, Karlsruhe, Germany

Chapter 5
Applications

The Organic Computing paradigm is particularly helpful in applications where the complexity, the autonomy, the requirements, the dynamics, or the a priori unknown structure of the systems under consideration make it hard to foresee their behaviour at design time. In these cases, it seems to be a good idea to defer decisions to runtime and to equip systems with the capability to organise, adapt, and optimise themselves. In Chap. 1 theoretical foundations for self-organising systems have been laid out, explaining the possibilities of how to construct self-x properties into technical systems. It is very plausible that in the above mentioned situations Organic Computing systems can outperform classical solutions. However, the proof of the pudding is in the eating.

Therefore, this chapter is devoted to applications of Organic Computing techniques. It is interesting to see some of the typical application areas chosen in the priority programme, the techniques that were applied, and to see comparisons to classical solutions. More applications can be found throughout the book serving there as illustrations of the presented results. The focus of this chapter is to look at some important application areas in more detail in order to improve the basis for assessing the potential impact of Organic Computing for computer science and society.

The 10 contributions of this chapter cover a variety of application areas for Organic Computing ranging from sensor networks to a six-legged walking robot. Two articles are applying OC principles to the coordination of robot ensembles: energy-aware teams exploring their surroundings, and a learning approach for societies of robots. Another topic covered in this chapter is self-organisation in the context of communication technology: bus-based communication on one hand, and multiple resource restricted transmitters on the other. Two contributions are devoted to investigating improvements to hardware devices such as smart real-time cameras and multi-core processor systems. And finally the chapter presents two articles on applying Organic Computing paradigms to improve our current traffic systems.

The contributions have at least one thing in common. They all apply self-organisation as a basic mechanism. However, looking closer at the different applications, the targets and the mechanisms for the organisational dynamics can vary quite considerably.

Traffic systems, especially connected intersections, are the focus of Chap. 5.1 (*Prothmann et al.: Organic Traffic Control*). A hierarchical self-organisation approach in which intersections learn the best way to deal with local traffic patterns, coordinate to achieve green waves and identify streams of traffic, alleviates the environmental and economical impact of traffic congestions caused by stop-and-go-traffic at intersections. The organic system decreases the average number of stops per vehicle significantly in comparison to current signalisation schemes.

A similar goal is tackled in Chap. 5.2 (*Fekete et al.: Methods for Improving the Flow of Traffic*). Self-organising local data clouds are used to improve the flow of congested traffic and to reduce the fuel consumption for cars in stop-and-go traffic. The data clouds are formed in reaction to different traffic situations and are based on methods for mobile ad-hoc networks using short-distance communication between cars. The article shows that the usage of these data clouds leads to fuel savings of up to 40 percent for vehicles in stop-and-go traffic, caused by less acceleration and fewer vehicles travelling at high speed, which are affected by increased air resistance.

The application of Organic Computing techniques to processors is presented in Chap. 5.3 (*Zeppenfeld et al.: Applying ASoC to Multi-Core Applications for Workload Management*). An Autonomic System on Chip self-organises by binding tasks to the cores of a multi-core processor at runtime based on a learning mechanism implemented in hardware. The techniques are applied to multi-core Internet-Protocol packet forwarding, they achieve nearly optimal results for dynamic workload balancing during runtime, and find an efficient parametrisation of the used hardware components.

Chapter 5.4 (*Sigg et al.: Efficient Adaptive Communication from Multiple Resource Restricted Transmitters*) introduces a protocol for distributed, adaptive transmit beamforming in wireless sensor networks, i.e. collaborative transmission of common messages in order to reach a remote receiver. As the performance of collaborative transmission is very sensitive to environmental changes, sensor nodes are able to self-adapt by sensing their environment and classifying the situation as well as to self-configure by learning locally optimal parameter settings.

To handle the complexity of communication in future distributed embedded systems and to increase its quality, Chap. 5.5 (*Ziermann et al.: OrganicBus: Organic Self-organising Bus-based Communication Systems*) discusses a decentralised bus-based approach. The introduced design methodology for an OrganicBus overcomes the drawbacks of offline design by considering the actual bus traffic and requirements of the system. Messages (called streams) are categorised and prioritised to guarantee hard-real-time behaviour and to optimise the usage of bandwidth by the dynamic offset adaptation algorithm (DynOAA). In comparison to classical approaches, this approach allows a significantly higher utilisation of the available communication resources.

A self-organising clustering mechanism is described in Chap. 5.6 (*Salzmann et al.: OC Principles in Wireless Sensor Networks*) that groups redundant sensor nodes to distribute the work load within the group in order to minimise resource usage in the individual node while maintaining full network functionality. The ability of the system to self-heal after a node is broken paired with a scale-free routing algorithm makes it robust against failures while ensuring energy efficiency and scalability.

The advantages gained from the use of self-organisation, self-reconfiguration, and self-healing principles for the implementation of a six-legged walking robot is shown in Chap. 5.7 (*Maehle et al.: Application of the Organic Robot Control Architecture ORCA to the Six-Legged Walking Robot OSCAR*). The application of the Organic Robot Control Architecture (ORCA) allows to master the complexity of engineering those robots. Reflexes are ensuring fast responses to unexpected events or faults and a high-level deliberative layer allows for adaptive planning based on the current situation.

Chapter 5.8 (*Degener et al.: Energy-Awareness in Self-organising Robotic Exploration Teams*) considers a team of autonomous mobile robots that have to build a short communication chain with local knowledge while minimising the energy spent for motion and sensing the environment. The robots autonomously adapt their positions by means of other robots' positions within view distance. The formation of short chains allows to minimise the number of necessary robots, which then can participate in other tasks.

A hierarchical learning approach for robots is presented in Chap. 5.9 (*Jungmann et al.: A Fast Hierarchical Learning Approach for Societies of Robots*). The architecture is based on three layers: The motivation layer which represents goals by drives (sub-goals), the strategy layer that uses reinforcement learning to learn a good strategy in order to satisfy the motivation layer, and the skill layer, which learns reactive skills that can be used by the strategy layer. The ability to learn strategies and skills enables the robots to adapt to changing tasks and environments.

A class of distributed, parallel algorithms for fast image vision detection processes (e.g. finding the centroid of an object) in restricted computing environments is discussed in Chap. 5.10 (*Fey et al.: Emergent Computing with Marching Pixels for Real-Time Smart Camera Applications*). Images are analysed by agents, called Marching Pixels, that explore the objects of an image. The propagation of agents, steered by a classic cellular automaton, is based on local information and interaction with other agents. Due to the capability to process multiple objects simultaneously, these algorithms are very scalable. Moreover, this approach is far more robust than classical techniques since it is able to compensate for failures during runtime.

The bandwidth of applications presented in this chapter shows impressively that Organic Computing principles are useful in diverse domains and applications and yield positive effects by providing more robust, flexible, and adaptive solutions to difficult and multi-faceted problems.

Wolfgang Reif
Institute for Software & Systems Engineering, Universität Augsburg, Augsburg, Germany
email: reif@informatik.uni-augsburg.de

Chapter 5.1
Organic Traffic Control

Holger Prothmann, Sven Tomforde, Jürgen Branke, Jörg Hähner, Christian Müller-Schloer, and Hartmut Schmeck

Abstract Urban road networks are an infrastructural key factor for modern cities. To facilitate an efficient transportation of people and goods, it is crucial to optimise the networks' signalisation and to route drivers quickly to their destination. As road networks are widespread and their traffic demands are dynamically changing, adaptive and self-organising (and therefore *organic*) control systems are required. This article demonstrates the potential benefits of organic traffic control: It presents an Observer/Controller that optimises an intersection's signalisation and introduces a self-organising coordination mechanism that allows for the traffic-responsive creation of progressive signal systems (or green waves). All presented mechanisms advance the state of the art and help to reduce the negative environmental and economical impact of traffic.

Keywords Observer/Controller architecture · Traffic signal control · Routing · Two-levelled learning · Learning classifier system · Evolutionary algorithm

1 Introduction

The road networks of modern cities are characterised by their numerous signalised intersections that regulate the traffic flows. Optimised signal plans, the improved coordination of neighbouring intersections, and intelligent vehicle routing mechanisms can help to utilise the existing road network more efficiently.

To keep the negative impact of traffic at a minimum, the network's signalisation needs to be continuously adapted to the dynamically changing traffic demands. Reoccurring changes, like those exemplified in Fig. 1a, can in principle be handled by a time-dependent switching of signal plans. However, pre-planned schedules are inflexible and cannot cope with irregularities like those illustrated by Fig. 1b. The figure compares the traffic demands of an arterial road at Karlsruhe, Germany, for two subsequent Sundays. While June 20, 2010 has been a regular Sunday, Germany played England in the round of sixteen in the FIFA World Cup in the afternoon of June 27, 2010. Although the game took place in Bloemfontein, South Africa, it affected the traffic at Karlsruhe. During the game, traffic counts are drastically reduced, while an increase can be observed for the rest of the day due to pre- or postponed trips.

C. Müller-Schloer et al. (eds.), *Organic Computing — A Paradigm Shift for Complex Systems*, 431–446
Autonomic Systems, DOI 10.1007/978-3-0348-0130-0_28,

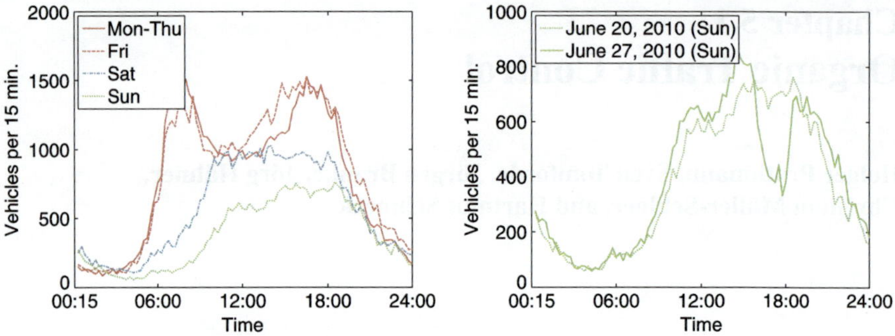

(a) Regular demand (b) 2010 FIFA World Cup

Fig. 1 Traffic demand of an arterial road at Karlsruhe, Germany (Data provided by Tiefbauamt, Stadt Karlsruhe)

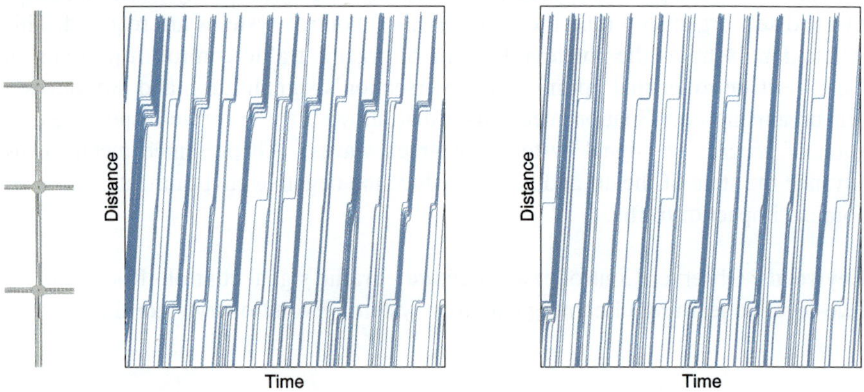

(a) Uncoordinated signals (b) Coordinated signals

Fig. 2 Time-distance diagram for an arterial road

As dynamic changes in traffic are difficult or even impossible to foresee at design time, it is necessary to move the optimisation and coordination of signalised intersections to the run-time of the signal system. This paradigm shift is a core idea of Organic Computing. At signalised intersections, it can be achieved by an Observer/Controller architecture that learns and optimises signal plans on-line. The architecture is based on the generic Observer/Controller pattern (see Chap. 4.1) and will be introduced in Sect. 2.

In urban areas, it is not sufficient to consider each intersection separately. As intersections are located in close vicinity, their traffic-responsive coordination is a crucial factor. This is illustrated by Fig. 2 which depicts vehicle trajectories at an arterial road. In the uncoordinated case (Fig. 2a), vehicles are stopped repeatedly which is indicated by horizontal lines in the time-distance diagram. By establishing a *progressive signal system* (or *green wave*), the number of stops can be significantly reduced once the vehicles have passed the first intersection where they arrive randomly (see Fig. 2b). The self-organised and traffic-responsive coordination of signalised intersections is in the focus of Sect. 3.

While signalised at intersections can adapt to the network's traffic flows, they cannot actively influence the drivers' route choice. Especially in the presence of disturbances (like traffic jams caused by high demands, accidents, or road works), it is beneficial to guide drivers around the affected areas. *Variable Message Signs* (VMS) can inform the drivers about the currently recommended routes. A decentralised routing mechanism can compute the necessary recommendations based on data available at the signalised intersections. The mechanism improves the robustness of the traffic system and is presented in Sect. 4. Finally, Sect. 5 concludes the article.

2 Adaptive Learning Intersections

Signalised intersections are the core component of urban road networks. This section briefly presents the state of the art in signal control and introduces an Observer/Controller architecture that allows for adaptive learning intersections.

2.1 State of the Art

Signalised intersections can be operated by fixed-time or traffic-actuated control: *Fixed-time signal plans* do not require detection hardware, but are designed to suit an expected average traffic demand (e.g., during the morning peak). Changes in demand are handled by a time-dependent switching among several predefined plans. An advantage of fixed-time control is its simplicity: The intersection's turnings are served by signal groups. Non-conflicting signal groups can be combined into phases that obtain the right of way in a reoccurring sequence. The duration of this sequence defines the cycle time. Phase sequence, phase durations, and cycle time are predefined and remain unaffected by the current vehicle arrivals.

In contrast to fixed-time operated intersections, *traffic-actuated controls* adapt their signalisation. Based on temporal and logical conditions that are specified by a traffic engineer, the intersections extend or truncate phase durations, change the phase sequence, or adjust the cycle time. Thereby, traffic-actuated signals operate demand-oriented and can prioritise public transport vehicles. However, they loose their adaptivity and behave like fixed-time controls in heavy traffic [2].

Neither fixed-time nor traffic-actuated controls evaluate their performance or optimise their signalisation at run-time. Signal plans have to be specified at design time and remain unchanged throughout the intersection's operation which causes a potential loss in performance.

2.2 An Observer/Controller Architecture for Signal Control

To allow for an on-line evaluation and optimisation of an intersection's signalisation, its fixed-time or traffic-actuated signal controller can be extended by an Observer/Controller component (see Chap. 4.1). While the signal controller forms the *System under Observation and Control* (SuOC), the *observer* monitors the local traffic demand and evaluates the active signal plan's performance. Based on these observations, the *controller* selects and optimises signal plans using a two-levelled learning mechanism. The resulting Observer/Controller framework for signal control is depicted in Fig. 3. Its working principle and experimental results are briefly presented in the following, a more extensive discussion is available in [9].

Observing the Traffic

It is the observer's task to monitor the intersection's traffic flows and to analyse the active signal plan's performance. The *monitoring* component collects data from the intersection's traffic detectors and stores the detection time series in a log file. During *preprocessing*, the stored time series are used to compute a traffic flow vector $M = (M_1, \ldots, M_n)$ that estimates the hourly flow for each of the intersection's n signal groups. If required, a forecast of future traffic developments can be obtained from the *predictor*. While short-term forecasts are based on recent vehicle detections

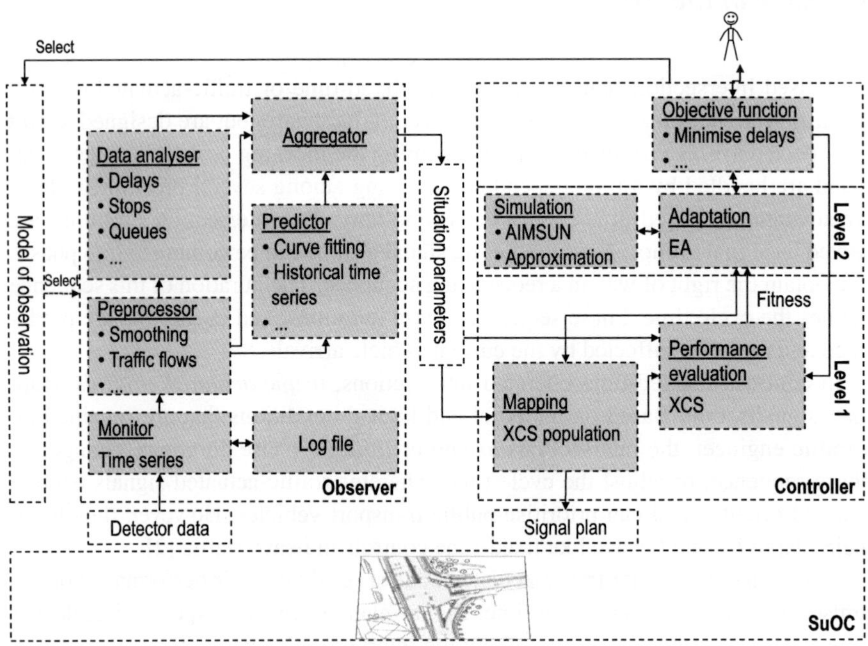

Fig. 3 An Observer/Controller architecture for signal control

(by computing a linear curve fitting of the last measured values and extrapolating it into the future), long-term predictions are obtained by an automatic matching of the latest measurements to historical time series (see [4]).

While preprocessor and predictor focus on current and expected demands, the *data analyser* derives performance measures for the intersection's active signal plan. Various measures can be evaluated, including vehicular delays, queue lengths, or stops. Unfortunately, most performance measures cannot be obtained directly from the monitored detection time series, but require supplementing traffic models or approximations. Traffic flows and performance measures are finally accumulated by the *aggregator* which provides them as *situation parameters* to the controller.

Controlling the Signalisation

Based on the observed situation parameters, the controller provides an appropriate signal plan to the SuOC by relying on a two-levelled learning and optimisation mechanism: On Level 1, suitable signal plans are selected on-line from a previously learnt mapping. On Level 2, new optimised signal plans are created whenever necessary. Learning and optimisation consider a user-specified objective (that is assumed to be the minimisation of vehicular delay in the following).

The on-line signal plan selection on Level 1 relies on a modified variant of the rule-based reinforcement learning system XCS [14]. Within the controller, XCS learns a *mapping* that assigns appropriate signal plans to observed traffic demands. The mapping is implemented as a population of rules (or *classifiers*). Each classifier consists of a condition, an action, and some quality indicators: The *condition* specifies to which traffic demands a classifier is applicable. It is a concatenation of n interval predicates $[l_i, u_i]$ that are said to match an observed traffic demand $M = (M_1, \ldots, M_n)$ if and only if $l_i \leq M_i \leq u_i$ for all $i \in \{1, \ldots, n\}$. A classifier's *action* corresponds to a signal plan that can be activated if the condition is matched. The quality indicators predict the vehicular delay resulting from the activation (*prediction*), keep track of the *prediction error*, and estimate the classifier's relative accuracy with respect to overlapping classifiers (*fitness*).

Given an observed traffic demand, XCS identifies the matching classifiers within its population and stores them in a *match set*. As the matching classifiers can advocate different signal plans, a selection has to take place. Using the match set, XCS computes a fitness-weighted prediction average for each signal plan and selects the best plan for activation in the SuOC. Matching classifiers advocating the selected plan are stored in an *action set* and the plan is activated. Once the plan has been active for a predefined number of cycles, XCS updates the quality indicators of all classifiers in the action set. The update is based on delays observed at the intersection and uses the update mechanism introduced in [14].

While signal plan selection and classifier updates are performed like in the original XCS, the classifier creation had to be modified. XCS creates matching classifiers randomly (*covering*) or modifies existing classifiers in the action set using genetic operators. The quality of the resulting classifiers is then explored in the system environment. In an adaptive traffic control system, this random exploration is likely to cause unacceptable delays at the controlled intersection.

The modified classifier system within the controller does not rely on random covering or genetic operators for classifier discovery. Classifiers are created on demand instead: Whenever an observed traffic demand is unmatched, the adaptation module on Level 2 is activated to obtain an optimised signal plan including an estimation of its performance. Once the optimised signal plan becomes available, a new classifier is added to the population for future use.

Unfortunately, the signal plan optimisation takes some time while the Observer/Controller is expected to react immediately. As an immediate reaction, the classifier exhibiting the minimal distance to the unmatched demand is selected from the population. Its condition is widened to match the unknown demand and, consequently, its signal plan is activated in the SuOC. The widening of closely located classifiers allows for an immediate response to unknown traffic demands, but increases the prediction error. The increase will be recognised when updating the widened classifier's quality estimates, such that more specific classifiers are preferred when available. The modifications to XCS are discussed in detail in [9].

Level 2 of the controller supports the on-line signal plan selection by providing optimised signal plans. Optimisations are performed by an *Evolutionary Algorithm* (EA) that implements the controller's *adaptation module*. EAs [5] are nature-inspired optimisation heuristics that mimic biological evolution: Starting with a set (called *population*) of initial signal plans, the EA selects plans with relatively low delays as parents. The parents are combined (*crossover*) and locally modified (*mutation*) to obtain new candidate plans as offspring. Based on their relative quality, some of the parent and offspring plans are selected to form the next generation's population. This process is repeated until a plan with a sufficiently low delay is found (or another stopping criterion is reached).

The quality of a plan (according to the specified *objective function*) is determined by the *simulation module* that can be implemented by a traffic simulator (like AIMSUN [1]) or an approximation formula: Traffic simulations allow to evaluate fixed-time and traffic-actuated controls with respect to a variety of objectives. Simulation-based evaluations are noisy and time-consuming, though. Depending on the signal controller and the considered objective, approximation formulas can be an alternative to simulations since they can be computed quickly and provide deterministic results. The average vehicular delay at a fixed-time controlled intersection can, e.g., be estimated by Webster's formula [12].

Experimental Results

To demonstrate the benefits of an on-line evaluation and optimisation of signal plans, an intersection located at Hamburg, Germany, has been investigated. The intersec-

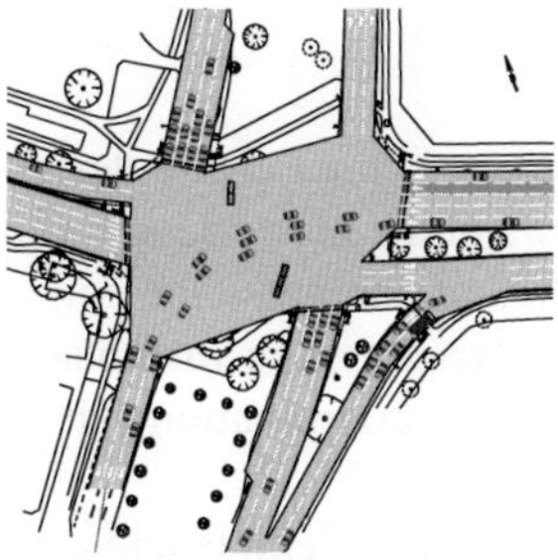

(a) Intersection model

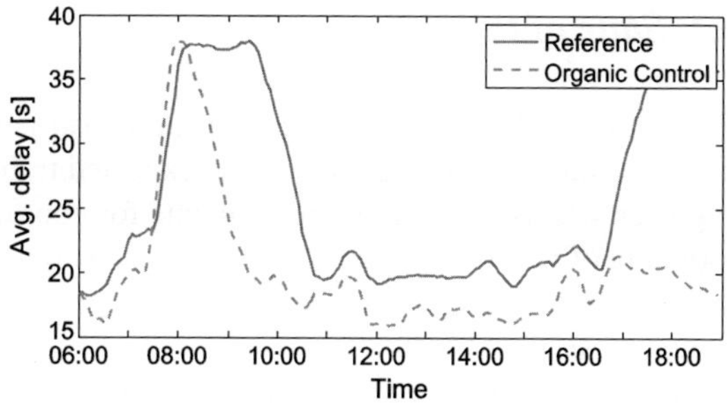

(b) Vehicular delays

Fig. 4 Adaptive signalisation for an intersection at Hamburg, Germany

tion (depicted in Fig. 4a) has been modelled in the microscopic simulation environment AIMSUN [1]. The model incorporates demand data taken from a recent traffic census and includes the signal plans used in the field as reference solution. Simulations are performed for a period of thirteen hours starting at 6 a.m. and consider the average vehicular delay as objective.

Figure 4b depicts the average vehicular delays obtained for the time-dependent switching of signal plans used in the field (labelled *reference*) compared to an organic on-line optimisation. The depicted delays are averaged over five repeated simulation runs and have to be minimised. The obtained results indicate that an on-line optimisation outperforms the time-dependent schedule of the reference solution for most of the simulation period. An exception is the beginning of the morning peak period, where the initially empty classifiers population cannot provide matching classifiers for the quickly rising demands. Despite this temporary increase, the vehicular delays for the whole simulation period are reduced by 22.5% compared to the reference solution used in the field.

3 Self-organised Coordination

In urban road networks, where intersections are located in close vicinity, the coordination of traffic signals becomes a crucial aspect. Badly coordinated signals induce unnecessary stops and increase travel times, fuel consumption, and emissions.

To coordinate signals along a traffic stream and obtain a progressive signal system, intersections have to be operated with carefully chosen offsets. An intersection's offset is defined as the difference between a reference time and the start time of the phase serving the coordinated traffic stream. By adjusting the offset difference among adjacent intersections to equal the expected travel time between those intersections, a coordination can be established. A prerequisite is a common cycle length for the coordinated intersections to ensure that the phases remain synchronised over time.

3.1 State of the Art

Traditionally, the signal coordination of a road network is specified at the design time of a signal system. Based on the expected traffic flows in the network, a traffic engineer selects some traffic movements for coordination and computes the corresponding offsets (and signal plans) with the help of a traffic engineering workplace. Often, several coordination schemes are prepared such that the coordination can be changed based on the time of day.

As the traditional coordination based on expected demands is inflexible and cannot cope with irregularities, *adaptive network control systems* evaluate and optimise a network's signalisation on-line. BALANCE [6] and MOTION [3] are two recently developed examples. Both systems exhibit a hierarchical architecture that combines a centralised network-wide optimisation with local traffic-actuated control. On the centralised level, detection and signalisation data is gathered from the network. The collected data is incorporated into a traffic model that is used to identify the network's dominant traffic streams and to optimise coordination and signal timings of the intersections. After the optimisation, the resulting signalisation constraints are provided to the network's signalised intersections as frame signal plans. The plans ensure a coordinated operation, but leave some freedom for a traffic-actuated adaptation of the signalisation.

Their hierarchical structure distinguishes BALANCE and MOTION from other network control systems that are mostly centralised. In [6], *Friedrich* motivates the hierarchical architecture of BALANCE by arguing that distributed intelligence can often handle the complexity of traffic better than strictly centralised systems that are error-prone and require a tremendous effort to communicate all relevant data.

3.2 Traffic-Responsive Decentralised Coordination

Following *Friedrich's* line of thought, the organic intersections introduced in Sect. 2 have been extended with a decentralised mechanism that coordinates the signalisation of neighbouring intersections in response to current traffic demands [9]. The mechanism is briefly introduced, before experimental results are presented.

Decentralised Progressive Signal Systems

The decentralised coordination mechanism for signalised intersections is called DPSS (for *Decentralised Progressive Signal Systems*) and works in three steps: In a first step, the intersections determine partners that collaborate to form a progressive signal system (PSS). Once the partnerships are established, the collaborating intersections agree on a common cycle time which is a prerequisite for coordination. In a third step, the partners select signal plans that respect the common cycle time, calculate offsets, and finally establish a coordinated signalisation. The three steps are briefly explained in the following:

Step D.1—Partnerships To obtain a sequence of intersections that should be coordinated when considering the network's current traffic demand, each intersection j determines which of its local turnings exhibits the strongest vehicle flow. Assuming that intersection j determines the turning from upstream intersection i to downstream intersection k as its strongest turning movement, intersection j informs its desired predecessor i that it would like to be i's successor in a PSS. After all intersections informed their desired predecessor, a local matching takes place: Each intersection j checks whether it was chosen by its downstream intersection k as k's desired predecessor. If this is the case, j acknowledges the partnership with k. Furthermore, reject messages are sent to other intersections that have selected j as desired predecessor, so no partnership is established with these intersections initially. Intersections that were not integrated in a PSS can repeat the above process with other less heavily used turning movements. When the first step is completed, intersections collaborating in a PSS know their partners.

Step D.2—Cycle time To determine a common cycle time that is suitable for all intersections in the PSS, the collaborating intersections determine their currently desired cycle length with the help of their controller's mapping. Afterwards, an echo algorithm starting at the first intersection of the PSS determines the maximum of the desired cycle times and distributes this information among the intersections. The maximum is selected since shorter cycles would reduce the capacity of the most heavily used intersection, thereby causing large delays.

Step D.3—Signal plans and offsets In the final step, signal plans need to be selected and offsets have to be calculated. The process starts at the first intersection of the PSS. The intersection uses its Observer/Controller to select a signal plan that is suitable for the current traffic demand and respects the common cycle time. Afterwards, the intersection communicates the relevant timing data for offset calculation to its successor. At the successor intersection, the process is repeated until the last intersection in the PSS is reached.

When Step D.3 is completed, the PSS is established. The coordination completely relies on locally available traffic data and on communication among neighbouring intersections. The DPSS mechanism is repeatedly executed every few minutes to continuously adapt the signal coordination to changing demands.

Experimental Results

To evaluate the benefits of a traffic-responsive decentralised coordination, the DPSS mechanism has been simulated for a small road network at Hamburg, Germany (see Fig. 5a). The network consists of eleven signalised intersections that are operated by an Observer/Controller. The simulated traffic demands are taken from a recent census.

Figure 5b compares the network-wide number of stops obtained for uncoordinated and coordinated intersections. Results are averaged over five simulation runs and indicate that a decentralised coordination is beneficial as stops are reduced

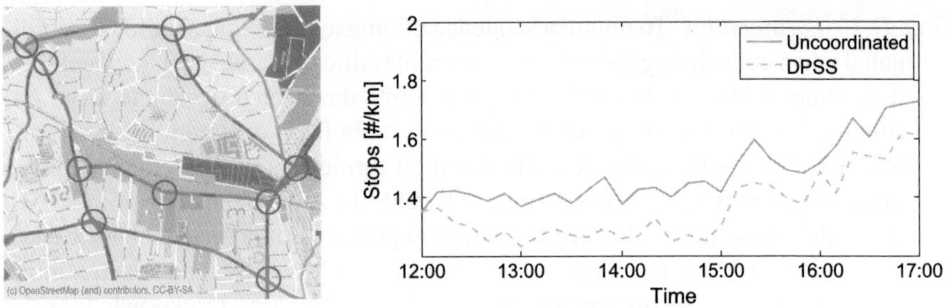

(a) Network map (b) Network-wide stops

Fig. 5 Signal coordination for a network at Hamburg, Germany

throughout the whole simulated period. On average, the reduction is about 8% compared to an uncoordinated signalisation.

3.3 *Limitations of Decentralised Control*

While a completely decentralised coordination of intersections has many advantages (like reduced effort and cost compared to centralised systems), the restriction to local information can result in a suboptimal coordination. When determining the partners for coordination in Step D.1, the DPSS mechanism prefers the strongest uncoordinated streams within the network. In most cases, this is a useful approach, but it does not necessarily lead to the best possible coordination.

A counterexample is depicted in Fig. 6a. The figure illustrates traffic streams in a Manhattan-type network. Streams are depicted as arrows, where the arrow width is proportional to the stream's traffic flow. The strongest stream is in the middle of the southbound streams, but in sum both eastbound streams encompass more vehicles than the three streams running from north to south. While the DPSS mechanism would establish a coordination for the three southbound streams, coordinating the eastbound streams would be more beneficial. To deal with such special cases, the DPSS mechanism has been extended with an additional hierarchical component [11].

Regional Manager

The *Regional Manager* (RM) executes a heuristic for finding the best possible combination of PSSs for the network's current traffic demand: It gathers traffic data from the signalised intersections to obtain a graph representation of the network's traffic flows. In a second step, the RM identifies strong traffic streams that serve as candidates for coordination. Unfortunately, PSSs cannot be installed for every candidate stream. Establishing PSSs for both directions of an arterial road is, e.g., possible only in special cases as restrictions are imposed on the intersections' distances, their cycle, and the local speed limits. Therefore, the RM combines non-conflicting

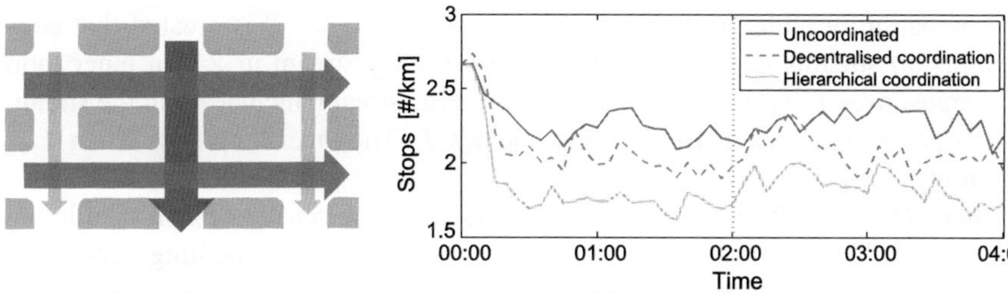

(a) Traffic flows (b) Network-wide stops

Fig. 6 Signal coordination in a Manhattan network

streams to stream systems and selects the most beneficial system in a third step. In the following, the three steps are briefly summarised. An extensive discussion is available in [11].

Step RM.1—Build the network graph To obtain a graph representation of the network's current traffic flows, each intersection creates a subgraph representing its topology and the local traffic flows. The subgraph contains one vertex for each outgoing section, one vertex for each incoming section, and one edge for each turning movement. Edges are directed and weighted, with the weight corresponding to the turning's current traffic flow. The resulting subgraphs are transmitted to the RM where they are combined to a network-wide representation.

Step RM.2—Determine the traffic streams To determine candidate streams for coordination, the RM iteratively connects graph edges (representing turnings) to an edge sequence (representing a traffic stream). To obtain a sequence, the RM selects an edge with maximal weight from the graph. The edge's predecessors and successors in the sequence are obtained by iteratively choosing the adjacent edges with highest weights. The sequence is ended when a selected adjacent edge is not the most heavily weighted edge in its subgraph. Once an edge sequence is completed, it is stored in a candidate set S and all of its edges are removed from the graph. Afterwards, additional sequences are created until the remaining edges' weights are below a predefined threshold.

Step RM.3—Determine the stream systems Finally, the most beneficial combination of non-conflicting streams has to be determined: Most beneficial means that the number of vehicles served by the coordinated streams should be maximal, while non-conflicting means that streams must not intersect each other or run in different directions on the same roads.

To create promising stream systems without generating the power set of streams, the RM applies a greedy approach that starts with some preprocessing: Pairs of conflicting streams are identified and stored in a table. The set S (containing the traffic streams created in Step RM.2) is sorted with respect to the number of benefiting vehicles. Furthermore, S is copied to a new set U that will contain unprocessed traffic streams which have not yet been included in a stream system. After preprocessing, stream systems are created in a loop that terminates when U is empty: The best unprocessed traffic stream s_1 is removed from U and included in the stream system Z that is currently under construction. Using the compatibility table created during preprocessing, a set T is created that contains all streams that are not in conflict with any stream in Z. An inner loop iterates over T, selects the best non-conflicting stream, includes it in the stream system Z, removes it from U, and updates T. The inner loop ends when T is empty.

Step RM.3 results in a set of stream systems that consist of non-conflicting streams. The stream system with the highest number of benefiting vehicles is selected for implementation in the road network.

All steps performed by the RM can be efficiently implemented (see [11] for details) such that Step D.1 of the DPSS mechanism can be substituted: After the partners for coordination have been identified by the RM, Steps D.2 and D.3 are reused to establish the coordination. Thereby, a hierarchical system architecture similar to that of BALANCE and MOTION is obtained, but with the difference that signal plans are selected and optimised locally at the intersections.

Experimental Results

To study the potential benefit of a hierarchical coordination, the DPSS mechanism has been compared to a hierarchical traffic control system that uses the RM. The investigated test case is a Manhattan-type road network consisting of six intersections (see Fig. 6a). Each intersection is equipped with an Observer/Controller responsible for the traffic-responsive selection and optimisation of signal plans. The simulated traffic demand resembles the motivational example discussed in the beginning of this section (see Fig. 6a), but all traffic flows change their direction after two simulated hours. The demands will cause the DPSS mechanism to create a suboptimal coordination by establishing PSSs for the southbound (or later northbound) streams. In contrast, the RM recognises the network-wide situation and coordinates the eastbound (or later westbound) streams.

Figure 6b depicts the network-wide number of stops obtained from the simulations. Compared to an uncoordinated operation, both coordination mechanisms are able to reduce the number of stops for (nearly) the complete simulation period. Reductions of 8.2% and 18.8% are obtained for the decentralised and the hierarchical approach, respectively. As expected, the hierarchical approach achieves a higher reduction of stops, but the DPSS mechanism still performs reasonably well.

4 Self-organised Routing

The previous sections have demonstrated the benefit of organic intersections that can adapt and coordinate their signalisation in response to changing traffic demands. In the following, the network's infrastructure will no longer only react to observed demands, but will actively guide drivers to prominent destinations. Thereby, the road network becomes more robust in the presence of disturbances or high demands.

4.1 State of the Art

While signalised intersections have been in the focus of research for decades, the traffic-responsive routing of drivers has gained attention only in recent years. Today, routing is mostly limited to highways where *Variable Message Signs* (VMS) at interchanges display the currently recommended routes to certain destinations.

A more fine-grained routing is performed by navigation systems installed in the vehicles. Early systems have not been traffic-responsive, but relied on a static road map. Today, systems additionally incorporate data of the radio's *Traffic Message Channel* (TMC, only for highways) or can access travel time databases that are provided by the system manufacturers. The databases contain travel times that have been recently experienced by drivers using a navigation system of the same manufacturer, but do not necessarily reflect the current situation correctly.

Recently, research investigates the use of *floating car data* (FCD) to determine the network's traffic situation. BeeJamA is an example of an FCD-based routing approach [13]. It assumes that vehicles transmit their position, speed, and destination to regional navigation servers. Using a modified Internet routing protocol, the servers update their routing tables and navigate vehicles from intersection to intersection on a next hop basis. While the use of FCD allows for a detailed view on current demands, it requires vehicles with special equipment.

4.2 Self-organised Routing

To extend the organic traffic control framework with routing capabilities, it is assumed that signalised intersections are equipped with *Variable Message Signs* (VMS) at their approaches. For some prominent destinations, these VMSs provide the next turn of a recommended route to the drivers. The route recommendations are computed in response to the current traffic demands by an adapted *Distance Vector Routing* (DVR) algorithm that is executed locally at the intersections.

Distance Vector Routing for Road Networks

At a signalised intersection, a *routing component* performs a DVR protocol that is adapted for road networks. The routing component maintains several routing tables for the intersection's approaches, updates them based on routing messages received from neighbouring intersections, and communicates changes to its own neighbours.

The implemented protocol resembles the DVR protocol known from the Internet (see, e.g., [10]): Initially, each intersection checks whether it is located in the vicinity of a prominent destination. In this case, it creates routing table entries for those approaches that can reach the destination via one of the intersection's turnings. The table entries contain the destination, the recommended turning, and its current delay. The delay corresponds to the travel time required to reach the destination. It can be

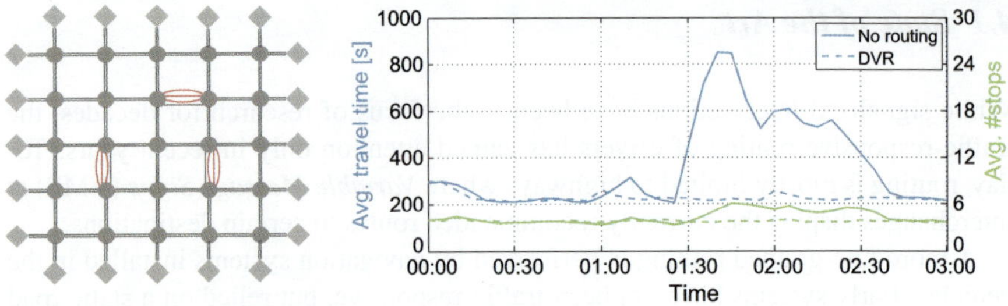

(a) Network with incidents (b) Network-wide travel times and stops

Fig. 7 Traffic-responsive routing in a Manhattan network

obtained from the intersection's Observer/Controller (where it is estimated from the current traffic flows and the turning's green time fraction).

Once a routing table entry has been created or updated, it is sent as a routing message to the corresponding upstream intersection where all routing tables are iteratively updated. To obtain the cost for the received route, the time that is required to reach the sender (i.e., the local turning delay plus the link travel time) is added to the travel time received with the message. Then, the routing table is checked: If the message's destination is unknown, a new table entry is created from the message. Otherwise, the existing routing entry is checked. If its recommended turning leads to the sender, its travel time is updated. If, on the other hand, the known route recommends a different turning, the travel times of the known and the novel route are compared and the novel route replaces the known one if it is faster.

The above steps are executed by each signalised intersection and result in a decentralised routing protocol that (when combined with an incident detection mechanism [7]) enables traffic-responsive routing in the presence of disturbances. The protocol can be refined to support an intra- and inter-regional routing that reduces the effort for computation and communication (see [8]) and does—in contrast to FCD-based approaches like those discussed in Chap. 2.7—not require specially equipped vehicles.

Experimental Results

The decentralised routing mechanism has been evaluated in a simulated Manhattan network. The network is illustrated in Fig. 7a and consists of 25 signalised intersections (depicted as circles) and 20 prominent destinations (depicted as diamonds). The simulated intersections are operated by an Observer/Controller and are equipped with VMSs on their approaches. Traffic flows are equally distributed among all destinations, but the encircled road segments are temporarily blocked due to incidents.

Figure 7b depicts the network-wide travel times and stops for the simulated network. When road segments are blocked and no route recommendations are provided, both measures are drastically increased due to uninformed drivers that try to use temporarily unavailable routes. When recommendations are available, neither travel times nor stops are significantly affected by the incidents although only 75% of all drivers consider the recommendations provided. In the investigated scenario, travel times and stops are reduced by 37.6% and 12.6%, respectively.

5 Conclusion

Urban road networks constitute an infrastructural key factor for modern cities. Their traffic demands are dynamically changing, while road works or accidents temporarily affect the availability of routes within the network. As dynamic environments, road networks benefit from Organic Computing concepts:

At signalised intersections, an Observer/Controller can reduce the average vehicular delay by adapting the signalisation to current demands. This on-line optimisation outperforms pre-specified signal plans and distinguishes organic intersections from traffic-actuated controls. In the road network, a self-organised coordination of signals reduces the network-wide number of stops. The coordination is traffic-responsive and relies on a distributed architecture such that it can be distinguished from pre-scheduled coordination schemes and most adaptive network control systems, respectively. Finally, a route recommendation system provides drivers with up-to-date routes. The recommendation system builds on existing infrastructure and can reduce travel times especially in the presence of incidents or heavy traffic.

In combination, Organic Computing concepts alleviate the negative economical and environmental impact of traffic by supporting the efficient use of the existing road network.

References

1. Barceló, J., Codina, E., Casas, J., Ferrer, J.L., García, D.: Microscopic traffic simulation: a tool for the design, analysis and evaluation of intelligent transport systems. J. Intell. Robot. Syst. **41**(2–3), 173–203 (2005)
2. Bretherton, R.D., Rai, G.I.: The use of SCOOT in low flow conditions. Traffic Eng. Control **23**(12), 574–576 (1982)
3. Busch, F., Kruse, G.: MOTION for SITRAFFIC—a modern approach to urban traffic control. In: Proc. 2001 IEEE Intelligent Transportation Systems Conf., pp. 61–64. IEEE, New York (2001)
4. Chrobok, R., Kaumann, O., Wahle, J., Schreckenberg, M.: Different methods of traffic forecast based on real data. Eur. J. Oper. Res. **155**(3), 558–568 (2004)
5. Eiben, A.E., Smith, J.E.: Introduction to Evolutionary Computing, 2nd edn. Springer, Berlin (2007)
6. Friedrich, B.: Steuerung von Lichtsignalanlagen: BALANCE – ein neuer Ansatz. Straßenverkehrstechnik **44**(7), 321–328 (2000)
7. Klejnowski, L.: Design and implementation of an algorithm for the distributed detection of disturbances in traffic networks. Masterarbeit, Institut für Systems Engineering—SRA, Leibniz Universität Hannover (2008)

8. Lyda, J.J.: Dezentrale adaptive Routingverfahren in selbst-organisierten Verkehrsnetzen am Beispiel von Organic Traffic Control. Masterarbeit, Institut für Systems Engineering – SRA, Leibniz Universität Hannover (2010)
9. Prothmann, H., Branke, J., Schmeck, H., Tomforde, S., Rochner, F., Hähner, J., Müller-Schloer, C.: Organic traffic light control for urban road networks. Int. J. Auton. Adapt. Commun. Syst. **2**(3), 203–225 (2009)
10. Tanenbaum, A.S.: Computer Networks, 4th edn. Pearson Education, Upper Saddle River (2002)
11. Tomforde, S., Prothmann, H., Branke, J., Hähner, J., Müller-Schloer, C., Schmeck, H.: Possibilities and limitations of decentralised traffic control systems. In: World Congress on Computational Intelligence, pp. 3298–3306. IEEE, New York (2010)
12. Webster, F.V.: Traffic signal settings. Road Research Technical Paper No. 39. Road Research Laboratory (1958)
13. Wedde, H.F., Lehnhoff, S., et al.: Highly dynamic and adaptive traffic congestion avoidance in real-time inspired by honey bee behavior. In: Holleczek, P., Vogel-Heuser, B. (eds.) Mobilität und Echtzeit – Fachtagung der GI-Fachgruppe Echtzeitsysteme, pp. 21–31. Springer, Berlin (2007)
14. Wilson, S.W.: Classifier fitness based on accuracy. Evol. Comput. **3**(2), 149–175 (1995)

H. Prothmann (✉) · H. Schmeck
Institute AIFB, Karlsruhe Institute of Technology (KIT), Bldg. 05.20, 76128 Karlsruhe, Germany
e-mail: holger.prothmann@kit.edu

H. Schmeck
e-mail: hartmut.schmeck@kit.edu

S. Tomforde · J. Hähner · C. Müller-Schloer
Institute of Systems Engineering, Leibniz Universität Hannover, Appelstr. 4, 30167 Hannover, Germany

S. Tomforde
e-mail: tomforde@sra.uni-hannover.de

J. Hähner
e-mail: haehner@sra.uni-hannover.de

C. Müller-Schloer
e-mail: cms@sra.uni-hannover.de

J. Branke
Warwick Business School, University of Warwick, Coventry CV4 7AL, UK
e-mail: juergen.branke@wbs.ac.uk

Chapter 5.2
Methods for Improving the Flow of Traffic

Sándor P. Fekete, Björn Hendriks, Christopher Tessars, Axel Wegener, Horst Hellbrück, Stefan Fischer, and Sebastian Ebers

Abstract We describe a distributed and self-regulated approach for the self-organisation of a large system of many self-driven, mobile objects, i.e., cars in traffic. Based on methods for mobile ad-hoc networks using short-distance communication between vehicles, and ideas from distributed algorithms, we consider reactions to specific traffic structures (e.g., traffic jams.) Building on current models from traffic physics, we are able to develop strategies that significantly improve the flow of congested traffic. Results include fuel savings up to 40% for cars in stop-and-go traffic; we present a number of simulation results illustrating the underlying mechanisms. In addition, we discuss the organic structure of urban traffic, and hint at how self-healing methods can lead to improvements in rush-hour traffic.

Keywords Vehicular traffic · Self-organising · Self-healing · Car-to-car communication · Traffic jam · Urban traffic · Rerouting

1 Introduction

1.1 Traffic

Traffic[1] is one of the most influential phenomena of civilisation. On a small scale, it affects the daily life of billions of individuals; on a larger scale, it determines the operating conditions of all industrialised economies; and on the global scale, traffic has a tremendous impact on the living conditions on our planet, both in a positive and in a negative way.

All this makes traffic one of the most important complex systems of our modern world, and an excellent application scenario for Organic Computing. It has several levels of complexity, reaching from individual actions of the drivers, over local phenomena like density fluctuations and traffic jams, traffic participants' choice of transport mode and time, regional and temporal traffic patterns, all the way up to long-range traffic development and regulation.

[1] The discussion of traffic in the introduction is based on parts of our survey paper [2].

C. Müller-Schloer et al. (eds.), *Organic Computing — A Paradigm Shift for Complex Systems*, 447–460
Autonomic Systems, DOI 10.1007/978-3-0348-0130-0_29,

In recent years, tremendous progress has been made in understanding the dynamics of traffic flow and traffic congestion. Arguably the most significant contribution has come from physics, interpreting traffic as a multi-particle system. These models explain how the complexity of traffic emerges from the behaviour of individuals that follow simple rules. They also meet the popular appeal of systems of interacting, autonomous agents as problem-solving devices and models of complex behaviour. In short, the "decentralised" view, which goes beyond attempts at centralised simulation and control, has improved our understanding of traffic.

Modern hardware has advanced to the point that it is technically possible to enable communication and coordination between traffic participants. However, combining mobility and communication for coordinated behaviour does not only solve problems, it also creates new ones, as it is a challenge in itself to maintain the involved ad-hoc networks, as well as the related information that is independent of individual vehicles.

1.2 Computing Methodologies in Traffic and Telematics

For an overview ("Traffic and related self-driven many-particle systems"), see the excellent survey [4]. Obviously, research on traffic as a whole is an area far too wide for a brief description in this short overview; we focus on a strain of research that is particularly relevant for our work.

Until the early 1990s, efforts for simulating traffic were based on complex multi-parameter models of individual vehicles, with the hope of extending those into simulations for traffic as a whole. Obvious deficiencies of this kind of approach are manifold:

1. Because the behaviour of even just an individual vehicle is guided by all sorts of factors influencing a driver, the attempt for a closed and full description of the complete system appears hopeless.
2. Determining the necessary data for setting up a simulation for a relevant scenario is virtually impossible.
3. Running such a simulation quickly hits a wall; even with today's computing power, simulating a traffic jam with a few thousand individual vehicles based on such a model is far beyond reach.

A breakthrough was reached when physicists started to use a different kind of approach: instead of modelling vehicles with ever-increasing numbers of hidden parameters, they tried to consider them as systems of many particles, each governed by a very basic set of rules. As Nagel and Schreckenberg managed to show [12], even a simple model based on cellular automata can produce fractal-like structures of spontaneous traffic jams, i.e., complex, self-organising phenomena. Over the years, these models [10] were generalised to two-lane highway traffic [15], extended for simulating commuter traffic in a large city [14], and have grown considerably; see the book chapter by Nagel [11] for an overview.

A closely related line of research uses an approach that is even closer to particle physics; see [13] for an excellent overview of models for traffic flow and traffic jams, with about 150 relevant references. Among many others, particularly remarkable is the approach by [7]: this model reproduces properties of phase transitions in traffic flow, focusing on the influence of parameters describing typical acceleration and deceleration capabilities of vehicles. This is based on the assumption that the capabilities of drivers to communicate and coordinate are basically restricted to avoiding collisions, which until now is frustratingly close to what drivers can do when stuck in dense traffic.

Parallel to the scientific developments described above, the interest in and the methods for obtaining accurate traffic data has continued to grow: for example, *floating car data* aims at keeping track of the movements of a suitable subset of vehicles (e.g., taxis in Berlin city traffic), with the hope of getting a more accurate overall image of traffic situations, both in time and space [9]. However, even this approach relies on the use of the central processor paradigm, and does not allow the use of ad-hoc networks for the active and direct interaction and coordination between vehicles.

1.3 Our Approach

In our project *AutoNomos*, we have pursued a distributed and self-regulated approach for the self-organisation of a large system of many self-driven, mobile objects, i.e., cars in traffic. Based on methods for mobile ad-hoc networks using short-distance communication between vehicles, and ideas from distributed algorithms, local data clouds are formed in reaction to specific traffic structures (e.g., traffic jams). These *Hovering Data Clouds* (HDCs) are described in detail in Chap. 2.7 of this compendium; see Fig. 1a. On the next level, the HDCs are used for forming *Organic Information Complexes* (OICs) that are functional entities within the traffic flow, hosted by—but independent of—the individual moving vehicles; see

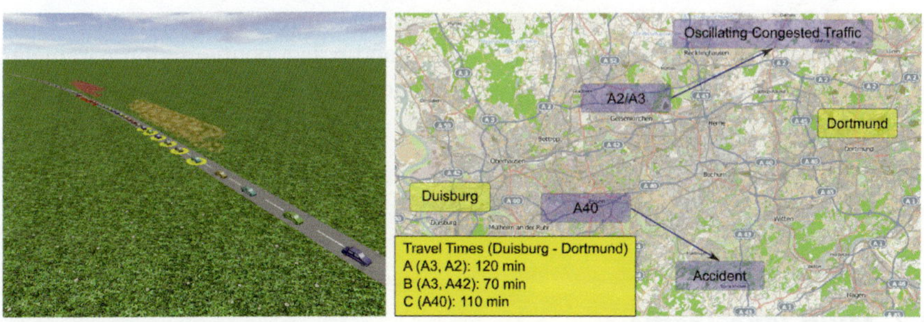

(a) HDCs marking beginning and end of a traffic jam.

(b) OICs with rerouting information. (map © www.OpenStreetMap.org contributors, CC-BY-SA)

Fig. 1 HDCs and OICs in traffic

Fig. 1b for an illustration. Using HDC-based OICs, we develop *Adaptable Distributed Strategies* (ADSs) for dealing with complex and changing traffic situations, such as the improvement of traffic congestions themselves.

In other words, HDCs and OICs produce structured data about the current traffic situation, which goes beyond the local information available by observing visible vehicles; furthermore, it is customised for each individual driver, which differs significantly from general traffic announcements.

The resulting ADSs lead to recommendations for the individual drivers of vehicles that are equipped with AutoNomos devices, e.g., to go slower than possible, to take a certain lane, or to use an alternative route.

2 Traffic Models

2.1 Single-Lane Traffic

Our approach for improving the flow of traffic is based on the well-known car-following model of Krauß [7], which derives position and velocity of a car from the gap to the predecessor and its velocity. Collisions are avoided based on a safe velocity v_{safe}:

$$v_{\text{safe}} = v_{\text{pred}} + \frac{g - \tau v_{\text{pred}}}{\frac{v+v_{\text{pred}}}{2b} + \tau}$$

v_{pred} is the velocity of the leading vehicle, g the gap to it, τ the reaction time (usually set to 1 s), and b the maximum deceleration. v_{safe} is proven by Krauß to prevent any collisions.

To compute the desired velocity for the next time step v_{next}, maximum velocity v_{max} and acceleration a must be considered, too; for a simulation time step Δt, this leads to:

$$v_{\text{des}} = \min\left[v_{\text{safe}}, v_{\text{max}}, v + a\Delta t\right]$$

Finally, actual velocities may be lower than the maximum possible one; this is accounted for by subtracting a random value from the current velocity. $rand$ returns a uniformly distributed value between 0 and 1, and σ is a configuration parameter between 0 and 1 to determine the degree of randomness within the laws of physics. Following [7], we set σ to 1. Thus, the velocity of the next timestep v_{next} is computed as

$$v_{\text{next}} = \max\left[0, \text{rand}(v_{\text{des}} - \sigma a\Delta t), v_{\text{des}}\right]$$

2.2 Multi-lane Traffic

Krauß has also suggested a lane-change model for two-lane roads (see [7], Chap. 8.2). According to this model, a change from the right to the left lane is favoured if v_{safe} on the right lane is less than v_{max}. A change from the left to the right lane is favoured if v_{safe} on the left lane, as well as a supposed $v_{safe,right}$ on the right lane are bigger than v_{max}. The asymmetry reflects the rule to use the right lane when not passing. If the velocity is lower than the congestion threshold v_{jam} (usually set to 60 km/h), no lane change will be favoured. In addition, a lane change will be favoured with a small probability p_{change} despite any other condition.

The favoured change will actually be carried out by moving the vehicle to the other lane, if and only if all safety conditions are satisfied after the change, i.e., the changed vehicle and its new follower are still within their v_{safe}.

To prevent passing on the right, a vehicle on the right lane is not allowed to go faster than the supposed $v_{safe,left}$ with respect to the predecessor on the left lane, as long as the vehicle is faster than the congestion threshold v_{jam}. Due to spatial discretisation, the gap to the predecessor on the left lane can be arbitrarily small after passing, causing the vehicle on the right to brake more than necessary. This is the case if the gap is smaller than the distance the passed car travels within a time step. In this case braking of the passed vehicle is limited to the velocity of the passing car.

2.3 Our Extensions to Krauß's Lane-Change Model

As Krauß's lane-change model is limited to two lanes, we extended it to an arbitrary number of lanes. The safety conditions become more complicated, because we have to consider the case in which we get a potential conflict when both a vehicle from the left and from the right intend to change to the same position of an interior lane.

In addition, we found that we need to allow vehicles in congested traffic to change back to the right lane. This will already be favoured if $v_{safe,right}$ is bigger than v_{safe} on the regarded vehicle's current lane, which is a far less strict condition for a change to the right than the original condition by Krauß for the uncongested state.

To prevent two vehicles of a slower type (e.g., trucks) to go side by side forever, blocking the road for faster vehicles (so-called elephant races), we add a decay of the maximum velocity from a more left to a more right lane, such that vehicles can go their defined v_{max} only on the leftmost lane. For each lane right to the leftmost lane a small decrement velocity (usually 5 km/h) will be subtracted. Hence, a vehicle on an empty street will go on the right with less than v_{max}, but such a case has no relevance for congested traffic.

As another extension we do not allow lane changes if the vehicle is very slow, usually less than 5 km/h. This prevents very slow or standing vehicles from changing their lanes.

2.4 Other Models

There are many traffic models at all levels of detail of traffic behaviour [4]. As a particular example we mention Helbing's Intelligent Driver Model (IDM, see [17]). Like the Krauß model, the IDM also falls into the category of car-following models.

Originally developed as a single-lane model, it was later extended to a multi-lane model by the lane-changing model MOBIL (Minimizing Overall Braking Induced by Lane-Changes, [18]). MOBIL allows to configure the left-/right-lane preferences by parameters so it can also simulate traffic without the obligation to drive on the right lane when possible.

3 Simulation

We have implemented the Krauß model with the above extensions in a simulator that allows complete control of a large variety of interactive aspects, including basic parameters like length of the road, number of lanes, vehicle density, and duration of a timestep. Furthermore, an arbitrary number of vehicle types can be configured, each with its own maximum velocity, length, acceleration, and parameters to compute its fuel consumption, and, of course, the fraction of vehicles to assign to each type.

The Krauß model is designed as a plugin, so, it can easily be replaced by another car-following model even during a running simulation. Of course, all of its parameters are configurable, too. We have also implemented the Intelligent Driver Model [17] as an alternative to the Krauß model, but without lane-changing capabilities.

The fuel consumption is computed in four parts: idle consumption, rolling resistance, air resistance and acceleration. If the acceleration is less than a fuel-cut-off threshold (set to $-0.38\ \mathrm{m/s^2}$) there is no consumption at all, accounting for the automatic complete fuel cut-off of modern cars. For a standing vehicle or acceleration between cut-off threshold and a coasting threshold (set to $-0.18\ \mathrm{m/s^2}$), there is only the (configurable) idle consumption of 1 l/h. In the remaining cases fuel consumption is the sum of a rolling resistance part (proportional to velocity), an air resistance part (proportional to the square of the velocity) and an acceleration part (proportional to the difference of current velocity and the velocity of the last timestep). All specific constants are derived for each vehicle type, such as front area, air resistance coefficient, and vehicle mass, as well as general parameters like air density, rolling resistance coefficient, the energy content of fuel and engine efficiency.

To test different strategies with respect to their improvement of the flow of traffic it is also possible to plug in so called influencers. Those are functions which are called on each timestep and car with the desired velocity and lane. Their output will be used as a new desired velocity and lane which will be further limited by the safety conditions. The *AutoNomos Strategy* (see Sect. 4) is implemented as such a pair of velocity- and lane-influencers.

To verify the results of our simulator we reproduced the results with SUMO [6]. SUMO (Simulator of Urban MObility) is an open-source microscopic traffic simulator developed by the German Aerospace Center DLR. Despite its name, it can also simulate highway traffic. SUMO is used by many different traffic researchers all over the world (see http://sourceforge.net/apps/mediawiki/sumo/index.php?title=Projects).

A variety of simulation results are shown in Sect. 5.

4 Improving the Flow of Highway Traffic

According to current research (see [17] and [16]), one of the major reasons for collapses of traffic flow on highways, known as traffic jams, is the wide velocity distribution of vehicles. This is well known to any experienced driver: even in dense traffic, high velocities are possible, as long as all vehicles move at almost the same speed; however, once a random fluctuation has occurred, the average speed drops considerably, and the overall pattern of motion becomes dominated by stop-and-go waves.

Overcoming non-uniform motion is a highly nontrivial matter. It is tempting to strictly enforce a uniform speed that seems to work so well before a collapse occurs. However, making large convoys of vehicles move in lockstep pushes the overall system into a highly unstable state; as a result, the catastrophic consequences of even a small failure or inaccuracy imply tremendous technical, legal, and psychological risks.

We have pursued a softer, and more self-regulating alternative, in which individual drivers are still responsible for avoiding collisions. Instead, our driving strategy tries to avoid excessive and unnecessary acceleration when it can be determined that deceleration is imminent. The overall objective was not only to conserve fuel, but also to homogenise the overall speed distribution; as it turns out, this does improve the average speed. One additional, but equally important requirement was not to depend on the participation of all drivers; instead, even a relatively small system penetration should lead to measurable individual benefits, which constitutes an incentive for using the system and comply with its recommendations.

After a large variety of different tests and simulations, we have developed a recommendation that is based on a convex combination of the desired velocity of the driver and the average velocity of the vehicles.

$$v_{\text{recommend}} = \lambda v_{\text{desired}} + (1-\lambda) v_{\text{average}} \tag{1}$$

We call this the *AutoNomos Strategy*. Here λ is a coefficient between zero and one. Setting $\lambda = 1$ means not to recommend any different velocity than without the strategy. On the other hand simulations (see Sect. 3) have shown that it is counterproductive to set $\lambda = 0$, i.e., to ignore the driver's desired velocity completely: whenever a vehicle ahead randomly slows down, vehicles following the strategy will also decelerate, implying that the average velocity converges to zero. Choosing the right compromise between these extremes works amazingly well. A patent [3] became effective in late 2010.

As simple as it is to state, the AutoNomos Strategy offers a variety of configuration options. Beside parameter λ, we have a range of choices how exactly the average of preceding vehicles should be taken. Taking the average over a particular number of vehicles ahead may run into problems in dense traffic, but can even encounter difficulties in intermittently sparse traffic, which happens in larger stop-and-go waves.

It is plausible to consider giving a higher weight to nearby vehicles; however, simulations have shown that this does not offer any clear advantage over the first choice of giving even weight to all considered preceding cars.

We also tried to account for current acceleration or deceleration, as we are more interested in the average velocity ahead when the guided vehicle will get there; we expected to have a better estimation for that moment, thus allowing even smoother motion. However, simulations resulted in huge oscillations, due to the feedback of local corrections. This is avoided by our simple recommendation.

Usually highways have more than one lane per direction. This offers more variations of the AutoNomos Strategy. The simplest strategy is not to distinguish between different lanes and take the average over all vehicles on all lanes. Simulations have shown that in the absence of a traffic jam, this will slow down vehicles on the passing lane, as they have to adapt to the average velocity of vehicles on the right lane, which are often slower trucks.

Therefore, we only consider the velocity average of vehicles on the same lane. In addition, we recommend not to change lanes if the average velocity of vehicles on the desired lane is similar to the average of the current lane.

5 AutoNomos Strategy Results

5.1 Single Lane

Our first results consider single-lane roads. For the vehicle-type parameters we chose typical passenger car values, which is the standard situation encountered by commuters in rush-hour traffic. We were particularly interested how well the AutoNomos Strategy works if not all vehicles are equipped so we tested different penetration rates.

We always started without the AutoNomos Strategy to let the system run into a natural stationary state and develop a traffic jam if the vehicle density is high enough. After this we turn the AutoNomos Strategy on to see its influence on the traffic situation. Clearly, being able to recover from a collapse of traffic flow is even harder than preventing it in the first place; as we will see in the context of urban traffic, this is not always possible when randomness is present.

Whenever our strategy was switched on, we observed a temporary drop in average velocity. This is because in the simulation the AutoNomos Strategy only recommends velocities up to the desired velocity, never above; it should be noted that this drop is an artifact of brutally switching on our strategy, so it does not constitute an actual disadvantage. After the initial velocity drop we observe that the average velocity recovers and in most cases even exceeds the average velocity without the AutoNomos Strategy.

Figure 2a shows the development of the average velocity. The low median before turning the AutoNomos Strategy on means that the majority of the vehicles is slower than the arithmetic average and only a few vehicles are faster. Figure 2b shows the development of the fuel consumption. Particularly the acceleration part and the air resistance part become significantly less, as the AutoNomos Strategy narrows the

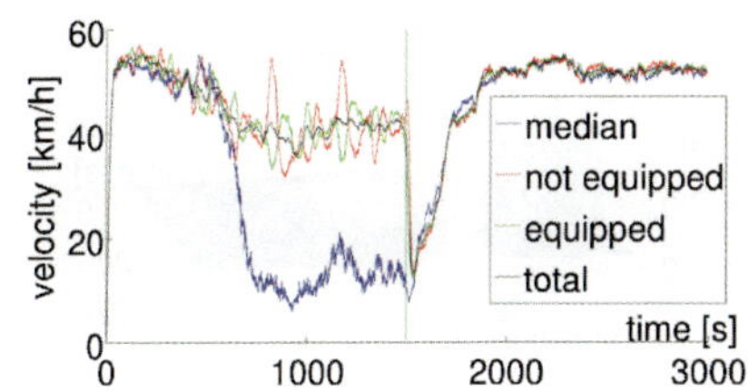

(a) Average velocities.

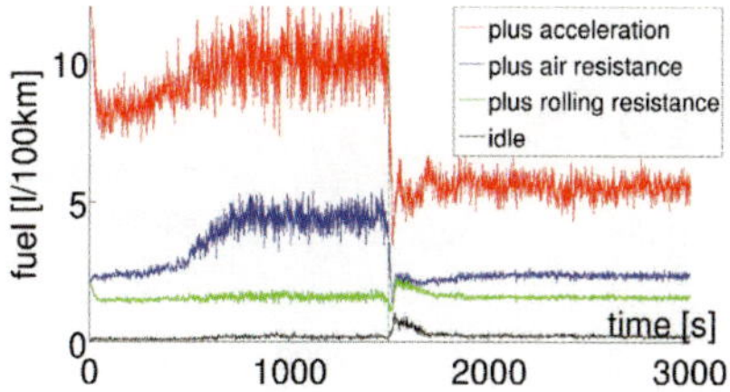

(b) Average fuel consumption, as described in Section 3.

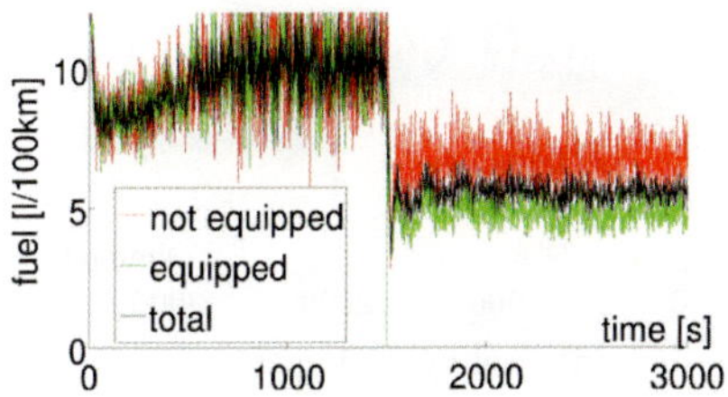

(c) Fuel by equipment.

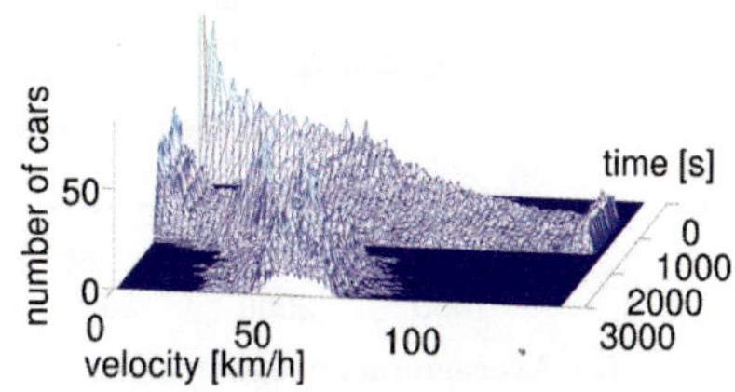

(d) Velocity histogram.

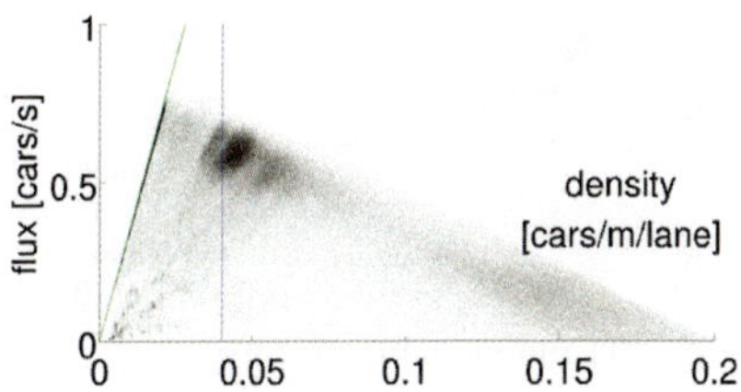

(e) Fundamental diagram without AutoNomos Strategy.

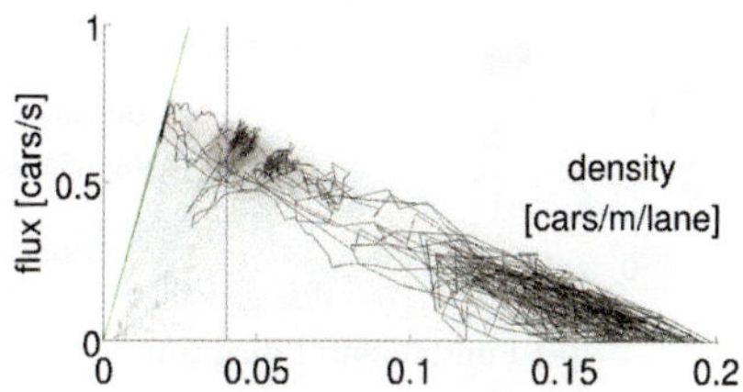

(f) Highlighted: a trajectory in the fundamental diagram.

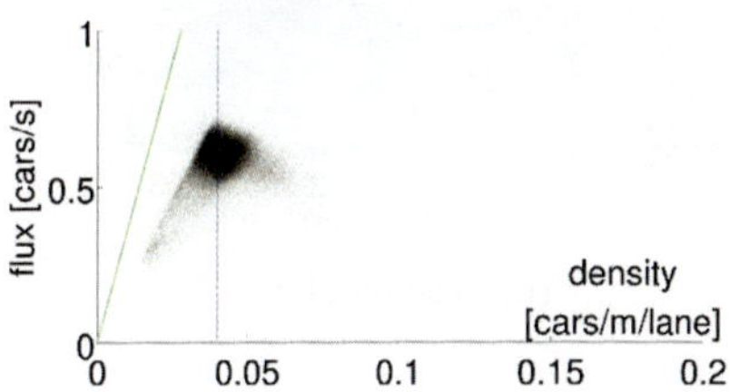

(g) Fundamental diagram with AutoNomos Strategy.

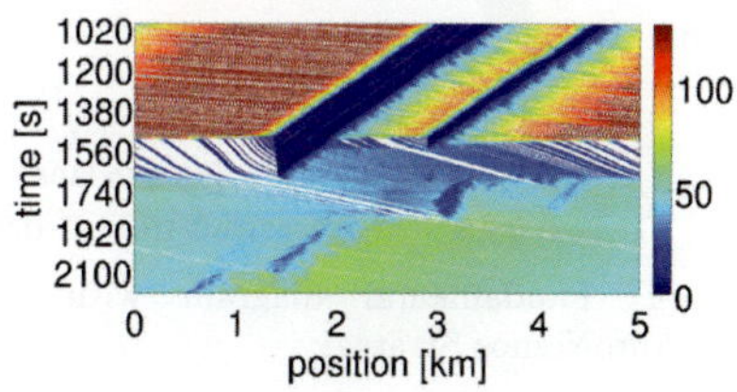

(h) Position plot coloured by velocity.

Fig. 2 A typical single-lane traffic scenario with 40 vehicles per kilometre, 60% equipment rate, $\lambda = 0.5$, averaging distance 500 m. Other values show similar behaviour of at least the same average velocity but still significant fuel saving. AutoNomos Strategy is switched on after 1500 s

velocity distribution, so that vehicles accelerate less, and there are fewer high-speed vehicles that are responsible for the high air resistance, which is proportional to the square of the velocity. Figures 2c to 2h are for comparison with the case of multiple lanes below (Fig. 3).

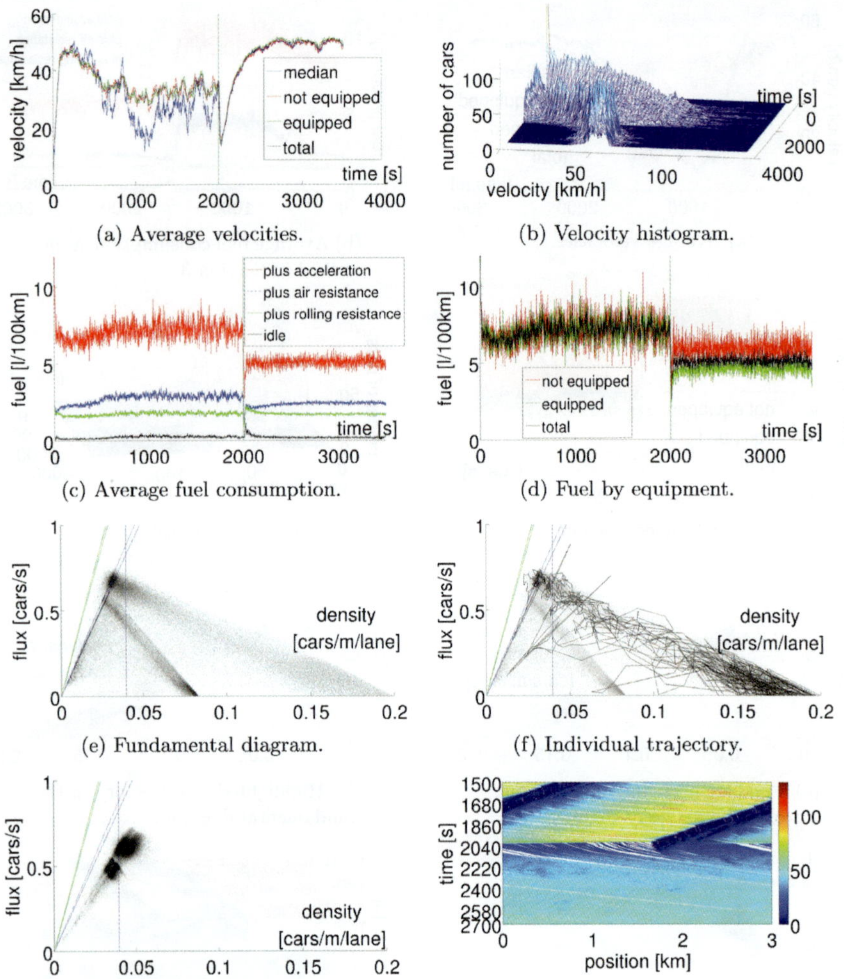

(a) Average velocities.

(b) Velocity histogram.

(c) Average fuel consumption.

(d) Fuel by equipment.

(e) Fundamental diagram.

(f) Individual trajectory.

(g) Fundamental diagram with AutoNomos Strategy.

(h) Position plot.

Fig. 3 Traffic on a multi-lane road with AutoNomos Strategy after 2000 s, 40 vehicles per km and lane, 60% equipment rate, $\lambda = 0.5$, averaging distance 700 m. In fundamental diagrams, lines mark maximum velocities of passenger cars (*green*) and trucks (*blue*)

5.2 Multiple Lanes

Multi-lane traffic is fundamentally different from single-lane traffic: non-equipped vehicles are now able to pass equipped vehicles, instead of indirectly obeying the AutoNomos Strategy by following an equipped vehicle. We also added 30% trucks as a different vehicle type with a lower maximum velocity of 80 km/h, bigger vehicle length and lower acceleration. In order to keep the fuel consumption comparable to the single-lane results, fuel consumption parameters are based on those of passenger cars. Equipment penetration did not distinguish between vehicle types, so the same fractions of passenger cars and trucks are equipped.

Figure 3a shows the same as Fig. 2a for a multiple-lane road. Like in the single-lane case, the average velocity recovers after some time and rises even beyond the average velocity without the AutoNomos Strategy. The tightening of the velocity distribution can clearly be seen in the velocity histogram in Fig. 3b. Without the AutoNomos Strategy we have a broad distribution with a few fast vehicles and a peak around 80 km/h from trucks. With the AutoNomos Strategy there is a narrow velocity distribution.

Figure 3c shows the fuel consumption split into its parts. Figure 3d shows the average fuel consumption of equipped and non-equipped vehicles beside the total average. The most fuel is saved by the equipped vehicles; but even the unequipped vehicles have a significant benefit from the AutoNomos Strategy.

The most important diagram in highway traffic research is the so-called fundamental diagram [8, 20]. It relates observed flows (vehicles per time) with their densities (vehicles per road stretch). Empirical fundamental diagrams have a triangular form with a sharply rising branch starting in the origin left of and a cloud of decreasing data points right of the maximum. The sharp branch on the left is known as the *free-flow branch*, the other branch as the *jam branch*. As the ratio of flow and density is the average velocity of a traffic state, the states in the free-flow branch all have similar velocity. The gradient of the jam branch shows the velocity of a traffic jam front [1, 4].

The fundamental diagram of our simulation (Fig. 3e) shows two jam branches. This is an artificial effect of the fact that we have only two vehicle types with fixed maximum velocities and vehicle lengths each; in reality, there is a distribution of maximum velocities and vehicle lengths. The intersections of our jam branches with the density axis at the reciprocals of the vehicle type lengths corresponds to the bumper-to-bumper density of standing vehicles. The gradient of the upper edge of the jam branch is approximately -15 km/h, which corresponds to the velocity of jam fronts observed all over the world.

Figure 3f marks the trajectory of a particular vehicle within the fundamental diagram. One can see that this vehicle—like all others—spends most of the time on the jam branch at low velocity and flow, and only a short time on the free-flow branch. Finally Fig. 3g shows the fundamental diagram after turning on the AutoNomos Strategy.

Figure 3h shows positions of all vehicles in each timestep coloured by their respective velocity. Traffic jams look like blue flanks between the flat areas of free flow. Again, the positive effects of our strategy can clearly be seen.

6 Urban Traffic

6.1 Traffic Collapse in an Urban Scenario

Most attempts that aim at improving the overall flow of traffic assume fixed capacities for roads in a street network. While classical optimisation methods consider

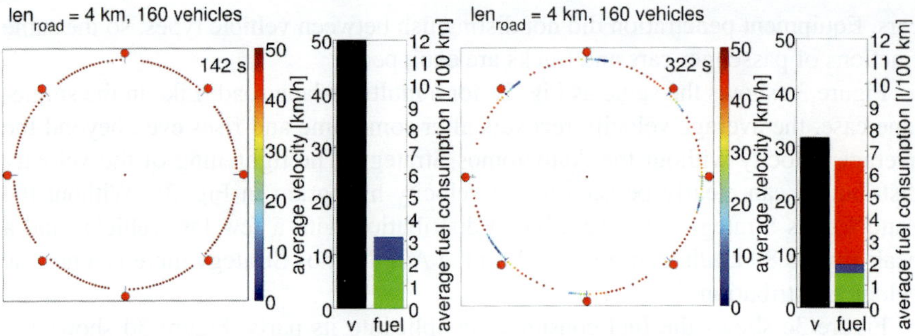

(a) A flow of vehicles optimally synchronised with a sequence of traffic lights.

(b) Synchronisation collapses in the presence of randomness.

Fig. 4 Optimisation of traffic-light synchronisation may yield unstable traffic flows. The *pictures* show screenshots of our simulator. The *left part* shows vehicles on a circular road coloured by their respective velocity together with eight traffic lights; also shown are current average velocity (*black, vertical bar*) and average fuel consumption (*multi-coloured, vertical bar*), which is split into parts, e.g., *red* for acceleration

only static flows that do not change over time, recent proposals are able to deal with dynamic flows; e.g., see [5]. However, even those methods do not account for phase transitions in the flow of traffic. Moreover, most attempts at improving the global flow of traffic tend to squeeze all slack out of the system; as it turns out, this is critical, as the resulting state becomes rather fragile, and leads to a situation in which minimal fluctuations lead to a drastic collapse of the flow capacity. This implies that the organic quality of traffic has to be taken into account when optimising its flow. So the goal is to optimise traffic flow such that usual fluctuations and capacity drops have as low impact as possible.

For a demonstration, consider the example in Fig. 4 in which traffic lights are tuned as a progressive signal system ("green wave"). With these settings we observed perfect platoons moving undisturbed by red lights (Fig. 4a). The average velocity is exactly at 50 km/h.

After switching on the usual randomness by setting σ to one, some vehicles do not make it through a green phase and have to stop and start, forcing the next platoon of cars to do the same (Fig. 4b). The average velocity breaks down to sometimes less than 30 km/h.

6.2 Flow Over Successive Traffic Lights

We have developed a method [19] for using local information to save fuel in an urban setting. From traffic lights, vehicles receive information about the next phase change. If the stop line cannot be reached during the green phase, the velocity during the approach will be optimised with respect to fuel consumption. In addition, if the vehicle already stands and has to wait more than a certain time for green, the engine will be turned off. Simulations have shown significant fuel savings with this algorithm.

Our next step will be to receive positions and velocities from other vehicles, too, and compute better estimates of the arrival time by accounting for vehicles between the guided one and the traffic light. Moreover, the optimisation can consider flow over several successive traffic lights.

6.3 Rerouting and Recovery

As mentioned in Sect. 6.1, urban road networks are optimised offline to fulfil the traffic flow demand as soon as possible. As stated above, this optimisation depends on fixed capacities between intersections, and cannot deal with the capacity drops caused by traffic collapses during rush-hour. Reoptimising flow through the whole network is too time consuming and not appropriate for the dynamics of real-world situations.

In current work, we are working on algorithmic methods to recover from such a collapse situation. A forthcoming paper by Fekete, Gray and Wiedenroth reassigns flows locally such that all flow is removed from the collapsed link and reassigned to links not at their respective capacity limit, so that the collapsed edge can recover, and then improve the total flow through the overall network. We envision this self-healing approach to be particularly useful for delaying the collapse at the onset of rush-hour traffic, and accelerate the recovery at the end of rush-hour.

This and further developments are left to current and future work.

References

1. Chowdhury, D., Santen, L., Schadschneider, A.: Statistical physics of vehicular traffic and some related systems. Phys. Rep. **329**(4–6), 199–329 (2000)
2. Fekete, S.P., Schmidt, C., Wegener, A., Hellbrück, H., Fischer, S.: Empowered by wireless communication: distributed methods for self-organizing traffic collectives. ACM Trans. Auton. Adapt. Syst. **5**(3), 439–462 (2010)
3. Fekete, S.P., Tessars, C., Schmidt, C., Wegener, A., Fischer, S., Hellbrück, H.: Verfahren und Vorrichtung zur Ermittlung Einer Fahrstrategie, September 2008. Patentnummer DE 10 2008 047 143 B4
4. Helbing, D.: Traffic and related self-driven many-particle systems. Rev. Mod. Phys. **73**, 1067–1141 (2001)
5. Köhler, E., Skutella, M.: Flows over time with load-dependent transit times. SIAM J. Optim. **15**(4), 1185–1202 (2005)
6. Krajzewicz, D., Hertkorn, G., Rössel, C., Wagner, P.: SUMO (Simulation of Urban MObility); an open-source traffic simulation. In: Al-Akaidi, A. (ed.) Proceedings of the 4th Middle East Symposium on Simulation and Modelling, pp. 183–187. SCS European Publishing House, Erlangen (2002)
7. Krauß, S.: Microscopic modeling of traffic flow: investigation of collision free vehicle dynamics. PhD thesis, Universität zu Köln (1998)

8. Kühne, R.: Das Fundamentaldiagramm – Grundlagen und Anwendungen. FGSV Merkblatt (Entwurf), p. 59 (2004)
9. Kwella, B., Lehmann, H.: Floating car data analysis of urban road networks. In: Pichler, F., Moreno-Díaz, R., Kopacek, P. (eds.) Proc. EUROCAST'99. LNCS, vol. 1798, pp. 357–367. Springer, Berlin (2000)
10. Nagel, K.: High-speed simulation of traffic flow. PhD thesis, Center for Parallel Computing, Universität zu Köln, Germany (1995)
11. Nagel, K.: Traffic Networks. In: Bornholdt, S., Schuster, H.G. (eds.) Handbook of Graphs and Networks—From the Genome to the Internet. LNCS. Wiley-VCH, Berlin (2003). Chap. 11
12. Nagel, K., Schreckenberg, M.: A cellular automation model for freeway traffic. J. Phys., I Fr. **2**, 2221–2229 (1992)
13. Nagel, K., Wagner, P., Woesler, R.: Still flowing: approaches to traffic flow and traffic jam modeling. Oper. Res. **51**(5), 681–710 (2003)
14. Rickert, M., Nagel, K.: Experiences with a simplified microsimulation for the Dallas/Fort Worth area. Int. J. Mod. Phys. C **8**, 133–153 (1997)
15. Rickert, M., Nagel, K., Schreckenberg, M., Latour, A.: Two-lane traffic simulation on cellular automata. Physica A **231**, 534–550 (1996)
16. Schönhof, M., Helbing, D.: Empirical features of congested traffic states and their implications for traffic modeling. Transp. Sci. **41**(2), 135–166 (2007)
17. Treiber, M., Helbing, D.: Microsimulations of freeway traffic including control measures. Automatisierungstechnik **49**, 478–484 (2001)
18. Treiber, M., Helbing, D.: Realistische Mikrosimulation von Straßenverkehr mit einem einfachen Modell. In: Tavangarian, D., Grützner, R. (eds.) ASIM 2002, Tagungsband 16. Symposium Simulationstechnik, pp. 514–520 (2002)
19. Wegener, A.: Organic-Computing-Konzepte und deren Umsetzung für dezentrale Anwendungen im Straßenverkehr. PhD thesis, Universität zu Lübeck (2009)
20. Wu, N.: Verkehr auf Schnellstraßen im Fundamentaldiagramm – Ein neues Modell und seine Anwendungen. Straßenverkehrstechnik **8**, 378–402 (2000)

S.P. Fekete (✉) · B. Hendriks · C. Tessars
TU Braunschweig, Braunschweig, Germany
e-mail: s.fekete@tu-bs.de

B. Hendriks
e-mail: b.hendriks@tu-bs.de

A. Wegener · S. Fischer · S. Ebers
Universität zu Lübeck, Lübeck, Germany

A. Wegener
e-mail: wegener@itm.uni-luebeck.de

S. Fischer
e-mail: fischer@itm.uni-luebeck.de

S. Ebers
e-mail: ebers@itm.uni-luebeck.de

H. Hellbrück
Fachhochschule Lübeck, Lübeck, Germany
e-mail: hellbrueck@fh-luebeck.de

Chapter 5.3
Applying ASoC to Multi-core Applications for Workload Management

Johannes Zeppenfeld, Abdelmajid Bouajila, Walter Stechele, Andreas Bernauer, Oliver Bringmann, Wolfgang Rosenstiel, and Andreas Herkersdorf

Abstract This article presents the use of decentralised self-organisation concepts for the efficient dynamic parameterisation of hardware components and the autonomic distribution of tasks in a symmetrical multi-core processor system. Using results obtained with an autonomic system-on-chip hardware demonstrator, we show that Learning Classifier Tables, a simplified XCS-based reinforcement learning technique optimised for a low-overhead hardware implementation and integration, achieve nearly optimal results for task-level dynamic workload balancing during run time for a standard networking application. Further investigations show the quantitative differences in optimisation quality between scenarios when local and global system information is available to the classifier rules. Autonomic workload management or task repartitioning at run time relieves the software application developers from exploring this NP-hard problem during design time, and is able to react to dynamic and unforeseeable changes in the MPSoC operating environment.

Keywords Autonomic · System on chip · Application · Network processor · Multi core · Learning classifier table

1 Introduction

Single-chip multi-processors have become the mainstream architecture template for advanced microprocessor designs across a wide spectrum of application domains. Intel, current market leader for general purpose computing, has abandoned its traditional strategy of primarily scaling frequency to increase microprocessor performance, instead introducing dual-, quad- and recently octal-core (Nehalem) Xeon processors [9]. In their research labs, Intel has even implemented the 80 core Tera-Scale processor [4] with a 2D network-on-chip and sophisticated 3D memory stack access. SUN Niagara, ARM MPCore, IBM Cell Broadband Engine, Nvidia GeForce, the CSX700 from ClearSpeed, and the TILE64 from Tilera are other examples of a non-exhaustive list of massively parallel multi-core architectures for use in mobile communications, graphics processing, gaming, industrial automation and high-performance scientific computing. While progress in deep sub-micron CMOS

C. Müller-Schloer et al. (eds.), *Organic Computing — A Paradigm Shift for Complex Systems*, 461–472
Autonomic Systems, DOI 10.1007/978-3-0348-0130-0_30,

technology integration enables the physical realisation of this vast amount of nominal processing capacity on a single chip, the application programmer community—across all of the above mentioned systems—is insufficiently supported by tools and methods to efficiently exploit the available parallel resources [5]. We regard this circumstance as a major challenge for a fast and efficient adoption of multi-core processors.

One of the key difficulties in efficiently executing an application composed of multiple, parallelisable tasks is to find an appropriate task distribution across available processing resources. Various approaches have been proposed to aid the designer in accomplishing this at design time, such as [11] which advocates the use of neural networks and support vector machines to map parallel programs onto multiple cores. Other approaches perform such a mapping at run time, but do not make use of learning techniques to improve mapping performance based on past experiences [10]. Combining the two, approaches such as [8] collect training data from function executions at run time, but perform any learning in additional design cycles offline. Numerous other publications propose similar solutions; however, we are not aware of any existing approaches that make use of machine learning techniques to optimise task distribution in hardware at run time. In this article we will therefore explore the applicability of the autonomic system on chip [1] paradigm (ASoC, presented in Chap. 4.7) to autonomously and dynamically partition individual tasks of a SW application among a set of homogeneous processor cores at run time [14]. Other optimisation goals achievable using ASoC, such as reliability gain, power reduction or performance improvements, can then be included to fashion a complete reliable and efficient system on chip.

2 System Overview

The application chosen for this article is multi-core Internet-Protocol (IP) packet forwarding for a variety of packet header- and payload-processing scenarios under variable traffic workloads. Packet forwarding lends itself well to the demonstration of autonomic enhancements, since variations in incoming packet rate, type and size can cause drastic changes in the system's workload, making it difficult to predict the system behaviour at design time. In addition, since individual packets are relatively small and quick to process, a large number of packets are processed in a relatively short amount of time. This shortens the required execution time needed to average out noise over a certain number of processed packets, and also allows for frequent changes in the system's workload to demonstrate the adaptivity and learning capabilities of the autonomic layer.

From a more general perspective, IP packet forwarding is a representative example for a class of applications with the following properties: the application consists of a set of N tasks, T_1 to T_N, with every task having different processing requirements in terms of instructions per task ($inst/T_i$). Tasks can be executed sequentially, conditionally and/or in iterative loops. The application is triggered by external events with varying parameter sets determining the system's overall workload.

Mapping N tasks to a set of M homogeneous processing elements P_1 to P_M with the objective of achieving equal workload distribution corresponds to a function partitioning problem of exponential complexity, $O(M^N)$, which cannot be solved with exact methods (e.g. integer linear programming) for real-world problem dimensions. Instead of applying conventional partitioning heuristics (e.g. hierarchical clustering [12], Kernighan-Lin [6], simulated annealing [7] or Tabu search [3]) to explore possible partitioning alternatives for various scenarios at design time, we propose adaptive run-time learning techniques to determine an appropriate partitioning for the specific scenario encountered by the system in the field. This saves a lot of guesswork during design time, since only very few of the potential partitionings will actually be needed by a certain application running in a specific environment, and both the application and the operating environment may not be completely defined until run time (especially in general purpose MP-SoCs destined for a wide variety of applications and operating conditions).

2.1 Functional Layer

The hardware system on which the network processing (NP) application presented here is executed is depicted in Fig. 1. The light shapes make up the functional layer, while the shaded components represent the autonomic elements—each containing monitors, actuators, a Learning Classifier Table (LCT [13]) evaluator, and a communication interface, as detailed in Chap. 4.7. The underlying functional layer is a fairly simple, bus-based MPSoC. Packets arrive and are transmitted over an Ethernet MAC (bottom centre), are stored in memory (bottom right), and are processed by one or more of the processing cores (top). Each core can query its associated autonomic element to determine the task(s) to be run on that core. Task execution is non-pre-emptive, i.e. once started a task will run until it completes. Migrating a task will cause a different core to run the task when new packet data arrives.

Due to the difficulty in obtaining repeatable results with real-world packet data, incoming packets are generated by a separate packet generation task, rather than from an external source. This allows for much finer control of the system's workload, and provides a mechanism by which various implementation alternatives (e.g. autonomic vs. static systems) can be compared given identical workload variations.

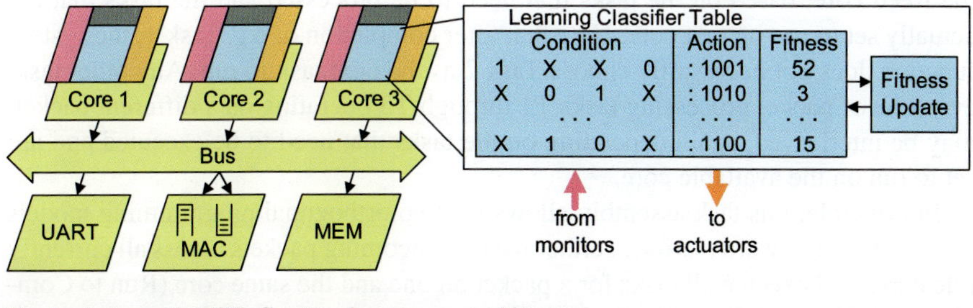

Fig. 1 Autonomic MPSoC network processor architecture

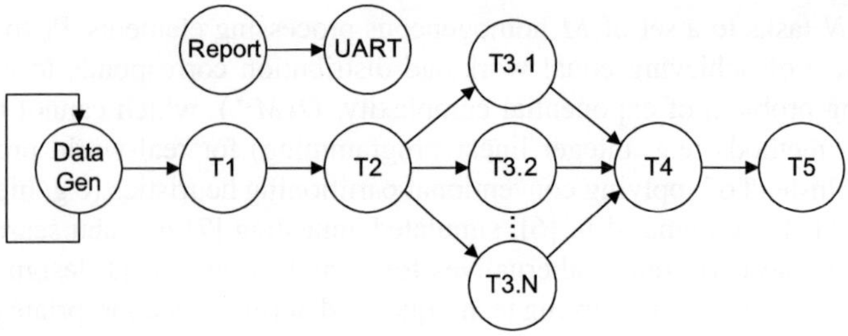

Fig. 2 Task graph of the network processing application

2.2 Application Software

The main focus of this article is to demonstrate the ability of an autonomic system to deal with workload management in a typical multi-core application. Since workload management becomes more complex and "interesting" with an increasing number of processing tasks, and to be able to run the application on any generic MPSoC platform (such as the system used here), as many packet processing functions as possible are implemented as software tasks that can be moved freely among the system's cores. This also includes tasks such as transferring data between the MAC and memory, which could also be accomplished easily—and perhaps more efficiently—by a dedicated hardware DMA controller.

In the representative network processing application used throughout this article, every packet passes through five stages of execution:

Task 1:	Transfer incoming packet to memory
Task 2:	Determine type of processing to be done on packet
Task $3.1 - N$:	Perform one of N packet header or payload processing tasks
Task 4:	Reorder packets for in-order transmission
Task 5:	Transfer packet from memory to MAC

While Tasks 1, 2, 4 and 5 are identical for every packet, Task 3 can be different for different packet types, as shown by the task graph in Fig. 2. Additional tasks are responsible for periodically reporting system monitor information over a RS232 console interface, or for generating packets as discussed above. Whenever a task completes, a system function determines the task that should be executed next on the freed core, based on the tasks that need to be processed and the tasks that are actually set to run on that core. Note that after completion of e.g. Task 1, the system function does not necessarily choose Task 2 as the next task to run. Any other task, even one of packet processing tasks T1 through T5 operating on a different packet, may be interleaved, again depending on the tasks that need to be executed and are set to run on the available core.

In principle, this task assembly allows for two orthogonal programming models on a generic hardware platform: either balance incoming packets across all currently idle cores and execute all tasks for a packet on one and the same core (Run to Completion (RTC) model), or distribute the tasks among all cores and let each packet traverse the cores in a sequential fashion (Pipelined model). Both models have unique characteristics, and both have advantages and disadvantages. RTC treats all cores as independent processing elements, thus eliminating the need for complex task partitioning among cores. RTC is easily scalable, but relies on little to no data sharing between successive packets, and that event/packet balancing in the application is an important factor. Pipelining requires equal workload sharing of tasks among processing elements in order to achieve optimised throughput and efficiency (avoiding pipeline "bubbles" and excessive slack). On the positive side, the pipelining model is applicable to a larger class of parallel applications, achieves a smaller instruction footprint per core and supports local state and data caching for each packet flow. We will show that, given an autonomic system to balance the workload at run time, a designer can treat tasks as if they were executed using a run to completion model, while additionally gaining the benefits provided by a pipelined system.

2.3 Autonomic Layer

The autonomic layer of the presented system contains an autonomic element for each of the system's three cores. Autonomic elements for the bus, memory, MAC and interrupt controller are not included, as the cores are currently our primary target for optimisation. However, given appropriate monitor and actuation interfaces in the remaining components, adding autonomic elements on top of these functional elements should not pose a major difficulty.

Monitors

Two local monitors keep track of the current core frequency and utilisation, and are multiplied together to produce a third monitor value—the core's workload:

$$load_{\text{core}} = util_{\text{core}} \cdot freq_{\text{core}} \tag{1}$$

Whereas the utilisation indicates the percentage of cycles that the core is busy (i.e. is processing a packet rather than waiting for new data in an idle loop), the workload indicates the actual amount of work that the core is performing per unit of time (useful cycles per second). Since the resulting value is helpful in comparing the amount of processing power being contributed by each core, it is shared with the other autonomic elements (AEs) over the AE interconnect. The workload information can then be averaged over all three cores, and by comparison with its own workload allows each core to determine whether it is performing more or less than its "fair share" of work. The difference between the core's and the average workload therefore provides another useful monitor value.

Actuators

In order to allow the AE to influence the operation of its core, two actuators are provided. The first of these adjusts the core's frequency by a certain value. Note that this is a relative change in frequency, i.e. the new frequency value depends on the old, which makes it easier to provide classifier rules that cover a larger range of monitor inputs. For example, with relative rules it is possible to express the statement "when utilisation is high increase the frequency" using a single classifier rule, which would not be possible with an absolute actuator that sets the frequency to a fixed value.

The second actuator triggers a task migration from the AE's core to one of the other cores. After migration of a task, any further requests to start the execution of that task will be performed by the target core instead. If a task migration is triggered while the task is already running, execution of the task is completed first (non-pre-emptive task migration). This minimises the amount of state information that needs to be transferred from one core to another, reducing the performance impact of migrating tasks. In the current implementation, the task chosen to be migrated is the task that most recently completed execution on that core. This ensures that only active tasks are migrated, since migrating an idling task would have no effect on the workload of either the source or destination core. Once a task is selected, a migration request is broadcast over the AE interconnect, where each of the other AEs can choose whether to accept the task or not by making a bid for that task. The task is then migrated to the highest bidder.

Evaluator

Choosing an appropriate action to be performed based on the incoming monitor signals is the responsibility of the LCT evaluator. By determining a fitness for each rule in the table, the LCT is able to learn which rules perform best in a given situation. Determining the fitness of a rule, both during the design-time rule creation phase described in Chap. 3.2 and during regular MPSoC operation, is accomplished by using a global objective function, which evaluates all available monitor signals to determine how well the system is currently functioning. First, a delta value is calculated for each monitor signal to indicate how close that signal is to its optimal value (low values are desirable):

$$\delta_{\text{freq}} \propto freq_{\text{core}} \tag{2}$$

$$\delta_{\text{util}} \propto (util_{\text{target}} - util_{\text{core}}) \tag{3}$$

$$\delta_{\text{load}} \propto |load_{\text{core}} - load_{\text{average}}| \tag{4}$$

The function used for each of these values is determined by the designer, and expresses the designer's optimisation goal. For our NP application, we would like each core to have as low a frequency as possible, which keeps the power consumption of the system low (voltage is assumed to scale in relation to the frequency). Likewise, we want the core's utilisation to reach a certain target value (in our case 75%), so that as few as possible processing cycles are wasted in an idle loop, while still allowing the system to clearly differentiate between an acceptably high load and an overload of the system. Finally, the workload of all cores should be similar to avoid temperature hot spots and to ensure similar aging across all components. Depending on the available monitor signals and the designer's optimisation goals, other delta functions may be chosen.

Combining the individual delta functions is accomplished by a simple weighting scheme, which yields the system's objective function:

$$f_{\text{objective}} = w_1 \cdot \delta_{\text{freq}} + w_2 \cdot \delta_{\text{util}} + w_3 \cdot \delta_{\text{load}} \tag{5}$$

The weights can be chosen according to which optimisation goal is most important; for the results presented below each delta function was weighted equally.

Although the objective function provides a value indicating how well the system is performing at some instant in time, in order to determine the worth of a certain rule we need to calculate a reward R that compares the objective value sampled before and after the rule was applied. If the previous objective value O_{T-1} is larger than the new objective value O_T, indicating that the system has gotten closer to its optimal state, a positive reward is chosen based on the magnitude of the change. Otherwise, a negative reward is returned. This reward is then used to update the fitness of the applied rule, which allows the LCT to make use of past experience when determining which rule should be applied during similar situations in the future.

3 Results

Given the NP packet forwarding application presented above, this section explores the differences between a classical implementation of the application on a purely functional system, and an implementation designed to make use of various autonomic enhancements.

In order to show the adaptive nature of the autonomic system, the packet generation task (remember that an internal packet generation task replaces the ingress path of the Ethernet MAC in order to generate identical traffic streams across multiple tests) creates a periodically changing traffic pattern to be processed by the system. Each input pattern is applied for 10 seconds, after which a new pattern is generated. For each pattern, the system must either adapt to meet the processing requirements, or maintain a sufficient reserve of processing power to be able to cope with even the largest and most processing intensive packets.

All systems used below are based on a common base system configuration. The hardware components and their functionality, as well as the software tasks used for packet processing were described in Sect. 2.1 and 2.2, respectively. Unless stated otherwise, the initial task distribution is such that one core is responsible for ingress path processing (Tasks 1 and 2), one core is responsible for all payload processing (Tasks 3.x), and the third core is responsible for egress path processing (Tasks 4 and 5). The initial core frequencies are chosen such that the system is able to handle all generated traffic patterns without requiring parameter adaptation. Those systems containing autonomic enhancements use the monitors, actuators and LCT evaluator described in Sect. 2.3 to autonomically adjust the frequency and migrate tasks.

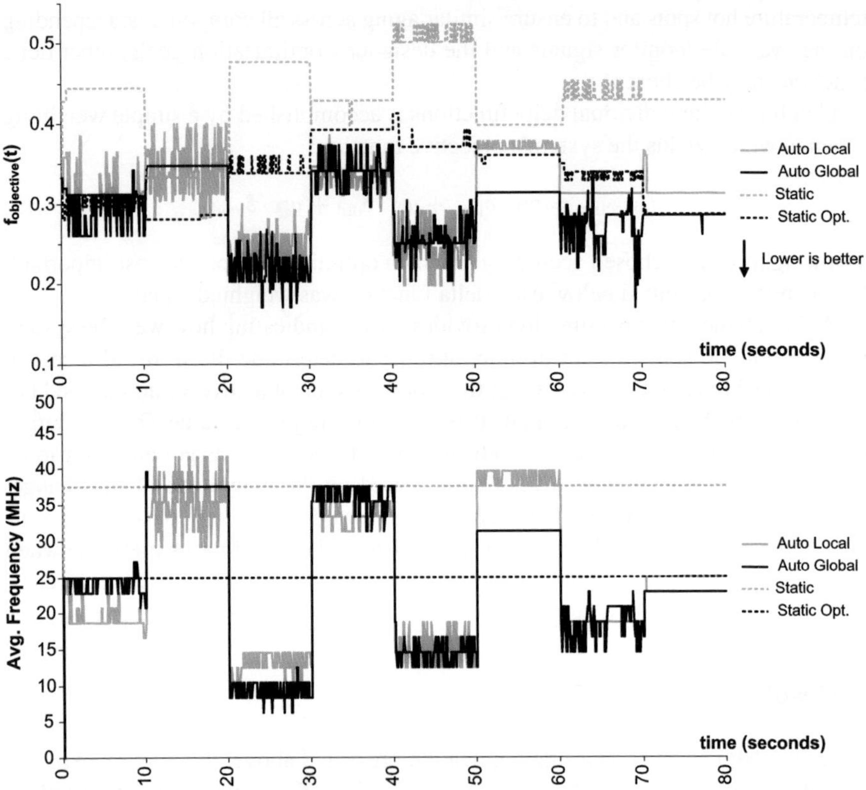

Fig. 3 Objective value and average core frequency of static and autonomic systems across various traffic patterns. Lower values are better

3.1 Comparison of Autonomic and Static Systems

The results shown in Fig. 3 compare the objective value and frequency adjustment of four different system configurations. The first two systems are static without any autonomic enhancements, where the first corresponds to the base system configuration described above. In the second static system, both the task distribution and core frequency were hand optimised with prior knowledge of the incoming traffic pattern. The results of this optimised system demonstrate how an ideally parameterised static system compares to a system with autonomic enhancements.

Both autonomic systems are based on the common system configuration, but are differentiated by the monitor signals available to them. While the second system uses the AE interconnect to share workload information globally among the core AEs, the first autonomic system must base its optimisation decisions solely on local monitor information. Both systems remain capable of migrating tasks, however.

In comparison to a static system, the trend of the objective value clearly shows the benefits of an autonomic system regarding fulfilment of the objective function chosen in Sect. 2.3. The autonomic system with global information is able to maintain system operation that is in most cases better than that achieved by the static system. Only for the second and most processing intensive traffic scenario, for which the static system was specifically hand optimised using prior knowledge of the traffic patterns, does the static system improve on the autonomic system's potential for optimisation. This is a direct consequence of the fact that a system optimised for one scenario (in this case for the traffic pattern starting at 10 seconds) is not necessarily optimised for other scenarios. Whereas the designer must choose a certain scenario for which the static system is optimised during design time, the autonomic system can optimise itself during run time to whatever scenario it is confronted with. Not only does this provide decent optimisation for a much larger set of foreseeable and unforeseeable scenarios, it also relieves the designer of having to optimise all aspects of the system at design time.

One benefit of the static system over the autonomic one is exhibited by the trend of the average core frequencies in the bottom plot of Fig. 3. Whereas the autonomic system must maintain a core utilisation below 100% in order to be able to determine that the system is not overloaded, the static system can be optimised to fully load the system, since no measurement of the utilisation is needed. Since the resolution of the utilisation monitor is relatively low, this results in a fairly large penalty in how far the frequency can be reduced by the autonomic systems. Nevertheless, the average frequency over all eight traffic patterns is roughly the same for the autonomic and optimised static system, whereas the non-optimised static system, whose initial parameters match those of the autonomic systems, maintains a high frequency over all traffic patterns.

Comparing the locally and globally optimised autonomic systems also yields some interesting results. As is to be expected, the autonomic system, given more information, is consistently able to more closely achieve optimal performance according to the objective function. The average frequency of the three cores shows a different trend, however, with an occasionally lower average core frequency in the locally optimised system, specifically for the first two traffic patterns. This is a result of the frequency directed optimisation occurring in the autonomic system given purely local information. Whereas the globally optimised system considers the performance of the system as a whole (resulting in a lower global objective value), the locally optimised system specifically tries to maintain as low a core frequency as possible. Although this does result in a lower frequency for specific cases, the global objective value suffers as a result.

3.2 Comparison of Autonomic and DVFS Systems

Figure 4 compares the globally optimised autonomic system presented in the previous section with a system that employs dynamic voltage and frequency scaling (DVFS). DVFS allows the non-autonomic system to adjust its operating frequency

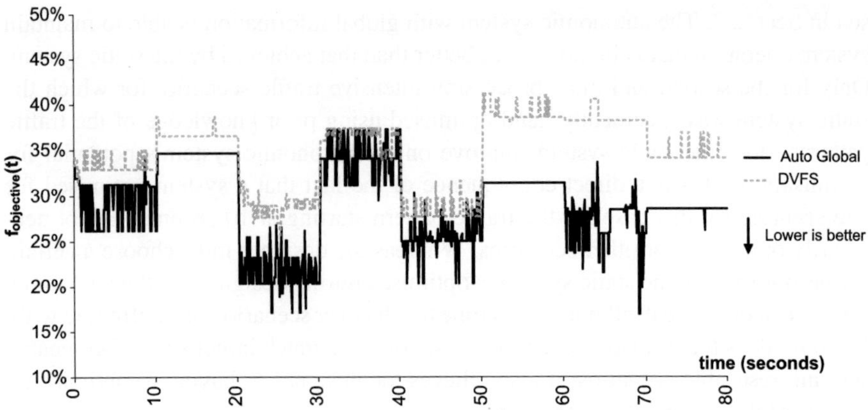

Fig. 4 Objective value of dynamic and autonomic systems across various traffic patterns. Lower values are better

in a fashion similar to the autonomic frequency actuator. The dynamic system corresponds to a DVFS-enhanced version of the hand-optimised static system from Sect. 3.1.

Comparing the objective values of the two systems, it can be seen that, although similar, the globally optimising autonomic system is still able to achieve somewhat better results than the DVFS system. This is due to the autonomic system's ability to migrate tasks in addition to adjusting the frequency of the cores. While the DVFS system can, as opposed to the static system from the previous section, maintain a consistently high utilisation by regulating the frequency, the distribution of tasks over various cores can be less than optimal in certain situations. The general increase in the objective value achievable by the DVFS system over the static systems from the previous section, noticeable especially during the second traffic pattern, is again explained by the necessity of a core utilisation below 100% to ensure that a decrease in frequency does not result in an overloaded system. This results in generally higher frequencies than achieved with the hand optimised system, on the other hand, the DVFS and autonomic systems do not need to know in advance what minimum frequency is needed by the cores to handle any workload presented to the system.

3.3 Area Overheads

The hardware overheads associated with various components of the autonomic layer are presented in Table 1. These results were obtained using a Gaisler Leon 3 [2] based FPGA implementation of the functional and autonomic system introduced in Sect. 2. The system was synthesised for a Xilinx Virtex 4 FPGA (XC4VLX100). The unmodified Leon 3 core provides a baseline comparison for hardware overheads used by the autonomic elements. With all modifications included (monitors,

Table 1 Hardware overheads for various AE components

	Flip-flops	LUTs	BRAMs	Mult.	Overhead
Leon3	1749	8936	28	1	–
Leon3 AE	2122	10213	29	2	14.3%
LCT	66	116	1	1	1.4%
Task Actuator	57	299	0	0	3.5%
Frequency Actuator	7	19	0	0	0.2%
Utilisation Monitor	35	74	0	0	0.8%
Load Monitor	20	40	0	0	0.5%
AE Interface	173	399	0	0	4.5%

actuators, LCT evaluator, AE communication infrastructure and additional supporting logic), the area overhead (number of slices) of the autonomic element is 14.3% of the core. Given the typical scaling between successive CMOS technology generations, we consider this a small price to pay for allowing the system to optimise itself during run time in regards to decisions that, due to a lack of information about the application or run-time environment, would have been difficult or even impossible to make during design time.

4 Conclusion

In this article we have demonstrated the applicability of self-organisation concepts and hardware-based machine learning techniques for the run-time binding of software tasks to homogeneous multi-core processors, specifically in the case of a generic packet forwarding network processor application. Software application developers can follow established design flows for functional partitioning of applications into sub-functions (tasks) without being forced to consider the underlying parallel MPSoC hardware architecture. LCT-based hardware evaluators take care of balancing the workload among available processing resources, without requiring a special programming language or fundamental OS modifications. It has been shown that the autonomic system is capable of parameterising a pipelined architecture such that it achieves similar or better performance when compared to static or DVFS-enabled implementations. By using run-time autonomic task migration it becomes possible to combine the benefits of pipelined and run-to-completion architectures, while delegating solution of the disadvantages, most notably the difficulty of efficiently distributing the workload of a pipelined system, to the autonomic layer.

References

1. Bernauer, A., Bringmann, O., Rosenstiel, W., Bouajila, A., Stechele, W., Herkersdorf, A.: An architecture for runtime evaluation of soc reliability. In: Informatik für Menschen. Lecture Notes in Informatics, vol. P-93, pp. 177–185. Köllen Verlag, Bonn (2006)

2. Gaisler: Leon3 processor. http://www.gaisler.com/
3. Glover, F., Laguna, M.: Tabu Search, pp. 70–150. Wiley, New York (1993)
4. Held, J., Bautista, J., Koehl, S.: From a few cores to many: a tera-scale computing research overview. Intel White Paper (2006)
5. Henkel, J.: Closing the soc design gap. Computer **36**, 119–121 (2003)
6. Kernighan, B.W., Lin, S.: An efficient heuristic procedure for partitioning graphs. Bell Syst. Tech. J. **49**(1), 291–307 (1970)
7. Kirkpatrick, S., Gelatt, C.D., Vecchi, M.P.: Optimization by simulated annealing. Science **220**(4598), 671–680 (1983)
8. Li, J., Ma, X., Singh, K., Schulz, M., de Supinski, B., McKee, S.: Machine learning based online performance prediction for runtime parallelization and task scheduling. In: IEEE International Symposium on Performance Analysis of Systems and Software, pp. 89–100 (2009)
9. Singhal, R.: Inside Intel next generation nehalem microarchitecture. http://blogs.intel.com/idf/2008/08/sample_idf_sessions_inside_neh.php (2008)
10. Streichert, T., Strengert, C., Haubelt, C., Teich, J.: Dynamic task binding for hardware/software reconfigurable networks. In: Proceedings of the 19th Annual Symposium on Integrated Circuits and Systems Design, SBCCI'06, pp. 38–43. ACM, New York (2006)
11. Wang, Z., O'Boyle, M.F.P.: Mapping parallelism to multi-cores: a machine learning based approach. In: Proceedings of the 14th ACM SIGPLAN Symposium on Principles and Practice of Parallel Programming, PPoPP'09, pp. 75–84. ACM, New York (2009)
12. Ward, J.: Hierarchical grouping to optimize an objective function. J. Am. Stat. Assoc. **58**, 236–244 (1963)
13. Zeppenfeld, J., Bouajila, A., Stechele, W., Herkersdorf, A.: Learning classifier tables for autonomic systems on chip. In: GI Jahrestagung (2), pp. 771–778 (2008)
14. Zeppenfeld, J., Herkersdorf, A.: Autonomic workload management for multi-core processor systems. In: Architecture of Computing Systems. Lecture Notes in Computer Science, vol. 5974, pp. 49–60. Springer, Berlin (2010)

J. Zeppenfeld (✉) · A. Bouajila · W. Stechele · A. Herkersdorf
Technische Universität München, München, Germany
e-mail: zeppenfe@tum.de

A. Bouajila
e-mail: a.bouajila@tum.de

W. Stechele
e-mail: stechele@tum.de

A. Herkersdorf
e-mail: herkersdorf@tum.de

A. Bernauer · W. Rosenstiel
Eberhard Karls Universität Tübingen, Tübingen, Germany

A. Bernauer
e-mail: bernauer@informatik.uni-tuebingen.de

W. Rosenstiel
e-mail: rosenstiel@informatik.uni-tuebingen.de

O. Bringmann
Forschungszentrum Informatik Karlsruhe, Karlsruhe, Germany
e-mail: Bringmann@fzi.de

Chapter 5.4
Efficient Adaptive Communication from Resource-Restricted Transmitters

Stephan Sigg, Michael Beigl, and Behnam Banitalebi

Abstract We present a protocol for distributed adaptive transmit beamforming in networks of wireless connected nodes and show that the performance of this protocol is sensitive to environmental changes. However, we show that it is possible to tune parameters of the protocol in order to compensate for these environmental aspects. We extend the protocol by Organic Computing principles to realise an adaptive, emergent behaviour so that optimum parameter settings for distributed environments are learned. For this organic behaviour, knowledge about the actual situation is required. To establish this situation awareness we present a novel approach to sense situations based exclusively on RF-channel measurements. We show that an awareness of the presence, position, count and even activity of persons can be established based on simple features from the RF-channel only. This situation awareness completes our proposal of an emergent protocol for collaborative transmission of distributed devices.

Keywords Distributed adaptive beamforming · Emergent radio · Adaptive transmission protocol · Situation awareness

1 Introduction

The superimposition of signals from various transmitters during communication allows for the exploitation of spatial diversity of nodes in order to improve the robustness and synchronisation quality [16]. This approach of distributed adaptive beamforming has been considered by various authors as summarised in [6]. From the multiple approaches presented, however, only the 1-bit feedback-based carrier synchronisation can be applied in practical instrumentations of wireless sensor networks since the other approaches are too resource demanding as they exploit, for instance, CDMA schemes or utilise further complex computations.

The former method iteratively traverses the search space spanned by the possible carrier phase offsets of transmit signals in a random manner. By this approach, iteratively a global optimum is approached [7, 9, 11]. General estimations on the expected synchronisation time have been published in [3, 15]. Also, algorithmic improvements of the general random search method have been presented in [2, 12, 13]. Finally, we derived an asymptotically optimal approach in [4, 14]. In

C. Müller-Schloer et al. (eds.), *Organic Computing — A Paradigm Shift for Complex Systems*, 473–487
Autonomic Systems, DOI 10.1007/978-3-0348-0130-0_31,

order to reach this optimality we reduced the amount of randomness in the traversal of the search space. In particular, the general function that describes the improvement of the carrier synchronisation during one iteration with respect to the change of the phase offset of a single node could be derived. With this knowledge it is possible to calculate the unknown parameters of this function (offset on both axis and amplitude) accurately from few feedback transmissions of a remote receiver. Each node then simply calculates its optimum phase offset according to this function in order to achieve a synchronisation among carrier phase offsets of transmit signals.

In this article, we present a protocol for distributed adaptive transmit beamforming in a network of not tightly synchronised nodes (Sect. 2). The performance of this protocol is, however, dependent on various parameters such as the probability P_{mut} to alter the phase offset, its probability distribution P_{dist} and variance P_{var}. We show in Sect. 2.2 that these parameters can be optimised for a given scenario. When, however, the environmental setting is altered frequently, a more advanced approach is required in order to prevent continuous learning and re-training to already observed settings. Consider, for instance, a conference room in which meetings are held, presentations are given or which is vacant at some times. In order to cope with such frequently changing scenarios, we present two implementations of simple learning classifier systems in Sect. 2.4 from which the first utilises an evolutionary learning approach and the second a metropolis learning component. Both approaches require the knowledge of environmental settings in order to match learned parameters to these situations. A novel approach to sense situations based on channel measurements is presented in Sect. 3. Section 4 draws our conclusion.

2 A Protocol for Distributed Adaptive Transmit Beamforming in Wireless Sensor Networks

A protocol for distributed adaptive beamforming in wireless sensor networks must contain an initial phase in which a remote receiver informs a set of nodes that it is expecting data transmission. Then, transmit nodes start the carrier synchronisation as, for instance, described in [5, 14]. When the synchronisation is sufficient, the receiver indicates this by broadcasting a transmission request. Transmitters then agree on a common transmit message and start transmission with the carrier phase offset established during the carrier synchronisation. This process was implemented for modulation schemes of several orders.

Figure 1 illustrates the process of modulation, transmission and demodulation for an amplitude modulation scheme that represents two bits per transmitted symbol.

We modulated a simple periodical symbol sequence on the transmit-carrier of each specific node. For this environmental setting we observe that the received superimposed signal is strengthened compared to a single transmit signal. Also, the symbol sequence is clearly visible from the received superimposed signal.

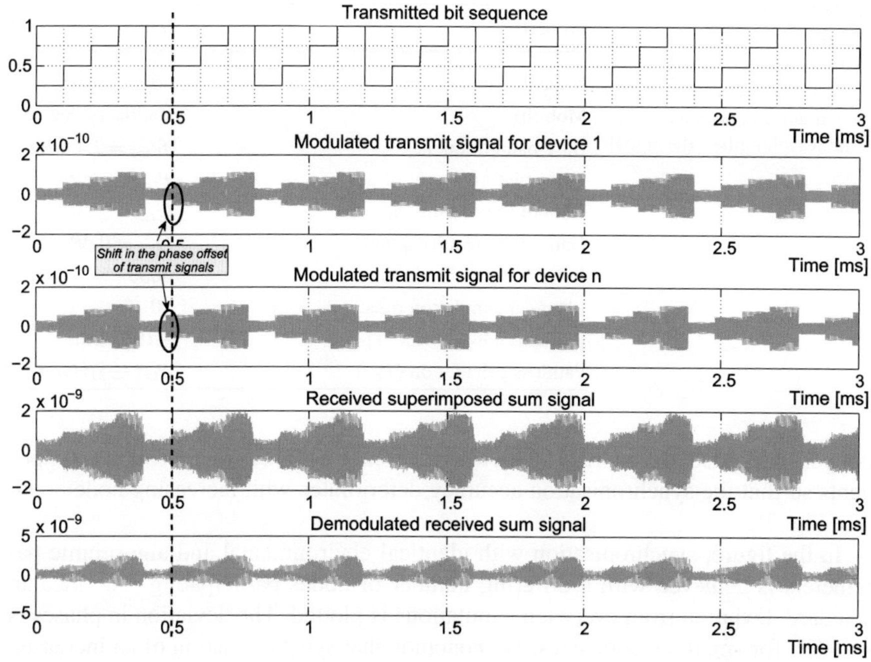

Fig. 1 Modulation and demodulation of a simple symbol sequence

2.1 Experimental Verification of the Protocol

We altered the count of participating devices, the distance between the transmitters and the receiver as well as the transmission bit-rate. The simulations were conducted in an Matlab environment. In each simulation run, first the carrier phases were synchronised for 6000 iterations of the iterative random synchronisation algorithm also utilised in [5, 14]. After synchronisation, devices simultaneously transmitted their data sequence. The simulation-environment is detailed in Table 1. In all simulations, 100 transmit-devices superimpose their carrier signals. Figure 2 shows the observed median performance of the protocol for various simulation settings. In these figures, the median BER for 10 independent synchronisations is calculated. It is derived from a binary sequence $s \in \mathbb{B}^n$ and a received binary sequence $s' \in \mathbb{B}^n$ as

$$\mathrm{BER} = \frac{\mathrm{ham}(s, s')}{n} \tag{1}$$

In this equation, $\mathrm{ham}(s, s')$ denotes the Hamming distance between s and s'.

We see that the BER is mainly impacted by the distance and the data rate. Observe from Figs. 2(a) and 2(b) that the BER increases for data rates of 10 kbps and 100 kbps with the transmission distance.

However, the lines in these figures cross for several network sizes. With increasing distance, also the noise level rises relative to the strength of the superimposed received sum signal. Due to noise and interference an optimum synchronisation is

Table 1 Configuration of the simulations. P_{rx} is the received signal power, d is the distance between transmitter and receiver and λ is the wavelength of the signal

Property	Value
Node distribution area	30 m × 30 m
Mobility	stationary devices
Base band frequency	$f_{\text{base}} = 2.4$ GHz
Transmission power of devices	$P_{tx} = 1$ mW
Gain of the transmit antenna	$G_{tx} = 0$ dB
Gain of the receive antenna	$G_{rx} = 0$ dB
Iterations per simulations	6000
Identical simulation runs	10
Random noise power [1]	−103 dBm
Pathloss calculation (P_{rx})	$P_{tx}(\frac{\lambda}{2\pi d})^2 G_{tx} G_{rx}$

not possible. Also, it is more difficult to synchronise a higher count of signal components so that the synchronisation accuracy deteriorates with increasing node count (see Fig. 3).

In the figure, synchronisation with identical environmental and algorithmic parameters is achieved with a differing number of nodes participating. The median standard deviation from every ten simulations is plotted. The deviation in phases is smallest for small network sizes. We conclude that synchronisation of an increased number of devices deteriorates in accuracy.

When both, network size and distance are altered simultaneously, the increased signal strength for greater network sizes acts adversely to the reduced synchronisation quality for increasing network sizes (cf. Fig. 2(c)).

When the transmit data rate is increased, we observe that also the BER increases as more bits are represented by one symbol (cf. Figs. 2(a), 2(b) and 2(d)). Figure 2(e) summarises that the BER increases with increasing transmission distance and datarate.

This observation is remarkable for collaborative transmission. Since the synchronisation-performance is directly impacted by the count of participating devices [13], it is beneficial when this count is just large enough to collaboratively reach a remote receiver. Greater numbers of transmitters do not lead to an improved BER but impair the synchronisation-performance.

2.2 *Environmental Impacts on the Performance of the Protocol*

The performance of the carrier synchronisation approach is impacted by the three parameters P_{mut}, P_{dist} and P_{var}. In [13] we observed that a steady optimisation is achieved when small changes are applied to the phase offset of nodes in each iteration. This means that either P_{mut} or P_{var} is small. In simulations, best results for $P_{\text{dist}} =$ ‘uniform’ have been achieved in a scenario with n nodes and $P_{\text{mut}} = \frac{1}{n}$ [13]. However, the environment may impact the optimum value for these synchronisation parameters. Next, we discuss three possible environmental impacts, namely the number of participating nodes, the noise figure and movement of devices.

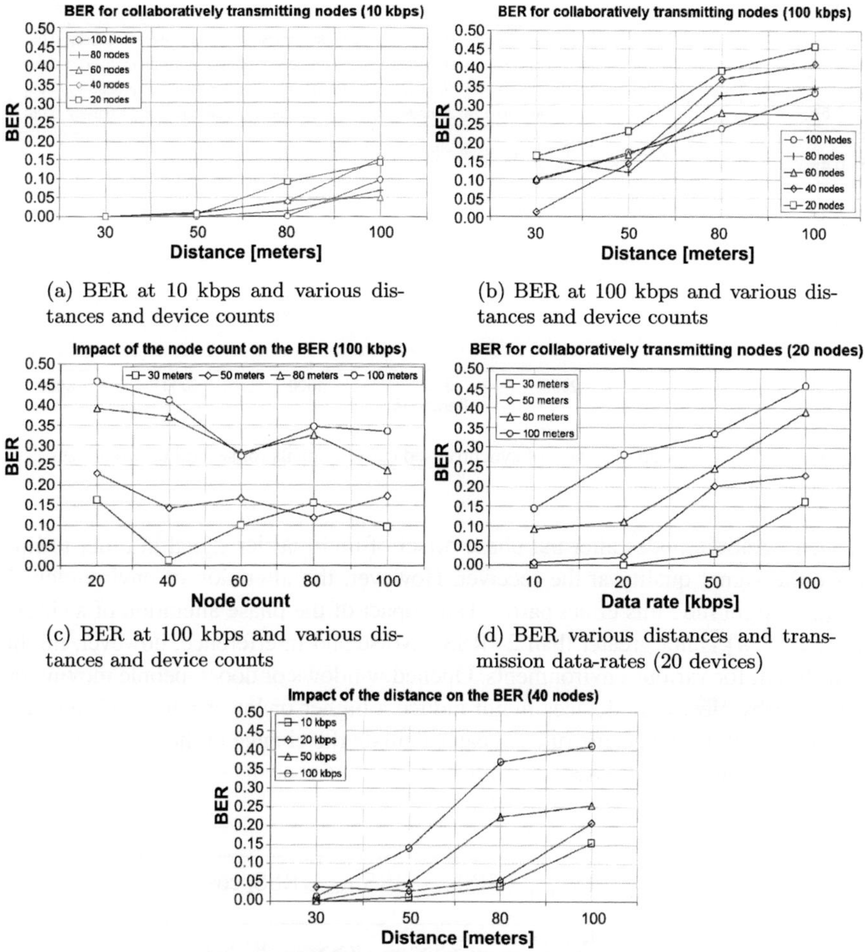

(a) BER at 10 kbps and various distances and device counts

(b) BER at 100 kbps and various distances and device counts

(c) BER at 100 kbps and various distances and device counts

(d) BER various distances and transmission data-rates (20 devices)

(e) BER at various distances and transmission data-rates (20 devices)

Fig. 2 BER for a distributed adaptive beamforming protocol among a set of wireless devices and various distances, data-rates and device counts

Impact of Noise and Interference

The signal observed by a remote receiver is composed of the superimposed sum signal and noise [8, 15]:

$$\zeta_{\text{sum}}(t) = \Re\left(m(t) e^{j2\pi f_c t} \sum_{i=1}^{n} \text{RSS}_i e^{j(\gamma_i + \phi_i + \psi_i)} + \mathcal{N} \right) \tag{2}$$

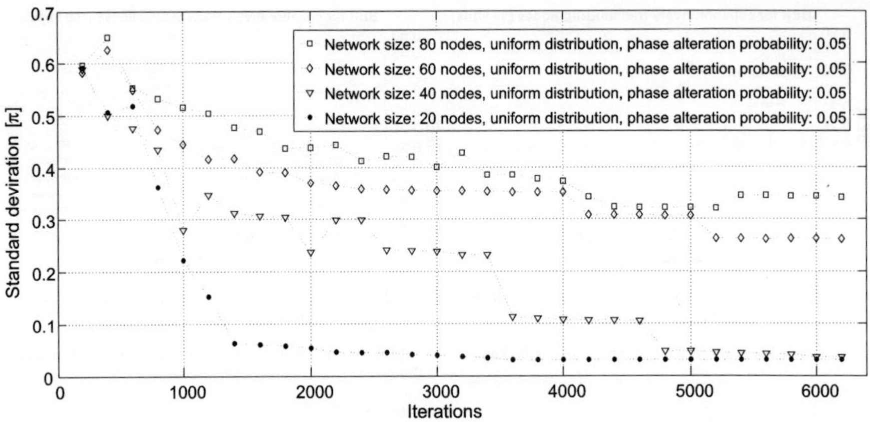

Fig. 3 Standard deviation of phases of synchronised carrier signals for several network sizes

In each iteration, nodes alter the phase offset of their carrier signal in order to improve the signal quality at the receiver. However, the alteration of environmental impacts might hide this effect partly. The impact of the phase alteration of a single link $i \in [1..n]$ is not greater than $2 \cdot \text{RSS}_i$. Noise and interference, however, might be different for various environments. Opened windows or doors, people moving or other nearby electronic devices might induce a higher or lower noise and interference figure. With the cosine rule we can calculate the change in the received signal strength of the superimposed signal at the event of a change of the carrier phase from γ_i to γ_i' as

$$\overline{\text{RSS}}(\gamma') = \sqrt{\text{RSS}_{\text{sum-}i}^2 + \text{RSS}_i^2 - 2\text{RSS}_{\text{sum-}i}\text{RSS}_i \cos(\gamma_i + \gamma_i')} - \sqrt{\text{RSS}_{\text{sum-}i}^2 + \text{RSS}_i^2 - 2\text{RSS}_{\text{sum-}i}\text{RSS}_i \cos(\gamma_i)} \tag{3}$$

as detailed in Fig. 4.

In this equation we denote the received signal strength achieved by the superimposition of all signals short of i by

$$\text{RSS}_{\text{sum-}i} = \sum_{j \neq i} \text{RSS}_j; \quad j \in [1..n] \tag{4}$$

Since the phase alteration is a random process, the actual gain of a single phase modification is typically smaller than $2 \cdot \text{RSS}_i$. When we assume that a single node can establish up to k equally distributed carrier phases physically, the average gain by the alteration of one carrier signal component is then

$$\frac{\sum_{i=1}^{k} \overline{\text{RSS}}(\frac{2\pi}{k} \cdot i)}{k} \tag{5}$$

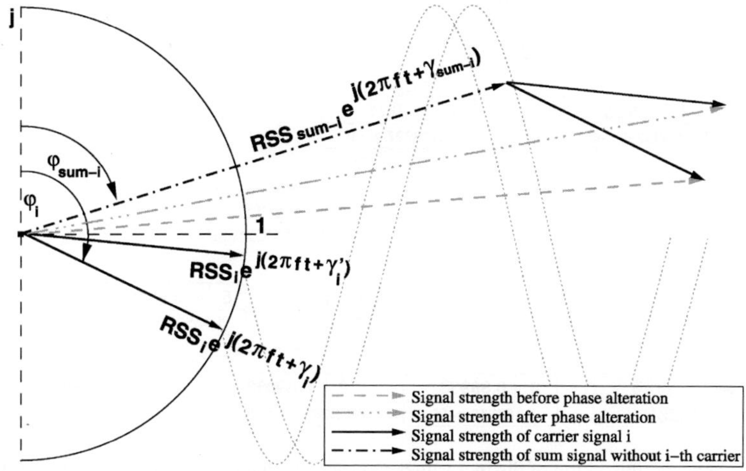

Fig. 4 Illustration of the impact of carrier phase alteration on the overall received signal strength

Consequently, when the noise figure is in the same order, the alterations of individual carriers has little effect on the observed signal quality.

Impact of the Network Size

The number of nodes that participate in collaborative transmission also impacts the synchronisation performance. At first, clearly, since the synchronisation is achieved by a random process, the synchronisation time is increased with the count of participating nodes [5].

In [13] we derived that the optimum synchronisation performance is achieved with a steady state configuration that is only slowly changed (e.g. with low P_{mut} and P_{var}). On the other hand, the impact of an individual node on the sum signal decreases with increasing node count (see (3)). The value $\overline{\text{RSS}}(\gamma')$ of the impact on the received signal strength decreases with increasing $\text{RSS}_{sum\text{-}i}$. With increasing node count, it is therefore beneficial to chose P_{mut} and P_{var} slightly higher than necessary for the steady case to increase the impact of modifications in one iteration.

2.3 Impact of Node Mobility

Movement has a dramatic impact on the synchronisation since phases will drift apart when receiver or transmit nodes move [4]. Synchronisation therefore has to be significantly faster than the velocity experienced. An increased P_{mut} or P_{var} might therefore be beneficial in the presence of node mobility.

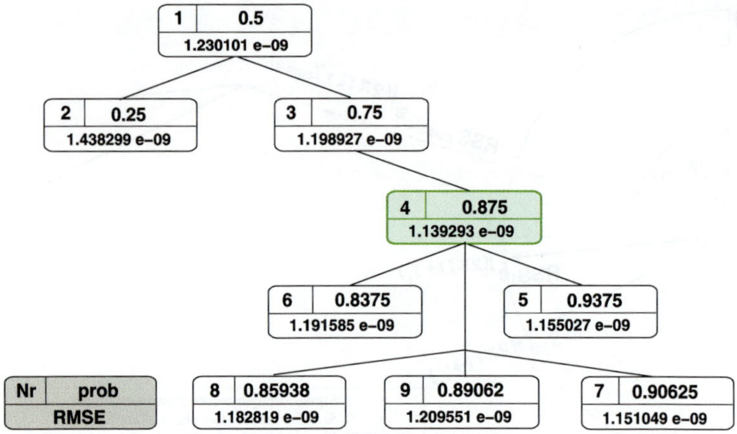

Fig. 5 Schematic illustration of distinct values requested by the binary learning approach. RMSE values depicted denote the mean RMSE after 10 synchronisations with identical P_{mut}

2.4 Adaptive Protocols for Distributed Adaptive Beamforming in Wireless Sensor Networks

In order to adapt the parameters of the transmission protocol to environmental impacts we propose to learn optimum values for a given environmental setting and to remember these at any re-occurrence of this setting. As a straightforward implementation we considered a binary search approach in order to find optimum algorithmic parameters. In Fig. 5, the application of this method is exemplarily detailed for the search of an optimum value of P_{mut}.

As shown in the figure, the approach divides the search space into a binary search tree and traverses the edges of this tree that lead to decreased RMSE values. The root of the Mean Square Error (RMSE) is calculated as

$$\mathrm{RMSE} = \sqrt{\sum_{t=0}^{\tau} \frac{(\zeta_{\mathrm{sum}} + \zeta_{\mathrm{noise}} - \zeta_{\mathrm{opt}})^2}{n}} \tag{6}$$

Here, τ is chosen to cover several signal periods.

The RMSE values depicted in the figure denote the mean RMSE over 10 distinct synchronisations. Although the synchronisation parameters can be adapted by this approach they had to be recalculated frequently in adaptive, changing environments.

2.5 Proposal of Two Adaptive Protocols

We propose two adaptive transmission schemes that both align learned parameters to scenarios in which they are learned. The first scheme is a learning classifier system [17]. In this approach, the parameter space spanned by the probability

$P_{\text{mut}} \in [0, 1]$ to alter the phase offset of a single node, the probability distribution $P_{\text{dist}} \in \{\text{uniform}, \text{normal}, \dots\}$ and the variance $P_{\text{var}} \in [-\pi, \pi]$ is explored by an evolutionary search algorithm.

Since this approach typically is slow to approach an optimum, we also consider a modification of this method in which the evolutionary search is replaced by a metropolis search algorithm [10].

An Evolutionary Learning Approach

The idea of a LCS is to combine a discovery and a learning approach [17]. In the former, a search algorithm iteratively improves the parameter settings applied. In the learning part, this performance achieved is matched with a given environmental situation. The method is therefore applied in combination with situation detection (see Sect. 3). It is then possible to implement a continuous learning to constantly adapt to a changing environment. This method will learn for each distinct environmental situation the optimum parameters P_{mut}, P_{dist} and P_{var}. Since we want to minimise the additional inter-node communication, this learning mechanism is achieved by individual nodes autonomously. Our implementation is detailed in Algorithm 1.

Algorithm 1 Adaptive learning of synchronisation parameters

1: populationSize = μ
2: offspringPopulationSize = λ
3: Init μ individuals with P_{mut}, P_{dist} and P_{var}
4: Synchronise nodes for each individual
5: **while** maximum population count not reached **do**
6:     create λ individuals by applying P_{mut}, P_{dist} and P_{var}
7:     **for** all individuals in the offspring population **do**
8:         synchronise nodes
9:         *% might be repeated for each individual*
10:     **end for**
11:     select μ best rated individuals for offspring population
12: **end while**

For the discovery part, we utilise an evolutionary algorithm. For simulation purposes the algorithm simultaneously maintains μ carrier synchronisation simulations with identical environmental settings for each scenario. A population of the algorithm consists of a complete network of distributed devices. In each iteration, all these simulations are synchronised in the course of 6000 iterations each. The new population consists of another λ simulations. For distinct individuals (simulations) the optimisation parameters P_{mut}, P_{dist} and P_{var} differ. In a changing environment, parameter settings are clustered for the situation at hand. After initialisation of the population in line 3, carrier phases are synchronised (line 4) and the evolutionary optimisation is executed in lines 5–12. We exemplarily detail the first 3 generations with a population size and offspring population size of 5 and selection in Fig. 6.

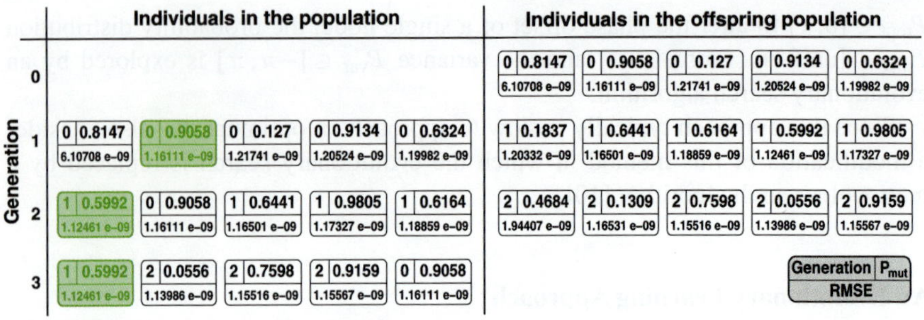

Fig. 6 Schematic of the optimisation process of the protocol utilising the LCS learning approach with an evolutionary algorithm. RMSE values depicted denote the mean RMSE after 10 synchronisations with identical P_{mut}

A Metropolis Learning Approach

Clearly, we can implement arbitrary search approaches for the LCS system. Since we expect the evolutionary search method to converge only at a moderate pace, in Algorithm 2 we detail a learning classifier system that features a metropolis search approach. This is basically a local search algorithm that also accepts search points with decreasing fitness value with a small probability. The probability is chosen inverse to the actual decrease in the fitness value.

Algorithm 2 Adaptive beamforming with metropolis learning

```
initialise search-point with P_mut, P_dist and P_var
synchronise nodes
while stop criterion not met do
    create new point by applying P_mut, P_dist and P_var
    synchronise nodes
    % might be repeated for each individual
    if RMSE of new point better than RMSE of old point then
        discard old search-point and continue with new one
    else
        if random decision positive then
            % probability inverse to fitness-decrease
            discard old point and continue with new one
        else
            discard new search-point
        end if
    end if
end while
```

3 Detection of Environmental Conditions in Wireless Sensor Networks

For the detection of situations we considered the possibility to detect these purely based on measurements of the wireless channel for the following reasons.

1. Since we consider situation awareness of wireless nodes, the one sensor each node trivially possesses is an RF-receiver. No additional sensors (increased cost, size and power consumption) are required.
2. The spectrum of activities that impact the wireless channel is immense. It ranges from the presence, position, number and movement of individuals to changed room conditions (e.g. opened or closed windows/doors, moved furniture, room size) to ongoing wireless activity to the detection of specific noise sources.
3. Detection based on the wireless channel comes virtually at no cost as the analysis of channel characteristics can be embedded in communication between nodes that is ongoing anyways.
4. Since the transmission protocol is impacted by wireless channel conditions, it is obvious to consider this wireless channel as the source of features for classification.

3.1 System

The situation recognition system is written in Python.[1] For signal processing with the USRP (Universal Software Radio Peripheral) nodes the GNU radio[2] framework was utilised. Learning and classification is conducted by the Orange data mining toolkit.[3] Interprocess communication is implemented using the messaging library QM[4] and its python bindings.[5]

The distinct modules of the situation recognition system are depicted in Fig. 7. The main component configures the transmit and receive components, records data and classifies situations. The transmitter component modulates a signal onto the RF carrier. The receiver component extracts features from a received signal and forwards them to the main component. The classification component evaluates the recognition rate.

3.2 Features and Classification

For classification we sampled the root mean square (RMS) of the received signal, the signal to noise ratio (SNR) and the average magnitude squared (AMS). These

[1] http://www.python.org/.

[2] http://www.gnuradio.org.

[3] http://www.ailab.si/orange/.

[4] http://www.zeromq.org/.

[5] http://www.zeromq.org/bindings:python.

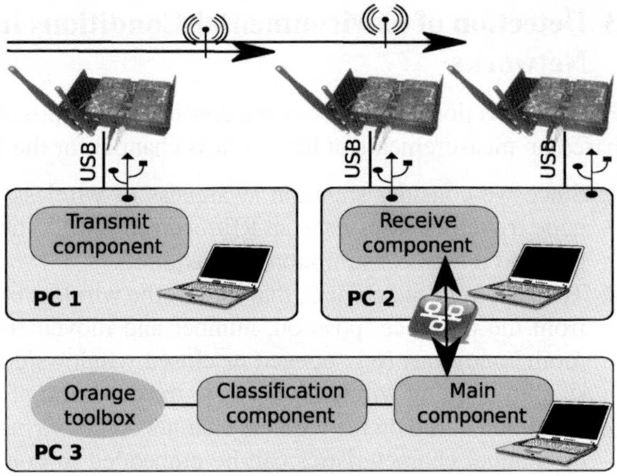

Fig. 7 Schematic illustration of the situation recognition system

simple features are typically available at the transceiver of nodes and do not require ambitious computation. These features are forwarded to the main component. A classification script is used for evaluation of the recognition accuracy by constructing classifiers from the learning data. We utilise the K-NN implementation of the Orange framework with $k = 1$ (count of neighbours) and weighting by ranks.

3.3 Experiment

The USRPs are distributed in a room with the sender connected to the first PC and the two receivers to the second. A third PC is set up outside the room to control the experiment. The system is trained depending on the situations to be recognised (around 500 value sets per situation). Depending on the situations, data is annotated (10 times per situation). This will be used to calculate the recognition rate. After all annotations are made, the created annotations file is converted from XML to a tab file format, which can be easily parsed with Orange. Finally the tab files created are run through the classify script.

Results

In our experiments we considered four distinct scenarios. In the first scenario we considered the pure detection of the presence of a person in a room and if the door to the room was opened or closed. In the second scenario, we tried to estimate the position of the person in the room. In the third scenario, the number of persons in a room was considered. Finally, in our fourth scenario we tried to distinguish between several activities of a single person in the room.

In 10 experiments, the classifier was applied in order to classify the trained situations. Figure 8 depicts our results.

	Door	Person	Empty
Door	**1.0**	0.0	0.0
Person	0.0	**0.7**	0.3
Empty	0.0	0.1	**0.9**

(a) Scenario 1: Detection of presence and opened/closed door

	P-1	P-2	P-3	Empty
P-1	**0.9**	0.0	0.0	0.1
P-2	0.36	**0.09**	0.45	0.09
P-3	0.0	0.0	**1.0**	0.0
Empty	0.1	0.0	0.4	**0.5**

(b) Scenario 2: Estimation of the position of a person in a room

	0	1	5	10
0	**1.0**	0.0	0.0	0.0
1	0.3	**0.5**	0.2	0.0
5	0.25	0.42	**0.08**	0.08
10	0.0	0.1	0.1	**0.8**

(c) Scenario 3: Estimation of the count of persons in a room

	Stand	Sit	Walk	Empty
Stand	**0.4**	0.3	0.1	0.2
Sit	0.0	**0.6**	0.0	0.4
Walk	0.1	0.1	**0.6**	0.2
Empty	0.2	0.0	0.0	**0.8**

(d) Scenario 4: Classification of activity

Fig. 8 Classification accuracy in four distinct scenarios

We observe that the diagonal of these confusion matrices contain values close to 1.0. However, for some of these scenarios the classification based on channel measurements seems to be better suited than for others. For instance, the detection if a person is present in a room or if the door is opened/closed in scenario 1 received generally a high classification accuracy (cf. Fig. 8(a)). The number of persons in a room can, however, only roughly be classified with our approach as depicted in Fig. 8(c). In particular, such a fine grained distinction between 0, 1, 5 and 10 persons is not feasible. While the distinction between 0 and 10 is sufficiently accurate, the distinction between 1 and 5 persons becomes confused. For the detection of the position of a person (see Fig. 8(b)), however, we receive quite accurate results for two of the three positions considered. Also, the detection of activity was surprisingly accurate given that only channel measurements are utilised to derive activities of persons (cf. Fig. 8(d)). Generally, we conclude that a rudimentary situation awareness is in fact possible based exclusively on cheap amplitude based channel measurements. With this situation awareness the LCS approaches described in Sect. 2.4 could be implemented without the need for additional sensors on wireless devices to derive an emergent radio protocol.

4 Conclusion

We have introduced a protocol for distributed adaptive transmit beamforming among wireless devices that is emergent in the sense that it is self-adaptive to changing environmental situations and that it is self configuring due to the ability to learn parameters for distinct environmental settings. The protocol was evaluated in simulations and experiments. A critical component of this protocol is the establishing of a situation awareness in order to link learned optimum parameter settings to situations. In order to provide this awareness we considered the classification of situations based exclusively on channel characteristics. We derived in a case study that we can, with only amplitude-based features already establish a rudimentary situation awareness regarding the presence, position, count and even activity of persons in a room.

References

1. 3GPP: 3rd generation partnership project; technical specification group radio access networks; 3g home nodeb study item technical report (release 8). Technical Report 3GPP TR 25.820 V8.0.0 (2008-03) (March 2008)
2. Barriac, G., Mudumbai, R., Madhow, U.: Distributed beamforming for information transfer in sensor networks. In: Proceedings of the Third International Workshop on Information Processing in Sensor Networks (2004)
3. Bucklew, J.A., Sethares, W.A.: Convergence of a class of decentralised beamforming algorithms. IEEE Trans. Signal Process. **56**(6), 2280–2288 (2008)
4. Masri, R.M.E., Sigg, S., Beigl, M.: An asymptotically optimal approach to the distributed adaptive transmit beamforming in wireless sensor networks. In: Proceedings of the 16th European Wireless Conference (2010)
5. Mudumbai, R., Barriac, G., Madhow, U.: On the feasibility of distributed beamforming in wireless networks. IEEE Trans. Wirel. Commun. **6**, 1754–1763 (2007)
6. Mudumbai, R., Brown, D.R., Madhow, U., Poor, H.V.: Distributed transmit beamforming: Challenges and recent progress. IEEE Commun. Mag. 102–110 (2009)
7. Mudumbai, R., Hespanha, J., Madhow, U., Barriac, G.: Scalable feedback control for distributed beamforming in sensor networks. In: Proceedings of the IEEE International Symposium on Information Theory, pp. 137–141 (2005)
8. Mudumbai, R., Hespanha, J., Madhow, U., Barriac, G.: Distributed transmit beamforming using feedback control. IEEE Trans. Inf. Theory **56**(1) (2010)
9. Mudumbai, R., Wild, B., Madhow, U., Ramchandran, K.: Distributed beamforming using 1 bit feedback: from concept to realization. In: Proceedings of the 44th Allerton Conference on Communication, Control and Computation, pp. 1020–1027 (2006)
10. Schwefel, H.-P.: Evolution and Optimum Seeking. Wiley-Interscience, New York (1995)
11. Seo, M., Rodwell, M., Madhow, U.: A feedback-based distributed phased array technique and its application to 60-GHz wireless sensor network. In: IEEE MTT-S International Microwave Symposium Digest, pp. 683–686 (2008)
12. Sigg, S., Beigl, M.: Algorithmic approaches to distributed adaptive transmit beamforming. In: Fifth International Conference on Intelligent Sensors, Sensor Networks and Information Processing—Symposium on Theoretical and Practical Aspects of Large-scale Wireless Sensor Networks (2009)
13. Sigg, S., Beigl, M.: Algorithms for closed-loop feedback based distributed adaptive beamforming in wireless sensor networks. In: Proceedings of the Fifth International Conference on Intelligent Sensors, Sensor Networks and Information Processing—Symposium on Adaptive Sensing, Control, and Optimization in Sensor Networks (2009)
14. Sigg, S., Masri, R., Ristau, J., Beigl, M.: Limitations, performance and instrumentation of closed-loop feedback based distributed adaptive transmit beamforming in WSNs. In: Fifth International Conference on Intelligent Sensors, Sensor Networks and Information Processing—Symposium on Theoretical and Practical Aspects of Large-scale Wireless Sensor Networks (2009)
15. Sigg, S., Masri, R.M.E., Beigl, M.: Feedback based closed-loop carrier synchronisation: a sharp asymptotic bound, an asymptotically optimal approach, simulations and experiments.

IEEE Trans. Mobile Comput. (2011). http://doi.ieeecomputersociety.org/10.1109/TMC.2011.21
16. Tse, D., Viswanath, P.: Fundamentals of Wireless Communication. Cambridge University Press, Cambridge (2005)
17. Urbanowicz, R.J., Moore, J.H.: Learning classifier systems: a complete introduction, review, and roadmap. J. Artif. Evol. Appl. (2009). doi:10.1155/2009/736398

S. Sigg (✉) · M. Beigl · B. Banitalebi
Karlsruhe Institute of Technology (KIT), Karlsruhe, Germany
e-mail: sigg@ibr.cs.tu-bs.de

M. Beigl
e-mail: michael@kit.teco.edu

B. Banitalebi
e-mail: behnan@kit.teco.edu

Chapter 5.5
OrganicBus: Organic Self-organising Bus-Based Communication Systems

Tobias Ziermann, Stefan Wildermann, and Jürgen Teich

Abstract Complexity in distributed embedded systems is growing and with that the complexity of the communication. One way to connect systems is to use priority-based communication such as the CAN protocol. In this article, we try to overcome drawbacks of today's pure offline design methods that are based on worst-case estimations, which are not expandable, and may easily degenerate when the environment or requirements change at run-time. In contrast, a decentralised approach using online *self-organisation* is able to monitor the actual traffic of the communication system and adapt at run-time individually. First, we introduce a communication model based on streams. Then, solutions are presented to schedule this communication.

Keywords CAN · Controller area network · Distributed embedded systems · Game theory · Real-time scheduling

1 Introduction

Distributed embedded systems distinguish themselves by a collaboration of multiple connected devices. These devices can be tiny, stand-alone, embedded micro controllers or networking devices, but also embedded PCs or robotics systems. Combining the local information with information from the other devices in the network enables new applications, for example, in automotive and avionics control systems, medical systems, and manufacturing and process automation. In these areas of application, the complexity of distributed embedded systems is very high. In addition, the demand for more and more functionality to incorporate has not yet eased [6]. A very important aspect of distributed embedded systems is the organisation of the communication. The communication can range from heterogeneous network-based systems, for example the Internet, to customised wireless radio networks. However, because reliability and timeliness are very important, a large number of distributed embedded systems relies on bus-based communication.

Today, one of the most popular bus-based protocols is the Controller Area Network (CAN) [1], which was originally designed for automotive applications and now is also used in other applications such as robotics and industrial automation. In CAN, a message-oriented approach is chosen in the data link layer. Each data

C. Müller-Schloer et al. (eds.), *Organic Computing — A Paradigm Shift for Complex Systems*, 489–501
Autonomic Systems, DOI 10.1007/978-3-0348-0130-0_32, © Springer Basel AG 2011

frame has an identifier that is unique for each message. This identifier defines the message priority by which the bus access is granted. After sending the identifier, only the message with the highest priority is left and has exclusive bus access. This arbitration mechanism enforces event-driven priority-based access to the bus. It is particular well suited for an organic communication approach, because on the one hand the event-driven aspect provides a dynamic change of the system, and on the other hand, the priority-based aspect allows determinism to fulfil real-time requirements. Due to these reasons, results are shown for this protocol. However, most of the methods introduced in this article can be also used with any priority-based bus-based communication protocol.

In real-time systems, communicating tasks often face hard deadlines. But also many tasks with soft deadlines exist that are not real-time critical and may also arrive some reasonable time after their deadline has expired. Due to these constraints, offline analysis and design is the common practise. Simulative and analytical methods are applied to check whether deadlines are satisfiable. However, a problem of offline design techniques is their lack of flexibility. A complete re-design may be needed when tasks are inserted or removed: even a small change in the architecture could require to restart the whole design process. Offline methods cannot consider system changes that happen at run-time. In addition, a precise knowledge of the whole system has to be available which is often not possible, because the network components are designed by different vendors or cannot be exactly predicted due to manufacturing variances. Finally, the complexity to design and analyse a system grows with the size of the communication network.

With our organic design methodology as described in the following, we want to overcome drawbacks of offline design by considering the actual bus traffic and requirements of the system. This has two advantages: (a) changes during the design phase of the system or different variants of a system can be tolerated, and (b) adaptability during run-time allows the system to react to situations not known at design time. These properties are achieved by using simple local rules and distributed lightweight algorithms.

The rest of the article is organised as follows. In Sect. 2, we describe a model of bus-based communication based on message streams. Section 3 shows ways to guarantee hard real-time behaviour. Section 4 explains how response times can be minimised using distributed online adaptation. Section 5 proposes an algorithm to share bandwidth equally on a priority-based medium. Finally, conclusions and future work are given in Sect. 6.

2 Model and Problem Definition

The system we are targeting can be described by a set of nodes communicating over a shared bus medium, as shown in Fig. 1. Here, one or more tasks on each node may initiate a communication, i.e. release messages. In our model, we abstract the tasks by considering only the mechanism used to release *messages* called a *stream*.

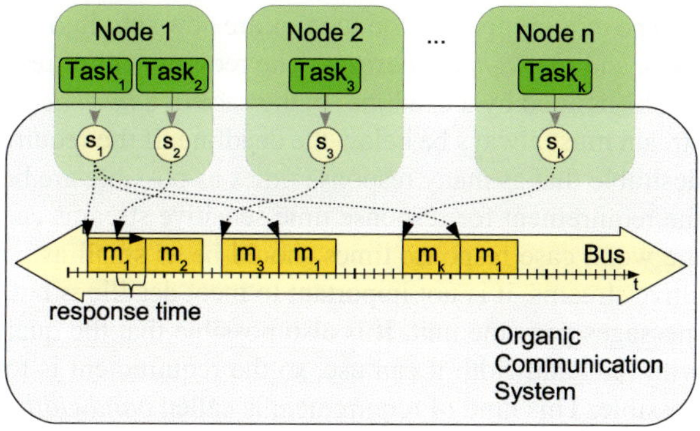

Fig. 1 CAN Bus-based system model

A stream s_i can be characterised by a tuple (N_p, U_i, R_i), that is, by a node N_p the stream is running on, an occurrence scheme U_i that describes when a message is released by stream s_i, and a requirement R_i of the amount of messages or the time until the generated message should or must be transmitted. A *scenario* consists of n streams. The priority by which access is granted to the bus is given by the numbering of the streams. In particular, the stream with the higher index, stream s_i, has higher priority than a stream with a lower index s_{i-1}. We assume the priorities are given by the designer according to application specific requisites.

A *message* is a single release such as a CAN frame [1] of the stream. The time between a message release and the start of its uninterrupted transfer over the bus is the *response time* of the message. We do not add the constant non-preemptive time to transfer the message to the response time, because then a response time of zero is always the best possible case for every message independent of its length. This will simplify later comparisons between different techniques to influence the scheduling of messages. In Fig. 1 for example, the response time of message m_2 is three time slots, because it is delayed by the running message m_1. The *worst case response time* ($WCRT_i(b, e)$) of a stream during a certain time interval starting at time b and ending at time e is the largest response time of the messages of the stream recorded during that time interval. An analytical approach calculates $WCRT_i(0, \infty)$.

In classical real-time models, the occurrence of a task, here stream, can have the following properties: *Periodic* and *sporadic*. Periodic messages are released with a fixed rate and are characterised by a period T_i and an offset O_i. The offset is the time until the first message is released relative to an imaginary global time reference. Because the local time reference differs from the global one, the offsets can drift over time. In our analysis, we assume an offset free system as defined in [7]. This means the offsets of individual streams are not bound by any constraints, but rather can freely be set. This approach potentially offers the avoidance of conflicts by setting the offsets appropriately [8]. Sporadic streams release messages irregularly but with a bounded interarrival period. This bounded interarrival period allows us to treat sporadic streams in the worst case as periodic streams. Therefore, we only consider periodic streams in the following. In addition to these classical models, we define the occurrence property *probabilistic*. It characterises the occurrence of a stream as a probability at which a message is generated during one transmission slot.

The requirements of the streams are now split into two groups: *response time sensitive* and *throughput sensitive*. The requirement of response time sensitive streams can be defined by a deadline. In the case of a *hard deadline*, the response times of a stream must always be below the deadline. If the requirement is a *soft deadline*, it is desirable that as many response times as possible are below the deadline. However, the requirement for response time sensitive streams can also be that the average or the worst case response times should be as small as possible. For throughput sensitive streams, it is not important to meet deadlines but to send a certain amount of messages per time unit. It is also possible that the quality of the application varies with the bandwidth it can use, so the requirement is to send as many messages as possible. This kind of requirement is called *bandwidth*.

2.1 Types of Streams

Any combination of occurrence and requirement of streams is imaginable. However, in our organic design methodology, we distinguish between three types of streams:

Hard Real-Time Streams

Safety critical applications often consist of hard real-time streams. These applications will lead to a system malfunction when the deadlines are not met, so the requirement is a hard deadline. The occurrence of hard real-time streams is assumed to be strictly periodic, otherwise it would be impossible to guarantee any behaviour. To keep this determinism, the organic system does not have any influence on the type of streams during run-time.

Soft Real-Time Streams

The majority of streams in typical distributed embedded systems are soft real-time. They are characterised by a periodic or sporadic occurrence. If a deadline exists, it is equal to the period or most often they do not have a deadline at all. Then, it is desirable to reduce the response time of individual messages to a minimum.

Bandwidth Streams

Streaming applications, such as in audio or video processing, are represented by bandwidth streams. On the one hand, their quality is depending on the amount of data they can send. For example, a video stream needs to be compressed less when there is more bandwidth available. On the other hand, these streams are very flexible in the amount of throughput they need. For example, using lossy compression the needed throughput can easily be reduced by an order of magnitude. This means, it is not so important to get a certain throughput, rather it is important that all bandwidth streams get the same amount of throughput. The requirement is bandwidth and the occurrence can either be probabilistic or periodic. The occurrence can be freely chosen (within reasonable bounds), so the objective of the organic system is to find the optimal occurrence.

2.2 Objectives of the Organic Communication System

In the following, we present the properties of our Organic Computing system. The objectives are defined as follows:

1. Satisfy all real-time requirements, if possible.
2. Increase the number of messages that fulfil imposed soft deadline constraints and bandwidth requirements.
3. Improve/maximise the bus utilisation without violating (1), and optimise (2) as a secondary goal.
4. Guarantee a certain amount of fairness, i.e., the resource allocation on the shared bus.

A solution to solve all four constraints, respectively objectives is to exploit the priority-based access mechanism of the CAN bus standard [1]. We divide streams into different priority classes depending on their constraints. As already mentioned in Sect. 2.1, three different types of streams are distinguished. On the one hand, CAN offers different priority classes intuitively by demanding one unique priority per stream. On the other hand, if the system (amount of streams) changes dynamically, we only have to ensure that each priority is only used once. Therefore, it is necessary to divide the available priorities beforehand during the planing phase of the system. This is not yet a problem, because in current systems the amount of streams is much lower than the amount of priorities (in a current Golf VI (2010) 140 streams per bus [12] compared to 2032 available priorities). In order to ensure maximal flexibility, we propose the following assignment strategy. First, the maximal amount of bandwidth streams has to be defined. With the bandwidth of CAN being limited to 1 Mbit/s, the number of bandwidth streams is also limited. Next, the amount of hard real-time streams has to be defined. This is a fixed number, because they require an offline analysis and will not change during runtime. The remaining priorities will be assigned to the soft real-time streams. This allows us to partition the available properties into three areas, one for each stream class. How the priorities within one priority class are assigned will be explained in the following sections.

3 Hard Real-Time Streams

The only possibility to guarantee that all deadlines are always met is to analyse all possible situations before the start of the system. Then, it has to be verified that all streams transmit their messages within their deadline under the given setting. In the following, we will present several approaches to analyse the response times of a set of streams.

The simplest and most used approach was first introduced in [15] and revised in [5]. To calculate the worst-case response time of each stream, we assume that CAN effectively implements asynchronous fixed priority non-preemptive scheduling of messages. The response time then consists of two elements: *blocking time* B, due to lower priority messages which cannot be preempted, and *interference* due to higher priority messages. The upper bound for blocking time B is the transmission time of the largest message in the system. The interference I can be calculated by the following construction: Start all streams concurrently with an offset of zero. The time until the first occurrence of the analysed stream is the interference. In summary, the worst-case response time $WCRT_i(0,\infty) = B + I$. By comparing the WCRTs to the deadline for each stream, we can decide whether the system can hold all constraints.

In addition to the above approach, a large variety of holistic approaches for distributed embedded systems exist, for example [4, 10]. They take into account the whole system including the timing of tasks running on resources and their communication. Similar to the specialised approaches, the timing analysis is done offline during the design phase and cannot adapt to system changes. A further disadvantage is that the computational complexity is very high, because the holistic approaches consider a lot of constraints.

One of these approaches is the compositional performance analysis methodology used in SymTA/S [9]. It solves the global timing analysis by decomposing the system into independently investigated components. The timing between the tasks (event streams) is then captured with event models that can efficiently be described by a small set of parameters. Based on this offline analysis, an online analysis was developed as described in [14]. To analyse the system distributed online, models of the application running on each embedded system as well as its architecture are stored on the embedded system itself. The reason for this is that the embedded system itself analyses the timing properties by communicating timing information between the embedded system components until the timing properties converge to a stable state. In order to allow online change of the system, a runtime environment is proposed in [11]. It establishes an abstraction layer between platform and application. Contracts are used to guarantee the real-time constraints of the tasks. These contracts are preverified using formal methods before admitting corresponding configurations to take effect in the system. Thus, the system will only transition between provenly safe configurations. Therefore, this approach perfectly suits the requirement for hard real-time streams. More details on this approach can be found in Chap. 4.6.

4 Soft Real-Time Streams

The main issue of soft real-time streams are too large response times. Large response times occur when many streams try to access the bus simultaneously. In the worst case, all streams try to access at the same time. This represents the analytical worst-case as described in Sect. 3 and occurs when all offsets are zero. So, the objective is to find a configuration of offsets, where as few streams as possible release messages at the same time and, in consequence, the response times are minimal. Algorithms to find the optimal static offsets are known to have exponential run-time [7]. In addition, calculating the offsets at design time does not make sense, because due to the asynchronous nature of the distributed embedded system, the offsets may change over time. Our solution to this problem is a new method of dynamic adaptation of offsets based on monitoring of the network traffic, where each task that releases messages changes offsets on-line.

4.1 DynOAA

The dynamic offset adaptation algorithm (DynOAA) [18] is run on each node independently and periodically. An illustration of the operation of the DynOAA for one stream is shown in Fig. 2. In the upper part of the figure, on the top of the time line, the periodically released messages of the stream are indicated by small arrows. The larger arrows on the bottom of the time line indicate the instances when the adaptations start or when DynOAA is executed. Each run consists of a traffic monitoring phase and a delay phase. During the monitoring phase, a list (*busy_idle_list*) is created. An example of it is shown in the lower part of Fig. 2. It contains for each time slot during the monitoring phase an idle element if the bus is idle and a busy element if the bus is busy. From the *busy_idle_list*, we can find the *longest_idle_time* and *longest_busy_time*, which are the maximum continuous intervals when the bus was idle or busy, respectively. The next message of the stream is then delayed, i.e., the offset is adjusted, so that a message in the next monitoring phase is released in the middle of the *longest_idle_time* interval (*next_position*).

In distributed systems, all streams are considered independent of each other. If more than one stream starts to execute the adaptation simultaneously, there is a high probability that the value of the *next_position* will be identical at more than one stream. Instead of spreading, the message release times would in that case be

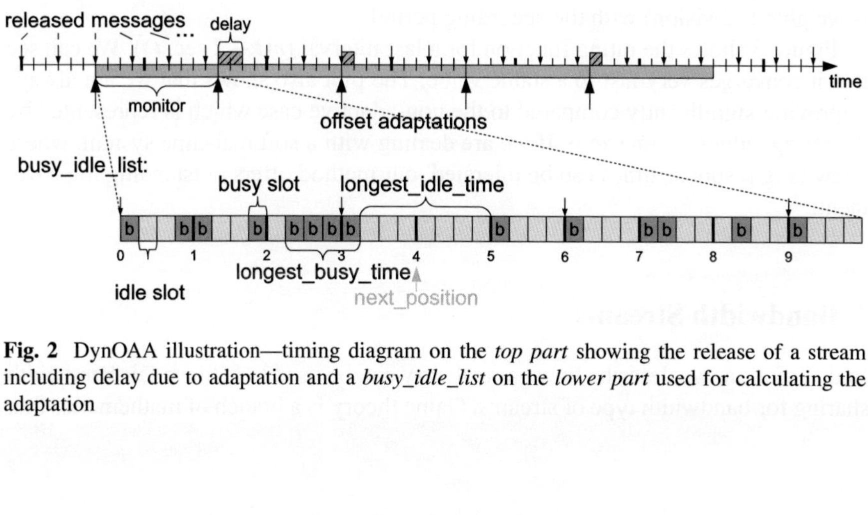

Fig. 2 DynOAA illustration—timing diagram on the *top part* showing the release of a stream including delay due to adaptation and a *busy_idle_list* on the *lower part* used for calculating the adaptation

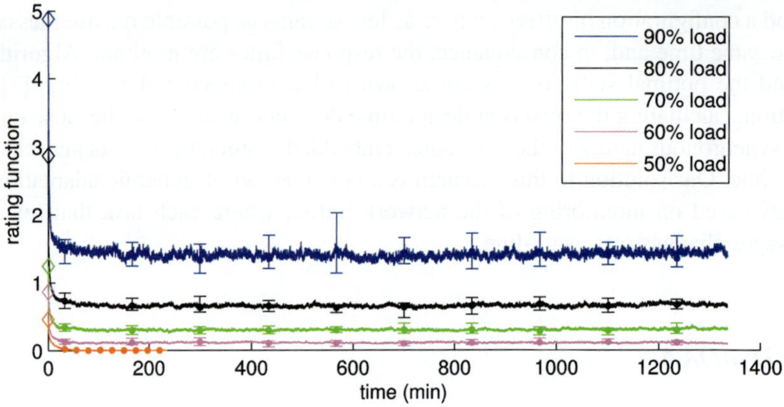

Fig. 3 Rating function as a function of time for different application and load scenarios—response times of the last adaptation interval taken into account: $r(t - 2\text{ sec}, t)$

clustered around the same time instance. Therefore, we need to ensure that only one stream is adapting its offsets at the same time. This is done by a unique criterion for all streams. Only the stream that is the first in the *longest_busy_time* interval will adapt its offset. All communication controllers adhere to this unique protocol individually.

4.2 Results

In order to evaluate our approach, we developed and used our own CAN bus simulator. The scenarios used for our experiments consist of synthetic scenarios generated by the Netcarbench [2] and are typical for the automotive domain with a bus load that can be freely adjusted. In order to compare different schedules we use the rating function: $r(b,e) = \sum_{i=1}^{k} \frac{WCRT_i(b,e)}{T_i}$. This function takes into account that streams with large periods are more sensitive to large response times, because each WCRT is weighted (division) with the according period.

Figure 3 shows the rating function for a last interval ($r(t - 2\text{ sec}, t)$). We can see that it converges very fast to a stable value. The plot also shows that we are always improving significantly compared to the non-adaptive case which is represented by the rating values at time zero. If we are dealing with a soft real-time system, where a few long response times can be tolerated, our method offers outstanding improvement.

5 Bandwidth Streams

In this section, we describe how game theory can be used to achieve fair bandwidth sharing for bandwidth type of streams. Game theory is a branch of mathematics that can be used to analyse multi-agent systems [3, 13]. In the following, we use the terms *stream* and *player* interchangeably.

5.1 Medium Access Game

The medium access in CAN can be modelled as a game:

Definition 1 A game G is defined as a tuple $G := (N, S, U)$ where

- $N = \{1, \dots, n\}$ is a set of n players.
- S is a set of mixed strategies available for all players.
- $u_i(\mathbf{p}) \in U$ is the utility function for player i dependent on the strategies chosen by all players $\mathbf{p} = (p_1, \dots, p_n)$.

A game can be displayed in *Normal Form* by showing for each strategy combination, the expected payoff of each player. For the two player priority-based medium access game, this Normal Form can be defined as a matrix as shown in Fig. 4. Each cell contains the payoff $(u_1(s_1, s_2), u_2(s_1, s_2))$ of the players' strategy dependent on the strategy selected by the other player. The strategy of player 1 is listed in the row, the one of player 2 in the column. If only one player wants to access the medium, it gets a grant and the payoff is 1. A collision occurs when both players want to send simultaneously. In this case, the player with the highest priority (here player 2) gets the grant. In the following, we assume that players are ordered by their priority: $\mathit{prior}(1) < \mathit{prior}(2) < \cdots < \mathit{prior}(n)$.

To achieve fair bandwidth sharing, a *mixed strategy* for each player has been introduced in [13]. Instead of having the two discrete strategies *send* and *wait*, a probability distribution on strategy space S_i is given. This means that the strategy of player i is to send with probability p_i and wait with probability $1 - p_i$.

The utility function for a player i is the probability to successfully transmit data. As described in [16], this Medium Access Game can be used to define and analyse fair bandwidth:

Definition 2 A strategy vector $\widetilde{\mathbf{p}}$ is called fair if each player has the same probability to get access to the shared medium. This means that the following constraint always holds:

$$u_1(\widetilde{\mathbf{p}}) = u_2(\widetilde{\mathbf{p}}) = \cdots = u_n(\widetilde{\mathbf{p}}) \tag{1}$$

We now want to consider a medium where $b \in [0, 1]$ determines the available bandwidth. Then, a strategy $\widetilde{\mathbf{p}}$ is fair if each player gets $\frac{b}{n}$ bandwidth ($u_i(\widetilde{\mathbf{p}}) = \frac{b}{n}$). To achieve this fairness, according to [16], each player i has to choose as sending probability:

$$\widetilde{p_i} = \frac{b}{n - (n - i) \cdot b} \tag{2}$$

Fig. 4 Two player priority-based medium access game in normal form

		player 2	
		wait	send
player 1	wait	0, 0	0, 1
	send	1, 0	0, 1

5.2 Enhanced Priority-Based Medium Access Game

Equation (2) describes the sending probability each player has to choose to reach fair bandwidth sharing. Its calculation requires global information of how many players are involved (n) and which priority the player has (i). To achieve self-organisation, the players need to be able to decide only on local information. Local information available for a stream can, e.g., be obtained by observing the bus traffic. In the proposed game theoretical model, this corresponds to the utility a player receives.

The problem in priority-based access is that streams with lower priorities cannot influence the strategies of higher-prior ones. Furthermore, there is no information available for a higher-prior stream that a lower-prior one tries to access the medium and gets blocked. Therefore, the medium access is extended by introducing a new constraint that claims that a minimal amount of bandwidth ϵ has to stay free. In case this constraint is not satisfied, the players are penalised, enforcing them to change their behaviours. A detailed description of the *enhanced priority-based medium access game* can be found in [16].

It can be shown that the fair strategy of (2) is a stable state. This means that as soon as the players reach this fair strategy, no player has an appeal to differ from its strategy. These results imply that it is possible to provide learning methods for the enhanced game to establish emergent and self-organising organic methods for bus-based communication architectures. In the following, an adequate learning algorithm is presented.

5.3 Penalty Learning Algorithm (PLA)

When playing the priority-based medium access game as described in Sect. 5.1, after *intervalLength* rounds the sending probability p_i of each player i is locally updated as illustrated in Fig. 5. This update is called a *learning step* and a pseudo code description is given in Algorithm 1. First, the *bus load* of the last *monitoring interval* has to be calculated (line 1) by dividing the number of rounds in which any player has sent a message by the total number of rounds of the monitoring interval. Additionally, the *success* (line 2) denotes the number of rounds player i has successfully transmitted in the monitoring interval divided by the total number of rounds. The learning algorithm works by adapting the sending probability according to the amount Δ calculated in lines 3 to 7. Depending on the load of the monitoring interval, the sending probability is lowered (*load* is greater $1-\epsilon$), or increased (*load* is less or equal to $1-\epsilon$). The amount depends on the success of the player. The more

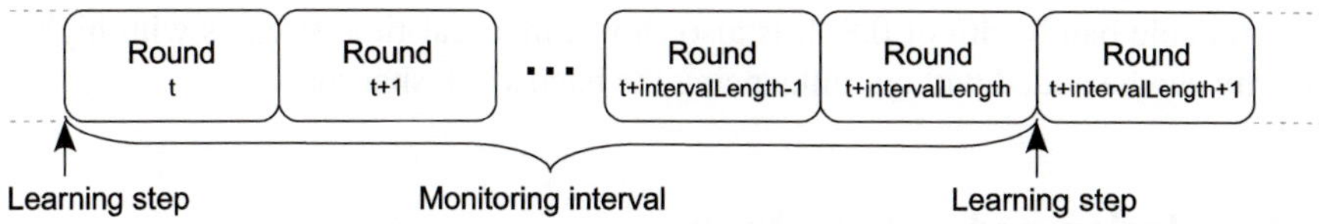

Fig. 5 PLA illustration—every *intervalLength* rounds a learning step is performed

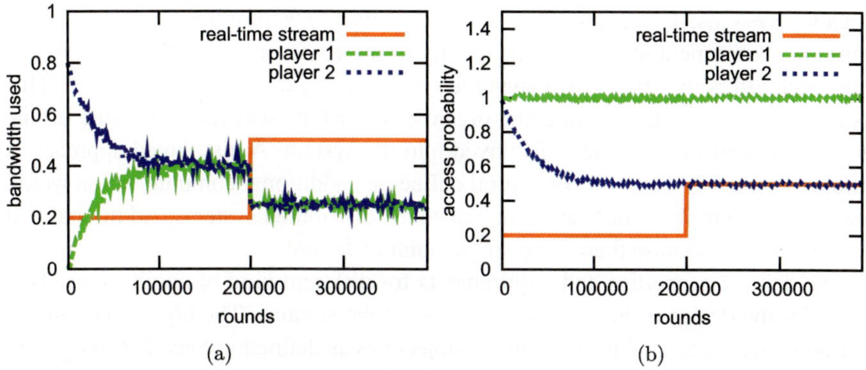

Fig. 6 Simulation results for PLA with mixed traffic, $intervalLength = 1000$ and $\eta = 0.02$

Algorithm 1 Learning step of PLA for each player

---

state $load = \frac{nOverallMsgSent}{intervalLength}$;
state $success = \frac{nSuccess}{intervalLength}$;
if $(load > 1 - \epsilon)$
state $\Delta = -success$;
else
state $\Delta = (1 - success)$;
endif
state $p_i^{(t+1)} = p_i^{(t)} + \eta \cdot \Delta$;

---

success the player had, the more it will decrease its sending probability in case of penalty and the less it will increase the sending probability in case of no penalty. This basic mechanism makes it possible that the players emerge to a fair bandwidth distribution. The change of the sending probability Δ is multiplied by a learning rate $\eta < 1$.

5.4 Results

Detailed results of the PLA are presented in [17]. Figure 6 shows the simulation results of the PLA for 2 players. After about 100 learning steps both players share the available bandwidth of 0.8. It is also shown that real-time streams with higher priority can be added without influencing the bandwidth sharing.

6 Conclusion and Future Work

In this article, we proposed a new methodology for bus-based priority based communication systems. This approach uses self-organisation and online adaptation to the current environment to improve the quality of the communication. In addition, this approach offers the ability to change the system configuration during runtime. This is possible because the planing and organisation, which was formerly done offline before the start of the system, is moved into the system. A distributed approach is chosen because in contrast to a central online one additional communication would be required, which is not feasible in embedded distributed systems. Additionally, it is more robust because there is no single point of failure.

We defined properties and requirements for different kind of applications. This provides the designer three classes to organise the streams. The organic communication system then enables to fulfil the objectives as defined in Sect. 2.2. By giving hard real-time streams the highest priority and using analytical methods, it is guaranteed that the hard deadlines are met. The analysis can be very pessimistic, i.e. very few streams are admissible and only a low average utilisation of the full bandwidth is possible. This is compensated by using lower priorities with less strict constraints. For soft real-time streams, we introduced DynOAA, which applies dynamic offset adaptation to reduce message response times and therefore may increase the number of fulfilled soft deadline constraints. In conclusion, compared to previous approaches, this adaptation allows a higher utilisation of the available resources. In general, the division into different types of applications provides a more efficient use. The bandwidth streams even allow a utilisation of 100%, because they can always fill up any remaining bandwidth.

Even though CAN uses a strictly priority-based access scheme, we can provide equal treatment for the streams. The PLA ensures that all bandwidth streams get equal amount of bandwidth. Although for soft real-time streams the priority finally decides the access to the bus, DynOAA distributes the access independent of the priorities, thus reducing response times for all messages equally.

Our future work will go into several directions. First, the predictability for soft real-time and bandwidth streams needs to be improved. One possible solution for soft real-time streams could be to introduce a fourth type of stream that guarantees (maybe with probability) a certain response time but is not as strict as the hard real-time. Secondly, we would to like to extend our methods for different architectures such as multi-segment buses or maybe even other types of communication protocols. Third, our model could be extended to incorporate tasks and their dependencies and their execution time. Finally, an implementation of the algorithms in real hardware needs to be done to prove the feasibility of the solution and to quantify its overheads.

References

1. CAN Specification 2.0 B: Robert Bosch GmbH, Stuttgart, Germany (1991)
2. Braun, C., Havet, L., Navet, N.: NETCARBENCH: a benchmark for techniques and tools used in the design of automotive communication systems. In: 7th IFAC International Conference on Fieldbuses and Networks in Industrial and Embedded Systems, Citeseer (2007)
3. Busoniu, L., Babuska, R., De Schutter, B.: A comprehensive survey of multiagent reinforcement learning. IEEE Trans. Syst. Man Cybern., Part C, Appl. Rev. **38**(2), 156–172 (2008)
4. Chakraborty, S., Künzli, S., Thiele, L.: A general framework for analysing system properties in platform-based embedded system designs. In: Proc. 6th Design, Automation and Test in Europe (DATE), Citeseer, pp. 190–195 (2003)
5. Davis, R.I., Burns, A., Bril, R.J., Lukkien, J.J.: Controller area network (CAN) schedulability analysis: refuted, revisited and revised. Real-Time Syst. **35**(3), 239–272 (2007)
6. Elmenreich, W., Kucera, M., Rinner, B., Seepold, R., Turau, V.: Challenges on complexity and connectivity in embedded systems. EURASIP J. Embed. Syst. (2009)
7. Goossens, J.: Scheduling of offset free systems. Real-Time Syst. **24**(2) (2003)
8. Grenier, M., Goossens, J., Navet, N., et al.: Near-optimal fixed priority preemptive scheduling of offset free systems. In: 14th International Conference on Real-time and Network Systems, Citeseer (2006)
9. Henia, R., Hamann, A., Jersak, M., Racu, R., Richter, K., Ernst, R.: System level performance analysis-the SymTa/s approach. In: Computers and Digital Techniques. IEE Proceedings, vol. 152, pp. 148–166 (2005)
10. Larsen, K.G., Pettersson, P., Yi, W.: UPPAAL in a Nutshell. Int. J. Softw. Tools Technol. Transf. **1**(1), 134–152 (1997)
11. Neukirchner, M., Stein, S., Schrom, H., Ernst, R.: A software update service with self-protection capabilities. In: Proc. of Design, Automation, and Test in Europe (DATE), Dresden, Germany (2010)
12. Racu, R.: The role of timing analysis in automotive network design. Talk, 4th Symtavision News Conference on Timing Analysis, Germany (2010)
13. Rakshit, S., Guha, R.K.: Fair bandwidth sharing in distributed systems: a game-theoretic approach. IEEE Trans. Comput. **54**(11) (2005)
14. Stein, S., Ernst, R.: Distributed performance control in organic embedded systems. In: IEEE 5th International Conference on Autonomic and Trusted Computing (ATC-08) Autonomic and Trusted Computing. Lecture Notes in Computer Science, vol. 5060, pp. 331–342. Springer, Berlin (2008)
15. Tindell, K., Hansson, H., Wellings, A.: Analysing real-time communications: controller area network (CAN). In: Proc. 15th IEEE Real-Time Systems Symposium, San Juan, Puerto Rico. IEEE Society Press, Los Alamitos (1995)
16. Wildermann, S., Ziermann, T., Teich, J.: Self-organizing bandwidth sharing in priority-based medium access. In: Proceedings of the Third IEEE International Conference on Self-Adaptive and Self-Organizing Systems (SASO'09), San Francisco, USA, pp. 144–153 (2009)
17. Ziermann, T., Mühleis, N., Wildermann, S., Teich, J.: A self-organizing distributed reinforcement learning algorithm to achieve fair bandwidth allocation for priority-based bus communication. In: 1st IEEE Workshop on Self-Organizing Real-Time Systems (SORT 2010), Carmona, Spain, pp. 11–20 (2010)
18. Ziermann, T., Salcic, Z., Teich, J.: DynOAA—dynamic offset adaptation algorithm for improving response times of CAN systems. In: Proceedings of Design, Automation and Test in Europe, 15–17 March. IEEE Computer Society, Grenoble, France (2011, to appear). ISBN 978-3-9810801-7-9 (http://www.date-conference.com/proceedings/)

T. Ziermann (✉) · S. Wildermann · J. Teich
Department of Computer Science, Hardware-Software-Co-Design,
University of Erlangen-Nuremberg, Nürnberg, Germany
e-mail: tobias.ziermann@informatik.uni-erlangen.de

Chapter 5.6
OC Principles in Wireless Sensor Networks

Jakob Salzmann, Ralf Behnke, and Dirk Timmermann

Abstract Large wireless sensor networks will play a major role in environmental observation of the near future. Due to limited resources and plenty of nodes, conventional control by use of a base station becomes infeasible. Organic principles derived from nature are able to bridge the gap between centralised and self-organised network control. This article proposes and evaluates the applicability of selected collaborative organic principles to wireless sensor networks.

Self-organised clustering schemes allow detection and deactivation of redundant network components, which are crucial for extending the lifetime of the network. The proposed self-healing schemes at node and cluster level allow rehabilitative as well as preventative healing. A superordinate routing tree algorithm interconnects the emerged clusters, and selected incorporated local rules enable a further increase in robustness and lifetime of the network.

Keywords Wireless sensor networks · Clustering · Scale-free routing · Self-organisation · Self-healing

1 Introduction

The proceeding miniaturisation of technical devices allows to join microcontroller, sensors and radio technology on tiny, battery driven devices, called sensor nodes. Several of these devices interconnected with each other establish a so called Wireless Sensor Network (WSN). Such WSNs are deployed in an area to fulfil their task often in environmental observation [2]. Potential scenarios are detection and tracking of objects, habitat monitoring, detection and prevention of forest fires, and laboratory automation. Here, nodes have to sense their environment permanently or periodically and transmit measured phenomena to a central station, called sink. Due to the size of the observed area, this is mostly done via multi-hop. Major problems in WSNs are the limited resources of the individual sensor nodes. By limiting the communication and sensing activities to a minimum without affecting the network functionality, a WSN is able to increase its lifetime dramatically.

To optimise this extension of lifetime, each node has to know whether it is required for network functionality or it is allowed to temporarily set itself into a sleep mode [6, 16]. The challenge here is to avoid periodic messages for a central control

C. Müller-Schloer et al. (eds.), *Organic Computing — A Paradigm Shift for Complex Systems*, 503–516
Autonomic Systems, DOI 10.1007/978-3-0348-0130-0_33,

of each sensor node, causing an unintentional waste of energy. Feasible solutions for this challenge are offered by clustering mechanisms [1]. By combining a self-organised intra cluster role changing algorithm with a redundancy detecting cluster structure, nodes of the WSN are able to save energy by reducing their transmission or sensing activities without global control. In addition, redundant nodes are able to replace defective or impaired nodes within their cluster. By the extension of the self-organised redundancy detection to cluster level, whole clusters are able to determine whether they are required for networks tasks or are allowed to be set into an energy-conserving sleep state. Next to an overall reduced node activity, several collaborating clusters are able to set themselves back to the active state and substitute failed clusters in their neighbourhood.

As a side-effect of clustering, differently populated clusters emerge due to a random node deployment. By considering the different strengths in an emerging routing tree, the network is able to increase overall robustness without any additional cost.

The remainder of the article is organised as follows: Sect. 2 investigates self-organised clustering mechanisms, Sect. 3 describes the healing mechanisms at node and cluster level. In Sect. 4, a routing tree for increased robustness is developed. The article concludes in Sect. 5.

2 Self-organisation in Wireless Sensor Networks

In large multi-hop networks, as given by WSNs, centralised control of each single node comes along with an infeasible amount of communication and hence wasted energy. To control such large distributed systems, self-organisation is a key-principle. Mechanisms like role assignment and role changing, redundancy detection, and clustering are used to achieve global control via local rules. Using role assignment and adaptive role changing, a self-organised hierarchy is applied to the network and lifetime increase at node level is achieved. Redundancy detection describes a scheme which allows identification of groups with similar abilities, while clustering presents different applicable mechanisms to analyse redundancy with the help of a top down view.

2.1 Role Assignment and Adaptive Role Change

Using role assignment, the flat hierarchy of a WSN, given in Fig. 1(A), is changed into a multilevel hierarchy, as given in Fig. 1(B). Here, the network is divided into sensor nodes and clusterheads. As shown in Fig. 1(C), the role of a clusterhead can be further divided into fine-grained tasks which can be assigned to additional nodes in the cluster. Although this distributed role assignment increases the amount of intra-cluster communication, it can be utilised by self-healing schemes, as explained in Sect. 3.

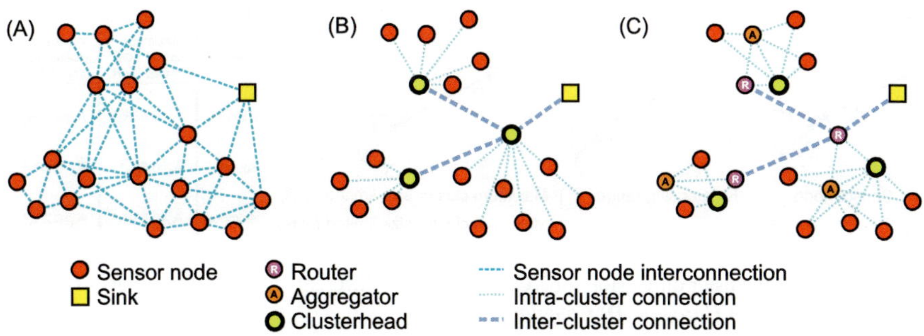

Fig. 1 Role assignment of sensor nodes. (**A**) Unstructured network (**B**) Hierarchy via clustering (**C**) Example with fine-grained role assignment in each cluster

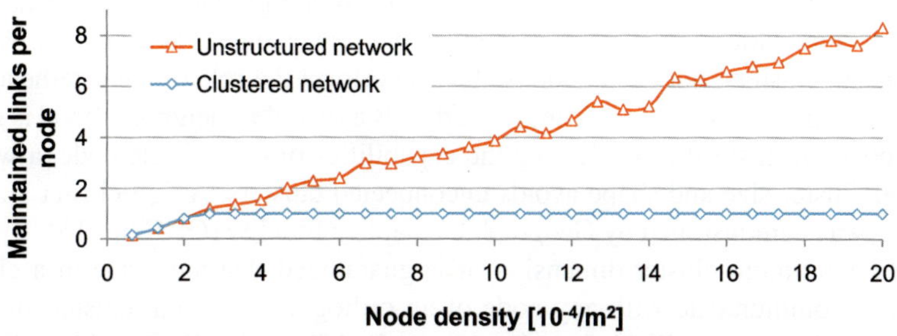

Fig. 2 Maintained links per node versus increased node density

At first glance, multilevel hierarchy only results in a structure which impedes the direct communication between two randomly selected sensor nodes. But this drawback, which is negligible for sink-oriented communication, comes along with the powerful advantage that the overall possible communication routes are reduced and the route maintenance per node is minimised, as shown in Fig. 2. This behaviour directly impacts the overall energy consumption and results in an increased network lifetime. Furthermore, a predetermined routing to a clusterhead or an aggregator allows a reduction of transmitted data, using aggregation as well as compression of measured sensor data.

Due to the different tasks of different roles, energy consumption of a hierarchic network is not balanced, as sensor nodes consume only a fraction of clusterheads' energy. To avoid a premature network breakdown, a role changing mechanism can be applied to the network. This task can be directly tackled by self-organisation. Instead of a permanent and cost-intensive central control of all clusterheads, each clusterhead is responsible for the nodes in its cluster. In particular, it is allowed to assign its own role to a node with a higher amount of remaining energy. This mechanism was implemented within a sensor network in [7] with fine-grained roles. Compared to a sink-initiated role-changing, the self-organised role-changing enables a lifetime increase of about 40%, as shown in more detail in [7].

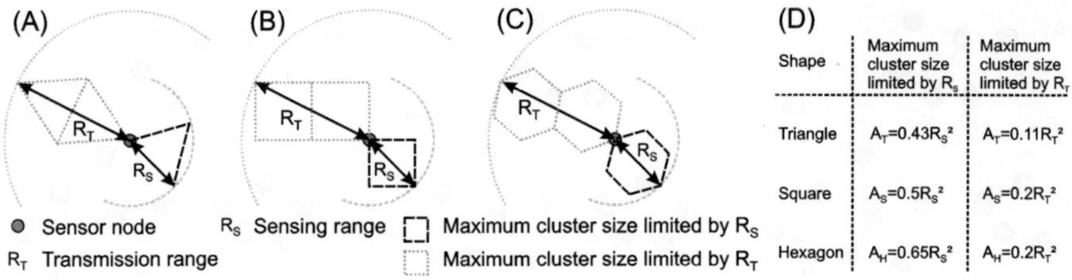

Fig. 3 Maximum cluster dimension for XGAF clusters of different tesselating cell shapes (**A**) *Triangle* (**B**) *Square* (**C**) *Hexagon* (**D**) *Calculated values*

2.2 Clustering Schemes

Although the proposed partition into clusters is an energy conserving mechanism, a feasible application to a WSN is only given if a role change is also applied. Within randomised clusters, not each node is able to perform the role of a clusterhead and maintain connections to all adjacent clusterheads as a node's neighbourhood is given by its position. Instead of evaluating the capabilities of each single node, a wisely selected cluster size and shape avoids unconnected clusters. A first cluster size restriction was demonstrated by Geographic Adaptive Fidelity (GAF) in [15]. By limiting the maximum cluster dimension, it is guaranteed that any node in a cluster is able to communicate with any node of an orthogonal adjacent cluster. In [13], this approach was extended by an algorithm called Extended Geographic Adaptive Fidelity (XGAF) in two directions. On the one hand, the applicability of different cell shapes was investigated. On the other hand, the resulting cluster sizes take also sensing range into account, i.e. the ability to detect phenomena. Compared to an unclustered sensor network, the lifetime was more than doubled as non active nodes could be switched to a hibernation state.

Different investigated tessellating cell shapes and their resulting maximum sizes are given in Fig. 3. As it can be seen, the hexagonal shape provides maximum possible cell area in terms of both, sensing and transmission range. The precondition of a mandatory spatial reference system for all those clustering schemes conflicts with the idea of a self-organising system. This problem was tackled in [10, 11] by self-organised clustering schemes called Localisation-Free Geographic Adaptive Fidelity (Free-GAF) and Localisation-Free Clustering with Approximation to Symmetric Hexagons (Free-CLASH), respectively. Both algorithms replace global cluster border detection by broadcasted messages from a starting clusterhead with adapted transmission range. Recipients are forced to join the cluster and their affirmation message is simultaneously utilised for the emergence of further clusterheads. As a result, a clustered network structure emerges with exchangeable clusterheads. Although the performance of this structure can usually not compete with a globally organised cluster structure in terms of required active nodes, the resulting clusters are independent from preceding global information and are unaffected by localisation errors. In the case of Free-CLASH, this performance increases with the number of deployed nodes due to an increased choice of feasible clusterheads.

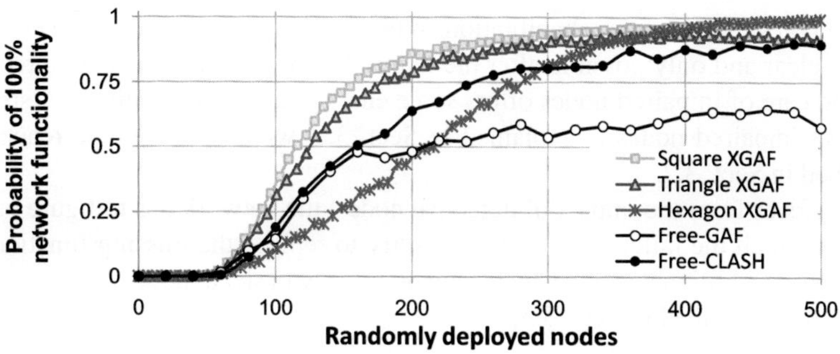

Fig. 4 Achieved probability for complete network functionality of a sensing range limited WSN. Network size: 162 m × 162 m, Sensing range: 27 m

In terms of achieved network functionality, Free-CLASH is even able to compete with hexagon based XGAF, as shown in Fig. 4.

3 Self-healing in Wireless Sensor Networks

The principle of self-healing allows a system to regenerate after a malfunction. This healing can either be done via spending additional resources or by transferring the tasks of an erroneous system component to different unutilised components. In the context of WSNs, the global task is usually continuous surveillance of a given area and a prompt message of a detected phenomenon from the acquiring node to the sink. Due to their random distribution in partly inaccessible areas, sensor nodes are affected by several error sources, e.g. atmospheric conditions in outdoor scenarios or mechanical stress during deployment. To handle the occurring errors, each node is assigned to one of the following three states.

Healthy nodes: These fully functional nodes have no impairments and shall be incorporated into the network after deployment.

Impaired nodes: Nodes which belongs to this group are only partly faulty and can be divided into Sensor Impaired Node (SINs) and Transmission Impaired Node (TINs). A SIN has a faulty sensor unit and is either never or seldom able to detect phenomena or detects phenomena misleadingly. The transmission range of TINs is limited to only some nodes in their neighbourhood. Reasons for both groups may be defective hardware or an inauspicious node deployment behind an obstacle.

Defective nodes: The third group are nodes which are completely broken, e.g. by a broken microcontroller or by exhausted batteries. In this case, the node is not able to be part of the network any longer but is also not able to affect the remaining network.

To apply self-healing, the network itself has to detect impaired or defective nodes and to react autonomously to retrieve its aspired performance. This can either be done via prevention or via rehabilitation. While the strategy of detecting a defective node is clear and only possible after the failure has occurred, a preventive recognition and care of impaired nodes offers some energy-saving possibilities. A strategy to detect impaired nodes is explained in Sect. 3.1, while a preventive reaction is described in Sect. 3.2.

To recover from the impact of defective nodes, the network has to figure out the function which the failed node served, and try to replace the missing functionality by activating additional nodes which were previously in sleep mode. A cluster-based approach is described in Sect. 3.3.

3.1 Impaired Node Detection

Conventional self-healing, also called rehabilitation, aims to retain a predefined functionality. In contrast, this section focuses on preventive healing, protecting the network from future malfunctions by reacting on early indications. Until now, just little attention has been paid in the literature to the detection and the ensuing exploitation of erroneous nodes, i.e. SINs and TINs.

While the occurrence of SINs does not necessarily affect energy consumption, it is possible to adapt network behaviour to save energy if additional knowledge about those erroneous nodes is available. As described in [4], in case of erroneous readings energy can be saved at various points. On the one hand, erroneous nodes neither need to acquire erroneous readings nor need to send erroneous values. On the other hand, routing nodes which do not need to forward dispensable messages are saving energy. In addition, erroneous nodes incapable of sensing the environment are well suited for other tasks like routing, reducing the load on healthy nodes.

A fundamental requirement for that kind of energy awareness is a detection of erroneous nodes. An algorithm which is especially designed to provide this information without use of any central instance is Efficient Localized Detection of Erroneous Nodes (ELDEN) [5]. Whenever a spatially correlated phenomenon, not limited to a single node, is to be monitored, e.g. vibration or sound in the case of vehicle tracking, ELDEN is capable to detect local outliers. Using ELDEN, each node uses measurement data from one-hop distant nodes to rate its own measurement trustworthiness. The detection algorithm is based on quartiles, which are robust against outlying data points. The quartiles can be found by sorting given data points in ascending order and dividing the ordered data into four equal-sized data subsets. The quartiles are given by the values of the data points, marking the borders between the subsets, i.e. 25%, 50% and 75% of the data points are less than or equal to the first ($Q1$), second ($Q2$) and third ($Q3$) quartile, respectively. Median ($Q2$) and Interquartile Range (IQR) ($IQR = Q3 - Q1$) are utilised by ELDEN to calculate a decision value y_i for node i, using the specific data of this node, as given in (1).

$$y_i = \frac{x_i - Q2_i}{IQR_i} \tag{1}$$

As given in (2), if the absolute value of y_i exceeds a threshold θ_{ELDEN}, the node is classified as faulty or erroneous, respectively. The threshold θ_{ELDEN} is found empirically. A well performing substitute of θ_{ELDEN} is 2, referring to [5]. That means a sensor reading is regarded as faulty if it differs more than twice the IQR from the median.

$$node_i = \begin{cases} \text{faulty} & \text{if } |y_i| > \theta_{ELDEN} \\ \text{normal} & \text{if } |y_i| \leq \theta_{ELDEN} \end{cases} \tag{2}$$

In case of TINs, the given malfunction directly affects the energy consumption of the node and its communication partners, due to increased retransmissions or even more hops, needed for a certain distance. While ELDEN takes any kind of input values, figuring out whether a specific value is likely to be faulty, it can be also used to detect TINs. Regarding the number of nodes within transmission range as input data, ELDEN detects if a node has a significant lower or higher number of nodes in its transmission range. By use of a modified decision, as given in (3), impaired nodes as well as highly connected nodes can be identified.

$$node_i = \begin{cases} \text{highly connected} & \text{if } y_i > \theta_{ELDEN} \\ \text{normal} & \text{if } |y_i| \leq \theta_{ELDEN} \\ \text{impaired} & \text{if } -y_i > \theta_{ELDEN} \end{cases} \tag{3}$$

3.2 Preventive Role Changing

If the network is structured into different roles, it is possible to utilise the remaining abilities of impaired nodes without affecting the remaining network. The more roles are assigned in a cluster, the less overall functionality is required from each node. Imagine a cluster with the roles environment observing sensor node, clusterhead, aggregator and router, as depicted earlier in Fig. 1(C). To achieve the most energy efficient and best performing behaviour, TINs should not be used as routing nodes, while they are well suited for local tasks like sensing, aggregation or managing a cluster as clusterhead. Similarly, SINs should be not used as sensor nodes, but can be used as aggregators, routers or clusterheads. If a node is a SIN and a TIN at the same time, it can be at least used as clusterhead or aggregator. The possible states of a node and the possible roles without restrictions are shown in Fig. 5.

3.3 Cluster-Based Rehabilitation

The detection of defective nodes and an appropriate reaction is essential for WSNs. Broken nodes due to exhausted energy or environmental impacts are usual for most WSN scenarios. Self-healing allows a lifetime prolongation by regenerating the network functionality after coverage or connectivity holes have been emerged. There are two basic requirements to apply a rehabilitation.

	Sensor node	Clusterhead	Aggregator	Router
Healthy node	✓	✓	✓	✓
Sensor impaired node	✗	✓	✓	✓
Transceiver impaired node	✓	✓	✓	✗
Defective node	✗	✗	✗	✗

Fig. 5 Nodes with different states and allowed roles

The first requirement is a couple of unused network resources. If a node breaks down, adjacent nodes must be able to assign its task. Although an active node offers some inherent healing abilities, e.g. by increasing the transmission power to reach nodes in two-hop distance, a more feasible solution is to utilise sleeping nodes in the same region. If a clustering as described in Sect. 2.2 is applied, each cluster contains several unused nodes, which can be utilised for self-healing. Moreover, due to the selected cluster size, each node in a cluster is proper to serve as backup for each other node in its cluster.

The second requirement is a timely detection of defective nodes. A sleeping node, which aims to serve as backup node, must wake up periodically and detect whether it is required as backup. If the network is unclustered, the detection is complex because an awaked node has to communicate with all nodes in its neighbourhood. In a clustered network, the detection of a defective clusterhead is simple and can be done via request and missed confirmation message of the corresponding clusterhead. This allows a fast and energy-aware intra cluster healing.

Next to a quickly possible intra cluster healing, a stronger overlap of adjacent clusters allows another kind of rehabilitation, the inter cluster healing, as done with the Mutual Assistance in a Cluster Environment (MASCLE) algorithm family in [8, 9, 14]. If the allowed cluster size is reduced to a fraction of the maximum one, it is guaranteed that a cluster is able to cover a part of the adjacent cluster and communicate to more distant clusters, as exemplarily shown for 2-MASCLE in Fig. 6(A). The reduced cluster size comes along with more clusterheads, which comes along with more active nodes at first glance. But due to the increase in neighbours and overlapping sensor coverage areas, only a fraction of cells is required to be active and sense the environment at each time point, as shown in Fig. 6(B). Here, only every second cluster has to be active for complete sensor coverage and connectivity. Instead of an increase, this ability allows an overall reduction of the active nodes.

As mentioned before, this reduced maximum cluster size allows for self-healing. If the last node of a cluster becomes defective, a network hole emerges. If the adjacent clusters recognise this failure, they are able to perform the task of the failed cluster jointly, as shown in Fig. 6(C). Due to the limited cluster size, neither the coverage nor the connectivity are impaired after this healing. In different MASCLE algorithms, this healing may be performed by less than all adjacent clusters, as further explained in [8, 14].

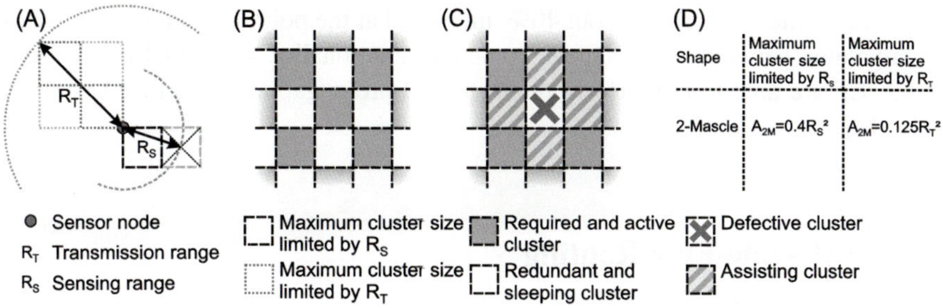

Fig. 6 Applied 2-MASCLE algorithm (**A**) Determination of maximum cluster dimensions (**B**) Regular operation with half of clusters active (**C**) Healing of a failed cluster with adjacent clusters (**D**) Calculated cluster size values

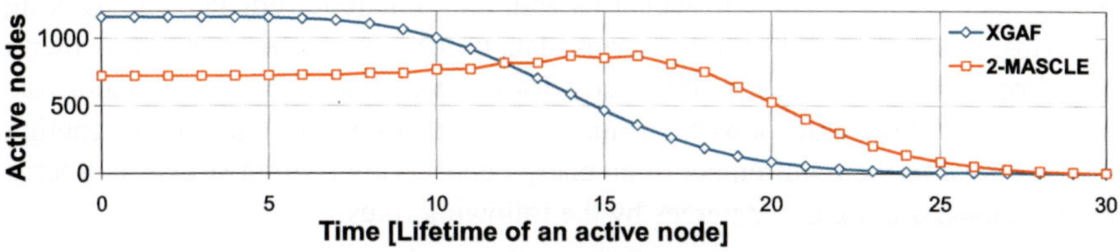

Fig. 7 Numbers of active nodes in an example scenario with applied XGAF and 2-MASCLE algorithm

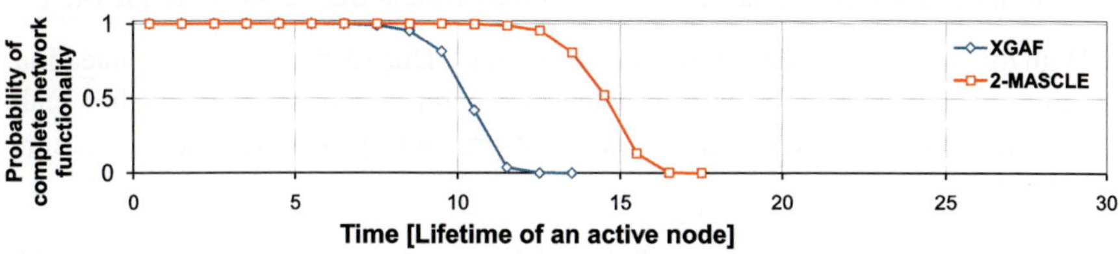

Fig. 8 Probability of complete network functionality over time of networks with applied clustering algorithms

Of course, this additional activation of adjacent clusters leads to a higher energy consumption during runtime, as shown in Fig. 7. One can see that the number of active nodes in XGAF starts with a certain number, depending on the size of the network. Over time, nodes and later clusters die due to their energy consumption and network holes emerge, which comes along with less active nodes in the network. In contrast, the number of active nodes increases over time using 2-MASCLE. Here, failed clusters are healed by surrounding clusters which leads to an increase of the overall number of active nodes. But after achieving a maximum, the network depletes its remaining energy faster compared to XGAF. However, this faster dying has no impact on the lifetime of a network as the complete network functionality is already lost before the number of active nodes achieves its maximum, as shown in Fig. 8. Here, one can see that the 2-MASCLE algorithm allows a lifetime prolongation of the network of about 40%, measured at the point of time when 50% of all simulated networks still achieve complete functionality. 4-MASCLE and Hex-MASCLE are able to outperform this achieved prolongation, as further described in [8, 14].

4 Robust Scale-Free Routing

While the last section explained principles and mechanisms for a self-organised and self-healing network at cluster level, an essential step to an efficient WSN is an intelligent routing structure, which is able to guarantee an undisturbed data flow in direction of the sink. Due to energy limitations in WSNs, idle listening of each routing node is only feasible in scenarios with real time requirements, e.g. tracking scenarios. In a large number of scenarios, data transmission is only required after a long time of data aggregation. Here, cost for route maintenance play a major role in relation to the overall power consumption. In this context, the idea of routing trees fits ideal to the requirement of an energy conserving routing structure. Such a self-organised routing tree emerges by the following rules:

1. A starting node, i.e. the sink, broadcasts a tree request.
2. Each node which is not part of the tree and is a recipient of the tree request, joins the tree with the broadcasting node as parent node.
3. After joining the tree, a node reproduces the original tree request broadcast.

With this simple scheme, a routing tree emerges. Due to the preferred connection, i.e. all nodes connect with the first requesting parent node, the network has scale free similar character, which means that the probability of having k connections follows (4).

$$P(k) \sim k^{-\gamma} \tag{4}$$

Usually, γ is a constant value in the range $2 < \gamma < 3$, which means that most nodes will have only few connections and few nodes, called hubs, have many connections. This behaviour is recognisable in Fig. 9. As known from the literature, this group of networks emerges after a network growth with preferred connections and are prevalent in several natural and artificial networks, e.g. social networks, air route networks, and networks of web links. As investigated in [3], the class of scale free networks is classified by several common properties. The most important properties are:

1. *Small World Networks:* Usually, the number of hops between two arbitrary nodes is relatively small in scale free networks.
2. *Robustness against random failures:* If a random failure occurs in such a network, there is only a low probability for a strong impact on the remaining network, as the failed node would mostly be a node, which is at a leaf of the routing tree, further referred to as leaf node.

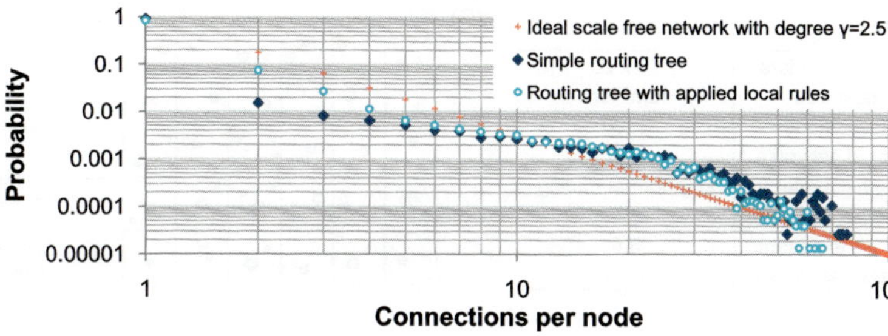

Fig. 9 Probability of numbers of connections in a routing tree, compared to an ideal scale free network

3. *Susceptibility to targeted attacks:* If some of the well connected nodes, called hubs, break down, the network functionality is extremely reduced.

Although a WSN benefits from the first two properties, the third one may lead to an undesirable and fast network breakdown. Fortunately, the already developed cluster structure offers the potential to strengthen the robustness of hubs by selecting appropriate clusters as hubs. Due to the fact that a centralised selection of hubs comes along with additional cost, it is more feasible to guide the routing tree emergence process by incorporating some local rules. In simulations, the following local rules have been applied to the routing tree:

1. *Range Reduction:* Sparsely populated clusters have the possibility to ignore tree request messages from distant clusters. Due to the fact that each cluster has to broadcast a tree request, it is likely that an adjacent cluster will respond to the tree request and re-broadcast it first. By waiting for this second chance to join the tree, a sparsely populated cluster can increase its chance of joining at a lower layer in the tree, becoming a node of less importance, probably a leaf.
2. *Wait and See:* Sparsely populated clusters are allowed to wait a certain time before re-broadcasting the tree request message. With this rule, the chance of well populated clusters to become hubs is increased dramatically.

To let a cluster determine by itself whether it is well populated or not, the mean value μ and the standard deviation σ of members in a cluster is broadcasted within the tree request message. The information either comes from the sink, or each receiving cluster adds the information about its own cluster to the values before re-broadcasting the tree request. To achieve robust routing, it is sufficient to classify the clusters based on the number of contained nodes N, as exemplarily given in (5).

$$\text{Cluster population} = \begin{cases} \text{Sparsely} & \text{if } N < \mu - \sigma \\ \text{Medium} & \text{if } \mu - \sigma \leq N \leq \mu + \sigma \\ \text{Well} & \text{if } N > \mu + \sigma \end{cases} \tag{5}$$

By this classification, a cluster detects its strength related to other clusters in the network, and appropriate local rules during the routing tree emergence can be applied. In a simulation environment, given in Fig. 10(A), the local rules were applied

(A)

Property:	Value:
Number of clusters	20 x 20
Cluster size	15m*15m
Deployed nodes	1000
Simulated networks	200
Channel model	Log normal Fading
Shadowing variance	8.4
Path loss exponent	3.6
Mean transmission range	70m

(B)

Cluster strength	Allowed communication range	Waiting period after connection
Weak cluster (Contained nodes = 1)	Limited to	8s
Medium cluster (1 < Contained nodes < 5)	No limit	4s
Strong cluster (Contained nodes > 4)	No limit	2s

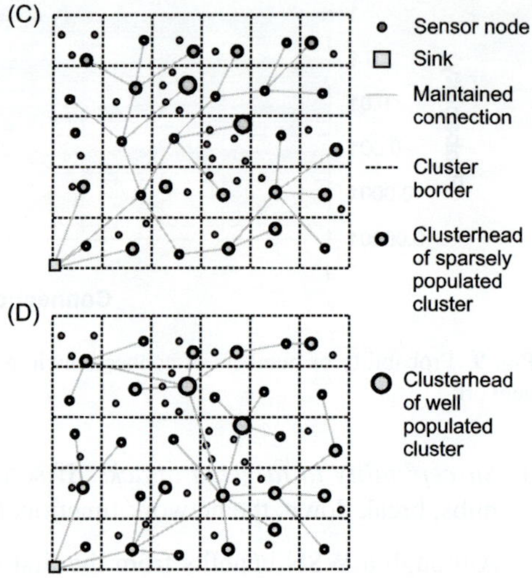

Fig. 10 Routing tree emergence (**A**) Simulation environment (**B**) Assignment of local rules (**C**) Simple routing tree (**D**) Emerged tree with applied local rules

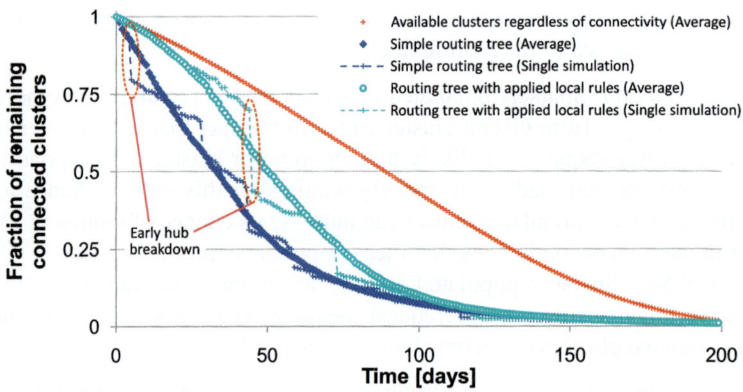

Fig. 11 Remaining connected clusters in a routing tree after error injection

as given in Fig. 10(B). For an example network, a conventional routing tree is shown in Fig. 10(C), while Fig. 10(D) shows the adapted routing tree. As shown in [12], such an adapted routing structure with strong clusters as hubs is able to increase the lifetime compared to a simple routing tree by more than 100%. The idea of a scale-free routing tree remains unaltered, as one can see in the connection probability in Fig. 9.

To investigate the robustness of this structure, an error rate of 2% per day was assumed in simulation. The error rate was selected arbitrarily and reflects environmental impacts of outdoor scenarios. Due to the short simulation duration, failed nodes regarding to usual energy exhaustion were not considered.

After applying the error rate, the number of clusters connected to the sink has been investigated as shown for two example networks in Fig. 11. Here, the breakdown of a hub is obvious. Furthermore, it is obvious that the routing tree with applied local rules easily outperforms the simple routing tree on average and achieves the maximum possible remaining connectivity.

5 Conclusion and Outlook

In this article, the applicability of organic principles in WSNs and their impacts have been discussed. By using clustering to establish a hierarchical structure in WSNs, organic principles offer optimised solutions for an energy-conserving and robust network. By using a self-organised intra cluster role changing algorithm, redundant nodes are able to detect and temporarily switch-off themselves. Due to a wisely chosen cluster size, the effort to detect and heal impaired or defective nodes is low and no central control is required. With cluster-based self-healing, a self-healing is assigned to cluster level, allowing a network to replace connectivity and coverage holes by an additional amount of energy. By an applied scale-free routing tree with assigned local rules, the clusters are interconnected to each other at routing level. Here, the strength of a cluster is determined via its population, and its task in the applied scale-free network is selected autonomously. Due to the application of organic principles at node, cluster and routing levels, the described mechanisms are able to be merged on a single WSN without disturbing each other.

Although the described principles yield robustness and lifetime enhancement of a general WSN, application specific mechanisms and parameters would allow to adjust the mechanism parameters to a given scenario. Moreover, the described algorithms offer only the application of selected Organic Computing principles to a WSN. The appliance of further principles is left open for future investigation.

References

1. Abbasi, A.A., Younis, M.: A survey on clustering algorithms for wireless sensor networks. Comput. Commun. **30**(14–15), 2826–2841 (2007)
2. Akyildiz, I.F., Weilian, S., Sankarasubramaniam, Y., Cayirci, E.E.: A survey on sensor networks. IEEE Commun. Mag. **40**(8), 102–114 (2002)
3. Albert, R., Jeong, H., Barabasi, A.-L.: Error and attack tolerance of complex networks. Nature **406**(6794), 378–382 (2000)
4. Behnke, R., Salzmann, J., Lieckfeldt, D., Thurow, K., Golatowski, F., Timmermann, D.: Exploiting malicious node detection for lifetime extension of sensor networks. In: 6th IEEE Consumer Communications and Networking Conference, CCNC 2009, pp. 1–2 (2009)
5. Behnke, R., Salzmann, J., Simanowksi, S., Timmermann, D.: Efficient localized detection of erroneous nodes (ELDEN). In: 2nd International Workshop on Sensing and Acting in Ubiquitous Environments, SEACUBE'10 (2010)
6. Cărbunar, B., Grama, A., Vitek, J., Cărbunar, O.: Redundancy and coverage detection in sensor networks. ACM Trans. Sens. Netw. **2**, 94–128 (2006)

7. Reichenbach, F., Bobek, A., Hagen, P., Timmermann, D.: Increasing lifetime of wireless sensor networks with energy-aware role-changing. In: In Proceedings of the 2nd IEEE International Workshop on Self-Managed Networks, Systems & Services, SelfMan 2006, Dublin, Ireland, pp. 157–170 (2006)
8. Salzmann, J., Behnke, R., Gag, M., Timmermann, D.: 4-MASCLE—improved coverage aware clustering with self healing abilities. In: The International Symposium on Multidisciplinary Autonomous Networks and Systems, MANS 2009, Brisbane, Australia, pp. 537–543 (2009)
9. Salzmann, J., Behnke, R., Lieckfeldt, D., Timmermann, D.: 2-Mascle—a coverage aware clustering algorithm with self healing abilities. In: 3rd International Conference on Intelligent Sensors, Sensor Networks and Information Processing, Melbourne, Australia (2007)
10. Salzmann, J., Behnke, R., Timmermann, D.: A self-organized localization-free clustering approach for redundancy exploitation in large wireless sensor networks. In: GI Jahrestagung, Workshop: Adaptive und organische Systeme, München, Deutschland, pp. 747–754 (2008)
11. Salzmann, J., Behnke, R., You, J., Timmermann, D.: Free-CLASH—improved localization-free clustering in large wireless sensor networks. In: International Conference on Ultra Modern Telecommunications Workshops, ICUMT'09, pp. 1–6 (2009)
12. Salzmann, J., Kubisch, S., Reichenbach, F., Timmermann, D.: Energy and coverage aware routing algorithm in self organized sensor networks. In: Proceedings of Fourth International Conference on Networked Sensing Systems, Braunschweig, Deutschland, pp. 77–80 (2007)
13. Salzmann, J., Behnke, R., Timmermann, D.: Tessellating cell shapes for geographical clustering. In: CIT, pp. 2891–2896 (2010)
14. Salzmann, J., Behnke, R., Timmermann, D.: Hex-MASCLE—hexagon based clustering with self healing abilities. In: IEEE Wireless Communications and Networking Conference, Cancun, Mexico (2011)
15. Xu, Y., Heidemann, J., Estrin, D.: Geography-informed energy conservation for ad hoc routing. In: ACM MOBICOM, pp. 70–84 (2001)
16. Zhou, Z., Das, S., Gupta, H.: Connected k-coverage problem in sensor networks. In: Computer Communications and Networks, Proceedings, 13th International Conference on, ICCCN 2004, pp. 373–378 (2004)

J. Salzmann (✉) · R. Behnke · D. Timmermann
Institute of Applied Microelectronics and Computer Engineering, University of Rostock, 18119 Rostock, Germany
e-mail: js228@uni-rostock.de

R. Behnke
e-mail: ralf.behnke@uni-rostock.de

D. Timmermann
e-mail: dirk.timmermann@uni-rostock.de

Chapter 5.7
Application of the Organic Robot Control Architecture ORCA to the Six-Legged Walking Robot OSCAR

Erik Maehle, Werner Brockmann, Karl-Erwin Grosspietsch, Adam El Sayed Auf, Bojan Jakimovski, Stephan Krannich, Marek Litza, Raphael Maas, and Ahmad Al-Homsy

Abstract Walking robots are complex machines, which are challenging to engineer and to program. In order to master this complexity, in this article Organic Computing (OC) principles in terms of self-organisation, self-reconfiguration and self-healing are applied to a six-legged walking robot named OSCAR (Organic Self-Configuring and Adapting Robot). The Organic Robot Control Architecture ORCA, developed in the same project, provides the architectural framework. OC principles are employed on all layers of the hierarchical robot control system starting at the reflexive layer with gait generation and reflexes over the reactive behavioural layer up to the deliberative planning layer. Many experimental evaluations with OSCAR have shown that the robot is able to flexibly adapt to internal faults as well as to unforeseen environmental situations and thus continues its mission in the best still possible way.

Keywords Walking robot · Robot control architecture · Organic computing · Fault tolerance · Anomaly detection · Reaction to anomalies · Robot leg amputation

1 Introduction

Walking is the most important propulsion method for mammals and many insects. Though not very fast compared to wheeled propulsion, it is very flexible and adaptable with respect to different grounds and surfaces. For mobile robots, wheeled drives still dominate, but there is also large interest in walking robots. One reason are applications with difficult environments like space missions, exploration of volcano craters or rescue scenarios in destroyed buildings, e.g. after earthquakes. Another reason is basic research for better understanding walking as such in biological as well as in technical systems.

Many different kinds of walking can be found in nature ranging from two legs (humans) over four legs (most other mammals), six (most insects), eight (spiders) up to several hundred legs (e.g. millipedes). In this article, we are mainly interested in walking robots which are robust and have a high potential for fault tolerance. Therefore, we have selected six legs, i.e. hexapod robots, because they are widespread in nature, provide static stability and offer enough redundancy.

C. Müller-Schloer et al. (eds.), *Organic Computing — A Paradigm Shift for Complex Systems*, 517–530
Autonomic Systems, DOI 10.1007/978-3-0348-0130-0_34,

There are many types of hexapod robots that have already been used for demonstration purposes for research on biologically inspired locomotion. Some of the current state-of-the-art hexapod robots include: "iSprawl" [15], "RHex" [2], "DLR Crawler" [9] and "AMOS-WD06" [22]. However, they all differ in the technology for their locomotion. For example, the leg design as well as the motion concept of the joints are different from one robot to the other.

While much research has been done on the locomotion of walking robots, much less attention has been paid to introducing organically inspired approaches within the robot itself, with the goal to make it fault-tolerant and give it the capability to remain functional even in unforeseen situations.

In this article, our hexapod robot OSCAR (Organic Self-Configuring and Adapting Robot) is introduced, which has been built at the University of Lübeck as a testbed for our Organic Robot Control Architecture ORCA. It is based on organic principles, in particular self-organisation, to achieve fault tolerance without making use of an explicit fault model. The general concepts and principles of ORCA are described in detail in a companion article (see Chap. 4.5). Since learning plays an important role in ORCA, it is subject of another companion article in this book (see Chap. 3.4). In the following, we will mainly concentrate on the implementation of ORCA on our walking robot OSCAR and the experiences with practical experiments gained from it so far.

2 Six-Legged Walking Robot OSCAR

OSCAR (Organic Self-Configuring and Adapting Robot) is a six-legged walking machine with three degrees of freedom in each leg. The legs are built largely from commercially available parts. The round body allows a symmetric construction and supports omni-directional movement of the machine. The joints are based on commercially available servos. Several generations of OSCAR have been built [5, 14].

The latest generation robot is OSCAR-X (Fig. 1) with the following main features:

- Diameter including stretched legs about 110 cm, weight about 7.5 kg
- Leg design that enables robot reconfiguration and physical leg detachment
- Digital servos with digital feedback for torque, current consumption, temperature, etc.
- Foot design for reliable ground sensing
- Movable head with camera, heat and distance sensors
- Orientation sensor (3D compass, accelerometer, gyro)
- Expandable design with interfaces for additional sensors
- Powerful controller based on Embedded Linux or Windows
- Programming in Java

A unique new feature of the OSCAR-X is the design of its legs, which aims for performing on-demand robot reconfiguration. Namely, the patent pending mechanism for Robot LEG AMputation, R-LEGAM [11], is integrated for each of

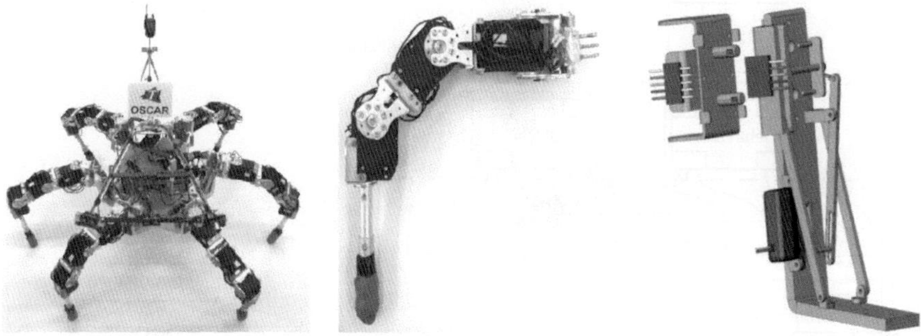

Fig. 1 *Left*: Walking robot OSCAR-X. *Center*: OSCAR's leg with three joints: alpha closest to body, beta in the middle, gamma closest to foot. *Right*: Leg detachment mechanism R-LEGAM

OSCAR-X's legs (Fig. 1, right). The robot's leg can be detached from the body under software control. Thus, instead of carrying a defective leg during the rest of the mission, it can be amputated to prevent any future negative influence on the functional remaining parts of the robot.

3 Robot Control Architecture ORCA

ORCA (Organic Robot Control Architecture) has a modular and hierarchical structure in order to be easily manageable and thus meets the requirements of a state-of-the-art control architecture for mobile robots. A major feature of ORCA is the strict distinction between Basic Control Units (BCUs) and Organic Control Units (OCUs), as depicted in Fig. 2. Each BCU encapsulates a specific functionality, which is required for the system's operation. It can, for example, implement controllers for servo motors, or may encapsulate sensors for perception and proprioception of the robot at lower system levels. At the reflexive system layer, gait pattern generation for walking machines or reflexes can e.g. be realised. Reactive robot behaviours like obstacle avoidance are provided in the medium behavioural layer. The highest deliberative layer is e.g. responsible for path and mission planning. The various BCUs interact by interchanging data (signals) upwards as well as downwards in the hierarchy. The connections between BCUs are also used to trigger activities in the receiving BCU when new data are sent.

OCUs supplement these BCUs. An OCU uses the same unified interface as a BCU to interchange data and to trigger activities. In contrast to BCUs, OCUs do not implement a predefined custom function for the robot, but rather monitor the signals generated by one or more BCUs. When these monitored signals show an anomaly, e.g. a substantial deviation from their normal range, the OCU can react by changing parameters of the BCUs. By defining e.g. "normal" or "good" ranges for some of the signals a BCU generates, an OCU can decide when to start changing parameters to bring the system back to a normal or "healthy" state (reasoner). The health state of a BCU indicating the degree of its deviation from its normal behaviour is

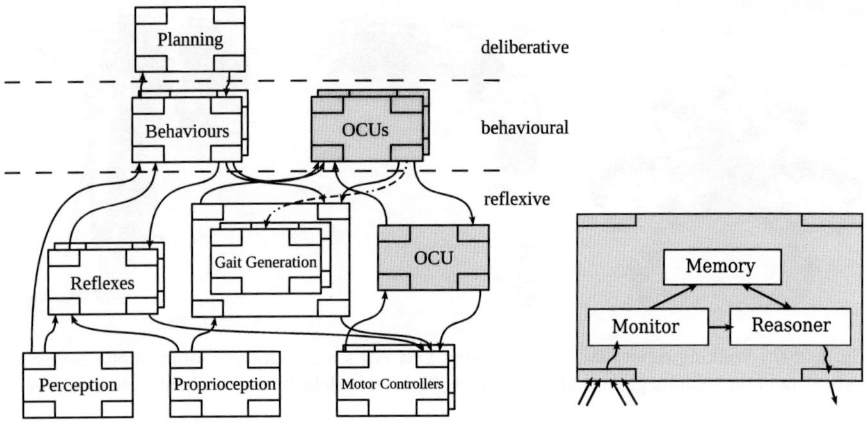

Fig. 2 *Left*: Principal ORCA architecture. *Right*: Organic Control Unit OCU

described by a so-called health signal. Various methods for anomaly detection and health signal generation have been developed (see Chap. 4.5 for more details). The selection of an appropriate method depends on the respective BCUs. Health signals can be combined and processed at the various levels of the architecture.

Our general view of anomalies comprises that the BCUs are able to learn in order to improve their reactions in similar situations in the future. This learning is controlled by OCUs which are provided with a memory for this purpose (see Chap. 3.4 for more details). The ORCA architecture can thus be viewed as a distributed and hierarchical variant of the popular Observer/Controller architecture [19] with OCUs encapsulating the observers and controllers.

4 Implementation of ORCA on OSCAR

Our walking robot OSCAR serves as a testbed for our robot control architecture ORCA. In the following section, the implementation of the various architectural layers is described in more detail from low-level gait generation over adaptive reflex-based walking up to primitive reactive behaviours and higher-level path planning. The corresponding functionalities are implemented in a modular and hierarchical way by various BCUs. On all levels OCUs are provided to continuously monitor the health state of the system and to react to anomalies in the best possible way.

4.1 Distributed Leg Control and Self-Organising Gait Patterns

Up to now, many approaches to gait pattern generation for walking robots were proposed. Centralised approaches use a central controller for all legs, distributed ones individual controllers for each leg which cooperate locally. For OSCAR we

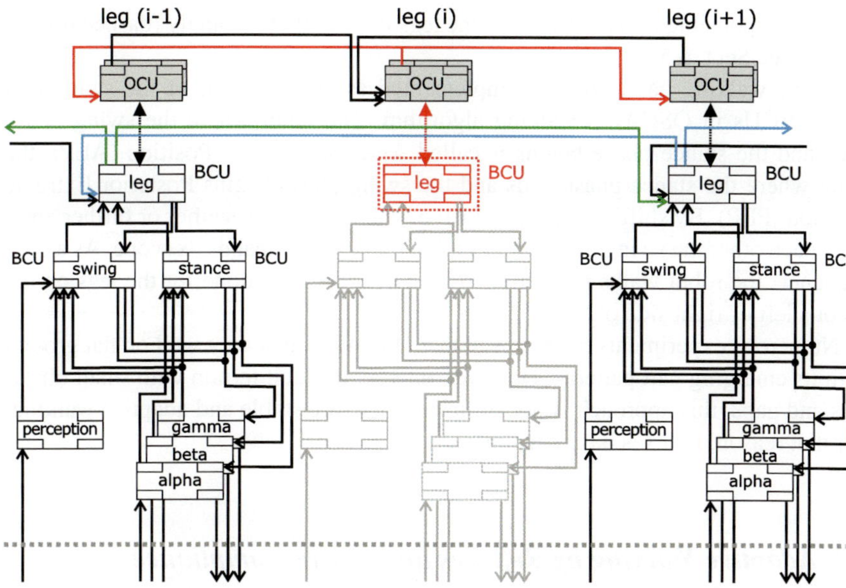

Fig. 3 Distributed leg control with ORCA. Leg(i) has been amputated and bypassed

use a distributed, biologically inspired method based on research results of H. Cruse on stick insects [4].

In order to achieve stable walking, the coordination of legs is based on the following simple rule: a leg is only allowed to swing if its two neighbouring legs perceive a ground contact detected by a binary sensor. With previously pre-coding the swing and stance phases for each of the robot's legs and implementing this rule, the hexapod robot shows emergent walking gaits [8]. The walking speed is increased by shortening the time for the stance phase while keeping the swing phase at a constant length. For a long stance phase (about 4 times the duration of the swing phase), the robot starts with a pentapod gait having only one leg in the air, then moves to tetrapod gait with two lifted legs and finally, with stance and swing phase having about the same length, it changes to the fastest gait: the tripod gait with three legs in the air at the same time. These emergent gait patterns have low engineering effort, since only the simple local coordination rule needs to be implemented by BCUs. As shown in Fig. 3, there is one BCU for each joint of a leg called alpha (closest to the body), beta (in the middle), and gamma (closest to the foot); see also Fig. 1. These BCUs are responsible for the control of the respective servos. On the next level, there is a BCU for the swing movement and one for the stance movement. The swing BCU is connected to the BCU for the ground contact sensor (perception). Each leg has the same structure. The leg BCUs cooperate with their left and right neighbour to implement the coordination rule and to control the stance and swing phases of their own leg accordingly. The neighbourhood relation is also extended between the two front and the two hind legs, i.e. all legs have exactly two neighbours. Each leg also has an OCU monitoring its health status. These OCUs are also responsible for reconfiguring the legs in case of an anomaly (dashed double arrow), see Sect. 4.5.

Curve walking has also been implemented as an extension of the swing and stance BCUs in OSCAR's walking algorithm. The point where the swing phase ends and the stance phase begins is called Anterior Extreme Position (AEP), the point where the stance phase ends and the swing phase begins Posterior Extreme Position (PEP). By shifting AEP and PEP of one leg closer together or further apart from each other the swing-stance-alternation reduces or expands its scope. As a consequence, the robot walks in a left or right curve, if this is done for the legs on the robot's left or right side only.

Numerous experiments have proven that OSCAR walks very well on flat ground with its emerging gait patterns. However, on very difficult terrain with small obstacles and gaps, this approach is not sufficient to assure stable and adaptive walking. For that reason, reflexes have been introduced.

4.2 Adaptive Walking by Reflexes and Active Compliance

Reflexes are an important means for living animals like insects to react and adapt to unforeseen situations, e.g. small obstacles or gaps. In OSCAR's control system five different reflexes were implemented improving its walking behaviour. The reflexes are elicited by defined stimuli like common disturbances of regular movements and can thus be regarded as a reaction to anomalies. The chosen reflexes are inspired by biological reflexes found in walking insects. Besides the *elevator reflex* [17] and *obstacle avoidance reflex* during the swing phase [21], a *ground detection reflex* and a *search reflex* after an uncompleted swing phase [3] were included into the control system.

In addition to the above mentioned reflexes performed by a whole leg, reflexes can also be executed by single joints in terms of active compliance, i.e. the robot modifies its behaviour based on the output of a force sensor. Our approach towards active compliance is implemented in the BCUs for each joint by reacting to an external force detected by exceeding a current consumption threshold of the servos. The joint then reacts by actively complying via correcting its servo position (*compliance reflex*). If the force is so strong that the joint leaves its optimal angular working range, a *short step reflex* can be triggered, i.e. the leg makes a short step into the direction of the applied force to move the joint back to its optimal angular working range [6].

The elevator reflex is triggered by an obstacle contact during the swing phase and allows the involved leg to overcome small obstacles by correcting the swing phase through lifting the leg higher (Fig. 4, left). By monitoring the alpha joint's current consumption during the swing phase, obstacle contact can be detected at any location of the leg. When pushing against an obstacle, the current consumption in the alpha servo increases significantly. This stimulus causes an interruption of the leg's swing phase and triggers the elevator reflex, lifting the leg higher trying

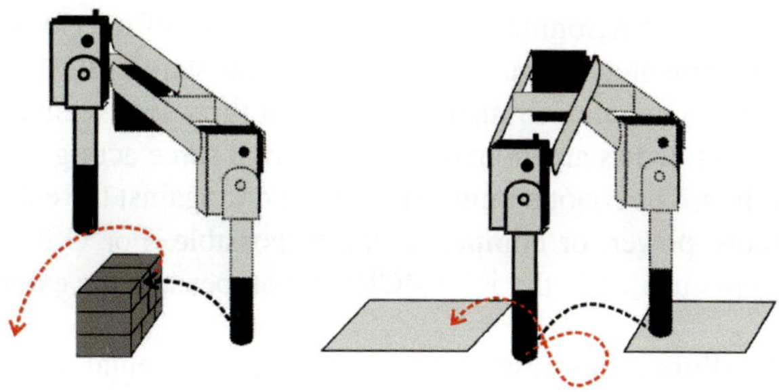

Fig. 4 *Left*: A leg hits an obstacle and an elevator reflex is triggered (*dashed trajectory*). *Right*: A leg detects no ground at the swing phase's end and a search reflex is triggered (*dashed trajectory*)

to overcome the obstacle that was hit. This reflex is implemented into the swing BCU and allows the leg to continue its swing phase even when bumping into small obstacles.

For the ground detection reflex, the ground detection is the stimulus and the stopping of the swing phase the reaction. The ground detection can be sensed by the current in the beta-servo exceeding a predefined threshold or by a mechanical ground detection sensor in the foot. This reflex makes the robot's walking adaptive to uneven terrain.

Searching for ground at the end of a swing phase is a reflex reaction triggered by the absence of ground at the leg's anterior extreme position. The stimulus of missing ground could indicate a stair, a gap or an end of the walkable ground. This search reflex extends the original swing phase to two search movements: one searching in the anterior direction and the other one in the posterior direction (Fig. 4, right). In case of a stair or a small gap in the ground, the search reflex can lead to the foot's ground contact and a stance phase trigger, although the leg at first was not detecting the ground. The search reflex is implemented as an extension of the swing phase.

The five reflexes mentioned above compensate most of the common disturbances during leg movements. The described reactions lead to a more adaptable walking and to more stability of the walking machine, as we have evaluated by letting OSCAR walk on rough artificial testing grounds [5]. But they can also be used to react to interior anomalies within the robot in order to achieve fault tolerance.

4.3 Reaction to Anomalies

Both environmental anomalies as well as interior anomalies can be roughly divided into three intensity levels: weak, medium, and strong. For all of them suitable reactions were developed, which allow the robot to continue its mission in the best still possible way.

Weak Anomalies A weak environmental anomaly could simply be a change of ground solidity while a weak interior own anomaly could e.g. be a change in the robot's weight or signs of wear in actuators or sensors. Reflexes such as ground contact reflexes or short step reflexes can manage the described weak anomalies in terms of environmental as well as in terms of interior anomalies. These reflexes are implemented in the existing swing and stance BCUs as default reflexes to handle small anomalies immediately. One of the weakest environmental influences that can be defined as an anomaly is an external force acting on one or several of the robot's joints. The robot's joint can either work against the external force, consuming much more power, or comply as far as possible. For OSCAR a compliance reflex was implemented in the joint BCUs to compensate the external force.

Medium Anomalies A medium environmental anomaly could be an uneven terrain with obstacles which is still manageable for walking in respect to step height and step length of the robot's legs. In terms of interior anomalies, a medium anomaly is given by light or small defects of the robot's body like a change in body segment length or sensor disturbances. Medium anomalies could cause the activation of additional reflexes like e.g. the elevator reflex for avoiding small obstacles or the ground search reflex to search for walkable ground in an environment with gaps. These simple reflexes are implemented in the corresponding swing and stance BCUs.

Strong Anomalies Strong anomalies within the environment could be a hardly passable obstacle like a large gap, a terrain with many small gaps or a strongly gradient ramp. Strong interior anomalies are severe defects like damaged joints, legs or other parts of the locomotion system, damaged sensor parts, groups of sensors or other parts of the perception system. They can no longer be handled by reflexes alone, but require the intervention of higher level OCUs which can e.g. react by amputating a leg or changing the robot's behaviour as described later.

4.4 Local Fault Masking by Means of Adaptive Filters

The gait of an autonomous robot usually depends on the complex, but nevertheless proper interaction of many of its components. For a hexapod robot like OSCAR, the influence of the components in the legs is most decisive. The function of a leg might be impaired e.g. due to ageing (because of mechanical friction, material degradation, loosening of screws etc.) of the servo motors associated to its joints, noise in communication signals from controllers, underperformance of batteries, or accidental situations.

In a detailed case study, for the example of OSCAR, it was systematically investigated how faults in the leg components can impair the leg function [1, 8]. This was carried out by simulation of OSCAR's trajectory behaviour in a robot simulator, which is based on a physical 3-dimensional movement model. A comprehensive spectrum of faults was considered, comprising, e.g. low response of joints, gradual degradation of joints, abnormal joint behaviour, jamming of joints, improper leg orientation or necessary leg amputation.

It was also studied how such faults in leg components can be masked at the local level of the impaired leg components. This was achieved by means of additional adaptive filters, i.e. filters that are self-adjusting according to an optimising algorithm, associated to the control of these components [10]. It was shown that a considerable number of the mentioned faults can be masked (i.e. corrected without needing operation interrupts) by filter actions modifying the local control. In some fault cases, faulty behaviour of one joint in the leg can be compensated with the help of a neighbouring joint, similar to some fault compensation strategies in biological beings.

It also turned out that the faults of the mentioned classes, by their different influence on the trajectories, can be sufficiently distinguished from each other. Thus, it was possible to integrate the adaptive filter solutions, developed for the different maskable faults that might appear in one leg, into one coherent fault correction algorithm.

In addition, adaptive filter schemes were developed which enable a combination of fault correction and fault monitoring [16]. Here, we derived two different architectural modifications:

- an architecture detecting the faults, which could not be corrected
- an architecture, which also reveals those faults that were successfully corrected; this may also be important, e.g. for being able to signal the occurrence of fault events to higher system levels.

4.5 Self-reconfiguration in Case of Amputated Legs

For six-legged insects, it can be observed that they are very well able to walk with missing legs [20]. Some species even sacrifice a leg to escape a predator. Biological research with stick insects has shown that they are able to immediately continue stable walking after a leg amputation. As already described in Sect. 4.1, our six-legged robot OSCAR is also capable of physical leg detachment. This feature can be used as a reaction to a strong anomaly in a leg, which can no longer be tolerated by adaptation based on reflexes or adaptive filters. However, in order to continue stable walking, the gait pattern generation has to be modified as well. To implement all this, OCUs are provided for each leg. They monitor the leg's health status and, in case of a strong anomaly like a broken servo, react by giving a signal to the leg BCU for amputation and disable all subordinate BCUs (shown in grey in Fig. 3 for leg (i)). The rule for self-organising gait pattern generation remains unchanged, i.e. a leg is only allowed to swing if its two neighbours are in their stance phase. Only the neighbourhood relation is changed by the OCUs such that the two neighbours of a failed leg now become direct neighbours (see Fig. 3 for legs $(i-1)$ and leg $(i+1)$ in case of a failure of leg (i)).

Experiments have shown that with this reconfiguration strategy OSCAR is able to continue walking after amputation of a single leg and of two middle legs, even for curve walking towards a goal [7]. However, the robot has to walk more slowly and stability suffers. To overcome this, an additional reconfiguration method of the remaining legs, Swarm Intelligence for Robot Reconfiguration (S.I.R.R.), was introduced inspired by swarm intelligence [12, 13].

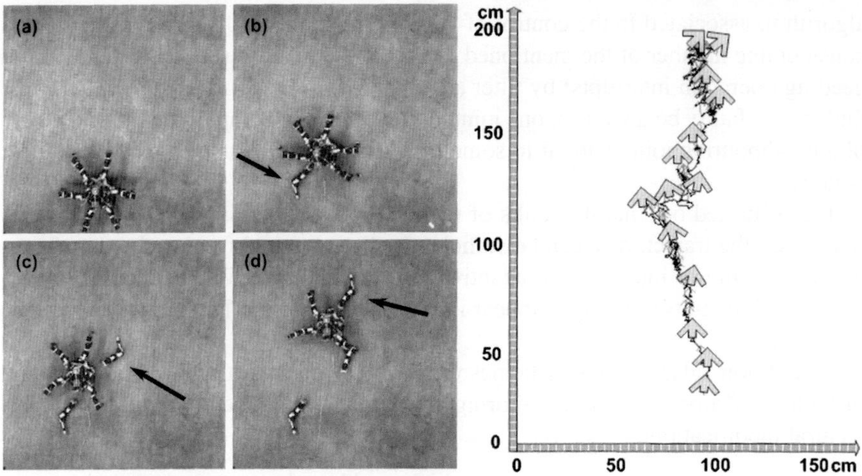

Fig. 5 *Left*: OSCAR performing leg amputations (see *arrows*) with R-LEGAM and using S.I.R.R. for self-configuration: fully functional (**a**), hind leg left (**b**), middle leg right (**c**), front leg right (**d**) with robot hitting the already ejected leg from (**c**). *Right*: Tracking of OSCAR's heading while ejecting legs during walking straight forward

S.I.R.R. reconfigures the posture of the remaining legs after a leg failure has occurred based on the intrinsic properties seen within swarms in nature. Each leg is considered as a swarm member and implements three local rules for flocking: separation, alignment and cohesion, according to Reynolds [18]. The legs are divided into two groups of three legs each, one on the left and one on the right side of the robot's symmetry line. The three rules are applied within each group. The S.I.R.R. approach additionally implements two rules: (1) avoid "overcrowding" by moving swarm members from one group to another and (2) change the parameters of the swarm members that were moved with respect to parameter characteristics of their new group.

The practical usability of S.I.R.R. has been demonstrated with various test scenarios on OSCAR which confirmed its robustness and efficiency for carrying out reconfiguration for up to three lost legs [14]. Figure 5 shows a typical example. The legs are amputated in the following order: hind leg left, middle leg right, front leg right. The resulting track of the robot using the self-organising gait pattern generation and self-reconfiguration as described above has been recorded (Fig. 5, right). It can be seen that the robot is still able to continue walking in the same direction, but shows some deviations from the straight path. This has to be taken into account at the higher behaviour and planning layers described next.

4.6 Primitive Reactive Behaviours

The reactive behaviour layer provides the robot with primitive behaviours, encapsulated in BCUs. *Wander* lets the robot e.g. move around randomly, *Escape* and

Avoid are used for moving around obstacles. Other primitive behaviours can easily be added. To monitor and reconfigure the BCUs, interconnected OCUs are provided for each one. Fluctuations in sensor data are evaluated in sensor OCUs and the generated health signals are propagated to the OCUs in the behaviour layer. Besides that information, the detection of anomalies is carried out by the behaviour OCUs by monitoring the performed actions of their underlying BCUs. Depending on the strength of the deviation from the normal behaviour the health signal of the BCU is adjusted. If the health signal drops below a certain threshold, the OCU disables its BCU and informs the other OCUs. For example, if an *Avoid* using sonar sensors is no longer possible, the robot can still move around obstacles by an *Escape* based on its infrared sensors.

In addition, an OCU with a BCU having a low health signal can request other OCUs to temporarily shut down their BCUs in order to initialise a self-test of the underlying BCU. If successful, the outcome of a self-test can lead to a rise of the health signal, or otherwise to further lowering of the health signal, up to complete termination of the BCU. Such tests also include gently approaching, even touching detected or presumably detected objects. This interplay of behaviour OCUs leads to a soft and temporal degradation of the reactive behaviour layer, as the health signals can be readjusted over time.

4.7 Path Planning Based on Health Signals

The health status of the robot can influence its capabilities to still fulfil its mission. One example is the propulsive power of OSCAR affected by the loss of a leg. Another example can be found in [7], where it is shown that OSCAR is capable of handling the loss of a leg and still continues its way toward a goal, but tends to drift toward the side of the defective leg. If these effects are not considered during the path planning the robot might get stuck in the environment. Therefore, a path planning algorithm is needed that takes the changing health status of the whole system into account and adapts the path accordingly.

The adaptive path planner consists of a BCU and an OCU. The BCU performs a path planning based on a wave front approach on a 2D grid map by searching the shortest path around obstacles towards a given goal and controls the robot to move along the resulting path. The robot itself is monitored on lower layers by corresponding OCUs that generate health signals. These signals are hierarchically processed and fed into the path planner OCU. The OCU may alter the map within the path planner BCU by adjusting the data of critical areas to represent obstacles. This results in alternative paths that lead the robot around areas that do not suit its actual fitness. All this happens transparently for the path planning BCU.

The effect of a decreasing health signal is shown in Fig. 6. The shades of grey specify the difficulty of a certain area, reaching from very difficult (black) to simple (white) terrain. The resulting path is shown as a bright line. Cells that are too difficult to be traversed are striped. As can be seen, the path planner is capable of planning a

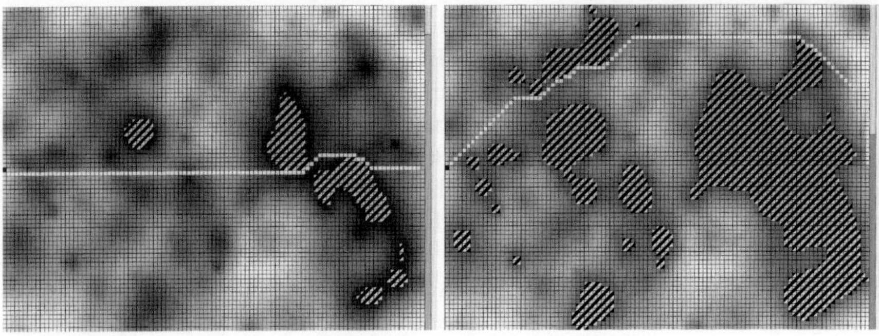

Fig. 6 Path for a health status of 91% (*left*) and the health status of 60% (*right*)

way through the terrain that avoids regions that might be too demanding for the robot at its current health state. Thereby the planner reacts to limitations of the robotic platform that might occur during its run time, caused by graceful degradation.

5 Conclusions and Outlook

In this article, we have demonstrated how Organic Computing (OC) principles like self-organisation, self-reconfiguration and self-healing can be used to make a mobile autonomous robot more robust and fault-tolerant. The robot is thereby able to flexibly adapt itself to its previously unknown environment as well as to internal faults in its body. The architectural framework was our Organic Robot Control Architecture ORCA. It provides a modular and hierarchical structure for the robot's functionality by its BCUs and the means for implementing OC principles by its OCUs, which are responsible for monitoring the health state of their assigned BCUs and reconfiguring them in case of anomalies. This approach was inspired by the natural immune system, which also monitors the body continuously and reacts in abnormal situations.

As common for many robot control architectures today, three layers are distinguished: reflexive, behavioural and deliberative. OCUs can be provided on all layers. As a testbed for ORCA, our six-legged walking robot OSCAR was introduced. On the reflexive layer, OC principles were realised in terms of self-organising gait patterns, reflexes in the legs and active compliance in the joints, which are able to tolerate small and medium anomalies. For strong anomalies, leg amputation and reconfiguration have been employed to allow the robot continue walking with fewer legs. On the higher layers, OCUs are provided for supervising elementary reactive behaviours and for path planning taking the robot's health status into account. While the reflexive layer has been investigated already in some detail, the two higher layers are still subject of ongoing research.

A remarkable feature of using OC approaches is the low engineering effort in contrast to conventional methods. There are no formal models required, neither for the kinematics of the machine, nor for faults. Instead, self-organisation is applied as widely as possible to let a solution evolve in place of explicitly designing and engineering it. Thereby, the robot becomes adaptive also to unforeseen situations—similar to living organisms.

References

1. Agarwal, R., Großpietsch, K.-E.: Fault tolerance for autonomous robots by means of adaptive filters. In: Proc. ARCS Workshop Dependability and Fault Tolerance. VDE Verlag, Berlin (2007)
2. Altendorfer, R., Moore, E.Z., Komsuoglu, H., Buehler, M., Brown, H., McMordie, D., Saranli, U., Full, D.E.K.R.: RHex: a biologically inspired hexapod runner. Auton. Robots **11**, 207–213 (2001)
3. Bläsing, B., Cruse, H.: Stick insect locomotion in a complex environment: climbing over large gaps. J. Exp. Biol. **207**, 1273–1286 (2004)
4. Cruse, H.: What mechanisms coordinate leg movements in walking arthropods? Trends Neurosci. **1**(13), 15–21 (1990)
5. El Sayed Auf, A.: Eine Organic Computing basierte Steuerung für einen hexpoden Laufroboter unter dem Aspekt reaktiver Zuverlässigkeit und Robustheit. Dissertation, University of Lübeck (2010)
6. El Sayed Auf, A., Dudek, N., Maehle, E.: Hexapod walking as emergent reaction to externally acting forces. In: Proc. ROBOTICA, pp. 67–72 (2009)
7. El Sayed Auf, A., Litza, M., Maehle, E.: Distributed fault-tolerant robot control architecture based on organic computing principles. In: Biologically-Inspired Collaborative Computing, IFIP International Federation for Information Processing, pp. 115–124 (2008)
8. El Sayed Auf, A., Mösch, F., Litza, M.: How the six-legged walking machine OSCAR handles leg amputations. In: From Animals to Animats 9, Rome, Italy (2006)
9. Görner, M., Wimbock, T., Hirzinger, G.: The DLR Crawler: evaluation of gaits and control of an actively compliant six-legged walking robot. Ind. Rob. **36**(4), 344–351 (2009)
10. Großpietsch, K.-E., Silayeva, T.A.: Organic computing—a new paradigm for achieving self-organized dependable behaviour of complex it systems. In: Proc. IDIMT Conference, pp. 127–138. Trauner, Linz (2006)
11. Jakimovski, B.: Patented robot leg amputation mechanism (R-LEGAM) for joint leg walking robots. DPMA-Az: 10 2009 006 934 (2009)
12. Jakimovski, B., Maehle, E.: In situ self-reconfiguration of hexapod robot OSCAR using biologically inspired approaches. In: Miripour, B. (ed.) Climbing and Walking Robots, pp. 11–332 (2010)
13. Jakimovski, B., Meyer, B., Maehle, E.: Swarm intelligence for self-reconfiguring walking robot. In: Proc. IEEE Swarm Intelligence Symposium (2008)
14. Jakimovski, B., Meyer, B., Maehle, E.: Self-reconfiguring hexapod robot OSCAR using organically inspired approaches and innovative robot leg amputation mechanism. In: Int. Conf. Automation, Robotics and Control Systems (ARCS) (2009)
15. Kim, S., Clark, J.E., Cutkosky, M.R.: iSprawl: design and tuning for high-speed autonomous open-loop running. Int. J. Robot. Res. **25**(9), 903–912 (2006)
16. Mladenov, M., Mock, M., Großpietsch, K.-E.: Fault monitoring and correction in a walking robot using LMS filters. In: Proc. Workshop on Intelligent Solutions in Embedded Systems (WISES), pp. 95–104 (2008)
17. Pearson, K., Franklin, R.: Characteristics of leg movements and patterns of coordination in locusts walking on rough terrain. Int. J. Robot. Res. **3**, 101–112 (1984)
18. Reynolds, C.: Flocks, herds, and schools: a distributed behavioral model. Comput. Graph. **21**(4), 25–34 (1987)

19. Richter, U., Mnif, M., Branke, J., Müller-Schloer, C., Schmeck, H.: Towards a generic observer/controller architecture for organic computing. In: Workshop Organic Computing—Status and Outlook, INFORMATIK 2006, Proc. 36. GI-Jahrestagung. Lecture Notes in Informatics, pp. 112–119. Köllen Verlag, Bonn (2006)
20. Schilling, M., Cruse, H., Arena, P.: Hexapod walking: an expansion to Walknet dealing with leg amputations and force oscillations. Biol. Cybern. **96**(3), 323–340 (2007)
21. Schmitz, J., Dean, L., Kindermann, T., Schumm, M., Cruse, H.: A biologically inspired controller for hexapod walking: Simple solutions by exploiting physical properties. Biol. Bull. **200**, 195–200 (2001)
22. Steingrube, S., Timme, M., Wörgötter, F., Manoonpong, P.: Self-organized adaptation of a simple neural circuit enables complex robot behaviour. Nature Physics (2010)

E. Maehle (✉) · A. El Sayed Auf · B. Jakimovski · S. Krannich · M. Litza · R. Maas · A. Al-Homsy
University of Lübeck, Lübeck, Germany
e-mail: maehle@iti.uni-luebeck.de

W. Brockmann
University of Osnabrück, Osnabrück, Germany
e-mail: werner.brockmann@uni-osnabrueck.de

K.-E. Grosspietsch
Fraunhofer Institute IAIS, Sankt Augustin, Germany
e-mail: karl-erwin.grosspietsch@online.de

Chapter 5.8
Energy-Awareness in Self-organising Robotic Exploration Teams

Bastian Degener, Barbara Kempkes, and Friedhelm Meyer auf der Heide

Abstract In the Smart Teams project, we consider a team of autonomous robots that have to cooperate in order to fulfil tasks like exploring an unknown terrain, assigning subtasks among each other in a fair way, or building and maintaining communication infrastructures. The challenge we face in this work is the locality constraint: a strategy for the team is not dictated by a global control, but consists of local strategies executed by the robots. "Local" means that the only input a robot can base its decision on is the information it can gather in its local environment.

In this article, we report about our research within the project related to the problem of building a short communication chain: consider two stations in the plane and a long, winding chain of mobile robots connecting the stations, i.e. consecutive robots are within viewing distance. How much energy must be spent in order to move the robots to the straight line between the stations using local strategies only? We present several algorithms to minimise the energy spent both for motion and for measurements (sensing the environment).

Keywords Mobile robots · Local algorithms · Robot formation problems · Geometry

1 Introduction

Technical innovations over the last decades have made it possible to build large numbers of small mobile robots at reasonable prices. Thus, teams of mobile robots have the potential to be increasingly used for tasks which are dangerous or even impossible for humans to fulfil. Examples for such tasks are explorations of terrains which humans cannot reach, like outer planets, or which are too dangerous for humans to enter, e.g. burning houses or contaminated terrains. If such an exploration task is fulfilled by a mobile robotic team, it is unrealistic to assume that there exists a central control which knows the positions and the current situation of all robots and guides them what to do next. Instead, the robots have to take their decisions on their own, knowing only their local surroundings. The challenge is to design such local protocols for the robots that lead to a globally desired behaviour of the whole team.

In this context our Smart Teams project deals with robotic collaboration under locality constraints. It focuses on the following four challenges.

C. Müller-Schloer et al. (eds.), *Organic Computing — A Paradigm Shift for Complex Systems*, 531–543
Autonomic Systems, DOI 10.1007/978-3-0348-0130-0_35, © Springer Basel AG 2011

Exploration One challenge is how an unknown terrain can be explored efficiently by a large group of mobile robots, provided that the robots can only interact and communicate with robots close by.

Communication Here we consider the communication infrastructure of the mobile robotic team, and ask questions like the following: If there is a mobile explorer robot which must permanently report to a base station despite a restricted communication range, how can a group of mobile relay robots build and maintain a communication chain between the explorer and the base station? How can such a group of mobile relays maintain the connectivity of a large group of explorers? How can mobile robotic teams build and maintain formations, like, in the simplest case, gather in some position?

Task assignment Robots can be required to fulfil tasks which they find in the terrain. The challenge is to assign the robots to such tasks maximising the number of tasks which can be processed.

Energy efficiency Mobile robots have to carry their energy in batteries, thus energy is a major limiting factor for their performance. Therefore analysing and optimising energy consumption of our strategies is an important issue.

In all four areas, the goal of the project is to provide algorithms which cope with robots that only have local information and limited energy. We provide thorough theoretical analyses and experimental evaluations via simulations of all developed algorithms.

Such local algorithms have several desirable properties of Organic Computing systems: they are self-organising in that they have no central control; they are self-optimising, as our analyses show; they are scalable and partly robust against inaccuracies of their actors and sensors. On the other hand, they are adaptive to their environment only to a limited extent. They can adapt in the sense of, e.g., modifying their positions as a reaction of movements of others or of obstacles, but they do not (yet) adapt to characteristics of their environment in the sense of learning.

1.1 Contents of the Article

In this article, we focus on a basic problem concerning communication among mobile robots, namely the transformation of an arbitrary, maybe long and winding, connected chain of mobile relay robots connecting two stationary units into a shortest one. Such a chain of n relay robots is said to be connected, if the first/last one has distance at most one to the first/second stationary unit, and the ith relay robot has distance at most one to the $(i+1)$st relay, for $i = 1, \dots, n-1$. This notion reflects the idea that, in order to maintain their connectivity, robots are restricted to optical sensors which are able to "see" their environment only within a fixed limited distance.

It makes sense in this setting to minimise the number of mobile robots which are used in the chain, presuming that the robots can also be used for other tasks. In order to achieve this goal, it is necessary to move the robots as close as possible to the line between the stations and thus to minimise the length of the chain. Afterwards, the robots can spread on the line and dispensable robots can return to one of the stations.

We will assume very restricted capabilities of the robot's sensors. The optical sensor will be able to provide a robot with the relative positions of other robots or stationary units within their viewing distance one. "Relative positions" means that a robot can measure the positions of the robots within the viewing distance, assuming a local coordinate system centred in its own position. We do not assume that the robots have the same sense of direction, i.e. they do not have a compass or even a GPS device. Thus, the coordinate systems of the robots may differ. Further, their algorithms do not need any information from the past, i.e., they are oblivious. This makes them simple and extremely memory efficient.

The algorithms executed by the robots consist of Look-Compute-Move (LCM) rounds: A robot observes the relative positions of its neighbours (Look). In our case we assume that these are only its successor and predecessor in the chain. Then it computes a target position based on these relative positions (Compute). Finally, it moves to this target position (Move). The LCM model is commonly used in the literature [7, 23, 25, 26]. In this article, we will concentrate on strategies that perform synchronous rounds. This means that all robots start a round at the same time. Our and other's algorithms in several asynchronous models are mentioned in the section on related work below.

Our goal is to devise energy-efficient strategies for the task to transform a communication chain into a shortest one. We identify two sources for energy consumption:

- the distance travelled by a robot (energy spent for motion) and
- the number of LCM rounds (energy spent for measurements).

As the communication chain breaks as soon as one robot runs out of energy, we are interested in bounding the maximum amount of energy spent, maximum taken over all robots. Note that our robots do not communicate, thus we do not consider the energy spent for communication.

Typically, the distance travelled is the major source of energy consumption. But also a round needs additional energy because of sensing and computing. It is intuitive and will be made explicit by our analyses that the two sources of energy consumption are conflicting: In order to find short paths to the destination, it seems favourable to sense the environment frequently and correct the direction based on these fresh observations of the neighbours. But this results in many LCM rounds. In this article, we will therefore consider first both sources for energy consumption separately, and then deal with good compromises between them.

In Sect. 2, we focus on the energy spent for measurements, i.e. the number of LCM rounds. For this, we describe the Go-To-The-Middle strategy. In a round of this strategy, each robot moves to the midpoint between its two neighbours. This and all other strategies presented in this article (and in most related work) fulfil the following invariant: if the initial chain is connected, then the chain will stay connected during the execution of the algorithm.

Table 1 Results overview

	1-bounded GTM	δ-bounded GTM	Continuous GTM	Continuous MOB
Number of rounds	$\Theta(n^2 \log \frac{n}{\epsilon})$ Theorem 1	$\Omega(n^2+\frac{n}{\delta})$ Theorem 5 $O(n^2 \log \frac{n}{\epsilon} + \frac{n}{\delta})$ Theorem 5	–	–
Maximum distance	$\Theta(n^2)$ Theorem 2	$\Theta(\delta n^2 + n)$ Theorem 6	$\Theta(n)$ Corollary 1	$O(\min\{n, (OPT + d) \log n\})$ Corollary 2

We will show that, after $\Theta(n^2 \log \frac{n}{\epsilon})$ rounds and a travelled distance of $\Theta(n^2)$ in the worst case, all robots are positioned in distance at most ϵ to the line between the stations.

In Sect. 3, we focus on minimising the energy spent for motion, i.e. the distance travelled. Here we study strategies where the robots permanently observe their neighbourhoods and immediately react to changes. We show that robots using such a "continuous" variant of the Go-To-The-Middle strategy travel a distance of only $\Theta(n)$ in the worst case. Then we introduce the so-called Move-On-Bisector strategy. With this strategy, the robots continuously move in direction of the angle bisector of the angle formed by themselves and their two neighbours. If they reach the line between them, they stay there for the remaining time. This strategy is analysed in more detail in this article; we compare the distance travelled by the robots with the distance the robots would have to travel with an optimal algorithm assuming an unbounded viewing range. We will show that the robots travel at most a distance of $O(\min\{n, (OPT + d) \log n\})$, where OPT is the distance needed by an optimal global algorithm and d the distance between the stations.

In Sect. 4, we consider both sources for energy consumption. For this, we introduce the δ-bounded variant of the Go-To-The-Middle strategy. In this strategy, a round consists of walking in direction to the middle between predecessor and successor, but for a distance of at most δ only, for some given $\delta > 0$.

Surprisingly, our results imply that, for $\delta = \Theta(1/n)$, both sources of energy consumption are minimised, resulting in $\Theta(n)$ travel distance and $O(n^2 + \log \frac{n}{\epsilon})$ and $\Omega(n^2)$ rounds in the worst case. An overview of the results can be seen in Table 1. While we present the results in Sect. 2 and 4 just briefly, we go into more depth in Sect. 3. Section 5 gives a conclusion of the stated results and presents future research directions.

1.2 Related Work

Our work on communication chains presented in this article is based on [4, 10, 12, 17]. The respective results will be explained in detail in the following sections. A more general problem than building a robotic *chain* for communication purposes is building a robotic *network* for communication. This problem is NP-hard to solve in general and has been considered from a global point of view in [14]. Here, efficient polynomial-time approximation algorithms are presented and lower bounds are given. The same problem is considered in [20] in a local setting. However, no bounds on the number of rounds are stated here. The problem of forming communication chains *locally* has also been considered using more powerful robots. In [13] the robots can act as beacons and the following robots know the positions of the beacons as well as the position of the base station. This way, the proposed strategy is even capable of handling obstacles in the environment and find an optimal communication chain with respect to the topology of the start configuration in linear time.

Closest to the work in this article is the Hopper strategy considered in [19]. The robots are slightly more powerful: They are able to distinguish their neighbours and can therefore base the decisions how to act depending on the action of their predecessors. In addition they are able to exclude robots if they are not needed any more to ensure connectedness. Furthermore, the goal is relaxed compared to this work, since the robots are only required to form a line that is at most a constant factor longer than the optimal line instead of converging towards the line. Also a timing model that is tailored to the strategy is applied. In this setting a more sophisticated strategy can be employed and a nice analysis yields a $O(n)$ bound for the number of rounds and the distance travelled, which is asymptotically optimal. For an overview of the problem of building communication chains or networks refer to [9]. Besides the theoretical approach to this problem, also some experimental work with physical robots is done [21, 22].

Building a short chain of robots can be seen as a formation problem. Other formation problems include forming a circle [5, 8] or just gathering in one point. Most work on the gathering problem or the closely related convergence problem where the robots only need to converge to a common point investigates which robot capabilities are inherently needed to achieve the task of meeting in a point in finite time [1, 11, 16, 24]. In [7] an upper bound of $O(n^2)$ for the number of rounds is provided for the algorithm that lets the robots iteratively move to the center of gravity of the robots positions. There is also work on local gathering, where the robots move as far as possible to the centre of the smallest enclosing circle of their neighbours without losing connectivity [2, 3].

In Sect. 3 we consider a continuous time model. A similar time model with a similar strategy has been considered before in [15]. Here the discrete step length converges to 0, while we take a direct approach. In contrast to our work, it is only shown that the robots gather in finite time.

1.3 Notation

We consider two stations and n relay robots $r_1, \ldots, r_n$. The stations are called r_0 and r_{n+1}. We denote the position of robot or station r_i by p_i, where the point of time will be clear from the context. The collection of the positions of the robots is called a *configuration*. In particular, we call the configuration at time 0 the start configuration. Each relay robot r_i is neighbour of r_{i-1} and r_{i+1} and thus it must be within viewing range of both neighbours. We set the viewing range to 1. The chain is *connected*, iff all neighbours are within viewing range of each other. We always assume that the start configuration is connected. Our algorithms will maintain the invariant that the chain stays connected at all times.

We call the line between the two stations the *target line*. Most of the algorithms for the robots considered in this article take as input the positions of the two neighbours and then compute a *target point*, towards which the robot then moves. Note that the robot might not reach the target point and that the target point changes over time. In contrast to this, a robot's *destination point* is the point on the target line on which the robot would be positioned, if the robots stood equally distributed on the target line. The time models will be described in the corresponding sections.

2 Energy Spent for Measurements

In this section, we describe a simple strategy, namely the Go-To-The-Middle strategy, and present results about its energy consumption for measurements. I.e., we present bounds on the number of LCM rounds needed.

We assume that the robots form a connected chain in the beginning and all robots know their respective two neighbours in the chain. (Our strategy ignores further robots that might be in their viewing range.) Note that we do not assume that the robots are able to distinguish between their two neighbours. Thus, at the beginning of a round, robot r_i is aware of the positions of the robots r_{i-1}, r_i and r_{i+1}. Based on these positions, it computes the position in the middle between the positions p_{i-1} and p_{i+1} of the robots r_{i-1} and r_{i+1}. Then it moves there (see Fig. 1). We assume a synchronous execution of rounds, that means that all robots start the next round at the same time. One can check that the concurrent execution of such a round by all robots maintains the connectivity of the chain.

It is easily seen that the robots do not always reach their destination point. Therefore we measure the number of rounds needed for an ϵ-approximation, i.e. until all robots are within a distance $\epsilon > 0$ of their destination point on the line between the stations.

Theorem 1 (See [12, 17, 18]) *Given an arbitrary connected chain of length n, the Go-To-The-Middle strategy ϵ-approximates the destination points of the robots within $\Theta(n^2 \log(\frac{n}{\epsilon}))$ rounds, in the worst case.*

The theorem consists of two parts, an upper and a lower bound. In order to show the upper bound $O(n^2 \log(\frac{n}{\epsilon}))$, we describe a round of an algorithm as a linear transformation of the spatial vectors connecting the positions of neighbouring robots. Applying techniques developed for bounding the mixing time of Markov-chains yields the upper bound. For details see [17]. A slightly different proof is shown in [18].

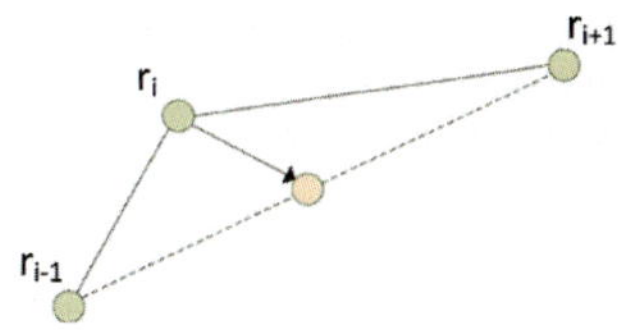

Fig. 1 Target point of robot r_i using Go-To-The-Middle

A direct proof from [18] shows a lower bound of $\Omega(n^2)$ for $\epsilon = 1$: the robots are arranged in a triangular shape, such that the robot $r_{\frac{n}{2}}$ is as far away as possible from its destination within the limits that are given by the fact that the chain has to be connected. Since the end points are on a horizontal line, the robots only move on the straight vertical line towards their destination point and this can be analysed explicitly. The improved lower bound of $\Omega(n^2 \log(\frac{n}{\epsilon}))$ was shown only recently in [17], again based on insights about the mixing time of Markov chains.

The above mentioned upper bound also yields the same bound on the travelled distance of the robots, because a robot travels at most distance 1 per round. Rather surprisingly, this bound is almost tight, although it seems that, especially in the final phase of the strategy, the distances travelled per round become smaller and smaller.

Theorem 2 (See [4]) *Given an arbitrary connected chain of length n, the Go-To-The-Middle strategy ϵ-approximates the destination points of the robots within a travel distance of $\Theta(n^2)$, in the worst case.*

3 Energy Spent for Motion

In this section we analyse the energy which the robots spend for motion. As mentioned already in the introduction, it seems plausible that the travelled distance decreases if the number of measurements is increased. Therefore we now consider an extreme situation in this context: We allow the robots to measure the positions of their neighbours continuously at all times. This makes sense if measuring positions does not need much energy compared to moving the robot. Iit shows the best motion energy costs for one strategy we can hope for if we reduce the number of measurements. We will now first describe the model in detail. Then we briefly state results for the continuous variant of the Go-To-The-Middle strategy, before we describe the Move-On-Bisector strategy [10] and its analysis.

We use a continuous time model instead of a discrete one to model the continuous observation of neighbours. Therefore, we are not given a classical round model, but rather all robots can perpetually and simultaneously measure the positions of their neighbours and adjust their movement paths to their observations. The robots are restricted to a maximum speed of 1. This leads to curves as trajectories for the robots. One can see this model as a "limit model" of the model from Sect. 4, where a discrete model like in Sect. 2 is used, but the step size of the robots in one round is reduced to δ. Letting $\delta \to 0$, we get the continuous time model. Note that the continuous model assumes that no time is needed to measure the positions of

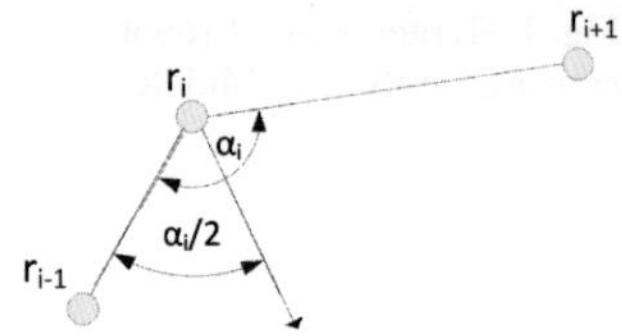

Fig. 2 Movement direction of robot r_i using Move-On-Bisector

neighbours and to execute the algorithm and so r_i can react to movements of its neighbours instantly. Note further that, since the speed of the robots is restricted to 1, the time it takes the robots to move to the line between the stations is an upper bound for the maximum travelled distance. Our strategy even guarantees that there is one robot which always moves with speed 1, which means that the time this robot needs to arrive at its destination point is equal to the distance it has travelled.

The continuous variant of the Go-To-The-Middle strategy works as follows. At time t, instead of moving to the midpoint between its neighbours, robot r_i moves in direction towards this point. Since the neighbours move continuously, the midpoint also moves resulting in a curve as movement path for r_i. The following result, which is a corollary from Theorem 6, shows that the travelled distance is optimal in the worst case.

Corollary 1 (See [4]) *When the continuous variant of the Go-To-The-Middle strategy is performed, the maximum distance travelled by a relay is* $\Theta(n)$ *for a worst-case start configuration.*

A more natural strategy in this setting is the Move-On-Bisector strategy. In this strategy, a robot r_i moves in direction of the angle bisector of the angle formed at r_i's position between the directions in which the two neighbours r_{i-1} and r_{i+1} are positioned. This angle is called $\alpha_i(t)$ when measured at time t. See Fig. 2 for a visualisation.

When using this strategy, the robots do not only converge to their destination points, but they actually reach them within finite time. We will see that the maximum distance travelled by the robots is also $\Theta(n)$ in the worst case, but we can analyse this strategy even in more detail: we do not only analyse worst case instances, but we will compare the maximum distance travelled when using Move-On-Bisector for each individual start configuration with the maximum distance *OPT* travelled when using an optimal global algorithm for the same start configuration.

Each algorithm, even if it has global information, must move all robots to the target line. Thus, the maximum distance between the target line and the robots is a lower bound for *OPT*. We call this maximum distance at time t the *height* $\mathfrak{h}(t)$ of a configuration at time t, see Fig. 3 for an illustration. The height can change over time, we denote the height of the start configuration by $h := \mathfrak{h}(0)$. As mentioned above, $h \leq OPT$.

A second property of a configuration is its *length* $\mathfrak{l}(t)$, which is the sum of the distances between neighbouring robots and stations: $\mathfrak{l}(t) := \sum_{i=0}^{n} |p(i+1) - p(i)|$. Again, we call $l := \mathfrak{l}(0)$ the length of the start configuration. Since the distance between two robots can be at most 1, it holds that $h \leq l/2$ and $l \leq n+1$. We will

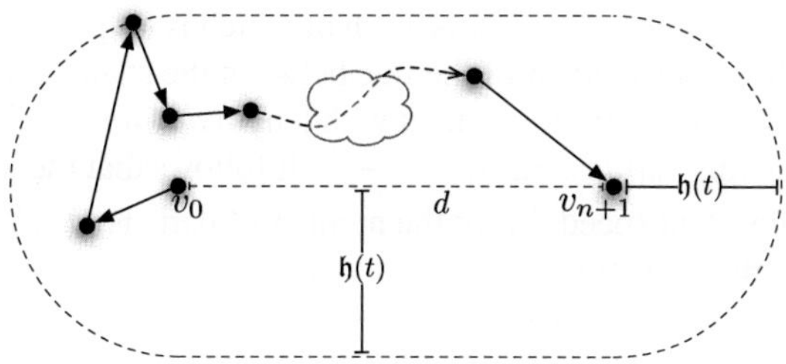

Fig. 3 Illustration of the height $\mathfrak{h}(t)$ of a configuration

now discuss two upper bounds for the distance that is travelled when using Move-On-Bisector. First, we will see that the travelled distance is $O(l)$ and therefore also $O(n)$. This bound is good for instances with a large height, but it does not tell anything about the travelled distance compared to an optimal global algorithm for instances with a small height and a large length. Therefore, we will then discuss a second upper bound which is better for these kinds of instances. Let d denote the distance between the two stations. We will see that the distance travelled by Move-On-Bisector can also be bounded by $O((h+d)\log l)$.

Theorem 3 (See [10]) *When using the Move-On-Bisector strategy, the maximum distance travelled by a robot is* $O(l)$.

In order to prove this bound, first observe that the height $\mathfrak{h}(t)$ as well as the length $\mathfrak{l}(t)$ are monotonously shrinking. Moreover, one can analyse how the distances between neighbouring robots or stations change over time by examining their derivatives with respect to the time t. The result is that the distance between r_i and r_{i+1} decreases with speed $\cos\alpha_i(t) + \cos\alpha_{i+1}(t)$. So the total length $\mathfrak{l}(t)$ decreases with speed $2\sum_{j=1}^{n} \cos\frac{\alpha_j(t)}{2}$. Now consider the case that there exists at least one robot r_i for which $\alpha_i(t) \le \frac{\pi}{2}$. In this case, $\mathfrak{l}(t)$ decreases with speed $2\sum_{j=1}^{n} \cos\frac{\alpha_j(t)}{2} \ge 2\cos\frac{\alpha_i(t)}{2} \ge 2\cos\frac{\pi}{4} = \sqrt{2}$. If on the other hand $\alpha_i(t) > \frac{\pi}{2}$ for all $1 \le i \le n$, then it can be shown that the robots which are in distance $\mathfrak{h}(t)$ from the target line move towards the target line with speed $\frac{1}{\sqrt{2}}$. Thus the height decreases with this speed. Since $\mathfrak{h}(t)$ can be lower bounded by 0 and $\mathfrak{l}(t)$ by d, and additionally $h \le \frac{l}{2}$, after time $\sqrt{2}h + \frac{\sqrt{2}}{2}l \le \sqrt{2}l$ all robots have reached their destination points. Thus, the maximum distance travelled by the robots is also $O(l)$.

Theorem 4 (See [10]) *When using the Move-On-Bisector strategy, the maximum distance travelled by a robot is* $O((h+d)\log l)$.

The main idea for the proof of this theorem is the following observation: the upper bound of $O(l)$ only needs one angle $\alpha_i(t)$ which is significantly smaller than π to decrease the length of the chain with constant speed. If there are more small angles, the speed with which the length of the chain decreases is higher. But if we are given a long chain with a small height, the sum of the angles cannot be arbitrarily close to πn, which would refer to a straight chain. The idea is now to divide the chain into several parts, each of which is longer than the boundary of the area which lies in distance h of the line between the stations. In particular, it can be shown that for each part of the chain with length $d + 4h$, the sum of the angles of the n_j robots in this part is at most $\pi n_j - \frac{\pi}{3}$. It follows that the length of each part decreases with constant speed. Since the number of parts is $O(\frac{l}{h+d})$ in the beginning and becomes smaller over time, this results in a number of rounds and therefore also a maximum distance travelled of $O((h + d)\log l)$.

The two upper bounds show on the one hand that Move-On-Bisector is asymptotically optimal for worst case instances. On the other hand, for $h \in \Omega(d)$, Move-On-Bisector is only by a factor of $O(\log l)$ worse than an optimal global strategy, although it only uses local information and very restricted robots. As $h \leq OPT$ and $l \leq n + 1$, we may conclude the following corollary.

Corollary 2 *The Move-On-Bisector strategy guarantees a travelled distance of* $O(\min\{n, (OPT + d)\log n\})$. *In particular, for every start configuration with* $d = O(OPT)$, *the travelled distance of this strategy is at most by a factor of* $O(\log n)$ *longer than the optimal distance.*

4 Energy Spent for Motion and Measurements

In this section we combine the results that we obtained in the previous sections.

We consider once again the Go-To-The-Middle strategy. However, it is adjusted such that a robot can travel at most a distance of δ in each round. We call this the δ-bounded variant of the Go-To-The-Middle strategy: Each robot r_i calculates the same target point as for the Go-To-The-Middle strategy (which is the midpoint of the straight line between p_{i-1} and p_{i+1}). In the subsequent move step, it either moves a distance of exactly δ or stops earlier if it reaches its target point (see Fig. 4). An important observation for the analysis is the fact that a robot that once reaches its target point will reach its target point in all following rounds. Note that the δ-bounded variant of the Go-To-The-Middle strategy fits the model of Sect. 2 when δ is set to 1 and that it fits the model of Sect. 3 for $\delta \to 0$.

We can prove the following bounds on the number of rounds.

Theorem 5 (See [4]) *Given an arbitrary connected chain of length n, the δ-bounded variant of the Go-To-The-Middle strategy ϵ-approximates the destination points of the robots within* $O(n^2 \log(\frac{n}{\epsilon}) + \frac{n}{\delta})$ $(\epsilon > 0)$ *and* $\Omega(n^2 + \frac{n}{\delta})$ *rounds.*

The definition of ϵ-approximation is exactly the same as in Sect. 2. The analysis, both for the upper bound of $O(n^2 \log(\frac{n}{\epsilon}) + \frac{n}{\delta})$ and the lower bound of $\Omega(n^2 + \frac{n}{\delta})$, uses the definition of two phases. In the first phase, at least one robot does not reach its target point and therefore travels a distance of exactly δ. In the second phase, all robots reach their target points in each round. The second phase can be analysed analogously to the Go-To-The-Middle strategy. Even a similar construction for the lower bound can be used.

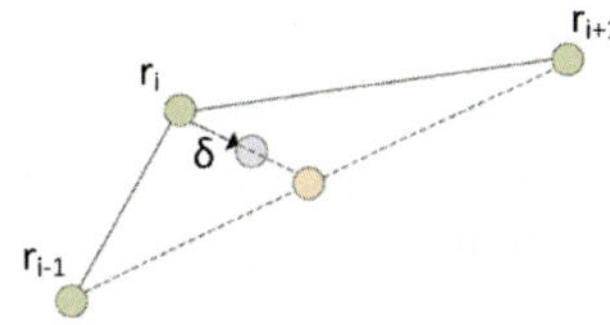

Fig. 4 Movement of robot r_i using the δ-bounded variant of Go-To-The-Middle

The fact that the robots travel exactly a distance of δ in the first phase can be used to prove the following theorem about the energy consumption due to motion.

Theorem 6 (See [4]) *Given an arbitrary connected chain, the δ-bounded variant of the Go-To-The-Middle strategy guarantees a maximum travel distance of $\Theta(\delta n^2 + n)$ in the worst-case.*

Again, the two phases are considered, but the analysis is more involved. Note that, as a direct consequence of the two theorems above, it follows that both energy consumers—measurement and motion energy—can be optimised at the same time (up to constant factors) when δ is set to $\Theta(\frac{1}{n})$:

Corollary 3 (See [4]) *Given an arbitrary connected chain of length n, the $\Theta(\frac{1}{n})$-bounded variant of the Go-To-The-Middle strategy ϵ-approximates the destination points of the robots within $O(n^2 \log(\frac{n}{\epsilon}))$ ($\epsilon > 0$) and $\Omega(n^2)$ rounds, travelling a distance of $\Theta(n)$ in the worst case.*

5 Conclusion and Outlook

We have presented simple strategies to form short chains of mobile robots. The two main energy consumers are studied, namely the energy spent for measurements and the energy spent for movement. We mainly looked at the Go-To-The-Middle strategy and its δ-bounded and its continuous variants. As a major insight, we showed that the $\frac{1}{n}$-bounded variant of the Go-To-The-Middle strategy is a good compromise for minimising the two sources of energy consumption. We also have presented a continuous strategy, the Move-On-Bisector strategy that improves upon the continuous variant of Go-To-The-Middle in so far, as it yields an—up to a logarithmic factor—best possible travel distance for each individual start configuration.

There are lots of open problems concerning forming chains. For example, the strategies presented here are the most efficient known ones in our synchronous LCM model. On the other hand, the Hopper-strategy from [19] only needs $O(n)$ rounds and travel distance. But it uses a very specific asynchronous model tailored to this strategy. It is still open whether better than our strategies exist for the synchronous model. Also, only little is known about dynamic versions where one or both stationary units are moving, or environments containing obstacles. Finally, our continuous models might become more realistic (and much harder to handle), if we incorporate a delay for the robots to react on measurements.

In addition, more general formation building problems open a huge range of research directions. An overview for the gathering problem can be found in [9], a very general, deep insight into the complexity of maintaining mobile flocks is given in [6].

References

1. Agmon, N., Peleg, D.: Fault-tolerant gathering algorithms for autonomous mobile robots. In: Proc. of the 15th annual ACM-SIAM symposium on Discrete algorithms, SODA'04, pp. 1070–1078 (2004)
2. Ando, H., Oasa, Y., Suzuki, I., Yamashita, M.: Distributed memoryless point convergence algorithm for mobile robots with limited visibility. IEEE Transactions on Robotics and Automation **15**(5), 818–828 (1999)
3. Ando, H., Suzuki, I., Yamashita, M.: Formation and agreement problems for synchronous mobile robots with limited visibility. In: Proc. of the IEEE Int. Symp. Intelligent Control, pp. 453–460 (1995)
4. Brandes, P., Degener, B., Kempkes, B., Meyer auf der Heide, F.: Building short chains of mobile robots locally with a bounded stepwidth. Preprint (2010). wwwhni.uni-paderborn.de/alg/publikationen
5. Chatzigiannakis, I., Markou, M., Nikoletseas, S.: Distributed circle formation for anonymous oblivious robots. In: Efficient and Experimental Algorithms, 3rd Int. Workshop, WEA'04, pp. 159–174 (2004)
6. Chazelle, B.: The convergence of bird flocking (2009). arXiv:0905.4241v1
7. Cohen, R., Peleg, D.: Convergence properties of the gravitational algorithm in asynchronous robot systems. SIAM Journal on Computing **34**(6), 1516–1528 (2005)
8. Défago, X., Konagaya, A.: Circle formation for oblivious anonymous mobile robots with no common sense of orientation. In: Proc. of the 2002 Workshop on Principles of Mobile Computing, POMC'02, pp. 97–104 (2002)
9. Degener, B., Fekete, S., Kempkes, B., Meyer auf der Heide, F.: A survey on relay placement with runtime and approximation guarantees. Computer Science Review **5**, 57–68 (2011)
10. Degener, B., Kempkes, B., Kling, P., Meyer auf der Heide, F.: A continuous, local strategy for constructing a short chain of mobile robots. In: Proc. of the 17th Int. Colloquium on Structural Information and Communication Complexity, SIROCCO'10, pp. 168–182 (2010)
11. Dieudonné, Y., Petit, F.: Self-stabilizing deterministic gathering. In: Proc. of the 5th Int. Workshop on Algorithmic Aspects of Wireless Sensor Networks, ALGOSENSORS'09, pp. 230–241 (2009)
12. Dynia, M., Kutyłowski, J., Lorek, P., Meyer auf der Heide, F.: Maintaining communication between an explorer and a base station. In: Proc. of IFIP 19th World Computer Congress, TC10: 1st IFIP Int. Conference on Biologically Inspired Collaborative Computing, BICC'06, pp. 137–146 (2006)
13. Dynia, M., Kutyłowski, J., Meyer auf der Heide, F., Schrieb, J.: Local strategies for maintaining a chain of relay stations between an explorer and a base station. In: SPAA'07: Proc. of the 19th Annual ACM Symposium on Parallel Algorithms and Architectures, pp. 260–269. ACM Press, New York (2007)
14. Efrat, A., Fekete, S.P., Gaddehosur, P.R., Mitchell, J.S.B., Polishchuk, V., Suomela, J.: Improved approximation algorithms for relay placement. In: Proc. of the 16th European Symposium on Algorithms, ESA'08, pp. 356–367 (2008)
15. Gordon, N., Wagner, I.A., Bruckstein, A.M.: Gathering multiple robotic a(ge)nts with limited sensing capabilities. In: Ant Colony, Optimization and Swarm Intelligence, 4th Int. Workshop, ANTS'04, pp. 142–153 (2004)

16. Izumi, T., Katayama, Y., Inuzuka, N., Wada, K.: Gathering autonomous mobile robots with dynamic compasses: An optimal result. In: Distributed Computing, 21st Int. Symposium, DISC'07, pp. 298–312 (2007)
17. Kling, P.: Unifying the analysis of communication chain strategies. Master's thesis, University of Paderborn (2010)
18. Kutyłowski, J.: Using mobile relays for ensuring connectivity in sparse networks. Dissertation, International Graduate School of Dynamic Intelligent Systems (2007)
19. Kutyłowski, J., Meyer auf der Heide, F.: Optimal strategies for maintaining a chain of relays between an explorer and a base camp. Theoretical Computer Science **410**(36), 3391–3405 (2009)
20. Meyer auf der Heide, F., Schneider, B.: Local strategies for connecting stations by small robotic networks. In: Proc. of IFIP International Federation for Information Processing, vol. 268; Biologically-Inspired Collaborative Computing, BICC'08, pp. 95–104. Springer, Boston (2008)
21. Nguyen, H.G., Farrington, N., Pezeshkian, N., Gupta, A., Spector, J.M.: Autonomous communication relays for tactical robots. In: Proc. of the 11th Int. Conference on Advanced Robotics, ICAR'03, pp. 35–40 (2003)
22. Nguyen, H.G., Pezeshkian, N., Gupta, A., Farrington, N.: Maintaining communication link for a robot operating in a hazardous environment. In: Proc. of the 10th Int. Conf. on Robotics and Remote Systems for Hazardous Environments. American Nuclear Society, La Grange Park (2004)
23. Prencipe, G.: Corda: distributed coordination of a set of autonomous mobile robots. Dissertation (2001)
24. Souissi, S., Défago, X., Yamashita, M.: Gathering asynchronous mobile robots with inaccurate compasses. In: Principles of Distributed Systems, 10th Int. Conference, OPODIS'06, pp. 333–349 (2006)
25. Suzuki, I., Yamashita, M.: Distributed anonymous mobile robots—formation and agreement problems. In: Proc. of the 3rd Annual Colloquium on Structural Information and Communication Complexity, SIROCCO'96, Waterloo, ON, Canada, pp. 313–330. Carleton Scientific, Kitchener (1996)
26. Suzuki, I., Yamashita, M.: Distributed anonymous mobile robots: Kitchener formation of geometric patterns. SIAM Journal on Computing **28**(4), 1347–1363 (1999)

B. Degener (✉) · B. Kempkes · F. Meyer auf der Heide
Heinz Nixdorf Institute and Computer Science Department, University of Paderborn,
33102 Paderborn, Germany
e-mail: bastian.degener@uni-paderborn.de

B. Kempkes
e-mail: barbaras@uni-paderborn.de

F. Meyer auf der Heide
e-mail: fmadh@uni-paderborn.de

Chapter 5.9
A Fast Hierarchical Learning Approach for Autonomous Robots

Alexander Jungmann, Bernd Kleinjohann, and Willi Richert

Abstract In this article we present an approach that enables robots to learn how to act and react robustly in continuous and noisy environments while not loosing track of the overall feasibility, i.e. minimising the execution time in order to keep up continuous learning. We do so by combining reinforcement learning mechanisms with techniques belonging to the field of multivariate statistics on three different levels of abstraction: the motivation layer and the two simultaneously learning strategy and skill layers. The motivation layer allows for modelling occasionally contradicting goals in terms of drives in a very intuitive fashion. A drive represents one single goal, that a robot wants to be satisfied, like charging its battery, when it is nearly exhausted, or transporting an object to a target position. The strategy layer encapsulates the main reinforcement learning algorithm based on an abstracted and dynamically adjusted Markovian state space. By means of state abstraction, we minimise the overall state space size in order to ensure feasibility of the learning process in a dynamically changing environment. The skill layer finally realises a generalised learning method for learning reactive low-level behaviours, that enable a robot to interact with the environment.

Keywords Robotics · Reinforcement learning · Markov decision process · Motivation system · Hierarchical learning

1 Introduction

Whenever a robot has to be programmed, its designer has to make many assumptions about the future environment to keep the task tractable. This holds even more, if the envisaged behaviour of the robot is so complex that it needs different levels of abstractions. The assumptions typically decrease the robot's autonomy and robustness in later application. Therefore, a learning robot architecture is desirable that places a minimum of assumptions into its algorithms in order to increase its robustness and autonomy, while not loosing track of the overall feasibility in terms of speed. This architecture should combine top-down goal specification with bottom-up exploration of its own capabilities. For that reason such an architecture should internalise the ability to learn and apply continuous actions (*skills*) in noisy domains

C. Müller-Schloer et al. (eds.), *Organic Computing — A Paradigm Shift for Complex Systems*, 545–558
Autonomic Systems, DOI 10.1007/978-3-0348-0130-0_36,

on the one hand. This so-called skill learner should find out by itself what types of capabilities are actually learnable. Additionally it should adapt already learned skills during execution. On the other hand this kind of architecture should provide a mechanism for finding state abstractions that are able to distinguish between sufficiently distinct states from the view of the learned skill set, while maintaining good generalisation.

This naturally leads to an architecture consisting of three layers: the motivation, the strategy, and the skill layer. The overall goal can be specified intuitively by different drives that make up the robot's motivation layer. Each drive is representing one sub-goal. The strategy layer has the task to group the infinitely large state space into a small number of abstract regions in order to escape the curse of dimensionality and determine the optimal sequence of actions for each one of those sub-goals. As the environment can change during runtime, the strategy layer also has to maintain a model about its behaviour in that environment. The low-level skills that are used to construct the overall behaviour are the responsibility of the skill layer. It has to find out, which actions the robot is actually capable of. It is not only in charge of exploring its own capabilities, but also in charge of optimising them while executing them.

Our approach provides a framework that combines learning mechanisms on different levels of abstraction in order to satisfy the different sub-goals of the overall motivation, namely enabling autonomous robots to learn in a robust manner while ensuring feasibility.

2 Overview of the ESLAS Architecture

We designed the ESLAS architecture along the *Triple Tower Architecture* developed by Nilsson [9]. Nilsson distinguishes towers for perception, modelling and action. Our approach focuses on the layered modelling tower, which is responsible for the control of the robot's behaviour, thus the overall learning process. Within the ESLAS architecture, the modelling tower is divided into three different levels of abstraction: the motivation layer, the behaviour layer and the skill layer (Fig. 1).

At the top level, a motivation layer provides a motivation function in terms of drives for the learning algorithm being the overall goal of the robot. This function determines which goal is the most profitable one to reach at each moment. With different drives, the learning algorithm is able to handle changes in the environment without the need of relearning everything. At the medium strategy layer is the Reinforcement Learning algorithm, which incorporates the state abstraction mechanism in parallel to Semi-Markov Decision Process (SMDP) policy calculation. A skill is described by a goal function and handled in the lowest layer. Skills can be simple, like driving forward, but also quite complex, depending on the goal function.

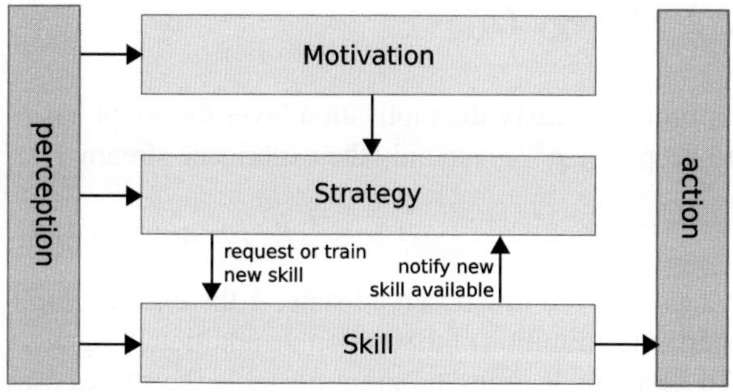

Fig. 1 ESLAS architecture, based on the well-known Triple Tower Architecture

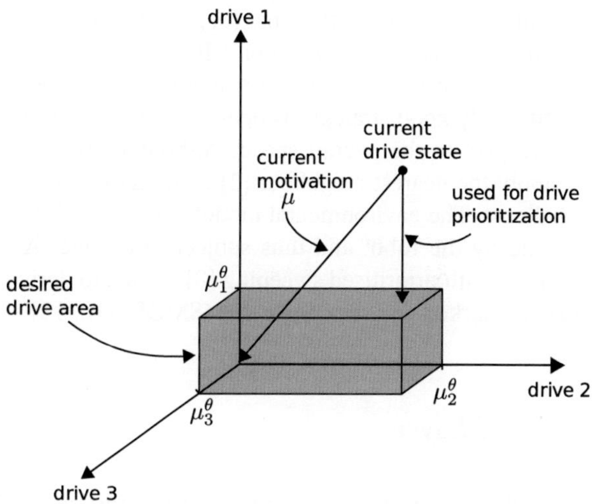

Fig. 2 The motivation system in terms of single, occasionally contradicting drives

2.1 Motivation Layer

For the evaluation of the robot's overall state, we use biologically inspired evaluation methods. We specify all high-level goals in the form of a drive system (Fig. 2):

$$\boldsymbol{\mu} = (\mu_1, \ldots, \mu_n)^T, \quad \mu_i \in \mathbb{R}^+ \tag{1}$$

Each motivation μ_i corresponds to one high-level goal, which is considered accomplished or satisfied if $\mu_i < \mu_i^\theta$, with μ_i^θ defining the threshold of the *well-being region*. By specifying $\mu_i : \mathbb{S} \to \mathbb{R}^+$ as a mapping from the strategy's state space to the degree of accomplishment of goal i and μ_i^θ as the satisfaction-threshold of that goal the designer is able to intuitively define the robot's overall goal, which it accomplishes by minimising each motivation's value. Furthermore, the motivation vector μ is used as a reward for the strategy learning algorithm.

2.2 Strategy Layer

In order to satisfy the motivation layer the robot has to learn a strategy that is able to keep $\boldsymbol{\mu} < \boldsymbol{\mu}^{\theta}$, given only the experience stream

$$\ldots, (o, a, d, \boldsymbol{\mu}, f)_{t-1}, (o, a, d, \boldsymbol{\mu}, f)_t, \ldots \quad (2)$$

where o is the raw observed state, a the executed action triggered in the last time step, d the duration of that action. In addition f signals whether the action has failed. To keep this learning program tractable the strategy layer is not trying to learn one strategy for the whole motivation system. Instead, it is generating one strategy for each motivation. The system then selects the strategy to follow dependent on the dynamic drive prioritisation $\max(\mathbf{0}, \boldsymbol{\mu} - \boldsymbol{\mu}^{\theta})$. In our approach it chooses the drive with the least satisfied motivation.[1] It will have to generalise the actual state observations into abstract regions on which it then uses Reinforcement Learning to find a sufficiently good strategy, as operating on the raw state space would be infeasible with respect to the overall speed. Although any form of abstraction can be used, we apply the nearest neighbour [2] method due to its simplicity and low computational cost. The environmental model is updated during runtime as new experience is made by the robot and thus subject to change. A model-based Reinforcement Learning with prioritised sweeping [8] is used to derive an optimal policy by means of semi-Markov decision processes (SMDP) [10, 12].

2.3 Skill Layer

The skill layer provides a generalised learning method for learning reactive low-level skills. Each skill is accompanied with a goal function which it tries to learn: $a_f : \mathbb{R}^d \rightarrow \mathbb{R}^+$, where d is the number of dimensions of the perception space. The function takes the state as an input and returns an error value which the skill has to minimise. It offers two working modalities to the strategy layer, one for training new skills and one for executing one of the learned skills. As long as no skills are available, the skill layer explores the space of the low-level actions by composing the output vector sent to the actuators with random values.

Each learned skill allows to control the perceived properties of the environment by continuously associating an error value to the input data. The learned skills are communicated to the strategy layer through the identifier that will be used to handle the skill and the definition of the skill. When the execution of a skill is requested, the skill layer reacts to the received inputs with low-level actions (output vectors) that minimise the error (cf. Sect. 4).

The skill layer starts with an exploration phase, where it generates random values for the output vector which determines the actions to be performed by the actuators.

[1] For the sake of simplicity we will restrict the description to one strategy in the following.

The skill layer records action/effect-traces to get information about how the goal function behaves in certain situations. It then creates likely hypotheses about dependencies between action and goal function changes. It uses these hypotheses in the execution phase to execute actions that minimise the goal function which in turn optimises the skill. For our strategy-learning algorithm, we assume that all skills have finished building hypotheses and are ready for execution.

There is one major benefit with the skill layer over using atomic actions: it gives the possibility to automatically create quite complex actions that evolve and adapt to changes.

3 Ensuring Feasibility by State Abstraction

The strategy layer consists of the three main components interacting with each other:

1. The processed, filtered and purified perception is stored as an sequence of interactions (2), previously called experience stream and thus denoted as *experience*.
2. By applying several heuristics, the raw state space is modified and abstracted to build a significantly simplified state space, which we call *model*.
3. Based on the model, a *policy* can then be generated much more efficiently then generating a policy on the raw state space.

At the beginning, all states belong to only one region, as the robot has no reason to believe otherwise. While interacting with the environment the model is modified by several heuristics, which are invoked recurrently to maintain a mapping of observations in the perception space $\mathbb{R}^d = \mathcal{I}_s$ (*state observation* in Fig. 3) to states in the abstracted region space $\mathbb{S}$ (*region* in Fig. 3), with d being the number of dimensions of the perception space. While applying the nearest neighbour method in

Fig. 3 State abstraction: raw states are clustered into abstract states

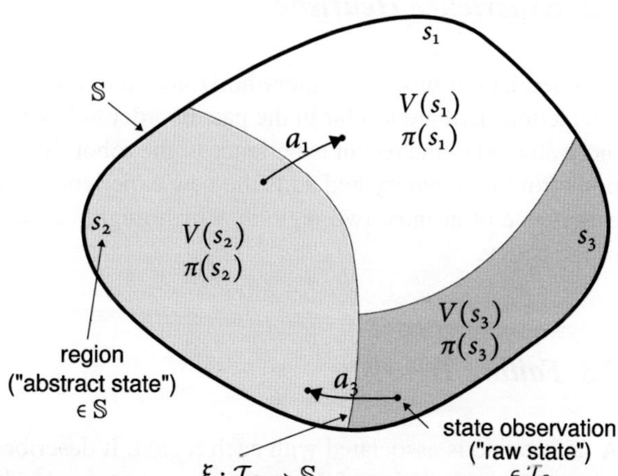

our approach, it can practically be replaced by any other abstraction mechanism that supports the tasks *add*, *remove* and *query* at runtime. The heuristics split or merge regions so that the model and underlying statistics reflect the world experience. For this approach, we identified the following heuristics to be necessary.

3.1 Transition Heuristic

As mentioned above, the continuous state space is split into regions so that for each raw state belonging to the same region executing the same action "feels" similar to the robot. That requires that $R(s,a,s')$ as the value for transitioning from s to s' with the greedy action $a = \pi(s)$ can be estimated with a sufficient confidence. This is calculated using interaction sequences starting in s and arriving in s' while only executing the greedy action a:

$$R(s,a,s') = \gamma(s,a,s')(r(s,a,s') + V(s')) \tag{3}$$

$$+ \lambda(s,a,s')\rho(s,a,s') \tag{4}$$

Let $\mathrm{succ}_a(s) = \{s' | P(s'|s,a) > 0\}$. If raw states are mistakenly grouped into the same abstract region the variance of the $R(s,a,s')$ values calculated for all the greedy traces belonging to the same region will increase. A high variance indicates that splitting that region will likely lead to better transition estimates in the split regions:

$$\mathrm{Var}(\{R(s,a,s')|s' \in \mathrm{succ}_a(s)\}) > \theta_{TV} \tag{5}$$

This is done by clustering the traces so that traces with similar $R(s,a,s')$ are grouped together. For each cluster, one region is created.

3.2 Experience Heuristic

This heuristic limits the memory horizon of the robot to θ_M interactions. It removes interactions that are too far in the past in order to keep the robot's model and policy more aligned to the recent experience of the robot. Basically, it removes old interactions from its memory and adds the new experience to it. Thus, it is modifying the experience of at most two regions, which might cause an update of the model and of the policy.

3.3 Failure Heuristic

A failure rate is associated with each region. It describes the ratio of failure signals when the greedy action of the corresponding region has been executed to the number of success signals. These signals are emitted by the strategy and skill layer. They are encoded as f in the interactions. Failure signals are scenario specific and can be emitted if e.g. the robot bumps into a wall or if it has not encountered something interesting for a longer period of time. The failure heuristic splits a region if its greedy action's failure rate is not homogeneous enough:

$$\theta_f < f < 1 - \theta_f \quad (0 < \theta_f < 1/2) \tag{6}$$

The lower the user defined threshold θ_f is, the more eager the failure heuristic is trying to split a region. This forces the state abstraction to arrive at regions that have failure rates with which a more deterministic strategy can be computed. For both resulting new regions individual greedy actions can then be determined by the reinforcement learning algorithm.

3.4 Simplification Heuristic

As splitting might lead to overly complex models a means is needed that again merges regions once the robot has gathered new experience that suggests a simpler model. This is the task of the simplification heuristic, which analyses sequences of regions connected by greedy actions. In our approach we consider chain and sibling merges. Let a behave nearly deterministically in s, then $\mathrm{succ}(s,a)$ denotes the region the execution of a leads to:

$$\mathrm{succ}(s,a) \equiv \begin{cases} s' & \text{if } P(s'|s,a) \approx 1 \\ \text{none} & \text{otherwise} \end{cases} \tag{7}$$

A chain merge of two regions s' and s'' is performed if

$$\mathrm{succ}(s',\pi(s')) = s'' \wedge \mathrm{succ}(s'',\pi(s'')) = s \wedge \pi(s') = \pi(s'') \tag{8}$$

In this case the region s'' is superficial and can thus be merged with s' into the new region $s''' = s' \cup s''$, with $\mathrm{succ}(s'''',\pi(s''')) = s$ and $\pi(s''') = \pi(s') = \pi(s'')$. All other regions that resulted into either s' or s'' are updated accordingly. In the same vein a sibling merge is triggered if

$$\mathrm{succ}(s',\pi(s')) = s \wedge \mathrm{succ}(s'',\pi(s'')) = s \wedge \pi(s') = \pi(s'') \tag{9}$$

In this case s' and s'' have similar expectations about the future region if the same action is executed.

3.5 Reward Heuristic

Especially in the beginning of the robot's lifetime, when there is not yet enough information for the transition and simplification heuristic to adapt the state space based on sufficient statistical data, the reward heuristic is of importance. It allows a region $s \in \mathbb{S}$ to be split if the reward rate variance is too high. This indicates that the action performed in that region gives a too diverse feedback. A split of that region will then lead to multiple regions, which are more consistent with regard to the expected reward rates. This also is vital in cases where the failure signal is too seldom, as it provides the only other possibility to initially split a region.

In particular, the reward heuristic is looking in the reward rate stream for a clear switch from low to high variance areas, where both areas are of sufficient length. Only such a switch in variance indicates clearly that a split is advisable. Therefore the reward heuristic considers the reward rates of the last n interactions made in the current region. The lump sum rewards in that time frame are not considered, as they will show non-zero values only in rare occasions. Let $\rho_{t_1}^{t_2} = (\rho_{t_1}, \ldots, \rho_{t_2})$ and t be the time at which the split is considered. The reward heuristic is searching for an index k that splits ρ_{t-n}^{t} into the two sequences ρ_{t-n}^{t-k-1} and ρ_{t-k}^{t}, such that the following condition holds:

$$\begin{aligned}&\left(\mathrm{Var}(\rho_{t-n}^{t-k-1}) \approx 0 \wedge \mathrm{Var}(\rho_{t-k}^{t}) > \theta_{RV} \wedge |\rho_{t-n}^{t-k-1}| > \theta_l\right)\\ \vee \quad &\left(\mathrm{Var}(\rho_{t-n}^{t-k-1}) > \theta_{RV} \wedge \mathrm{Var}(\rho_{t-k}^{t}) \approx 0 \wedge |\rho_{t-k}^{t}| > \theta_l\right)\end{aligned} \tag{10}$$

The minimum variance threshold θ_{RV} is dependent on the motivation system design. Recall that the reward received by the motivation system can be interpreted as a reward rate, e.g. if $|\dot{\mu}_i| \leq \rho_\theta$. With $\theta_{RV} = k \cdot \rho_\theta$ $(0 < k < 1)$, a switch is easily detected by the reward heuristic. The minimum low variance sequence length θ_l ensures that the reward heuristics does not find trivial splits. Naturally it is set to be a fraction of the considered time horizon n.

4 Learning Skills at the Lowest Level

So far we have assumed that a set $\mathbb{A}$ of actions is always provided beforehand and that the strategy simply has to choose the right action at each state. For real-world scenarios it would be advantageous if also $\mathbb{A}$ could be learned at run-time.

The input of the skill-learning algorithm is given by the strategy layer in terms of an error function e. Figure 4 shows a camera image that has been taken from one of our robots in a real-world scenario. Within the camera image two different coloured balls were detected by a computer vision algorithm. Now let d be the Euclidean distance of the yellow ball to the image centre and Δs the difference between the maximum and current size in terms of number of corresponding pixels. One possible error function that formulates the goal to maximise the yellow ball in the middle of the camera image would be:

$$e(d, \Delta s) = \sqrt{c_0 * d^2 + c_1 * \Delta s^2} \tag{11}$$

The first step is then to get a set of training examples that will later be generalised. During this initial exploration the algorithm gathers information about the relationship between the actuators and the effects. The changing of the actuators

Fig. 4 Camera images of one of our robots, detecting a yellow and a magenta ball in a real world scenario

is called an *action* A. An *effect* is the perceived result of an action. The actions are generated randomly and are applied for some time. In this phase we call the actuator values the input I and the perceived effect the output O, as they are seen from the skill learning algorithm's perspective. This information comprises the components of a trace T with the length t: $T = (A, \{(I_0, O_0), \ldots, (I_{t-1}, O_{t-1})\})$. Several traces are recorded. Now the error function e is used to extract the good traces forming the training set. To get as many traces as possible, every trace is cut at the position i $(0 \leq i \leq t-1)$ of the lowest error $e(O_i)$. Then every trace not leading to a decrease of the error function e is discarded.

Previous to the generalisation, the number of traces and the dimensions have to be refined to reduce the generalisation complexity. The most important attributes of a trace are A, I_0, O_0 and O_{t-1}. If the actuator configuration I_0 and the sensor vector O_0, which describe the current situation, are given, A has to be used to reach the effect O_{t-1}. To reduce the number of traces, we perform an agglomerative hierarchical clustering. Only the mentioned attributes of a trace are used. The distance measure between two traces is the Euclidean distance of the attribute values. The distance between two clusters is defined by average-linkage. The dimensions of I_0 and O_0 depend on the number of actuators and effect properties. Actuators that don't influence the effect can be ignored in the generalisation step. Another side-effect of the dimension reduction is the noise reduction, because also the data dimensions with no significant effect to the action-effect can be ignored. We use PCA [1, 4] for this and specify the number of principal components to be kept by the fraction of variance to be explained. In our experiments we were able to reduce the dimensions from six to two while maintaining 95 percent of the data's accuracy.

The last two steps reduce the trace data to the basic properties. In the PCA step a mapping from the data into a new artificial space is done. To generalise the data, a mapping from the principal components. $x_0, \ldots, x_n$ to the individual actor elements $a_i \in A$ is calculated. We use a polynomial regression for every a_i. To get the simplest polynomial of $\sum_{i=0}^{d} \prod_{j=0}^{n} p_{ij} {x_j}^i$ that fits the data sufficiently the algorithm starts with $d = 0$ and increments it until the prediction error drops below a predefined error threshold. This process can be seen in Fig. 5. There is also a thresh-

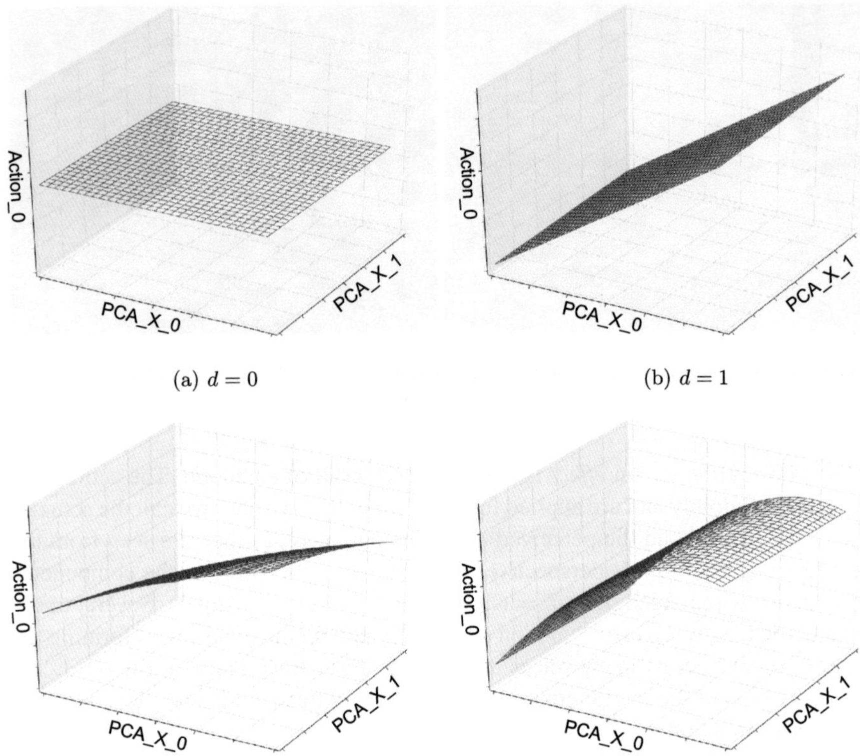

(a) $d = 0$ (b) $d = 1$

(c) $d = 2$ (d) $d = 3$

Fig. 5 Finding the simplest reasonable hypothesis for the first actor element in PCA space. The graphs show the fitted function for one actor dimension dependent on the two calculated PCA dimensions. The degree of the polynomial is incremented from $d = 0$ up to $d = 3$. In (**d**) the final function can be seen. The increase of d has been stopped because the fitting error falls below a defined threshold

old for the complexity's degree. To avoid over-fitting a maximal possible degree can be specified. Finally, a function $f_i(x_0,\ldots,x_n) = a_i$ is calculated for each a_i. When applying the learned skill I_0 and O_0 are known as the current parameter values of the actuators and sensors. A mapping to the PCA space has then to be done before using the calculated f_is to build the next action A. With this approach the robot can reach maximal adaptivity and robustness with regard to sudden breaks or graceful degradation [11].

5 Exploration vs. Exploitation

When learning behaviour at two different layers, the exploration-exploitation dilemma—naturally found in learning problems—is amplified. On the one hand, the strategy layer has to explore strategies over actions that are assumed to reliably yield the same result over time. On the other hand, the skill layer has to explore skills in order to provide the strategy layer with a sufficient set of usable skills.

This is solved by letting the strategy layer decide when the skill layer is allowed to explore. If the skill layer is exploring, the strategy does not interfere by commanding, which skill to execute next. It waits for the skill layer to signal new skills that it deems to be reliable. The strategy layer updates its own action space each time accordingly. It triggers the skill layer to be in exploitation mode again if it deems its current action space as sufficient. From then on, the strategy layer is again in control and allowed to request the skill layer to execute skills.

6 Discussion

Hasselt and Wiering devised the *Continuous Actor Critic Learning Automaton* approach, which empowers reinforcement learning to operate on continuous state and action spaces [13]. They calculate real valued actions by interpolating the available discrete actions based on their utility values. Therefore, the performance is highly dependent on initial assumptions about the value function.

It is obvious that a full search in continuous state and action spaces is infeasible. For reinforcement learning approaches to be applied in realistic domains, it is therefore vital to limit the search to small areas in the search space. One approach to do that is the *Actor-Critic* method [6]. It separates the presentation of the policy from the value function. The actor maintains for each state a probability distribution over the action space. The critic is responsible for providing the reward from the actions taken by the actor, which in turns modifies its policy. As this relieves the designer from assumptions about the value function, it introduces new assumptions about the underlying probability distribution. To overcome this problem Lazaric et al. devised *Sequential Monte Carlo Learning* [7], which combines the Actor-Critic method with a non-parametric representation of the actions. After initially being drawn from a prior distribution, they are re-sampled dependent on the utility values learned by the critic.

The *Adaptive Modelling and Planning System* (AMPS) by Kochenderfer [5] maintains an adaptive representation of both the state and the action space. In his approach, the abstraction of the state and action space is combined with policy learning in a smart way: states are grouped into abstract regions, which have the common property that perception-action-traces, previously performed in that region, "feel" similar in terms of failure rates, duration, and expected reward. It does so by splitting and merging abstract states at runtime. AMPS not only dynamically abstracts the state space into regions, but also the action space into action regions. This is, however, done in a very artificial way that could not yet be shown to work in real world domains.

Although our strategy layer is inspired by AMPS, we differ from it in the following important points: AMPS applies the splitting and merging also to the action space, which works fine in artificial domains but will not cope with the domain dependency one is typically faced with in real environments. In contrast to that, we use the skill layer's goal functions (cf. Sect. 2.3) as the strategy's actions. This leads to a perfect separation of concerns: the task of the strategy layer is to find sequences of actions and treats actions as mere symbols. The skill layer by means of data driven skill functions then grounds these symbols.

Another aspect is the supported number of goals. Take for example a system, which has to fulfil a specific task while paying attention to its diminishing resources. While, on the one hand, accomplishing the task, the resources might get exhausted. If it, on the other hand, always stays near the fuel station, the task will not be accomplished. Approaches like AMPS, which do not support multiple goals by multiple separate strategies, have to incorporate all different goal aspects in one reward function. This leads to a combinatorial explosion in the state space and implicates a much slower learning convergence.

In our approach, we use abstract motivations, which the designer has to specify. These motivations may also contain competing goals. The major advantage of our approach is that the robot can learn a separate strategy for each motivation. Depending on the strength of each motivation, it has now a means to choose the right strategy for the actual perception and motivation state.

Except for our approach, all the discussed approaches have the following underlying restricting assumptions. First, they assume that optimal actions are either possible to be predefined or effectively learnable within the reinforcement learning framework. That means that prior to using these approaches a careful analysis of all occurring events in the environment has to be carried out by the designer. Except for AMPS, they are all based on Markov Decision Processes (MDP). Time varying actions, which are the norm in realistic scenarios, however require a semi-Markov Decision Process (SMDP), which complicates the search in continuous action spaces. Arguing that models are difficult to approximate at runtime the model-free approaches do not learn a model on which the policy is approximated but only the value function. Furthermore, they always solve only one goal and it is not intuitively clear, how multiple possibly contradicting goals could be integrated using the same state and action space for all goals.

7 Conclusion and Future Work

In this article we presented a fast hierarchical learning approach in terms of a framework that is able to handle system and environmental changes by learning autonomously at different levels of abstraction while pursuing multiple, possibly contradicting goals. It is able to do so in continuous and noisy environments by (1) an active strategy-learning module that uses reinforcement learning and (2) a dynamically adapting skill module that proactively explores the robot's own action capabilities and thereby provides actions to the strategy module. Our proposed modular architecture for autonomous robots enables them to adapt selfishly to real-world

Fig. 6 The miniature robot BeBot, having an orange ping-pong ball in its passive gripper

environments while paying attention to the robots' needs. By using time- and perception-dependent drives to endow the robots with dynamic motivations, the robots are enabled to proactively reorganise the task allocation among themselves and trigger corresponding learning processes, if they are not yet able to process the allocated tasks.

We are currently assembling a real-world scenario for evaluating our framework under more realistic conditions. As a mobile robotic platform, we want to apply the miniature robot *BeBot* [3], which was developed at the University of Paderborn. As depicted in Fig. 6, the BeBot supplies a video camera, which is attached to the front side of its case and thus delivers images from a subjective point of view. Furthermore, the robot can push small objects by means of a passive gripper. In order to overcome the common problem of self-localisation, we intend to use a set of external cameras for the time being. With the help of these cameras, we will be able to observe the whole setting and track one or more BeBots in order to determine their absolute positions.

References

1. Alpaydin, E.: Introduction to Machine Learning. MIT Press, Cambridge (2004)
2. Cover, T., Hart, P.: Nearest neighbor pattern classification. IEEE Trans. Inf. Theory **13**(1), 21–27 (1967)
3. Herbrechtsmeier, S., Witkowski, U., Rückert, U.: BeBot: a modular mobile miniature robot platform supporting hardware reconfiguration and multi-standard communication. In: Progress in Robotics. Communications in Computer and Information Science, vol. 44, pp. 346–356. Springer, Berlin (2009)
4. Jolliffe, I.T.: Principal Component Analysis. Springer, Berlin (2002)
5. Kochenderfer, M.J.: Adaptive modelling and planning for learning intelligent behaviour. PhD thesis, School of Informatics, University of Edinburgh (2006)
6. Konda, V.R., Tsitsiklis, J.N.: On actor-critic algorithms. SIAM J. Control Optim. **42**(4), 1143–1166 (2003)
7. Lazaric, A., Restelli, M., Bonarini, A.: Reinforcement learning in continuous action spaces through sequential Monte Carlo methods. In: Advances in Neural Information Processing Systems, vol. 20, pp. 833–840. MIT Press, Cambridge (2008)

8. Moore, A., Atkeson, C.: Prioritized sweeping: Reinforcement learning with less data and less time. Mach. Learn. **13**(1), 103–130 (1993)
9. Nilsson, N.J.: Artificial Intelligence: A New Synthesis. Morgan Kaufmann, San Mateo (1998)
10. Puterman, M.L.: Markov Decision Processes: Discrete Stochastic Dynamic Programming. Wiley-Interscience, New York (1994)
11. Richert, W., Lüke, O., Nordmeyer, B., Kleinjohann, B.: Increasing the autonomy of mobile robots by on-line learning simultaneously at different levels of abstraction. In: Proceedings of the Fourth International Conference on Autonomic and Autonomous Systems, pp. 154–159. IEEE Computer Society, Los Alamitos (2008)
12. Sutton, R.S., Barto, A.G.: Reinforcement Learning: An Introduction. MIT Press, Cambridge (1998)
13. van Hasselt, H., Wiering, M.A.: Reinforcement learning in continuous action spaces. In: IEEE International Symposium on Approximate Dynamic Programming and Reinforcement Learning (ADPRL), pp. 272–279 (2007)

A. Jungmann (✉) · B. Kleinjohann · W. Richert
C-LAB, University of Paderborn, Paderborn, Germany
e-mail: global@c-lab.de

Chapter 5.10
Emergent Computing with Marching Pixels for Real-Time Smart Camera Applications

Dietmar Fey and Marcus Komann

Abstract In this article the concept of an emergent algorithm class called Marching Pixels is presented. These algorithms are intended for a direct implementation in hardware. Marching Pixels can be understood as a cluster of hardware agents that are crawling directly within a two-dimensional pixel grid in order to analyse pixels with regard to certain features. These locally acting entities communicate either directly or indirectly with each other by leaving information on grid points. In cooperation they solve after a certain number of time steps a task, e.g. to find the centroids of objects contained in a binarised image. In this article we present a series of Marching Pixels algorithms for centroid detection—the so-called Marching Pixels toolbox—that differ in complexity and in their capability to analyse diversely complex shaped objects.

Keywords Emergent computing · Marching pixels · Smart camera chips

1 Introduction

Marching Pixels denote a class of emergent computing algorithms which were developed in [7] to realise fast image vision detection processes for embedded hardware like e.g. smart sensors. In modern factories, machine vision is an elementary part of the production process. Machines often need vision capabilities in order to grab, drill, grind, or in general, handle tools and work pieces. The machines are thereby more and more required to cope with complex tasks and have to gather and compute data about the items they work on. Originally Marching Pixels focused on the detection of known objects on an assembly line, whose exact locations are unknown. The goal of this work was to find parallel and distributed algorithms for fast detection of multiple objects, their centroids and their rotation according to emergent computing principles.

This detection process must be executed in milliseconds since many industrial vision tasks require to process more than 100 images per second. This time of 10 ms per image includes image capturing as well as transfer, processing, and output of the data. For the handling of megapixel resolution a simple calculation yields that only 10 ns would remain for the processing of a pixel. To carry out a reasonable number

C. Müller-Schloer et al. (eds.), *Organic Computing — A Paradigm Shift for Complex Systems*, 559–572
Autonomic Systems, DOI 10.1007/978-3-0348-0130-0_37,

of operations within this time range would require a clock frequency of several GHz on a serial processor. However, in embedded systems one would like to avoid such high frequencies due to the high energy dissipation.

Our answer to both the strict real-time requirements and the scalability with regard to increasing pixel resolutions are appropriate emerging computing principles. Such algorithms work in parallel to be fast but also in a distributed way to be robust as well. Emergent computing denotes a computing paradigm that was characterised by Hordijk, Crutchfield and Mitchell [9]. Emergent computing is based on local operators which can be executed by small entities. These entities operate on a discretised grid data structure, e.g. an image. The entities operating locally on a micro level cooperate in direct and also in indirect way, e.g. by deposing information on a grid point. After a certain number of discretised time steps the desired result emerges. This number of steps is sometimes not easy to recognise in advance. It depends primarily on the input information and can only be observed on the macro level.

The local entities in our algorithms are implemented as a kind of autonomous hardware we denote as *Marching Pixels* or short MPs. We consider them to be agents because they are instructed with the task to travel virtually within a pixel grid, which corresponds to the image, in order to find the centroids' of objects in the image. After the agents have found them the corresponding coordinates are given to a robot control to enable the robot to grasp the objects.

The propagation of MPs is steered by classic cellular automata (CA) [19]. A MP has an own state as well as a pixel it currently visits. Depending on these states and the states of neighbouring pixels one or several instructions have to be applied to the data the MP is carrying with it, or the data stored on the grid point it is visiting.

During our research on MPs we designed different algorithms which differ in their complexity and their capability of handling different complex shaped objects. Therefore these algorithms strongly differ in capability and cost, too. We collected all these algorithms in a so-called *Marching Pixels Algorithm Toolbox*. Some of the algorithms can be used for few object types only but do neither need long computation time nor large memory. Others are able to handle complex situations but require much more effort to be successful. For a series of such application-specific MP toolbox algorithms we designed application-specific architectures as ASICs.

The rest of this article is organised as follows. Section 2 shows work related to our MP approach. The algorithmic base for those architectures is laid in Sect. 3. Furthermore we explain in that chapter the principle of a MPs' algorithm exemplarily for an algorithm called *Flooding*. In Chap. 2.4 in this compendium we demonstrate how such algorithms can be implemented with application-specific and programmable hardware for realising arbitrary MP and other emergent computing algorithms. In that chapter we will show that the centroids of an arbitrary number of objects given in a 1000×1000 pixel image can be found within 8 ms using a system-on-chip of reasonable size. Finally, we conclude this article with a summary and a brief outlook how we used MPs successfully for further applications like path planning and as algorithm scheme in GPGPUs (General Purpose Graphics Processing Units).

2 Related Work

The *cellular neural networks* (CNN) model from Chua, Yang and Roska [2–4, 15] provides a concept for massively-parallel visual microprocessors. The CNN architecture consists of an array of so-called *analogic* cells. In each cell, a partial differential equation of second order is solved iteratively until the output and the states of the cells are running in equilibrium. This corresponds to a hard-wired implementation of a reaction-diffusion-architecture.

The different generations of CNN vision chips [14] are primarily based on analogue electronics. Various filter operators for image preprocessing can be executed very flexibly by means of so-called templates. A template defines a matrix of weights to control the inputs and the state of a CNN cell for the calculation of the cell's next state.

Beside other models, CNNs were used to implement so-called *Pixel Level Snakes* (PLS) by Vilarino and Dudek [17, 18]. PLS belong to the contour altering techniques where contour shapes evolve until they reach a final desired location. Their evolution is influenced by the underlying image information like edges or intensities and the desire to keep the contour smooth. With proper altering functions, PLS are able to compute a lot of complex vision tasks. Different to classic contour techniques, the contours of PLS are modified by local functions, which create a global image of snakes moving on the objects. This locality allows for massively-parallel implementation on CNN machines [18].

With CNNs, it is possible to realise movement of single pixels within a grid. However, the movement of whole virtual state-based agents with assigned memory as needed for Marching Pixels or sophisticated cellular automata is not possible. Furthermore, the integration of integer arithmetic logic in each cell is not in the scope of the CNN concept. Marching Pixels need that capability to determine attributes of multiple objects contained in a binary image, i.e., to calculate local moments (see Sect. 3.3). For this, with the exception of the analogue signal capturing of the optical signals, a purely digital realisation for Marching Pixels in an array of processing elements (PE) is favoured. The price to pay for that is a higher effort concerning the required chip area per pixel as compared to CNNs.

Another scientific area that is related to Marching Pixels are *lattice gas* simulations [13]. There, flows of gases and fluids can be simulated by calculating the movement of particles on microscopic level. The particles move in a regular 2-D or 3-D grid and locally interact with other particles. Lattice gases form complex global behaviour as a result of this local particle interaction. Hence, like Marching Pixels, they are emergent.

In the meantime the concept of Marching Pixels has also found international recognition. Geese and Dudek used MPs in [8] for the realisation of autonomous long-distance data transfers in cellular SIMD arrays.

3 The Principle of Marching Pixels Algorithms

The input for the MP algorithms in a real-time embedded system like a smart camera arises e.g from an array of photo detectors. Then, some classic fast image preprocessing algorithms can be applied to improve image quality, e.g. transformation of a grey-valued image into a binary one via thresholding, erosion and dilation, or noise reduction. The starting point for an MP algorithm is usually to detect the edge pixels of objects in an image. Edge detection can be carried out with classic image processing schemes like the Sobel operator or a combination of Sobel and a subsequent Gaussian filter as in the Canny Edge detector [1]. In the following, edges are identified by applying very simple local masks to object pixels, which compare the values of the neighbouring pixels to an object pixel. If any of the neighbouring pixels is a background pixel, the current pixel is identified as an edge pixel. All these operations of the preparing phase are applied locally and thus can be performed in parallel on each pixel. Hence, they only require constant time independent of the image size. A binarised image in which the edges of the objects are recognised is then the starting point for the MP algorithms.

3.1 The Basic Procedures of Marching Pixels Algorithms

MP algorithms for objects' attributes detection in images generally consist of three major phases. The first phase is called the MPs' birth phase. After edge detection, MPs are born at certain detected edge pixels. E.g., the edge pixels on the left side of an object can be the birth place of an MP agent, meaning it will start its march from that position. After the MPs are born the second phase starts in which MPs move from the edge pixels and collaboratively gather information about the objects while running on object pixels and possibly on pixels in the neighbourhood of these object pixels. This is why we call this second phase the marching phase. During their march across the pixel array, MPs perform specific operations depending on data values stored in the pixels where they currently reside, and data stored in the directly neighbouring pixels as well as their own state. Hence, the behaviour of an MP is completely defined by local rules. As consequences of these local rules, different things can happen. MPs may lay data on the pixel, and a possibly new direction in which the MP will be moving in the next step is calculated. It is also possible that the march is stopped when the MP has found its final destination, e.g., the centre point of an object. Furthermore, an MP might meet other MPs and interact with them causing a change of its own state and stored data values, or even the death of the MP can occur.

Entering a situation in which the condition for the stopping of the MPs' march is fulfilled corresponds to the third phase. This phase is called the stop phase. In the stop phase, the state of halted MPs is collected and evaluated by an outside processor we denote as *observer*. The observer is some authority outside the pixel grid with higher capabilities. Its task is to carry out post-processing of the data preprocessed by the MPs. Furthermore the observer can determine which object is positioned at the location where the MP stopped its march. This can be done by evaluating the number of object pixels a MP has counted during its march and retrieving the corresponding object from a data base containing the area or number of pixels for certain objects. Finally, the task of the observer is to transmit the calculated results to the outside world, e.g., a robot control.

3.2 The Local Calculation Tasks of Marching Pixels

The propagation direction of MPs is steered by a CA. During this propagation the MPs have to calculate data values in order to fulfil their ordered task, namely the calculation of attributes of the objects contained in the image. This is performed by calculating the moments of objects by MPs as a local calculation procedure.[1] For that, each MP agent contains the following 6-tuple which is brought from one pixel to another: $S = (m, s_x, s_y, s_{xx}, s_{yy}, s_{xy})$. All of the component values are integers. At the beginning of the march, each of these values is zero with the exception of m, which is equal to 1. A tuple of a MP located at pixel (x, y) has the following meaning: m counts how many object pixels were collected by the MP so far. Assuming the MP passed pixels at $(x_1, y_1), (x_2, y_2), \ldots, (x_m, y_m)$, the other variables have the values shown in (1). This means, that e.g. s_x contains the sum of the horizontal distances of the current pixel value x to all other pixels x_i, $1 \leq i \leq m$, the MP has visited so far. The same holds for y_i concerning the y coordinate. The values s_{xx}, s_{yy} are the sums of the quadratic distances and s_{xy} is the sum of the product of the distances in x- and y-direction. We will use these values for the determination of the moments. The moments allow us to calculate the features we want in a post-processing step, namely centroid, orientation and area of the investigated object.

$$
\begin{aligned}
s_x &= \sum_{i=1}^{m}(x - x_i), \qquad & s_y &= \sum_{i=1}^{m}(y - y_i) \\
s_{xx} &= \sum_{i=1}^{m}(x - x_i)^2, \qquad & s_{yy} &= \sum_{i=1}^{m}(y - y_i)^2 \\
s_{xy} &= \sum_{i=1}^{m}(x - x_i)(y - y_i)
\end{aligned}
\tag{1}
$$

During the propagation of an MP in the pixel field it must be assured that (1) is always true. Generally, all values in the tuple are local moments. Hence, a pixel

[1] These algorithms have been partially developed together with A. Kröller, C. Schmidt and S. Fekete from Braunschweig University of Technology during a cooperation within the Priority Programme 1183 *Organic Computing*.

receiving a MP with the tuple $S = (m, s_x, s_y, s_{xx}, s_{yy}, s_{xy})$ will adjust the values to make itself the centre. It therefore computes a new tuple S':

- If S is received from the left:

$$S' = (m, s_x + m, s_y, s_{xx} + 2s_x + m, s_{yy}, s_{xy} + s_y)$$

- If S is received from the right:

$$S' = (m, s_x - m, s_y, s_{xx} - 2s_x + m, s_{yy}, s_{xy} - s_y)$$

- If S was sent from below:

$$S' = (m, s_x, s_y + m, s_{xx}, s_{yy} + 2s_y + m, s_{xy} + s_x)$$

- If S was sent from above:

$$S' = (m, s_x, s_y - m, s_{xx}, s_{yy} - 2s_y + m, s_{xy} - s_x)$$

As we will see later, there is not only one MP but a wave of them, each carrying moment values relative to their current position. If some MPs of that wave meet in a certain pixel, i.e., multiple MPs arrive from different sources, their tuples can simply be added component-wise. If the pixel itself should be considered for counting in this MP, m' is increased by one. An object pixel must be counted only once, multiple counts of the same pixel would invalidate the final result. To avoid that, a flag is attached to each pixel when a MP meets the pixel the first time.

Only basic arithmetic operations are needed here (addition, subtraction and left shift). For a pixel field of size $2^k \times 2^k$, the memory requirement is as follows. For m, the number of object pixels holds $m \leq 2^k \times 2^k$. Hence we need $2k+1$ bits to store m. For each of the 2^k rows it holds that $|s_x| \leq \sum_{i=1}^{2^k} i$. Since $2^k \sum_{i=1}^{2^k} i < 2^{3k}$ we need $3k + 1$ bits including sign (analogous for s_y) at most. The remaining three entries have an upper bound of $2^k \sum_{i=1}^{2^k} i^2 < 2^{4k}$. Hence, they require at most $4k + 1$ bits including sign.

In [10] it is shown that these invariants S' fulfil (1). Consequently, an MP can easily determine the position of the centre of gravity of the traversed pixels in relation to its own location: $sumx/m$ steps to the left and $sumy/m$ downwards, if the propagation direction of the MPs determined by the steering CA leaded the MP swarm from left to right and bottom to top.

In the end of an object computation, all gathered information is accumulated at a single pixel (x, y). The values of the tuple S at this final pixel provide all information that is necessary to compute the characteristic values.

1. The centre of gravity (μ_x, μ_y) can be computed with the first moment and lies at

$$\mu_x = x - \frac{s_x}{m}, \qquad \mu_y = y - \frac{s_y}{m} \tag{2}$$

   This is s_x/m pixel steps to the left and s_y/m downwards from the final pixel. From that pixel either a marker MP can be sent to the centroid position or the observer processor collects the values together with the position (x, y) and computes centroid and orientation from the above formula.

2. The second moments are obtained by first calculating the variances

$$\sigma_x^2 = \frac{s_{xx}}{m} - \frac{s_x^2}{m^2} \quad \text{and} \quad \sigma_y^2 = \frac{s_{yy}}{m} - \frac{s_y^2}{m^2} \tag{3}$$

as well as the covariance

$$\sigma_{xy} = \frac{s_{xy}}{m} - \frac{s_x s_y}{m^2} \tag{4}$$

Using the variances and the covariance the orientation θ of an object can be determined according to (5). This complicate operation should be carried out on the observer processor in the post-processing step.

$$\theta = 0.5 \times \arcsin \frac{2\sigma_{xy}}{\sqrt{4\sigma_{xy}^2 + (\sigma_x^2 - \sigma_y^2)^2}} \tag{5}$$

3.3 Example

For clarification, look at the example in Fig. 1. The right upper part shows the coordinate system, the object pixels in dark grey, and background pixels in white. An exemplary MP movement is depicted by dark arrows. Two MPs start their march from the left edge pixels at coordinates (6, 14) and (6, 15). After the second discrete time step the upper MP turns to the right about 90° and moved downward. At the position (7, 14) it hits on a pixel position where the lower MP was already one time step before. In the last time step a MP arrives at (8, 14). The rest of the figure zooms the object pixels and presents the local moments' values stored in the 6-tuple S at each pixel. It displays how the initial 6-tuple value $S = (1, 0, 0, 0, 0, 0)$ is changed according to the formulas for S' shown above. Note, that at position (7, 14) the 6-tuple $S = (2, 1, -2, 1, 2, -1)$ of the upper MP is added to the 6-tuple S of the lower MP whose tuple values had to be stored at that coordinate before.

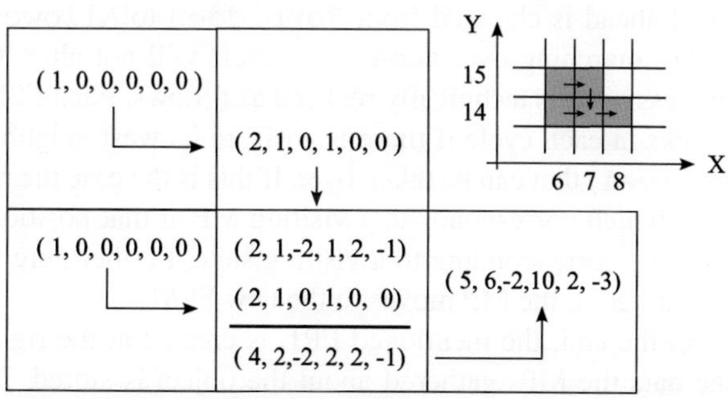

Fig. 1 An example for MPs collecting LMs in a 5-pixel object

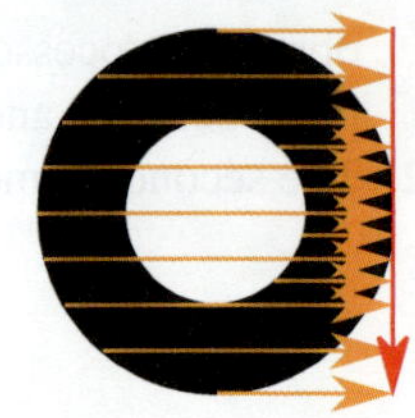

Fig. 2 Scheme of MP flooding algorithm

After the MP movement has stopped, the outside observer processor collects the tuple $(5, 6, -2, 10, 2, -3)$ from the pixel at (8, 14). It computes the center at $(8 - \frac{6}{5}, 14 + \frac{2}{5})$ according to (2), the variances σ_x^2, σ_y^2 as $\frac{14}{25}$ resp. $\frac{6}{25}$, and the covariance σ_{xy} as $-\frac{3}{25}$ according to (3) and (4). These values allow to determine the orientation θ with (5).

3.4 Flooding as an Example of a MP Algorithm

In Fig. 2 the marching scheme of a MP algorithm called *flooding* can be seen. The MPs start from the western edge pixels and run eastwards. After a certain time all MPs in one row stop their march along a vertical line through the easternmost object pixel. This line will be called Flooding Reduction Line (FRL) from now. Figure 3 shows the CA that steers the MPs in the flooding algorithm to achieve that. The moving of a MPs is realised in such a way that each pixel position is modelled by a Finite State Machine. Background pixels are in the state `Passive0:Forbidden`, meaning they are not allowed to host a MP, object pixels which have a western object pixel as neighbour are in the state `Passive1:Allowed`, meaning they host no MP but they are allowed to do that in principle, and object pixels which have a western non-object pixel as neighbour are in the state `Active1:Run eastwards`, meaning they are the birth place of a MP. A kind of virtual diffusion process allows MPs to march over all pixels enclosed by the object. This diffusion process works as follows. A non-object pixel being in the state `Forbidden` can become `Allowed`, if a MP passed along its northern or southern neighbour pixel `North=Active1 OR South=Active1`. A MP is moving eastwards until it meets a non-object pixel ahead `West=Active1`, elsewhere it is blocked. The blocking is dissolved when a pixel ahead is changed from `Forbidden` to `Allowed` by the diffusion process. A MP marching over non-object pixels will not alter its 6-tupel S. The marching process itself is technically realised as follows. Each FSM, denoted as current FSM, checks in each cycle if there is a MP in its west neighbouring FSM (being in state `Active1`) that can be taken over. If this is the case the neighbouring FSM clears its state which corresponds to a visiting MP at that position and the current FSM sets the state corresponding to a visiting state. Furthermore all information gathered by and stored in the MP moves to the new FSM.

In the end, the mentioned FRL is created at the right side of the object, where the data the MPs gathered about the object is stored. It corresponds to horizontal

Fig. 3 CA that steers MPs in the flooding algorithm

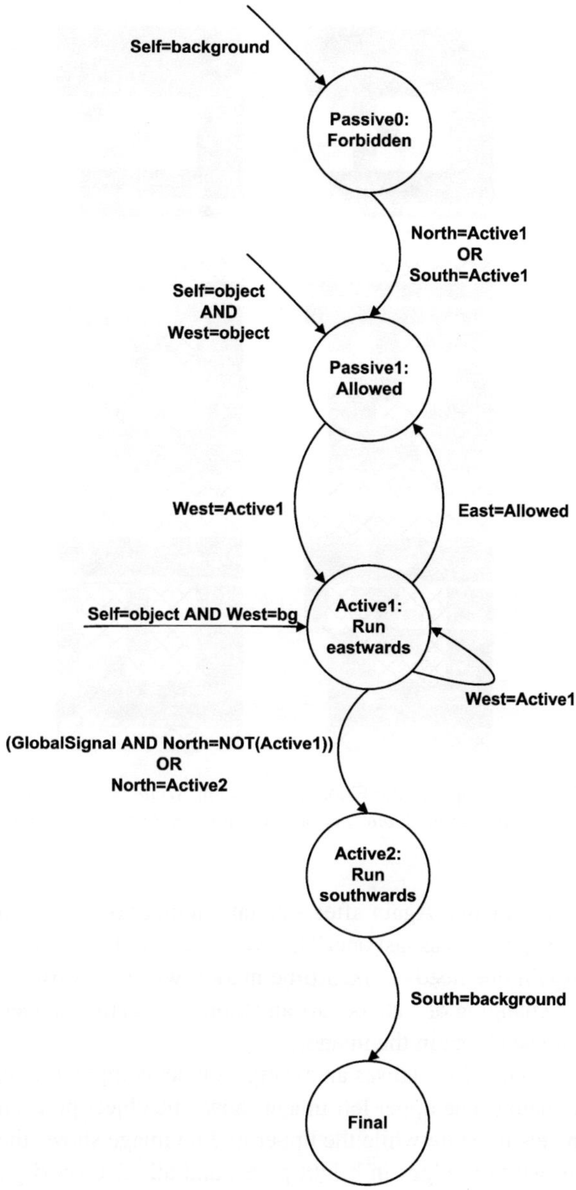

projections applied row-wise. Then a global signal becomes active and starts a new MP that runs southwards along the FRL and accumulates the information that is stored in the FRLs pixels (red arrow). The accumulated data is used to compute the desired attributes of the object. The global signal is switched to Active by the observer processor, e.g. after a fixed amount of time. Alternatively all MPs can also march to the south all the time. These MPs will carry only information about a part

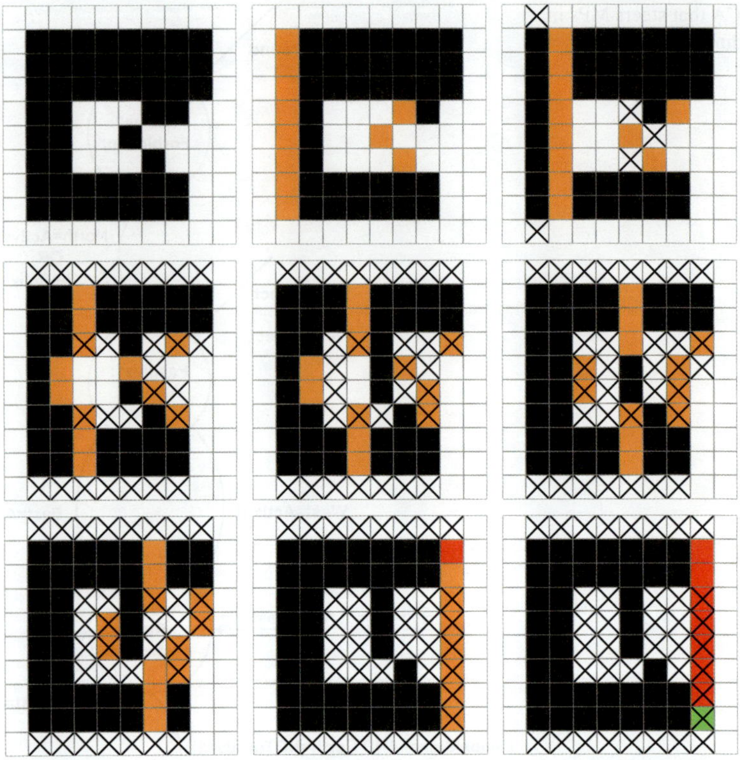

Fig. 4 Example of the Flooding algorithm. *White* = background, *black crosses on white* = `Allowed` background, *black* = object, *orange* = `Active1`, *red* = `Active2`, *green* = `Final`

of the image. Again after a certain number of time steps one MP will arrive at the final position as last one. That MP possesses the information about the whole image. Again one needs a fixed time marker when the observer asks for final results, i.e., no changing of state occurs any more. When this happens depends on the size of the largest object in the image.

Figure 4 visualises an example object computation of the presented flooding automaton. The upper left image shows the object pixels in black and the background pixels in white while the upper middle image shows the starting `Active1` MPs at all western edges in bright pixels and all `Allowed` pixels in black. The first step can be seen in the upper right image. There, all active MPs have been taken by their previously eastern neighbours. In addition, all northern and southern background neighbours of active MPs turned to state `Allowed` which is depicted by a black cross for background pixels. The left image in the middle row shows the second step. Some of the MPs have stopped their march because they do not find `Allowed` pixels in their eastern neighbours. One MP moved outside the object to an `Allowed` which is depicted by a background pixel (bright pixel with black cross). The next images illustrate the following steps. MPs are marching as long as they can.

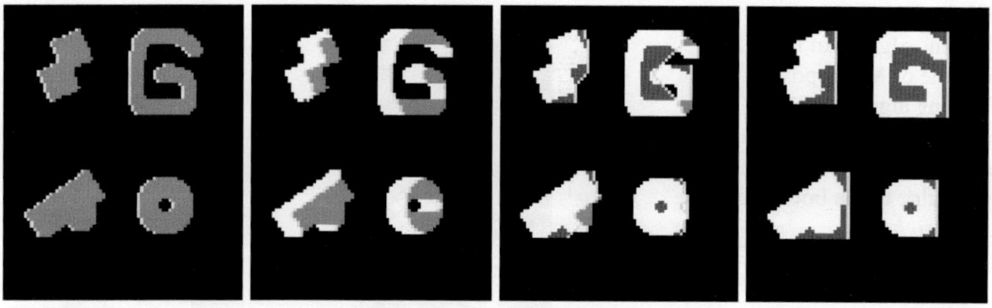

Fig. 5 SWARM2.2 simulation of Flooding

The operation of the diffusion process, which controls the MPs leaving the object, can be seen by the increasing amount of black crosses. After the 18th step (lower middle image) the straight FRL is built at the x-coordinate of the easternmost pixel and the first pixels at the FRL became `Active2`. The corresponding MP marches southwards and accumulates the data gathered by the `Active1` MPs. Finally, the lower right image presents the situation after step number 26 when the southern `Active2` MP has turned to the state `Final` (see brighter colour at bottom pixel).

Figure 5 illustrates a sample simulation of flooding applied to more than one object. The simulation has been carried out with the SWARM2.2 environment [11]. The leftmost image shows the starting MPs at the western edges. Sets of MPs flooding as a wave over the objects can be seen in the second image, while the third image also shows MPs moving outside of objects. The final image presents the FRLs at the eastern sides of all objects.

3.5 Limits of Flooding and Further MP Algorithms

Figure 6 (left) shows a scenario where flooding fails. This is the case if two objects lie so close together that their convex hulls intersect. Then their FRLs will merge to one line. This situation can be avoided by another MP algorithm, in which two opposite waves of MPs are propagating from the western and the eastern edge pixels starting at the same time. Therefore we called this MP algorithm as *opposite flooding*. Figure 6 (right) illustrates the principle. A MP running from left to right may only proceed over non-object pixels, if they are marked by the diffusion procedure and in addition if a MP running in the opposite direction has passed them. The MPs running in the opposite direction are steered by an automaton realising the simple flooding algorithm. This prevents MPs from running over the right most object pixel in each row. As result not a straight RFL is formed but a RFL which is running exactly along the right most pixels in each row. The steering CA is much more complex compared to the flooding MP algorithm. For lack of space the details are not shown here, they can be found in [10]. Opposite flooding costs much more realisation effort on one side, but on the other side the algorithm is much more efficient than flooding. However, there is also a situation where opposite flooding fails.

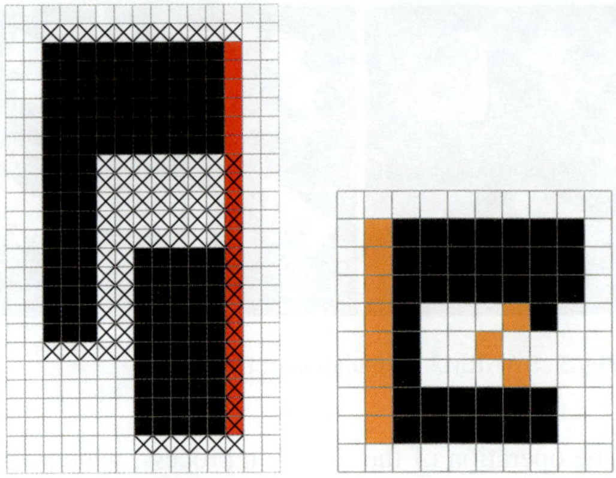

Fig. 6 Scenario where MP flooding algorithm fails (*left*); Scheme of MP opposite flooding algorithm. The reduction line runs from the top object pixel along the right edge to the bottom edge pixel (*right*)

This is given for objects with comb-like structures which are interleaved to each other. A further extended MP algorithm can manage such a situation. Its called the Pearl Chain algorithm (see Fig. 7). In this algorithm each MP is not more than one pixel ahead from its direct upper and lower neighbouring MP belonging to the same swarm of MPs which investigates a single object. It looks as if the MPs belonging to an object are chained together. This allows that MPs from different objects can interact with each other. This MP algorithm is functionally the most powerful one, but it causes the highest effort for the realisation. Details on the Pearl Chain algorithm can be found in [5], animations for all MP toolbox algorithms can be found in [6].

4 Outlook and Summary

In this article we presented distributed and parallel algorithms based on an emergent computing scheme we denoted as Marching Pixels. The application of MP

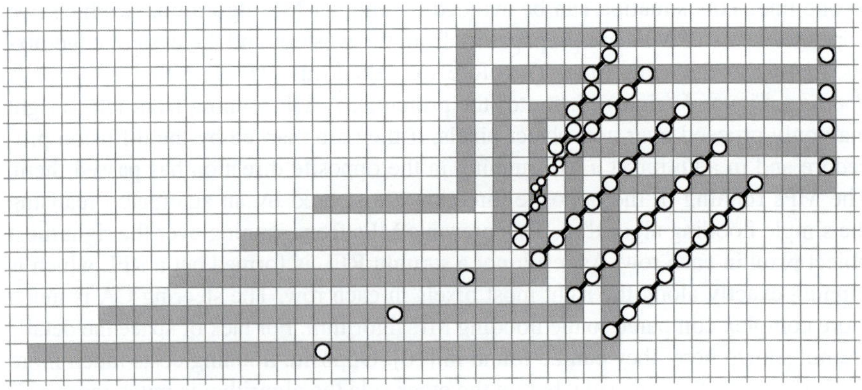

Fig. 7 Scheme of MP Pearl chain algorithm

algorithms lies in the field of smart camera chips to fulfil real-time requirements in industrial vision tasks. In Chap. 2.4 we show that response times in the ms range can be achieved with those algorithms. Their strength lies in their capability to process multiple objects simultaneously. Thanks to the emergent computing scheme their run time does not depend on the size of the image. Therefore, the algorithms are very scalable. In addition they offer robust solutions when one MP fails during its march, there will be other MPs that can take over the task of centroid detection. Furthermore the principle of MPs was also exploited for a fast path planning scheme in FPGAs by us [16]. It was also shown that these algorithms can be implemented very well on task parallel architectures like a GPGPU, too [12]. With an MP algorithm on a GPGPU it was possible to improve the runtime against other path planning algorithms running on a standard multi core processor by two orders of magnitude.

References

1. Canny, J.F.: A computational approach to edge detection. IEEE Trans. Pattern Anal. Mach. Intell. **8**(6), 679–698 (1986)
2. Chua, L.O., Roska, T.: Cellular Neural Networks and Visual Computing: Foundations and Applications. Cambridge University Press, New York (2002)
3. Chua, L.O., Yang, L.: Cellular neural networks: applications. IEEE Trans. Circuits Syst. **35**(10), 1273–1290 (1988)
4. Chua, L.O., Yang, L.: Cellular neural networks: theory. IEEE Trans. Circuits Syst. **35**(10), 1257–1272 (1988)
5. Fekete, S.P., Fey, D., Komann, M., Kröller, A., Reichenbach, M., Schmidt, C.: Distributed vision with smart pixels. In: Hershberger, J., Fogel, E. (eds.) Symposium on Computational Geometry, pp. 257–266. ACM, New York (2009)
6. Fey, D.: Marching pixels website demos @http://www3.informatik.uni-erlangen.de/Research/MarchingPixels/ (June 2009)
7. Fey, D., Schmidt, D.: Marching pixels: a new organic computing paradigm for smart sensor processor arrays. In: Bagherzadeh, N., Valero, M., Ramírez, A. (eds.) Proceedings of the Second Conference on Computing Frontiers, Ischia, Italy, 2005, pp. 1–9. ACM, New York (2005)
8. Geese, M., Dudek, P.: Autonomous long distance transfer on SIMD cellular processor arrays. In: 12th International Workshop on Cellular Nanoscale Networks and Their Applications, CNNA 2010, pp. 1–6 (2010)
9. Hordijk, W., Crutchfield, J.P., Mitchell, M.: Mechanisms of emergent computation in cellular automata. In: Eiben, A.E., Bäck, T., Schoenauer, M., Schwefel, H.-P. (eds.) PPSN. Lecture Notes in Computer Science, vol. 1498, pp. 613–622. Springer, Berlin (1998)
10. Komann, M.: Manual engineering and evolution of emergent algorithms for agents on two-dimensional grids. PhD thesis, University Erlangen-Nuremberg (2010)
11. Minar, N., Burkhart, R., Langton, C., Askenazi, M.: The Swarm simulation system: a toolkit for building multi-agent simulations. Technical report, Santa Fe Institute (1996)
12. Seidler, A.S.R., Schmidt, M., Fey, D.: Comparison of selected parallel path planning algorithms on GPGPUs and multi-core processors. In: Proceedings of the ADPC, pp. A133–A139 (2010)
13. Rivet, J.-P., Boon, J.: Lattice Gas Hydrodynamics. Cambridge University Press, Cambridge (2001)
14. Rodriguez-Vazquez, A., Linan-Cembrano, G., Carranza, L., Roca-Moreno, E., Carmona-Galan, R., Jimenez-Garrido, F., Dominguez-Castro, R., Meana, S.: ACE16k: the third generation of mixed-signal SIMD-CNN ACE chips toward VSoCs. IEEE Trans. Circuits. Syst. I, Regul. Pap. **51**(5), 851–863 (2004)

15. Roska, T., Rodriguez-Vazguez, A.: Towards visual microprocessors. Proc. IEEE **90**, 1244–1257 (2002)
16. Schmidt, M., Fey, D.: An optimized FPGA implementation for a parallel path planning algorithm based on marching pixels. In: International Conference on ReConFigurable Computing and FPGAs, ReConFig 2010 (2010)
17. Vilarino, D.L., Dudek, P.: Evolution of pixel level snakes towards an efficient hardware implementation. In: ISCAS, pp. 2678–2681 (2007)
18. Vilario, D.L., Rekeczky, C.: Implementation of a pixel-level snake algorithm on a CNNUM-based chip set architecture. IEEE Trans. Circuits Syst. I **51**, 885–891 (2004)
19. Wolfram, S.: A New Kind of Science. Wolfram Media Inc., Champaign (2002)

D. Fey (✉)
Department Computer Science, Chair for Computer Architecture,
Friedrich-Alexander-Universität Erlangen-Nürnberg, Nürnberg, Germany
e-mail: dietmar.fey@informatik.uni-erlangen.de

M. Komann
Institute of Computer Science, Chair for Computer Architecture and Communication,
Friedrich-Schiller-Universität Jena, Jena, Germany
e-mail: marcus.komann@googlemail.com

Chapter 6
Status and Outlook

From the humble beginnings with the first Organic Computing (OC) workshops in 2003 to the presentation of the results of the DFG Priority Programme (SPP) in 2011, OC has made considerable progress. This compendium reports scientific results achieved within the SPP. But beyond that, the SPP has also been very successful as an enabler to spark a plethora of additional activities in OC and neighbouring disciplines. It was a special wish of the initiators to build or integrate into international research networks interested in the technical applications of biological and nature-analogous paradigms. Organisational participations in international conferences and workshops are one of the main indicators of such common activities. OC has become a standard thread in a variety of international conferences such as

- ARCS (GI/ITG Conference on Architecture of Computing Systems)
- ICAC (International Conference on Autonomic Computing)
- SASO (IEEE International Conference on Self-adaptive and Self-organising Systems)
- IEEE Symposium Series on Computational Intelligence
- BICC (IFIP Conference on Biologically-Inspired Collaborative Computing)
- NaBIC (IEEE World Congress on Nature and Biologically Inspired Computing).

Three Dagstuhl seminars on various aspects of OC between 2006 and 2011 have been key in the discussion of long-term goals and visions for the further development of OC. It is our hope that OC will continue to thrive and develop. Fortunately, SPP participants have already started a variety of projects with OC background. Eleven such projects are briefly introduced in the first section of this chapter.

In the final article, the initiators of the SPP try to sketch out the future of OC: Organic Computing—Quo vadis? There we discuss some of the lessons learnt from the research done so far, and try to point out future directions of OC research.

Christian Müller-Schloer
Institute for Systems Engineering – System and Computer Architecture,
Leibniz Universität Hannover, Hannover, Germany
email: cms@sra.uni-hannover.de

Chapter 6.1.1
OC Techniques Applied to Solve Reliability Problems in Future 1000-Core Processors

Arne Garbade, Sebastian Weis, Sebastian Schlingmann, and Theo Ungerer

Keywords Reliability · Many-core systems · Future and emerge technologies · Organic computing · Self-healing

Autonomic and Organic Computing draws the attention to self-X properties to simplify management of complex, distributed systems. Typically an observer (system monitoring and analysis) and a controller (planning and execution) on top or integrated into the distributed system provide a high degree of autonomy to the system itself. Within the EC project TERAFLUX the research group at the University of Augsburg adapted the "Lazy Heartbeat" failure detector [3] developed in Priority Programme project OCμ (see Chap. 4.2) to detect faulty cores and interconnections of a future 1000-core processor. Furthermore, self-X techniques of Autonomic/Organic Computing [2] are applied to observe and control the reliability of cores and interconnections.

As shown in [4], shrinking the architecture size from 180 nm to 65 nm and downscaling of the supply voltage to 1.0 V leads to increasing failure rates up to 400%. The sheer number of installed components coupled with the shrinking transistor size will lead to error-prone processors. Due to the complexity of the processor's structure and economical constraints, design tests cannot completely ensure its functional correctness. This means that physical defects may be present at the shipping time of the chip. Moreover, ageing (caused by electromigration) and overheating can also lead to dynamically occurring hardware defects, which can partially be recognised by including self-testing units. Such units check the functionality of individual hardware blocks for correctness. But limited space per core restricts the use of self-testing units, which hence cannot ensure the functional correctness in all cases. Therefore, faults are likely to arise during the lifetime of the processors. Accordingly, errors caused by hardware defects can be less and less prevented in future processors. From the reliability perspective, it seems natural to apply self-healing

The research leading to these results has received funding from the European Community's Seventh Framework Programme (FP7/2007-2013) under grant agreement n°249013 (TERAFLUX).

C. Müller-Schloer et al. (eds.), *Organic Computing — A Paradigm Shift for Complex Systems*, 575–577
Autonomic Systems, DOI 10.1007/978-3-0348-0130-0_38,

principles to enable the construction of a reliable system out of unreliable components. We propose distributed Fault Detection Units (FDU), which are designed to detect faulty cores and interconnections of a chip by using heartbeats [3]. Heartbeats are messages that each core of a cluster has to send periodically to its affiliated FDU. A cluster is dynamically created and contains a number of cores and a supervising FDU. Within the FDU, the arrival time of the messages will be evaluated and compared with earlier arrival times from previous heartbeats. These interarrival times give some indication about the condition of the processor's interconnection network. Additionally, the heartbeats can encapsulate core state information about the current error rate, heat, and performance of a core. The information collected by the FDU will be used to manage the cluster and will also be made available to the Thread Scheduling Unit (TSU) that coordinates the thread distribution to the cores.

Besides the processing and disclosure of information, the reclustering of the cores is another task of the FDUs. An ideal cluster has a symmetric shape without holes. That eases the communication, because all messages will arrive at their destination device in a deterministic manner. However, due to faults on individual cores or interconnections, a cluster may become irregular. In such a case one or more cores of that cluster may be separated from the remaining cores and may be hard to reach due to congestions caused by bottlenecks in the interconnection network. The FDU will assign the separated cores to a neighbouring cluster. Since the code of an FDU is executed on one core of the cluster, an FDU itself may also become faulty. Therefore, the FDUs also have to monitor each other.

The gathered information of the cores cannot only be exploited to maintain a cluster's reliability, but also to support an optimised processor utilisation by the TSU. The FDU implementation is inspired by the MAPE-cycle [1] as it monitors cores, analyses the provided information, and draws first conclusions. Furthermore, it plans changes regarding the cluster structure and executes all actions needed in order to satisfy the plans. The execution includes adjusting the cores' configuration, reporting core states to the TSU, and setting up a new cluster structure.

References

1. Kephart, J.O., Chess, D.M.: The vision of autonomic computing. Computer **36**(1), 41–50 (2003)
2. Richter, U., Mnif, M., Branke, J., Müller-Schloer, C., Schmeck, H.: Towards a generic observer/controller architecture for Organic Computing. In: GI Jahrestagung (1), pp. 112–119 (2006)
3. Satzger, B., Pietzowski, A., Trumler, W., Ungerer, T.: A lazy monitoring approach for heartbeat-style failure detectors. In: ARES, pp. 404–409 (2008)
4. Srinivasan, J., Adve, S.V., Bose, P., Rivers, J.A.: The impact of technology scaling on lifetime reliability. In: International Conference on Dependable Systems and Networks, pp. 177–186 (2004)

A. Garbade (✉) · S. Weis · S. Schlingmann · T. Ungerer
Department of Computer Science, University of Augsburg, Universitätsstr. 6a,
86159 Augsburg, Germany
e-mail: garbade@informatik.uni-augsburg.de

S. Weis
e-mail: weis@informatik.uni-augsburg.de

S. Schlingmann
e-mail: schlingmann@informatik.uni-augsburg.de

T. Ungerer
e-mail: ungerer@informatik.uni-augsburg.de

Chapter 6.1.2
Dynamic Classification for Embedded Real-Time Systems

Florian Kluge, Jörg Mische, and Theo Ungerer

Keywords Embedded system · Hard real-time system · Organic computing

This article gives a brief overview of our architecture for *Dynamic Classification for Embedded Real-Time Systems* (DCERT) [1, 3]. The goal of DCERT is to introduce the concepts of Autonomic and Organic Computing (AC/OC) into the domain of embedded hard real-time systems. Such systems are usually resource-constrained and thus do not have much computing performance available beneath their main application. Also, the timing behaviour of this main application is crucial: it must never miss a deadline due to the hard real-time requirements. So any optimisation technique applied to such a system must not introduce any unpredictable influence on the real-time application's timing behaviour.

The basic idea of AC/OC is to embed a productive system into a control loop of observation and control. In most systems there is a vast range of parameters possibly relevant for such a control loop to optimise the system. However, in many cases it is sufficient to evaluate only few parameters to find a solution to a problem. Hence, the DCERT architecture is separated into two layers that are placed upon the productive system, similar to a multi-level Observer/Controller architecture (see Chap. 4.1). On the lower layer there are *module managers* which are bound to specific aspects or components of the productive system. They monitor the few parameters of their aspect and can react locally based on a small and static problem-specific rule set. Only if this local reaction cannot overcome a nuisance, a *global manager* is notified. For this purpose the module managers pre-process their current parameter values and transform them into a domain common to all module managers.

The global manager can access the monitored and pre-processed parameters of all available module managers. Thus it has a comprehensive view of the underlying system's current state. Using a generic algorithm, the global manager classifies the current situation and selects a reaction which is executed in one of the module managers. Thus the global manager creates a kind of interaction between the module managers.

The algorithm of the global manager is based on the ideas of automatic planning and of Learning Classifier Systems (LCS). As mentioned above, we have to ensure

C. Müller-Schloer et al. (eds.), *Organic Computing — A Paradigm Shift for Complex Systems*, 579–581
Autonomic Systems, DOI 10.1007/978-3-0348-0130-0_39,

that any reaction applied to the productive real-time system will not have a negative influence on the system's behaviour. Such guarantees can be gained by using actions with known preconditions and effects, like in automatic planning. However, the planning algorithms themselves are usually very complex. We represent preconditions and effects of actions using strings of *true, false and don't-care* symbols known from LCS. This enables a very efficient implementation of the classification. During one classification cycle DCERT selects only one action for reaction. In most situations this is sufficient and thus DCERT achieves an improvement of the system's state at a low overhead. Also, the DCERT algorithm allows to dynamically change the available monitors and actors during runtime.

We implemented the DCERT architecture on the simultaneous multithreaded CarCore processor [4], which is available as SystemC simulation and as well as FPGA prototype. The CarCore's integrated hard real-time scheduler allows to run helper threads in parallel to a hard real-time application without influencing the application's timing behaviour. DCERT is implemented as such a helper thread. The operation system CAROS [2] ensures a further separation of applications regarding memory and resource usage. By integration with the CARISMA middleware [5] it is possible to fit embedded real-time networks with organic capabilities. Further research in applying OC concepts in real-time embedded systems will be conducted.

Acknowledgements This research work was funded by the German Science Foundation (DFG), project "CAR-SoC".

References

1. Kluge, F.: Autonomic- und Organic-Computing-Techniken für eingebettete Echtzeitsysteme. PhD thesis, Universität Augsburg, July 2010
2. Kluge, F., Mische, J., Uhrig, S., Ungerer, T.: An operating system architecture for organic computing in embedded real-time systems. In: 5th Int. Conf. on Autonomic and Trusted Computing (ATC 2008), Oslo, Norway, June 2008, pp. 343–357. Springer, Berlin (2008)
3. Kluge, F., Uhrig, S., Mische, J., Ungerer, T.: A two-layered management architecture for building adaptive real-time systems. In: 6th IFIP Workshop on Software Technologies for Future Embedded & Ubiquitous Systems (SEUS 2008), Capri Island, Italy, pp. 126–137. Springer, Berlin (2008)
4. Mische, J., Guliashvili, I., Uhrig, S., Ungerer, T.: How to enhance a superscalar processor to provide hard real-time capable in-order SMT. In: 23rd Int. Conf. on Architecture of Computing Systems (ARCS 2010), Hanover, Germany, Feb. 2010, pp. 2–14. Springer, Berlin (2010)
5. Nickschas, M., Brinkschulte, U.: Guiding organic management in a service-oriented real-time middleware architecture. In: 6th IFIP Workshop on Software Technologies for Future Embedded & Ubiquitous Systems (SEUS 2008), Capri Island, Italy, pp. 90–101. Springer, Berlin (2008)

F. Kluge (✉) · J. Mische · T. Ungerer
Department of Computer Science, University of Augsburg, Universitätsstraße 6a, 86159 Augsburg, Germany
e-mail: kluge@informatik.uni-augsburg.de

J. Mische
e-mail: mische@informatik.uni-augsburg.de

T. Ungerer
e-mail: ungerer@informatik.uni-augsburg.de

Chapter 6.1.3
On the Future of Chemistry-Inspired Computing

Peter Kreyssig and Peter Dittrich

Keywords Self-organisation · Chemical computing · Unconventional computing · Rule-based models · Membrane computing · Protocells · Molecular computing

Various research projects have successfully demonstrated how (bio-)chemical information processing principles can inspire the design of novel self-organising computing systems with various self-* properties (e.g., [8, 9]). In particular, within the Priority Programme "Organic Computing", artificial hormones have been applied for the formation of computational "organs" in task allocation (Chap. 4.4), for load balancing within the middle ware OCμ, and in hierarchical monitoring infrastructure for heterogeneous adaptive computing systems (Chap. 2.3). The chemical metaphor allows to easily import theory from chemistry and systems biology (Chap. 2.6). This theory can guide the design of the rule-system and provides insight into the dynamics. Apart from the advancement made so far many issues and challenges remain:

- Develop suitable programming languages and environments (especially for rule-based chemical programming languages using structured molecules, e.g., Fraglets [9] and Chap. 2.3), efficient run-time systems, and therefore make the approach accessible to the professional in a coherent way.
- Explore further application areas, like knowledge technologies and work flow systems [8], artificial developmental systems (e.g., distributed robotics and Chap. 4.4) showing advantages, and differences with other approaches.
- Develop a theory for the emergence of function from local chemical program rules, which encompasses quantitative elements (Chap. 1.4).

The chemical metaphor is not only used for the design of self-organising computing systems, but has also inspired abstract models of parallel computation (e.g., the Chemical Abstract Machine and P-systems). Moreover there is a highly active research field where novel chemical, bio-chemical, and biological computational substrates are developed (e.g., the current EU FET initiative CHEM-IT [1, 4–6]). Very likely, the molecular computing approaches developed in this field will not compete with silicon based conventional computers in solving hard computational

C. Müller-Schloer et al. (eds.), *Organic Computing — A Paradigm Shift for Complex Systems*, 583–585
Autonomic Systems, DOI 10.1007/978-3-0348-0130-0_40,

tasks. Their innovative potential is rooted in the integration of information processing and material processing at a micro or even nano scale, e.g., self-assembling and self-healing materials, molecular factories, smart drugs, protocells, and artificial immune system cells [1].

Although the aims of in-silico and real chemical computing are fundamentally different, the mechanisms of information processing are basically the same from a structural science point of view [7], i.e., they share the same mechanisms for design, analysis, and programming. Future work on the theoretical base of in-silico and real chemical IT should consider this overlap in order to obtain general integrating theories and formal methods. This should, for example, lead to methods allowing to describe a "chemical program" independent of its substrate, whether it is executed as an artificial hormone system in-silico or within a bunch of lipid-based protocells. Two advantages will be possessing a formal science bridging different domains and providing a common ground when in-silico technologies and molecular IT converge [3].

This convergence implies further questions for future research: How can wet chemical IT benefit from formal chemical-like approaches, e.g., what are possible applications of concepts like P-systems to real chemical computing systems? How can in-silico chemical approaches like membrane computing or hormone systems benefit from chemical IT, e.g., can we find ideas or problems from real chemical IT interesting for a theoretical method of resolution? Could it make sense to hybridise in-silico and real chemical computing? This endeavour requires the collaboration of experts from many different disciplines, whose coordination is on its way, e.g., COBRA [2].

Acknowledgements We acknowledge financial support by the DFG (grant Di 852/4, CHEMORG) and the EC (NEUNEU: ICT-2009.8-3: FET Proactive 3, CHEM-IT and COBRA: ICT-2009.8-9).

References

1. BACTOCOM: Bacterial Computing with Engineered Populations (2010). http://www.bactocom.eu/
2. COBRA: Coordination of Biological and Chemical It Research Activities (2010). http://www.cobra-project.eu/
3. EECell: Electronic Chemical Cells (2010). http://www.bactocom.eu/
4. FET Proactive Initiative 7: CHEM-IT (2010). http://cordis.europa.eu/fp7/ict/fet-proactive/chemit_en.html
5. MATCHIT: Matrix for Chemical IT (2010). http://fp7-matchit.eu/
6. NEUNEU: Artificial Wet Neuronal Networks from Compartmentalised Excitable Chemical Media (2010). http://www.bactocom.eu/
7. Artmann, S.: Historische Epistemologie der Strukturwissenschaften. Wilhelm Fink (2010)
8. Fernández, H., Priol, T., Tedeschi, C.: Decentralized approach for execution of composite web services using the chemical paradigm. In: 8th International Conference on Web Services (ICWS 2010), Miami, USA, July 5–10. IEEE Press, New York (2010)
9. Tschudin, C.: Fraglets-a metabolistic execution model for communication protocols. In: Proc. 2nd Annual Symposium on Autonomous Intelligent Networks and Systems (AINS), Menlo Park, USA (2003)

P. Kreyssig · P. Dittrich (✉)
Bio Systems Analysis Group, Jena Centre for Bioinformatics and Department of Mathematics and Computer Science, Friedrich Schiller University Jena, Ernst-Abbe-Platz 1-3, 07743 Jena, Germany
e-mail: peter.dittrich@uni-jena.de

P. Kreyssig
e-mail: peter.kreyssig@uni-jena.de

Chapter 6.1.4
Agent-Based Thermal Management for Multi-core Architectures

Thomas Ebi, Janmartin Jahn, and Jörg Henkel

Keywords Dynamic thermal management · Agent-based systems · Multicore architectures

Technology scaling has resulted in a decrease in feature size which comes at the cost of higher power densities on a given area of a chip (e.g., on average, an increase from 2 W/mm^2 in 65-nm technology to 7.2 W/mm^2 for 45-nm technology [4]). These power densities and their resulting increase in temperature variations directly have a negative impact on system reliability by resulting in both intermittent and permanent faults (e.g. due to temperature-dependent electromigration), which reduces the reliability of the designed chips. For example, reducing the feature size by 66% (e.g. from 180 nm to 65 nm) increases the temperature of the hottest structure from 342 K to 356 K assuming a fixed number of transistors. This results in a reduction of the mean-time-to-failure (MTTF) by 76% [5].

In our dynamic thermal management (DTM) approach [1], we cope with the complexity incurred by large-scale multi-processor systems—expected to reach a processor count in the hundreds or even thousands in the near future—by introducing an agent hierarchy which reduces the problem space of thermal management from a global problem to one limited to *clusters*, thereby keeping the approach scalable [2]. We define a cluster as a connected subset of tiles which can be dynamically resized, created, and destroyed depending on application requirements. All tasks of an application are limited to one cluster, but there may be multiple applications per cluster. Our agent hierarchy is comprised of three layers whose functionality is summarised in Fig. 1.

The key for providing Self-X properties in agent-based systems lies in their negotiation techniques. Through these, several agents working together exhibit emergent global behaviour while relying on simple local rules. In our approach, the cluster agent (CA) is triggered at runtime by a tile agent when a tile located inside its cluster reaches a threshold temperature. The CA first tries to reduce the temperature by remapping tasks inside the cluster. However, if this is not possible, the CA begins to negotiate with the CAs of adjacent clusters.

During negotiation, all neighbouring CAs are asked if they can give up a tile which is adjacent to the current cluster. Once these tiles have been determined, the

C. Müller-Schloer et al. (eds.), *Organic Computing — A Paradigm Shift for Complex Systems*, 587–588
Autonomic Systems, DOI 10.1007/978-3-0348-0130-0_41,

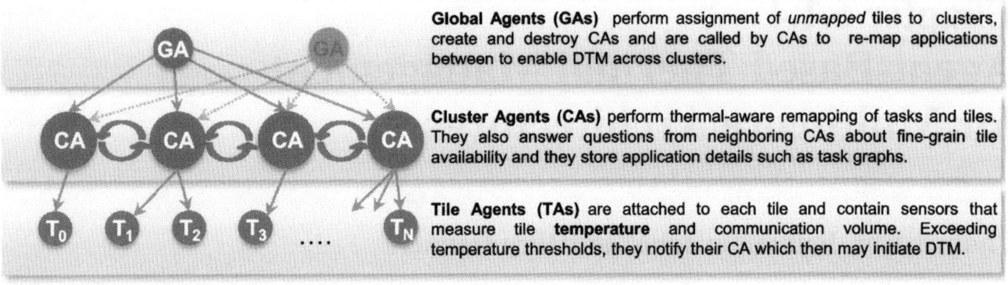

Fig. 1 Agent hierarchy composition

one with the lowest remapping cost is chosen, and the tile is then removed from its cluster and added to the current set of tiles of the new cluster, realising reclustering. This new tile allows the CA to offload computation onto it using a lightweight context-aware task migration mechanism [1], which results in lower power densities through the spreading of power consumption. In the case where this is not enough to deal with the thermal hotspot, the CA must inform one of the global agents (GA) which then negotiates with all CAs to determine which cluster is best suited for the application. The GA chooses the CA that reports the lowest cost and the application is then migrated to the respective cluster. By exploiting the locality of clusters, we are able to reduce the remapping communication volume by a factor of 4.2 compared to a centralised approach.

References

1. Al Faruque, M.A., Jahn, J., Ebi, T., Henkel, J.: Runtime thermal management using software agents for multi- and many-core architectures. IEEE Des. Test Comput. **27**(6), 58–68 (2010)
2. AlFaruque, M.A., Krist, R., Henkel, J.: ADAM: run-time agent-based distributed application mapping for on-chip communication. In: ACM/EDAC/IEEE Design Automation Conference (DAC), pp. 760–765 (2008)
3. Ebi, T., Al Faruque, M.A., Henkel, J.: TAPE: thermal-aware agent-based power economy for multi/manycore architectures. In: IEEE/ACM Int'l Conf. Computer-Aided Design (ICCAD), pp. 302–309 (2009). Best Paper Award
4. Link, G.M., Vijaykrishnan, N.: Thermal trends in emerging technologies. In: Int'l Symp. Quality Electronic Design (ISQED), pp. 625–632 (2006)
5. Srinivasan, J., et al.: Lifetime reliability: toward an architectural solution. IEEE MICRO **25**(3), 70–80 (2005)

T. Ebi (✉) · J. Jahn · J. Henkel
Karlsruhe Institute of Technology, Kaiserstr. 12, 76131 Karlsruhe, Germany
e-mail: thomas.ebi@kit.edu

J. Jahn
e-mail: jahn@kit.edu

J. Henkel
e-mail: henkel@kit.edu

Chapter 6.1.5
Trust Management—Handling Uncertainties in Embedded Systems

Werner Brockmann, Andreas Buschermöhle, Jens Hülsmann, and Nils Rosemann

Keywords Organic computing · Trust management · Dynamic uncertainties

Modern embedded systems increasingly permeate applications, in which a safe and trustworthy operation is a must, e.g. robots in human interaction or car driver assistance systems. These applications do not only increase in complexity, but also in uncertainties they have to cope with. These uncertainties arise during the design process, e.g. because of missing (formal) models or a lack of concrete knowledge of system properties and of its environment. But uncertainties arise also especially at runtime operation. The sources of uncertainty are therefore manifold and can be categorised as follows:

- perception uncertainties, e.g. sensor noise, drift, outliers, and faults,
- informational uncertainties, e.g. unknown or conflicting information and dynamically changing systems parameters,
- insufficient specification, e.g. suboptimal or incomplete parameterisation at design time or due to online learning, and
- interaction uncertainties, e.g. unforeseeable interaction patterns of an actuator with its environment like slippage or blockage.

Because all these different sources of uncertainty have to be dealt with at once, trust or trustworthiness, respectively, becomes a central concept which has to be addressed both at design time as well as at runtime. The aim of our *trust management* approach is thus to build up a framework to increase dependability and engineerability by making uncertainties explicit and by taking them dynamically into account during runtime operation.

We therefore build on the ORCA-architecture [3] which was introduced within the Organic Computing initiative as a control architecture for dependable robotic systems. Besides online learning, one key feature of the ORCA architecture is monitoring the system operation and detecting as well as signalling anomalies and disturbances by health signals. During the investigations of the ORCA architecture it turned out that the health signal principle is more generally applicable to other kinds of application, e.g. control and automation in general. And what is more, a

C. Müller-Schloer et al. (eds.), *Organic Computing — A Paradigm Shift for Complex Systems*, 589–591
Autonomic Systems, DOI 10.1007/978-3-0348-0130-0_42,

separation into specific observing modules, called OCUs, and functional modules (BCUs), which allow full system functionality without the OCUs, is no longer appropriate.

Within the trust management framework, the notion of health signals is hence generalised to so-called *trust signals* because health signals only address a subset of the above mentioned uncertainties, namely abnormal deviations of the normal operation (see Chap. 4.5). Another key feature is that trust signals are now integrated intrinsically into the normal system operation. This allows addressing not only anomalies, but also uncertainties in general at design time and at runtime in an intuitive way. The trustworthiness of the operation of an embedded system is hence modelled explicitly at a very general level. That is why we call this way *trust management* [2].

Up to now, health signals were applied only within the ORCA architecture for robust, behaviour-based robot control and its framework of controlled self-optimisation. In order to lift the health signal mechanism to the more general level of trust management, its concepts have to be generalised and investigated also in the light of different classes of algorithms. The aim is to get a deeper understanding, especially of the trust level semantics in different application domains, and to improve dependability and engineerability issues further. Current work hence addresses more diverse control applications, especially industrial ones, and data analysis.

Building appropriate tools and deriving suitable engineering guidelines requires a deeper understanding and formalisation of the health signals semantics and processing. Another aim is hence to work out general theories and formal proofs in order to give warranties of certain system properties at design time, e.g. robustness, dependability, or anomalies the system is capable to counteract. Here more information-theoretical concerns get relevant, e.g. to answer the question how far and under which requirements the concepts and algorithms of trust management are applicable in a generic way. Thus, theories like general information theory [4] and info-gap theory [1] will be investigated in depth. Finally, our aim is to build a framework for trust management.

References

1. Ben-Haim, Y.: Info-Gap Decision Theory: Decisions Under Severe Uncertainty. Academic Press, Oxford (2006)
2. Brockmann, W., Buschermöhle, A., Hülsmann, J.: A generic concept to increase the robustness of embedded systems by trust management. In: Proc. IEEE Conf. Systems, Man, and Cybernetics, SMC, pp. 2037–2044 (2010)
3. Brockmann, W., Maehle, E., Mosch, F.: Organic Fault-Tolerant Control Architecture for Robotic Applications. Technical report, 4th IARP/IEEE-RAS/EURON Workshop on Dependable Robots in Human Environments, Nagoya, Japan, June 16–18, 2005, Nagoya University, Japan (2005)
4. Klir, G.J.: Generalized information theory. Fuzzy Sets Syst. **40**(1), 127–142 (1991)

W. Brockmann · A. Buschermöhle · J. Hülsmann · N. Rosemann (✉)
Institute of Computer Science, University of Osnabrück, Albrechtstrasse 28,
49069 Osnabrück, Germany
e-mail: nils.rosemann@uni-osnabrueck.de

Chapter 6.1.6
OC-Trust: Towards Trustworthy Organic Computing Systems

Jan-Philipp Steghöfer and Wolfgang Reif

Keywords Trust · Self-organisation · Usability · Safety · Functional correctness · Reliability · Credibility

The development of Organic Computing systems towards more open environments, towards an increasing heterogeneity of participating entities, into safety- or mission-critical settings, and into domains that require involvement of human users necessitates a perspective on these systems that incorporates different facets of trust. Functional correctness, safety, security, reliability, credibility, and usability go hand in hand to ensure that openness, complex interactions of many entities, and user involvement can be dealt with [1]. In some cases, trust can be ensured a priori by certifying adherence to certain standards or giving formal proofs with regard to a system's functional correctness, security, safety, or reliability. Other facets however, like credibility of agents, actual reliability, or ease of use and effectiveness of the interface require repeated interactions between the agents or between the user and the system. Trust can thus only be built gradually. Additionally, due to the dynamic nature of Organic Computing systems, even those facets considered at design time need to be appraised during runtime to ensure correct, safe, and reliable operation under any conditions that may occur.

The aim of the DFG research group OC-Trust is to create provisions that help develop systems that can be deemed trustworthy by their users, by other systems interacting with them, and by authorities and organisations that certify and deploy systems in safety- or mission-critical environments. The means to achieve this are fourfold:

Control of emergent behaviour: One of the main advantages of Organic Computing systems is that they exhibit emergent behaviour. However, sometimes such behaviour is detrimental to a system and thus has to be limited by analysing, verifying, and restraining the interactions between agents during design or runtime and fostering cooperative behaviour.

Trustworthy self-x algorithms: Self-x mechanisms in open systems need to prevent malicious or unintentional exploitation or defection. Furthermore, values for credibility and reliability can help in dealing with uncertainty in the data the algorithms' decisions are based on.

C. Müller-Schloer et al. (eds.), *Organic Computing — A Paradigm Shift for Complex Systems*, 593–595
Autonomic Systems, DOI 10.1007/978-3-0348-0130-0_43,

Models and metrics for trust in OC systems: To be able to autonomously react to changes in the relationships between the entities in a system or the user and a system, trust has to be monitored and measured continuously during runtime. The measurements will have to be evaluated and decisions will have to be made to restore the trust relationships.

Design of trustworthy OC-systems: Trust and its facets have to be regarded as principal issues during the software engineering process for trustworthy Organic Computing systems. Additionally, a focus on user interface design is necessary to ensure usability of a self-organising system.

The techniques developed in the course of the project are demonstrated with three case studies. All applications are based on the Trust-Enabling Grid, an underlying middleware framework that provides interfaces for communication as well as access to trust models and metrics.

The *Trusted Computing Grid* deals with open, social agent environments. Cooperation of an agent is rewarded by inclusion in a Trusted Community, a self-organising group of agents known to be trustworthy from previous interactions. Observations of the other agents' behaviour regarding credibility and reliability are recorded in a reputation management system. Egoists are effectively excluded from the community, making the system robust against uncooperative agents and increasing performance within the community.

In the *Trusted Energy Grid*, a large-scale, safety-critical system, power plants are controlled in an autonomous, decentralised fashion. The major challenge is uncertainty introduced by energy sources based on natural resources like wind and sun and by a fluctuating energy demand. The system has to plan and operate under these conditions, maintain safety, and cover demand by dynamically adapting and planning the controllable power plants' output.

Multi-user multi-device environments let users interact with a system on public and private devices. When several participants are using the same device, privacy concerns as well as usability come into play. The system has to decide on which device controls and information have to be displayed and which interaction modes should be available to the user. User preferences guide these decisions which are evaluated at runtime on a dynamic user model.

Achieving the goals of OC-Trust will be a big step towards taking Organic Computing systems out of the research laboratories and bringing them to innovative software companies and finally into real world applications.

References

1. Steghöfer, J.-P., Kiefhaber, R., Leichtenstern, K., Bernard, Y., Klejnowski, L., Reif, W., Ungerer, T., André, E., Hähner, J., Müller-Schloer, C.: Trustworthy organic computing systems: Challenges and perspectives. In: Xie, B., Branke, J., Sadjadi, S., Zhang, D., Zhou, X. (eds.) Autonomic and Trusted Computing. Lecture Notes in Computer Science, vol. 6407, pp. 62–76. Springer, Berlin (2010)

J.-P. Steghöfer (✉) · W. Reif
Institute for Software & Systems Engineering, Universität Augsburg, Universitätsstrasse 6a, 86159 Augsburg, Germany
e-mail: steghoefer@informatik.uni-augsburg.de

W. Reif
e-mail: reif@informatik.uni-augsburg.de

Chapter 6.1.7
Emergence in Action

Matthias Werner, Jan Richling, and Hans-Ulrich Heiss

Keywords Cyber-physical systems · Distributed systems · Emergence · Constraints · Real space-time

1 Cyber-physical Systems

Cyber-physical systems (CPS) are distributed systems that incorporate physical elements of the environment into cooperating computing systems, cf. [2]. Designing cooperating systems that have to consider physical processes entails the requirement to deal with real space and real time, i.e., the system has to be location/motion-aware and real-time capable.

In general, there are many approaches to design distributed systems. We can identify two extreme cases: (a) *Purely protocol-based*: The global behaviour is described by protocols, i.e., by enumeration of interactions between the components. Such protocols tend to become very complex (and thus error-prone) and/or to limit the freedom of the system, possibly avoiding useful behaviour. (b) *Swarm intelligence*: The designer sets rules for the local elements. The global behaviour *emerges* from the sum of local behaviours. This approach gives the system more degrees of freedom, but usually requires a rather high degree of redundancy. In addition, it is a tough challenge to decompose a desired behaviour to local rules. We have chosen to strike a balance between these two extreme cases. We assume a distributed system, composed of a number of (possibly mobile) computing nodes where each node may be equipped with sensors and/or actuators. In cyber-physical systems, the objects of interest are non-computing objects of the physical world. In classical embedded system design, these objects would be called "controlled objects".[1]

2 Actions

The aim of an application is to control one or more controlled objects in a desired way. This is done by actions of actuators that have a (direct or indirect) impact on

[1] *Kopetz* uses the term *real-time entities* (RTE) [1].

C. Müller-Schloer et al. (eds.), *Organic Computing — A Paradigm Shift for Complex Systems*, 597–599
Autonomic Systems, DOI 10.1007/978-3-0348-0130-0_44, © Springer Basel AG 2011

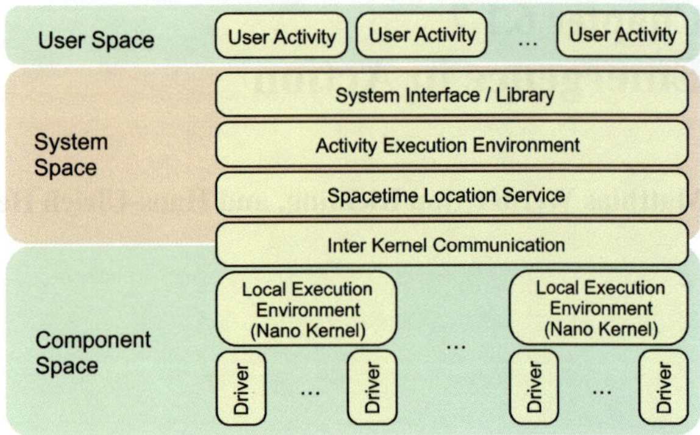

Fig. 1 Run-time system

the controlled object. Since we are dealing with real time and space, it is important that the actions are performed in a timely manner and at certain places in the space. However, it does *not* matter which node performs a specific action as long as it is capable to do so, i.e., it owns the proper actuators and does it at proper space-time.

For that reason, our approach allows to specify actions in the usual imperative way, but we constrain these actions by *spatiotemporal conditions*. In that way, we add a virtual layer. In our model, the programmer is able to keep a systemic view (i.e., *what* has to be done with the controlled object, but not *who* does it) and can abstract from, e.g., the actual number of computing nodes or their corresponding capabilities. The actual behaviour is computed at run time with the help of a constraint-driven spatiotemporal scheduler. At run time, the flow of actions may migrate from one node to another, and actions may even happen concurrently, without any explicitly specified concurrency. Thus, the actual behaviour *emerges* at run time.

3 Run-Time System

To support our model, we are developing a run-time system that allows considerations of spatiotemporal constraints at run time. The architecture of the run-time system consists of three layers (cf. Fig. 1): a component space that exists in single instances at every computing node; a shared system space that provides services as constraint-based scheduling; and the user space where the applications (called activities) reside and which is shared as well. We are also developing a tool chain that parses the constraint-enriched code and allows to create individual instances from non-distributed and non-concurrent code. More details as well as application examples can be found, e.g., in [3].

References

1. Kopetz, H.: Real-Time Systems—Design Principles for Distributed Embedded Applications. Kluwer Academic, Norwell (1997)

2. Lee, E.A.: Cyber-physical systems—are computing foundations adequate? In: Position Paper for NSF Workshop On Cyber-Physical Systems: Research Motivation, Techniques and Roadmap, October 2006
3. Werner, M., Richling, J., Müller, D., Mühl, G.: Towards a holistic approach for distributed real space-time systems. In: International Workshop on Dependable Network Computing and Mobile Systems at the 27th International Symposium on Reliable Distributed Systems, pp. 63–67 (2008)

M. Werner
Operating Systems Group, Chemnitz University of Technology, Strasse der Nationen 62, 09111 Chemnitz, Germany
e-mail: mwerner@informatik.tu-chemnitz.de

J. Richling (✉) · H.-U. Heiss
Communication and Operating Systems, Berlin University of Technology, Einsteinufer 17, 10587 Berlin, Germany
e-mail: richling@cs.tu-berlin.de

H.-U. Heiss
e-mail: heiss@cs.tu-berlin.de

Chapter 6.1.8
Organic Computing in Off-highway Machines

Sanaz Mostaghim, Hartmut Schmeck, Micaela Wünsche, Marcus Geimer, and Timo Kautzmann

Keywords Generic Observer/Controller architecture · Application · Mobile machine · Learning

This article is about an application of Organic Computing (OC) on machine management systems that enables an off-highway machine (such as a tractor and a wheel loader) to adapt to new situations and optimise the fuel consumption accordingly. Fuel consumption is one of the major issues in vehicle technology and particularly in mobile machinery. The state of the art machine management systems are developed to reduce fuel consumption by providing strategies and parameter settings that ensure efficient coordination of the machine's components in different operating modes [2, 3]. However, these systems are designed to work in specific pre-optimised modes and if they are used in situations for which they have not been specifically designed, fuel consumption may increase dramatically. In addition, off-highway machines like tractors are intended for multi-purpose use, making it especially difficult to optimise them for all eventualities. Our goal is to illustrate how concepts from OC may be used to realise a self-organising, reliable, adaptive, and robust machine management system that is capable of adjusting to new situations.

Different from the existing adaptive control systems used in mobile machineries such as Model Reference Adaptive Control (MRAC), Self Tuning Regulators (STR) and gain scheduling, in OC we consider the mobile machine as a collective of many subsystems (each individually controlled) which work together to achieve a common goal. We adapt the generic Observer/Controller (O/C) architecture from OC [1] (Chap. 4.1) and present the off-highway machine as the System under Observation and Control (SuOC) as shown in Fig. 1. In contrast to other typical SuOCs to which the generic O/C architecture has already been successfully applied (e.g., Chap. 5.1), an off-highway machine does not consist of several separate entities. Instead, there are the components of a single machine whose interactions are closely interrelated. In off-highway machines, those components are the traction drive, the power take-off, the hydraulic system, as well as varying auxiliary components. These units are mechanically or fluidically connected or they communicate and interact, e.g., via bus systems.

C. Müller-Schloer et al. (eds.), *Organic Computing — A Paradigm Shift for Complex Systems*, 601–603
Autonomic Systems, DOI 10.1007/978-3-0348-0130-0_45, © Springer Basel AG 2011

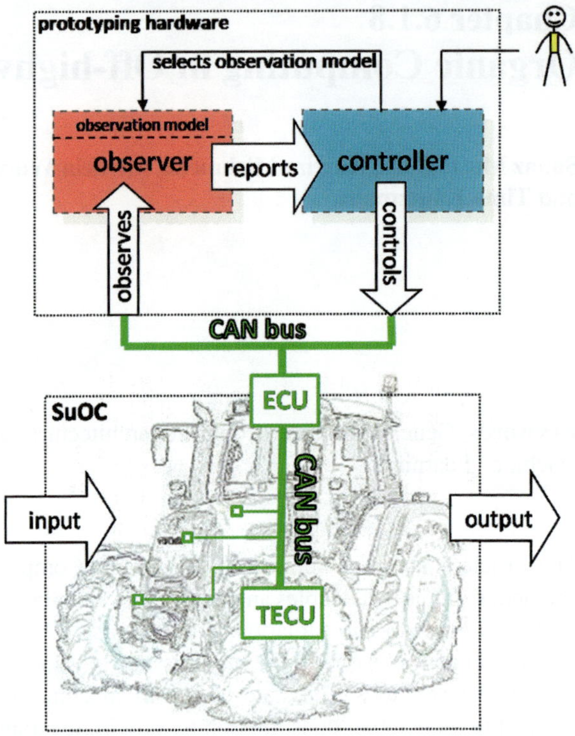

Fig. 1 Off-highway machine as SuOC in O/C architecture (ECU refers to Electronic Control Unit and TECU indicates the Tractor ECU)

The input to the SuOC consists of a series of input variables such as fuel, driver interactions and environmental conditions like changing subsoil or varying peripheral components connected to the machine. The main output of the system which can be measured is its fuel consumption. Sensors in the SuOC measure input and output of the system, as well as selected parameters of its components such as torque of rear and front axles. The possibilities for the controller to interact with the system consist of the setting of reference values for conventional controllers within the machine, or communication with the driver by providing suggestions how to increase efficiency.

This application of OC opens a full spectrum of challenges. In the observer, we deal with a large amount of large-scale data streams from which a valid characterisation of the system state with respect to the target function (fuel consumption) must be derived. The identification of suitable sensor streams for this task is done in a simulation model of the tractor and all its subsystems. The vector representing the system state is clustered online by the observer. As the system must be able to react to previously unknown situations, the number of clusters is not known a priori and the clustering algorithm must be able to adapt accordingly. The major issues in the controller are on-line and off-line learning, specifically the design and integration of a Simulink simulation model within the controller, that will serve as a basis for evaluating the rules in the controller's situation/action mapping. Beyond the aspects in the O/C architecture, it is required to separately or simultaneously handle alternative and conflicting objective functions such as fuel consumption and exhaust emission NO_x.

References

1. Branke, J., Mnif, M., Mueller-Schloer, C., Prothmann, H., Richter, U., Rochner, F., Schmeck, H.: Organic computing—addressing complexity by controlled self-organization. In: Post-Conference Proceedings of the 2nd International Symposium on Leveraging Applications of Formal Methods, Verification and Validation (ISoLA 2006), pp. 185–191. IEEE Press, New York (2006)
2. Forche, J.: Management hydraulischer Antriebe in mobilen Arbeitsmaschinen. In: Landtechnik, pp. 239–244 (2003)
3. Mariutti, H.: Lastkollektive fuer die Fahrantriebe von Traktoren mit Bandlaufwerken (Fortschritt-Berichte VDI). VDI Verlag (2003)

S. Mostaghim (✉) · H. Schmeck · M. Wünsche · M. Geimer · T. Kautzmann
Karlsruhe Institute of Technology, Kaiserstr. 89, 76133 Karlsruhe, Germany
e-mail: sanaz.mostaghim@kit.edu

H. Schmeck
e-mail: hartmut.schmeck@kit.edu

M. Wünsche
e-mail: micaela.wuensche@kit.edu

M. Geimer
e-mail: marcus.geimer@kit.edu

T. Kautzmann
e-mail: timo.kautzmann@kit.edu

Chapter 6.1.9
Decentralised Energy Management for Smart Homes

Florian Allerding, Birger Becker, and Hartmut Schmeck

Keywords Energy management · Smart-home · Smart-grid · Observer/controller architecture

The climate protection targets of the German government aim to cover 35% of the electricity consumption from renewable sources by 2020. This growing share of fluctuating renewable energy sources will cause a higher demand for flexible power suppliers and consumers. The German national development plan for electric mobility predicts one million electric vehicles on German roads by 2020 which may significantly contribute to satisfying this demand.

Within the research project *MeRegioMobil* an energy management system (EMS) has been developed to adapt the electric power demand of smart homes supporting the challenge of balancing supply and demand in the electric grid. External signals, reflecting the state of the low voltage grid, are sent to smart homes, which are able to adapt their energy demand automatically without restraining the smart home's resident. For this purpose, the load of single devices is shifted within certain degrees of freedom complying with specific constraints specified by the appliance or by the user [2].

To manage the smart home's appliances and to efficiently integrate electric vehicles as energy storages, decentralised measure and control systems are investigated. The EMS is based on a hierarchical Observer/Controller (O/C) architecture. The aim is a mostly self-organised system which reduces the interaction between the smart home resident and his appliances to a minimum, while still allowing explicit interference by the resident.

Each power consumer in the smart home has a local O/C unit that measures its relevant data, filters the data, and provides it to the global observer. The global observer analyses the collected data and generates a prediction of the future energy consumption. The results are communicated to the global controller, which additionally receives external signals (like price signals or user constraints). Finally, the global controller calculates the optimal operation time of each appliance in the smart home and sends control signals to the local O/C units. These units can have their own control functions, e.g. the local O/C unit of the electric vehicle starts the

C. Müller-Schloer et al. (eds.), *Organic Computing — A Paradigm Shift for Complex Systems*, 605–607
Autonomic Systems, DOI 10.1007/978-3-0348-0130-0_46,

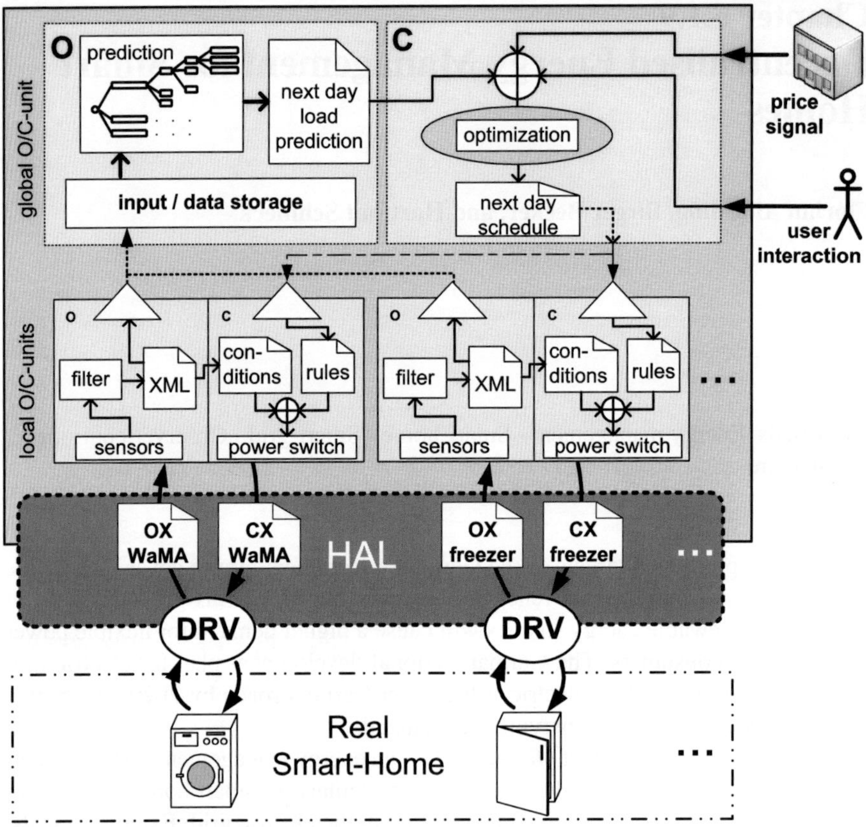

Fig. 1 O/C architecture for smart homes

charging process autonomously, if the battery's state of charge (SOC) is below a specific threshold value.

Within the architecture, a hardware abstraction layer (HAL) adapts the different communication protocols of the appliances which are connected via individual drivers. For simulating the smart home environment, the drivers can be substituted with simulation agents. In that way, the presented O/C architecture can be used for simulation aspects as well as for managing real hardware components [1].

The EMS for electric appliances has already been extended to manage thermal energy consumers in the smart home (e.g. heating system) as well as suppliers for thermal energy (e.g. CHP) [3]. Altogether, the O/C architecture is particularly suitable to realise energy management systems.

References

1. Allerding, F., Becker, B., Schmeck, H.: Integration intelligenter Steuerungskomponenten in reale smart-home-Umgebungen. In: GI Jahrestagung 2010. LNI, pp. 455–460. GI (2010)

2. Becker, B., Allerding, F., Reiner, U., Kahl, M., Richter, U., Pathmaperuma, D., Schmeck, H., Leibfried, T.: Decentralized energy-management to control smart-home architectures. In: Proceedings of the 23rd International Conference on Architecture of Computing Systems (ARCS 2010). LNCS, vol. 5974, pp. 150–161. Springer, Berlin (2010)
3. Graessle, S., Becker, B., Knapp, T., Allerding, F., Schmeck, H., Wagner, A.: Intelligent control system for chp-equipment in smart-homes. In: Proceedings of the 2nd International Conference on Microgeneration and Related Technologies (2011)

F. Allerding · B. Becker (✉) · H. Schmeck
Institute of Applied Informatics and Formal Description Methods (AIFB),
Karlsruhe Institute of Technology (KIT), Karlsruhe, Germany
e-mail: birger.becker@kit.edu

F. Allerding
e-mail: florian.allerding@kit.edu

H. Schmeck
e-mail: hartmut.schmeck@kit.edu

Chapter 6.1.10
Self-organising Distributed Smart Camera Systems

Michael Wittke and Jörg Hähner

Keywords Smart camera · Self-optimisation · Active vision network · Runtime optimisation · Distributed vision

A new generation of surveillance systems relies on Smart Cameras (SCs) and overcomes drawbacks of today's systems in terms of privacy protection, cost efficiency and robustness towards security threats. The research project PRISMATICA [3], which has been funded by the European Union, started in 2005 to investigate SCs for building intelligent surveillance systems. A SC consists of a CCD sensor acquiring images, a computing unit that allows to analyse the collected images automatically, and a pan/tilt/zoom drive to adjust its field of view. No image data leaves the SC as long as no predefined incidents are detected. Advances in the research area of computer vision allow for object detection, tracking and recognition and thereby enable SCs to analyse scenes autonomously. Large camera systems (as e.g. used at international airports) are comprised of thousands of cameras. For instance, almost half a million surveillance cameras were in operation in London in 2002, Heathrow Airport and London Underground using more than 5,000 cameras each [3]. The research project *Distributed Smart Cameras (DISC)*—funded by the German Ministry of Education and Research (BMBF) through the CamInSens project (www.caminsens.org)—investigates how *Self-organising Distributed Smart Camera Systems* can be created and how image analysis and camera alignment benefit from a self-organising system architecture.

In self-organising distributed SC systems, SCs analyse scenes cooperatively and anticipate dangerous situations or aggregate useful statistics. Apart from collaboration in terms of computational image analysis, management tasks need to be carried out. For example, cameras need to cooperatively adjust their fields of view by panning and tilting their CCD sensors in order to observe areas [2] or track objects efficiently [1]. These management tasks have to—as the analysis of video streams—be carried out in a distributed and self-organising fashion and should not rely on central components that might be prone to errors and attacks and even lead to system failure. For this, the following self-* properties have been translated into specific distributed algorithms (i.e., ROCAS [2] and DMCtrac [1]) that have been implemented as a part of the DISC project:

C. Müller-Schloer et al. (eds.), *Organic Computing — A Paradigm Shift for Complex Systems*, 609–610
Autonomic Systems, DOI 10.1007/978-3-0348-0130-0_47, © Springer Basel AG 2011

- **Self-Organisation:** SCs can reorganise their fields of view by turning their heading. These SCs self-organise their fields of view and achieve good tracking performance in spite of disturbances such as communication failure. The user can set up constraints (priority regions, blind spots) but does not need to take care of the process in detail.
- **Self-Optimisation:** SCs are further able to arrange their fields of view so that an optimal surveillance coverage is achieved. This process of optimising the arrangement of the cameras' fields of view is an NP complete derivative of the art gallery problem. The heuristics ROCAS and DMCtrac are examples for distributed heuristics that bring self-optimisation properties to camera networks.
- **Self-Healing:** SCs are not only able to detect a system failure but also to react in order to compensate its effects. Failure detection mechanisms allow for a detection of fail-stop errors of single nodes. An appropriate reaction to a fail-stop error of a single SC is to re-arrange the cameras' fields of view so that the area covered by the failing node is covered as well as possible by neighbouring nodes.

Future work of the DISC project will investigate how node mobility (e.g. by mounting SCs to mobile entities such as ground or air vehicles) can be used to further decrease the number of SCs needed in surveillance applications [4]. These cameras extended by position drives are called *Active Smart Cameras (ASCs)*. ASCs shall enable, for example, efficient observation by reconfiguring themselves (i.e., self-configuration) to capture images of specified targets, i.e. locations, periodically. They imply the continuous feedback between the sensing device, i.e. the camera, and its processing unit.

References

1. Hoffmann, M., Wittke, M., Bernard, Y., Soleymani, R., Hahner, J.: Dmctrac: Distributed Multi Camera Tracking, pp. 1–10 (2008)
2. Hoffmann, M., Wittke, M., Hähner, J., Müller-Schloer, C.: Spatial partitioning in self-organising camera systems. IEEE J. Sel. Top. Signal. Process. **2**, 480–492 (2008)
3. Velastin, S.A., Boghossian, B., Lo, B., Sun, J., Vicencio Silva, M.A.: Prismatica: Toward ambient intelligence in public transport environments. IEEE Trans. Syst. Man Cybern., Part A, Syst. Hum. **35**(1), 164–182 (2005)
4. Wittke, M., Grenz, C., Hähner, J.: Towards organic active vision systems for visual surveillance. In: 24th International Conference on Architecture of Computing Systems 2010—Workshop Proceedings (ARCS '11—24th International Conference on Architecture of Computing Systems 2011) (2011)

M. Wittke (✉) · J. Hähner
Institute of Systems Engineering, Leibniz Universität Hannover, Appelstr. 4,
30167 Hannover, Germany
e-mail: wittke@sra.uni-hannover.de

J. Hähner
e-mail: haehner@sra.uni-hannover.de

Chapter 6.1.11
Organic Network Control

Björn Hurling, Sven Tomforde, and Jörg Hähner

Keywords Network protocols · Parameter adaptation · Runtime optimisation · Machine learning · Self-organising · Self-learning

In recent years technical systems have shown increasing complexity and, at the same time, an increasing demand for communication. Additionally, the number of interconnected devices and the corresponding transfer load of data communication networks is steadily increasing. Traditional network protocols have to compete with newly developed protocols and applications for limited resources. A demonstrative example can be found in the domain of cellular networks: usage scenarios are switching from pure voice communication to data-oriented services. This shift leads to much heavier load than researchers and engineers expected when the current protocols were put into operation years ago.

Modern network protocols are characterised by a high degree of flexibility in terms of different configurations—especially considering the increasing number of variable parameters like, e.g., delay times or buffer sizes. This flexibility allows for tuning the behaviour of the algorithm in particular operating conditions. One of the major challenges is to carefully choose appropriate settings for the numerous parameters. New solutions are needed to deal with this flexibility and to guarantee an optimal behaviour in every possible situation.

In [2] *Tomforde et al.* introduce a novel concept that allows for the dynamic adaptation of a network protocol's settings, the Organic Network Control (ONC) System. The ONC System is based on the ideas of the Organic Computing Initiative (cf. [1]), a recent research area, which focuses on designing self-organising and self-adaptive systems.

From a technical perspective the ONC System is based on the three-layered Observer/Controller architecture as described in Chap. 4.1. ONC regards network protocols as black box systems that provide an interface to change variable parameters. To learn an appropriate mapping from locally observed situations to optimal parameter values ONC uses standard machine learning techniques. Being generic in nature, ONC has been successfully applied to several protocols from different domains. In [4] the ONC concept has been used to increase the performance of a commonly used Peer-to-Peer filesharing protocol.

C. Müller-Schloer et al. (eds.), *Organic Computing — A Paradigm Shift for Complex Systems*, 611–613
Autonomic Systems, DOI 10.1007/978-3-0348-0130-0_48,

In a second investigation we applied ONC to the Reliable Broadcast Protocol in mobile ad-hoc networks [3]. Since nodes are moving, the situation surrounding a particular node is steadily changing. By properly choosing parameters like packet forwarding delays, the ONC-enhanced system increased the delivery ratio while reducing the overhead caused by retransmissions.

Most recently, we applied the ONC concept to the domain of wireless sensor networks (WSN) [5]. WSN scenarios are characterised by high uncertainties regarding the deployment of sensors in the field. The topology of the deployed network and the detailed geological characteristics of the environment are typically unknown at design-time. Therefore, these systems demand solutions to dynamically adapt their behaviour to the situation at hand.

In our investigation we focused on the Adaptive Distributed Resource Allocation Scheme (ADRA). Sensor nodes may switch from an active to a passive state to reduce energy consumption. ADRA assigns priorities to different aspects like coverage or battery saving that control whether a node will be active or passive in the next cycle. The ONC system adapts these priorities to the characteristics of the locally observed neighbourhood. In our experiments we were able to increase both the surveillance quality in terms of detected events and the system's lifetime.

Besides an application to further protocols, current work focuses on cross-layer and learning aspects. Additionally, we investigate different solutions for the simulation-based rule generation mechanism (cf. Chap. 4.1). Beyond parameter tuning our group will focus on strategies to allow for structurally changing and optimising existing protocols.

References

1. Müller-Schloer, C.: Organic Computing: On the feasibility of controlled emergence. In: Proc. of the 2nd Int. Conf. on Hardware/Software Codesign and System Synthesis, pp. 2–5 (2004)
2. Tomforde, S., Cakar, E., Hähner, J.: Dynamic control of network protocols—A new vision for future self-organised networks. In: Proc. of the 6th Int. Conf. on Informatics in Control, Automation, and Robotics (ICINCO'09), pp. 285–290 (2009)
3. Tomforde, S., Hurling, B., Hähner, J.: Dynamic control of mobile ad-hoc networks—network protocol parameter adaptation using organic network control. In: Proc. of the 7th Int. Conf. on Informatics in Control, Automation, and Robotics (ICINCO'10), Setubal, PT, vol. 1, pp. 28–35 (2010). INSTICC
4. Tomforde, S., Steffen, M., Hähner, J., Müller-Schloer, C.: Towards an Organic Network Control System. In: Proc. of the 6th Int. Conf. on Autonomic and Trusted Computing (ATC'09), pp. 2–16. Springer, Berlin (2009)
5. Tomforde, S., Zgeras, I., Hähner, J., Müller-Schloer, C.: Adaptive control of sensor networks. In: Proc. of the 7th Int. Conf. on Autonomic and Trusted Computing (ATC'10) (2010)

B. Hurling (✉) · S. Tomforde · J. Hähner
Institute of Systems Engineering, Leibniz Universität Hannover, Appelstr. 4,
30167 Hannover, Germany
e-mail: hurling@sra.uni-hannover.de

S. Tomforde
e-mail: tomforde@sra.uni-hannover.de

J. Hähner
e-mail: haehner@sra.uni-hannover.de

Chapter 6.2
Organic Computing: Quo vadis?

Christian Müller-Schloer and Hartmut Schmeck

Abstract Over and above the many results of the single projects, the intensive discussions during the colloquia and mini workshops of the SPP have revealed a multitude of commonalities, insights and lessons learnt. In this concluding chapter we want to discuss these findings and point towards possible future research questions in the context of Organic Computing.

Keywords Organic Computing · Challenges · Outlook

1 Design Time to Runtime

Classical system design follows a top-down path from the system level to the lower implementation levels. Different aspects of the design (software, hardware, mechanical parts, etc.) have to be kept in line and consistent. We might subsume the Organic Computing (OC) research initiative under the motto: "*Move design time decisions to run time!*" Today, the designer has the job to anticipate all future situations his final system will run into during lifetime. This is, of course, only partially possible. To a certain extent it is possible, if the environment of the system under design is carefully controlled, and if the external variations can be anticipated. Beyond this, the system behaviour might get unpredictable and is bound to fail. A higher degree of freedom at runtime has to be combined with increasing the robustness of a system, which means to extend the range of acceptable disturbances. Taking the above motto seriously defines a long-term research effort. We want to list a few consequences in the following.

- **Manual (re)configuration becomes self-organisation:** In the classical design process it is evident that the designer has all the power to change the system under design in any conceivable way. It is an objective of OC to let the system reconfigure without manual intervention during runtime in response to external demands or disturbances. This requires some active mechanism to take over the task of the designer, and a well-defined adaptability of the system under design. Adaptability can be realised in a relatively simple way for software systems, especially if mode-switching techniques are used. For hardware, this is much more complicated and requires special reconfigurable hardware, e.g. FPGAs. Reconfiguration with no or only minimal manual intervention is a form of self-organisation. In

C. Müller-Schloer et al. (eds.), *Organic Computing — A Paradigm Shift for Complex Systems*, 615–627
Autonomic Systems, DOI 10.1007/978-3-0348-0130-0_49, © Springer Basel AG 2011

OC, this is intended to be a decentralised process. Hence, each of the subsystems has partial responsibility for triggering a reconfiguration, but the whole reconfiguration is an orchestrated effort to be agreed upon by the subsystems.

- **Design space exploration becomes runtime optimisation:** The classical designer's task is design space exploration. In order to come up with sensible alternative solutions, candidates are evaluated e.g. by simulation or by real model building. The results of these evaluations are compared with the requirements and lead to acceptance or rejection. The design space of a real system is huge. Therefore, the designer follows template designs, introduces delta-modifications, and, above all, uses experience. At the end of the design phase, the blueprint goes into production, which is relatively simple for software, and possibly difficult and expensive for hardware. Doing the same job at runtime means above all to define the configuration space of the system under observation and control (SuOC). Software modifications are relatively simple if they can be controlled by strategic parameters. Hardware modifications become increasingly feasible by using FPGA technology. Modifications of the mechanical structure of a system (such as the wing extension of an aeroplane) are only possible with very high effort.
  But before the modifications can be activated, the OC system has to explore the alternatives. This can be done by selection from predefined candidates. This approach, however, limits the design space to a few anticipated solutions. In the more general case, we want to allow the system to come up with new solutions at runtime. This entails the usage of learning and optimisation technologies at runtime (such as Evolutionary Algorithms). We will discuss the implications of runtime learning and optimisation—especially in terms of performance and errors—below.
- **Model-based simulation becomes runtime modelling:** During the last decades model-based design has become standard. Instead of testing a real prototype we can derive certain properties of the future system from modelling and simulation. The simulation can be done at significantly different levels of detail, from microscopic modelling and simulation of individual behaviours of components to a macroscopic view based on mathematical models of aggregated effects (e.g. the *Webster* formula [14] in the area of urban traffic control). What happens to such models when the design process is moved to runtime? For one, the models are still useful as a basis for runtime simulation. But now we have to solve the additional problem of model adaptation at runtime. The changing environment has to be reflected in a changing environmental model, and the technical system itself, being adaptive, will result in a changing system model. Furthermore, at runtime, there will be time constraints with respect to decisions on system adaptations. As a consequence, quite often, one has to go for alternatives that are "good enough", since there will not be sufficient time to generate optimal configurations. Models of the environment should resemble the actual state of the environment as closely as possible. In this sense they are descriptive. If they deviate from reality the simulation-based system adaptation will lead to incorrect conclusions. Therefore, mechanisms are needed that keep model and real environment in line. The system model plays a double role: It is *prescriptive* (in the sense of a top-down

constraint flow), and at the same time it is *descriptive* (in the sense of a true description of the real state of the system). A contradiction between the two has to be resolved and may lead to what is called "yoyo design" in the classical design process. Yoyo design at runtime means a permanent adjustment process between *prescriptive* top-down goals and constraints with *descriptive* bottom-up models. For some interesting work in the area of software adaptation see e.g. [13].

- **Test:** Testing is similar to simulation with the exception that the real system is used rather than a model. The system is stimulated with selected inputs, and its performance is evaluated by observing its response and comparing the outputs to the specification or to the outputs of a "golden device" if available. In contrast to the classical design process, due to the shift of design decisions to runtime, in the case of OC systems the real system is available. However, in most cases it is not possible to use it for testing purposes. A sandbox, i.e. a hermetically closed test environment, might be formed temporarily if effects on the real environment can be modelled (inside the sandbox), or if side effects of test outputs can be tolerated. In any case it is necessary to roll the system state back to the state before testing.
- **Certification and verification:** In particular in critical application areas (like traffic control, automotive, avionics, or energy systems) there is a definite need to derive verifiable statements on the expected behaviour of a system (software and/or hardware). This is a severe challenge for systems operating in static environments and becomes much more necessary and challenging, if environmental conditions and behavioural requirements are changing at runtime. This is even more true, if a system adapts autonomously to changing requirements. This is also closely related to the issue of trust: To establish trust, you may perform initial tests, extend them by formal statements towards certification and in this way reduce the uncertainty on the essential properties of a system. Such guarantees have to be renewed at runtime.

2 Cautious Configuration Space Design

The configuration space defines the interface available to the higher-level manager (the Observer/Controller) for modifying the system under observation and control. It is crucial to define this interface explicitly and cautiously in order to provide the necessary capability of controlling the behaviour of the system while avoiding unwanted side effects. This interface has to provide the potential for explicitly influencing the system, even if most of the time there might be no need for external control. One has to take care that different control actions are not interfering with each other. An ill-defined multitude of such control activities would probably interact and produce unexpected results. A clear definition of this configuration space interface will also enforce a separation between the manager and the managed system. This is particularly important in systems having multiple levels of control, and it is crucial for adhering to the principle of "non-critical complexity" (see below). Another pitfall could be a loose coding of the parameters, which constitute the configuration space. The configuration parameters span the axes of the configuration space. In order to keep the configuration space as small as possible, and to exclude meaningless configurations from the beginning, care should be taken to encode the possible states as densely as possible. There is a definite trade-off between the need for an abstract representation of the essential characteristics of the system and the intention to provide a more detailed view to allow for potential explicit interference. This corresponds to the coding (or representation) problem frequently encountered during the design of Genetic Algorithms.

3 Self-organisation is not Magic

Self-organisation is a tempting concept because it promises to get something for nothing: "I don't know how to solve a certain problem, therefore I let the system somehow self-organise to do the job for me." Unfortunately, this does not work, not even in nature when we observe how an anthill is miraculously built by self-organisation. The elements of a self-organising system must be endowed with certain basic capabilities as preconditions for self-organisation. They must be able to communicate, to sense the environment and other agents, and to trigger a reconfiguration if changing conditions require it. For such managerial tasks we have to provide explicit resources. This leads to the dichotomy between the managed system (SuOC) and the manager (the Observer/Controller and the external "user" providing system objectives and evaluation criteria). Self-organisation is a task to be done based on resources dedicated to this task. It looks "magic" only when observed from outside the black box.

And who manages the manager? A separation into a lower-level managed (i.e. passive) part of the system and a higher level actively managing part might not be sufficient. It has been shown that the Observer/Controller itself can be adapted by a next-level managing system, and so on. Such a multi-level architecture supports a separation of concerns where different aspects of adaptation are handled by different management levels. This is also advantageous because the adaptation mechanisms on the different levels have to react in different time scales: While the lowest level is basically a stimulus-response system, the higher levels have more time to "deliberate". Although it is not clear whether an ISO/OSI-like stacked architecture is appropriate, a 4-level Observer/Controller system could be devised (Fig. 1), where the two lower levels are responsible for local adaptations, level 3 handles peer-to-peer negotiations, and level 4 interacts with higher-level institutions.

This extended structural view of an Observer/Controller architecture has some similarity with the Viable System Model as introduced by *Stafford Beer* [1] as a recursively applicable template for structuring the management of large enterprises.

4 Overhead and Complexity

With the introduction of the (multi-level) Observer/Controller architecture future systems will develop into a dichotomy. In addition to the production system, which

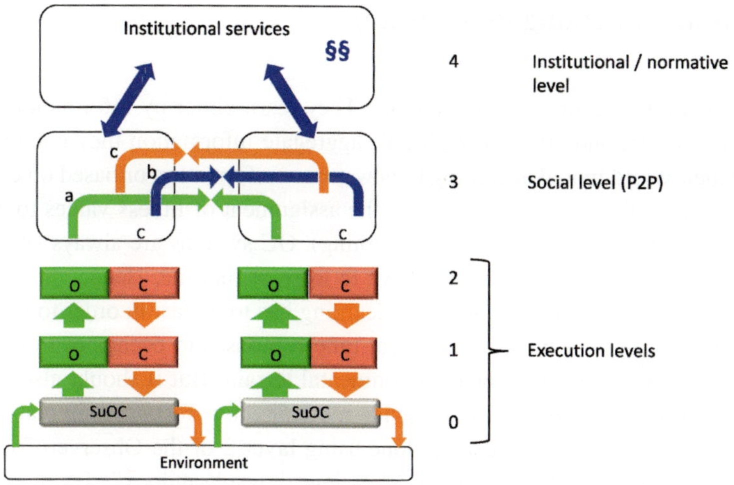

Fig. 1 4-level Observer/Controller architecture

serves to fulfil the productive tasks (such as responding to sensor inputs by a temporary release of the brakes in an ABS system), we will have to spend additional effort for the higher control levels. This means that we are adding complexity and overhead. Additional complexity is contradictory to the original goals of OC if it increases the risk of failure. This complexity is acceptable only if it is non-critical in terms of the basic functionality of the production system. This "**principle of non-critical complexity**" means that the production system must remain fully functional even in case the Observer/Controller system fails. In this case, however, the production system cannot be adapted any more, it will stay in the last state it has been adapted to, unless it has some adaptivity potential on its own or some other external entity takes over. Observer/Controller levels mean additional effort in terms of lines of code and/or transistors.

How much additional cost is acceptable? And: Can we call this additional effort "overhead"? Observer/Controller levels are meant to improve the robustness and flexibility of the SuOC. As long as there are no severe variations in the execution environment the additional code for achieving robustness and flexibility will definitely be unnecessary overhead. But whenever situations beyond normal operation have to be dealt with, these additional capabilities are essential. If they do the job, they can't be called "overhead". But undoubtedly they produce additional cost. We predict that this additional effort will be an integral and necessary part of future highly complex systems. How much effort will be devoted to these control levels remains speculation. It is interesting to note, however, that in the human brain the ratio of "overhead" (like the limbic system responsible e.g. for emotional reactions) and the productive system (i.e. the sensor-motor subsystems) is estimated to be $10^5 : 1$![1]

[1] *Gerhard Roth* (Institut für Hirnforschung, Bremen), personal communication.

5 Runtime Learning (Sandboxing)

Learning is a key ability of OC systems. They learn either by information aggregation or by trial and error (or both). To aggregate information they continuously update their world model according to new sensor information or based on communication input. Trial and error relies on the assignment of fitness values to actions performed in the past (reinforcement learning). OC systems are always situated in a real environment. Therefore, they have to learn at runtime. Hence, their actions directly influence the environment, and learning has to be fast in order to be meaningful. An OC system must provide real-time responses to events, even if it runs into (at least partially) unknown environmental terrain. But it should also be able to "think in the back-ground" and provide more effective responses in real-time or just "some time later". This can be done using layer 2 of the Observer/Controller architecture (Fig. 1), which could also be viewed as a "sandbox". Trying out new actions in a safety-critical environment is acceptable only if the actions have been verified against constraints set by the designer, or if the actions have been validated in a simulated environment. This latter approach is also called sandbox learning: The sandbox constitutes a hermetically closed environment where actions cannot cause harm. A sandbox approach was developed within the Organic Traffic Control (OTC) project [10], it became the template for "off-line learning" in the generic Observer/Controller architecture.

A sandbox with a simulated environment can also solve the performance problem. In order to find a legal new action with a high fitness value (e.g. via an Evolutionary Algorithm) a very large number of trials is necessary. Each trial has to be evaluated (by applying it to the real environment where the evaluation time constant is dictated by the typical delay of this environment). For a traffic system this delay (i.e. the time until the action's effect can be measured) might be as long as 5 minutes. If we assume just 1,000 trials, this amounts to an evaluation time of 5,000 minutes! If learning takes place within the sandbox using a simulation model, the evaluation time can be reduced to a few minutes, or, if we apply an approximation formula, to milliseconds. Of course, this does not come without a price to be paid: The learning will adapt to the model (or the formula), including possible errors.

Sandbox learning is a potentially powerful technique. Although first steps have been taken there remains much to do. So far the sandbox and the included model are strictly local. Influences from neighbouring nodes can be taken into account only indirectly, e.g. in the traffic example by assuming traffic loads as expected from the outside world. It would be interesting to investigate the possibilities for a distributed sandbox where the sandboxes of all nodes cooperate. A second problem of sandboxing, at least, if it is simulation-based, is performance. A solution might be to use fast parallel optimisation algorithms running on multi- or many-cores. Finally, for real-time optimisation we need algorithms, which can deliver results with scalable quality within a predefined time frame (as opposed to a quality-determined runtime).

6 OC Devices Can Be Interpreted as Cognitive and Self-optimising Systems

OC devices are situated in a real environment. They must observe at least certain aspects of the world via their sensory equipment, derive an aggregated situation description (some kind of world model), map this situation description into a plan, and execute this plan through actuators. This control loop has been described in a similar way in Autonomic Computing (as MAPE cycle [6]), in robotics as sense–plan–act cycle (or as a subsumption architecture), in agent technology as belief–desire–intention (BDI) cycle, in Management Science as Viable System Model, and in OC as Observer/Controller architecture. These are all realisations of cognitive systems. While for OC it is not the goal to model cognitive (or living) systems, it is worthwhile to use such architectures as a template to improve the Observer/Controller mechanism. However, care must be taken not to overstress the analogy. Classical reasoning mechanisms as proposed in agent technology (BDI) are not feasible in practical applications where real-time response is required. So far, OC "agents" employ a rather simple stimulus-response scheme, which is well suited for a Learning Classifier implementation. But with this kind of decision mechanism we cannot handle multi-step planning and plan execution. Here an extension is necessary.

Adaptive agents situated in a real environment are playing two roles at the same time: They act within the environment, and they try to optimise their behaviour in accordance with some fitness function. In other words: OC agents navigate in a fitness landscape. Unfortunately, usually this fitness landscape is not known, and, worse, it changes dynamically, and even worse, it changes in response to the actions of the agents. We call such a situation a "self-referential fitness landscape" (see also Chap. 3.1 of this compendium). Examples for such behaviours are self-modifying robots or trading agents in the stock market. It could prove useful to model OC problems in terms of self-optimising agents. In this context, similar to classic approaches of game theory, solutions have to be found for a balance between an egoistic behaviour (i.e. an optimisation of an egoistic utility function) and an altruistic behaviour (optimising the system utility).

7 Definition of Emergence Leads to Analysis of Distribution Functions

Emergence has been defined as self-organised order [8]. Although emergence has played a major role in the discussions of the OC community, it was not the goal to verbally define a somewhat elusive term. It was rather the objective to find a quantitative (and hence computable) definition and use it within real OC systems. It has been shown that a quantitative definition of emergence can be used to characterise multi-element systems in terms of order (or structure) in feature space. This allows conclusions not only regarding the present system state but also a prediction of future states. Such an early indicator is especially useful if it triggers countermeasures in case of unwanted future system states. The measurement of emergence results is a fast but quite unspecific characterisation of the system. It can be used as a saliency monitor, which then can lead to a more detailed analysis of the situation e.g. through a cluster analysis.

Recent work [4] (see also Chap. 1.3 of this compendium) has generalised the basic idea of measuring emergence in terms of an entropy difference to a comparative analysis of distribution functions. This technique can be used as a detector for new but so far unknown processes in a system. An important application is intrusion detection where generally attacks of unknown character have to be detected (cf. Chap. 3.3).

8 No Decentralisation at Any Cost!

Decentralisation has developed into the holy grail of systems architectures. Any central authority is to be avoided, and the arguments are always the same: No single point of failure, no central bottleneck! After all, self-organisation in its pure meaning has been defined as decentralised. However, experience shows that neither a purely centralised nor a purely decentralised system architecture is optimal. Moreover, the optimal architecture might change dynamically during runtime between the two extremes.

It is necessary to make a more subtle distinction between different functions and their respective most adequate architecture in terms of a deep or a flat hierarchy or no hierarchy at all. Of course, a large hierarchical communication architecture with all messages being routed through one root node would not make sense. On the other hand, in many cases, purely local decisions made by many autonomous agents without higher-level knowledge are a bad solution. We have shown (in the case of self-organised progressive signal systems [12]) that a moderate regional coordination results in better performance than the decentralised solution. However, this must not be confused with a central manager for the coordination of all progressive signal systems in a city! Central (or better: regional) coordination by a regional manager is an advantage if

1. the coordinating entity is involved only occasionally and not during each micro-decision of one of the subsystems, and/or if
2. the coordinating entity has more data from a wider sensory horizon than the single subsystems, and if
3. the work load of the central entity remains moderate.

Generally, the system performance will be better with a more decentralised coordination during periods of high workload [2]. Conflict resolution between many parties can be handled more efficiently with support from a central authority than by a collection of peers. The observation that central entities can be helpful is not limited to technical systems. Institutional economics (*W. Hamilton*, *W. Hodgson*) has investigated the relation of parts of an economic system to the whole. *Silva and Lima* have applied these ideas to technical systems (Institutional Robotics [11]).

9 Human-Centric OC

It has been an objective of the Organic Computing Initiative to improve the relationship between user and system. Unfortunately, little progress has been made so far in this context. Additional research in Human Computer Interaction is all the more necessary since adaptive systems put higher demands on the user interface (UI) design. Two research aspects can be identified: We need (1) better UIs for adaptive systems, and (2) we need better adaptive UIs.

1. A UI for a permanently changing system must take the crucial decision when to hide internals and when to inform the user. A system might reconfigure internally in order to keep the functionality—as experienced by the user—constant. This is e.g. the case when failures of subsystems can be covered by replacement with spare parts. As long as the functionality is not endangered the user should not be bothered. However, there might be a critical situation where a reconfiguration is not possible any more. In this case it will be necessary to warn the user, even in advance. The more general question to be answered is: How much of the internals of a self-organising system have to be communicated to the user?
2. The idea of an adaptive UI is to learn from the interaction with the user and to adapt accordingly by offering frequently used menus and buttons more conveniently. Infamous examples show that such a UI can become very awkward to use if the placement of inter-action elements is not predictable any more.

Both aspects have one thing in common: The user has to be always in control, and control is based on predictability and transparency. Predictability means that the system behaves as expected (therefore predictability is closely related with trust). Transparency means that the user can always gain sufficient insight into the internals of the system, without being overwhelmed by details. Different research approaches are being followed in this context.

1. The concept of guarantees and corridors (see e.g. *Güdemann et al.* [5]) lets the user define boundaries, which the system is guaranteed not to exceed. Within the boundaries the system acts autonomously.
2. A relatively new research field is the use of emotions to transport mental states between human user and system. This can be done in both directions. It is especially interesting to define an emotional state of a machine. *Kleinjohann et al.* [3] use the concept of contentedness of a robot team to express the collective degree of goal achievement.

10 Social OC

One of the key insights about self-organising systems is the role of interaction between multiple entities as the major cause for emergent global phenomena. Hence, it is reasonable to investigate the influence of "social" activities on organic computing systems. This has also been outlined in [9]. Complex systems in nature often develop in a bottom-up way due to the self-organising capabilities of their components. Each component and the system as a whole respond to the demands of the environment. In doing so they are guided by two principles (cf. *Arthur Koestler's* holon model [7]): Their drive to survive as an individual (selfishness) and the necessity to cooperate (altruism). Successful individuals are able to balance both drives.

In a similar sense, OC is two-faced. It has an inward as well as an outward view: On the one hand, the Observer/Controller is concerned with observing and controlling its own system (the system under observation and control) in order to keep it operational even in the presence of external or internal disturbances, or to reconfigure it in case the user sets a new goal. On the other hand, every OC system is part of a major system, including the environment. In this role, it has to interact with systems on the same level (peer to peer) or with super- or subordinated systems in a hierarchy. While this two-faced character of OC systems has been recognised early in the research community, it is only now (i. e. increasingly in the third phase of the Special Priority Programme) that the emphasis has been shifting towards the aspect of "computing societies". This new orientation does not make the inside view obsolete. However, it poses interesting new questions for future OC research. In this role, OC subsystems behave like agents, and it is not surprising that OC can borrow techniques from the research area of MultiAgent Systems (MAS—[15]). We want to mention four MAS areas worthwhile to be investigated more closely also from the OC point of view:

- **Cooperation and coordination**: Agents within a MultiAgent system are autonomous (in OC they are semi-autonomous being controlled by goals and constraints from the user) and therefore potentially self-interested (or egoistic). This can lead to conflicts or—in the more desirable case—to cooperation. Game Theory has extensively investigated the different situations and the choices made by the so-called "rational agent". Especially interesting is the fact that rational agents, although they act entirely comprehensible, will not reach a state of maximal system utility. If we want to achieve maximal system utility we must endow the agents with a wider sensory horizon, with the ability to negotiate and, most importantly, with trust mechanisms. The problem has been investigated theoretically in game theory. It will be one of the OC challenges to make game theory operational (at runtime) within a community of agents.
- **Normative and institutional systems**: Societies (of humans, animals or technical systems) have never developed on a purely decentralised or peer-to-peer basis. They have tended to install special agents playing the role of coordinators, conflict resolvers or legislators. If OC is to mimic such forms of organisation, it must define institutions and interaction protocols, which serve to regulate a constructive living together. Institutions are responsible for stating rules and enforcing them. Rules can be e.g. hard (deontic) or soft norms, constraints, recommendations, commitments or obligations. Norms are not fixed. We must also devise ways of ("democratically") finding and modifying them.
- **Agent architectures**: The Observer/Controller architecture existing so far is a relatively simple (however efficient) stimulus-response type of "brain". As soon as the agents become social, this architecture must be augmented with at least

two more layers of capabilities (Fig. 1): a social and a normative layer. The social layer is responsible for interactions on the P2P level, e.g. for observation exchange, goal reconciliation and joint action planning. The normative layer takes care of adherence to (or disobedience of) norms, and of negotiations to modify norms. It is interesting to note that these two additional layers can generally not be organised in a pure top-down fashion: We will have to find mechanisms to resolve internal conflicts resulting from contradictions between top-down goals and bottom-up (e.g. physical) constraints.

- **Methodologies and tools**: The MAS community offers a variety of methods and tools, which could be useful also in social OC. Among them are communication languages based on speech-act theory (e.g. FIPA ACL), development frameworks (such as JADE or Jadex) or simulation environments (e.g. RePast). However, in contrast to pure software agents an OC agent is always physically embodied, and it "lives" in a real environment. This leads to additional problems like resource limitations and hard real-time constraints that have to be addressed.

11 Technical Applications?

One of the key requirements of the Priority Programme on Organic Computing has been the need for considering technical applications as the driving motivation for introducing self-organisation and adaptivity. While so far, the major application scenarios have been traffic on highways and in cities and production lines, the short papers later in this chapter indicate further areas where the concepts of Organic Computing may provide essential contributions for innovation, in particular in data communication networks and in smart control of power grids. In general, the current thrive for "green by IT" provides a multitude of scenarios where it is essential to utilise intelligent adaptive feedback and control mechanisms as developed in Organic Computing and in related initiatives.

12 Organisational Sciences

The term Organic Computing has been derived from the analogy with living systems. In a somewhat wider context, we could also redefine Organic Computing—as proposed by *Kirstie Bellmann*—as part of the overarching discipline of **Organisational Sciences**. Questions of organisation play a crucial role in a variety of otherwise quite unrelated research fields like

- Technical systems
- Social systems
- Economic systems, Operations Research
- Biological systems
- Physical systems.

Topics of common interest are e.g.

- Adequate or even optimal organisational structures (or architectures)
- Institutions and organisations as structuring elements
- Control vs. self-organisation
- Dynamic behaviour
- Metrics like robustness, stability, flexibility
- Modelling of system behaviour in state space.

It remains to be seen if such diverse disciplines will be able to find a common terminology and methodology.

13 Conclusion

Initiated by the Priority Programme, a range of projects has addressed various aspects of the visionary concept of Organic Computing. As outlined in this contribution, our research has led to interesting results with respect to many related challenges for research and development, but there is still a long way to go. In particular, the trustworthiness and safety of self-organising adaptive systems and the role of social interactions between technical components and with human users are of essential importance for future technical applications. Hence we might claim that good research answers some questions while—and this is more important—posing new ones.

References

1. Beer, S.: Diagnosing the System for Organizations. Managerial Cybernetics of Organization. Wiley, Hoboken (1994)
2. Cakar, E., Hähner, J., Müller-Schloer, C.: Creating collaboration patterns in multi-agent systems with generic observer/controller architectures. In: Manzalini, A. (ed.) Proceedings of the 2nd International Conference on Autonomic Computing and Communication Systems, Autonomics 2008, September 23–25, Turin, Italy, p. 6 (2008)
3. Esau, N., Kleinjohann, L., Kleinjohann, B.: An adaptable fuzzy emotion model for emotion recognition. In: Montseny, E., Sobrevilla, P. (eds.) Proceedings of the Joint 4th Conference of the European Society for Fuzzy Logic and Technology and the 11th Rencontres Francophones sur la Logique Floue et ses Applications, Barcelona, Spain, September 7–9, 2005, pp. 73–78. Universidad Polytecnica de Catalunya, Catalonia (2005)
4. Fisch, D., Jänicke, M., Sick, B., Müller-Schloer, C.: Quantitative emergence – A refined approach based on divergence measures. In: Proc. SASO 2010, Budapest, Sept. 2010, pp. 94–103 (2010)
5. Güdemann, M., Nafz, F., Ortmeier, F., Seebach, H., Reif, W.: A specification and construction paradigm for organic computing systems. In: Second IEEE International Conference on Self-Adaptive and Self-Organizing Systems, 2008. SASO '08, pp. 233–242 (2008)
6. Kephart, J.O., Chess, D.M.: The vision of Autonomic Computing. IEEE Comput. **36**(1), 41–50 (2003)
7. Koestler, A.: Die Wurzeln des Zufalls. Scherz Verlag, München (1984). ISBN 3502153868

8. Mnif, M., Müller-Schloer, C.: Quantitative emergence. In: Proc. 2006 IEEE Mountain Workshop on Adaptive and Learning Systems (SMCals 2006), pp. 78–84. IEEE Press, New York (2006)
9. Müller-Schloer, C., Schmeck, H.: Organic computing: A grand challenge for mastering complex systems. Inf. Technol. **52**(3), 135–141 (2010)
10. Prothmann, H., Branke, J., Schmeck, H., Tomforde, S., Rochner, F., Hähner, J., Müller-Schloer, C.: Organic traffic light control for urban road networks. Int. J. Auton. Adapt. Commun. Syst. **2**(3), 203–225 (2009)
11. Silva, P., Lima, P.U.: Institutional robotics. In: Proceedings of the 9th European Conference on Advances in Artificial Life. ECAL'07, pp. 595–604. Springer, Berlin (2007)
12. Tomforde, S., Prothmann, H., Branke, J., Hähner, J., Müller-Schloer, C., Schmeck, H.: Possibilities and limitations of decentralised traffic control systems. In: WCCI 2010 IEEE World Congress on Computational Intelligence, pp. 3298–3306. IEEE Press, New York (2010)
13. Vogel, T., Giese, H.: Adaptation and abstract runtime models. In: Proceedings of the 5th Workshop on Software Engineering for Adaptive and Self-Managing Systems (SEAMS 2010) at the 32nd IEEE/ACM International Conference on Software Engineering (ICSE 2010), Cape Town, South Africa, pp. 39–48. ACM, New York (2010)
14. Webster, F.: Traffic Signal Settings - Technical Paper No 39. Road Research Laboratory, London, UK (1959)
15. Wooldridge, M.J.: An Introduction to Multiagent Systems, 2nd edn. Wiley, Hoboken (2009)

C. Müller-Schloer (✉)
Institute of Systems Engineering, System and Computer Architecture, Leibniz Universität Hannover, Appelstr. 4, 30167 Hannover, Germany
e-mail: cms@sra.uni-hannover.de

H. Schmeck
Institute AIFB, Karlsruhe Institute of Technology (KIT), Kaiserstr. 89, 76128 Karlsruhe, Germany
e-mail: hartmut.schmeck@kit.edu

